Lecture Notes in Computer Science

Vol. 88: Mathematical Foundations of Computer Science 1980. Proceedings, 1980. Edited by P. Dembiński. VIII, 723 pages. 1980.

Vol. 89: Computer Aided Design - Modelling, Systems Engineering, CAD-Systems. Proceedings, 1980. Edited by J. Encarnacao. XIV, 461 pages. 1980.

Vol. 90: D. M. Sandford, Using Sophisticated Models in Resolution Theorem Proving.
XI, 239 pages. 1980

Vol. 91: D. Wood, Grammar and L Forms: An Introduction. IX, 314 pages. 1980.

Vol. 92: R. Milner, A Calculus of Communication Systems. VI, 171 pages. 1980.

Vol. 93: A. Nijholt, Context-Free Grammars: Covers, Normal Forms, and Parsing. VII, 253 pages. 1980.

Vol. 94: Semantics-Directed Compiler Generation. Proceedings, 1980. Edited by N. D. Jones. V, 489 pages. 1980.

Vol. 95: Ch. D. Marlin, Coroutines. XII, 246 pages. 1980.

Vol. 96: J. L. Peterson, Computer Programs for Spelling Correction: VI, 213 pages. 1980.

Vol. 97: S. Osaki and T. Nishio, Reliability Evaluation of Some Fault-Tolerant Computer Architectures. VI, 129 pages. 1980.

Vol. 98: Towards a Formal Description of Ada. Edited by D. Bjørner and O. N. Oest. XIV, 630 pages. 1980.

Vol. 99: I. Guessarian, Algebraic Semantics. XI, 158 pages. 1981.

Vol. 100: Graphtheoretic Concepts in Computer Science. Edited by H. Noltemeier. X, 403 pages. 1981.

Vol. 101: A. Thayse, Boolean Calculus of Differences. VII, 144 pages. 1981.

Vol. 102: J. H. Davenport, On the Integration of Algebraic Functions. 1–197 pages. 1981.

Vol. 103: H. Ledgard, A. Singer, J. Whiteside, Directions in Human Factors of Interactive Systems. VI, 190 pages. 1981.

Vol. 104: Theoretical Computer Science. Ed. by P. Deussen. VII, 261 pages. 1981.

Vol. 105: B. W. Lampson, M. Paul, H. J. Siegert, Distributed Systems – Architecture and Implementation. XIII, 510 pages. 1981.

Vol. 106: The Programming Language Ada. Reference Manual. X, 243 pages. 1981.

Vol. 107: International Colloquium on Formalization of Programming Concepts. Proceedings. Edited by J. Diaz and I. Ramos. VII, 478 pages. 1981.

Vol. 108: Graph Theory and Algorithms. Edited by N. Saito and T. Nishizeki. VI, 216 pages. 1981.

Vol. 109: Digital Image Processing Systems. Edited by L. Bolc and Zenon Kulpa. V, 353 pages. 1981.

Vol. 110: W. Dehning, H. Essig, S. Maass, The Adaptation of Virtual Man-Computer Interfaces to User Requirements in Dialogs. X, 142 pages. 1981.

Vol. 111: CONPAR 81. Edited by W. Händler. XI, 508 pages. 1981.

Vol. 112: CAAP '81. Proceedings. Edited by G. Astesiano and C. Böhm. VI, 364 pages. 1981.

Vol. 113: E.-E. Doberkat, Stochastic Automata: Stability, Nondeterminism, and Prediction. IX, 135 pages. 1981.

Vol. 114: B. Liskov, CLU, Reference Manual. VIII, 190 pages. 1981.

Vol. 115: Autcmata, Languages and Programming. Edited by S. Even and O. Kariv. VIII, 552 pages. 1981.

Vol. 116: M. A. Casanova, The Concurrency Control Problem for Database Systems. VII, 175 pages. 1981.

Vol. 117: Fundamentals of Computation Theory. Proceedings, 1981. Edited by F. Gécseg. XI, 471 pages. 1981.

Vol. 118: Mathematical Foundations of Computer Science 1981. Proceedings, 1981. Edited by J. Gruska and M. Chytil. XI, 589 pages. 1981.

Vol. 119: G. Hirst, Anaphora in Natural Language Understanding: A Survey. XIII, 128 pages. 1981.

Vol. 120: L. B. Rall, Automatic Differentiation: Techniques and Applications. VIII, 165 pages. 1981.

Vol. 121: Z. Zlatev, J. Wasniewski, and K. Schaumburg, Y12M Solution of Large and Sparse Systems of Linear Algebraic Equations. IX, 128 pages. 1981.

Vol. 122: Algorithms in Modern Mathematics and Computer Science. Proceedings, 1979. Edited by A. P. Ershov and D. E. Knuth. XI, 487 pages. 1981.

Vol. 123: Trends in Information Processing Systems. Proceedings, 1981. Edited by A. J. W. Duijvestijn and P. C. Lockemann. XI, 349 pages. 1981.

Vol. 124: W. Polak, Compiler Specification and Verification. XIII, 269 pages. 1981.

Vol. 125: Logic of Programs. Proceedings, 1979. Edited by E. Engeler. V, 245 pages. 1981.

Vol. 126: Microcomputer System Design. Proceedings, 1981. Edited by M. J. Flynn, N. R. Harris, and D. P. McCarthy. VII, 397 pages. 1982.

Voll. 127: Y.Wallach, Alternating Sequential/Parallel Processing. X, 329 pages. 1982.

Vol. 128: P. Branquart, G. Louis, P. Wodon, An Analytical Description of CHILL, the CCITT High Level Language. VI, 277 pages. 1982.

Vol. 129: B. T. Hailpern, Verifying Concurrent Processes Using Temporal Logic. VIII, 208 pages. 1982.

Vol. 130: R. Goldblatt, Axiomatising the Logic of Computer Programming. XI, 304 pages. 1982.

Vol. 131: Logics of Programs. Proceedings, 1981. Edited by D. Kozen. VI, 429 pages. 1982.

Vol. 132: Data Base Design Techniques I: Requirements and Logical Structures. Proceedings, 1978. Edited by S.B. Yao, S.B. Navathe, J.L. Weldon, and T.L. Kunii. V, 227 pages. 1982.

Vol. 133: Data Base Design Techniques II: Proceedings, 1979. Edited by S.B. Yao and T.L. Kunii. V, 229–399 pages. 1982.

Vol. 134: Program Specification. Proceedings, 1981. Edited by J. Staunstrup. IV, 426 pages. 1982.

Vol. 135: R.L. Constable, S.D. Johnson, and C.D. Eichenlaub, An Introduction to the PL/CV2 Programming Logic. X, 292 pages. 1982.

Vol. 136: Ch. M. Hoffmann, Group-Theoretic Algorithms and Graph Isomorphism. VIII, 311 pages. 1982.

Vol. 137: International Symposium on Programming. Proceedings, 1982. Edited by M. Dezani-Ciancaglini and M. Montanari. VI, 406 pages. 1982.

Vol. 138: 6th Conference on Automated Deduction. Proceedings, 1982. Edited by D.W. Loveland. VII, 389 pages. 1982.

Vol. 139: J. Uhl, S. Drossopoulou, G. Persch, G. Goos, M. Dausmann, G. Winterstein, W. Kirchgässner, An Attribute Grammar for the Semantic Analysis of Ada. IX, 511 pages. 1982.

Vol. 140: Automata, Languages and programming. Edited by M. Nielsen and E.M. Schmidt. VII, 614 pages. 1982.

Vol. 141: U. Kastens, B. Hutt, E. Zimmermann, GAG: A Practical Compiler Generator. IV, 156 pages. 1982.

Lecture Notes in Computer Science

Edited by G. Goos and J. Hartmanis

176

Mathematical Foundations of Computer Science 1984

Proceedings, 11th Symposium
Praha, Czechoslovakia
September 3–7, 1984

MFCS '84

Edited by M. P. Chytil and V. Koubek

Springer-Verlag
Berlin Heidelberg New York Tokyo 1984

Editors

Michael P. Chytil
Václav Koubek
Charles University
Malostranské nám. 25, 11800 Praha 1, Czechoslovakia

CR Subject Classifications (1982): F. 1, F. 2, F. 3, F. 4, G. 2

ISBN 3-540-13372-0 Springer-Verlag Berlin Heidelberg New York Tokyo
ISBN 0-387-13372-0 Springer-Verlag New York Heidelberg Berlin Tokyo

Printed in Germany

Printing and binding: Beltz Offsetdruck, Hemsbach/Bergstr.
2145/3140-543210

FOREWORD

MFCS '84

This volume contains 11 invited papers and 44 short communications contributed for presentation at the 11th Symposium on Mathematical Foundations of Computer Science - MFCS´84, held at Prague, Czechoslovakia, September 3 - 7, 1984.

The symposium was organized by the Committee of Applied Cybernetics of the Czechoslovak Scientific and Technological Society in cooperation with the Faculty of Mathematics and Physics of Charles University, Prague, Computing Science Department of the Purkyně University, Brno, the Faculty of Natural Sciences of the Šafárik University, Košice, and the Federal Ministry for Technical and Investment Development.

The contributions in these Proceedings were selected from 107 papers submitted in response to the call for papers. The selection was made on the basis of originality and relevance to theoretical computer science by the international program committee.

The program committee of MFCS´84 consisted of J. Bečvář /Prague/, J. Berstel /Paris/, A. Blikle /Warsaw/, R. Book /St. Barbara/, W. Brauer /Hamburg/, M. Chytil /Prague/ - chairman, P. van Emde Boas /Amsterdam/, R. Freivalds /Riga/, J. Hartmanis /Ithaca/, V. Koubek /Prague/, W. Lipski /Warsaw/, A. Meyer /Cambridge/, A. Salwicki /Warsaw/, A. Slisenko /Leningrad/, J. W. Thatcher /Yorktown Heights/, L. Valiant /Cambridge/, G. Wechsung /Jena/, J. Wiedermann /Bratislava/, M. Yannakakis /Murray Hill/.

The organizing committee of MFCS´84 consisted of M. Benešovský, M. Chytil, A. Goralčíková, J. Gregor, Š. Hudák, V. Koubek, L. Kučera, B. Miniberger, Z. Renc - organizing secretary, B. Sekerka, I. Vavrysová, M. Vlach - chairman.

In evaluating the submitted papers, the following referees assisted to the program committee members: J. Adámek, A. Andžans, K. R. Apt, W. Bartol, J. M. Barzdin, M. Ben-Or, S. L. Bloom, L. Boasson, A. Brandstädt, B. Chlebus, C. Choffrut, J. Dassow, O. Demuth, P. Enjalbert,

K. Fischer, M.-C. Gaudel, P. Goralčík, A. Goralčíková, H. Grabowski, D. Yu. Grigor´ev, M. Gross, J. Gruska, I. Guessarian, P. Hájek, H. D. Hecker, A. Hemmerling, F. Hoffmann, J. Hromkovič, L. Janiga, A. Kelemenová, E. B. Kinber, I. Korec, D. Kozen, L. Kučera, M. Kudlek, P. Lamačka, M. Latteux, A. I. Litwinink, J. Mandel, A. Mazurkiewicz, G. L. Miller, G. Mirkowská, J. C. Mitchell, A. W. Mostowski, W. Nehrlich, E. Nelson, C. F. Nourani, F. J. Oles, P. Orponen, L. Pacholski, D. Perrin, J. E. Pin, J. Pittl, I. Privara, P. Pudlák, J. C. Raoult, J. H. Reif, W. Reisig, C. Reutenauer, P. Růžička, L. Staiger, J. Steiner, P. Štěpánek, O. Štěpánková, J. Šturc, J. Tiuryn, J. Úlehla, R. Valk, E. G. Wagner, K. Wagner, R. Wiehagen, J. Winkowski.

The editors wish to thank all those who submitted papers for consideration, and the program committee members with the referees for their meritorious work on evaluating the papers. Finally, the excellent cooperation of Springer-Verlag in the publication of this volume is highly appreciated.

Prague, May 1984 Michal Chytil, Václav Koubek

CONTENTS

INVITED LECTURES

COMMUNICATIONS

APPENDIX

INVITED LECTURES

COMMUNICATIONS

SEPARATING, STRONGLY SEPARATING, AND COLLAPSING RELATIVIZED COMPLEXITY CLASSES

José Luis Balcázar
Facultad de Informática de Barcelona
Jordi Girona Salgado 31, Barcelona 34
SPAIN

Le Choeur: Créon, ses oracles ne se trompent jamais.
(J. Cocteau: Antigone.)

0. Introduction.

Relativization of complexity classes is currently a field of interest in the research on foundations of computer science. After a period spent looking for widely accepted definitions, the study of the most interesting complexity classes became surprisingly difficult. Most of the central problems -- whether some of these classes coincide, are today open problems, and the real reasons of these difficulties are still far from being known.

Originally, one of the reasons justifying the undecidability of some problems was the possibility of describing machine computations, and hence diagonalize over them; whether the difficulty in deciding some seemingly "hard" problems arises from a similar cause (the possibility of diagonalizing over "easy" problems) is one of the central points under research.

The interest of relativizing complexity classes is that diagonalization techniques and simulation techniques can be used for separating and equalizing, respectively, complexity classes whose exact relationship in the unrelativized case is unknown. The first important results in this direction were obtained by Baker, Gill, and Solovay [BGS], and his paper is the basis of the results to be presented here. Ways of separating in different manners relativized complexity classes will be studied, and we will try to relate these results to the general framework of the research in relativizations of complexity classes.

We hope with this research to obtain an idea of the reasons that make these relativizations possible, in order to classify as "relativizing" or "unrelativizing" the different techniques widely used in complexity theory.

We assume the reader familiar with the general concepts of complexity theory. Our model of algorithm is the multitape Turing machine, possibly nondeterministic or alternating. Machines are allowed to query oracles, obtaining in one step an 1-bit answer about the membership of a word to a fixed oracle set. Machines will usually be bounded in time and/or space in order to define complexity classes. The central classes under consideration in this work are, among others, the relativizations to oracles A of polynomially time-bounded deterministic machines, P(A), and nondeterministic machines, NP(A); of polynomially space-bounded machines, PSPACE(A); and of the polynomial hierarchy, defined by polynomially bounded alternating machines with a fixed constant number of alternations. The complements of sets in NP(A) form the class coNP(A), and the intersection of both is biNP(A). Relativizations of the probabilistic classes of Gill [G], PP(A), BPP(A), R(A), coR(A), and ZPP(A) will also be considered. The concepts of polynomially time-bounded reducibility and completeness with respect to this reducibility are assumed to be known.

1. Separating.

The original paper [BGS] includes two fundamental separations: P from NP and NP from coNP. Other more complex cases are studied as well. The construction of an oracle A for which P(A) $\neq$ NP(A) is based in the following idea: at stage n we add information to the oracle about whether the n-th polynomial time-bounded deterministic machine accepts or not a word of the form 0^m for a convenient m. This information is "hidden" in the fact that there is in A some word of length m; the word is chosen so that the deterministic machine does not query it. A nondeterministic machine can reach this information by guessing the word and querying it to the oracle, thus knowing whether the deterministic machine accepts 0^m and behaving contrarily on this word. As the deterministic machine can query only polynomially many strings of length m but there are exponentially many of them, for a large enough m there is a word to add to the oracle.

A similar idea separates NP from coNP. A word is added to the oracle if the n-th nondeterministic polynomial time-bounded oracle machine accepts 0^m for the convenient m, preserving the accepting computation, which depends on only polynomially many words. In this case a nondeterministic machine can behave on 0^m exactly as the n-th one does, and hence its complement can not coincide with any of the sets accepted by the NP machines.

(Observe by the way that these relativizations also separate P from PSPACE and NP from PSPACE, respectively.)

This bound leads to an interesting study: sometimes it is possible to decompose a relativized proper inclusion $C(A) \subset D(A)$ between two relativized classes C and D into two inclusions $C(A) \subseteq D'(A) \subset D(A)$, where D' is a restriction of D defined by bounding the acces to the oracle in the nondeterministic machines, in such a way that a similar construction allows the separation between D' and D but a separation between C and D implies a separation of the unrelativized classes C and D. This is the idea behind the so-called "positive relativizations". (See for examples [B], [BW], [BLS], and the references there.) As an example of these, define the class NPO#Q(A) as the class of the sets accepted by nondeterministic polynomially time-bounded machines that only query to the oracle A polynomially many strings in its whole set of computations. Then there is an oracle that separates NPO#Q from NP (even a strong separation as defined later), but it can be shown that $P \neq NP$ (without relativization) iff there exists an oracle separating P from NPO#Q. (See [BLS].)

Also classes higher in the polynomial hierarchy can be considered. Heller [H] shows different oracles for which Σ_2 equals Π_2. NP differs from coNP, and Δ_2 equals Σ_2 in one of them but is different from it in the other. A variant of his construction will be indicated later. Baker and Selman [BS] prove the existence of an oracle separating Σ_2 from Π_2. This is the highest separation achieved. The counting arguments necessary are very elaborated, and show a heavy combinatorial contents. It seems very difficult to find similar ideas in order to separate higher classes of the polynomial hierarchy.

2. Strongly separating.

Recently, some investigators are studying some structural properties of the complexity classes. In particular, complexity classes as lattices under inclusion (modulo finite variations) is one of the subjects of research. Immunity and simplicity have been considered by Homer and Maass [HM], and "slow diagonalizations" are being used. We will survey this kind of diagonalization.

We say that an infinite set L is immune with respect to a class of sets C if no infinite subset of L belongs to the class C. A coinfinite set L is simple with respect to class C if $L \in C$ and the complement of L is immune with respect to C. We will say simply "C-immune" and "C-simple", and omit C if it is clear from the context.

In [BGS] separation is achieved by a a set in one class which is not in the other. Sometimes it is possible to separate two classes by a set in one of them which is immune with respect to the other. This kind of separation has been called "strong separation". A variation of the diagonalizations of [BGS] proves the following:

Theorem. [SB] There is a recursive oracle A such that a P(A)-immune set exists in NP(A).

Proof (sketch). We will construct A such that the set

$$L(A) = \{ 0^n \; / \text{ there exists a word of length } n \text{ in } A \}$$

becomes P(A)-immune. To do this, we do not diagonalize over one machine at each stage: we keep a list of "not yet spoiled" machines whose accepted sets could be included in L(A), and at each stage we try to "spoil" one of these machines. We try to find a word of a fixed form, 0^m, accepted by some machine in the list, and leave it out of L(A) by adding no word to A. If we are able to do it, we delete the machine from the list. We ensure that if a machine accepts under oracle A an infinite set then it is spoiled at some stage. On the other hand, if at some stage it is not possible to diagonalize over a machine then we add a not queried word to A, so that A (and hence L(A)) becomes infinite. The length m is chosen so that there is such a word.

□

Observe that the key hypothesis are again the same as in the relativizations of [BGS]: the possibility of adding to the oracle a word not queried by the deterministic machines. In the reference [SB] the reader can find a generalization of this theorem to classes specified by nondeterministic machines under very weak hypothesis that describe, essentially, this limitation on the number of queries allowed. This generalization applies, for example to the class NPO#Q defined above, proving that this class can be separated from NP, in the corresponding relativization, by a immune set.

(Different proof techniques, namely priority methods, are used in [HM] in order to obtain similar results.)

Following similar guidelines, we can prove the following:

<u>Theorem.</u> There is a recursive oracle B such that an NP(B)-simple set exists.

<u>Proof</u> (sketch). The construction is a combination of the separation of NP from coNP in [BGS] with the slow diagonalization of [SB], and the central ideas should be credited to Uwe Schöning. We keep a similar list of "not yet spoiled" machines, and we try to spoil some of these at each stage. We do it by looking for an accepted word of the form 0^m and, if found, we add to A a word of the same length not queried in the accepting computation. Hence the set L(A) defined just as before intersects every set in NP(A), and its complement can not contain none of these. Infinity of the complement of L(A) is achieved in the stages where no diagonalization is possible.

□

(A nonrecursive recursively enumerable oracle set with similar properties is constructed in [HM].)

Again the crucial point is the bound on the number of queries, although in this case we only need to bound the number of queries in some accepting computation. A generalization much like the above indicated holds, and many relativized complexity classes can be shown to possess simple sets. (See [Bal] if interested.) Moreover, both diagonalizations being similar, it is possible to merge them alternating stages creating a P(A)-immune set with stages creating a NP(A)-simple set; the resulting oracle has both types of sets in NP(A).

As we will see later, this slow diagonalization can be combined with the proofs of the other theorems in [BGS].

Variants of this diagonalization can be used for dealing with probabilistic classes. The following property of some words is needed:

<u>Definition.</u> Given a probabilistic "R" (it also applies to "BPP") machine M working under oracle A, w is a critical word for x if M is an "R" machine no matter whether w is added to or deleted from A, but M accepts or rejects x depending on whether w is added to or deleted from A.

The interest of this definition is that it can be shown that the number of critical words is polynomial in the length of x, and that only critical words are worth to be controlled when diagonalizing: a noncritical word either leaves the acceptance or rejection of x unchanged, or else breaks the probabilistic character of the machine under consideration.

So similar constructions allow to diagonalize slowly over the probabilistic classes R and BPP, constructing oracles that strongly separate them from P and NP and from each other. All these constructions can be found in [BR]. Different uses can be found for the definition of critical, and we will see later another one.

Still another approach to these separations relies in the observation that large sets of potential queries can be generated only by nondeterministic steps; generating k potential queries needs at least log k such steps. Bounding them implies also a bound on the number of queries. In [K] it is shown that oracles can be constructed separating simultaneously infinitely many levels in a hierarchy between P and NP obtained by bounding in several ways the number of nondeterministic steps. A very general separating theorem was stablished in [XDB] including the results in [K] as particular cases, and it was shown in [SB] that using slow diagonalizations all the separations can be shown strong. The diagonalization constructing simple sets also works for these cases, and it can be shown that under the appropriate relativization all the classes specified by these different bounds on the nondeterminism have simple sets.

A remark concerning the possibility of lifting this kind of results to the second level of the polynomial hierarchy. The proof of [H] can be modified in order to admit a slow diagonalization strongly separating Δ_2 from Σ_2. An idea similar in its philosophy to the technique of "forcing" allows to diagonalize over deterministic machines using as oracle an NP-complete set relative to the constructed oracle, thus diagonalizing over Δ_2. The proof can be found in the author's dissertation. We do not know how to apply these ideas to Σ_2, and it is an open problem the existence of Σ_2-simple sets in some relativization. New techniques are currently being investigated in order to attack this problem.

3. Collapsing to NP.

The basic idea to be developed in this section is presented in [BGS] for constructing an oracle such that NP = coNP but P ≠ NP. It goes this way: simultaneously, a diagonalization separating P from NP is performed using only odd length words, and a coding process "hids" in the oracle information about the complement of the complement of an NP-complete set using words of even length. The information is coded in such a way that a NP machine can recover it, and so the coNP-complete set coded is in NP; this implies NP = coNP under this oracle.

We want to study two ways of extending these ideas. First, we would like to know whether a slow diagonalization could be substituted for the usual one, so that the separation between P and NP is strong. Secondly, we are interested in characterizing the sets that could be coded in this way, trying to find the fundamental property of the coNP-complete set used by [BGS] that allows the coding to be performed.

We propose the following result as a solution to these problems. We state it as a lemma. We denote by $|w|$ the length of the word w.

Collapsing Lemma. Let M be a (possibly nondeterministic) oracle machine that always halts, and such that on input x every word w queried in any computation verifies $|w| < q(|x|)$, for some fixed polynomial q and for every oracle set. Then there is a recursive oracle C such that the set accepted by M under C is in NP(C) and NP(C) has a P(C)-immune set.

Proof (sketch). We slowly diagonalize as in the previous section using only words of odd length; with this diagonalization we achieve a P(A)-immune set in NP(A). Simultaneosly we code the set accepted by M under the currently constructed oracle as follows:

$$\forall u,\ u \in L(M,A) \iff \exists v\ |v| = q(|u|)\ uv \in A.$$

Clearly a nondeterministic machine can guess v and check the second part of the equivalence in polynomial time. If an initial part of the oracle is already constructed until length $q(|u|)^{i}$, then we can decide without error whether M accepts u with oracle A, and if this is the case then we add a word of the form uv (longer than $q(|u|)$) to A. Notice that if the polynomial $q(n)$ is of the form n^{i} (we can assume it without loss of generality) then the length of uv is even.

Each stage consists of two parts. After chosing a large enough odd length for the next diagonalization part, we scan all the words u whose coding words uv are of intermediate lengths between the last place where we diagonalized and the new one just chosen. For each such u we decide whether M accepts it and if necessary we add an appropriate uv not queried in previous diagonalization parts (the lengths can be taken large enough to ensure this, because the number of "dangerous" words can be polynomially bounded, and we have exponentially many v's available).

Once this coding is done, the second part of the stage consists of a diagonalization very much like the ones in section 2. This part ensures he immunity of a set similar to the L(A) defined before (the only change is that only odd length words are to be considered).

□

As corollaries of this lemma we can obtain:

Corollary. [HM] There exists a recursive C such that NP(C) = coNP(C) but NP(C) contains a P(C)-immune set.

Proof. The usual universal set for coNP can be accepted without querying words longer than the input. Hence the Collapsing Lemma applies.

□

Corollary. There exists a recursive D such that NP(D) = PSPACE(D) but NP(D) contains a P(D)-immune set.

□

Corollary. There exists a recursive E such that NP(E) = PP(E) but NP(E) contains a P(E)-immune set.

□

Two remarks before closing this section. First, it seems to be possible to prove very similar lemmas for classes other than NP; for example, for the probabilistic classes of Gill. See [R]. Oracles relative to which similar properties hold for these classes could be constructed in this way. And, secondly, it is worth noting that a partially similar coding technique has been used by Stuart Kurtz [Ku] for collapsing to P the sparse sets in NP relative to an oracle (without collapsing, of course, all of NP). The coding we will use for collapsing particular kinds of sets to P is somehow different, and is the subject of the next section.

4. Collapsing to P.

Among the theorems of [BGS], there is a very complicated construction collapsing biNP to P but keeping P different from NP. This is the basis of this section. Our questions are, as in the last section, whether it is possible to get a strong separation and whether other classes admit this collapse to P. For example, Rackoff [R] shows that such a result holds for R with a similar construction. We will try to isolate the common part of these constructions. Warning: we consider that this line of research is far from being completed.

Let A be a PSPACE-complete set having only words of odd length. Let $e(n)$ denote the number resulting from the evaluation of a "stack" of $2n+1$ exponentiated 2's; i.e., we obtain $e(0) = 2$, $e(1) = 2^{2^2} = 16$, and so on. Assume that a set D has been defined consisting only of words of lengths $e(n)$, and at most one word per length; let E be the union of A and D. We will see later how to construct such a set D.

Assume that a definition has been given of "machine M behaves well on input x under oracle A"; we say then that machine M is well-behaved under A if it is on every input. (This assumption will become clearer later.) Further, assume that the definition is such that for a properly constructed oracle E the following "lemma" holds; we state it as a "pseudo-lemma" whose proof is to be provided together with the definition of "well-behaved".

Pseudo-lemma. If M is a well behaved machine under oracle E, then there is a function f_M computable in polynomial time with oracle E such that for every x, $f_M(x)$ is a machine M′ running under the same time bounds as M and satisfying:

$$x \in L(M,E) \iff x \in L(M',A).$$

We prove that in this case every machine well-behaved under E accepts a set belonging to P(E). Essentially, this pseudo-lemma implies that the differences between oracles E and A relevant to M′s computation are only a few and small enough to be computed in polynomial time and incorporated to the finite control of machine M, thus constructing M′. Once under A, we are able to eliminate the nondeterminism. A more formal proof follows.

Lemma. If the definition of "to behave well" and the oracle E satisfy the pseudo-lemma above, then for every polynomial time-bounded nondeterministic oracle machine M well-behaved under E we have L(M,E) ∈ P(E).

Proof. First, observe that P(A) = PSPACE(A), because A is PSPACE-complete; secondly, observe that A ∈ P(E) because they coincide on the odd length words and A has no even length words. Hence there are NP(A)-complete sets in P(E) that allow to decide with oracle E whether x ∈ L(M′,A). By the pseudo-lemma, such a M′ can be constructed in polynomial time with an oracle for E. Now, in order to decide whether x ∈ L(M,E), compute M′ and then decide whether x ∈ L(M′,A). Both computations can be made in polynomial time with oracle E.

□

We now sketch how to construct the oracle E such that the pseudo-lemma holds for particular classes of machines and P(E) is strongly separated from NP(E). At stage n we try to satisfy two kinds of goals: (a) diagonalizing over a P machine which accepts a word $0^{e(n)}$ by letting this word out of L(E) (which is defined as usual), and (b) proving that with the oracle as currently constructed some machine is not well behaved on some input word. In both cases no word is added to the oracle. If none of these goals can be satisfied at the current stage (it can be shown that this happens infinitely often), then a nonqueried word of length e(n) is added to E so that L(E) becomes infinite.

We ensure with the construction the P(E)-immunity of L(E) with a slow diagonalization identical to the one of section 2. The construction also ensures that a machine which is well-behaved under the final oracle E is also well-behaved on all but finitely many inputs under each partially constructed part of E. This property is used below in proving the corresponding instantiation of the pseudo-lemma. The construction is recursive in the predicate "M behaves well on x".

In this way it is enough to define "to behave well" and to prove the pseudo-lemma for showing that there is an oracle for which P-immune sets exist in NP but for which the class defined by this property is collapsed to P. Let us apply it to biNP and R. An analogous result for BPP can be found in [BR].

Define for the first one "to behave well on x" for a nondeterministic machine as follows: either some computation accepts and no computation rejects, or some computation rejects and no computation accepts. (We assume a computation can have three outcomes: accept, reject, or question-mark.) It is well known that these machines (sometimes called "strong") define the class biNP. For these machines the pseudo-lemma holds.

<u>Lemma.</u> If M is a well behaved machine under oracle E, then there is a function f_M computable in polynomial time with oracle E such that for every x, $f_M(x)$ is a machine M′ running under the same time bounds as M and satisfying:

$$x \in L(M,E) \iff x \in L(M',A).$$

<u>Proof</u> (sketch). Given a machine M, the hypothesis on A allow to compute in polynomial time from a word $x \in L(M,A)$ the sequence of words queried by M on x with oracle A and the final state of some accepting computation.

We can use this for computing the needed function f_M as follows: First, compute the larger n such that M can query a word of length e(n). (Here a finite number of particular cases must be solved by table-lookup.) Then we can compute all the words in E of lengths e(k) for k < n by "brute force", querying E in a systematic way for all of them; observe that these words are small: no longer than $\log(|x|)$ (because their length is at most e(n-1), which equals log log e(n)). We know now all the differences between E and A relevant to the computation on x but at most one: the one of length e(n).

Now compute the function above indicated for the machine that with oracle A simulates M on x under E using the computed differences, and finally accepts if and only if M does not output question-mark. This gives the words that are queried by M in some accepting or rejecting computation. If any of these is of length e(n), query it to E. Either one of them is in E, and then we know the remaining difference between A and E, or none of them is in E, in which case the simulated computation is faithful and the final state of the simulation tells us whether M accepts x under E or not. (Recall that M is strong under E.) In both cases we can output a machine that accepts x with A if and only if M does so with E.

□

Define now "to behave well on x" as follows: either no computation accepts, or at least half the computations accept. The well-behaved machines define the class R relativized to the considered oracle. Also for these machines the pseudo-lemnma holds. Recall that the notion of "critical" was defined in the section 2.

Lemma. If M is a well behaved machine under oracle E, then there is a function f_M computable in polynomial time with oracle E such that for every x, $f_M(x)$ is a machine M′ running under the same time bounds as M and satisfying:

$$x \in L(M,E) \iff x \in L(M',A).$$

Proof (sketch). The idea is very similar to the one for biNP. The "magic" auxiliar function is the one that gives the critical strings for M under A. As PSPACE(A) = P(A), it is computable in polinomial time with oracle A, and hence with oracle E.

Now the machine is constructed as follows. First compute in the same way as before all the differences between A and E but the one of length e(n) (the longer the machine can query on x). Then consider the machine M′ that with oracle A simulates M under E on x using the computed differences, and compute the critical words for x and for this new machine M′. Query E for the ones of length e(n). If one is found to be in E, then all the relevant differences are known. If not, the remaining difference (the possible word of length e(n)) is not critical, and hence the machine M′ accepts x under A if and only if M accepts x under E. In both cases we can output a machine satisfying the equivalence above.

□

We consider that a work yet to be done is to characterize the classes to which this kind of construction can be applied.

5. A look to the diagonalizing set.

We strongly believe that the deep study of known techniques could help us in understanding the "true" intuitive reasons of the difficulties we find in our main open problems. Generalizing known results as far as possible may not yield splendorous results, but sheds new light into some of the fundamental aspects of our actual knowledge.

In this sense, there is at least one of the particularities of the complexity theoretic diagonalizations that we have not analyzed yet: the set we use to diagonalize out of a class. We do not know the exact kind of sets that can be used in the diagonalizations, nor the suitable ones for slow diagonalizations. Are they the same?

An interesting result in this direction is used, without explicit mention, by Baker and Selman [BS]. In the separation of Σ_2 from Π_2, they only prove that "there is room to diagonalize". The implicit result was established by Angluin [A]. We present it here in a more general way.

Lemma. Let $\{ M_i \mid i \in N \}$ be a "reasonable" class of oracle machines. Let M be an oracle machine such that, for every word w, the fact of whether w is or not in the oracle is only relevant for the acceptance by M of a finite number of input words. Assume that $\forall i \exists A\ L(M_i,A) \neq L(M,A)$. Then $\exists A \forall i\ L(M_i,A) \neq L(M,A)$ (and hence L(M,A) does not belong to the class specified by the machines M_i).

The hypothesis we impose on M could be called "finite dependence of the oracle". The hypothesis of "reasonability" on the M_i is, besides halting on every input, closure under two kinds of finite "patching", or coding in the control of the machine of a finite amount of information. Namely, we ask for (a) closure under finite variations: if a set L differs only in finitely many words from some $L(M_i,A)$, then L is some $L(M_j,A)$; and (b) closure under finite oracle variations: the class specified by the M_i is the same if the oracle changes in finitely many words.

If this is the case, then for each finite initial part of the oracle under construction we can find an extension B that makes L(M,B) different from $L(M_i,B)$. (The hypothesis of finite dependence of the oracle is needed here.) Pick a word in which they differ, and fix into A or out of A the words that are relevant to the computations of M and M_i on this word (only finitely many, obviously). The final result of the construction makes L(M,A) different from every $L(M_i,A)$.

We ask now the following question: is it possible to state and prove a similar result yielding immune sets? We do not know the answer... Not yet.

6. Open problems and new ideas.

We have presented already two open problems to investigate, related to the presented results: the existence of simple sets in the second level of PH and the modification of Angluin's results to yield immune sets. We believe that the solutions of these problems can be closely related.

Other directions of research we would like to mention are: first, variants of the positive relativizations; second, analysis of the oracles collapsing NP to P.

In the first one, "low sets" have been defined. It has been shown that P = NP if and only if for every tally set T, P(T) = NP(T); and that $\Sigma_2 = \Sigma_3$ if and only if for every sparse set S, $\Sigma_2(S) = \Sigma_3(S)$. (See [BBS] and the references there.) These results differ from the presented positive relativizations in that the restriction is no longer over the relativized classes but over the kind of oracles allowed. Comparisons between them have been studied by Long [L]. More information on this subject can be found in [BBS] and [BBLSS].

For the second one, in which very little is known, the starting position is: well, positive relativizations intuitively explain why it is possible to construct A such that $P(A) \neq NP(A)$; the key hypothesis is the polynomial bound on the number of queries. This indicates where a proof of P = NP should fail to relativize. But, why is it possible to construct B such that P(B) = NP(B)? Does this happen only for PSPACE-complete sets? How to characterize this class of oracles? Where a proof of $P \neq NP$ should fail to relativize?

An idea proposed by Ken Regan may be useful: define the following class:

$$H := \bigcap_{P(A)=NP(A)} P(A)$$

It can be easily seen that $PH \subset H$ and $H \subset PSPACE$. H is a candidate to a class not Σ_3 arithmetic. If it is not in Σ_3, then it differs from PH and PSPACE, which implies a lot of wonderful results (like PH is infinite, and hence $P \neq NP \neq coNP$, and NP-complete sets do not have polynomial size circuits [KL], and...). The relation of this class with the above indicated question is that H = PSPACE if and only if every oracle in PSPACE relative to which P = NP must be PSPACE-complete. A whole family of classes between PH and PSPACE can be defined taking the intersection over oracles relative to which PH extends i levels, $i \in N$, and we do not know whether these classes are the same. On the other hand, Heller [H] presents two different ways of defining w-jumps of PH, and only one of these yields a PSPACE-complete set; could the other have any relation with H?

More information on the class H can be found in the recent work of Regan [Re].

References:

[A] D. Angluin: On counting problems and the polynomial-time hierarchy, Theor. Comp. Sci. 12 (80) 161-173.

[BGS] T. Baker, J. Gill, R. Solovay: Relativizations of the P =? NP question, SIAM J. Comp. 4 (75), 431-442.

[BS] T. Baker, A. Selman: A second step toward the polynomial hierarchy, Theor. Comp. Sci. 8 (80) 177-187.

[Bal] J. L. Balcázar: Simplicity, relativizations, and nondeterminism, SIAM J. Comp. (84), to appear.

[BBLSS] J. Balcázar, R. Book, T. Long, U. Schöning, A. Selman: Sparse oracles may as well be empty, unpublished manuscript (83).

[BBS] J. Balcázar, R. Book, U. Schöning: Sparse oracles, lowness, and highness, this conference.

[BR] J. Balcázar, D. Russo: Immunity and simplicity in relativizations of probabilistic complexity classes, submitted for publication.

[B] R. Book: Bounded query machines: on NP and PSPACE, Theor. Comp. Sci. 15 (81), 27-39.

[BLS] R. Book, T. Long, A. Selman: Quantitative relativizations of complexity classes, submitted for publication.

[BW] R. Book, C. Wrathall: Bounded query machines: on NP( ) and NPQUERY( ), Theor. Comp. Sci. 15 (81), 41-50.

[G] J. Gill: Computational complexity of probabilistic Turing machines, SIAM J. Comp. 6 (77), 675-695.

[H] H. Heller: Relativized polynomial hierarchies extending two levels, Dr. R. Nat. dissertation, Inst. für Informatik, München (82).

[HM] S. Homer, W. Maass: Oracle dependent properties of the lattice of NP sets, Theor. Comp. Sci 24 (83), 279-289.

[K] C.M.R. Kintala: Computations with a restricted number of nondeterministic steps, Ph.D. dissertation, Penn. St. Univ. (77)

[KL] R. Karp, R. Lipton: Some connections between nonuniform and uniform complexity classes, 12th STOC-ACM (80), 302-309.

[Ku] S. Kurtz: On sparse sets in NP - P: relativizations, submitted for publication.

[L] T. Long: On restricting the size of oracles compared with restricting access to oracles, submitted for publication.

[R] C. Rackoff: Relativized questions involving probabilistic algorithms, J. ACM 29 (82), 261-268.

[Re] K. Regan: On diagonalization methods and the structure of language classes, FCT 83.

[SB] U. Schöning, R. Book: Immunity, relativizations, and nondeterminism, SIAM J. Comp. (84), to appear.

[XDB] Xu Mei-rui, J. Doner, R. Book: Refining nondeterminism in relativized complexity classes, J. ACM 30 (83), 677-685.

COMPLEXITY OF QUANTIFIER ELIMINATION IN THE THEORY OF ALGEBRAICALLY CLOSED FIELDS

A.L.Chistov, D.Yu.Grigor'ev

Leningrad Scientific Research Computer Centre
of the Academy of Sciences of the USSR,
Mendeleevskaya 1, Leningrad, 199164, USSR

Leningrad Department of V.A.Steklov Mathematical
Institute of the Academy of Sciences of the USSR,
Fontanka 27, Leningrad, 191011, USSR

Abstract

An algorithm is described producing for each formula of the first order theory of algebraically closed fields an equivalent free of quantifiers one. Denote by N a number of polynomials occuring in the formula, by d an upper bound on the degrees of polynomials, by n a number of variables, by a a number of quantifier alternations (in the prefix form). Then the algorithm works within the polynomial in the formula's size and in $(Nd)^{n^{(2a+2)}}$ time. Up to now a bound $(Nd)^{n^{O(n)}}$ was known ([5] , [7] , [15]).

1. Fast algorithms for factoring multivariable polynomials and for solving systems of algebraic equations

Lately the considerable progress in the polynomial factoring problem was achieved. Lenstra A.K., Lenstra H.W., Lovasz L. [12] have designed an ingenious polynomial-time algorithm for factoring onevariable polynomials over $\mathbb{Q}$. Independently Kaltofen E. [8] , [9] has constructed a reduction of multivariable factoring over $\mathbb{Q}$ to onevariable factoring, running within the polynomial-time provided that the number of variables is fixed. The authors [1] , [4] , have suggested a polynomial-time algorithm for factoring multivariable polynomials over $\mathbb{Q}$ and over finite fields. Later another polynomial-time algorithm for the case of finite fields was exhibited in [13] spreading the method [12] .

Also an essential progress has taken place in another important problem of the commutative computer algebra, namely in the problem of solving systems of algebraic equations. Earlier a complexity bound of the order d^{2^n} was known for it, e.g. from [5] , [7] , [15] . Lazard D. [11] has designed an algorithm for solving homogeneous systems of algebraic equations in the case when the variety of roots in the projective space of the system is null-dimensional, i.e. finite, working within the time $d^{O(n)}$ if the coefficients of the input system are taken from a finite field (certainly, provided that we are supplied with a polynomial-time algorithm for polynomial factoring). The authors [2] , [3] , [4] involving the polynomial-time algorithm for polynomial factoring [1] , [4] and the method from [11] have constructed an algorithm for solving an arbitrary system of algebraic equations, running within a polynomial in the size L_2 of the input data (system) and in d^{n^2} time. Moreover, the algorithm finds all the irreducible compounds $W_\alpha \subset \mathbb{P}^n(\bar{F})$ of the variety of roots of the homogeneous system within the polynomial time in d^{nc} and in L_2 where $c = 1 + \max_\alpha \dim W_\alpha$ (the general case is reducible here to homogeneous one). Finding W_α allows to answer the principle questions, e.g. emptiness, dimension of the variety of roots.

Now we turn ourselves to the exact formulations of the mentioned results. Let a ground field $F = H(T_1, \dots, T_\ell)[\eta]$ where either $H = \mathbb{Q}$ or $H = \mathbb{F}_{q^\varkappa}$, $q = char(H)$, the elements $T_1, \dots, T_\ell$ be algebraically independent over H ; the element η is separable and algebraic over a field $H(T_1, \dots, T_\ell)$, denote by $\varphi = \sum_{0 \le i < deg_Z(\varphi)} (\varphi_i^{(1)}/\varphi^{(2)}) \cdot Z^i \in H(T_1, \dots, T_\ell)[Z]$ its minimal polynomial over $H(T_1, \dots, T_\ell)$ with the leading coefficient $lc_Z(\varphi) = 1$, herewith $\varphi_i^{(1)}, \varphi^{(2)} \in H[T_1, \dots, T_\ell]$ and the degree $deg(\varphi^{(2)})$ is the least possible. Any polynomial $f \in F[X_0, \dots, X_n]$ can be uniquely represented in a form $f = \sum_{0 \le i < deg_Z \varphi;\, i_0, \dots, i_n} (a_{i, i_0, \dots, i_n}/b) \eta^i X_0^{i_0} \dots X_n^{i_n}$ where $a_{i, i_0, \dots, i_n}, b \in H[T_1, \dots, T_\ell]$, the degree $deg(b)$ is the least possible; the polynomials $a_{i, i_0, \dots, i_n}, b$ are determined uniquely up to a factor from H^* . Set $deg_{T_j} f = \max_{i, i_0, \dots, i_n} \{deg_{T_j}(a_{i, i_0, \dots, i_n}), deg_{T_j}(b)\}$. By a length of description $l(h)$ in the case $h \in \mathbb{Q}$ we mean its bitwise length, and in the case $h \in \mathbb{F}_{q^\varkappa}$ we mean $\varkappa \log_2(q)$. By $l(f)$ denote the maximum of the lengths of descriptions of the coefficients from H in the monomials in $T_1, \dots, T_\ell$ of the polynomials $a_{i, i_0, \dots, i_n}, b$.

Let $deg_{X_j}(f) < r$, $deg_{T_j}(f) < r_2$, $deg_{T_j}(\varphi) < r_1$, $deg_Z(\varphi) < r_1$, $l(f) \le M_2$, $l(\varphi) \le M_1$. As a size $L_1(f)$ of the polynomial f we consider in the theorem I a value $r^{n+1} r_2^{\ell} r_1 M_2$ and analogously $l(\varphi) = r_1^{\ell+1} M_1$.

THEOREM I. ([1] , [4]). One can factor the polynomial f over F within the polynomial in $L_1(f)$, $L_1(\varphi)$, q time.

Remark that it is possible within the same time to obtain also the absolute factorization of f i.e. the factors irreducible over the algebraic closure $\bar{F}$ of the field F ([2] , [4]).

Proceed to the problem of solving systems of algebraic equations. Let an input system of algebraic equations $f_0 = \ldots = f_K = 0$ be given (we can assume w.l.o.g. that $f_0, \ldots, f_K$ are linearly independent). As a matter of fact we suggest an algorithm which decomposes an arbitrary projective variety on the irreducible compounds, so one can suppose w.l.o.g. that $f_0, \ldots, f_K \in F[X_0, \ldots, X_n]$ are homogeneous relatively to $X_0, \ldots, X_n$ polynomials. Let $deg_{T_1,\ldots,T_\ell,Z}(\varphi) < d_1$, $\ell(f_i) \leq M_2$, $deg_{X_0,\ldots,X_n}(f_i) < d$, $deg_{T_1,\ldots,T_\ell}(f_i) < d_2$ for all $0 \leq i \leq K$ and in the theorem 2 a size $L_2(f_i) = \binom{d+n}{n} d_1 d_2^\ell M_2$ and $L_2(\varphi) = d_1^{\ell+1} M_1$. Denote $L = L_2(f_0) + \ldots + L_2(f_K)$.

The projective variety $\{f_0 = \ldots = f_K = 0\} \subset \mathbb{P}^n(\bar{F})$ of roots of the system $f_0 = \ldots = f_K = 0$ is decomposable on the compounds $\{f_0 = \ldots = f_K = 0\} = \bigcup_\alpha W_\alpha$, herewith each compound W_α is defined and irreducible over the maximal purely inseparable extension $F^{q^{-\infty}}$ of F . Moreover $W_\alpha = \bigcup_\beta W_{\alpha\beta}$ where the (absolutely irreducible) compounds $W_{\alpha\beta}$ are defined and irreducible over $\bar{F}$. Denote $c = 1 + \max_\alpha \dim W_\alpha$. The algorithm designed in [2], [3], [4] finds all W_α and thereupon $W_{\alpha\beta}$ (actually, W_α, $W_{\alpha\beta}$ are defined over some finite extensions of the field F which are also constructed by the algorithm). We (and the algorithm) represent every compound W_α or $W_{\alpha\beta}$ in two following manners: by its general point [16] and on the other hand by a certain system of algebraic equations such that the compound under consideration coincides with a variety of the roots of this system, in the similar case we say that the system determines the variety.

For functions $g_1, g_2, h_1, \ldots, h_s$ a relation $g_1 \preccurlyeq g_2 \mathcal{P}(h_1, \ldots, h_s)$ denotes further that $g_1 \leq g_2 P(h_1, \ldots, h_s)$ for an appropriate polynomial P .

Let $W \subset \mathbb{P}^n(\bar{F})$ be a closed projective variety, $codim_{\mathbb{P}^n}(W) = m$, defined and irreducible over some field F_1 being a finite extension of F , denote by F_2 the maximal subfield of F_1 which is a separable extension of F . Let $t_1, \ldots, t_{n-m}$ be algebraically independent over F . A g e n e r a l p o i n t of the variety W can be given by the following fields isomorphism

$$F(t_1,\ldots,t_{n-m})[\theta] \cong F_2(X_{j_1}/X_{j_0}, \ldots, X_{j_{n-m}}/X_{j_0}, (X_0/X_{j_0})^{q^\nu}, \ldots, (X_n/X_{j_0})^{q^\nu}) \subset F_1(W) \quad (1)$$

for suitable q^{ν} (here and further $\nu \geq 0$ when $q > 0$ and we set $q^{\nu} = 1$ when $char(F) = 0$), index $0 \leq j_0 \leq n$ and an element θ is algebraic separable over a field $F_2(t_1, \ldots, t_{n-m})$; denote by $\Phi(Z)$ its minimal polynomial such that $lc_Z(\Phi) = 1$. The elements X_j / X_{j_0} are considered herein as the rational functions on the variety W , herewith W is not situated in a hyperplane $\{X_{j_0} = 0\}$, under the isomorphism (1) $t_i \to X_{j_i}/X_{j_0}$, $1 \leq i \leq n-m$. The algorithms further represent the isomorphism (1) by the images of rational functions $(X_j / X_{j_0})^{q^{\nu}}$ in the field $F_2(t_1, \ldots, t_{n-m})[\theta]$. Sometimes, when there is no misunderstanding, we identify a rational function with its image.

THEOREM 2. ([2] , [3] , [4]). a) An algorithm is suggested which for every compound W_{α} produces its general point and constructs a certain family of homogeneous polynomials $\psi_1^{(\alpha)}, \ldots, \psi_N^{(\alpha)} \in F[X_0, \ldots, X_n]$ such that a system $\psi_1^{(\alpha)} = \ldots = \psi_N^{(\alpha)} = 0$ determines the variety W_{α} . Denote $m = codim\, W_{\alpha}$, $\theta_{\alpha} = \theta$, $\Phi_{\alpha} = \Phi$. Then $q^{\nu} \leq d^{2m}$, $deg_Z(\Phi_{\alpha}) \leq deg\, W_{\alpha} \leq d^m$, for all i, j the degrees $deg_{T_1,\ldots,T_{\ell},t_1,\ldots,t_{n-m}}(\Phi_{\alpha})$, $deg_{T_1,\ldots,T_{\ell},t_1,\ldots,t_{n-m}}(X_j/X_{j_0})^{q^{\nu}}$ (the latter two degrees are defined according to the isomorphism (1) analogously to how $deg_{T_i}(f)$ was defined above) are less than $d_2 \mathcal{P}(d^m, d_1)$, apart that $l(\Phi_{\alpha}), l((X_j/X_{j_0})^{q^{\nu}}) \leq (M_1 + M_2 + (n+\ell)\log d_2)\mathcal{P}(d^m, d_1)$. A number of equations $N \leq m^2 d^{4m}$, the degrees $deg_{X_0,\ldots,X_n}(\psi_s^{(\alpha)}) \leq d^{2m}$ and the degrees $deg_{T_1,\ldots,T_{\ell}}(\psi_s^{(\alpha)}) \leq d_2 \mathcal{P}(d^m, d_1)$; besides that the algorithm represents each $\psi_s^{(\alpha)}$ in a form $\psi_s^{(\alpha)} = \bar{\psi}_s^{(\alpha)}(Z_{s,0}, \ldots, Z_{s,n-m+2})$ for suitable linear forms $Z_{s,j}$ in the variables $X_0, \ldots, X_n$ with the coefficients from H and the polynomials $\bar{\psi}_s^{(\alpha)} \in F[Z_{s,0}, \ldots, Z_{s,n-m+2}]$, thereto $l(\bar{\psi}_s^{(\alpha)}) \leq (M_1 + M_2 + (n+\ell)\log d_2)\mathcal{P}(d^m, d_1)$, lastly the size $L_2(Z_{s,j}) \leq \mathcal{P}(n, \log d d_1 d_2)$ for all s, j . The total running time of the algorithm can be bounded from above by $\mathcal{P}(M_1, M_2, (d^n d_1 d_2)^{c+\ell})$ Obviously, the latter value is less than $\mathcal{P}(L^{c+\ell}(q+1)) \leq \mathcal{P}(L^{\log L}(q+1))$ if $n = \mathcal{O}(d)$.

b) An algorithm is suggested which for every absolutely irreducible compound $W_{\alpha\beta}$ finds the maximal separable subfield $F_2 = F[\xi_{\alpha\beta}]$ of the minimal field of definition F_1 (containing F) of the variety $W_{\alpha\beta}$. The algorithm produces a general point of $W_{\alpha\beta}$ and some system of equations with the coefficients from the field F_2 determining the variety $W_{\alpha\beta}$. For the parameters of the general point and the system of equations hold the same bounds as in the item a) of the theorem. Denote by $\varphi_{\alpha\beta} \in F[Z]$ the minimal polynomial for $\xi_{\alpha\beta}$ such that $lc_Z(\varphi_{\alpha\beta}) = 1$, then $deg_Z(\varphi_{\alpha\beta}) \leq deg\, W_{\alpha\beta}$ and the degrees $deg_{T_1,\ldots,T_{\ell}}(\varphi_{\alpha\beta}) \leq d_2 \mathcal{P}(d^m, d_1)$, lastly $l(\varphi_{\alpha\beta}) \leq (M_1 + M_2 + (n+\ell)\log d_2)\mathcal{P}(d^m, d_1)$. The time bound is the same as in the item a).

REMARK. If we are supplied with a general point (with the same bounds on its parameters as in the theorem 2) of a closed irreducible variety $V_1=\overline{\pi(W_\alpha)}$ where $\pi(X_0:\ldots:X_n)=(X_0:\ldots:X_m)$ is a lenear projection $\pi:\mathbb{P}^n\to\mathbb{P}^m$ and W_α is some compound of the variety $\{f_0=\ldots=f_K=0\}\subset\mathbb{P}^n(F)$, then we can produce a system of equations determining V_1 with the same bounds on the parameters as for the family $\psi_s^{(\alpha)}$ in the theorem 2 within the same time bound.

In conclusion of the section 1. The authors make a conjecture that one can find the compounds within time $\mathcal{P}(d^{(c'+\ell+1)n},(d_1d_2)^{n+\ell},L)$ where $c'=\max_\alpha \min\{\dim W_\alpha+1,\operatorname{codim} W_\alpha\}$.

2. Projecting a constructive set

Let an input formula $\exists X_1\ldots\exists X_s\,(\&_{1\leq j\leq K}(f_j=0)\,\&\,(g\neq 0))$ be given, herein the parameters of the polynomials $f_j,g\in F[Z_1,\ldots,Z_{n-s},X_1,\ldots,X_s]$ satisfy the same bounds as of f_j in the section 1. The goal in the present section is to produce an equivalent quantifier-free formula $\vee_{1\leq i\leq N}(\&_{1\leq j\leq \varkappa_i}(f_{ij}^{(1)}=0)\,\&\,(g_i^{(1)}\neq 0))$ where $f_{ij}^{(1)},g_i^{(1)}\in F[Z_1,\ldots,Z_{n-s}]$.

The input formula is equivalent to $\exists X_0\exists X_1\ldots\exists X_s\exists X_{s+1}((X_0\neq 0)\,\&\,\&_{1\leq j\leq K}(\bar f_j=0)\,\&\,(\bar f_0=X_{s+1}\bar g-X_0^{1+\deg g}=0))$, therein X_0, X_{s+1} are new variables and $\bar f_j=X_0^{\deg_{X_1\ldots X_s}(f_j)}f_j(Z_1,\ldots,Z_{n-s},X_1/X_0,\ldots,X_s/X_0)$, $\bar g=X_0^{\deg_{X_1\ldots X_s}(g)}g(Z_1,\ldots,Z_{n-s},X_1/X_0,\ldots,X_s/X_0)$ (cf. [7]). The desired projection, i.e. the constructive set consisting of all the points $(z_1,\ldots,z_{n-s})\in\mathbb{A}^{n-s}(\bar F)$ satisfying the latter formula, we denote by Π. One can assume further w.l.o.g. that $\deg_{X_0,\ldots,X_{s+1}}\bar f_j=d-1$, $0\leq j\leq K$, replacing $\bar f_j$ by the family of polynomials $\{\bar f_j X_i^{d-1-\deg \bar f_j}\}_{0\leq i\leq s+1}$.

Introduce a variety $U=\{(z_1,\ldots,z_{n-s};(x_0:\ldots:x_{s+1}))\in(\mathbb{A}^{n-s}\times\mathbb{P}^{s+1})(\bar F):\&_{0\leq j\leq K}(\bar f_j=0)\}$ and a natural linear projection $\pi:\mathbb{A}^{n-s}\times\mathbb{P}^{s+1}\to\mathbb{A}^{n-s}$, then the desired $\Pi=\pi(U\cap\{X_0\neq 0\})$. For each point $z=(z_1,\ldots,z_{n-s})\in\mathbb{A}^{n-s}(\bar F)$ consider the variety (the layer) $U_z=\pi^{-1}(z)\cap U\subset\{z\}\times\mathbb{P}^{s+1}\simeq\mathbb{P}^{s+1}$. The condition $z\in\Pi$ is true iff for an appropriate $0\leq m\leq s+1$ the layer U_z has at least one compound W with the dimension $s+1-m$ such that $W\not\subset\{X_0=0\}$.

Fix a point z in the following speculations for some time. It is not difficult (see e.g. §2 [2]) to indicate a family of $N'=Kd^m+1$ vectors $u^{(1)},\ldots,u^{(N')}\in H^{K+1}$ any $K+1$ from which are linearly independent (we suppose here and below that H contains sufficiently many element, extending it if necessary). Denote $h_i=\sum_{0\leq j\leq K}u_j^{(i)}\bar f_j$, herewith $u^{(i)}=(u_0^{(i)},\ldots,u_K^{(i)})$. The relevant compound W of U_z exists iff there are such indices $1\leq i_1\leq\ldots<i_m\leq N'$

that W is a compound of the variety $\{h_{i_1}(z)=\ldots=h_{i_m}(z)=0\}\subset\mathbb{P}^{s+1}$, herein the coordinates of the point z are substituted instead of $Z_1,\ldots,Z_{n-s}$, i.e. $h_{i_j}(z)\in\bar{F}[X_0,\ldots,X_{s+1}]$ (cf. §4a [2]).

One can construct (see §2 [2]) a family $\mathfrak{M}=\mathfrak{M}_{s,s-m,d^m}$ consisting of $(s-m+1)$-tuples of linear forms in variables $X_1,\ldots,X_{s+1}$ with the coefficients from H such that for every variety $W_1\subset\mathbb{P}^s$ satisfying the inequalities $\dim W_1\leqslant s-m$, $\deg W_1\leqslant d^m$ there is $(s-m+1)$-tuple $(Y_1,\ldots,Y_{s-m+1})\in\mathfrak{M}$ for which $W_1\cap\{Y_1=\ldots=Y_{s-m+1}=0\}=\emptyset$. Thereto $\operatorname{card}(\mathfrak{M})\leqslant\binom{(s+1)d^m+1}{s-m}$. Let us take a variety $W\cap\{X_0=0\}$ as W_1. Supplement linear forms $Y_0=X_0, Y_1,\ldots,Y_{s-m+1}$ up to a basis $Y_0,\ldots,Y_{s+1}$ with the coefficients from H of the space of linear forms in $X_0,\ldots,X_{s+1}$ (in arbitrary manner). Replacing variables denote $\hat{h}_i(z,Y_0,\ldots,Y_{s+1})=h_i(z)$ and $\tilde{h}_i(z)=\hat{h}_i(z,Y_0,0,\ldots,0,Y_{s-m+2},\ldots,Y_{s+1})$. Thus, the condition under consideration about the existence of W is equivalent to that there are indices $1\leqslant i_1<\ldots<i_m\leqslant N'$ and linear forms $Y_1,\ldots,Y_{s-m+1}$ for which the variety $\{\tilde{h}_{i_1}(z)=\ldots=\tilde{h}_{i_m}(z)=0\}\subset\mathbb{P}^m$ as one of its compounds has a certain point $\tilde{\Omega}=(\xi_0:\xi_{s-m+2}:\ldots:\xi_{s+1})$ such that the point $\Omega=(z,(\xi_0:0:\ldots:0:\xi_{s-m+2}:\ldots:\xi_{s+1}))\in U_z\cap\{Y_0\neq 0\}$ (in force of the theorem about the dimension of intersection [14]).

Introduce a system of homogeneous algebraic equations

$$\tilde{h}_{i_j}(z)-YY_{s-m+j+1}^{d-1}=0;\quad 1\leqslant j\leqslant m \tag{2}$$

in the variables $Y_0,Y_{s-m+2},\ldots,Y_{s+1}$ with the coefficients from $\bar{F}[Y]\subset\bar{F}(Y)=K$ where Y is algebraically independent over F. One can prove (see also lemma 11 §5 [3]) that the set of roots in $\mathbb{P}^m(\bar{K})$ of the system (2) is finite. The variety of roots is decomposable on the irreducible and defined over K nulldimensional compounds $V_{\mathfrak{p}_K}$ corresponding to the minimal prime ideals $\mathfrak{p}_K\subset K[Y_0,Y_{s-m+2},\ldots,Y_{s+1}]/(\{\tilde{h}_{i_j}(z)-YY_{s-m+j+1}\}_{1\leqslant j\leqslant m})$. The system (2) can be considered apart that as the system in the variables $Y,Y_0,Y_{s-m+2},\ldots,Y_{s+1}$ with the coefficients from F which determines a variety $\tilde{U}_z^{(F)}\subset\mathbb{A}^{m+2}(\bar{F})$. It is not difficult to show (cf. lemma 12 §5 [3]) that there is a bijective correspondence between the points $V_{\mathfrak{p}_K}$ and on the other side such compounds $V_{\mathfrak{p}_F}$ of the variety $U_z^{(F)}$ that $V_{\mathfrak{p}_F}$ is not contained in any union of finite number of hyperplanes of the kind $\{Y-c_1=0\}\subset\mathbb{A}^{m+2}$ for $c_1\in\bar{F}$, notice that $\dim V_{\mathfrak{p}_F}=2$.

Now we exhibit an important auxiliary device from [11] (see also §3 [2]). Let $g_0,\ldots,g_{k-1}\in F[X_0,\ldots,X_n]$ be homogeneous polynomials of degrees $\delta_0\geqslant\ldots\geqslant\delta_{k-1}$ respectively. Introduce new variables

$U_0,\dots,U_n$ algebraically independent over $F(X_0,\dots,X_n)$. Set $g_K = X_0U_0+\dots+X_nU_n \in F(U_0,\dots,U_n)[X_0,\dots,X_n]$ and $D=\sum_{0\le i\le n}\delta_i - n$, herein $\delta_j = 1$ if $K \le j \le n$. Consider linear over $F(U_0,\dots,U_n)$ mapping $\mathcal{O}\!\mathcal{L}:\mathcal{B}_0\oplus\dots\oplus\mathcal{B}_K\to\mathcal{B}$ where $\mathcal{B}_i$ (correspondingly $\mathcal{B}$) is the space of homogeneous polynomials in $X_0,\dots,X_n$ over the field $F(U_0,\dots,U_n)$ of degree $D-\delta_i$ (correspondingly D) for $0\le i\le K$, namely $\mathcal{O}\!\mathcal{L}(b_0,\dots,b_K)=\sum_{0\le i\le K} b_i g_i$. Any element $b=(b_0,\dots,b_K)\in \mathcal{B}_0\oplus\dots\oplus\mathcal{B}_K$ can be written in the form $b=(b_{0,1},\dots,b_{0,s_0},b_{1,1},\dots,b_{1,s_1},\dots,b_{K,1},\dots,b_{K,s_K})$ where $s_i=\binom{n+D-\delta_i}{n}$ and $b_{i,1},\dots,b_{i,s_i}$ are the coefficients of the polynomial b_i provided that a certain numeration of all the monomials of the degree $D-\delta_i$ is fixed. Analogously one can write the elements of the space $\mathcal{B}$. In the chosen system of coordinates the mapping $\mathcal{O}\!\mathcal{L}$ has a matrice A of the size $\binom{n+D}{n}\times\left(\sum_{0\le i\le K} s_i\right)$. One can represent $A=(A',A'')$ where A' (call it the number part of A) contains $\sum_{0\le i\le K-1} s_i$ columns and A'' (call it the formal part) contains s_K columns, besides that the entries of A' belong to F, the entries of A'' are linear forms over F in variables $U_0,\dots,U_n$ (cf. [6]). There is proved in [10] that the system $g_0=\dots=g_{K-1}=0$ has no roots in $\mathbb{P}^n(\bar F)$ iff the ideal $(g_0,\dots,g_{K-1})\supset(X_0,\dots,X_n)^D$. Besides that, the following proposition is ascertained in [11].

PROPOSITION. ([11]). 1) The system $g_0=\dots=g_{K-1}=0$ has a finite number of roots in $\mathbb{P}^n(\bar F)$ iff the rank $rg\,A=\binom{n+D}{n}=r$;

2) all $r\times r$ minors of A generate a principal ideal whose generator $R\in F[U_0,\dots,U_n]$ is their g.c.d.;

3) the homogeneous form $R=\prod_{1\le i\le D_1} L_i$ where $L_i=\sum_{0\le j\le n}\xi_j^{(i)}U_j$ is a linear form over $\bar F$, moreover $(\xi_0^{(i)}:\dots:\xi_n^{(i)})$ is a root of the system and the number of occuring of the forms proportional to L_i for each i in the product equals to the multiplicity of the corresponding root. Apart that $\deg R = D_1 = r - rg(A')$.

The algorithm designes the matrix A with the entries from the ring $F[Y,Z_1,\dots,Z_{n-s},U_0,U_{s-m+2},\dots,U_{s+1}]$ corresponding to the modified system (2) in which $Z_1,\dots,Z_{n-s}$ are considered as variables (instead of $z_1,\dots,z_{n-s}$) according to the just exhibited device. Denote by A_z the matrix obtained from A by means of substituting the coordinates of the point z instead of $Z_1,\dots,Z_{n-s}$. Let the polynomial $R_z\in\bar F[Y,U_0,U_{s-m+2},\dots,U_{s+1}]$ correspond to the matrix A_z as in the proposition. One can suppose w.l.o.g. that $Y\nmid R_z$ (dividing R_z on the greatest possible power of the variable Y).

Regard a certain representation of the union $\cup_{\mathfrak{p}_F} V_{\mathfrak{p}_F} = \{S_0 = \ldots = S_{K'-1} = 0\}$ for suitable polynomials $S_i \in \bar{F}[Y, Y_0, Y_{s-m+2}, \ldots, Y_{s+1}]$ homogeneous relatively to $Y_0, Y_{s-m+2}, \ldots, Y_{s+1}$. Considering a system $S_i(0, Y_0, Y_{s-m+2}, \ldots, Y_{s+1}) = 0$; $0 \leq i \leq K'-1$ and basing on the proposition (see also lemma 16 §5 [3]), one proves that $R_z(0, U_0, U_{s-m+2}, \ldots, U_{s+1}) = \prod_i L_i^{c_i}$ and moreover the linear forms $L_i = \sum_j \xi_j^{(i)} U_j$ correspond bijectively to the points $(\xi_0^{(i)} : \xi_{s-m+2}^{(i)} : \ldots : \xi_{s+1}^{(i)}) \in W_z' \subset \mathbb{P}^m$ where the cone $con(W_z') = (\cup_{\mathfrak{p}_F} V_{\mathfrak{p}_F}) \cap \{Y = 0\}$. Thereupon it is not difficult to check that $\Omega \in W_z'$ (cf. lemma 13 §5 [3]). Summarizing and utilizing the notations introduced above, we have ascertained the following.

LEMMA 1. The formula $\exists X_1 \ldots \exists X_s (\&_{1 \leq j \leq K} (f_j = 0) \& (g \neq 0))$ is valid in a point $z \in \bar{F}^{n-s}$ iff for appropriate $0 \leq m \leq s+1$ there exist such indices $1 \leq i_1 < \ldots < i_m \leq N'$, a set of linear forms $(Y_1, \ldots, Y_{s-m+1}) \in \mathfrak{M}$ and a point $\Omega = (z, (\xi_0 : 0 : \ldots : 0 : \xi_{s-m+2} : \ldots : \xi_{s+1})) \in U_z \cap \{X_0 \neq 0\}$ (in the coordinates $Y_0, Y_1, \ldots, Y_{s+1}$) that the linear form $(\xi_0 U_0 + \xi_{s-m+2} U_{s-m+2} + \ldots + \xi_{s+1} U_{s+1}) | R_z(0, U_0, U_{s-m+2}, \ldots, U_{s+1})$.

Now make more precise the definition of a version of Gaussian algorithm (v.G.a) for reducing the matrices to the generalized trapezium form (cf. [7]).V.G.a. is determined by a succession of pairs of indices (pivots) $(i_0, j_0), (i_1, j_1), \ldots, (i_p, j_p)$. Herewith $i_\alpha \neq i_\beta$ and $j_\alpha \neq j_\beta$ if $\alpha \neq \beta$. For any initial matrix $A^{(0)}$ v.G.a. yields the chain of matrices $A^{(0)}, A^{(1)}, \ldots, A^{(p+1)}$. Introduce a notation $A^{(\alpha)} = (a_{ij}^{(\alpha)})$. Apart that $a_{i_\alpha j_\alpha}^{(\alpha)} \neq 0$ and $a_{ij}^{(\alpha+1)} = a_{ij}^{(\alpha)} + a_{i j_\alpha}^{(\alpha)} a_{i_\alpha j}^{(\alpha)} / a_{i_\alpha j_\alpha}^{(\alpha)}$ for all i distinguished from $i_0, \ldots, i_\alpha$, lastly $a_{i_\beta j}^{(\alpha+1)} = a_{i_\beta j}^{(\alpha)}$ where $0 \leq \beta \leq \alpha$. The matrix $A^{(p+1)}$ is in the generalized trapezium form, namely, $a_{ij}^{(p+1)} = 0$ when either i differs from $i_0, \ldots, i_p$ or $i = i_\alpha, j = j_\beta$ and $\alpha > \beta$, besides that $a_{i_\alpha j_\alpha}^{(p+1)} = a_{i_\alpha j_\alpha}^{(\alpha)} \neq 0$.

Denote by $\Delta_{ij}^{(\alpha)}$ the determinant of $(\alpha+1) \times (\alpha+1)$ matrix formed by the rows with the indices $i_0, \ldots, i_{\alpha-1}, i$ and the columns with the indices $j_0, \ldots, j_{\alpha-1}, j$ provided that $i \neq i_0, \ldots, i \neq i_{\alpha-1}$ and $j \neq j_0, \ldots, j \neq j_{\alpha-1}$. Then $a_{ij}^{(\alpha)} = \Delta_{ij}^{(\alpha)} / \Delta_{i_{\alpha-1} j_{\alpha-1}}^{(\alpha-1)}$ (see e.g. lemma 7 [7]).

Now we turn ourselves to considering an arbitrary point $z \in A^{n-s}$. Fix for some time $0 \leq m \leq s+1$ indices $1 \leq i_1 < \ldots < i_m \leq N'$ and a set of linear forms $(Y_1, \ldots, Y_{s-m+1}) \in \mathfrak{M}$ (see lemma 1). By r denote the number of rows of the matrix A. Produce a certain succession of v.G.a.s $\Gamma_1, \Gamma_2, \ldots$ over a field $F(Y, Z_1, \ldots, Z_{n-s}, U_0, U_{s-m+2}, \ldots, U_{s+1})$ and a succession of polynomials $P_1, P_2, \ldots \in F[Y, Z_1, \ldots, Z_{n-s}, U_0, U_{s-m+2}, \ldots, U_{s+1}]$ thereto v.G.a. Γ_i can be applied correctly to the matrix A_z for all points $z=(z_1,\dots,z_{n-s})$ of (possibly empty) quasiprojective variety ([14]) $W_i \subset \mathbb{A}^{n-s}$ which is defined by the following conditions: inequality $0 \neq P_i(Y, z_1, \dots, z_{n-s}, U_0, U_{s-m+2}, \dots, U_{s+1}) \in \bar{F}[Y, U_0, U_{s-m+2}, \dots, U_{s+1}]$ and equalities $0 = P_j(Y, z_1, \dots, z_{n-s}, U_0, U_{s-m+2}, \dots, U_{s+1})$ for $1 \leq j \leq i-1$ are fulfilled. Apart that the variety $\{(z_1,\dots,z_{n-s}) : P_i(Y, z_1, \dots, z_{n-s}, U_0, U_{s-m+2}, \dots, U_{s+1}) = 0$ for all $i\} = \emptyset$, henceforth $\cup_i W_i = \mathbb{A}^{n-s}$. Exposed below construction is close to the proof of the lemma 9 [7].

Later on we apply the v.G.a.s $\Gamma_1, \Gamma_2, \dots$ to the initial matrix A. As Γ_1 one can take an arbitrary v.G.a. Set a polynomial $P_1 = \prod_{0 \leq \alpha \leq \rho_1} \Delta^{(\alpha)}_{i_\alpha j_\alpha}$ (for v.G.a. regarded at the current step the same notations as above are utilized). Assume that $\Gamma_1, \dots, \Gamma_i$; $P_1, \dots, P_i$ are already produced. Then as Γ_{i+1} we take v.G.a. in which for every $0 \leq \alpha \leq \rho_{i+1}$ the column index j_α of the pivot in the matrix $A^{(\alpha)}$ is the least possible, moreover $j_\alpha > j_{\alpha-1}$ and the polynomials $P_1, \dots, P_i, \prod_{0 \leq \beta \leq \alpha} \Delta^{(\beta)}_{i_\beta j_\beta}$ are linearly independent over F. Finally, put $P_{i+1} = \prod_{0 \leq \alpha \leq \rho_{i+1}} \Delta^{(\alpha)}_{i_\alpha j_\alpha}$. The algorithm stops producing v.G.a.s $\Gamma_1, \Gamma_2, \dots$ when it is impossible to produce Γ_{i+1} satisfying formulated above requirements (if $\rho_{i+1} < \tau - 1$ then $W_{i+1} = \emptyset$).

One can ascertain that if $W_i \neq \emptyset$ then for each $z \in W_i$ the polynomial R_z (see proposition) is obtained as the value in the point z of the polynomial $\det \Delta_i$ (up to a factor Y^ε for a suitable ε), where $\tau \times \tau$ submatrix Δ_i of the matrix A is generated by the columns with the indices $j_0, \dots, j_{\tau-1}$ corresponding to v.G.a. Γ_i. This follows from the fact that in the matrix $(A^{(\alpha)})_z$ an entry $a^{(\alpha)}_{\beta j} = 0$ when $\beta \neq i_0, \dots, i_{\alpha-1}$ and $j < j_\alpha$ in force of the choice of j_α. Therefore, if for an appropriate α a cell $(i_{\alpha-1}, j_{\alpha-1})$ belongs to the number part A' of A and a cell (i_α, j_α) belongs to the formal part A'' of A then $\mathrm{rg}((A')_z) = \alpha$ that implies the mentioned representation of R_z.

Write $\det \Delta_i = \sum_\varepsilon \Delta^{(\varepsilon)}_i Y^\varepsilon$, herewith $\Delta^{(\varepsilon)}_i(Z_1,\dots,Z_{n-s}) \in F[Z_1,\dots,Z_{n-s}, U_0, U_{s-m+2},\dots,U_{s+1}]$. Introduce varieties $W^{(\varepsilon)}_i = \{(z_1,\dots,z_{n-s}) \in W_i : \Delta^{(0)}_i(z_1, \dots, z_{n-s}) = \dots = \Delta^{(\varepsilon-1)}_i(z_1,\dots,z_{n-s}) = 0; \Delta^{(\varepsilon)}_i(z_1,\dots,z_{n-s}) \neq 0\}$ for $\varepsilon \geq 0$. The variety $W^{(\varepsilon)}_i$ is quasiprojective as the intersection of two quasiprojective varieties, namely, if $\Xi^{(j)} = \{\&_\beta (G^{(j)}_\beta = 0) \& \vee_\gamma (C^{(j)}_\gamma \neq 0)\}$; $j = 1, 2$ then $\Xi_1 \cap \Xi_2 = \{\&_{\beta(1), \beta(2)} (G^{(1)}_{\beta(1)} = 0) \& G^{(2)}_{\beta(2)} = 0) \& \vee_{\gamma(1), \gamma(2)} (C^{(1)}_{\gamma(1)} C^{(2)}_{\gamma(2)} \neq 0)\}$. Moreover $W^{(\varepsilon_1)}_i \cap W^{(\varepsilon_2)}_i = \emptyset$ for $\varepsilon_1 \neq \varepsilon_2$ and $\cup_\varepsilon W^{(\varepsilon)}_i = W_i$.

Thereupon represent $\Delta^{(\varepsilon)}_i = \sum_{0 \leq j \leq D_2} e^{(\varepsilon, j)}_i U_0^{D_2 - j}$ where $e^{(\varepsilon,j)}_i (Z_1,\dots,Z_{n-s}) \in F[Z_1,\dots,Z_{n-s}, U_{s-m+2}, \dots, U_{s+1}]$. Consider quasiprojective varieties $W_i^{(\varepsilon,j)}=\{(z_1,\dots,z_{n-s})\in W_i^{(\varepsilon)}: e_i^{(\varepsilon,\varkappa)}(z_1,\dots,z_{n-s})=0,\ 0\leqslant\varkappa<j;\ e_i^{(\varepsilon,j)}(z_1,\dots,z_{n-s})\neq 0\}$, then $W_i^{(\varepsilon,j_1)}\cap W_i^{(\varepsilon,j_2)}=\emptyset$ when $j_1\neq j_2$ and $\bigcup_{0\leqslant j\leqslant D_2} W_i^{(\varepsilon,j)}=W_i^{(\varepsilon)}$. Observe that the proposition and the ascertained earlier entail that $(\Delta_i^{(\varepsilon)})_z=\Delta_i^{(\varepsilon)}(z_1,\dots,z_{n-s},U_0,U_{s-m+2},\dots,U_{s+1})=\prod_{\varkappa} L_{\varkappa}^{c_\varkappa}$ is a product of linear forms for $z\in W_i^{(\varepsilon)}$. This implies that for $z\in W_i^{(\varepsilon,j)}$ the polynomial $(e_i^{(\varepsilon,j)})_z$ equals to the product of powers $L_{\varkappa}^{c_\varkappa}$ of all linear forms $L_\varkappa$ in which the coefficient at U_0 vanishes. Henceforth $(e_i^{(\varepsilon,j)})_z \mid (\Delta_i^{(\varepsilon)})_z$ in the ring $\bar{F}[U_0,U_{s-m+2},\dots,U_{s+1}]$.

Our nearest purpose is to calculate the quotient $(\Delta_i^{(\varepsilon)})_z/(e_i^{(\varepsilon,j)})_z$ for $z\in W_i^{(\varepsilon,j)}$. If $I=(I_{s-m+2},\dots,I_{s+1})$ is a multiindex then denote $U^I=U_{s-m+2}^{I_{s-m+2}}\dots U_{s+1}^{I_{s+1}}$, apart that by $I<J$ denote the lexicographical order on multiindices. Write $e_i^{(\varepsilon,j)}=\sum_I \gamma_I U^I$ and let $0\neq\gamma_I\in F[Z_1,\dots,Z_{n-s}]$ for a certain I (fixed in further speculations). Introduce a quasiprojective variety $W_{i,I}^{(\varepsilon,j)}=\{(z_1,\dots,z_{n-s})\in W_i^{(\varepsilon,j)}:$ $\gamma_J(z_1,\dots,z_{n-s})=0$ when $J>I$ and $\gamma_I(z_1,\dots,z_{n-s})\neq 0\}$. Evidently $W_{i,I_1}^{(\varepsilon,j)}\cap W_{i,J_1}^{(\varepsilon,j)}=\emptyset$ if $I_1\neq J_1$ and $\bigcup_{I_1} W_{i,I_1}^{(\varepsilon,j)}=W_i^{(\varepsilon,j)}$. For any point $(z_1,\dots,z_{n-s})\in W_{i,I}^{(\varepsilon,j)}$ the quotient $(\Delta_i^{(\varepsilon)})_z/(e_i^{(\varepsilon,j)})_z$ can be obtained by means of the described below process of dividing polynomial on polynomial and after that substituting the coordinates $z_1,\dots,z_{n-s}$ instead of variables $Z_1,\dots,Z_{n-s}$.

Let $0\neq\Psi\in F(Z_1,\dots,Z_{n-s})[U_{s-m+2},\dots,U_{s+1}]$. Denote by $lex(\Psi)\neq 0$ the monomial of Ψ in variables $U_{s-m+2},\dots,U_{s+1}$ for which in $\Psi-lex(\Psi)$ occur only the monomials less than $lex(\Psi)$, set $\overline{\overline{\Psi}}=\Psi(U_{s-m+2}^m,U_{s-m+3}^{m-1},\dots,U_{s+1})$ and $\sigma(\Psi)=deg(\overline{\overline{\Psi}})$. Delete from $e_i^{(\varepsilon,j)}$ all the monomials $\gamma_J U^J$ (except $\gamma_I U^I$) with $\sigma(U^J)\geqslant\sigma(U^I)$ and denote obtained polynomial by $\widetilde{\widetilde{e}}_i^{(\varepsilon,j)}$. Then $(e_i^{(\varepsilon,j)})_z=(\widetilde{\widetilde{e}}_i^{(\varepsilon,j)})_z$, when $z\in W_{i,I}^{(\varepsilon,j)}$ since $(e_i^{(\varepsilon,j)})_z$ is the product of linear forms. For any index $j<\varkappa\leqslant D_2$ the algorithm designs a succession of nonzero polynomials $\Psi_0=e_i^{(\varepsilon,\varkappa)},\Psi_1,\dots,\Psi_\rho$. Represent uniquely $\Psi_t=\Psi_t^{(1)}+\Psi_t^{(2)}+\Psi_t^{(3)}$, herewith $\overline{\overline{\Psi_t^{(1)}}},\overline{\overline{\Psi_t^{(2)}}}$ are homogeneous, $\sigma(\Psi_t^{(3)})<\sigma(\Psi_t)=\sigma(\Psi_t^{(1)})=\sigma(\Psi_t^{(2)})$ and $\Psi_t^{(1)}/U^I\in F(Z_1,\dots,Z_{n-s})[U_{s-m+2},\dots,U_{s+1}]$, lastly each monomial from $\Psi_t^{(2)}$ is not divided by U^I. Then $\Psi_{t+1}=\gamma_I(\Psi_t-\Psi_t^{(2)})-\Psi_t^{(1)}\widetilde{\widetilde{e}}_i^{(\varepsilon,j)}/U^I$ for every $0\leqslant t\leqslant\rho-1$ (obviously, $\sigma(\Psi_{t+1})<\sigma(\Psi_t)$). Regard a polynomial $\Psi_{i,I}^{(\varepsilon,j,\varkappa)}=\sum_{0\leqslant t\leqslant\rho-1}\Psi_t^{(1)}\gamma_I^{\rho-t-1}/U^I\in F[Z_1,\dots,Z_{n-s},U_{s-m+2},\dots,U_{s+1}]$ and set $\Psi_{i,I}^{(\varepsilon,j)}=\gamma_I^\rho U_0^{D_2-j}+\sum_{j<\varkappa\leqslant D_2}\Psi_{i,I}^{(\varepsilon,j,\varkappa)}U_0^{D_2-\varkappa}$. One can check that $(e_i^{(\varepsilon,\varkappa)})_z/(\gamma_I^{-\rho}e_i^{(\varepsilon,j)})_z=(\Psi_{i,I}^{(\varepsilon,j,\varkappa)})_z$ for $z\in W_{i,I}^{(\varepsilon,j)}$ and therefore $(\Delta_i^{(\varepsilon)})_z/(e_i^{(\varepsilon,j)})_z=(\gamma_I^{-\rho}\Psi_{i,I}^{(\varepsilon,j)})_z$ equals to the product of $L_\mu^{c_\mu}$ for all linear forms L_μ in which the coefficient at the variable $\mathcal{U}_0$ does not vanish.

Thereupon remind that $\operatorname{con} W'_z = U_{\wp_F} V_{\wp_F} \cap \{Y=0\}$ and introduce $W' = \bigcup_{z\in W_{i,I}^{(\varepsilon,j)}} (\{z\}\times(W'_z \cap \{Y_0 \neq 0\}))$ (as above we fix $i,\varepsilon,j,$ I). Observe that $W' = \{(z_1,\dots,z_{n-s},(y_0:y_{s-m+2}:\dots:y_{s+1}))\in W_{i,I}^{(\varepsilon,j)} \times \mathbb{A}^m(\bar{F}) \subset W_{i,I}^{(\varepsilon,j)} \times \mathbb{P}^m(\bar{F}): 0 = (\varphi_{i,I}^{(\varepsilon,j)}(-\sum_{s-m+2\leq\alpha\leq s+1} \mathcal{U}_\alpha y_\alpha, y_0\mathcal{U}_{s-m+2},\dots,y_0\mathcal{U}_{s+1}))_z \in \bar{F}[\mathcal{U}_{s-m+2},\dots,\mathcal{U}_{s+1}]\}$. Representing the polynomial $\psi_{i,I}^{(\varepsilon,j)}(-\sum_{s-m+2\leq\alpha\leq s+1} \mathcal{U}_\alpha Y_\alpha, Y_0\mathcal{U}_{s-m+2},\dots,Y_0\mathcal{U}_{s+1}) = \sum_J E_J \mathcal{U}^J$ leads to an equality $W' = \{\&_J(E_J=0)\}\cap(W_{i,I}^{(\varepsilon,j)}\times\mathbb{A}^m)$. Because of that the subset W' is closed in the quasiprojective variety $W_{i,I}^{(\varepsilon,j)}\times\mathbb{A}^m$.

Consider the natural linear projection $\pi_2: \mathbb{A}^{n-s}\times(\mathbb{P}^m\cap\{Y_0\neq 0\})\to\mathbb{A}^{n-s}$ defined by the formula $\pi_2(Z_1,\dots,Z_{n-s},(Y_0:Y_{s-m+2}:\dots:Y_{s+1})) = (Z_1,\dots,Z_{n-s})$. Let a morphism $\pi_1: W'\to W_{i,I}^{(\varepsilon,j)}$ be the restriction of π_2 on W'. Our nearest goal is to show that π_1 is finite ([14]). Obviously, the inverse image $\pi_1^{-1}(V)\subset W'$ of any open affine subset $V\subset W_{i,I}^{(\varepsilon,j)}$ is isomorphic to $(V\times\mathbb{A}^m)\cap W'$, henceforth $\pi_1^{-1}(V)$ is open in W' and besides that $\pi_1^{-1}(V)$ is affine since $\pi_1^{-1}(V)$ is closed in the open affine set $V\times\mathbb{A}^m$ ([14]). Now we check that every coordinate function $Y_\varkappa/Y_0$ on the variety $\pi_1^{-1}(V)$ satisfies a suitable relation of integral dependence over the ring $\bar{F}[V]$ where $s-m+2\leq\varkappa\leq s+1$. Let $\psi_{i,I}^{(\varepsilon,j)} = \psi_{i,I}^{(\varepsilon,j)}(\mathcal{U}_0,\mathcal{U}_{s-m+2},\dots,\mathcal{U}_{s+1})$. Then $\psi_{i,I}^{(\varepsilon,j)}(Y_\varkappa/Y_0,0,\dots,0,-1,0,\dots,0)=0$ on W', herein -1 is substituted instead of the variable $\mathcal{U}_\varkappa$. Taking into account that $(\gamma_I)_z\neq 0$ when $z\in W_{i,I}^{(\varepsilon,j)}$ this yields an equation of integral dependence. So, we infer that the morphism π_1 is finite.

Utilizing the notations from the lemma 1 one concludes that a set $V_{i,I}^{(\varepsilon,j)}$ consisting of all such points $z=(z_1,\dots,z_{n-s})\in W_{i,I}^{(\varepsilon,j)}$ that there exists a point $\Omega=(z,(\xi_0:0:\dots:0:\xi_{s-m+2}:\dots:\xi_{s+1}))\in U_z\cap\{X_0\neq 0\}$ is closed in $W_{i,I}^{(\varepsilon,j)}$ as $V_{i,I}^{(\varepsilon,j)}$ coincides with the image under projection π_1 of the closed in the domain of definition of π_1 (i.e. in W') set $\pi_1^{-1}(W_{i,I}^{(\varepsilon,j)})\cap\{\tilde{f}_0=\dots=\tilde{f}_K=0\}$ where $\tilde{f}_\varkappa(Y_0,Y_{s-m+2},\dots,Y_{s+1}) = \hat{f}_\varkappa(Y_0,0,\dots,0,Y_{s-m+2},\dots,Y_{s+1})$ and $\hat{f}_\varkappa(Y_0,Y_1,\dots,Y_{s+1}) = \bar{f}(Z_1,\dots,Z_{n-s},X_0,\dots,X_{s+1})$ for $0\leq\varkappa\leq K$ and since the image of the closed set under a finite morphism is again closed ([14]).

Now we describe a procedure for constructing the required $V_{i,I}^{(\varepsilon,j)}$. Let the quasiprojective variety $W_{i,I}^{(\varepsilon,j)} = \{\&_\beta(G_\beta=0)\&(V_\gamma(C_\gamma\neq 0))\}$, herewith the polynomials $G_\beta, C_\gamma\in F[Z_1,\dots,Z_{n-s}]$ were actually produced earlier. Denote the closure of the projection $\pi_2\{\&_\beta(G_\beta=0)\&\&_J(E_J=0)\&\&_{0\leq\varkappa\leq K}(\tilde{f}_\varkappa=0)\} = \tilde{V}_{i,I}^{(\varepsilon,j)}$. On the other hand in force of the aforesaid the equalities hold $V_{i,I}^{(\varepsilon,j)} = \tilde{V}_{i,I}^{(\varepsilon,j)}\setminus\{\&_\gamma(C_\gamma=0)\}$

$= \mathcal{V}_{i,I}^{(\varepsilon,j)} \setminus \{ \&_{\gamma} (C_{\gamma} = 0)\}$. Thus, it remains only to design the affine variety $\mathcal{V}_{i,I}^{(\varepsilon,j)}$.

Involving the theorem 2 (see section 1) the algorithm finds the general points of the compounds $\mathcal{V}$ of the variety $\{\&_{\beta}(G_{\beta}=0)\&\&_{J}(E_{J}=0)\&\&_{0\leq\varkappa\leq K}(\tilde{f}_{\varkappa}=0)\}$. It is sufficient for each $\mathcal{V}$ to construct the closure of its projection $\overline{\pi_2(\mathcal{V})}$. Notice that there is an imbedding of the fields of functions $F^{q^{-\infty}}(\overline{\pi_2(\mathcal{V})}) = F^{q^{-\infty}}(Z_1,\dots,Z_{n-s}) \in F^{q^{-\infty}}(Z_1,\dots,Z_{n-s}, Y_1/Y_0,\dots,Y_{s+1}/Y_0) = F^{q^{-\infty}}(\mathcal{V})$. Therefore, the algorithm can produce the general point of $\overline{\pi_2(\mathcal{V})}$ yielding firstly a trascendental basis and after that a primitive element (cf.(1), section 1). Searching a transcendental basis and also a primitive element is based on the procedure for calculating a polynomial relation over F (if it exists) between the elements $a_1,\dots,a_{p+1} \in F(t_1,\dots,t_{n-m_1})[\theta] \subset F^{q^{-\infty}}(\mathcal{V})$ provided that $a_1,\dots,a_p$ are algebraically independent over F, the procedure in its turn is reducible to solving a linear system whose indeterminates are the coefficients of the relation (cf. § 1 [2], §§ 4b, 6 [3]). Thereupon with the help of the remark just after the theorem 2 the algorithm computes a representation $\overline{\pi_2(\mathcal{V})} = \{\&_{\delta}(B_{\delta}=0)\}$ where the polynomials $B_{\delta} \in F[Z_1,\dots,Z_{n-s}]$.

We summarize the results of the present section in the following lemma, in which bounds are obtained making use of the theorem 2.

LEMMA 2. An algorithm is suggested which outputs the constructive set $\Pi = \pi(U \cap \{X_0 \neq 0\}) = \{(z_1,\dots,z_{n-s}) \in \mathbb{A}^{n-s}(\bar{F}) : \exists X_1 \dots \exists X_s (\&_{1\leq\varkappa\leq K}(f_{\varkappa}(z_1,\dots,z_{n-s}, X_1,\dots,X_s)=0) \& g(z_1,\dots,z_{n-s},X_1,\dots,X_s)\neq 0)\}$, i.e. the projection in the form $\{\bigvee_{0\leq m\leq s+1} \bigvee_{1\leq i_1<\dots<i_m\leq N'} \bigvee_{(Y_1,\dots,Y_{s-m+1})\in \mathfrak{M}} \bigvee_{i,\varepsilon,j,I} \bigvee_{\mathcal{V},\gamma} (\&_{1\leq\delta\leq N_1}(B_{\delta}=0)\&(C_{\gamma}\neq 0))\} = \{\bigvee_{\mu}(\&_{\delta}(B_{\delta}^{(\mu)}=0)\&(C^{(\mu)}\neq 0))\}$. Thereat $\deg_{Z_1,\dots,Z_{n-s}}(B_{\delta}^{(\mu)}) \leq d^{4(n+2)(2s+3)}$, $\deg_{T_1,\dots,T_\ell}(B_{\delta}^{(\mu)}) \leq d_2\mathcal{P}(d^{(s+1)n}, d_1)$, lengths of descriptions $\ell(B_{\delta}^{(\mu)}) \leq (M_1+M_2+(n+\ell)\log d_2)\times \mathcal{P}(d^{(s+1)n}, d_1)$. Apart that $\deg_{Z_1,\dots,Z_{n-s}}(C^{(\mu)}) \leq (3d)^{(2s+3)}$, $\deg_{T_1,\dots,T_\ell}(C^{(\mu)}) \leq d_2\mathcal{P}(d^{(s+1)}, d_1)$ and $\ell(C^{(\mu)}) \leq (M_1+M_2+(n+\ell)\log d_2)\mathcal{P}(d^{(s+1)}, d_1)$.
Besides that, $\delta \leq (s+1)^2(3d)^{4(2s+3)(n+2)}$, $\mu \leq d^{12(s+2)(n+s+3)}$. The running time of the algorithm can be estimated by $\mathcal{P}(M_1+M_2, d^{sn(n+\ell)}, (d_1+d_2)^{n+\ell}, q)$.

3. Subexponential-time deciding the first order theory of algebraically closed fields

Let a Boolean formula Q with N atoms of the kind $f_i = 0$ where $f_i \in F[X_1,\dots,X_n]$ satisfies the same bounds as in the section 1, be given, $L_2(Q)$ denotes the size of Q. Firstly we exhibit a procedure reducing Q to a disjunctive normal form.

Following [7] name $(g_1,\dots,g_\rho)$-cell for $g_1,\dots,g_\rho \in F[X_1,\dots,X_n]$ any nonempty quasiprojective variety of the kind $\{\&_{j\in \mathcal{J}_1}(g_j=0)\,\&\,\&_{j\in\mathcal{J}_2}(g_j\neq 0)\}\subset \mathbb{A}^n(\bar F)$, herewith $\mathcal{J}_1\cup\mathcal{J}_2=\{1,\dots,\rho\}$, $\mathcal{J}_1\cap\mathcal{J}_2=\emptyset$. By means of the Bezout inequality [14] it is ascertained in [7] that a number of all $(g_1,\dots,g_\rho)$-cells is less or equal to $(1+\deg g_1+\dots+\deg g_\rho)^n$. We shall describe the method for decomposing the space $\mathbb{A}^n$ on $(g_1,\dots,g_\rho)$-cells by recursion on ρ. Assume that we are supplied with all $(g_1,\dots,g_{\rho-1})$-cells $(\rho\geq 1)$. Every $(g_1,\dots,g_\rho)$-cell is of the form either $K\cap\{g_\rho=0\}$ or $K\cap\{g_\rho\neq 0\}$ for a pertinent $(g_1,\dots,g_{\rho-1})$-cell K. Henceforth it is sufficient to pick out (involving the theorem 2 from the section 1) all nonempty sets among quasiprojective varieties of the forms $K\cap\{g_\rho=0\}$ and $K\cap\{g_\rho\neq 0\}$.

Applying the just described method the algorithm yields all $(\{f_i\}_{1\leq i\leq N})$-cells. Again repeatedly making use of the theorem 2 by induction on the number of logical signs in Q the algorithm for each $(\{f_i\}_{1\leq i\leq N})$-cell checks, whether this cell is contained in the constructive set $\Pi_Q=\{Q\}\subset\mathbb{A}^n$ determined by the formula Q, and thereby represents Π_Q as a union of $(\{f_i\}_{1\leq i\leq N})$-cells $K^{(\mu)}$ that means reducing Q to a disjunctive normal form $\vee_\mu(\&_{\delta\geq 1}(f_\delta^{(\mu)}=0)\&(f_0^{(\mu)}\neq 0)))$. Moreover $1\leq\mu\leq(1+Nd)^n$, $1\leq\delta\leq N$, any polynomial $f_\delta^{(\mu)}=f_i$ for a relevant i and $f_0^{(\mu)}=\prod_{j\in\mathcal{J}} f_j$ for an appropriate $\mathcal{J}\subset\{1,\dots,N\}$. The working time of the exhibited procedure can be estimated according to the theorem 2 by $\mathcal{P}(L_2(Q), N^n, (d^n d_1 d_2)^{n+\ell}, q)$.

Finally we pass to the general case. Let an input formula of the first order theory

$$\exists Z_{1,1}\dots\exists Z_{1,s_1}\forall Z_{2,1}\dots\forall Z_{2,s_2}\dots\exists Z_{a,1}\dots\exists Z_{a,s_a} Q \tag{3}$$

be given where the formula Q is of the kind as at the beginning of the section, $f_i\in F[Z_1,\dots,Z_{s_0},Z_{1,1},\dots,Z_{a,s_a}]$, herein $Z_1,\dots,Z_{s_0}$ occur free, $n=s_0+s_1+\dots+s_a$, by L_2 denote the size of (3). Applying to (3) alternatively the just exhibited procedure for reducing to a disjunctive normal form and the lemma 2 (section 2) the algorithm arrives after performing $\varkappa$ steps at an equivalent to (3) formula

$$\exists Z_{1,1}\dots\exists Z_{1,s_1}\neg\dots\exists Z_{a-\varkappa,1}\dots\exists Z_{a-\varkappa,s_{a-\varkappa}}\neg(\vee_{1\leq i\leq N^{(\varkappa)}}(\&_{1\leq j\leq K^{(\varkappa)}-1}(f_{ij}^{(\varkappa)}=0)\&(f_{i0}^{(\varkappa)}\neq 0))).$$

Denote $d^{(\varkappa)}=\max_{ij}\deg_{Z_1,\dots,Z_{s_0},Z_{1,1},\dots,Z_{a-\varkappa,s_{a-\varkappa}}}(f_{ij}^{(\varkappa)})$; $d_2^{(\varkappa)}=\max_{ij}\deg_{T_1,\dots,T_\ell}(f_{ij}^{(\varkappa)})$; $q^{(\varkappa)}=N^{(\varkappa)}K^{(\varkappa)}d^{(\varkappa)}$; $M_2^{(\varkappa)}=\max_{ij} l(f_{ij}^{(\varkappa)})$; $\sigma=s_{a-\varkappa+1}$. Then in force of the theorem 2 and the lemma 2 the inequalities hold: $d^{(\varkappa)}\leq$

$(q^{(\varkappa-1)})^{8(\sigma+2)(n+2)}$, $N^{(\varkappa)} \leq (q^{(\varkappa-1)})^{n+12(\sigma+2)(n+\sigma+3)}$, $K^{(\varkappa)} \leq (\sigma+1)^2 (3q^{(\varkappa-1)})^{8(\sigma+2)(n+2)}$. Therefore $q^{(\varkappa)} \leq (q^{(\varkappa-1)})^{48n(\sigma+8)} \leq (Nd)^{(48n\sum_{a-\varkappa+1\leq j\leq a}(s_j+8)/\varkappa)^{\varkappa}}$. Apart that $d_2^{(\varkappa)} \leq d_2^{(\varkappa-1)} \times \mathcal{P}(q^{(\varkappa-1)}, d_1) \leq d_2 \mathcal{P}(q^{(\varkappa)}, d_1^{(\varkappa)})$, $M_2^{(\varkappa)} \leq (M_1 + M_2 + \ell \log d_2)\mathcal{P}(q^{(\varkappa)}, d_1^{\varkappa})$. Lastly the running time of the algorithm (after $\varkappa$ steps) is less than $\mathcal{P}(M_1+M_2, (Nd^n)^{(48n/\varkappa)^{\varkappa}(\sum_{a-\varkappa+1\leq j\leq a}(s_j+8))^{\varkappa}(n+\ell)}, (d_1^{\varkappa} d_2)^{n+\ell}, q)$.

Performing a steps completes the proof of the following

THEOREM 3. An algorithm is proposed which for a formula (3) outputs an equivalent to it a quantifier-free one $\bigvee_{1\leq i\leq \mathcal{N}}(\&_{1\leq j\leq \mathcal{K}} (g_{ij}=0) \& (g_{i0}\neq 0))$ where $g_{ij} \in F[Z_1,\ldots,Z_{s_0}]$, herewith $\deg_{Z_1,\ldots,Z_{s_0}}(g_{ij}) \leq (Nd^n)^{(48n(n+8a)/a)^a} = \mathcal{D}$, $\deg_{T_1,\ldots,T_\ell}(g_{ij}) \leq d_2 \mathcal{P}(\mathcal{D}, d_1^a)$; besides that $\ell(g_{ij}) \leq (M_1+M_2+\ell \log d_2)\mathcal{P}(\mathcal{D}, d_1^a)$. The integers $\mathcal{N}, \mathcal{K} \leq \mathcal{D}$. Finally, the algorithm works within the time $\mathcal{P}(L_2, L_2(\varphi), (Nd^n)^{(48n(n+8a)/a)^a (n+\ell)}, (d_1^a d_2)^{n+\ell}, q)$.

REFERENCES

1. Chistov A.L., Grigor'ev D.Yu. Polynomial-time factoring of the multivariable polynomials over a global field. - LOMI preprint E-5-82, Leningrad, 1982.
2. Chistov A.L., Grigor'ev D.Yu. Subexponential-time solving systems of algebraic equations. I. - LOMI preprint E-9-83, Leningrad, 1983.
3. Chistov A.L., Grigor'ev D.Yu. Subexponential-time solving systems of algebraic equations. II. - LOMI preprint E-10-83, Leningrad, 1983.
4. Chistov A.L., Grigor'ev D.Yu. Polynomial-time factoring of polynomials and subexponential-time solving systems and quantifier elimination. - Notes of Scientific seminars of LOMI, Leningrad, 1984, vol.137.
5. Collins G.E. Quantifier elimination for real closed fields by cylindrical algebraic decomposition. - Lect.Notes Comput.Sci., 1975, vol.33, p.134-183.
6. Grigor'ev D.Yu. Multiplicative complexity of a bilinear form over a commutative ring. - Lect.Notes Comp.Sci., 1981, vol.118, p.281-286.
7. Heintz J. Definability and fast quantifier elimination in algebraically closed fields. - Prepr.Univ.Frankfurt, West Germany, December, 1981.

8. Kaltofen E. A polynomial reduction from multivariate to bivariate integral polynomial factorization. - Proc.14-th ACM Symp.Th. Comput., May, N.Y., 1982, p.261-266.
9. Kaltofen E. A polynomial-time reduction from bivariate to univariate integral polynomial factorization. - Proc.23-rd IEEE Symp.Found Comp.Sci., October, N.Y., 1982, p.57-64.
10. Lazard D. Algébre linéaire sur $k[X_1,\dots,X_n]$ et élimination. - Bull.Soc.Math.France, 1977, vol.105, p.165-190.
11. Lazard D. Résolutions des systèmes d'équations algébriques. - Theor Comput.Sci., 1981, vol.15, p.77-110.
12. Lenstra A.K., Lenstra H.W., Lovasz L. Factoring polynomials with rational coefficients. - Math.Ann., 1982, vol.261, p.515-534.
13. Lenstra A.K. Factoring multivariate polynomials over finite fields. - Preprint Math.Centrum Amsterdam, IW 221/83, Februari, 1983.
14. Shafarevich I.R. Basic algebraic geometry. - Springer-Verlag, 1974.
15. Wüthrich H.R. Ein Entscheidungsverfahren für die Theorie der reellabgeschlossenen Körper. - Lect.Notes Comput.Sci., 1976, vol.43, p.138-162.
16. Zariski O., Samuel P. Commutative algebra, vol.1, 2. - van Nostrand, 1960.

SYSTOLIC AUTOMATA - POWER, CHARACTERIZATIONS, NONHOMOGENEITY

Jozef Gruska
VUSEI-AR, Dúbravská 3, [1]
84221 Bratislava, Czechoslovakia

1. INTRODUCTION

1.1 From cellular automata to systolic automata. Cellular automata (CA) of von Neumann have been the first model of highly concurrent automata. They have been created to model activities of cells and to study problems of selfreproducibility. They are, however, such a natural model of parallel computations that very soon also computational problems on CA and on their various modifications, for example on iterative arrays (IA), have been investigated ([He],[Co]).

For a long time numerous attempts to implement some variants of CA or IA failed because of the lack of proper technology. However, the recent advent of new technologies (VLSI), which make implementations of very large multiprocessor systems feasible, significantly rearoused the interest in highly concurrent computations. New technologies allow to design parallel systems with 10^6 or more elements. It has become clear that in order to make design and verification of such systems feasible, sufficiently formal methods have to be used. This in turn implies that from the viewpoint of design highly regular and modular parallel systems are desirable. The VLSI technology makes interconnections area- and time-consuming and therefore requires regular and short interconnections also from the cost and performance point of view. The possibility of using new technologies for hardware implementation of important algorithms, to be used as external devices, brings two additional requirements. It got important to minimise the data flow between the main computer and external devices and it is therefore desirable that once data enter a multiprocessor system, they flow through the system until they enter all computations where they are needed. Secondly, it got important to maximise the throughput of parallel systems and therefore the period becomes a new and important complexity measure.

Multiprocessor systems satisfying these requirements have been called *systolic systems* by Kung and Leiserson [KL]. Since then a large variety of often surprisingly simple and fast systolic systems have been designed.

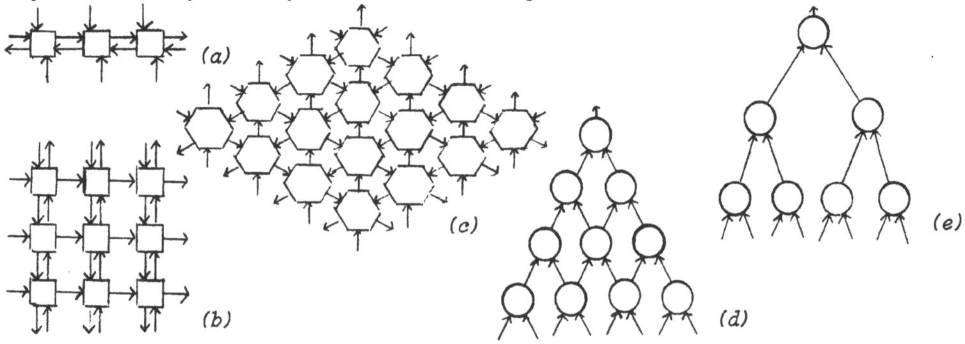

Fig. 1.1

1) This work has been finished during the author's visit at the Universität Hamburg, Fachbereich Informatik, in the Spring of 1984.

The main characteristics of systolic systems are as follows: very short and regular interprocessor communications, very regular data flow and very regular processor distributions. The most frequently used communication topologies and some data flows are shown in Fig. 1.1.

One attempt to get results bringing a deeper insight into possibilities and limitations of systolic systems with various topologies has been made through the study of systolic automata which has been initiated in [CSW, CGS-1].

The systolic automata are highly concurrent acceptors in the form of networks of very simple processors (combinational logics) with uni-directional data flow. (Transducers are obtained by a suitable modification of the model). So far mainly tree and trellis topologies (Fig. 1.1 e,d) have been investigated.

There are two main differences between systolic automata and CA + IA: the first one is uni-directional data flow and no-memory processors in systolic automata. (Therefore they have minimal period of computation). However, see Section 3, they are also a very useful tool to study networks with memory processors and multidirectional data flow. The second difference is that systolic automata are a better framework than CA and IA for studying the power of nonhomogeneity and of regular distributions of processors in networks. Homogeneity has been quite a natural requirement for CA and IA. However, for systolic systems homogeneity is a too strong requirement. Various regular but nonhomogeneous systolic systems to perform important computations have been designed. Moreover, new technologies and design trends [DFKM] make design of nonhomogeneous systems possible and thus the study of their properties desirable. (For further motivation see [CSW, CGS-1, Sa].

1.2 Summary The present paper presents an overview of results and research directions in the area of systolic automata including new results concerning trellis automata.

The paper is structured as follows. In Section 2 systolic tree automata are discussed: the power and limitations including the power of nonhomogeneity and of nondeterminism. A special attention is given to regular language recognition by programmable systolic tree automata (suitable for VLSI implementation).

Section 3 contains quite a general treatment of systolic trellis automata. Various characterizations of CA, IA and parallel Turing machines in terms of systolic trellis automata and of systolic trellis automata in terms of sequential Turing machines are shown. A speed-up theorem for systolic trellis automata is discussed which is stronger than that for Turing machines. Finally it is shown that computation processes for systolic trellis automata can be represented by the so-called generalised Pascal triangles and this gives a rise to a variety of new problems and results which can be applied also to processors distributions in nonhomogeneous trellis automata.

In the last two sections real-time systolic trellis automata are discussed: the power, limitations and processors distributions. In Section 4 homogeneous automata are investigated and in Section 5 two approaches to non-homogeneity are considered: in "regular trellis automata" processor distributions are "top-down programmable" and in "modular trellis automata" a trellis network can be assembled using modules of arbitrarily large size.

2. SYSTOLIC TREE AUTOMATA

The systolic tree automata are a natural abstraction of a very natural idea to use a network of tree connected processors to process an input. Tree processing is "context-free" and can be done theoretically in logarithmic time in the case of balanced trees.

2.1 Systolic tree automata (tree automata or StA for short). A *systolic tree automaton* is specified by an *infinite rooted tree* T with no leaves, with infinitely many nodes with at least two sons (*linear growth condition*), and with only finitely many different infinite subtrees (*uniformity condition*), by an *input alphabet* Σ, an *operating alphabet* Γ with $\lambda \in \Gamma$ (λ is the blank symbol), a *terminating alphabet* $\Gamma_0 \subseteq \Gamma$

and by two functions - an *input function* f_n: $\Sigma\cup\{\lambda\}\to\Gamma$, and an *output function* $g_n:\Gamma^n\to\Gamma$ for any natural number n such that T has a node with n sons. (The uniformity condition implies that there are only finitely many such n's). Formally

$$A=\langle T,\Gamma,\Sigma,\Gamma_0,f,g\rangle$$

is said to be a systolic tree automaton on T .

Initial parts of two infinite trees which satisfy linear growth and uniformity conditions are depicted in Fig. 2.1. An internal node of the tree in Fig. 2.1(a) has three (two) sons iff it has one (two) brother. Each node of the tree in Fig. 2.1(b) has two sons. This is a so-called *binary balanced tree*. In general a tree is a *balanced*

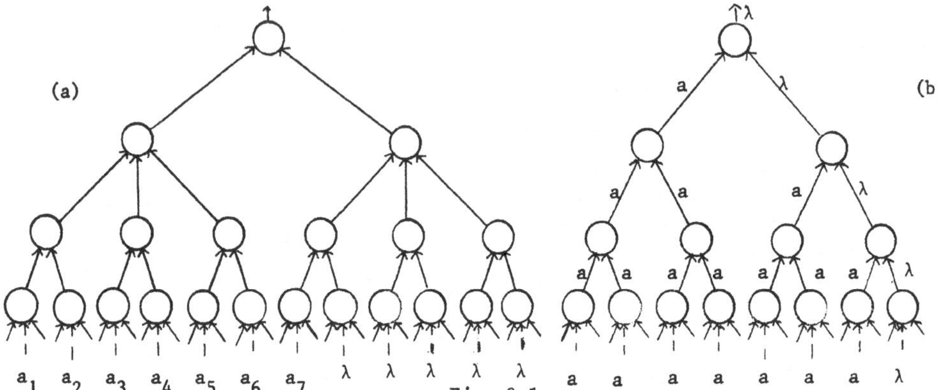

Fig. 2.1

tree of the order p>1 if each its node has p sons.

All nodes of a tree T at the distance n≥0 from the root form *"the n-th level of nodes of T"* and $N_T(i,j)$ will denote the i-th node (from the left, starting with 0) at the j-th level of T.

A tree automaton specification $A=\langle T,\Gamma,\Sigma,\Gamma_0,f,g\rangle$ specifies an infinite network of processors interconnected to form the tree T. A processor of A in a node with k sons has one external input (depicted by dashed arrows in Fig. 2.1), one output and k internal inputs connected with outputs of its son-processors. The activity of such a processor on an external input is specified by the input function f_k and its activity on internal inputs by the transition function g_k.

An input word $w=w_0w_1\ldots w_{n-1}\in\Sigma^n$ is processed by A as follows. Let j_0 be the smallest integer such that there are m≥n nodes at the j_0-th level of T. To input w, the i-th symbol of w-w_i-is fed to the external input of the i-th processor at the j_0-th level and λ is fed to external inputs of the remaining processors (if there are any) at the j_0-th level. All these processors get their external inputs simultaneously and immediately process these inputs using their input functions, and simultaneously (in no time units) send results to their fathers. It is assumed that it takes exactly one time unit to transmit data along any edge. Thus all processors at the (j_0-1)-th level receive simultaneously data on their internal inputs. All these processors simultaneously process their inputs, using their transition functions, and immediately send results to their fathers. This process continues, and after j_0 parallel computation steps,the root-processor outputs a symbol. If this symbol is in Γ_0, w is accepted, otherwise w is rejected.

In this way with every tree automaton A the language L(A) of all words accepted by A is associated. Moreover, by L_T(StA) we denote the class of languages accepted by tree automata on T. L(bStA) [L(BStA)] denotes the class of languages accepted by tree automata on binary [balanced] trees.

<u>Example</u>: Let $A=\langle T,\{a,\lambda\},\{a\},\{a\},f,g\rangle$ where T is the binary tree, f(a)=a, g(λ,a)= =g(a,λ)=g(λ,λ)=λ, g(a,a)=a. Obviously $L(A)=\{a^{2^n}|n\geq 0\}$. (The processing of $w=a^7$ is shown in Fig. 2.1b).

<u>Remark</u> It has been shown [St] that there is a very close relation between the systolic tree automata and "classical" tree automata.

2.2 Power and Limitations The systolic tree automata are clearly not too powerful. If the underlying tree T of a StA A satisfies *"exponential growth condition"* ([CSW], i.e. there exists an $\alpha>1$ such that there are at least α^j nodes at the j-th level of T), as it is in the case of balanced trees, then A accepts L(A) in logarithmic parallel time. This fast acceptance is of special interest and therefore we shall pay here particular attention to StA on balanced trees. The main results concerning the power of StA are summarised as follows.

Theorem 2.1 (1) If T is an infinite tree satisfying linear growth and uniformity conditions, then the class L_T(StA) contains all regular languages and is closed under Boolean operations.

(2) If T is a balanced tree, then the class L_T(StA) is also closed under right concatenation with regular sets, but it is not closed under the following operations: left concatenation with regular set, reversal, inverse homomorphism, ε-free homomorphism and Kleene closure. Δ

The closure under right concatenation with regular sets and no-closure results have been shown in [IK-2] for the case of binary trees but can be generalized for all balanced trees. The proof in [IK-2] is based on the following theorem [CSW], where a language L is called *subexponential* if for every n there is an integer k such that there are at least n words in L whose length is in the interval $(2^{k-1},2^k]$ and L is said to have *bounded tail ambiguity* if there is an integer t such that, for every k and every word w of length 2^{k-1}, there are at most t words w' with length $\leq 2^{k-1}$ such that ww' is in L.

Theorem 2.2 No subexponential language with bounded tail ambiguity is in L_T(bStA). Δ

The structure of classes L(BStA) and L(StA) is not well understood. L(BStA) contains, e.g. properly context-free and context-sensitive languages [Pt].

Theorem 2.3 The emptiness, finiteness and equivalence problems are decidable for StA on balanced trees. Δ

It is, however, an *open problem* whether the emptiness problem decidability (which implies the equivalence problem decidability) holds true for all StA. This problem seems to be quite difficult because of the following result [CGS-4] where (1) and (2) are well-known open decision problems in the theory of formal power series.

Theorem 2.4 If the emptiness problem for StA is decidable, then the following decision problems are also decidable: (1) Does a negative number appear in a given Z-rational sequence of integers? (2) Does zero appear in a given Z-rational sequence of integers? Δ

2.3 Modifications We shall consider now several modifications of StA. In all cases we get devices with the same recognition power. This indicates that the notion of StA is well chosen. We commence with a simple normal form for StA on balanced trees, but only the case of binary trees is discussed in details.

Theorem 2.5 To every StA $A=\langle T,\Gamma,\Sigma,\Gamma_o,f,g\rangle$ on a balanced tree T one can construct an equivalent StA $\bar{A}=\langle T,\bar{\Gamma},\Sigma,\bar{\Gamma}_o,\bar{f},\bar{g},\rangle$ such that $\bar{f}(a)=g(a,\lambda,\dots,\lambda)$ for each $a\in\Sigma$. Δ

This implies that without loss of generality we can view StA on balanced trees as having all processors with only internal inputs and that an input symbol is fed into a processor if it is fed into its leftmost internal input and λ's into all remaining inputs (Fig. 2.2). This means that StA on binary trees can be considered in the form $A=\langle\Gamma,\Sigma,\Gamma_o,g\rangle$ where $\Sigma\cup\{\lambda\}=\Gamma\supseteq\Gamma_o$ and $g(a,\lambda)=a$ for $a\in\Gamma$. If we now define for $w=w_1\dots w_n\in\Gamma^n$

$$\gamma(w_1\dots w_n)=\begin{cases} g(w_1,w_2)g(w_3,w_4)\dots g(w_{n-1},w_n) & \text{if n is even} \\ g(w_1,w_2)g(w_3,w_4)\dots g(w_n,\lambda) & \text{if n is odd} \end{cases}$$

then $L(A)=\{w \mid w\in\Sigma^+ \text{ and } \gamma^{\lceil\log_2|w|\rceil}(w)\in\Gamma_o\}$

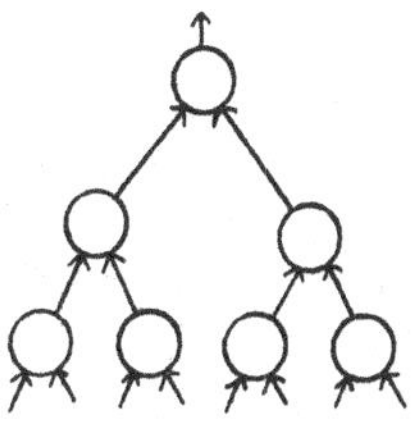

Fig. 2.2

Let us now relax input regulations for StA. A StA is called *stable* [*superstable*] if the accepted language remains invariant, if we drop the requirement that an input word w has to be fed to the first possible level of processors [and the requirement that the i-th symbol of w is fed into the i-th leftmost processor, and we require only that, for $1 \le i < j \le |w|$, the i-th symbol of w is fed into a processor which is to the left from the processor into which the j-th symbol is fed]. Informally this means that in the case of a stable [superstable] StA a "fixed-size chip" can be used to process all inputs up to a certain size [and no fixed input connections are necessary].

Theorem 2.6 (1) To every StA on a balanced tree T one can effectively construct an equivalent stable StA on T.

(2) A language L is recognisable by a superstable StA on a balanced tree if and only if L is regular. $\triangle$

Let us now generalise the notion of StA by dropping out our homogeneity requirement that all processors with the same number of sons have to be identical. Instead of that we shall consider StA on node labelled trees and we shall assume that nodes of a tree T are labelled by symbols from a finite alphabet Δ (of processors names). For a StA

$$A=\langle T,\Delta,\Gamma,\Sigma,\Gamma_o,f,g\rangle$$

over a labelled tree T we shall assume that all nodes labelled by the same symbol, say δ, have the same number of sons and contain the same processor, with the input function f_δ and the transition function g_δ.

A StA over a labelled tree process an input in a similar way as a homogeneous StA. The only difference being that the way inputs are processed in a node does not depend only on the number of sons of the node but also on its label.

This generalization of StA is obviously too strong. If no restrictions are put on processor distribution, then also nonrecursive languages can be accepted.

A natural way to deal with this problem is to put some regularity condition on tree labelling. Two such conditions will be considered now. A labelling of a tree T is said to be *uniform* (or *globally regular*) if T, as a labelled tree, has only finitely many non-isomorphic infinite labelled subtrees. A labelling is said to be *top-down* (or *locally regular*) if the label of each node uniquely determines labels of its sons. Finally a StA is said to be *regular systolic tree automaton* if the labelling of the corresponding tree is globally or locally regular.

It is natural to consider also non-deterministic StA where a set of input and transition functions is associated with each processor and these functions are non-deterministically chosen to process given inputs.

It follows from the next theorem [CSW,IK-2] that neither regularity nor non-determinism increase the power of StA. Inspite of that the result if of importance for the design of StA because regular StA are sometimes more easy to construct. (See also [CY] where very powerful regular tree automata with two directional data flow are investigated).

Theorem 2.7 To every nondeterministic regular StA one can construct an equivalent deterministic homogeneous StA (over the same tree). $\triangle$

2.4 Programmable systolic tree automata for regular language recognition. Fast recognition of regular languages is important in many applications. StA can recognize regular languages in logarithmic time. It is therefore worthwhile to study regular language recognition by StA in more details.

It is easy to design a homogeneous StA on the binary tree to recognise a given regular language $R \subset \Sigma^+$. Indeed, let us consider the syntactical monoid of R with the set $\{[w] \mid w \in \Sigma^+\} = \Gamma$ of syntactical classes. Then $R=L(A_R)$, where $A_R=\langle T,\Gamma,\Sigma,\Gamma_o,f_R,g_R\rangle$ is the StA on the binary tree T with $\Gamma_o=\{[w]\ w\in R\}$, $f_R(a)=[a]$ for $a \in \Sigma \cup \{\lambda\}$ (with λ also as the empty word), $g_R([w_1],[w_2])=[w_1w_2]$ for $w_1,w_2 \in \Sigma^*$. (Note also that A_R is superstable because g_R is transitive).

This construction of a StA to recognise R is very simple. However, is it the best possible? If $R \neq \Sigma^*$, then $\Gamma_o \neq \Gamma$ and therefore the root-processor of A_R need not be able to produce any symbol from Γ. A "smaller size" processor may be sufficient - it is enough if this processor is able to produce any symbol from $\bar{\Gamma}_o=\Gamma_o \cup \{NO\}$ where NO is a new symbol not in Γ.

In [SJ] this problem is studied in more details and it is shown how to construct a regular StA $\bar{A}_R$, to recognise R, which is for some R "smaller" than A_R.

The main disadvantage of both constructions is that the resulting StA A_R and $\bar{A}_R$ are infinite. However, from the way A_R is designed it is clear that R is recognizable also by the following - Fig. 2.3 - *finite StA with feedback* [CJ] where the processor in a node labelled f_R (g_R) computes the function $f_R(g_R)$. To accept a $w \in \Sigma^+, |w|=4k$, w is divided into subwords of the length 4, $w=w_1 \dots w_k, |w_i|=4$, and, with the period one, the subwords $w_1, \dots, w_k$ are fed into external inputs. It is easy to see how to generalise this construction that it holds [CJ]:

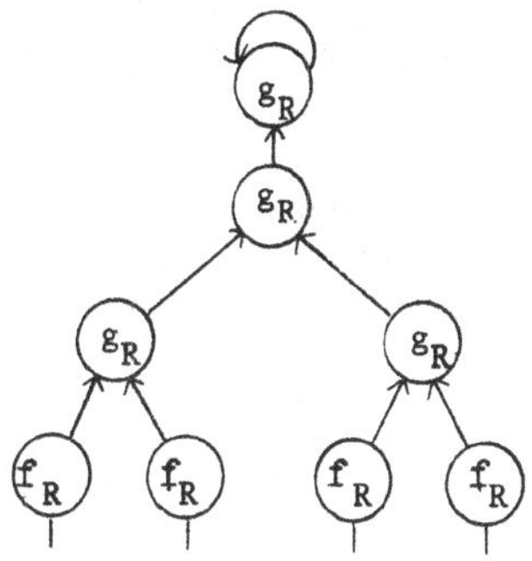

Fig. 2.3

Theorem 2.8 To any regular language R and any k>0 there is a finite systolic tree automaton with feedback which has k leaves, accepts R and needs $\lceil \frac{n}{k} \rceil + \lceil \log k \rceil$ steps to recognise a word of the length n.

A very interesting hardware realization [CJ], suitable for VLSI, of finite StA with feedback is based on the fact [SS] that for any regular language $R \subset \Sigma^+$, which can be recognised by a nondeterministic finite automaton with n states, there is a representation $\phi_R : \Sigma \to \beta_n$ (nxn Boolean matrices), Boolean column and row vectors η_R and π_R, both of the length n, such that

$$w \in R \quad \text{iff} \quad \pi_R \phi_R(w) \eta_R = 1$$

If this representation is used, the input function evaluation-$f_R(a)$-reduces to the creation of the transition matrix corresponding to a and the transition function g_R realisation reduces to Boolean matrix multiplication. Finally, the output evaluation reduces to one matrix-vector and one vector-vector multiplication. The important point now is that all these modifications can be done using well-known systolic arrays.

This realization has a big advantage. It is build for a fixed n and it is easily programmable to accept any language R which is accepted by a nondeterministic finite automaton with n states. (Only vectors π_R, η_R and the input function are to be fixed).

2.5 Turing machine characterization It is not easy to design systolic automata because it requires thinking about a large number of simultaneous operations. It is therefore useful to have their characterizations in terms of sequential machines - they are more easy to program. A characterization of StA on binary trees, which is easy to generalize to all balanced trees, is given in [IK-2]:

Theorem 2.9 Let p>1. StA on p-ary balanced trees are effectively equivalent to the so called deterministic p-ary counter synchronized Turing machines.

For p=2 such a Turing machine M has an input alphabet Σ, an operating alphabet Γ, a set of states with the initial state q_1. An input word $w=a_1 \dots a_n\$$ is presented to the external input of M - Fig. 2.4 - with the endmarker \$. The activity of M consists of n right-to-left and left-to-right sweeps. During left-to-right sweeps M is in a special state q_0 and the tape contents is not changed. During right-to-left sweeps tree computations are simulated using a special p-ary counter - starting from a leaf and "climbing" the tree along right edges only, as far as possible - as shown in Fig. 2.5. For details se [IK-2].

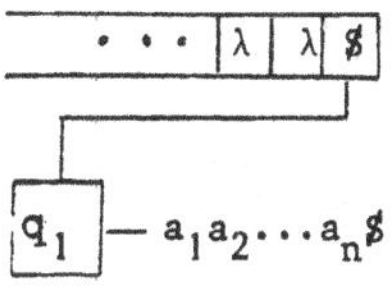

Fig. 2.4

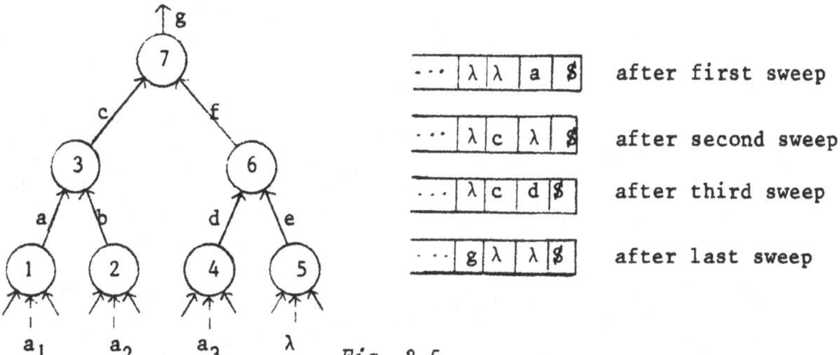

Fig. 2.5

3. SYSTOLIC TRELLIS AUTOMATA

The systolic automata we shall deal with in this section are a natural abstraction of systolic systems with two-dimensional arrays of processors and with uni-directional flow of data. They are also a natural framework for getting insight into computations on various other parallel devices as cellular automata, iterative arrays and parallel Turing machines. On the other side, systolic trellis automata have simple characterizations in terms of sequential Turing machines what allows to use well understood sequential techniques to program and to study various parallel systems. Moreover, there is a close relation between computation processes in trellis automata and generalized Pascal triangles over finite algebras what allows to use algebraical and combinatorial tools to get a deeper insight into parallel computations. Our model of systolic trellis automata is a natural unification and generalization of models discussed in [CGS-1, IKM,CC].

3.1 Homogeneous systolic trellis automata (trellis automata or STA for short) are devices $A=\langle\Gamma,\Sigma,\Gamma_0,g\rangle$ consisting of an infinite triangularly shaped array of identical processors (combinational circuits) with unit propagation delay between neighbouring

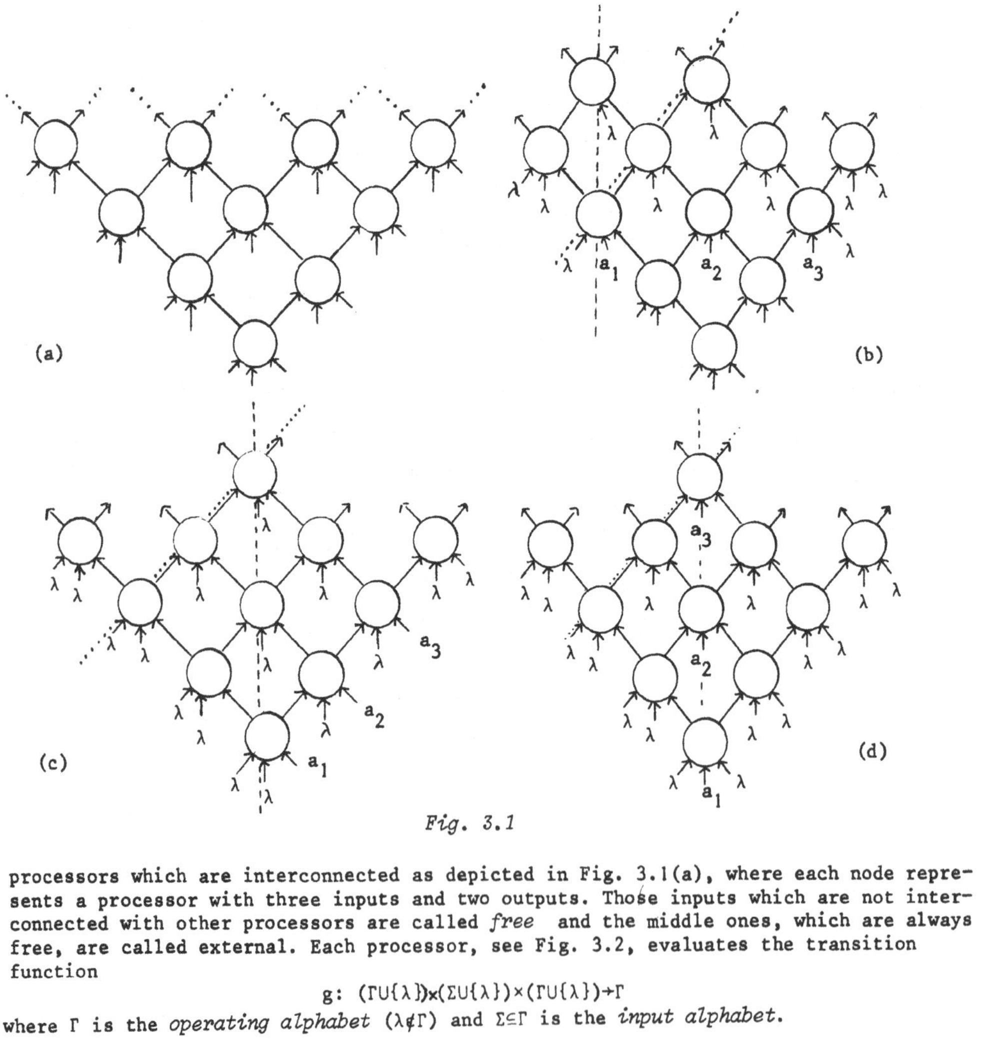

Fig. 3.1

processors which are interconnected as depicted in Fig. 3.1(a), where each node represents a processor with three inputs and two outputs. Those inputs which are not interconnected with other processors are called *free* and the middle ones, which are always free, are called external. Each processor, see Fig. 3.2, evaluates the transition function

$$g:\ (\Gamma\cup\{\lambda\})\times(\Sigma\cup\{\lambda\})\times(\Gamma\cup\{\lambda\})\to\Gamma$$

where Γ is the *operating alphabet* ($\lambda\notin\Gamma$) and $\Sigma\subseteq\Gamma$ is the *input alphabet*.

Three input modes will be considered for STA. In *parallel input mode*, to input a word $w=a_1\ldots a_n\in\Sigma^n, a_i\in\Sigma$ for $1\leq i\leq n$, all processors at levels with less than $|w|$ processors are "cut off" and the consecutive symbols of w, symbol by symbol, are fed, in parallel, to consecutive external inputs of processors at that level of processors which has exactly $|w|$ processors - Fig. 3.1(b)- all other free inputs get λ's. In the case of *skewed input mode* (*sequential input mode*) symbols of w are fed to the external inputs of the first $|w|$ processors of the right diagonal (middle column) of processors - Fig. 3.1(c,d) - other free inputs get λ's.

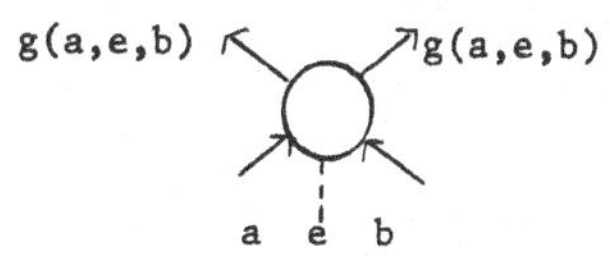

Fig. 3.2

For all three input modes a computation process starts at that row of processors which contains the processor activated by a_1 and proceeds from one level of processors to the next one as in the case of tree automata. During a computation process at any moment processors at exactly one level are active. They compute in parallel values of their transition functions and transmit them synchronously to their fathers.

There are several ways how to define an acceptance for STA and we shall consider three of them. In all of them the alphabet $\Gamma_0\subseteq\Gamma$ of *accepting symbols* is used. A $w\in\Sigma^+$ is said to be *vertically accepted* (*diagonally accepted*) by a STA A if during the computation process a symbol from Γ_0 is produced by a processor in the column (in the right diagonal) of processors which starts with the processor activated by a_1 (with the $|w|$-th processor of the middle column of processors). See dashed lines for vertical acceptance and dotted lines for diagonal acceptance in Fig. 3.1. Moreover, w is said to be *totally accepted* if in a computation step all processors produce symbols from Γ_0.

For short we shall often use symbols *p*, *sk* and *se*, respectively, to denote parallel, skewed and sequential input, respectively, and symbols *v*, *d* and *t*, respectively, to denote vertical, diagonal and total acceptance, respectively. For example, for $i\in\{p,sk,se\}$ and $a\in\{v,d,t\}$ and a STA A, $L_i^a(A)$ will denote the language a-accepted by A for i-input. Moreover, a language L is said to be a-accepted in time T(n) by A with i-input if $L=L_i^a(A)$ and not more than $T(|w|)$ levels of processors are needed to accept w. Finally, let $STA_i^a(T(n))$ denote the class of languages a-accepted by a STA with i-input in time T(n).

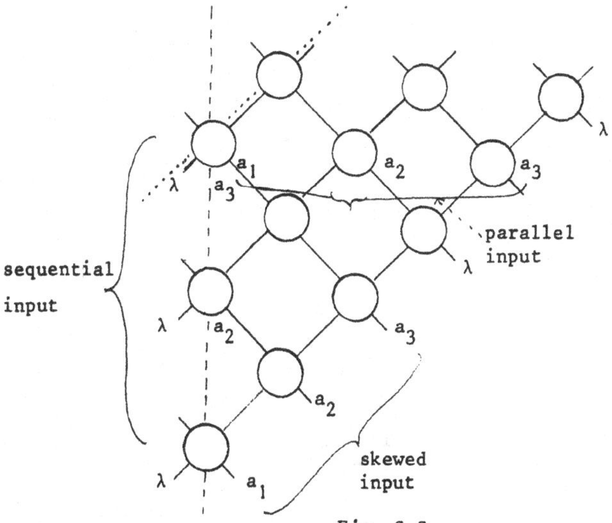

Fig. 3.3

Remark 1 It is easy to see that for all our input modes and acceptance rules STA are real-time equivalent to systolic automata with the network of processors as depicted in Fig. 3.3, where also all input modes are shown and where all processors have only two inputs and their transition function

$$g:(\Gamma\cup\{\lambda\})\times(\Gamma\cup\{\lambda\})\to\Gamma$$

has two arguments. The dashed (dotted line) shows where *v*-acceptance (*d*-acceptance) is checked. The network in Fig. 3.3 is obtained from the network in Fig. 3.1(a) by *folding* around the middle column. The equivalence for skewed and sequential input mode easily follows from this. To see that for the parallel input, fold the network in Fig. 3.1(b) around the dashed line. (Folding [CP] is a very useful technique to produce and transform systolic systems.) [The following modification of the parallel input mode is natural: two symbols are fed into each node. TA with such input have the same power but they are about twice as fast [IKM].]

Remark 2 In order to study acceptance and time complexity of STA with the parallel input we can restrict ourselves, without loss of generality, to such STA $A=\langle\Gamma,\Sigma,\Gamma_o,g\rangle$ that $g(\lambda,a,\lambda)=a$ for each $a\in\Sigma$. If we now define, for $a,b\in\Gamma$, $l(a)=g(\lambda,\lambda,a)$, $r(a)=g(a,\lambda,\lambda)$ and $\bar{g}(a,b)=g(a,\lambda,b)$, then computation processes and acceptances can be defined in a very simple way. Indeed, let $\pi:\Gamma^+\to\Gamma^+$ be the mapping defined by $\pi(a)=l(a)r(a)$ for $a\in\Gamma$ and for $w=a_1\ldots a_n\in\Gamma^n, n>1$,

$$\pi(w)=l(a_1)\bar{g}(a_1,a_2)\bar{g}(a_2,a_3)\ldots\bar{g}(a_{n-2},a_{n-1})\bar{g}(a_{n-1},a_n)r(a_n)$$

Then $\{\pi^i(w)\}_{i=0}^{\infty}$ is an infinite sequence of words produced by processors at the consecutive levels of A when w is being processed by A. Moreover

$$L_p^v(A)=\{w\mid w\in\Sigma^+,\pi^{2i}(w)\in\Gamma^i\Gamma_o\Gamma^* \text{ for some } i>o\}$$
$$L_p^d(A)=\{w\mid w\in\Sigma^+,\pi^{i}(w)\in\Gamma^i\Gamma_o\Gamma^* \text{ for some } i>o\}$$
$$L_p^t(A)=\{w\mid w\in\Sigma^+,\pi^{i}(w)\in\Gamma_o^+ \text{ for some } i>o\}$$

Observe also that in this way with each STA $A=\langle\Gamma,\Sigma,\Gamma_o,g\rangle$ we can associate the algebra $\mathcal{A}=(\Gamma,l,r,\bar{g})$ of the signature (1,1,2). However, also the opposite holds.With any algebra $\mathcal{A}=(\Gamma,l,r,g)$ of the signature (1,1,2) we can associate, for each $\Sigma\subseteq\Gamma$ and $\Gamma_o\subseteq\Gamma$ a STA in a natural way. Moreover, if $\Sigma=\Gamma$ and acceptance is not of importance, then there is actually one-to-one correspondence between STA and finite algebras of the signature (1,1,2). For any such algebra $\mathcal{A}=(\Gamma,l,r,g)$ and $w\in\Gamma^+$ we have defined the sequence $\{\pi^i(w)\}_{i=0}^{\infty}$ which will be called the *generalized Pascal triangle* over $\mathcal{A}$ and w and denoted GPT($\mathcal{A}$,w). (Observe that if Γ is the set of natural numbers, l and r are the identity functions and g is the plus operation, then GPT($\mathcal{A}$,1) is exactly the set of rows of the usual Pascal triangle).

These relations between STA and algebras of the signature (1,1,2) on one side and between computation processes of STA and GPT on the other side, may help to use algebraical and combinatorial methods to study parallel computations - see Section 3.6.

It is possible to show that as far as acceptance is concerned one can consider without the loss of generality only such STA that l and r are the identity functions and therefore algebras of the signature (2) are sufficient to consider.

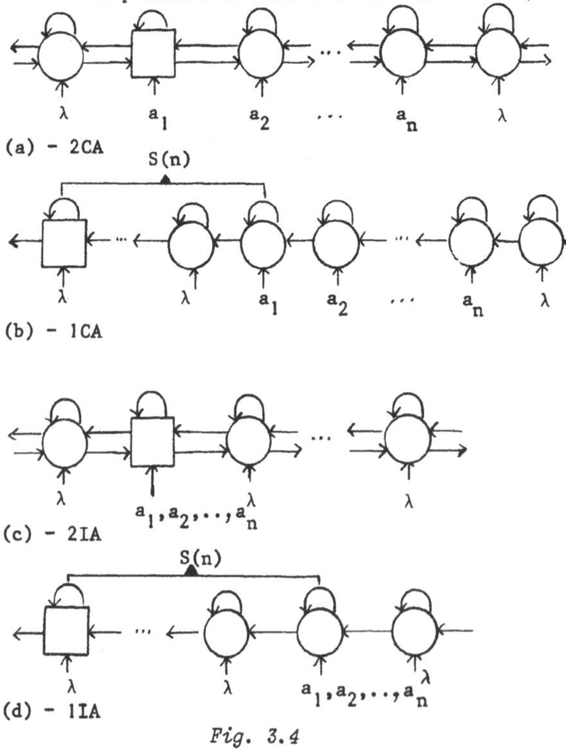

Fig. 3.4

3.2 Tellis automata characterizations of cellular automata and iterative arrays. In order to make relations between trellis automata, cellular automata and iterative arrays more transparent and the overall treatment of parallel automata sufficiently uniform (see also [CC]) we shall consider processors of CA and of IA as combinational circuits and not as sequential machines as it is usually the case.

A *two-way cellular automaton* (2CA) $A=\langle\Gamma,\Sigma,\Gamma_o,g\rangle$ is an infinite linear array of identical processors with unit propagation delay along all interconnections. Each processor is two-way connected with both neighbours, one-way connected with itself, and has one external input - see Fig. 3.4(a). Each processor computes the transition function $g:(\Gamma\cup\{\lambda\})^4\to\Gamma\cup\{\lambda\}$ such that $g(\lambda,\lambda,\lambda,\lambda)=\lambda$. The first (second) argument comes from the left (right) neighbour, the third from the feedback input and the last one from the external input. To start a computation an input word $w=a_1...a_n \in \Sigma^n$ is fed, symbol by symbol, in parallel, to external inputs of some n consecutive processors and all other inputs get λ's. In all the next time moments all external inputs get λ's. w is accepted in time t if after t parallel steps the processor activated by a_1-depicted by the square in Fig. 3.4(a)- produces a symbol from $\Gamma_0 \subseteq \Gamma$. 2CA(T(n)) denotes the class of languages accepted by 2CA in time T(n).

A *one-way cellular automaton* (1CA - Fig. 3.4(b)) $A=\langle\Gamma,\Sigma,\Gamma_0,g\rangle$ is defined in a similar way as a 2CA only g is a ternary function and $g(\lambda,\lambda,\lambda)=\lambda$. A is said to accept a language L in time T(n) and space S(n) if the following condition is satisfied: $w \in L$ if and only if after T(|w|) parallel steps the S(|w|)-th processor to the left from the processor activated by a_1 produces a symbol from Γ_0. Let 1CA(T(n),S(n)) be the classof languages accepted by 1CA in time T(n) and space S(n).

Two-way and *one-way iterative arrays* (2IA and 1IA - Fig. 3.4(c,d) and their acceptance are defined in a similar way with the only difference that an input word is fed, symbol by symbol, sequentially to the external input of one fixed processor during |w| consecutive moments - afterwards that input receives λ's.

A very useful technique to deal with parallel computations on CA or IA is to *unroll* their computations in time into computations in space (the so-called *time-space transformation*) where each processor of CA or IA is simulated either by a column of processors (let us call it c-unrolling) or by a diagonal of processors (d-unrolling).

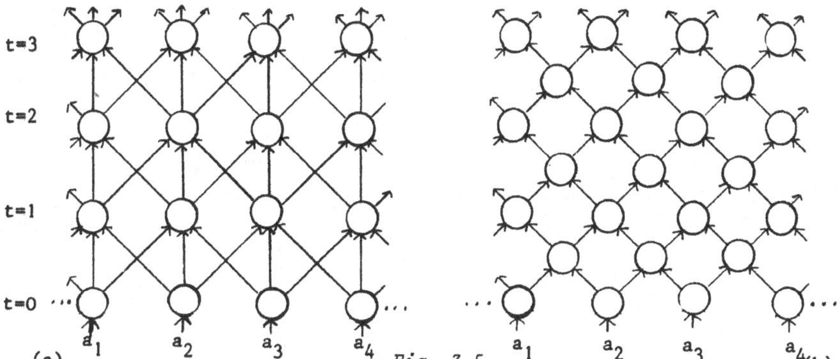

Fig. 3.5

C-unrolling of computations of 2CA gives the network shown in Fig. 3.5(a). As far as the column acceptance is concerned, this network is simulated by the network from Fig. 3.5(b) which is obtained from the network in Fig. 3.5(a) by removing vertical connections and by replacing all crossings by new processors. From the above it follows that if a language L is accepted in time T(n) by a 2CA, then it is accepted by a STA with parallel input in time 2T(n). Obviously also the opposite is true. On the other hand, d-unrolling of 1CA gives exactly STA with parallel input. In a similar way, d-unrolling of 2IA and of 1IA leads to STA with sequential and skewed input. Using these relations the following results can be shown [IK-3].

Theorem 3.1 For any $T(n)\geq 0$ it holds

(1) $STA_p^v(2T(n)-1)=STA_{sk}^v(2T(n)-1)=2CA(T(n)=1CA(2T(n),T(n))=1IA(2T(n),T(n))$

(2) $STA_p^d(T(n))=1CA(T(n),1)$

(3) $STA_{se}^v(2T(n)-1)=2IA(T(n))$

(4) $STA_{sk}^d(T(n)+n-1)=1IA(T(n),n)$ if $T(n)\geq n$. △

3.3 Trellis automata characterizations of parallel Turing machines There is also a close relation between STA and a new model of parallel computations - parallel Turing machines (PTM).

Only a brief description of one-tape PTM will be given here, for details see [Wi]. A specification $\mathcal{S}$ (the initial state, final states, next move relation) of a PTM M is the same as of a nondeterministic TM, only the interpretation is different. M starts each computation with exactly one processor which is in the initial state and its read-write head scans the leftmost symbol of a given input word, which is written on consecutive squares of the tape and all other squares of this two-way infinite tape are filled with blanks. During each computation process M may have several processors active, each of them having its head on a square of the tape and more than one head on a square is allowed, Each processor of M acts independently and according to the specification $\mathcal{S}$ with the exception that the next move relation of $\mathcal{S}$ allows more than one - say k - possible moves. In such a case the processor under consideration creates k-1 new copies of itself and each of these copies performs one of the k possible moves and acts further independently (according to the specification $\mathcal{S}$). If, during a parallel step, several processors want to write on the same square, this simultaneous writing is allowed if all try to write the same symbol - otherwise the computation becomes illegal. An input is said to be accepted if, after some parallel steps, each processor of M is in an accepting state.

Theorem 3.2 One-tape parallel Turing machines and systolic trellis automata with skewed input and total acceptance are real-time equivalent. $\triangle$

Sketch of the proof (1) Let a PTM be given and let tape squares be numbered from $-\infty$ to ∞ with zero for the square with the leftmost input symbol. The processors of a STA simulating M will output at a level j triples (S_i,S_{i+1},S_{i+2}) - see Fig. 3.6 - where $S_i=(b_i,Q_i)$, b_i is a symbol on the square number i after the j-th parallel step and Q_i is a set of states of those processors heads of which scan the i-th square during the j-th step.

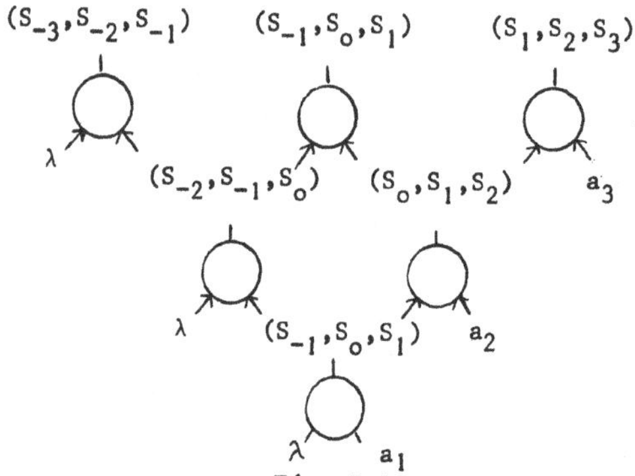

Fig. 3.6

(2) Let a STA $A=\langle\Gamma,\Sigma,\Gamma_0,g\rangle$ be given. A PTM simulating A will have an initial state q_0 and a pair of states $q_a,\bar{a}$ for each $a\in\Gamma$. Final states of M will be states q_b and $\bar{b}$ for $b\in\Gamma_0$. A processor of M functions as follows. If it reads an $a\in\Sigma$ in the state q_0, then writes $b=g(\lambda,a,\lambda)$, enters the state q_b and sends to the right one new processor in the state $\bar{b}$. If a processor is in a state q_b, $b\in\Gamma$, then writes $c=g(\lambda,\lambda,b)$, enters the state q_c and sends to the right a new processor in the state $\bar{c}$. If a processor is in a state $\bar{c}$, $c\in\Gamma$, and reads d, then it writes $e=g(c,d,\lambda)$ and moves right in the state $\bar{e}$. Clearly, M simulates A for the case of skewed input.

Corollary 1 Any one-tape PTM of time complexity T(n) can be simulated by a sequential Turing machine in time $O(T^2(n))$.

Corollary 2 There exists a universal STA simulating in linear time any other STA with skewed input and total acceptance.

Proof The corollary follows from Theorem 3.2 and from the fact that there is a universal one-tape PTM simulating any one-tape PTM in the linear time [Wi].

The existence of universal STA can be shown also for other input modes, vertical and total acceptance.

3.4 Sequential Turing machine characterizations of trellis automata In [IK-3] nice characterizations of STA in terms of the so-called full scan Turing machines (STM) are given. There is actually several such machines, one - STM_i^a - for each input mode i and vertical and diagonal acceptance rules. They differ in the initial position of their heads, in the positions they read new input symbols, and how they react on the endmarker.

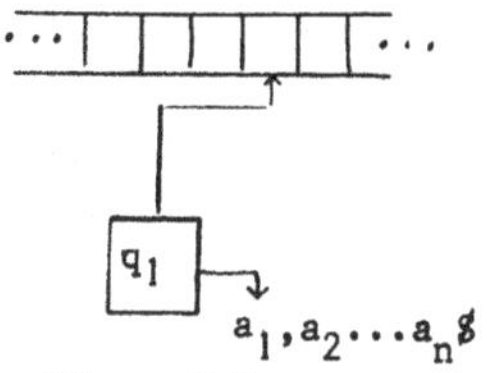

Fig. 3.7

A STM_i^a M is a one-tape one-head Turing machine with the external input - Fig. 3.7 - to receive input words $a_1 \dots a_n \$$, $a_i \varepsilon \Sigma, \$ \notin \Sigma$ - the input alphabet. M starts a computation in an initial state q_1 and performs right-to-left and left-to-right sweeps. An input is accepted if M writes an accepting symbol (from $\Gamma_o \subseteq \Gamma$ - the tape alphabet). Fig. 3.8 shows for various STM their initial position and complete sweeps. "R" shows where reading of the external input takes place. "W" stands for writing and W($) means that in the indicated place, after the endmarker $ is read, $ is written. During their left-to-right sweeps all STM stay in the state q_o and do not change tape contents and move right until they find $ or λ. During right-to-left sweeps STM can change both tape contents and their states but they enter q_o, when coming to a square with $ or λ, only to start a left-to-right sweep.

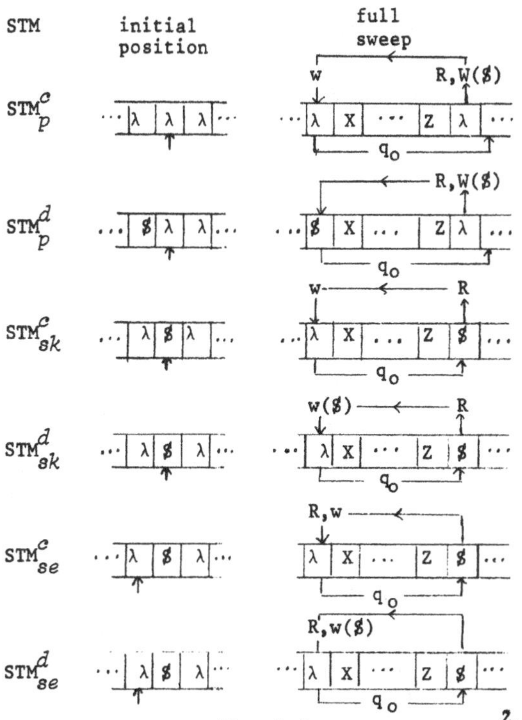

Fig. 3.8

The sweep complexity of an STM on an input is the least number of complete sweeps needed to accept this input. The following theorem follows from the results in [IK-3] and assumes $T(n) \geq 0$.

<u>Theorem</u> 3.3 (1) Let $i \varepsilon \{p, sk, se\}$. A language L is accepted by a STA_i^v in time $2T(n)-1$ if and only if it is accepted by a STM_i^v with sweep complexity $T(n)$.

(2) A language L is accepted by a STA_p^d in time $T(n)$ if and only if L is accepted by a STM_p^d with sweep complexity $T(n)$.

(3) Let $i \varepsilon \{sk, se\}$. For $T(n) \geq n$, a language L is accepted by a STA_i^d in time $T(n)+n-1$ if and only if L is accepted by a STM_i^d with sweep complexity $T(n)$. Δ

Fig. 3.9(a) shows a computation on a STM_p^d A for the input abcd and Fig. 3.9(b) shows a profile of a STM_p^d simulating A.

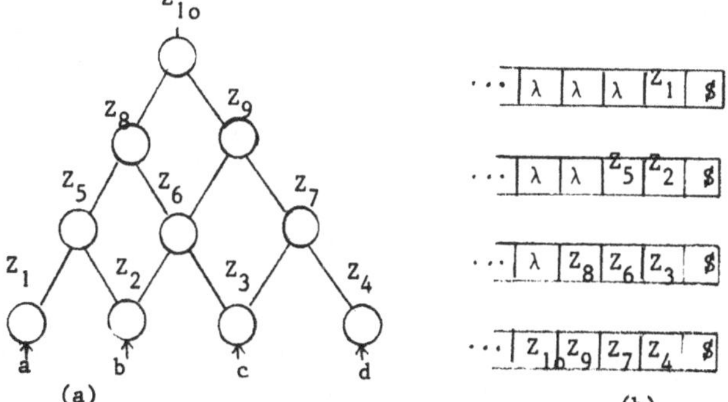

Fig. 3.9

3.5 Speed-ups and hierarchies The following theorem has been shown in [IKM] for vertical and diagonal acceptance using characterizations of STA in terms of STM. This proof can be rewritten in terms of STA and then extended to cover total acceptance, too.

Theorem 3.4 If $T(n)\geqq 0, k>0$, then

(1) $STA_p^d(n+T(n))=STA_p^d(n+\lfloor T(n)/k\rfloor)$ and $STA_p^d(n+k)=STA_p^d(n+1)$

(2) $STA_i^a(2n-1+T(n))=STA_i^a(2n-1+2\lfloor T(n)/k\rfloor)$ and $STA_i^a(2n-1+k)=STA_i^a(2n+1)$

for $i\varepsilon\{p,sk,se\}$, $a\varepsilon\{v,t,d\}$. Δ

In view of Theorems 3.1 and 3.2 this result implies that for CA, IA and STM a stronger speed-up is possible - from $n+T(n)$ to $n+T(n)/k$ - than it is known for deterministic TM - from $n+T(n)$ to $2n+18+\lceil (n+T(n))/k\rceil$ [LP]. For example, the speed-up from $n+\log^* n$ is possible for CA and such a speed-up is not known for TM.

The following hierarchy result has been shown [IKM] for STM_p^v but it holds also for some other models of STM. From that also hierarchy results for CA, IA and PTM follow.

Theorem 3.5 If $i\varepsilon\{p,sk,se\}$, $a\varepsilon\{v,t\}$, then $STA_i^a(T_2(n))-STA_i^a(T_1(n))\neq 0$ if T_2 is time constructible on a one tape Turing machine and

$$\inf_{n\to\infty} \frac{T_1^2(n)\log T_1(n)}{T_2(n)} = 0 \quad \Delta$$

3.6 Trellis automata and generalized Pascal triangles It has been shown in Section 3.1 that there is one-to-one correspondence between infinite computations of STA with parallel input and generalized Pascal triangles $GPT(\mathcal{A},w)$ over algebras $\mathcal{A}=(A,l,r,g)$ of the signature (1,1,2) and $w\varepsilon A^+$. (See Fig. 3.10 for the case $A=\{0,1,2\}$, $l=r=id$, $g=$ the addition modulo 3). This gives rise to new problems which arise naturally if one thinks in terms of GPT and which may throw light on infinite computations of STA (and therefore also of CA, IA and PTM). For example:

1 2 1 2 1 2 1 2 1
1 1 0 2 2 0 1 1
1 0 0 2 0 0 1
1 2 1 1 2 1
1 1 0 1 1
1 0 0 1
1 2 1

Fig. 3.10

(1) What kind of structure have diagonals and columns of GPT?
(2) With every $GPT(\mathcal{A},w)=\{\pi^i(w)\}_{i=0}^{\infty}$ one can associate the language $\{\pi^i(w)\mid i\geq 0\}$. What kind of languages are they?
(3) Let $T=\{\pi^i(w)\}_{i=0}^{\infty}$ be a GPT and let $T(i,j)$ denote the $(j+1)$-th symbol of $\pi^i(w)$. T is said to have self-embedding property if there are nonnegative p,q,a,b,c such that $T(i,j)=T(i+p,j+q)$ for $i\geq a, i+|w|-c>j\geq b$. What can be said about such GPT?
(4) Decision problems. Is it decidable whether a given symbol (word) appears - as a subword in at least one (in infinitely many) [in almost all] words of a given GPT? Is it decidable whether a given word (infinitely many words [almost all words] of a given GPT are words over a given subalphabet {of one symbol}?

Theorem 3.6 (i) All decision problems in (4) - and many others - are undecidable for GPT over finite algebras with a commutative binary operation.

(ii) All decision problems in (4) - and many others - are decidable for GPT such that the corresponding language is a simple semilinear language.*

(iii) If a GPT has a self-embedding property, then the corresponding language is simple semilinear and context-free.

(iv) If $\mathcal{A}=(\Gamma,l,r,g)$ and g is idempotent and associative (or some other identities hold), then all GPT over $\mathcal{A}$ have self-embedding properties. Δ

This theorem has been shown in [Ko-1] where many other results concerning GPT have been obtained. The first part of the next theorem is due to [Ko-3], the second is from [IKM].

---

* A language is simple semilinear if it is union of finitely many languages of the type $\{U_oV_1^iU_1\ldots V_k^iU_k \mid i\geq o; U_o,V_1,\ldots,U_k \text{ fixed}\}$.

Theorem 3.7 (1) If $\mathcal{A}=(A,l,r,g)$ is a finite algebra of the signature (1,1,2), then all diagonals of GPT over $\mathcal{A}$ are ultimately periodic and all their periods are divisible only by primes smaller than cardinality of A.

(2) It is undecidable (decidable) for a given column (diagonal) of a GPT and a given symbol whether this symbol occurs in that column (diagonal). $\triangle$

4. HOMOGENEOUS REAL-TIME TRELLIS AUTOMATA

Real-time trellis automata are of a special interest from theoretical as well as from implementation point of view. Actually there exist several types of real-time trellis automata - for different input modes and acceptance rules. However, we shall deal here only with automata of the form depicted in Fig. 4.1. They seem to be more important and to have more interesting properties than others. In the notation of Section 3.1 they are actually real-time SPA_p^d and they have a similar characterization in terms of STM [IK-1] - see Fig. 3.9. We shall call them homogeneous real-time STA - HRTSTA for short. (There is also another way of looking at HRTSTA - the basic model is as in Section 3.1 but data flow in the opposite direction).

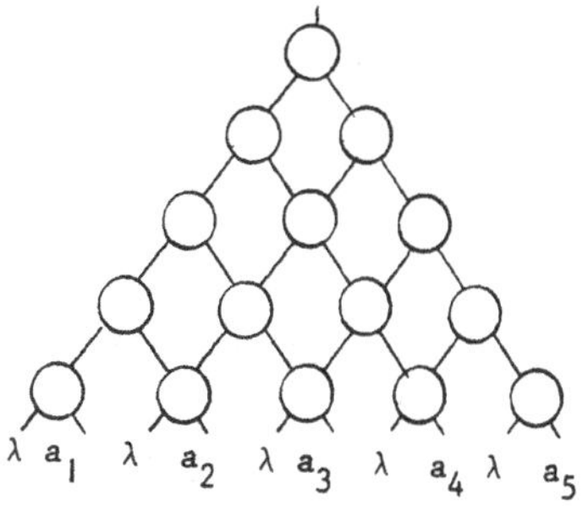

Fig. 4.1

4.1 Power and limitations Several techniques to design HRTSTA have been discussed in [CGS-1] and another follows from TM-characterization of HRTSTA. Using these techniques one can show that the following languages are in L(HRTSTA): $\{x\#x^R/x\epsilon\Sigma^+,\#\notin\Sigma\}$, $\{xx^R/x\epsilon\Sigma^+\}$, $\{x\#x/x\epsilon\Sigma^+,\#\notin\Sigma\}$. On the other hand it is an open problem whether the language - which is known [IKM] to be in $L(STA_p^v)$-$\{xx/x\epsilon\Sigma^+\}$ is in L(HRTSTA), too.

Theorem 4.1 [CGS-1,IK-1] L(HRTSTA) is an abstract family of deterministic languages which contains all linear ε-free context-free languages (but only regular languages over one-letter alphabets), some languages complete for polynominal time with respect to log-space reducibility, and it is closed under Boolean operations **and injective length multiplying morphisms, but not under letter-to-letter morphisms.** $\triangle$

It is an *open problem* whether L(HRTSTA) is closed under concatenation and Kleene closure. Theorem implies that the language $\{a^{2^n}/n\geq o\}$ which is in L(BStA) is not in L(HRTSTA) and that the language $\{a^nb^n/n\geq 1\}$ which is not in L(BStA) is in L(HRTSTA). Hence these two families are incomparable. (Theorem 4.1 has been used in [BC] to show some closure results for families of languages accepted by various CA.) The following theorem shows that also decidability results for HRTSTA and StA on balanced trees differ [CGS-2].

Theorem 4.2 The emptiness and equivalence problems for HRTSTA are undecidable. $\triangle$

Theorems 4.1 and 4.2 indicate that HRTSTA are quite powerful. However, they are not too powerful:

Theorem 4.3 Any language accepted by HRTSTA can be accepted in $O(n^2)$ time and O(n) space on one-tape Turing machines. $\triangle$

It has been shown [CGS-2] that this time bound cannot be improved and, likely, neither for multitape Turing machines. It follows from Theorem 4.1 [IK-1] that even space bounds cannot be improved significantly for multitape Turing machines.

The only known direct technique to show a language is not in L(HRTSTA) is the following pumping lemma [Hr].

Lemma 4.4 Let $L=\langle\Gamma,\Sigma,\Gamma_o,g\rangle$ be a HRTSTA and $k=|\Gamma|$. If $uw^rv\epsilon L(A)$ for a $r>|w|k^{|uwv|}+|uv|+1$, then there is a z>o such that $uw^{r\pm z}v\ \epsilon L(A)$. $\triangle$

Corollary $L=\{0^m 1 2^{m^2} / m\geq 1\}$ is not in L(HRTSTA).

4.2 Modifications Stability, superstability and nondeterminism are defined for HRTSTA in a similar way as for StA. This time, however, we have quite different results [CGS-2, IK-1].

Theorem 4.5 (1) To every HRTSTA one can effectively construct an equivalent superstable HRTSTA.

(2) Languages accepted by nondeterministic HRTSTA form an AFL which contains some NP complete languages and is contained in DSPACE (n log n).

5. NONHOMOGENEOUS REAL-TIME TRELLIS AUTOMATA

For CA and IA it has been natural to require homogenity. However, a different situation seems to be for systolic trellis automata. There are no big problems to design and implement nonhomogeneous trellis networks provided only few types of processors are used and they are distributed through the network in a sufficiently regular and/or modular way. It is therefore interesting to find out how much one gains with nonhomogeneous trellis networks with different processor distributions.

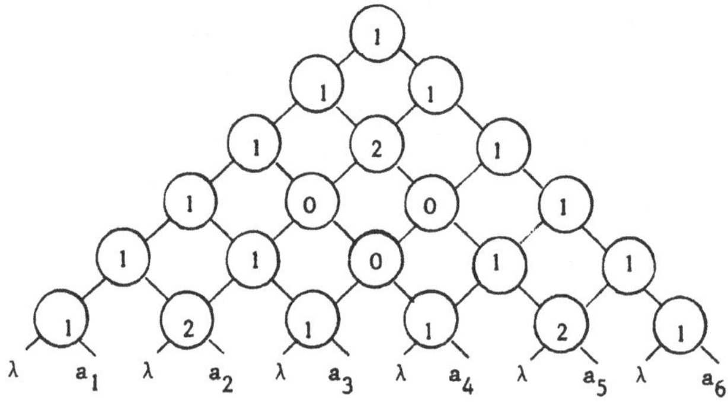

Fig. 5.1

5.1 Top-down regular real-time trellis automata (RRTSTA for short), their computations and acceptance are defined in a similar way as for homogeneous ones. The underlying network is again the infinite triangularly shaped trellis - Fig. 5.1 - of combinational circuits, each with two inputs. This time, however, several types of processors are used but they are distributed in the network in such a way that their names, depicted as labels of nodes in Fig. 5.1, form a GPT($\mathcal{A}$,a) over an algebra $\mathcal{A}=\langle\Delta,l,r,g\rangle$, and a symbol $a\in\Delta$ - see Sections 3.1 and 3.6. (This means that labelling of nodes satisfies a similar local regularity condition as in Section 2.3).

Formally, a RRTSTA is a device $A=\langle T,\Delta,\Gamma,\Sigma,\Gamma_0,G\rangle$ where T is a GPT($\mathcal{A}$,a) over an algebra $\mathcal{A}=(\Delta,l,r,g)$; Γ,Σ,Γ_0 are alphabets of operating, terminal and accepting symbols, $\Sigma\subseteq\Gamma,\Gamma_0\subseteq\Gamma,\lambda\notin\Gamma$ and $G=\{g_\delta \mid \delta\in\Delta,\ g_\delta:(\Gamma\cup\{\lambda\})\times(\Gamma\cup\{\lambda\}\to\Gamma\}$ is a family of transition functions.

RRTSTA can be viewed as having a top-down flow of control (fixing a program in each node), and a bottom-up flow of information (when an input is processed). In order to process its inputs a processor labelled by $\delta\in\Delta$ uses the transition function g_δ. Otherwise RRTSTA process inputs in the same way as HRTSTA!

From the next theorem [Gr] it follows that the family of languages accepted by RRTSTA is richer than the family of languages accepted by HRTSTA and, in a reasonable sense, also richer than the family of languages accepted by StA on balanced trees.

Theorem 5.1 To any StA $A=\langle\Gamma,\Sigma,\Gamma_0,g\rangle$ on a p-ary balanced tree there exists effectively a RRTSTA $\bar{A}=\langle T^p,\Delta_p,\Sigma\cup\{\#\},\bar{\Gamma}_A,\Gamma_0,G_A\rangle$ which uniformly simulates A in the following sense: to any node N of the p-ary balanced tree corresponds a node $\bar{N}$ of the infinite trellis such that the following holds: if a $w\in\Sigma^+$ is processed by A, and $w\#^i, i<|w|$, $|w\#^i|=p^k$, is processed by A, then the processor of A in the node N and the processor of $\bar{A}$ in the node $\bar{N}$ produce the same output.Δ

On the other hand, the following theorem indicated that RRTSTA are not much more powerful than HRTSTA [CGS-1].

Theorem 5.2 If a language L is accepted by a HRTSTA, then L is accepted by a deterministic multitape [one-tape] Turing machine in time $O(n^2)$ [$O(n^3)$] and space $O(n)$.

There is, however, no simple way to find out whether the language accepted by a RRTSTA can be accepted by a HRTSTA [Če].

Theorem 5.3 It is undecidable, given a RRTSTA A, whether $L(A) \in \mathcal{L}(\text{HRTSTA})$.△

RRTSTA can also be characterized in terms of full scan Turing machines [IK-1] - this time with a push-down store.

Three descriptional complexity measures are quite natural for RRTSTA $A=\langle T,\Delta,\Gamma,\Sigma,\Gamma_o,G\rangle$: the number of processor names (i.e. $|\Delta|$), the number of types of processors (i.e. the number of different functions in G), and the size of the auxiliary alphabet $\Gamma-\Sigma$.

Theorem 5.1 implies that having more than one type of processors we gain in the recognition power. But is it true that the more processors we have the more powerful automata we get? The following result [Ko-2] gives the negative answer.

Theorem 5.4 To every RRTSTA one can effectively construct an equivalent RRTSTA which uses only two types of processors.△

This result may be of importance for systolic system implementations because it shows that quite a large class of nonhomogeneous systems can be replaced by systems with only two types of processors. On the other hand, no restriction on the size of auxiliary alphabets can be made without restricting the power of RRTSTA [Ko-2]. However, it is an *open problem* whether there is a k such that any language $L \in \mathcal{L}(\text{RRTSTA})$ can be accepted by a RRTSTA $A=\langle T,\Delta,\Gamma,\Sigma,\Gamma_o,G\rangle$ with $|\Delta|<k$.

One can also show that any language $L \in \mathcal{L}(\text{RRTSTA})$ can be accepted by a RRTSTA which is "almost homogeneous" in the following sense.

Theorem 5.5 If $L \in \mathcal{L}(\text{RRTSTA})$, then for any $o<\varepsilon<1$ there exists a RRTSTA A and a processor P of A such that $L(A)=L$ and for any $n \geq 1$ $N_A^P(n)>(1-\varepsilon)\frac{n(n+1)}{2}$ (where $N_A^P(n)$ denotes the number of occurences of P at the first n levels of processors).△

Theorems 5.2, 5.4 and 5.5 and also the next theorem indicate that the regular RTSTA are quite a natural and not too strong generalization of HRTSTA [IK-1].

Theorem 5.6 A language L is accepted by a nondeterministic RTSTA if and only if L is accepted by a nondeterministic HRTSTA.△

Undecidability and decidability results for GPT - Theorem 3.6 - imply undecidability and decidability of various problems concerning processors distribution in RRTSTA.

Theorem 5.7 The following problems are undecidable for all RRTSTA but decidable for those RRTSTA the underlying GPT of which has the self-embedding property: (1) Does a given processor appear in a given network (i) at least once, (ii) infinitely often, (iii) at almost all levels? (2) Is it true that all but finitely many processors are the same?△

5.2 Semihomogeneous real-time trellis automata (SRTSTA). The first idea for a global regularity condition for processor distribution is to take a uniform distribution defined similarly as in the case of StA: a RTSTA is called semihomogeneous if the corresponding labelled trellis graph has only finitely many infinite nonisomorphic labelled subtrellises. The following theorem shows that as far as the recognition power is concerned we do not gain anything [CGS-1]. However, inspite of that this theorem represents an important design methodology for HRTSTA because SRTSTA are quite often more easy to construct (see [CGS-1] for examples).

Theorem 5.8 To any semihomogeneous RTSTA one can construct an equivalent homogeneous RTSTA.

5.3 Modular real-time trellis automata (MRTSTA for short) are another type of non-homogeneous RTSTA with global regularity condition for processor distribution. The basic idea is to consider trellis automata the underlying network of which can be assembled in a fixed way using modules of arbitrarily large size.

Formally, a MRTSTA is a device $A=\langle T,\Delta,\Gamma,\Sigma,\Gamma_o,G\rangle$ where Γ,Σ,Γ_o and G have the same meaning as in the case of RRTSTA and T is a trellis network, labelled by labels from Δ in such a way that the following modularity condition is satisfied: there are integers $p>1, q>1, \delta^o \in \Delta$ and for each $\delta \in \Delta$ a pxq matrix M_δ^1 of elements from Δ (see Fig. 5.2(a) such that $\delta_{oo}{}^o = \delta^o$ and $T(i,j) = M_{\delta^o}^j(i,j)$, where $T(i,j)$ denotes the label of the i-th node at the (i+j)-th level of T, and for each $\delta \in \Delta$ and $i \geq 1$, M_δ^{i+1} - Fig. 5.2 - is a $p^{i+1} x q^{i+1}$ matrix.

$$M_\delta^1 = \begin{vmatrix} \delta_{oo} \cdots \delta_{p-1,o} \\ \vdots \qquad \vdots \\ \delta_{o,q-1} \cdots \delta_{p-1,q-1} \end{vmatrix} \qquad M_\delta^{i+1} = \begin{vmatrix} M^i_{\delta_{oo}} \cdots M^i_{\delta_{p-1,o}} \\ \vdots \qquad \vdots \\ M^i_{\delta_{o,q-1}} \cdots M^i_{\delta_{p-1,q-1}} \end{vmatrix}$$

(a) *Fig. 5.2* (b)

MRTSTA are more powerful than HRTSTA. Indeed, Theorem 5.1 can be proven in such a way that all T^p are also modular. This implies that MRTSTA can simulate StA on balanced trees and therefore they accept all languages $L_k = \{a^{k^n} \mid n \geq o\}, k \geq 2$.

It could seem that it is more difficult to determine labels of nodes of modular trellises than of regular trellises. However, the opposite is true.

Theorem 5.9 A labelled trellis T is modular if and only if there is a finite automaton M such that if M receives on its input a pair $(i,j), 1 \leq j \leq i$, in parallel and digit by digit, then M produces the label T(i,j) of T.

A better upper bound are also known for Turing machine recognition of languages accepted by MRTSTA than for languages accepted by RRTSTA.

Theorem 5.10 If a language L is accepted by a MRTSTA [where the underlying trellis is also to-down regular], then L can be accepted in time $O(n^2)$ and space O(n) on multitape deterministic Turing machines and in time $O(n^2 \log n)$ $[O(n^2/\log n)]$ and space O(n) on one-tape deterministic Turing machines.

Modular trellises have, in a sense, a more transparent structure as it follows from the following theorem. (All results of this section are from [CG]).

Theorem 5.11 The following problems are decidable for MRTSTA:

(1) Does a given processor appear in a given network (i) at least once, (ii) infinitely often?

(2) Is it true that almost all processors of a given MRTSTA are the same?

Open problems: (1) Is it true that to every MRTSTA there exists an equivalent MRTSTA which uses only two types of processors? (2) Which, if any, of the two classes L(RRTSTA) and L(MRTSTA) is larger? (But we know there exists a RTSTA which is modular but not regular and vice versa). (3) Have nondeterministic MRTSTA different power than nondeterministic RRTSTA?

R E F E R E N C E S

[Če] Černý, A.: Personal communication, 1984.

[ČG] Černý, A. - Gruska, J.: Modular trellis automata (in preparation), 1984.

[Co] Cole, S.N.: Real-time computation by n-dimensional iterative arrays of finite-state machines, IEEE Trans. on Comp., V18, 1969, 349-365.

[CC] Choffrut, C.-Culik, K.II.: On real-time cellular automata and trellis automata, Res. Rep. F114, Inst. für Informationsverarbeitung, TU Graz, 1983.

[CB] Culik, K.II.- Bucher, W.: On real-time and linear time cellular automata, Res. Rep. F115, Inst. für Informationsverarbeitung, TU Graz, 1983.

[CGS-1] Culik, K.II. - Gruska, J. - Salomaa, A.: Systolic trellis automata (for VLSI), Res. Rep. CS-81-34, Dept. of Comp. Sci., University of Waterloo, 1981.

[CGS-2] Culik, K.II. - Gruska, J. - Salomaa, A.:Systolic trellis automata: stability, decidability and complexity, Res. Rep. CS-82-04,Dept. of Comp. Sci., University of Waterloo, 1982.

[CGS-3] Culik, K.II. - Gruska, J. - Salomaa, A.: Systolic automata for VLSI on balanced trees, Acta Informatica 18, 1983, 335-344.

[CGS-4] Culik, K.II. - Gruska, J. - Salomaa, A.: On a family of L languages resulting from systolic tree automata, TCS 23, 1983, 231-242.

[CJ] Culik, K.II. - Jürgensen, H.: Programmable finite automata for VLSI, Intern. J. Computer Math., V14, 1983, 259-275.

[CP] Culik, K.II. - Pachl, J.: Folding and unrolling systolic arrays, Proc. ACM SIGACT - SIGOPS Symp. on Principles of Distributed Computing, 1982, 254-261.

[CSW] Culik, K.II. - Salomaa, A. - Wood, D.: VLSI systolic trees as acceptors, RAIRO, Theoretical Informatics, 1984, to appear.

[CY] Culik,K.II. - Yu, S.: Iterative tree automata, Res. Rep. CS-84-03, Dept. of Comp. Sci., University of Waterloo, 1984.

[DFKM] Dohl, Y. - Fischer, A.L. - Kung, H.T. - Monier, L.M.: The programmable systolic chip: project overview, Proc. Workshop on Algorithmically Specialized Computer Organisations, Purdue Univ., Sept. 1982.

[Gr] Gruska, J.: Simulation of systolic tree automata by systolic trellis automata, (in preparation) 1984.

[He] Hennie, F.C.: Iterative arrays of logical circuits, MIT Press, 1961.

[Hr] Hromkovič, J.: Personal communication, 1984.

[IK-1] Ibarra, O. - Kim, S.: Characterizations and computational complexity of systolic trellis automata, TCS, V29, N1,2, 1984.

[IK-2] Ibarra, O. - Kim, S.: A characterization of systolic binary tree automata and applications, to appear in Acta Informatik, 1984.

[IK-3] Ibarra, O. - Kim, S.: Characterizations of linear iterative arrays, Res. Rep., Dept. of Comp. Sci., University of Minnesoty, 1983.

[IKM] Ibarry, O. - Kim, S. - Moran, S.: Trellis automata: characterizations, speed-up, hierarchy, decision problems, Res. Rep.,Dept. of Comp. Sci., University of Minnesota, 1983.

[JS] Jürgensen, H. - Salomaa, A.: Syntactic monoids in the construction of systolic tree automata, TI-5/83, Inst. für Theoretische Informatik, Tech. Hochschule Darmstadt, 1983.

[Ko-1] Korec, I.: Generalized Pascal triangles: decidability results, to appear in Acta Math., Univ. Comen, Bratislava, 1985.

[Ko-2] Korec, I.: Two kinds of processors are sufficient and large operating alphabets are necessary for regular trellis automata languages,Bulletin EATCS, 1984.

[Ko-3] Korec, I.: Personal communications, 1984.

[Ku] Kung, H.T.: Why systolic architecture, Computer Magazine, 1982, 37-46.

[KL] Kung, H.T. - Leiserson, C.E.: Systolic arrays (for VLSI), Sparse Matrix Proc. 1978, Society f. Industrial and Applied Mathematics, 1979, 256-282.

[LP] Lewis, H.R. - Papadimitriou, C.H.: Elements of the theory of computation, Prentice-Hall, 1981.

[Pa] Patterson, M.: Solution to P8, EATCS Bulletin, 18, 1982.

[Sa] Salomaa, A.: Systolic tree and trellis automata, Proc.Colloq ACLCS, Györ, 1984.

[SS] Salomaa, A. - Soittola, M.: Automata - theoretic aspects of formal power series, Springer-Verlag, Berlin, 1978.

[St] Steinby, M.: Systolic trees and systolic language recognition by tree automata, TCS, V22, 1983, 219-232.

[Wi] Wiedermann, J.: Parallel Turing machines and systolic computations, Res. Rep., VÚSAI-AR Bratislava. 1984.

A NOTE ON UNIQUE DECIPHERABILITY*

Christoph M. Hoffmann
Department of Computer Science
Purdue University
West Lafayette, Ind. 47907

We consider the following problem: Given a set $\Gamma = \{c_1, ..., c_n\}$ of nonempty strings over a fixed, finite alphabet Σ, is every string in Γ^+ uniquely decipherable, or does the equation

$$cx = dy,$$

where $c, d \in \Gamma$, $c \neq d$, and $x, y \in \Gamma^*$, have a solution? We give an $O(n\,L)$ algorithm for this problem, where $L = |c_1| + ... + |c_n|$, and use this algorithm to investigate the impact of structural properties of Γ on the complexity of testing unique decipherability. We then give an $O(L \log(n))$ unique decipherability test for sets Γ which may be linearly ordered by the prefix relation.

1. Introduction

Let Σ be a fixed finite alphabet, $\Gamma = \{c_1, ..., c_n\}$ a set of nonempty, finite words over Σ. We say that Γ is a *uniquely decipherable code* if the equation

$$cx = dy$$

where $c, d \in \Gamma$, $x, y \in \Gamma^*$, and $c \neq d$, has no solution.

Many algorithmic solutions to the problem have been given, e.g., in [SP 53], [Mar 62], [Blu 65], [Spe 75], [Cap 79], [Eve 79], [Rod 82], [AG 84]. The two fastest algorithm for the problem are due to Rodeh [Rod 82] and Apostolico and Giancarlo [AG 84], and take $O(n\,L)$ steps, where $L = |c_1| + ... + |c_n|$ is the sum of the lengths of the words in Γ. Conceptually, these algorithms are based on the Sardinas-Patterson algorithm [SP 53], and their superior performance is based primarily on the use of advanced pattern matching techniques.

We add to this another $O(n\,L)$ algorithm to test whether Γ is uniquely decipherable, which is conceptually related to Spehner's algorithm [Spe 75]. The advantage of our algorithm over

* This work was supported in part by the National Science Foundation under Grant MCS 80-21066

the others is that it retains the simplicity of the Apostolico-Giancarlo algorithm while making explicit structural properties of codes, such as synchronizability and finite deciphering delay. Our algorithm also permits exhibiting a deciphering ambiguity in case Γ is not uniquely decipherable, which cannot be done with the other two.

However, the purpose of this paper is not to add another $O(n\,L)$ method to the repertoire of available unique decipherability tests. Rather, we wish to investigate the relationship between the complexity of testing unique decipherability and combinatorial properties of the words in Γ. Indeed, apart from the well-known fact that all prefix codes are uniquely decipherable, it seems to be unknown why the worst case running time of these algorithms is $O(n\,L)$ rather than, say, $O(L)$: Are there specific structural properties of the words in Γ which have to be present for the above algorithms to attain the $O(n\,L)$ upper bound? And if so, can we find special techniques to circumvent the difficulty? The investigation of these questions is the focus of this paper.

Our $O(n\,L)$ algorithm happens to be the ideal vehicle for making explicit the connection between the repetitive structure of code words and the complexity of testing unique decipherability. This algorithm is developed in Sections 2 and 3 of the paper. From then on, we investigate the impact of word periodicity on the algorithm and show that high periodicity, in conjunction with a high degree of prefix and suffix relationship among code words, causes the worst case bound. Without proof, we mention that the other $O(n\,L)$ tests attain their worst case for the same reason.

We investigate the question in detail in the special case where the set Γ may be linearly ordered by the prefix relation. For this case, as it were, the worst case time for testing unique decipherability is still $O(L \log(n))$, even in the absence of high periodicity. We demonstrate this situation in Section 4.

In Sections 5 and 6 we investigate in detail how high periodicity influences the graph structure underlying the $O(n\,L)$ algorithms. We show that the graph size becomes large precisely on very regular strings, and that its structure consequently can be understood and exploited, so that large portions of the graph never need to be constructed. The remaining subgraph is sufficiently small, permitting an $O(L \log(n))$ unique decipherability test for all nonbranching sets Γ, regardles of the periodicity of the words in the set. Section 7 presents the structure of the resulting algorithm, which is essentially a very simple modification of our $O(n\,L)$ algorithm. It is possible to similarly modify the Apostolico-Giancarlo algorithm.

At the outset of this investigation, it appeared that the influence of the word structure on a unique decipherability test was so complicated that the resulting, faster test would not have practical merits. This turned out to be false. The faster algorithm is still simple and can be easily implemented. At this time it is unclear, however, how this approach carries over to the general case. We believe that an $O(L \log(n))$ unique decipherability test is possible for the general case, but exactly how the analytical methods developed here are to be generalized remains to be seen.

2. Terminology, Notation and Basics

In the following, Σ is a fixed, finite alphabet of letters, Σ^* the set of all finite words over Σ, $\lambda \in \Sigma^*$ the empty word, and $\Sigma^+ = \Sigma^* - \{\lambda\}$. The length of $w \in \Sigma^*$ is denoted by $|w|$.

Let $x = uv$ be in Σ^*. Then u is a *prefix* of x, written $u < x$, and is *proper* if $v \neq \lambda$. Similarly, v is a *suffix* of x, written $x > v$, and is *proper* if $u \neq \lambda$. Let $x = uv = wu$. Then u is a *divisor* of x and is *proper* if $v \neq \lambda$. Clearly there is a unique longest proper divisor of x, and the relation *longest proper divisor of* has been called the *failure function* in [AC 75].

Some of the techniques to be developed for analyzing the unique decipherability problem depend on combinatorial properties of strings. Recall that a word w is *primitive* if $w = u^k$ implies $u = w$ and $k = 1$. Equivalently, w is primitive iff, for all partitions $w = pq$, where $p \neq \lambda$ and $q \neq \lambda$, $w \neq qp$. That is, w is different from all its *conjugates* qp.

Every string w can be written $w = x^i y$, where x is primitive, y is a proper prefix of w, not necessarily nonempty, and $i \geq 1$. We make this decomposition unique by insisting that x be the shortest among all possible primitive x. Let $w = x^i y$ be this unique decomposition of w. Then x is the *root* of w, $|x|$ is the *period* of w, and i is the *exponent* of w.

In the following, Γ is a finite set of nonempty words, and Γ^+ is the set of $w \in \Sigma^+$ which may be obtained by concatenating elements of Γ. We define the sets

$$Prefix(\Gamma) = \{\, u \in \Sigma^* \mid u < c,\ c \in \Gamma\}$$
$$Suffix(\Gamma) = \{\, u \in \Sigma^* \mid c > u,\ c \in \Gamma\}$$
$$Divisor(\Gamma) = Prefix(\Gamma) \cap Suffix(\Gamma)$$

A constructive proof that Γ is not a code might proceed by finding $c, c' \in \Gamma$ such that $c' < c$, followed by an attempt to construct $w, w' \in \Gamma^*$ such that $cw = c'w'$. At any point in this a construction, we have found prefixes p and p' of $cw = c'w'$ which are in Γ^+, and, without loss of generality, the difference between the two prefixes is a suffix of some code word. These suffixes have been called *tails*. Initially, $p = c$ and $p' = c'$, so the first such tail is s, where $c's = c$.

Consider the prefixes p and p' of $cw = c'w'$ with the current tail s, assuming $p' < p$ without loss of generality. By extending p' with a code word c'' we obtain the prefixes p and $p'c''$ with the new tail s'. Two cases arise:

1. The word c'' is prefix of s, i.e., $s = c''s'$. Then we say that s has been *contracted* to s'.
2. The tail s is prefix of c'', i.e., $c'' = ss'$. Then we say that s has been *extended* to s'.

When the two cases coincide we have obtained a proof that Γ is not a code.

It is easy to map this constructive approach into a graph traversal problem. Consider the following graph $U(\Gamma)$ with vertex set $Suffix(\Gamma)$: There is an edge (u, v) in $U(\Gamma)$ if there is a $c \in \Gamma$ such that $c = uv$ or $u = cv$. We identify the subset $J(\Gamma) = \{\, s \mid cs \in \Gamma, c \in \Gamma, s \neq \lambda \,\}$ of initial tails among the vertices. The following is obvious:

Proposition 2.1
Γ is not uniquely decipherable iff there is a path in $U(\Gamma)$ from a vertex in $J(\Gamma)$ to the vertex λ.

It is more convenient to construct a related graph $R(\Gamma)$ whose vertex set is $Prefix(\Gamma)$. There is an edge (u, v) in $R(\Gamma)$ if there is a $c \in \Gamma$ such that $c = uv$ or $uc = v$. We associate with the suffix s all prefixes u such that $us \in \Gamma$. Clearly, this induces the association of the set Γ with λ, and of the set $S(\Gamma) = \{\, c \in \Gamma \mid cs \in \Gamma, s \neq \lambda \,\}$ with $J(\Gamma)$. It is now easy to see the following

Corollary 2.2
Γ is not uniquely decipherable iff there is a path in $R(\Gamma)$ from a vertex in $S(\Gamma)$ to a vertex in Γ.

The specific advantage of using $R(\Gamma)$ instead of the more traditional graph $U(\Gamma)$ is that the pattern matching methods of [AC 75] can be applied directly to the set Γ rather than applying them to the set of reversed words as is done in [AG 83]. We also remark, without proof, that the edges of $R(\Gamma)$ may be suitably labelled so that the label sequence of a path from $S(\Gamma)$ to Γ permits specifying a deciphering ambiguity for Γ. Thus our data structure not only permits testing unique decipherability, it also allows finding explicitly ambiguous messages if Γ is not uniquely decipherable. In [CH 84] we have explored the suitability of $R(\Gamma)$ for a number of semigroup algorithms.

The construction of $R(\Gamma)$ begins by finding the trie of all prefixes of words in Γ. This trie forms the graph $Tree(\Gamma)$, which has been called the *graph of the goto-function* in [AC 75]:

(1) The vertex set of $Tree(\Gamma)$ is $Prefix(\Gamma)$.
(2) For $u, v \in Prefix(\Gamma)$, $a \in \Sigma$, if $ua = v$ then (u, v) is an edge of $Tree(\Gamma)$.
(3) Nothing else is an edge of $Tree(\Gamma)$ unless required by (2).

Now $R(\Gamma)$ has the same vertex set as $Tree(\Gamma)$, so it is obtained by adding edges to $Tree(\Gamma)$. We distinguish in $R(\Gamma)$ *reach* and *divisor* edges:

$$E_{\mathrm{Reach}} = \{ (u, v) \mid uc = v,\ u \neq \lambda,\ u, v \in Prefix(\Gamma),\ c \in \Gamma \}$$
$$E_{\mathrm{Divisor}} = \{ (w, u) \mid wu \in \Gamma,\ u \in Divisor(\Gamma),\ u \neq \lambda,\ w \neq \lambda \}$$

Example 2.1
Figure 2.1(a) shows an edge in E_{Reach} and Figure 2.1(b) an edge in E_{Divisor}. Wiggly lines indicate paths in $Tree(\Gamma)$.

In general, E_{Reach} may contain $O(n\,L)$ edges, where $L = |c_1| + \ldots + |c_n|$, as can be seen by considering the sets

$$\Gamma_n = \{ a^i \mid 1 \leq i < n \} \cup \{ a^{n^2} \},$$

whose graphs $R(\Gamma)$ have approximately $\frac{2}{3}(n-1)L$ reach edges. But since $R(\Gamma)$ has at most $L+1$ vertices, it is clear that Γ is not a code iff there is a path of at most $L+2$ vertices in $R(\Gamma)$. Hence most of the edges in a large set E_{Reach} are not needed. After giving the basic $O(n\,L)$ algorithm in Section 3, therefore, we will subsequently concentrate on the problem of identifying a suitably small subgraph of $R(\Gamma)$ which still will permit testing unique decipherability.

Without proof, we remark that $R(\Gamma)$ permits testing a number of special properties a uniquely decipherable code may have. Specifically, we mention the properties of *finite decipherability delay* and of *finite synchronizability*. Intuitively, Γ has finite decipherability delay if there is a constant d such that, after reading the first d consecutive symbols of $x \in \Gamma^+$, we can uniquely decipher the leftmost code word in x. Moreover, Γ is finitely synchronizable, if after reading s consecutive symbols x in a message $yxz \in \Gamma^+$, s a constant, we can split $x = x_1 x_2$ such that the split is on a correct code word boundary, i.e., such that $yx_1 \in \Gamma^*$ and $x_2 z \in \Gamma^*$.

Formally, a uniquely decipherable code Γ has *finite decipherability delay* if there exists an integer d such that, for all $x \in \Sigma^+$, $|x| \geq d$, there is $c \in \Gamma$ such that, for all y with $xy \in \Gamma^+$, we have $xy = cz$ and $z \in \Gamma^*$.

A uniquely decipherable code Γ is *finitely synchronizable* if there is an integer s such that, for all $x \in \Sigma^+$, $|x| \geq s$, there is a split $x = x_1 x_2$ such that, for all y and z with $yxz \in \Gamma^+$, we have $yx_1 \in \Gamma^*$ and $x_2 z \in \Gamma^*$.

Note that every synchronizable code has finite decipherability delay, but not necessarily vice versa. Moreover, codes with finite decipherability delay can be deciphered by a sequential transducer. The following proposition is easy to prove; compare also [Cap 79] and [CH 84]:

Proposition 2.3
A uniquely decipherable code Γ is synchronizable iff $R(\Gamma)$ is acyclic. Moreover, Γ has finite decipherability delay if no cycle in $R(\Gamma)$ can be reached from a vertex in $S(\Gamma)$.

3. An $O(n\,L)$ Unique Decipherability Test

In the following, $\Gamma = \{c_1, ..., c_n\}$, and $L = |c_1| + ... + |c_n|$. We assume that the reader is familiar with the pattern matching algorithm of [AC 75]. The unique decipherability test to be given is based on Corollary 2.2 above. Specifically, it first constructs the directed graph $R(\Gamma)$, and then, by a breadth-first search of $R(\Gamma)$, it determines whether there exists a vertex $u \in S(\Gamma)$ from which, by a nonempty path in $R(\Gamma)$, a vertex $v \in \Gamma$ can be reached.

The algorithm for testing whether Γ is a uniquely decipherable code has the following major steps:

Algorithm 1

1. Construct a pattern matching machine M with Γ the set of patterns.
2. Construct the edge set E_{Divisor}.
3. Construct the edge set E_{Reach}.
4. Search $R(\Gamma)$ breadth-first to locate a path from $S(\Gamma)$ to Γ. If such a path exists, then Γ is not uniquely decipherable; otherwise Γ is uniquely decipherable.

Step (1) is implemented using the algorithm of [AC 75]. A routine modification enables it to locate all matches in a subject, rather than the first match only. As shown in [AC 75], Step (1) can be implemented in $O(L)$ steps. Note that M contains $Tree(\Gamma)$ in its description. Steps (2) and (3) will have to perform the following operations on $Tree(\Gamma)$:

(a) Given a vertex v of $Tree(\Gamma)$, find the length $|v|$ of v.
(b) Given a vertex v of $Tree(\Gamma)$, find the vertex u of a prefix u of v of prescribed length.

It is clear that both operations can be implemented in constant time assuming a preprocessing step creating index structures requiring $O(L)$ steps. Briefly, with every leaf v of $Tree(\Gamma)$ is associated a vector V such that $V[i]$ points to the prefix of length i of v. Each interior vertex u uses the vector associated with an arbitrarily selected leaf in the subtree rooted at u.

Recall from [AC 75] the notation $f(u) = v$ which means that v is a maximal proper suffix of u which is in $Prefix(\Gamma)$, and recall that M contains the graph of f for all strings in $Prefix(\Gamma)$. Consider $c \in \Gamma$, and assume that $f(c) = y_1$, $f(y_1) = y_2$, ..., $f(y_k) = \lambda$, where $y_k \neq \lambda$. Note that $k < |c|$. We add the arcs (u_i, y_i) to E_{Divisor}, $1 \leq i \leq k$, where u_i is the prefix of c of length $|c| - |y_i|$. Tracing the graph of f from c, this can be implemented in $O(|c|)$ steps, hence Step (2) requires $O(L)$ steps in all.

Consider the determination of E_{Reach}, Step (3). Here we run M on $Tree(\Gamma)$. Each time we have a match of the set $\{d_1, ..., d_k\} \subset \Gamma$ at vertex u, we add to E_{Reach} the arcs (u_i, u), $1 \leq i \leq k$, where u_i is the prefix of length $|u| - |d_i|$ of u. Now a traversal of $Tree(\Gamma)$ for matching purposes requires at most $O(L)$ steps, since the sum of the lengths of all root-to-leaf paths is bounded by L. Therefore, Step (3) is $O(L + |E_{\text{Reach}}|)$ which is $O(n\,L)$.

Step (4) is a standard breadth-first search of R(Γ). We have a "current" vertex set and all arcs originating in the set are explored. Newly reached vertices become the next current vertex set. Exploration ends when a vertex in Γ is reached or all edges have been examined. If a vertex in Γ is reached, then Γ is not a code, otherwise it is. The initial current vertex set is $S(\Gamma)$ which is easily identified in Step (1). Clearly the time required for Step (4) is $O(n\,L)$, proportial to the number of edges in $R(\Gamma)$. Correctness of the algorithm is evident from Corollary 2.2, so that we have established:

Theorem 3.1
Let $\Gamma = \{ c_1, ..., c_n \}$ be a subset of Σ^+, where Σ is a fixed alphabet, and let $L = |c_1| + ... + |c_n|$. Then we can test in $O(n\,L)$ steps whether Γ is a uniquely decipherable code.

4. Nonbranching Sets

Algorithm 1 fails to be $O(L)$ in the worst case because of the number of reach edges which are possible. We wish to investigate whether it is possible to eliminate certain reach edges without losing the ability to discriminate between ambiguous and uniquely decipherable codes with the smaller graph. In obtaining a suitable subgraph, we separate the question of how many reach edges are needed, based on local considerations, from the question of how to identify these edges efficiently. Our approach to this problem tries to account for structural properties of the code words which influence the density of reach edges.

It is clear that in order to achieve a high density of reach edges, one needs both a high in- and a high out-valence of the reach edge subgraph. This implies that there must be a relatively large subset of code words whose elements are all prefixes of some code word, as well as a relatively large subset of code words whose elements are all suffixes of some code word. In order to simplify the situation, we assume that Γ can be linearly ordered by the prefix relation.

Definition
A subset $\Gamma = \{c_1, \ldots, c_n\}$ of Σ^+ is *nonbranching* if, for $1 \leq i < n$, $c_i < c_{i+1}$.

Clearly Γ is nonbranching iff $Tree(\Gamma)$ is a nonbranching tree. For a nonbranching set Γ, we will now estimate the total number of matches of c_i in a word w, assuming that consecutive matches of c_i in w do not overlap. In particular, if each c_i is divisor free, then consecutive c_i matches (and hence consecutive c_i-edges) cannot overlap in any word w.

Proposition 4.1
Let Γ be nonbranching, w a word in Σ^+. If, for all $c \in \Gamma$, no two c-edges overlap, then there are at most $|w|(\log(n) + 1)$ reach edges on w, where n is the cardinality of Γ.

Proof Since Γ is nonbranching, the number of reach edges is bounded by $|w|\,H_n$, where $H_n = \sum_{i=1}^{n} i^{-1}$. ∎

Corollary 4.2
Let $\Gamma = \{c_1, \ldots, c_n\}$ be nonbranching such that, for $1 \le i < n$, the exponent of c_i is bounded by m. Then there are at most $(m+1)\,|c_n|\,(\log(n)+1)$ reach edges in $R(\Gamma)$.

Proof Note that the set of reach edges can be partitioned into $m+1$ blocks such that no two c_i-edges of the same block overlap. ∎

It should be noted that the bound of Corollary 4.2 cannot be tight, since $m\log(n)$ may exceed n. However, we can show that the bound of Proposition 4.1 is optimal up to a constant factor: Consider the alphabet $\Sigma = \{0, 1, 2\}$, and let b_i be the number i in binary; e.g., $b_3 = 11$. We define strings s_i recursively by

$$\begin{aligned} s_1 &= 2 \\ s_{i+1} &= s_i\, b_i\, s_i \end{aligned}$$

Hence $s_2 = 21210212$. The length $|s_i|$ is governed by the recursion

$$\begin{aligned} |s_1| &= 1 \\ |s_{i+1}| &= 2|s_i| + \lfloor\log(i)\rfloor + 1 \end{aligned}$$

Let $s_i = q_i\, 2^i$, and set $a_{i+1} = q_{i+1} - q_i = (\lfloor\log(i)\rfloor + 1)/2^{i+1}$. Since $\sum_{i=2}^{\infty} a_i$ converges, we have $2^{i-1} < |s_i| < q\, 2^i$, for some constant $q < 5$.

We consider sets Δ_i, where

$$\begin{aligned} \Delta_1 &= \{s_1\} \\ \Delta_{i+1} &= \Delta_i \cup \{\, s_i b_i c \mid c \in \Delta_i \,\} \end{aligned}.$$

Here $|\Delta_i| = 2^i$. The number e_i of reach edges in $R(\Delta_i)$ is obtained from

$$\begin{aligned} e_1 &= 0 \\ e_2 &= 1 \\ e_{i+1} &= 2e_i + |\Delta_{i-1}| \end{aligned}$$

from which we obtain $e_i = (i-1)\, 2^{i-2}$. Note that e_i is the number of reach edges on $s_i b_i s_i$.

Proposition 4.3
For infinitely many positive integers n, there are nonbranching, uniquely decipherable sets Γ of cardinality n such that $R(\Gamma)$ contains $O(L\log(n))$ reach edges and, for all $c \in \Gamma$, no two c-edges overlap. Here L is the sum of the lengths of the words in Γ.

Proof Let $n = 2^i$ and let $\Gamma = \Delta_i - \{s_i\} \cup \{w = (s_i b_i)^R s_i\}$ where $R = |s_i|$. Then Γ is nonbranching and the sum of the lengths of strings in Γ is $O(|s_i|^2) = O(|w|)$. Γ is uniquely decipherable since each $c \in \Gamma$ begins and ends with the letter 2, and no c contains 22 as substring. It is easy to see that if $c \in \Gamma$, then no pair of c-edges is overlapping. Since $e_i = O(i\,|s_i|)$, there are $O(i\,|w|)$ reach edges in $R(\Gamma)$. ∎

Corollary 4.2 implies that Algorithm 1 has worst case performance $O(L\log(n))$ on nonbranching sets with bounded exponent. Proposition 4.3 implies that the estimate of this worst case performance is tight. We summarize the situation as follows:

Theorem 4.4
Let Γ be a nonbranching set and assume that there is a constant m such that every word in Γ has exponent not exceeding m. Then Algorithm 1 tests unique decipherability of Γ in $O(L \log(n))$ steps.

5. Nests on Repeated Roots

Consider the case where $R(\Gamma)$ contains more than $O(L \log(n))$ reach edges. Then there must be a word $c \in \Gamma$ whose exponent is large, and c must occur in the longest string $c_n \in \Gamma$ with a high degree of overlap. It follows that c has a relatively short root x and c_n contains a relatively long substring x^t. It is useful to consider all words in Γ which have the same root x:

Definition
A *nest* N is a maximal subset of words in Γ which have a common root, and the longest string in N has exponent greater than 3.

Definition
Let x be primitive and $c_n = \alpha x^t z \beta$, where x is not suffix of α, z is a proper prefix of x, the prefix z' of x with length $|z| + 1$ is not a prefix of $z\beta$, and $t > 2$. Then the substring $x^t z$ of c_n is called an x^* *region*.

Note that the reach edges generated by the words in N, hereafter referred to as *N-edges*, have high density only on the x^* regions of c_n. The number of N-edges not entirely contained in such a region cannot exceed the $O(L \log(n))$ bound, as they cannot overlap more than $|x| - 1$ characters.

We study the structure of $R(\Gamma)$ on x^* regions in c_n and investigate which N-edges are needed for the purpose of deciding the existence of a path from $S(\Gamma)$ to Γ. Here an edge can be useless either if it leads to a dead-end, that is, following it we know we cannot reach a vertex in Γ, or else because it is redundant in the sense that after deleting it the same subset of Γ is reachable from $S(\Gamma)$. Ignoring for the moment how to detect that an edge is useless, we will be able to show that for all uniquely decipherable sets Γ there cannot be too many useful edges, and this ultimately leads to an efficient detection of the presence of such edges. We distinguish certain prefixes of $\alpha x^t z$:

Definition
A prefix γ of $\alpha x^t z$ is an *exit point* of the x^* region if $|\alpha x^t z| - |\gamma| < |x|$. A prefix αx^i, $0 \leq i \leq t$, is a *division point*. A prefix γ of $\alpha x^t z$ is a *critical point* if there is a $c \in \Gamma$ such that $|c| - |\gamma| < |x|$.

We will make use of the following observation which is a simple consequence of the primitivity of x:

Lemma 5.1
In an x^* region, all c-edges with $x < c$ originate at division or exit points. If the edge originates at a division point which is not an exit point and $c \in N$, then it must terminate in the x^* region. All divisor edges originate at division, exit, or critical points.

Consider an edge due to $c = x^i y \in N$, originating at some division point u and ending at the point $p = uc$ within the x^* region. If we wish to reach a vertex in Γ following this edge, then there must be a path from p to a division, exit, or critical point and the shortest such path must consist entirely of reach edges generated by words which are prefixes of x. (Note that the critical points include those words in Γ which are situated in the x^* region). We consider, therefore, whether from a point u in the x^* region one of these points may be reached.

Definition
A point u in the x^* region *reaches* p if there is a path from u to p in $R(\Gamma)$ consisting entirely of reach edges.

Lemma 5.2
Let u be a division point, c and c' in N, uc and uc' prefixes of αx^t. Assume that uc reaches p and uc' reaches p'. If both p and p' are division points, or if $p = p'$, then Γ is not uniquely decipherable.

Proof Let $p = ucw$ and $p' = uc'w'$. By definition, $w, w' \in \Gamma^*$. If both p and p' are division points, then $cw = x^i$ and $c'w' = x^j$. Since $c \neq c'$, x^{ij} is not uniquely decipherable. Otherwise, we have $cw = c'w'$. Hence cw is not uniquely decipherable. ∎

Since x is primitive, two consecutive x^* regions cannot overlap more than $|x| - 1$ characters. In conjunction with the definition of region, this leads to the following length estimate:

Lemma 5.3
If x is primitive, then the sum of the lengths of all x^* regions in a word w is bounded by $\frac{3}{2}|w|$.

Theorem 5.4
Let Γ be a nonbranching set where all words $c \in \Gamma$ either have an exponent bounded by 3 or have the same root x. Let X be the subset of vertices in $R(\Gamma)$ from which a vertex in Γ can be reached. If Γ is uniquely decipherable, then the subgraph of $R(\Gamma)$ induced by X has no more than $7L + 4L\log(n)$ reach edges.

Proof Let N be the nest of words with root x. By Corollary 4.2, there are at most $4L(\log(n) + 1)$ reach edges not entirely contained in some x^* region of c_n or due to words not in N. We count the number of N-edges entirely contained within the x^* regions of c_n: Consider all N-edges originating at a division point u which is not an exit point of the region. By Lemma 5.2, at most $|x| + 1 \leq 2|x|$ of them are in the subgraph because of reaching either a division point or an exit point. By Lemma 5.3, there are at most $3|c_n|$ of those edges in the subgraph. Now for each c contained in the region, there are up to $|x|$ additional edges possible at all those division points preceding c, hence there are at most $\frac{3}{2}|c|$ additional N-edges for each such word c. The total number of N-edges on all x^* regions of c_n is therefore bounded by $3L$. ∎

Theorem 5.4 implies that a simple variation of Algorithm 1 can test unique decipherability of nonbranching sets containing a single nest in $O(L\log(n))$ steps. In the more special case where the nest includes every word of the set, a special $O(L)$ algorithm is possible because of the following:

Theorem 5.5 (Wrathall)
If there is a primitive x such that, for all $c \in \Gamma$, $x < c$ and $c < x^t$, for sufficiently large t, then Γ is not uniquely decipherable iff there exist $x^i, x^j \in \Gamma$, $i \neq j$, or there exist $x^i, x^j u, x^{ki+j} u \in \Gamma$.

Proof If there are strings of the required form, it is clear that Γ is not uniquely decipherable. So, we assume that Γ is not uniquely decipherable. Then there are $c, d \in \Gamma$ and $w_1, w_2 \in \Gamma^*$ such that $cw_1 = dw_2$ and $cy = d$ for a nonempty y. If $c \neq x^i$, then x is not prefix of $w_1 \neq \lambda$, contradicting that $w_1 \in \Gamma^*$; hence $c = x^i \in \Gamma$. We may split $cw_1 = v_1 c_1 v_2$, where $v_1, v_2 \in \Gamma^*$, $c_1 \in \Gamma$, and such that $v_1 y_1 = d$, and $dy_2 = v_1 c_1$. Let $d = x^r u$. By induction, $v_1 \in x^+$. If $u \neq \lambda$, then $y_2 = \lambda$, for otherwise x cannot be prefix of w_2. The conclusion now follows. ∎

6. Interaction of Nests

Suppose N_2 is a nest with root x_2, and there is a nest N_1 with root x_1, where $x_1 < x_2$. Then, by the definition of nests, x_2 must contain $x_1{}^*$ regions. Consider an $x_2{}^*$ region in c_n. We wish to count how many N_2 edges, originating at an x_2 division point are in the subgraph induced by the set X of Theorem 5.4, provided that Γ is uniquely decipherable. As before, we make use of Lemma 5.2, but we will have to count more carefully the number of x_2 exit points and x_2 critical points. Here the primitivity of x_1 is important.

Intuitively, an exit point p is inessential if from p no edge leads outside the x^* region. In counting the number of N-edges on an x^* region, the reachability of inessential exit points should not matter. In the following definition, we account for this possibility, and we also assure that certain exit and critical points are not multiply counted:

Definition
An exit point p of an x^* region is *essential* if at p a reach edge originates which terminates outside the x^* region, or if at p a divisor edge originates and p is not a critical point. A critical point p is *essential* if a divisor edge originates at p and p is not a division point.

We derive structural statements about the position of essential exit and critical points:

Lemma 6.1
Let x be the root of the nest N in Γ, and assume that $c_n = \alpha x^t z \beta$ is an x^* region in c_n, and p an essential exit point of the region. Then there is a prefix u of x such that $pu = \alpha x^t z$. Moreover, if p is an essential critical point, then there is $c \in \Gamma$ and $u < x$ such that $pu = c$.

Proof Define u by $pu = \alpha x^t z$, where p is an essential exit point. If there is a reach edge from p to pc where pc is not in the region, then $\alpha x^t z < pc$, hence $u < c$. Since $|\alpha x^t z| - |p| < |x|$, $u < x$. Similarly, if there is a divisor edge from an essential critical point p to a point u, then there is a $c \in \Gamma$ such that $pu = c$. Since p is not a division point, $|c| - |p| < |x|$, hence $u < x$. ∎

Lemma 6.2
Let x_2 be the root of the nest N_2, x_1 the root of the nest N_1, and assume that $x_1 < x_2$. Then $|x_2| > 3|x_1|$.

Proof Let c be the longest word in N_1. Then $x_1{}^4 < c$. Since $x_1 < x_2$ and $x_2{}^2 < c_n$, by primitivity, $|c| - |x_2| < |x_1|$. ∎

Lemma 6.3
Let $N_1, \ldots, N_r$ be the nests in the nonbranching set Γ, where x_i is the root of the nest N_i and $x_1 < x_2 < \cdots < x_r$. Let $c_n = \alpha x_i^t z \beta$ be an $x_i{}^*$ region in c_n, and p an essential exit point of the region, where $\alpha x_i^t z = pu$. If $|u| \geq 3|x_j|$, then p is a division point in an $x_j{}^*$ region.

Proof By Lemma 6.1, if $|u| \geq 3|x_j|$, then $x_j{}^3 < u$. The conclusion now follows from the primitivity of x_j. ∎

Lemma 6.4
Let $N_1,...,N_r$ be the nests in the nonbranching set Γ, where x_i is the root of the nest N_i and $x_1 < x_2 < \cdots < x_r$. Let $c_n = \alpha x_i^t z \beta$ be an $x_i{}^*$ region in c_n, and assume that the region contains $c \in \Gamma$. Assume there is a divisor edge originating at the essential critical point p which terminates at u, where $pu = c$. If $|u| \geq 3|x_j|$, then p is a division point in an $x_j{}^*$ region.

Lemma 6.5
Let $\{l_i\}$, $1 \leq i$, be an infinite sequence of natural numbers such that, for all i, $l_{i+1} > 3l_i$. Then, for all $k > 2$,

$$L_k = \sum_{i=k}^{\infty} l_k/(l_i l_{k-2}) < \frac{3}{(2l_{k-2})}$$

Proof Since $l_{i+1} > 3l_i$, we have $L_k < l_{k-2}^{-1}(\sum_{i=0}^{\infty} 3^{-i}) = 3/(2l_{k-2})$ ∎

Theorem 6.6
Let Γ be a nonbranching set. Let X be the subset of vertices in $R(\Gamma)$ from which a vertex in Γ can be reached. If Γ is uniquely decipherable, then the subgraph of $R(\Gamma)$ induced by X has no more than $16L + 4L \log(n)$ reach edges.

Proof Let $N_1,...,N_r$ be the nests in Γ, where x_i is the root of the nest N_i, and such that $x_1 < x_2 < \cdots < x_r$. Let l_i be the length of x_i, and observe that $l_{i+1} > 3l_i$. By Theorem 5.4, there are at most $7L + 4L \log(n)$ reach edges in the subgraph, not counting the N_k-edges, $k > 1$, which are entirely contained in some $x_k{}^*$ region.

Consider the number of N_k-edges originating at the same division point of an $x_k{}^*$ region. By Lemma 5.2, the number of edges which are also in the subgraph is then bounded by the reachable essential exit points, essential critical points, and one additional edge which may reach a division point. By Lemma 6.3, the number of essential exit points may be estimated by e_k where

$$e_2 = 3l_1 + (l_2 - 3l_1)/l_1$$

$$e_k = e_2 + \sum_{i=3}^{k} (l_i - l_{i-1})/l_{i-2}.$$

By Lemma 6.4, the number of essential critical points due to some $c \in \Gamma$ which is in the region is also bounded by e_k. Since the division points in an $x_k{}^*$ region are l_k characters apart, the total number of N_k-edges on all $x_k{}^*$ regions is bounded by $\frac{3}{2} L (e_k + 1)/l_k$. Since $\sum_{k=2}^{\infty} e_2/l_k \leq 3$, we have $\sum_{k=2}^{r}(e_k/l_k) < 3 + \sum_{k=3}^{\infty} L_k$ which, by Lemma 6.5, is bounded by $3 + 9/4$. Moreover, $\sum_{k=2}^{r}(1/l_k) < 1/2$. Therefore, the number of all N_k-edges, $k \geq 2$, which are contained in all $x_k{}^*$ regions and are in the subgraph cannot exceed $9L$, unless Γ is not uniquely decipherable. ∎

7. An $O(L \log(n))$ Algorithm for Nonbranching Sets

By Theorem 6.6, we may test unique decipherability of nonbranching sets as follows: Construct the subgraph of $R(\Gamma)$ induced by the set of vertices X from which a vertex in Γ is reachable. This subgraph may be constructed by first placing all divisor edges as in Algorithm 1, followed by placing reach edges from a subset of X. Initially, this subset is Γ. As a reach edge terminating in a subset vertex is placed, the origin of the edge is added to the subset. Moreover, the edges are counted as they are placed. If the bound of Theorem 6.6 is exceeded, then the graph construction terminates as Γ cannot be uniquely decipherable. In this case it is possible to exploit the proof of Lemma 5.2 to construct explicitly a deciphering ambiguity.

Suppose that the bound of the theorem is not exceeded. Then we have constructed a subgraph with $O(L \log(n))$ reach edges, and may search it in $O(L \log(n))$ steps for a path from $S(\Gamma)$ to Γ. In the event that such a path exists, Γ is not uniquely decipherable and the path defines a deciphering ambiguity. Otherwise, Γ is uniquely decipherable.

Algorithm 2

1. Construct a pattern matching machine M with Γ the set of patterns, and test whether Γ is nonbranching.
2. Construct the edge set E_{Divisor}
3. Run M on c_n: For each match, deposit the *name* of the match set at the vertex.
4. Compute $m := 16L + 4L \log(n)$. Initialize $X := \Gamma$, marking each element of X *unprocessed*.
5. While there is another *unprocessed* vertex u in X do Steps 6–9. Thereafter, the algorithm continues with Step 10.

    6. Mark u as *processed*.
    7. If there is a match set name for $\{d_1, ..., d_k\}$ at u, add the appropriate reach edges (v_i, u), where $v_i d_i = u$, $1 \le i \le k$.
    8. Decrement m by the number of reach edges placed. If $m \le 0$, then stop—Γ is not uniquely decipherable.
    9. Add the vertices v_i to X, unless already in the set, and mark them as *unprocessed*.

10. Search the subgraph of $R(\Gamma)$ so constructed for a path from $S(\Gamma)$ to Γ. If such a path exists, then Γ is not uniquely decipherable; otherwise it is.

By Theorem 6.6, Algorithm 2 correctly determines whether Γ is uniquely decipherable. It is clear that the time required is proportional to the number of edges placed (plus L), hence it is $O(L \log(n))$ in the worst case. Note, however, that the constructed subgraph is no longer sufficient to determine whether Γ has a finite decipherability delay or is synchronizable.

Acknowledgements

It is a pleasure to acknowledge many useful discussions with Alberto Apostolico, Mikhail Atallah, Renato Capocelli, Michael Drazin, Rao Kosaraju, and Walter Schnyder. Shimon Even pointed out an error in an earlier version of this algorithm. Celia Wrathall communicated Theorem 5.5 to me in November of 1983.

References

[AC 75] A. V. Aho and M. J. Corasick
"Efficient string matching: an aid to bibliographic search", *CACM 18:6* (1975) 333–343

[AG 84] A. Apostolico and R. Giancarlo
"Pattern matching machine implementation of a fast test for unique decipherability" *Inf. Proc. Letters*, forthcoming in 1984

[Blu 65] E. K. Blum
"Free subsemigroups of a free semigroup", *Mich. Math. J. 12* (1965) 179–182

[Cap 79] R. M. Capocelli
"A note on uniquely decipherable codes", *IEEE Trans. on Inf. Thy. IT-25* (1979) 90–94

[CH 84] R. M. Capocelli and C. M. Hoffmann
"Algorithms for factorizing semigroups", *Proc. NATO Workshop on Combinatorial Algorithms on Words*, Maratea, Italy, 1984

[Eve 79] S. Even
"Graph Algorithms", Comp. Sci. Press, Potomac, Md., (1979)

[KMP 77] D. Knuth, J. Morris, V. Pratt
"Fast pattern matching in strings", *SIAM J. on Comp. 6:2* (1977) 323–350

[Lal 79] G. Lallement
"Semigroups and Combinatorial Applications", J. Wiley & Sons, New York (1979)

[Mar 62] A. A. Markov
"Non-recurrent coding", *Problem. Kybern. 8* (1962) 169–189

[McC 76] E. M. McCreight
"A space-economical suffix tree construction algorithm", *J. ACM 23:2* (1976) 262–272

[Rod 82] M. Rodeh
"A fast test for unique decipherability based on suffix trees", *IEEE Trans. on Inf. Thy. IT-28:4* (1982) 648–651;

[SP 53] A. A. Sardinas and C. W. Patterson
"A necessary and sufficient condition for the unique decomposition of coded messages", *IRE Intl. Conv. Rec. 8* (1953) 104–108

[Spe 75] J. C. Spehner
"Quelques constructions et algorithmes relatifs aux sous-monoides d'un monoide libre", *Semigroup Forum 9* (1979) 334–353

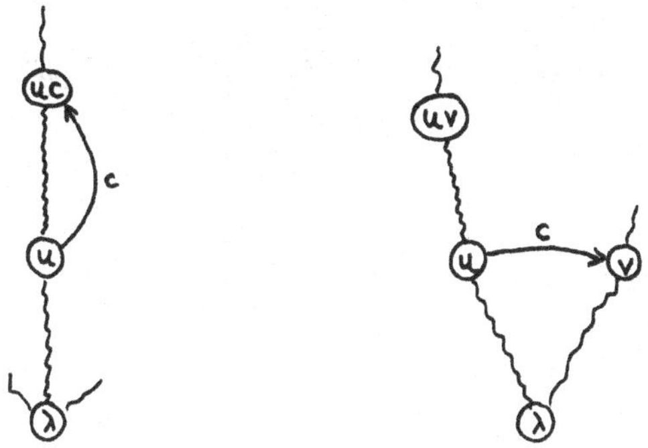

Figure 2.1(a) Figure 2.1(b)

Figure 2.1(a) shows a reach edge; the code word c occurs as proper substring in some other code word, beginning at u. Figure 2.1(b) shows a divisor edge; the code word c is partitioned as uv, where the proper suffix v occurs as prefix of some code word.

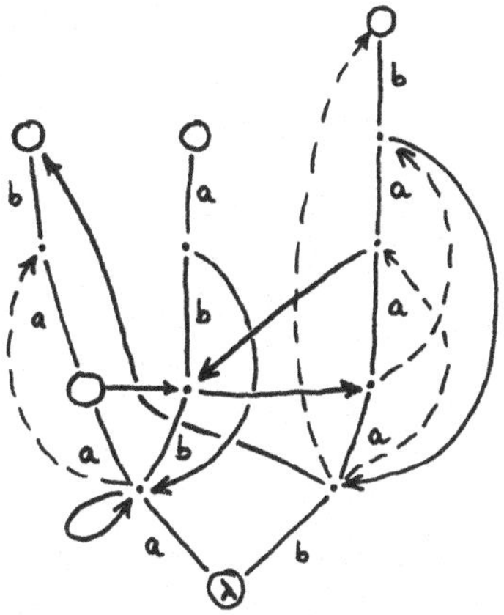

Figure 2.2

Figure 2.2 shows the graph $R(\Gamma)$ for the set $\Gamma = \{aa, aaab, abba, baaab\}$. Γ is not uniquely decipherable, as evidenced by the path from aa to $aaab$. Divisor edges are shown as solid arrows, reach edges are dashed.

OUTLINE OF AN ALGEBRAIC LANGUAGE THEORY

Günter Hotz
Informatik, Universität des Saarlandes
6600 Saarbrücken

Abstract: The algebraic theory we present here continues the early work of Chomsky-Schützenberger [Ch,Sch], Shamir [Sh] and Nivat [N]. The leading idea is to develop a machine and production free language theory. The interest in such a theory is support by the hope that the proofs in such a theory don't need so much case discussions which often lead to errors and that a view which is free from non-essentials of language theory will lead to a progress in the direction of our problems. Even if the theory is in an early stage the attempt pays out in a machine free definition of LL(k) and LR(k) languages which leads easily to generalisations of non-deterministic LL(k) and LR(k) languages with the same space and time complexity behaviour. Too, we are able to show that this theory is not restricted to the context free languages but also concerns the whole Chomsky hierarchy. Our theory is in a sense a dual to the theory of formal power series as introduced by M. Schützenberger.

Introduction: Let X be a set and X^* be the free monoid generated by X. The empty word is $1 \in X^*$ and $|u|$ means the length of $u \in X^*$. For monoids M and semirings R is $R\langle M\rangle$ the semiring of the finite sums

$$p = \sum_{m \in M} \alpha_m \cdot m \qquad \text{where } \alpha_m \in R.$$

We often write $\alpha_m = \langle p,m\rangle$. Only for finitely many elements $m \in M$ it holds $\langle p,m\rangle \neq 0$. We always assume that R has a multiplicative unit which we identify with $1 \in M$.
Of special importance for our theory is the syntactical monoid $X^{(*)}$ of the Dyck language D(X) over X. This monoid called polycyclic monoid by Perrot [Pe] can also be defined as follows: One takes an bijectively equivalent set X' to X such that $X \cap X' = \emptyset$. The bijection be $x \rightarrow \bar{x}$. We take further a symbol $0 \notin X \cup X'$ and form $(X \cup X' \cup 0)^*$; then we take the quotient of this free monoid by the relation system

$$x \cdot \bar{x} = 1,\ x \cdot \bar{y} = 0,\ 0 \cdot z = z \cdot 0 = 0 \text{ for } x,y \in X, z \in X \cup X' \cup \{0\}.$$

For $\bar{x}$ we also write x^{-1} and $x^1 = x$.

We further make use of context free grammars $G=(X,T,P,S)$ with $X \cap T = \emptyset$, $P \subset X \times X^2 \cup X \times T$ and $S \in X$. From this it follows that we have no ε-productions and $1 \notin L(G)$, if L(G) is the language generated by G. We assume further G to be free from superfluous variables. This means, that for $x \in X$ there exist derivations f and g such that

$$S \xrightarrow{f} u x v \xrightarrow{g} w \text{ and } w \in T^*.$$

The last assumption on G is, that S does not appear in the right hand side of any production $q \in P$.

It is usual to write P too as an equation system

$$x = \sum \alpha_{x,u} \circ u \text{ for } x \in X$$

and $\alpha_{x,u} = 1$ if $(x,u) \in P$ and $\alpha_{x,u} = 0$ in all other cases.

Schützenberger has shown, that this makes sense in the following way: The equation system can be solved by a system of formal power series. L(G) can be looked at as the support of the power series belonging to S. The coefficient of the word w in the series gives the multiplicity of w relative to G, this means the number of essentially different derivations of w from S.

We assign to the grammar an equation system in a dual way by writing the quadratic terms on the left side and the corresponding linear terms as sums on the right side. This means that we study equation systems of the form

$$x \circ y = \sum_{z \in X} \alpha^{z}_{x,y} \cdot z, \quad t = \sum_{z \in X} \alpha^{z}_{t} \circ z$$

with $\alpha^{z}_{x,y}, \alpha^{z}_{t} \in \{0,1\}$ and

$$\alpha^{z}_{x,y} = 1 \Longleftrightarrow (z,xy) \in P,$$

$$\alpha^{z}_{t} = 1 \Longleftrightarrow (z,t) \in P.$$

These relations are similar to the multiplication rules of finite dimensional algebras over a ring R. In general such an equation system does not define an associative algebra. But with a simple trick we get an associative algebra from this idea.

We assign to G a new alphabet $\overline{X}$ by setting

$$X_l = \{(x,l) \mid (z,xy) \in P\},$$

$$X_r = \{(y,r) \mid (z,xy) \in P\},$$

$$\overline{X} = X_l \cup X_r.$$

For (x,l) resp. (y,r) we write often shorter x_l resp. y_r. Now we define the grammar $\overline{G} = (\overline{X}, T, \overline{P}, S_r)$ with

$$\overline{P} = \{(x_i, y_l z_r) \mid (x,yz) \in P, x \neq S, i \in \{l,r\}\}$$

$$\cup \{(S_r, x_l z_r) \mid (S,xz) \in P\}$$

$$\cup \{(x_i, t) \mid i \in \{l,r\}, (x,t) \in P\}$$

Obviously $L(G) = L(\overline{G})$ holds.

We assign to G now the following equation system

$$x \cdot y = \sum \alpha^{z}_{x,y} \cdot z \qquad \text{for } x \in X_l, y \in Y_r \qquad (\mathcal{R}_G)$$

and

$$\alpha^{z}_{x,y} = \begin{cases} 1 \text{ if } (z,xy) \in \overline{P} \\ 1 \text{ if } z = S_r, x = S_l, y = S_r \\ 0 \text{ in all other cases.} \end{cases}$$

We generalize $\mathcal{R}_G$, because our proofs will not become harder by this, to the following situation: X_1 and X_r are any two alphabets with $X_1 \cap X_r = \emptyset$. We put $\overline{X} = X_1 \cup X_r$. Further there are given two mappings

$$\delta' : X_1 \times X_r \to R\langle \overline{X}^* \rangle$$

and

$$\eta' : T \to R\langle \overline{X}^* \rangle$$

with

$$\delta'(x,y) = \sum_{z \in \overline{X}} \alpha^z_{x,y} \cdot z \quad \text{for } (x,y) \in X_1 \times X_r,$$

$$\eta'(t) = \sum_{z \in \overline{X}} \alpha^z_t \circ z \quad \text{for } t \in T.$$

We extend δ' to $\overline{X}^*$ by defining

$$\delta(u) = \begin{cases} u & \text{for } u \in X_r^* \cdot X_1^* \\ \delta'(x,y) & \text{for } u = xy \in X_1 \circ X_r \\ u_1 \delta(xy) u_2 & \text{for } u_1 \in X_r^* \cdot X_1^* \text{ and } xy \in X_1 \circ X_r. \quad \text{where } u = u_1 xy u_2 \end{cases}$$

Now we extend δ linear to $R\langle \overline{X}^* \rangle$. η is the corresponding extension of η' to $R\langle T^* \rangle$. The equation system

$$xy = \delta(xy) \text{ for } xy \in X_1 \cdot X_r \qquad (\mathcal{R})$$

is the generalisation of the system $(\mathcal{R}_G)$.

We assign now an associative algebra $\mathcal{A}_R(\delta)$ to $(\mathcal{R})$. For this reason we iterate δ and finitely form the transitive closure δ^* of δ. This means it holds $\delta \bullet \delta^* = \delta^*$.

Now one easily proofs

LEMMA 1: $\delta^*(uv) = \delta^*(\delta^*(u) \circ \delta^*(v))$.

Now we define the operation '$\bullet$' on $R\langle \overline{X}^* \rangle$ by setting

$$u \bullet v := \delta^*(uv).$$

From our lemma it follows that:

$$(u \bullet v) \bullet w = \delta^*(\delta^*(uv) \circ w) = \delta^*(\delta^*(uv)\delta^*(w)) = \delta^*(uvw),$$
$$u \bullet (v \bullet w) = \delta^*(u \cdot \delta^*(vw)) = \delta^*(\delta^*(u)\delta^*(vw)) = \delta^*(uvw).$$

Therefore the following theorem holds.

THEOREM 1: $\mathcal{A}_R(\delta) := (R\langle \overline{X}^* \rangle, +, \bullet)$ is an associative algebra and

$$\delta^* : (R\langle \overline{X}^* \rangle, +, \circ) \to (R\langle \overline{X}^* \rangle, +, o)$$

is an algebra homomorphism.

For the algebra we so assigned to our grammar G we write $\mathcal{A}_R(G)$. We extend this algebra to include the terminals too. For this reason we use the defined mapping η and extend η to $(\overline{X} \cup T)^*$ by setting $\eta(x) = x$ for $x \in \overline{X}$.

Now for $u,v\in(\overline{X}\cup T)^*$ we define

$$u\bullet v := \delta^*(\eta(uv)).$$

The associative algebra we get by this construction we call $\bar{\mathcal{A}}_R(G)$.
For $u_1\bullet u_2\bullet\ldots\bullet u_n$ we write again $u_1u_2\ldots u_n$.
In a case where it is not clear which product we mean
we write

$$u_1u_2\ldots u_n\ [\mathcal{A}_R(G)]$$

if the product is in $\mathcal{A}_R(G)$. Analogously we proceed with other algebras.
The following concerns the questions.

How are the algebras $\mathcal{A}_R(G)$ structured?
Which information contains $\mathcal{A}_R(G)$ about $L(G)$?
How is the structure of $\mathcal{A}_R(G)$, if G is deterministic?
Is it possible to generalize the theory to non-c.f. languages?

The following section is dedicated to the first question.

A representation theorem for $\mathcal{A}_R(\delta)$

We are going to show, that for each algebra $\mathcal{A}_R(\delta)$ there exist a non-trivial representation $\varphi:\mathcal{A}_R(\delta)\to R\langle X^{(*)}\rangle$. We will show that the algebra $R\langle X^{(*)}\rangle$ for our algebras and for the finite dimensional algebras plays a similar role as the matrix ring in the finite dimensional case. The following lemma shows that $R\langle X^{(*)}\rangle$ has a very simple algebraic structure.

LEMMA 2: $\mathcal{A}_D = R\langle X^{(*)}\rangle$ contains only trivial two-sided ideals. Ideals $\mathfrak{a}$ of $\mathcal{A}_D$ here are considered to be trivial, if there exists an ideal $\mathfrak{a}'$ of R such that $\mathfrak{a}=\mathfrak{a}'\langle X^{(*)}\rangle$.
Let be Z a finite basis of $\mathcal{A}$ over R and $\mathcal{A}$ being given by the relations

$$x\cdot y = \sum_{z\in Z}\alpha^z_{x,y}\cdot z,\quad \alpha^z_{x,y}\in R.$$

We define

$$\varphi:\mathcal{A}\to\mathcal{A}_D \text{ by defining.}$$

$$\varphi(y) := \sum_{z,u\in Z}\bar{z}\cdot\alpha^u_{z,y}\cdot u \quad \text{for } y\in Z.$$

This defines φ uniquely. ($\bar{z}$ is the inverse of z in $Z^{(*)}$).

THEOREM 2: φ is an algebra homomorphism. If $\mathcal{A}$ contains a multiplicative unit, then φ is injective.
For the case of matrix rings we give a second representation.

THEOREM 3: Let $\mathcal{A}$ be a finite dimensional ring of quadratic matrices $(a_{z,y})_{z,y\in Z}$. Then

$$\varphi(a) = \sum_{z,y\in Z}\bar{z}a_{z,y}\cdot y$$

is a monomorphism from $\mathcal{A}$ into $\mathcal{A}_D$.

Now we come to the main result of this section. To construct the representation

$\varphi : \mathcal{A}_R(\delta) \to \mathcal{A}_D$ we first define a suitable alphabet $\mathcal{A}_D$.
For $u \in \overline{X}$ and $x \in X_r$ (remember $\overline{X} = X_1 \cup X_r$), we define

$$[u:x] = \begin{cases} 0 \text{ if for all } w \in \overline{X}^* \text{ it holds } \langle \delta^*(uw), x \rangle = 0, \\ 1 \quad \text{for } u = x \\ \text{free variable in all other cases.} \end{cases}$$

Clearly it follows from $[u:x] \neq 0$ and $u \in X_r$ that $u=x$.
We set

$$Z = \{[u:x] \mid [u:x] \neq 1,0;\ u \in \overline{X},\ x \in X_r\}$$

and $\mathcal{A}_D = R\langle Z^{(*)}\rangle$.

For $z \in \overline{X}$ we define

$$\varphi'(z) = \sum_{\substack{y,v,u,x \\ [y:x] \in Z}} \alpha^{u}_{y,v} \ \overline{[y:x]}[u:x][z:v]$$

<u>THEOREM 4</u>: There exists an uniquely defined extension of φ' to an algebra homomorphism $\varphi : \mathcal{A}_R(\delta) \to \mathcal{A}_D$

<u>PROOF:</u> $\overline{X}$ generates $\mathcal{A}_R(\delta)$ and therefore there exists not more than one homomorphic extension of φ' onto $\mathcal{A}_R(\delta)$. To show that such an extension exists, it is sufficient to show that for the linear extension φ of φ' it holds

$$\varphi(z_1) \circ \varphi(z_2) = \varphi(z_1 z_2) \qquad \text{for } z_1 \in X_1,\ z_2 \in X_r.$$

By straight forward calculation one gets

$$\varphi(z_1) \cdot \varphi(z_2) =$$

$$\sum_{\substack{y_1,v_1,u_1,x_1, \\ y_2,v_2,u_2,x_2}} \alpha^{u_1}_{y_1,v_1} \ \overline{[y_1:x_1]}[u_1:x_1][z_1:v_1] \ \alpha^{u_2}_{y_2,v_2} \ \overline{[y_2:x_2]}[u_2:x_2][z_2:v_2]$$

$$= \sum_{\substack{y_1,v_1,u_1,x_1 \\ v_2,u_2}} \alpha^{u_1}_{y_1,v_1} \ \alpha^{u_2}_{z_1,v_2} \ \overline{[y_1:x_1]}[u_1:x_1][u_2:v_1][z_2:v_2].$$

For $z_2 \neq v_2$ we have $[z_2:v_2] = 0$ because $z_2 \in X_r$. Therefore there remain only the cases $z_2 = v_2$, that means $[z_2:v_2] = 1$. We use the commutativity of R and have

$$\varphi(z_1) \cdot \varphi(z_2) = \sum_{u_2} \alpha^{u_2}_{z_1,z_2} \cdot \sum \alpha^{u_1}_{y_1,v_1} \ \overline{[y_1:x_1]}[u_1:x_1][u_2:v_1]$$

$$= \sum \alpha^{u_2}_{z_1 z_2} \ \varphi(u_2) = \varphi(z_1 \cdot z_2).$$

<u>Historical remark:</u> Nivat uses in his thesis a homomorphism which formally looks like our homomorphism φ. But ψ is a mapping $\psi : R\langle X^*\rangle \longrightarrow R\langle H(B)\rangle$,

where H(B) is the free half group generated by B. The main difference comes from the different domains of φ and ψ. Nivat uses ψ to proof the representation theorem of Shamir. But he needs for this proof the normal form theorem of Greibach, which follows as the theorem of Shamir from the existence of φ. The reason is, that $\mathcal{A}_R(G)$ contains a lot of information on G, but R<X*> not at all. More detailed informations on this subject the reader may find in the book [Sa] of Salomaa.

As we will show later one can derive from φ a representation of L(G) by a grammar in Greibach normal form. The size of the grammar corresponds to the size of φ. We define

$$|\mathcal{A}_R(\delta)| = \sum_{x,y,z\in\overline{X}} |\alpha^z_{x,y}|$$

with

$$|\alpha| = \begin{cases} 1 \in \mathbb{N} & \text{for } \alpha \neq 0 \\ 0 \in \mathbb{N} & \text{otherwise} \end{cases}$$

For $p \in \mathcal{A}$ we put

$$|p| = \sum_{w\in Z^{(*)}} |\langle p,w\rangle| .$$

We define the size $|\varphi|$ of φ by

$$|\varphi| = \sum_{z\in\overline{X}} |\varphi(z)| .$$

One easily proofs

LEMMA 3: $|\varphi| \leq |\mathcal{A}_D| \circ |\overline{X}|^2$,
where $|\overline{X}|$ is the number of elements of $\overline{X}$.

Invariants of the Transformation $G \to \overline{G}$.

We return to grammars and study which properties of G remain unchanged when passing from G to $\overline{G}$ as we did in section 1. The set of derivations of words into other words using G we call $\mathcal{F}$. If $f\in\mathcal{F}$ the Q(f) is the word on which the derivation starts and Z(f) is the result of the derivation f. If f, $g\in\mathcal{F}$ and Q(f) = Z(g), then $f\bullet g$ is the derivation, which one gets by first applying g and then applying f. Obviously $Q(f\bullet g) = Q(g)$ and $Z(f\bullet g) = Z(f)$ and "o" is associative. The empty derivation belonging to the word w is 1_w. We have $1_{Z(f)}\bullet f\bullet 1_{Q(f)} = f$. In the case Q(f) = w, Z(f) = v we write too $w \xrightarrow{f} v$.
If we have

$$w_1 \xrightarrow{f_1} v_1 \quad \text{and} \quad w_2 \xrightarrow{f_2} v_2$$

we may form the derivation

$$w_1 \circ w_2 \xrightarrow{f_1 \times f_2} v_1 \circ v_2 .$$

This leads to a further associative operation on $\mathcal{F}$. The unit belonging to 'x' is 1_1.
Both operations are connected by the property

$$(f_1\bullet g_1)\times(f_2\bullet g_2) = (f_1\times f_2)\bullet(g_1\times g_2)$$

if the left side is defined. ($\mathcal{F}$,(XUT)*, Q,Z,o,x) forms a free monoidal category, which in [Ho.O] has been called free x-category and syntactical category in [Be]. The elements of $\mathcal{F}$ are trees or words over the derivation trees in the case of context free grammars. The trees of the production set P generate $\mathcal{F}$. $\overline{\mathcal{F}}$ is the category belonging to $\overline{G}$. The structure preserving mappings are called x-functors. An x-functor consists of two mappings (φ_1,φ_2), the first one is a monoid homomorphism from the monoid of the source category into the monoid of the target category. φ_2 maps the derivation set into the derivation set. We use further the abbreviations

$$\mathrm{Mor}_{\mathcal{F}}(w,v) = \{f\in\mathcal{F} \mid Q(f) = w, Z(f) = v\},$$

$$\mathrm{mult}_G\,(w) = \mathrm{card}\ \mathrm{Mor}_{\mathcal{F}}\,(S,w).$$

The multiplicity of w over G tells us in how many essentially different ways w may be derived from S using G.

LEMMA 4: For $w\in T^*$ it holds $\qquad \mathrm{mult}_G(w) = \mathrm{mult}_{\overline{G}}(w)$

Now we show that the LL(k) and LR(k) properties of G do not change, when passing from G to $\overline{G}$. [Kn], [H.S.]. For this reason we introduce the following notations.

$f\in\mathcal{F}$ we call u-left-prime for $u\in(X\cup T)^*$,

iff from

$f = (1_u\times h)\circ g$ it follows $g = f$.

The definition u-right-prime is symmetric to the former definition.
One easily shows

LEMMA 5: For each $f\in\mathcal{F}$, u prefix of Z(f), there exists exactly one decomposition $f = (1_u\times h)\circ g$ such that g is u-left-prime
Relating to the notion in this lemma, we call g the u-left-prime factor of f and h the v-right-base of f if $Z(f) = u\cdot v$. We write

$$g = \text{left-prim}\,(u,f), \quad h = \text{right-base}\,(v,f).$$

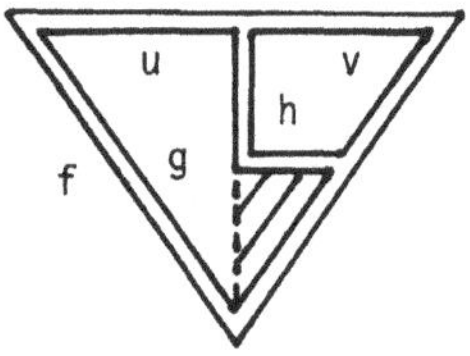

This figure should explain the definitions. We use the notions too, which we get from this definition by changing 'left' into 'right' and right into 'left'.

We give now the definition of LR(k) which is equivalent to the definition [Har.p.502] and of LL(k) equivalent to the one given by Hennie and Stearns [H.S.]. The reader should remember, that we assume G to be in Chomsky NF and G without ε-productions.
G is a LR(k) grammar resp. LL(k+1) grammar for k = 0,1, ... , if the following holds:

For all $f,f'\in\mathcal{F}$ with $Z(f) = u\cdot v$ and $Z(f') = u\cdot v'$

we have

left-base(u,f) = left-base(u,f')
for $Q(f) = Q(f') = S$ and $\mathrm{First}_k(v) = \mathrm{First}_k(v')$

resp.: left-prim(u,f) = left-prim(u,f') for $Q(f) = Q(f')\in X$ and $\mathrm{First}_k(v) = \mathrm{First}_k(v')$.

Remember that we assume that S never appears on the right hand side of any production. From this it follows [Har.p.525], that our LR(o) grammars produce only ALR(o)-languages i.e. strictly determin. languages. Now we are able to prove

LEMMA 6.: If G is a LL(k) resp. LR(k) grammar then $\overline{G}$ is a LL(k) resp. LR(k) grammar.

LEMMA 7: If f is v-right-prime then it holds $u \in X_1^*$ for $z(f) = u \cdot f$. If h is u-left-base of f, then is $Q(h) \in X_1^*$. The lemma remains true if we exchange the words left and right.

Connections between L(G), $\mathcal{A}_R(G)$ and φ

In this section we work out the general relations between L(G) and $\mathcal{A}_R(G)$ and our representation φ. A first information is given by

THEOREM 5: $w \in L(G) \Leftrightarrow \langle \eta(w), S_r \rangle \neq 0$ for $\chi(R)=0$ ($\chi(R)$=characteristic of R),

$$\text{mult}_G = \langle \eta(w), S_r \rangle [\mathcal{A}_N(G)].$$

(remember: '[]' contains the algebra in which the relation is to be understood.

PROOF: As we have shown in lemma 4 we may use $\overline{G}$ instead of G. The proof is by induction on the length |w| of w. For the proof we show a little more general result:

$$\text{mult}_{\overline{G}}(x_i, w) = \langle \eta(w), x_i \rangle \quad \text{for } w \in T^*,\ x_i \in \overline{X}$$

where

$$\text{mult}_{\overline{G}}(x_i, w) = \text{card Mor}_{\mathcal{F}}(x_i, w).$$

The theorem is obvious for |w| = 1.

Let be $f: x_i \to w$ a derivation and |w|>1. Then we may decompose

$$f = (f_1 * f_2) \circ p, \quad p \in \overline{P}.$$

Therefore we have

$$\text{mult}_{\overline{G}}(x_i, w) = \sum_{\substack{w_1 \cdot w_2 = w \\ w_1 \neq 1, w_2 \neq 1 \\ \langle y_1 z_r, x_i \rangle = 1}} \text{mult}_{\overline{G}}(y_1, w_1) \cdot \text{mult}_{\overline{G}}(z_r, w_2).$$

By the induction hypothesis we get

$$\text{mult}_{\overline{G}}(x_i, w) = \sum_{\substack{w_1 \cdot w_2 = w \\ w_1 \neq 1, w_2 \neq 1}} \langle \eta(w_1), y_1 \rangle \cdot \langle \eta(w_2), z_r \langle \cdot \rangle y_1 z_r, x_i \rangle = \langle \eta(w), x_i \rangle,$$

what has to be proved.

In the following we use the definition

$$(u) = u + \mathcal{A}_R(G).$$

(u) is the additive residual-class of u.

Corollary of THEOREM 5: For $R = \mathbb{B}$ = boolean-ring with two elements we have

$$L(G) = \eta^{-1}(S_r).$$

The following lemma is crucial. It shows how the representation φ transforms the residual-class (S_r).

LEMMA 8: For $z_0 z_1 \ldots z_n \in \bar{X}^*$ it holds for $z_0 \neq s$

$$<z_0 z_1 \ldots z_n, s> = <\varphi(z_1 \ldots z_n), \overline{[z_0 : s]}>.$$

LEMMA 9: Using the notation of lemma 8 it holds

$$<z_1 \ldots z_n, S_r> = <\varphi(z_1 \ldots z_n), \overline{[S_1 : S_r]}>.$$

PROOF: From Lemma 8 it follows

$$<S_1 z_1 \ldots z_n, S_r> = <\varphi(z_1 \ldots z_n), \overline{[S_1 : S_r]}>.$$

By definition of $\mathcal{A}_R(G)$ we get

$$<S_1 z_1 \ldots z_n, S_r> = <z_1 \ldots z_n, S_r>$$

and from this directly our lemma.

If we now concatenate the homomorphisms η and φ in this sequence we get a homomorphism $h = \varphi \circ \eta$ from T^* into $R<X^{(*)}>$. This leads us to a representation theorem for c.f. languages, which is nearly the theorem of Shamir ([Sh] see too [N_i]). Shamir uses instead of $X^{(*)}$ the half group $H(X)$, that means he does not make use of the relations $x \cdot \bar{y} = 0$ for $x \neq y$.

THEOREM 6: (Shamir) To each c.f. language $L \subset T^*$ there exists a monoidhomomorphism $h : T \longrightarrow R<Z^{(*)}>$ and an additive residual class $(\$)$ such that $L = h^{-1}((\$))$ holds.

PROOF: The proof follows from lemma 9 and theorem 5 by choosing $\$ = \overline{[S_1 : S_r]}$.

Each polycyclic monoid $Z^{(*)}$ can be embedded by a monomorphism into $\{x_1, x_2\}^{(*)}$. This embedding even can be done such that $\overline{[S_1 : S_r]}$ in all cases will be mapped onto the same element $a_0 \in \{x_1, x_2\}^{(*)}$. We extend this embedding to a ring homomorphism from $R<Z^{(*)}>$ into $R<\{x_1, x_2\}^{(*)}>$ and put it behind h. The resulting homomorphism let be $\bar{h}$. Then the following holds.

COROLLARY TO THEOREM 6: For each c.f. language $L \subset T^*$ there exists a homomorphism

$$\bar{h} : T \longrightarrow R<\{x_1, x_2\}^{(*)}>$$

such that

$$L = \bar{h}^{-1}((a_0)).$$

In this form this theorem was first given in [Ho.1], where it was derived from the theorem of Chomsky-Schützenberger as an algebraic version of the theorem of Greibach about a hardest language under homomorphic reduction [Gr]. This language one gets from the representation given above by forming the c.f. language of the expressions consisting of products of polynomials of $R<\{x_1, x_2\}^*>$. The theorem of Greibach and the representation above have been found independently from the theorem of Shamir. A long time one has not payed attention to the theorem of Shamir outside of the French School, because its complexity theoretic aspects had not been seen. As shown in [Ho.4] one can construct similar representations for r.e., c.s., d.c.s. and other classes of languages. It seems to be possible to construct for each complexity class given by a time bound $T(n)$ a language which is hardest in the categorie of homomorphic reductions.

We show that it is as easy as in the case of the theorem of Shamir to prove the theorem of Chomsky-Schützenberger from our theorem 5 and lemma 9. For this reason we change a little bit the definition of h, but such that lemma 9 remains applicable.

We define a homomorphism $g : T^* \longrightarrow R\langle(\overline{Z}\cup\overline{T})^*\rangle$ by setting for $t \in T$

$$g(t) = \sum_{z \in X} \sum \alpha_t^z \sum \alpha_{y,v}^u \; \overline{[y:x]}t\bar{t}[u:x][z:v].$$

We notice that the difference of g and h consists in two things: The codomain is different and between $\overline{[y:x]}$ and $[u:x][z:v]$ the product $t\bar{t}$ has been inserted.

Let $\bar{g}$ be the prolongation of g to a homomorphism from T^* into $R\langle(\overline{Z}\cup\overline{T})^{(*)}\rangle$, by applying the canonical mapping from $R\langle(\overline{Z}\cup\overline{T})^*\rangle$ into $R\langle(\overline{Z}\cup\overline{T})^{(*)}\rangle$ behind g. Obviously then it holds

Corollary to LEMMA 9:

$$\langle\bar{g}(w),\overline{[S_1:S_r]}\rangle = \langle h(w),\overline{[S_1:S_r]}\rangle .$$

We define now a regular set over $\overline{Z}\cup\overline{T}$.

$$REG = [S_1:S_r] \cdot \{v \mid \exists(w \in T)\langle g(w),v\rangle \neq 0\}^*.$$

Let $D(\overline{Z}\cup\overline{T})$ be the Dyck-language over $\overline{Z}\cup\overline{T}$ and $\sigma:(\overline{Z}\cup\overline{T})^* \longrightarrow T^*$ the monoidhomomorphism with

$\sigma(z) = \varepsilon$ for $z \in \overline{Z}$,

$\sigma(\bar{t}) = \varepsilon$ for $t \in T$,

$\sigma(t) = t$ for $t \in T$,

then it holds because of lemma 9 the

THEOREM 7 (Chomsky-Schützenberger): $L(G) = \sigma(REG \cap D(\overline{Z}\cup\overline{T}))$.

In conclusion of this section we construct a grammar for $L(G)$ in Greibach normal form. We define

$$\tilde{P} = \{[y:x] \rightarrow t[z:v][u:x] \mid \alpha_t^z \cdot \alpha_{y,v}^u \neq 0\}$$

and

$$\tilde{G} = (Z,T,\tilde{P},[S_1:S_r]).$$

Obviously $\tilde{G}$ is in Greibach normal form. It holds

THEOREM 8: It is $L(G) = L(\tilde{G})$ and more precisely it holds

$$\mathrm{mult}_G(w) = \mathrm{mult}_{\tilde{G}}(w) \text{ for } w \in T^*.$$

The size of $|G|$ and $|\tilde{G}|$ are related as follows

$$|\tilde{G}| \leq 32 \cdot |P_N| \cdot |P_T| \cdot |X|,$$

where $P = P_N \cup P_T$, P_N the set of non-terminals and P_T the set of terminal productions.

REMARK: From this theorem it follows immediately

$$|\tilde{G}| \leq \frac{16}{3}|G|^2 \cdot |X| < \frac{8}{3}|G|^3.$$

For large production systems, this means

$$|P_T| = O(|T|\cdot|X|),\ P_N = O(|X|^3)$$

it holds for $|T|<|X|$ and $\varepsilon > 0$

$$|\tilde{G}| \leq O(|T|\circ|X|^5) \leq O(|G|^{2+\varepsilon}).$$

Syntactical congruences

In this section we transfer the syntactical congruences on our algebra $\mathcal{A}_R(G)$ and we study how this congruences are related under our representation $\varphi : \mathcal{A}_R(G) \rightarrow R\langle Z^{(*)}\rangle$. In this connection the following lemma is entral.

LEMMA 10: If $w\in\overline{X}^*$ and $u\in\overline{Z}^*$ such that $\langle\varphi(w), [z_0:x_0]u\rangle = \alpha\neq 0$. Then there exists $w'\in X_r^*$ such that $\langle z_0ww', x_0\rangle = \alpha$.

For $L \subset T^*$ we define as usually

$$u =_r v(L) \Leftrightarrow \forall_w (uw\in L \Leftrightarrow vw\in L).$$

$=_r(L)$ is the syntactic right congruence.

For an easy formulation of the following results we extend our alphabet Z by one new element $\dashv$. The new alphabet is again Z. And we use the abbreviation $\$ = \dashv \circ[S_1:S_r]$.

The idea is to annulate words, which have not the form $\overline{[S_1:S_r]}\circ Z^*$ and which are in $\varphi(\mathcal{A}(G))$ by multiplying it from the left with \$. Remember $\$ \circ \overline{z} = 0$ for $z\in Z$ and $z \neq [S_1:S_r]$ and $\$[S_1:S_r]\cdot\overline{z} = 0$ for all $z\in Z$.

THEOREM 9: $w =_r O(L) \Leftrightarrow \$h(w) = 0$. Here h is the homomorphism of theorem 6.

PROOF: We assume $\$\cdot h(w) \neq 0$. Applying lemma 10 we find w' such that $\langle S_1\eta(ww'), S_r\rangle \neq 0$, and by lemma 9 we have $ww'\in L$. Therefore it holds $w \neq_r O(L)$.

On the other hand there exists w' for w such that $w\circ w'\in L$, then by lemma 9 is $\langle S_1\eta(ww'), S_r\rangle \neq 0$ and therefore $\$\cdot h(w) \neq 0$ too. This proves our theorem.

This theorem describes a procedure to decide $w =_r O(L)$, L being a c.f. language.

Now we transfer the right congruence to $\mathcal{A}_R(G)$ by defining for $p,p'\in\mathcal{A}_R(G)$

$$p =_r p'(L) \Leftrightarrow \forall_{q\in\mathcal{A}_R(G)} (\langle p\cdot q, S_r\rangle = 0 \Leftrightarrow \langle p'\cdot q, S_r\rangle = 0).$$

In a symmetrical way one defines the left congruence $=_l(L)$.

One easily seesthat these definitions for R=B or R=N define congruence relations, but this is not true for R=Z or R being a field. The same holds for the following definition of the syntactical equivalence modulo L:

$$p = p'(L) \Leftrightarrow \forall_{q',q\in\mathcal{A}_R(G)} (\langle q\cdot p\cdot q', S_r\rangle = 0 \Leftrightarrow \langle q\cdot p'\cdot q', S_r\rangle = 0).$$

The quotient of $\mathcal{A}_R(G)$ by the syntactical congruence yealds the syntactical algebra $\mathcal{A}_R(G)/(L)$. Because the syntactical monoid even for c.f. languages is hard to compute, this holds for $\mathcal{A}_R(G)/(L)$ too. Therefore it is of interest to look for algebras between $\mathcal{A}_R(G)$ and $\mathcal{A}_R(G)/(L)$.
We put

$$\mathfrak{a}_r(L) = \{p \in \mathcal{A}_R(G) \mid p =_r 0(L)\}$$

and

$$\mathfrak{a}(L) = \{p \in \mathcal{A}_R(G) \mid p = 0(L)\}.$$

obvioulsy it holds

LEMMA 11: $\mathfrak{a}_r(L)$ is a right ideal.
$\mathfrak{a}(L)$ is a two-sided ideal.

Immediately one has the

Corollary of THEOREM 9: The word problem $w \in \mathfrak{a}_r(L)$ will be decided by $\mathfrak{S}\cdot\varphi(w)$ for $R = \mathbb{N}$ or $R = \mathbb{B}$.
Now $\varphi^{-1}(0)$ is a two-sided ideal of $\mathcal{A}(G)$ and it is $\varphi^{-1}(0) \subset \mathfrak{a}_r$. Therefore one may ask if $\varphi^{-1}(0)$ has an interesting syntactical property. Obviously it holds $\varphi^{-1}(0) \subset \mathfrak{a}(L)$ too. One may ask if it is possible to prolongate φ to a homomorphism $\psi : \mathcal{A}(G) \rightarrow R\langle Y^{(*)}\rangle$ with a suitable Y, such that $\psi^{-1}(0) = \mathfrak{a}(L)$ holds. Because of lemma 2 one this by a homomorphism from $R\langle Z^{(*)}\rangle$ into $R\langle Y^{(*)}\rangle$. But it could be that such a prolongation from $\varphi(\mathcal{A}(G))$ into a suitable $R\langle Y^{(*)}\rangle$ exists, because $1 \notin \varphi(\mathcal{A}(G))$.
Presumably such a homomorphism does not exist, because each semigroup homomorphism from $Z^{(*)}$ in $R\langle Y^{(*)}\rangle$, which is induced by transformations $[y:x] \rightarrow \Sigma q[y:x]q'$ maps the elements $[y:x]\cdot[z:v]$ for $v \neq x$ into 0. Therefore it remains an

Open question: Do there exist non trivial representations of $\mathcal{A}_R(G)/\mathfrak{a}(L)$ in $R\langle Y^*\rangle$? Answering this question is of practical interest too, because a section u of a program of a language L is syntactically incorrect if $u = 0(L)$. By evaluating of $\mathfrak{S}\cdot\varphi(w)$ we are able to find the shortest syntactically incorrect prefix of a program u.

The representation of $\mathcal{A}_R(G)/\mathfrak{a}(L)$ we are looking for would do the same for the shortest syntactically incorrect sections of a program. One could object that the evaluation of our ring homomorphisms is not trivial. This is indeed so, if we wish to do it in a most efficient way. But there are several other important problems that are reducible to this problem. We take the occasion and point out some further problems which seem to be important. The syntactical congruence of a language L(G) does not reflect the structure of G very strongly. As the week equivalence of two languages L(G) = L(G') does not say much about relations between G and G'. One of the most important applications of language theory is to describe the syntax of programming or natural languages. The semantics of these languages depends strongly on the grammars G, which generate the syntax. Therefore I think that the grammars deserve more interest than the languages. Languages are just one under different properties of a grammar. If the grammars G and G' describe the syntax of two programming languages and if L(G) = L(G') then these languages as programming languages are not necessarily equal. This leads to the question to formulate structural equivalences between grammars. Different such equivalences have been defined but only one of them, the "strong" equivalence is well known. This equivalences will be reflected by the existence of certain homomorphisms and products between our algebras $\mathcal{A}_R(G)$. We will come back to this problem at another place. Here we give only a definition of a finer syntactical congruence, which is identical with the normal one in the case of unambiguous grammars.

For $p,p' \in \mathcal{A}_R(G)$ we define p syntactically congruent p' modulo G:

$$p = p'(G) \Leftrightarrow \underset{q,q' \in \mathcal{A}_R(G)}{\forall} (\langle qpq', S_r\rangle = \langle qp'q', S_r\rangle).$$

We see that the 0-classes in both congruences (L) and (G) are the same.

The word-problem for the quotient algebra $\mathcal{A}(G)/(G)$ is closely related to the equivalence problem in the case of unambiguous grammars. Therefore as one may assume, these algebras are hard to compute. It is clear that in this connection lot of interesting questions arise.

For R being a field we have

$$p = p'(G) \Leftrightarrow p = p' \, \mathcal{A}_R(G)/\mathfrak{a}.$$

Therefore in this case $\mathcal{A}_R(G)/(G)$ is the syntactical algebra Reutenauer [Re] associated to the formal series belonging to the grammar G. We think it very important to study each of these cases. Restricting to $R = \mathbb{Z}$ or R a field, important practical questions disappear from the theory.

Unambiguous grammars, LL(k) grammars

In this section we always assume $R = \mathbb{N}$ and we write therefore just $\mathcal{A}(G)$ for $\mathcal{A}_R(G)$.

By definition it holds for unambiguous grammars

$$\langle w, S_r\rangle \leq 1 \text{ for } w \in T^*.$$

Because of lemma 10 this is equivalent to

$$\langle \mathfrak{S} \cdot \varphi(u), a\rangle \leq 1 \quad \text{for } u \in \overline{X}^* \text{ and } a \in Z^*.$$

If one goes through the proof of lemma 10 again, one sees that the following lemma is true.

LEMMA 12: Let G be an unambiguous c.f. grammar and $w \circ w' \in L(G)$. Then there exists exactly one monom $a \in Z$ such that

$$\alpha = \langle S \cdot h(w), a$$

$$\alpha' = \langle h(w), \overline{a}\rangle$$

and $\alpha = \alpha' = 1$ holds. Here is $\overline{x_1 \dots x_k} = \overline{x_k \dots x_1}$.

We assume in the following G to be a LL(k) grammar if the converse isn't not explicitly will stated.

We are interested here to study $\mathcal{A}(G)$ and our representation for LL(k) grammars. As we have shown in lemma 7, it follows from f u-left prime and $Z(f) = u \cdot v$, that $v \in X_r^*$. In $\mathcal{A}(G)$ we then have $\langle uv, Q(f)\rangle = 1$ if $Q(f) \in \overline{X}$. We call $v \in X_r^*$ as <u>almost inverse</u> to u <u>from the right</u> if there exists $z \in \overline{X}$ such that $\langle uv, x\rangle \neq 0$.

<u>LEMMA 13:</u> For each $u \in (\overline{X} \cup T)^*$ card X = m there exist at most $2\,m^{k+2}$ elements $v \in X_r^*$, which are almost inverse to u from the right, if G is LL(k).

We define for $p \in R\langle Z^{(*)}\rangle$

$$|p| = \sum_{Zu \in Z^*} \langle p, u\rangle$$

|p| is the sum of the coefficients of the monoms of p which contain only at the first place an inverse of Z.

<u>LEMMA 14:</u> For all $u \in \overline{X}^*$ it holds

$$|\varphi(u)| \leq m^{k+3}$$

<u>LEMMA 15:</u> Let be $u \in (\overline{X} \cup T)^*$ and $[y_0 : x_0] \in Z$.
If $\dashv \cdot [y_0 : x_0]\varphi(u) \neq 0$, then there exists a decomposition $u = u_1 \cdot u_2$ and $w \in Z^*$ such that

$$\dashv \cdot [y_0 : x_0]\varphi(u) = w \cdot (u_2), \quad |u_2| \leq k.$$

From this directly follows

<u>THEOREM 10:</u> The word problem $w \in L(G)$, $G \in LL(k)$ can be decided in linear space and linear time by multiplying out $\$\ \varphi(u)$ sequentially from left to right.
The method described in this theorem applied to LR(k) languages would generally lead to exponentially growing space complexity.
The converse of our theorem 10 is not true. There exist c.f. grammars G for non-deterministic languages such that their word problem can be decided by sequentially multiplying out from left to right in linear space and linear time.

<u>Definition:</u> We call this class of c.f. languages SMLR(N), iff

$$|\$ \cdot \varphi(n)| \leq N \text{ for all } u \in T^*.$$

Obviously it holds because of lemma 15 and these remarks the

<u>THEOREM 11:</u> 1) The word problem for SMLR(N) can be decided in linear time and linear space.
2) $LL(k) \subset SMLR(m^{k+3})$
3) $LRSM = \bigcup_N SMLR(N)$ is closed under "$\cup$".

<u>Open problems:</u> 1. Is it decidable for $G \in$ c.f. if $G \in SMLR(N)$ for fixed N?
2. Is it decidable, if $L(G) = L(G')$ for $G, G' \in SMLR(N)$?

This section shows that in our theory we get a pure algebraic definition of the LL(k) languages. We will show in the next section that this remains true for LR(k) languages.

THEOREM 12: $\varphi \in SMLR(N)$ is recursively undecidable.

LR(k) - Grammars: Here we derive similar results as in the section before. The only difference comes is the substitut on $R\langle X^{(*)}\rangle$ by $\mathcal{A}_R(G)$ mod $\mathcal{A}_r(L)$ in the characterisation of LR(k). A first information we get by the following

LEMMA 16: Let G be a LR(k) - grammar and $u \in T^*$. If $u = \tilde{u}_1 + \ldots + \tilde{u}_m$ $\mathcal{A}_r(L)$ with $\tilde{u}_i \neq_r 0(L)$, then $m \leq (|\bar{X}|+1)^k$ holds.

This lemma not yet characterizes LR(k)-grammars. But going a second time through the proof of lemma 17, we see that $\tilde{u}_1, \ldots \tilde{u}_m$ have a common prefix, which is uniquely determined by u. We see this from the decomposition $u = u_1 \cdot u_2$ such that $|u_2| = k$. Therefore it holds one direction of the

THEOREM 13: The c.f. grammar G is of type LR(k) iff for each $u \in T^*$ it holds $u = \tilde{u} \cdot p$, $\tilde{u} \in X_T^*$, $p = \tilde{u}_1 + \ldots + \tilde{u}_m$, $u_i \in X_T^*$ and $|\tilde{u}_i| \leq k$.

To proof this theorem completely it is sufficient to show, that the word problem $w \in L(G)$ can be decided by a deterministic pda. It is a simple consequence of the following theorem, which concerns a more general class of c.f. languages.
We generalize LR(k) as before LL(k) in the following

Definition: The c.f. grammar G is in the class BSLR(N) iff for all $u \in T^*$ holds:

from $u = \alpha_1 \cdot \tilde{u}_1 + \ldots + \alpha_m \cdot \tilde{u}_m$ $(\mathcal{A}_r(L)$, $L = L(G)$, $\tilde{u}_i \in X_T^*$, $\alpha_i \in R$

$R = \mathbb{N}$ it follows $\sum_{i=1}^{m} ||\alpha_i|| \leq N$, $||\alpha_i|| = \begin{cases} 1 & \text{if } \alpha_i \neq 0 \\ 0 & \text{if } \alpha_i = 0. \end{cases}$

The letters BS come from bounded size and LR from the use of the right congruence $=_r(L)$.

THEOREM 14: The word problem $w \in L(G)$ for $G \in BSLR(N)$ can be decided sequentially in time $O(|w|)$.

Acknowledgments: The autor wishes to thank J. Berstel, M. Harrison, C. Reutenauer and J. Sakarovitch for helpful discussions. The comments of B. Becker concerning an earlier version of this paper lead to improvements of several proofs.

Literatur:

[Be] Benson, D.: "Syntac and Semantic: A Categorial View". Information and Control (1970), p 145-160

[Ch.-Sch] Chomsky, N. and Schützenberger, M.P.: "The algebraic theory of context-free languages". P. Braffort and S. Hirschberg eds., Computer Progr.and Formal Systems, North Holland, Amsterdam, p 116-161 (1970).

[Gr.1] Greibach, S.A.: "The hardest context-free language". Siam, J. comp.2, (1970)

[Gr.2] Greibach, S.A.: "Erasable context-free languages". Information and Control 4 (1975).

[Har] Harrison, M.A.: "Introduction to Formal Language Theory", Addison-Wesley, 1978.

[L.S.] Lewis, P.M. and Stearns, R.E.: "Syntax-directed Transduction", J. Assoc. Comp. Mach., 1968, 465-488.

[Ho] Hotz, G.: "Eine Algebraisierung des Syntheseproblemes von Schaltkreisen". EIK 1965, p. 185-231.

[Ho.1] Hotz, G.: "Der Satz von Chomsky-Schützenberger und die schwerste kontextfreie Sprache von S. Greibach." Astérisque 38-39(1976) p. 105-115, Société Mathématique de France.

[Ho.2] Hotz, G.: "Normal-form transformations of contextfree grammars." Acta Cybernetica, Tom., Fasc. 1, p 65-84 (1978)

[Ho.3] Hotz, G.: "A representation theorem for infinite dimensional assoziatives algebras and applications in language theory." Extended abstract in Proceedings de la 9ième Ecole de Printemps d'informatique théorétique, Murol 1981.

[Ho.4] Hotz, G.: "About the Universality of the Ring $R<X^{(*)}>$." Actes du Seminaire D'Informatique Theoretique,Université Paris VI et VII, p 203-218, ANNE 1981-1982

[Ho-Est] Hotz, G. und Estenfeld, K.: "Formale Sprachen", Mannheim: Bibliographisches Institut (1981)

[Kn.1] Knuth, D.E.: "On the translation of languages from left to right". Inform. and Control (1965) p 607-639.

[Kn.2] Knuth, D.E.: "Top-down-Syntax Analysis". Acta Informatica (1971)p79-110

[Ni] Nivat, M.: "Transductions des Languages des Chomsky". Ann. de L'Institut Fourier 18 (1968)

[Pe] Perrot, J.F.: "Contribution à l'étude des monoides syntactiques et des certains groupes associés aux automates finis. Thèse Sc. Math., Université Paris VI (1972)

[Re] Reutenauer, C.: "Series rationelles et algèbres syntactiques". Thèse de Doctorat D'Etat à Université Paris VI p 1-209 (1980).

[Sa] Salomaa, A. and Soittola, M.: "Automata-Theoretic Aspects of Formal Power Series". Springer Verlag (1978)

[Sh] Shamir, E.: "A representation theorem for algebraic and context-free power series in non commuting variables". Information and Control 11, 239-254 (1967).

THUE SYSTEMS AND THE CHURCH-ROSSER PROPERTY

Matthias Jantzen

Fachbereich Informatik
Universität Hamburg
Rothenbaumchaussee 67
D-2000 Hamburg 13
F.R.G.

1. Introduction

Transformation systems are of great importance in various parts of computer science: they are used for constructing theorem provers, algebraic simplifiers, program optimizers, and are the basis for describing program transformations, abstract data types, formula manipulations, program specifications, graph grammars, and many other formal systems.

In many cases the transformation systems take the form of term rewriting systems. However, just as much important are string rewriting systems which have been studied in mathematics and computer science since the pioneering work of Axel Thue in 1914, [63]. Thue was also interested in the more general problem of rewriting of other combinatorial objects such as graphs or trees. Semi-Thue systems have been the basis for Chomsky to define formal grammars and languages, whereas Thue systems as symmetrical rewriting systems are used as presentations of monoids and groups.

One of the questions of primary interest in all transformation systems is whether two syntactically different objects do have the same meaning, i. e., are congruent modulo the closure of the transformation relation. In some, but not all, cases the completion method of Knuth and Bendix produces a decision algorithm for this question. As it is well known this question posed for Thue systems is the so-called word problem and is in general undecidable. This suggests that this problem may be undecidable for more sophisticated systems, too. In order to understand other systems with may be more practical appeal it is worthwhile to learn from string rewriting systems as much as one can.

Therefore we consider finite semi-Thue systems having the Church-Rosser property in its original meaning, where reductions are defined by the direction of the rules and not solely based on the length of the strings involved. This has the advantage that one can use Thue systems that are Church-Rosser with respect to the special type of reduction, as done in [5, 7-17, 32, 34, 46, 47, 50, 53, 54, 59-61] and many others cited there. On the other hand this approach rules out many interesting problems and useful considerations, such as using different orderings as in [23, 24, 56] or questions that are already solved by definition. Using thus semi-Thue systems as the basic object of consideration we simply speak about Thue congruences, Thue systems, and their word problem as being induced or presented by the semi-Thue system.

Having defined the necessary notation we shall discuss results about the relation between congruence classes induced by certain types of semi-Thue systems and formal languages.

Apart from the word problem in Thue systems there are many other interesting problems which are undecidable, too. We include many of those results in order to get a feeling about the complexity of simple questions. We shall also include some of the recent results about a special Thue system studied by the author the last four years. We shall omit proofs and give in most cases only the reference.

2. General Transformation Systems

We only give the absolute minimum of definitions in this section and want to ask the interested reader to consult the excellent work of Huet [29], Huet and Oppen [30], and the work cited there (166 entries).

Definition: Let B be a set of objects and $\Rightarrow$ be a binary relation on B. Then the structure $(B,\Rightarrow)$ is a transformation (or reduction) system and $\Rightarrow$ is called the transformation (or reduction) relation. $\overset{*}{\Rightarrow}$ (and $\overset{*}{\Leftrightarrow}$) is the reflexive transitive (resp. symmetric, reflexive and transitive) closure of $\Rightarrow$. If $x \in B$ and there is no $y \in B$ such that $x \Rightarrow y$, then x is called irreducible, and if in addition $x \overset{*}{\Leftrightarrow} y$ for some $y \in B$ then x is called a normalform for y and for the whole equivalence class $[y] := \{x \in B \mid y \overset{*}{\Leftrightarrow} x\}$.

Usually normal forms are not unique. One condition that guarantees the existence of unique normal forms in transformation systems is the Church-Rosser property.

Definition: Let $S := (B,\Rightarrow)$ be a transformation system. Then (a) S is Church-Rosser if for all $x, y \in B$, $[x] = [y]$ implies $x \overset{*}{\Rightarrow} z$ and $y \overset{*}{\Rightarrow} z$ for some $z \in B$. (b) S is confluent if for all $w, x, y \in B$, $w \overset{*}{\Rightarrow} x$ and $w \overset{*}{\Rightarrow} y$ imply $x \overset{*}{\Rightarrow} z$ and $y \overset{*}{\Rightarrow} z$ for some $z \in B$. (c) S is locally confluent if for all $w, x, y \in B$, $w \Rightarrow x$ and $w \Rightarrow y$ imply $x \overset{*}{\Rightarrow} z$ and $y \overset{*}{\Rightarrow} z$ for some $z \in B$.

The next result is from Newman and was rediscovered by other authors for special cases.

Theorem,[49]: A transformation system $(B,\Rightarrow)$ is confluent iff it is Church-Rosser.

Theorem,[49]: A transformation system is Church-Rosser iff normal forms are unique and exist.

In order to force a transformation system to actually have normal forms one should not have infinite chains of transformations.

Definition: The transformation system $(B,\Rightarrow)$, as well as $\Rightarrow$, is called terminating (uniformly terminating or Noetherian) if there is no infinite chain $x_o \Rightarrow x_1 \Rightarrow x_2 \Rightarrow \ldots$ with $x_i \in B$ for all $i \geq 0$. The relation $\overset{*}{\Rightarrow}$ is then called well founded.

If a transformation system is terminating, then normal forms do always exist and thus are unique if the system is Church-Rosser. Hence it would be appropriate to call such systems 'uniquely terminating'. Often uniquely terminating systems are called 'canonical' or 'complete', giving credit to the completion method of Knuth and Bendix, [36]. In general it is undecidable whether a transformation system is Church-Rosser, but if it is terminating then the following result by Newman may be of help. The proof by Newman is very complicated but Huet has given a very elegant proof in [29].

Theorem: Let $S := (B, \Rightarrow)$ be a terminating transformation system. Then S is confluent iff S is locally confluent.

The property of a transformation system being locally confluent is decidable for terminating semi-Thue systems, and thus it will be important to decide whether a semi-Thue system is terminating or to guarantee termination by definition.

3. Semi-Thue and Thue Systems

Definition: If X is an alphabet, then e shall denote the empty string which is the identity in the free monoid X^* generated by X under the usual concatenation as operation. The length of a string $w \in X^*$ is denoted by $|w|$ and for sets $A \subseteq X^*$ their cardinality is written $|A|$.

Definition: Let X be an alphabet.

(a) A semi-Thue system (abb.: STS) S on X is a subset $S \subseteq X^* \times X^*$. Each element $(u,v) \in S$ is called a rule and X is always the smallest alphabet possible. Usually X will be finite.

(b) Given S the one step derivation relation $\Rightarrow_{(S)}$ is defined for all strings $x,y \in X^*$ by $x\,u\,y \Rightarrow_{(S)} xvy$ iff $(u,v) \in S$ is a rule.

(c) $\overset{*}{\Rightarrow}_{(S)}$ (or $\overset{+}{\Rightarrow}_{(S)}$) is the reflexive transitive (transitive, resp.) closure of $\Rightarrow_{(S)}$.
Based on these relations we define for each string $w \in X^*$:

(d) $\Delta^+(w)_{(S)} := \{v \in X^* \mid w \overset{+}{\Rightarrow}_{(S)} v\}$ and

$\Delta^*(w)_{(S)} := \{v \in X^* \mid w \overset{*}{\Rightarrow}_{(S)} v\}$ as the set of all

descendants of w (with respect to S).

(e) $\langle w \rangle^+_{(S)} := \{v \in X^* \mid v \overset{+}{\Rightarrow}_{(S)} w\}$ and

$\langle w \rangle^*_{(S)} := \{v \in X^* \mid v \overset{*}{\Rightarrow}_{(S)} w\}$ as the set of all

ancestors of w (with respect to S).

This notation is used for sets $A \subseteq X^*$ in the obvious way by taking the union of the sets generated by the elements of A.

Now $(X^*, \Rightarrow_{(S)})$ is a transformation system and we call the STS S on X terminating if the derivation relation $\Rightarrow_{(S)}$ is terminating. Using the well known Theorem of Rice it is easy to show:

Theorem: It is undecidable whether a finite STS is terminating.

This situation is avoided if one restricts a STS in such a way that the length of the strings is decreasing in every derivation step, as done by Nivat, Book, and others.

We shall use a special notation for this in order to be able to reference results without confusing the reader. Compare also the recent work [34].

Definition: A STS S is called lg-terminating if $(u,v) \in S$ implies $|u| > |v|$. If $>$ is a total ordering of X^* such that $x\,u\,y > x\,v\,y$ for any $(u,v) \in S$ and all $x,y \in X^*$, then S is called $>$-terminating.

According to what has been said about general transformation systems we shall define irreducible strings and normal forms.

Definition: Let S be a STS on X.

(a) A string $w \in X^*$ is irreducible with respect to S if $\Delta^+(w)_{(S)} = \emptyset$. By IRR(S) we denote the set of all strings $w \in X^*$ that are irreducible with respect to S.

(b) For each $w \in X^*$ let $IRR(w) := \{v \in X^* \mid v \in IRR(S) \cap \Delta^*(w)_{(S)}\}$ and for sets $A \subseteq X^*$ we let $IRR(A)_{(S)} := \bigcup_{w \in A} IRR(w)_{(S)}$.

(c) If $w \overset{*}{\Leftrightarrow}_{(S)} v$ and $v \in IRR(S)$, then v is called a normal form of w.

(d) If there is no string w such that $w \overset{*}{\Leftrightarrow} v$ and $|w| \leq |v|$, then v is called minimal.

Now, in order to speak about the Church-Rosser property of a STS we need also the definition of the induced Thue system or Thue congruence.

Definition: Let S be a STS on X.

(a) The Thue congruence $\overset{*}{\Leftrightarrow}_{(S)}$ induced by S is the symmetric, transitive, and reflexive closure of $\Rightarrow_{(S)}$. The Thue system (abb.: TS) induced by S is $T(S) := S \cup \{(v,u) \mid (u,v) \in S\}$.

(b) The monoid M(S) presented by S or equivalently by T(S) is the quotien monoid $X^*/\overset{*}{\Leftrightarrow}_{(S)}$, the elements of which are the congruence classes $[w]_{(S)} := \{v \in X^* \mid w \overset{*}{\Leftrightarrow}_{(S)} v\}$ with the usual multiplication $[u]_{(S)} \cdot [v]_{(S)} = [uv]_{(S)}$. M(S) is finitely presented (or generated) if S (resp. X) is finite.

(c) S is Church-Rosser (and thus confluent) if for all $w,v \in X^*$, $[w]_{(S)} \cap [v]_{(S)} \neq \emptyset$ implies $\Delta^*(w)_{(S)} \cap \Delta^*(v)_{(S)} \neq \emptyset$.

(d) If S is Church-Rosser and terminating (lg-terminating, $>$-terminating, resp.) then S will be called uniquely terminating (uniquely lg-terminating, uniquely $>$-terminating, resp.)

In what follows we shall omit the subscript (S) whenever it is possible. If a TS is used as a presentation for a group, then the Thue congruence is usually written as = meaning the equality in the group which will be clear from the context. We shall use this simpler notation for a STS only in case the presented monoid is in fact a group.

Notice that a Thue system is called Church-Rosser in [7-17, 22, 32, 34, 46, 50, 52-54, ...] while we want to write it as a uniquely lg-terminating STS.

The first positive result is due to Berstel.

Theorem: If S is a finite STS on X, then IRR(S) is a regular subset of X^*, a regular expression for which can be constructed effectively from S.

From the result that the set IRR(S) is regular for a finite STS S it is clear that it is decidable whether a string is irreducible. However, the situation is different for minimal strings.

Theorem, [16]: For arbitrary finite STS S on X and $w \in X^*$ it is undecidable whether w is minimal with respect to S.

One of the basic questions for Thue systems is the word problem which is known to be undecidable by the results of Markov and Post. Using the Post correspondence problem one can show that the common ancestor problem is undecidable, see [15] and Narendan, cited in [12].

Theorem: For arbitrary uniquely lg-terminating STS S on X and $w, v \in X^*$, the common ancester problem, i. e., the question "$\langle w \rangle^* \cap \langle v \rangle^* \neq \emptyset$?" is undecidable.

The common descendant problem for finite STS's which are uniquely terminating, however, is decidable. One only has to compute the unique irreducible normal forms which will be equal if the strings one starts with are congruent and thus have a common descendant.

Theorem: Let S be a finite uniquely terminating STS on X. Then the common descendant problem, i. e., the question "$\Delta^*(w) \cap \Delta^*(v) \neq \emptyset$?", is decidable for arbitrary $w,v \in X^*$.

If S is a STS then the common descendant problem is in general a different question than the word problem for the Thue system T(S) induced by S. But if S is uniquely terminating, the common descendant problem, the word problem for T(S), or the monoid M(S), presented by S, i. e., "[w] = [v]?" coincide.

In order to get some information about the complexity of the common descendant problem for uniquely terminating STS's one has to look at the length of the derivations finding the unique normal forms, compare also [16]. Using a technique from Book, [7], one obtains

Theorem, [33]: Let S be a uniquely terminating STS on X and f be a function such that no derivation $u \overset{*}{\Rightarrow} v$ has more than f(u) steps. Then the common descendant problem for $w,v \in X^*$ is solvable in time $c \cdot (f(w) + f(v) + |wv|)$.

Corollary, [7]: The word problem for TS's induced by finite, uniquely lg-terminating STS's is decidable in linear time.

This, however, is only an upper bound, since the word problem for a TS T may be decidable by using a different method and not by resorting to some uniquely terminating STS which induces T. There are some more results of interest in connection with the complexity of the word problem. The following result is minor modification of a result of Bauer, cited in [33].

Theorem: For any Gregorzyk class E_n there exists a finite, uniquely terminating STS S such that the word problem for T(S) is E_n-decidable, but not E_{n-1}-decidable.

It is easily seen that for each finite STS S there is an infinite uniquely terminating STS S' equivalent to S, and S' is recursive iff S has a decidable word problem, compare [33] and [16]. However, such a result is not of great help, since there exist infinite Thue systems with a decidable word problem for which there does not exist any equivalent finite Thue system. And even worse, it is undecidable whether for a given finite STS S there exists an equivalent finite STS S' which is uniquely terminating or which has a decidable word problem. To formulate those and other important results we need a precise definition of equivalence for STS's which actually describes the equivalence of the induced Thue systems.

Definition: Let S and S' be two STS's on X, then S and S' (likewise T(S) and T(S')) are equivalent iff $\overset{*}{\Leftrightarrow}_{(S)} = \overset{*}{\Leftrightarrow}_{(S')}$.

This definition is much stronger than simply to require that M(S) and M(S') are isomorphic, since S and S' have to operate on the same alphabet X.

Let $S := \{(a\, b^n\, c^n\, d,\, e) \mid n \geqq 1\}$ be an infinite STS on $X := \{a, b, c, d\}$ then S is uniquely lg-terminating and T(S) has a decidable word problem, but there is no finite STS on X equivalent to S. There are results stating that certain groups with recursively enumerable presentation and solvable word problem do not have a finite presentation, so that the above mentioned result also holds if our notion of equivalence is weekened to the requirement that M(S) and M(S') are isomorphic.

Theorem, [54, 16]: It is undecidable whether an arbitrary STS S is equivalent to some uniquely lg-terminating STS S'.

This result holds if both systems S and S' are infinite, if one of the two systems is finite, or if both are finite. Even if we force S' to be a certain infinite STS induced by S we cannot do better,compare [52, 53].

Theorem, [54]: It is undecidable whether for an arbitrary finite STS S on X the following infinite system $P(S) := \{(u,v) \mid [u] = [v] \wedge |u| > |v|\}$ is uniquely lg-terminating.

Narendan, communicated in [54], has shown that $S := \{(aba, ab)\}$ is a STS such that P(S) is uniquely lg-terminating but there is no finite STS equivalent to S which also is uniquely lg-terminating.

The question whether any finite STS with decidable word problem for T(S) has an equivalent STS which is uniquely terminating and which could then be used as a basis for the decision algorithm for the word problem is still open. What one can do, however, is to effectively construct an equivalent uniquely terminating (or lg-terminating) STS if such a system exists. This can be done using the Knuth-Bendix completion method, [36] and well described in [31], and has been applied for other transformation systems for instance in [39, 40, 43, 48, 62]. For Thue and semi-Thue systems one gets

Theorem, [16, 34, 43]: There is an algorithm that on input a finite STS S will eventually halt and give as output a finite STS S' which is uniquely terminating and equivalent to S if such a system exists. Moreover S' can be made to be reduced, see [34, 43].

Due to Nivat's criteria we can decide whether a given terminating STS is uniquely terminating (or lg-terminating). If this fails one can not do much more because the completion method might not hold and it is undecidable whether it eventually will hold.

Theorem, [16, 34]: There is an algorithm that decides whether a given finite STS S is uniquely terminating. The algorithm works in time $o(|S|^3)$ if S is lg-terminating.

For arbitrary STS's the equivalence problem was undecidable. It becomes decidable if the both systems in question are uniquely lg-terminating, as shown in [17].

4. Congruential Specification of Formal Languages

One motivation for studying Thue and semi-Thue systems in connection with formal languages is the observation that in case the given STS has certain structural properties and is uniquely terminating, then the induced congruence classes are context-free languages and not just complicated recursively enumerable sets. One of the most promising approaches was the choice of restricting attention to lg-terminating STS's.

Definition: A language $L \subseteq X^*$ is congruential if there exists a finite STS on X such that $L = [A] := \bigcup_{w \in A} [w]$ for some finite set $A \subseteq X^*$.

It is easily seen, [5], that the linear context-free language $\{ww^R \mid w \in \{a, b\}^*\}$ is not congruential. On the other hand there are uniquely terminating STS's where for example the class [e] is not a context-free set.

For example, let S:= {(abba, e)} and S':= {(bbaa, e), (aab, baa), (abb, bba)} be two STS then S and S' are equivalent and [e] is not context-free. M(S) = M(S') is actually a group and hence not a context-free group, see [44]. S' is a preperfect system, but results in [13] show that there cannot exist an uniquely lg-terminating STS which is equivalent to S. There are some syntactical restrictions, originally formulated for Thue systems, that can be adapted for STS's and will be important to formulate very often positive results.

Definition: Let S be a STS on X.

(a) S is special, if $(u,v) \in S$ implies $v = e$ and $|u| \geq 1$.

(b) S is homogenous of degree k, if it is special and $(u, e) \in S$ implies $|u| = k$.

(c) S is homomorphic of there exists a homomorphism $h: X^* \to X^*$ such that $(u, v) \in S$ implies $v = h(u)$.

(d) S is monadic, if $(u, v) \in S$ implies $v \in X \cup \{e\}$ and $|u| \geq |v|$.

(e) S is regular (or context-free) if $R(S) := \{v \mid (u, v) \in S\}$ is finite and for each $v \in R(S)$ the set $\{u \mid (u, v) \in S\}$ is a regular (resp. context-free) language.

(f) S is preperfect if $(u, v) \in S$ implies $|u| > |v|$ and in addition S is Church-Rosser.

Notice that we have changed the definition (e) of regular (or context-free) STS's slightly by not forcing them to be lg-terminating, compare [52-54]. On the other hand we have weakened the definition (f) of preperfect STS's compared with that used for Thue systems in [7-17], since a STS still defines a direction for the application of lg-preserving rules. But special, monadic, and homogenous STS's all are lg-terminating.

The word problems for Thue systems induced by STS's of the above forms are all undecidable. The same holds for the common descendant problem of the STS's themselves.

Since monadic STS's are by definition lg-terminating and the rules are context-free rewrite rules in opposite direction, they are much easier to deal with and allow the formulation of a number of interesting results. We shall use [12] as the basic source. From results about iterated context-free substitutions one immediately gets

<u>Theorem, [15]:</u> Let S be a context-free monadic STS on X and $L \subseteq X^*$ be any context-free language, then the set of its ancestors $\langle L \rangle^*$ is again a context-free language.

This is a nice and strong result about certain infinite systems, but can obviously be used for finite systems as well. A weaker version of this result has been established by Nivat [50]. If L and the context-free monadic STS are specified by finitely many context-free grammars, then a context-free grammar for $\langle L \rangle^*$ can be constructed effectively.

A similar result for descendants can only be proved for regular sets.

<u>Theorem, [15]:</u> Let S be a context-free monadic STS on X and $R \subseteq X^*$ be a regular set, then the set of its descendants $\Delta^*(R)$ is again a regular set.

From the finite specifications of R and S one can effectively construct the finite automaton accepting $\Delta^*(R)$. The result remains true for arbitrary infinite monadic STS's but the effectiveness of the construction depends on some more properties of S.

The undecidability of the common ancester problem remains valid if only finite monadic systems are considered. There are negative results for finite monadic STS's even if they are uniquely terminating.

<u>Theorem, [15]:</u> For any recursively enumerable set $L \subseteq X^*$ there exists a finite special STS S on X and a context-free language $C \subseteq X^*$ such that S is uniquely terminating and $\Delta^*(C) \cap X^* = [C] \cap X^* = L$.

Hence, L may be non-recursive, and the result about the descendants of regular sets cannot be generalized to context-free languages. The situation may even be worse because of the following 'birdy-result'.

<u>Theorem [15]:</u> Let S be a finite special STS which is uniquely terminating, and let $C := L(G)$ be a context-free language specified by the context-free grammar G. Even knowing that [C] (or $\Delta^*(C)$) is context-free there is no algorithm that will construct a context-free grammar for the latter set.

However, in many other cases the results are not so disappointing if we look at uniquely terminating monadic systems. One of the nice results that strengthened one of Nivat and Cochet, [22], is the following

Theorem, [15]: Let S be a finite monadic STS on X which is uniquely terminating, then $\{[R] \mid R \subseteq X^*$ is regular$\}$ is a Boolean algebra of deterministic context-free languages.

This not only shows that the congruence classes induced by some finite, monadic, and uniquely terminating STS are context-free languages, but also that their complements, intersections and unions can be written as congruence classes of the very same system by using a different regular set as a generator. From a regular expression for R one can construct the deterministic pushdown automaton for [R]. Unfortunately we know of no simple uniform method to do so, see the discussion in [12]. If we consider context-free monadic and uniquely terminating systems, then from the mentioned results we can conclude that in this case, too, [R]is context-free for any regular set R, but [R] is not necessarily deterministic. In order to get deterministic context-free congruence classes we only have a result for regular monadic STS's which is slightly weaker .

Theorem, [52, 53]: Let S be a regular monadic STS on X which is uniquely lg-terminating then for every regular set $R \subseteq X^*$ the congruence class [R] is a deterministic context-free language.

As the property of being uniquely lg-terminating was useful for finite systems, it is so for infinite systems that have a simple specification like regular monadic STS's.

Theorem, [52, 53]: There is an algorithm to decide whether an arbitrary regular monadic STS is uniquely lg-terminating.

This question becomes undecidable for context-free monadic STS's,[15], and for regular lg-terminating STS's that are not monadic, [52, 53].

Once a regular STS is known to be uniquely lg-terminating, then the common descendant problem is solvable in linear time, [53], and this is a surprising generalization of the same result for finite lg-terminating systems. For context-free STS's the problem is a bit more complicated, since it uses the membership problem for context-free languages.

Theorem, [12, 15, 53]: If S is a context-free (or regular) STS which is uniquely lg-terminating, then the common descendant problem is solvable in polynominal (resp. linear) time.

Again we encounter a positive result which applies to regular monadic STS's but not to context-free monadic STS's and is a generalization of the similar result for finite STS's.

Theorem, [53, 15]: It is decidable (resp. undecidable) whether two given regular (resp. context-free), monadic, and uniquely terminating STS's are equivalent.

Sometimes the strong requirement of being uniquely terminating can be dispensed if the other syntactical restrictions are strong enough to still yield positive results.

Theorem, [13]: Let S be a finite homogenous STS of degree 2, then there is a linear time algorithm for the word problem for T(S).

This is a stronger version of a result by Adjan, [1], who also showed that there exists a finite Thue system with undecidable word problem which is induced by some homogenous STS of degree 3.

There are still more results in connection with formal languages, see [12], and others that were open quite a while and solved quite recently in [46] and [54].

Theorem, [46, 54]: It is undecidable, whether an arbitrary finite STS is preperfect.

There are other concepts for combining Thue systems with formal languages that have been considered.

Boasson and his students have studied a class of deterministic context-free languages, called 'nonterminal separated' or shortly NTS, [6, 18, 25, 58] and may be others.

Narendan and McNaughton extended the notion of lg-terminating STS's and called their class of languages obtained in this way Church-Rosser languages', [47].

Another interesting area is the connection between monoids or groups and questions which can be answered from their uniquely terminating presentations. This is a relatively young field of interest, [2, 3, 4, 9, 10, 11, 16, 39, 40, 44]. Generally undecidable questions like "is the monoid M(S) presented by the STS S a group?" become decidable if S has nice properties, see [9, 16] for more results in this direction.

We shall close this paper by adding a section with results about an example of a special STS, studied by the author for the last 4 years.

5. A Class of Special Semi-Thue Systems

The class of STS's we shall introduce now all are finite special systems which have only single rule. The example studied in [32, 56] as well as by Squier and Wrathall, will turn out to be equivalent to one STS out of this class. In what follows we shall use the fixed alphabet $X := \{a, b\}$.

Definition: For each $n \geq 1$ let $S_n := \{((ab)^n ba, e)\}$ be a special STS on X. Let $=_n$ be the Thue congruence induced by S_n.

As it turns out $M(S_n)$ is actually a group which we shall denote by G_n, and thus $=_n$ is the equality in G_n. Let $J := M(S)$ be the monoid studied in [32] and presented by $S := \{(abbaab, e)\}$, then $J = G_2$. G_1 is the example of the group which is not context-free and has a preperfect presentation but as all the groups G_n no uniquely lg-terminating presentation. This result follows from [10]. In contrast to G_1 it has been shown in [32] that G_2 does not even have a finite preperfect presentation and it would be interesting to see whether this is true for all the other groups G_n, $n \geq 2$, too.

The first interesting result follows from the characterization of context-free groups in [44] and a characterization of the groups G_n as semi-direct products of nice and well known groups.

Theorem: G_n is isomorphic to $\mathbb{Z} \ltimes \mathbb{Z}(\frac{1}{n})$, the semi-direct product of $\mathbb{Z}(\frac{1}{n}) := \{p\cdot n^q \mid p,q \in \mathbb{Z}$ by $\mathbb{Z}$.

The homomorphism $\phi\colon \mathbb{Z} \to \mathrm{Aut}(\mathbb{Z}(\frac{1}{n}))$ determines the product we have to use in $\mathbb{Z} \ltimes \mathbb{Z}(\frac{1}{n})$ by $(x,p)\cdot(y,q) := (x\cdot y, \phi(y)(p)\cdot q)$, where $x,y \in \mathbb{Z}$ and $p,q \in \mathbb{Z}(\frac{1}{n})$. Now one can conclude that G_n is torsion-free but not free and therefore, [44], G_n cannot be a context-free group, a result that was established for $G_2 = J$ in [32] by a relatively direct method. The groups G_n are isomorphic to the groups presented by $\langle a,x;\ a^{-1}xa = x^{-n}\rangle$, where the Tietze-transformation is determined by $ab = x$. For each $n \geq 1$, G_n has a faithful matrix representation in GL_2, that can be used for many interesting results in a more direct manner.

Theorem: Each group G_n has a faithful matrix representation given by

$$a := \begin{pmatrix} -\frac{x}{n} & 0 \\ 0 & x \end{pmatrix} \quad , \ b := \begin{pmatrix} -\frac{n}{x} & -\frac{n}{x} \\ 0 & \frac{1}{x} \end{pmatrix}$$

where $x \in \mathbb{Z}$, $x \neq 0$, and $|x| \neq 1$ if $n = 1$.

Many results that have been proved for G_2 in [32] can be generalized for G_n, $n \geq 2$. We shall only mention a few, and some only for G_2 since all the possible generalizations have yet not been worked out in exact detail, and are quite recent results.

Theorem: Each G_n has infinitely many proper subgroups isomorphic to G_n itself.

Remark: For all $n \geq 1$ we have the following equations in G_n:

(a) $baab =_n abba$

(b) $(ab)^n =_n bbaa(ba)^{n-2}$

These equations can be proved directly from the presentation of G_n by the STS S_n or more easily using the matrix representation.

Theorem: $\mathbb{Z}(\frac{1}{n})$ is a normal subgroup of G_n, which is abelian and is, as a subgroup, not finitely generated if $n \geq 2$.

In [32, 33] this normal subgroup is abbreviated by D_1^* which is the set of all strings in X^* that have an equal number of b's and a's. We shall now recall some results for which we have detailed proofs for the group G_2, than can partly be found in [33], and can very likely be generalized for all the groups G_n.

The first result which is quite interesting is the so-called 'Reflection-Theorem', which has nice geometrical interpretation, see [33].

Definition: The reflection refl(w) of a string $w \in X^*$ is defined by refl(e) := e, and inductively refl(wa) := b refl(w), refl(wb) := a refl(w).

Theorem, [33]: For each string $w \in D_1^*$ one has $w =_2 \text{refl}(w)$.

The normal subgroup D_1^* of G_2 (or G_n) is of primary interest and thus the next result, [33], is important and quite helpful.

Theorem: For each string $w \in D_1^*$ there exists some $q \in \mathbf{Z}(\frac{1}{2}) \subseteq \mathbb{Q}$ such that $w =_2 (ab)^q$.

From this result one can easily deduce that D_1^* is commutative, but there may be other strings in G_2 that commute and can be characterized.

Theorem: Let $u, v \in G_2$, then $uv =_2 vu$ iff there exist $w \in G_2$, and $r, s \in \mathbf{Z}$ such that $u =_2 w^r$ and $v =_2 w^s$.

Theorem: Let $u, v \in G_2$, then $u^m = v^m$ for some $m \neq 0$ implies $u =_2 v$.

This statement is stronger than to say that G_2 (or likewise G_n, for which this result is true, too) is torsion-free, since G_n is not abelian.

As regards the specific example G_2 we still have no proof for the conjecture that there is no finite uniquely terminating presentation for G_2, equivalent to S_2 even though the word problem is decidable and can be solved deterministically in logarithmic space. There are only a few results in this direction: Potts, [56], has given an infinite STS S on {a,b,c,d} which is uniquely terminating and can be used to solve the word problem for G_2 since it has a simple finite specification. Unfortunately S is not equivalent to the STS S_2 since it uses a different alphabet, but the groups that both systems define are isomorphic. Also there may be finite presentations for G_2 which are uniquely terminating but not equivalent to S_2 according to our strong definition. On the other hand we know that $S := \{(baa, a(ab)^4), (ababb, babba), (ababba, e)\}$ is equivalent to S_2 and Church-Rosser but not terminating. S becomes a uniquely terminating STS by imposing a certain regular control upon the application of its rules.

6. Conclusion

The results we surveyed showed that:

- Simple questions are in general undecidable for finite semi-Thue and Thue systems.
- Some, but not all, of these problems become decidable if the systems under consideration are uniquely terminating.
- Very often uniquely lg-terminating systems provide efficient algorithms for the solvable questions.
- Infinite but finitely specified systems sometimes produce negative results even in cases where they satisfy those properties that yielded the positive results for finite systems.
- As often in formal language theory the really best results are obtained for regular systems, while context-free systems do produce more undecidability results.

The connection to group theory still has to be explored and there are other questions related to formal language theory worth to be studied. Then there is the question which types of total orderings $\succ$ behave nice such that uniquely $\succ$-terminating systems do allow positive results similar to the uniquely lg-terminating systems, which apparently turned out to be a fruitful but very restricted approach.

References

1. Adjan, S., Defining Relations and Algorithmic Problems for Groups and Semigroups, Proc. Steklov Inst. Math. 85, (1966) (English version published by the American Mathematical Society, 1967.)

2. Avenhaus, J., Book, R., and Sqier, C., On Expressing Communtativity Monoids, R.A.I.R.O. Informatique Thèorique, to appear.

3. Avenhaus, J., and Madlener, K., Algorithmische Probleme bei Einrelatorgruppen und ihre Komplexität, Arch. math. Logic 19, (1978), 3-12.

4. Avenhaus,J.,and Madlener, K., String Matching and Algorithmic Problems in Free Groups, Revista Columbiana de Mathematicas 14, 1-16(1980).

5. Berstel, J., Congruences plus que parfaites et langages algèbriques, Sèminaire d`Informatique Thèorique, Institute de Programmation, (1976-77), 123-147.

6. Boasson, L., Dèrivations et réductions dans les grammaires algèbriques, Automata, Languages, and Programming, Lecture Notes in Computer Science, Vol. 85, Springer-Verlag (1980), 109-118.

7. Book, R., Confluent and Other Types of Thue Systems, J. Assoc. for Comput. Mach. 29, (1982), 171-182.

8. Book, R., The Power of the Church-Rosser Property in String-Rewriting Systems, Proc. 6th Conf. on Automated Deduction, Lecture Notes in Computer Science, Vol. 85, Springer-Verlag (1982) 360-368.

9. Book, R., When is a Monoid a Group? The Church-Rosser Case is Tractable, Theoret. Comp. Sci. 18, (1982), 325-331.

10. Book, R., A Note On Special Thue Systems with a Single Defining Relation, Math. Syst. Theory, 16, (1983), 57-60.

11. Book, R., Decidable Sentences of Church-Rosser Congruences, Theoret. Comput. Sci. 24 (1983) 301-312.

12. Book, R., Thue Systems and the Church-Rosser Property: Replacement Systems, Specification of Formel Languages, and Presentations of Monoids, in Combinatorics on Words, J.L. Cummings (Ed.), Academic Press Canada (1983) 1-38.

13. Book, R., Homogenous Thue Systems and the Church-Rosser Property, Discrete Math. 48 (1984) 137-145.

14. Book, R., Jantzen, M., Monien, B., ÓDúnlaing, C., and Wrathall, C., On the Complexity of Word Problems in Certain Thue Systems, Math. Found. of Comput. Sci., Lecture Notes in Computer Science 118, Springer-Verlag (1981), 216-223.

15. Book, R., Jantzen, M., and Wrathall, C., Monadic Thue Systems, Theoret. Comput. Sci.19, (1982), 231-251.

16. Book, R., and ÓDúnlaing, C., Thue Congruences and the Church-Rosser Property, Semigroup Forum 22, (1981), 367-379.

17. Book, R., and ÓDúnlaing, C., Testing for the Church-Rosser Property, Theoret. Comput. Sci. 16, (1981), 223-229.

18. Butzbach, P., Une famille de congruences de Thue pour lesquelles le Problème de l'équivalence est décidable. Application à l'équivalence des grammaires séparés, in Nivat, M., (ed.) Automata, Languages, and Programming, North Holland (1973), 3-12.

19. Church, A., and Rosser, J.B., Some Properties of Conversion, Trans. Amer. Math. Soc. 39 (1936) 472-482.

20. Cochet, Y., Sur l'algébricité des classes de certains congruences définies sur le monoide libre, Thèse 3ème cycle, Rennes, 1971.

21. Cochet, Y., Church-Rosser Congruences on Free Semigroups, Colloq. Math. Sco. Janos Bolyai: Algebraic Theory of Semigroups 20, (1976), 51-60.

22. Cochet, Y., and Nivat, M., Une généralization des ensembles de Dyck, Israel J. Math. 9, (1971), 389-395.

23. Dershowitz, N., Orderings for Term-Rewriting Systems, Theoret. Comput. Sci. 17 (1982) 279-301.

24. Dershowitz, N., and Manna, Z., Proving Termination with Multiset Orderings, Commun. Assoc. Comput. Mach. 22 (1979) 465-476.

25. Frougny, C., Une famille de langages algébriques congruential: les langages a nonterminaux séparés, Thése 3éme cycle, Rennes (1980).

26. Greendlinger, M., Dehn's Algorithm for the Word Problem, Comm. on Pure and Applied Math. 13 (1960) 67-83.

27. Hindley, R., An Abstract Form of the Church-Rosser Theorem I, J. of Symbolic Logic 34 (1969) 545-560.

28. Hindley, R., An Abstract Form of the Church-Rosser Theorem II: Applications, J. of Symbolic Logic 39 (1974) 1-21.

29. Huet, G., Confluent Reductions: Abstract Properties and Applications to Term Rewriting Systems, J. Assoc. Comput. Mach. 27, (1980), 797-821.

30. Huet, G., and Oppen, D., Equations and Rewrite Rules, in R. Book, Ed., Formal Language Theory: Perspectives and Open Problems, Academic Press, (1980), 349-405.

31. Huet, G., A Complete Proof of Correctness of the Knuth-Bendix Completion Algorithm, J. Comput. Syst. Sci. 23 (1981) 11-21.

32. Jantzen, M., On a Special Monoid with a Single Defining Relation, Theoretical Computer Science 16, (1981), 61-73.

33. Jantzen, M., Semi-Thue Systems and Generalized Church-Rosser Properties, Proc. Fêtes des mots, Rouen, June (1982) 60-75.

34. Kapur, D., and Narendan, P., The Knuth-Bendic Completion Procedure and Thue Systems, Proc.3rd Conf. on Found. of Software Techn.and Theoret. Comput. Sci., Bangalore, India, (1983) 363-385.

35. Kasincev, E.V., On the Word Problem for Special Semigroups, Math. USSR Izvestija, 13 (1979) 663-676

36. Knuth, D., and Bendix, P., Simple Word Problems in Universal Algebras, in J. Leech, Ed., Computational Problems in Abstract Algebra, Pergamon Press, (1970), 263-297.

37. Lallement, G., On Monoids Presented by a Single Relation, J. Algebra 32, (1974), 370-388.

38. Lallement, G., Semigroups and Combinatorial Applications, Wiley, 1979.

39. Lankford, D.S., A Unification Algorithm for Abelian Group Theory, Report MTP-1, Math. Dept., Louisiana Techn. Univ., Jan (1979).

40. Lankford, D.S., and Ballantine, A.M., Decision Procedures for Simple Equational Theories with Commutative Axioms: Complete Sets of Commutative Reductions, Report ATP-35, Depts. of Math. and Comp. Sci., Univ. of Texas at Austin, March (1977)

41. Markov, A., On the Impossibility of Certain Algorithms in the Theory of Associative Systems, Dokl. Akad. Nauk 55, (1947), 587-290; II, 58 (1947), 353-356 (in Russian).

42. A. Markov, Impossibility of Algorithms for Recognizing some Properties of Associative Systems, Dokl. Akad. Nauk SSSR 77 (1951) 953-956. A. Mostowski, Review of (41), J. Symbolic Logic 17 (1952) 151-152.

43. Metivier, Y., About the Rewriting Systems Produced by the Knuth-Bendix Completion Algorithm, Info. Proc. Letters 16 (1983) 31-34.

44. Muller, D.E., and Schupp, P.E., Groups, the Theory of Ends, and Context-free Languages, J. Comput. Syst. Sci. 26 (1983) 295-310.

45. Musser, D.R., On froving Inductive Properties of Abstract Data Types, 7th ACM Symp. on Principles of Programming Languages, Jan (1980) 154-162.

46. Narendran, P., The Undecidability of Preperfectness of Thue Systems, Abstracts, Amer. Math. Soc. 82T-68-435, 3, Oct. (1982), 546.

47. Narendran, P., and McNaughton, R., Church-Rosser Languages, in preparation.

48. Nelson, C.G., and Oppen, D.C., Fast Decision Algorithms Based on Congruence Closure, J. Assoc. Comput. Mach, to appear.

49. Newman, M.H.A., On Theories with a Combinatorial Definition of "Equivalence", Ann. Math. 43, (1943), 223-243.

50. Nivat, M. (with Benois, M.), Congruences Parfaites, Seminaire Dubreil, 25e Année, 1971-72, 7-01-09.

51. Ó'Donnell, M., Computing in Systems Described by Equations, Lectures Notes in Computer Science 58 (1977).

52. Ó'Dúnlaing, C., Finite and Infinite Regular Thue Systems, Ph.D. dissertation, University of California at Santa Barbara, 1981.

53. Ó'Dúnlaing, C., Infinite Regular Thue Systems, Theoret. Comput. Sci. 25, (1983), 171-192.

54. Ó'Dúnlaing, C., Undecidable Questions of Thue Systems, Theoret. Comput. Sci. 23, (1983) 339-346.

55. Post, E., Recursive Unsolvability of a Problem of Thue, J. Symb. Logic 12, (1947), 1-11.

56. Potts, D., Remarks on an Example of Jantzen, Theoret. Comput. Sci., to appear.

57. Rosen, B., Tree-Manipulating systems and Church-Rosser Theorems, J. Assoc. Comput. Mach. 20, (1973), 160-187.

58. Senizergues, G., A New Class of C.F.L. for Which the Equivalence Problem is Decidable, Info. Proc. Letters 13, (1981), 30-34.

59. Squier, C., The Group of Units of a Monoid with a Church-Rosser Presentation, in preparation.

60. Squier, C., and Book, R. Almost All One-Rule Thue Systems Have Decidable Word Problems, in preparation.

61. Squier, C., and Wrathall, C., The Freiheitssatz for One-Relation Monoids, Proc. Amer. Math. Soc., to appear.

62. Steckel, M.E., Unification Algorithms for Artificial Intelligence Languages, Ph. D. Thesis, Carnegie-Mellon Univ. (1976).

63. Thue, A., Probleme über Veränderungen von Zeichenreihen nach gegebenen Regeln, Skr. Vid: Kristiania, I Mat. Naturv. Klasse 10, (1914), 34 pp.

64. Yasuhara, A., Some Non-Recursive Classes of Thue Systems with Solvable Word Problem, Zeitschr. f. Math. Logik und Grundlagen d. Math. 20 (1974) 121-132.

LIMITS, HIGHER TYPE COMPUTABILITY AND TYPE-FREE LANGUAGES

G. Longo
Dip. Informatica, Università di Pisa
Corso Italia 40, I-56100 Pisa

Content:

Introduction. Denotational semantics of programming languages has been invented in order to interpret computer programs in a way which is independent of their representation mechanism (denotations). Meaning is actually added to syntactic definitions if the intended interpretation is given into mathematical structures defined by essentially different tools. Because of our "historical acquaintance" with Set Theory, Algebra, Category Theory, say, we know more about programs when we may look at them as elements of structures defined by these theories. Moreover, the more "interpretations" are known, in Set Theory, Algebra ..., the better programming languages are understood. A greater understanding may lead to improve the design of programming languages (e.g. Gordon et al. (1978), Albano et al. (1983)) or may give an insight in some of their properties (see §.5 and Longo (1984) for a discussion).

The key issue is that programming languages operationally define computable functions. Thus the interpretation of programs, in the chosen mathematical structure, needs to be a "computable" element, in some sen-

Work partially supported by C.N.R.(Comitato per la Matematica: GNSAGA)

se possibly independent of the given operational mechanism (see references). If the given programming language only deals with (denotations for) ω, the natural numbers, then PR, the partial recursive functions, may suffice. Most (high level) programming languages, though, in particular most functional languages, possess some more refined type discipline, by which they may compute with various types of data, including representations of functions. A fully expressive typed language, say, may deal with functions of functions of functions,.., in any finite higher type, for programs are strings of arbitrary finite lenght. Sometimes functional languages may be type-free and this makes life more complicated.

(Convention: Given a set At of atomic type symbols, $T \supseteq At$ is the (least) set of cartesian type-symbols, i.e. $\sigma, \tau \in T \Rightarrow \sigma \times \tau, \sigma \to \tau \in T$).

The purpose of this lecture is to outline a few issues in computability in abstract structures and its relation to the semantics of a "minimal" functional language: typed and type-free λ-calculus.

First, independently of this language, properties of total and partial elements in some interesting type structures are surveyed. Then results on the operational and denotational behaviours are briefly mentioned.

§.0 Computable elements in limit structures

There is a simple unifying way to understand the various approaches proposed in the late 60's and 70's to generalized computability. The connecting point is the construction of categories of sets where a suitable notion of limit gives an abstract notion of computability, independently of the specific type one deals with. This may be done with enough generality by using the Moore-Smith net-convergence or, equivalently, the filter-convergence, see 0.6 below. For the sake of simplicity, we look at the simplest notion of limit.

0.1 Definition (Kuratowski (1952)) A limit-space (L-space) $(X, \downarrow)$ is a set X and a relation (convergence) between countable sequences $\{x_i\}_{i\in\omega} \subseteq X$ and elements $x \in X$ (notation: $\{x_i\} \downarrow x$) s.t.:

1) if all but finitely many x_i are x, i.e. $\{x_i\}$ is eventually x, then $\{x_i\} \downarrow x$;
2) if $\{x_i\} \downarrow x$ and $k(0) < k(1) < \ldots < k(n) < \ldots$, then $\{x_{k(i)}\} \downarrow x$;
3) if $\{x_i\} \not\downarrow x$, then there is $k(0) < k(1) < \ldots < k(n) < \ldots$ s.t. for no subsequence $\ell(0) < \ell(1) < \ldots < \ell(n) < \ldots$ one has $\{x_{\ell(i)}\} \downarrow x$.

0.2 Definition A L-space $(X, \downarrow)$ has a countable basis (is separable) iff for some given countable $X_o \subseteq X$ $\forall x \in X$ $\exists \{x_i\} \subseteq X_o$ $\{x_i\} \downarrow x$.

From now on we use assume that each countably based L-space $(X, \downarrow)$ comes toghether with a given enumeration $e: \omega \to X_o$ (onto) of the base.

A well known example of separable L-space is the set of real numbers endowed with the usual notion of sequence convergence (Cauchy).

As required for our purposes in generalized computability and denotational semantics, limit structures are inherited at higher types. This is trivial for cartesian products, by defining limits componentwise.

0.3 Definition Let $(X, \downarrow)$ and $(Y, \downarrow)$ be L-spaces. Define the set of continuous functions $L(X,Y)$ by

$$f \in L(X,Y) \quad \text{iff} \quad \forall x \in X \quad \forall \{x_i\} \downarrow x \quad \{f(x_i)\} \downarrow f(x),$$

in the intended spaces.

Define also $(L(X,Y), \downarrow)$ as L-space by

$$\{f_i\} \downarrow f \quad \text{iff} \quad \forall x \in X \quad \forall \{x_i\} \downarrow x \quad \{f_i(x_i)\} \downarrow f(x)$$

With some work one can actually show that if $(X, \downarrow)$ and $(Y, \downarrow)$ are separable, then also $(L(X,Y), \downarrow)$ is separable. By a further condition on the enumeration of the bases of $(X, \downarrow)$ and $(Y, \downarrow)$, one can canonically define and enumerate the basis of $(L(X,Y), \downarrow)$ (see Hyland (1979), via filter spaces). Moreover, L-spaces and separable L-spaces, with continuous maps as morphisms, form Cartesian Closed Categories (CCC's). That is, one also has that the operations of product and function space nicely commute (see MacLane (1971), say).

0.4 Remark $(L(X,Y), \downarrow)$, defined as in 0.3, is the coarsest (i.e. with more converging sequences) limit structure such that eval: $L(X,Y) \times X \to Y$, with eval $(f,x) = f(x)$, is continuous.

0.5 <u>Definition</u> 1) Let $(X,\downarrow)$ be an L-space. Define the <u>induced topology</u> top on X by: $O \in top$ iff $\forall x \in O$ $\forall\{x_i\} \downarrow x$ $\{x_i\} \subseteq O$ eventually.

2) Let (X,top) be a topological space. Define $(X,\downarrow)$ by: $\{x_i\} \downarrow x$ iff $\forall O \in top\ (x \in O \Rightarrow \{x_i\} \subseteq O$ eventually).

Definition 0.5 relates topological spaces to L-spaces by an obvious "injection" (see Hyland (1976/9) for a precise categorical approach: it may suffice to know that by applying first 2 then 1 in 0.5 one obtains the identity on structures). There exist, though, L-spaces whose limit structure is not topological, i.e. it doesn't derive from a topology in the sense that by applying first 1 then 2 in 0.5 one doesn't recover the original L-space. The point is that some of these L-spaces are among those needed for the study of the total computable functionals (see 1.2 below).

Let R be the total recursive functions and $(X,X_o,e,\downarrow)$ a separable L-space. Define then, in an informal way (see 0.6), the collection $X_c \subseteq X$ of <u>computable elements</u> by:

(1) $$x \in X_c \text{ iff } \exists f \in R\ \{e_{f(i)}\} \downarrow x.$$

In other words, given a countably based limit structure, an element is computable (or recursive) when it is the limit of a countable sequence indexed over an r.e. set. This is the underlying idea for generalized computability in Scott (1970), Scarpellini (1971), Ershov (1972), Hyland (1979), Scott (1982).

0.6 <u>Comment</u> L-spaces actually carry too little structure to yield a good definition of "recursive" just by taking arbitrary limits of r.e. converging sequences. Their simplicity and generality, though, should give an immediate intuition of what's going on. One of the weakness of these structures is that limits are far away from being uniquely determined and that there is no obvious way to characterize "interesting" sequences and limit points (see later). Thus, say, Scarpellini's notion of computable functional seems too wide.

The point then is to take only "some" limits. This is done in Hyland (1979) by coded filter spaces and in Scott (1970-1982) by directed sets in algebraic cpo's (see §.2, below). In short, for the reader acquainted with the Moore-Smith convergence or the filter spaces (see Kelley (1955)), this may be sketched as follows. (X,F) is a filter space iff $\forall x \in X\ F(x)$ is a filter of filters such that the ultrafilter generated by x is in $F(x)$. Any (X,F) induces (X,top) by: $O \in top$ iff $\forall x \in O\ \forall \Phi \in F(x)\ O \in \Phi$; conversely, (X,top) defines (X,F) by $F(x) = \{\Phi \mid \Phi$ is a filter and $\forall O \in top\ (x \in O \Rightarrow O \in \Phi)\}$. Define

(s) $\Phi \overset{s}{\downarrow} x$ iff $\Phi \in F(x)$ and $\forall O \in \Phi \cap top\ \ x \in O$.

Note that the induced topology is T_0 iff s-convergent filtres have a unique limit. Then, for a separable (X,F) with base $\{U_i\}_{i \in \omega}$, define

(1_s) $x \in X$ is computable iff $\exists \Phi \overset{s}{\downarrow} x\ \{i/U_i \in \Phi\}$ is r.e.

Note that, with the obvious names for the intended categories, TOP $\Rightarrow$ FIL by the above injection (analougously to TOP $\Rightarrow$ L-spaces in 0.5); moreover, FIL $\Rightarrow$ L-spaces in a similar way. Thus filter spaces are "half-a-way" structured.

The definition in (s) loosely summarizes the variant in Ruggeri (1984) of coded filter spaces in Hyland (1979). The related notion of computability, (1_s), gives Hyland's over coded filter spaces which are Hausdorff (i.e. the induced topology is T_2) as well as Scott's over effective domains (§.2).

§.1 Total recursive functionals

In the '50s several approaches were proposed for higher type Recursion Theory (see Gandy & Hyland (1976), Normann (1980) for surveys). The various notions were essentially aimed at the semantics of intuitionistic logic and arithmetic. Namely, for the purposes of a constructive approach to the foundation of Mathematics, given a formula such as $\forall x\, \exists y\, A(x,y)$, one wants to recover y effectively from x, i.e. its functional interpretation is $\exists f\, \forall x\, A(x,f(x))$, for some $f \in R$. More alternations of quantifiers clearly require a notion of higher type total and computable functional.

In Hyland (1979) an elegant approach is given to the classical work done in Kleene (1959/a) and Kreisel (1959). The technical implications are beyond the scope of this lecture and give an insight into recursion on the higher type total functionals (see also Gandy & Hyland (1976),

Normann (1980)).

As far as the classical recursion theoretic hierarchy is concerned, i.e. the type structure generated by ω in the category of L-spaces, Hyland's stronger condition (Hausdorff filter spaces) is automatically satisfied. Thus it may be sufficient to consider the informal notion of computability in (1) above, with reference to 0.6, Hyland (1979), Ruggeri (1984).

1.1 Definition Let $(\omega,\downarrow)$ be the natural numbers with the limit structure induced by the discrete topology. Set $(0) \in T$ and $H^{(0)} = (\omega,\downarrow)$. Define then $H^{\sigma\times\tau} = (H^{\sigma} \times H^{\tau},\downarrow)$ and $H^{\sigma\to\tau} = (L(H^{\sigma},H^{\tau}),\downarrow)$, for $\sigma,\tau \in T$.

H_c^{σ} are the total continuous and computable functionals of type σ. Moreover, $\{H^{\sigma}\}_{\sigma\in T}$ are exactly the Kleene-Kreisel countable functionals. They also turn out to be closed under computation via Kleene's schemes S1-S9 (Kleene (1959)).

1.2 Theorem (Hyland (1979)) $H^{(0\to 0)\to 0}$ is not topological.

Moreover no topology, preserving continuity of functions, yields for any element a countable neighbourhoods' filter (i.e. the first countability axiom fails). By this and by 1.2, already at pure type 2 (i.e. for $2 = 1\to 0$, where $1 = 0\to 0$), there is no way to characterize the computable functionals by a topological notion of convergence (of r.e. -indexed sequences). Thus, the full generality of L-spaces (more precisely: of filter spaces, see 0.6) is required to construct Hyland's model of the classical recursion-theoretic total functionals.

§.1½ From total to partial maps

There are at least two good reasons to look at partial functions and functionals. First, all intersting (high level) programming languages formally compute partial functions. Second, in general there are no "effective enumerations" of classes of total maps, i.e. the "universal function" lies outside the class under examination.

With reference to Hyland's $\{H^{\sigma}\}_{\sigma\in T_x}$ a counterexample is immediately found, for $H_c^{(1)} = R$ and, hence, it cannot be effectively enumerated. A similar fact holds also for the recursive reals, i.e. the limits of r.e. indexed sequences (nested intervals) of rational numbers (see Moschovakis (1963)), for there is no way to enumerate effectively all recursive functions whose range are the indexes of a Cauchy sequence of rational numbers.

In general this is due to the following fact. Consider a separable L-space $(X,X_o,e,\downarrow)$; then there is no uniform effective way to turn an arbitrary r.e. set A into an r.e. set B s.t. $\{e_i\}_{i\in B}$ is convergent and, if $\{e_i\}_{i\in A}\downarrow x$, then $\{e_i\}_{i\in B}\downarrow x$. The decidability properties of the basis in Scott's effectively given domains easily give such a uniform procedure.

§.2 Domains as partial objects

By a (Scott) domain $(X,X_o,<)$ we mean a bounded (or consistently) complete algebraic cpo, with base X_o, endowed with the Scott topology top_s. That is, $\underline{X} = (X,X_o,<)$ is a complete partial order where all bounded subsets have a least upper bound (notation: $\sqcup$); moreover, for $\check{x} = \{y \in X/x \leqslant y\}$ and $\hat{x} = \{x_o \in X_o/x_o \leqslant x\}$, one has $x = \sqcup\hat{x}$ and $X_o = \{x \in X/\check{x} \in top_s\}$.

The partial order and top_s are nicely related by

(2) $$x \leqslant y \text{ iff } \forall 0 \in top_s\ (x \in 0 \Rightarrow y \in 0);$$

moreover $\leqslant$ collapses to = iff top_s is T_1. (By (2), top_s is always T_o).

2.1 Remark Since in non trivial posets top_s is T_o and not even T_1, each converging sequence may have lots of limit points. More precisely, if $(X,\downarrow)$ is induced by top_s on $(X,X_o,<)$ as in 0.5.2 (i.e. $\downarrow$ is just topological convergence), then $\{x_i\}\downarrow x$ and $y \leqslant x$ imply $\{x_i\}\downarrow y$, by (2). Thus one has to choose "some" sequences and limits in order to define computability in a reasonable way. This will be done by taking only directed sequences and their l.u.b's. (Of course, if D is directed, $D\downarrow \sqcup D$).

A domain $\underline{X} = (X,X_o,e,<)$ is effectively given (effective) iff $e:\omega \to X_o$ (one-one, onto) and

(3) "$\exists z \in X\ \ e_n,e_m \leqslant z$" and "$e_k = e_n \sqcup e_m$" are decidable in k,n,m.

2.2 <u>Definition</u> Let $(X,X_o,e,<)$ be an effective domain. Define then $x \in X$ is <u>computable</u> iff $\exists f \in R\ \{e_{f(i)}\}_{i \in \omega}$ is directed <u>and</u> $\sqcup\{e_{f(i)}\} = x$. (cf. (1) in §.0).

Note that, by (3), x is computable iff $\{n/e_n \leq x\}$ is r.e.; moreover, no other limit point of the given r.e. directed set needs to be computable.

2.3 <u>Comment</u> We are now in the position to look at the general notion of computability given in 0.6. Let $(X,X_o,<)$ be a domain and (X,F) the induced filter structure by top_s. It is then easy to prove that $\Phi \downarrow^s x$ iff $D = \{x_o \in X_o/\check{x}_o \in \Phi\}$ is directed and $\sqcup D = x$.

Thus, also in effective domains, computable elements are defined exactly as in (1_s) in 0.6, since, for $e:\omega \to X_o$, $\{\check{e}_i/i \in \omega\}$ is a base for the induced (X,F).

As well known, see Scott (1982), for (effective) domains $\underline{X}$ and $\underline{Y}$, also $\underline{X \times Y}$ and $\underline{D}(X,Y)$ are (effective) domains, where $\underline{D}(X,Y)$ are the morphisms in the category of domains, i.e. the continuous functions. Actually, (effective) domains form sub-CCC's of L-spaces, by 0.5.2. Thus, the notion of partial computable element (functional) is inherited at any finite higher type.

Notice now that, given sets A and B, the collection of partial maps from A to B form a domain, partially ordered by $f \leq g$ iff $\mathrm{dom}\ f \subseteq \mathrm{dom}\ g$ <u>and</u> $\forall x \in \mathrm{dom}\ f\ (f(x) = g(x))$. The morphisms in $\underline{D}(X,Y)$ above are total maps, though. Why can we soundly consider them as partial functionals?

In Longo & Moggi (1984 an) (see also Heller (1983)) an obvious notion of category with partial morphisms (pC) is given. Clearly, in a pC, any $f:x \to y$ (partial) corresponds to a total $f:x \to y^{\perp}$, where $y^{\perp}$ is obtained from y by adding an "undefined" element.
(<u>Notation</u>: Let C be a pC. Then $C_T(x,y)$ are the total morphisms from x to y and C_T the corresponding category. $x \triangleleft y$ (x is a <u>retract</u> of y) iff $\exists in \in C(x,y), \exists out \in C(y,x)\ out \circ in = id_x$).

2.4 <u>Definition</u> (Longo & Moggi (1984 an)) Let C be a pC. Define then

1) $-^{\perp}: C \to C_T$ is a <u>bottom functor</u> if $C(x,y) \cong C_T(x,y^{\perp})$.

2) x is a <u>partial object</u> if $x \lhd x^{\perp}$ in C_T.

With some work one can naturally define partial Cartesian Closed Categories (pCCC's: see Longo & Moggi (1984 an)). In a pCCC, x_p^y is the object which represents $C(x,y)$ (and x^y, if it exists, represents $C_T(x,y)$). The following fact should fully relate partial and total morphisms to partial objects.

2.5 <u>Theorem</u> (Longo & Moggi (1984 an)) Let C be a pCCC and t a terminal object (i.e. $\forall x\ \exists! f \in C_T(x,t)$). Then

(i) $x_p^t \cong x^{\perp}$, i.e. $-_p^t$ is a bottom functor,

(ii) x_p^y is a partial object.

Assume also that C_T is a CCC and x a partial object. Then

(iii) $x^y \lhd x_p^y$.

The point is that domains (or even cpo's) are indeed partial objects, in suitable categories (e.g. numbered sets). The intuition is that domains "already possess a least (underfined) element", thus they trivially satisfy $x \lhd x^{\perp}$. 2.4 and 2.5 may be regarded as a precise way to understand why total maps on arbitrary domains (e.g. higher type functionals) may be actually considered as partial maps (see Longo & Moggi (1984 an) for applications). 2.2 then gives a sound notion of partial computable functional in any higher finite type.

2.6 <u>Remark</u> There is a natural way to get back from partial to total maps. At type $1 = 0 \to 0$, just consider the maximal elements. Similarly at higher types, modulo the equivalence relation which equates functions with the same behaviour on the previously defined "total" maps (see Ershov (1974), Longo & Moggi (1983) for details). Of course, at type 1, i.e. for the domain P of the partial number theoretic functions, one obtains exactly the total functions. The induced topology turns ω^{ω} into the Baire's metric space and corresponds to the limit structure in $H^{(1)}$, with $H_c^{(1)} = R$. At higher types one recovers the Kleene-Kreisel countable functionals. By 1.2, the quotient operation gives L-spaces and all the required topological properties for a direct definition of computability are lost. This is why for defining a smaller class of functionals, the total ones, a category larger than domains is needed, i.e. the category of L-spaces. It would be interesting to characterize which L-spaces one can obtain by this technique.

In §.1½ we mentioned the intrinsic impossibility of enumerating effectively the class of total computable functionals in any finite type. In contrast to this, for any effective domain $(X,X_o,e,<)$, by (3) in 2.2 it is easy to define $f_X \in R$ such that, for any r.e. set W_i,

i) $W_{f_X(i)}$ is directed (an abuse of language for "$e(W_{f_X(i)})$ is directed")

i)) if W_i is directed, then W_i and $W_{f_X(i)}$ are cofinal.

Then $c_i = \bigsqcup e(W_{f_X(i)})$, for $i \in \omega$, enumerates exactly the computable elements in X.

2.7 Remark There are at least two more reasons, besides the effectiveness of the construction, to regard $\{c_i\}_{i\in\omega}$ above as an effective enumeration of the computable elements. First, relatively to $e:\omega \to X_o$, since $\exists g \in R\ \forall i\ c_{g(i)} = e_i$. Second, for any $\sigma \in T$, if $\{c_i^\sigma\}_{i\in\omega}$ enumerates the computable functionals of type σ, then, for $f(<n,m>) = c_n^{\sigma\to 0}(c_m^\sigma)$, one has $f \in PR$. This is an easy consequence of the elementary approach to the partial computable functionals in Longo & Moggi (1983) (see Longo (1984) for a discussion).

2.8 Comment The technique used in L-spaces, and their sub-categories, for defining the computable functionals in "extenso" should now be clear. Given for granted the notion of recursive functions and r.e. set, by "local", i.e. convergence or topological, properties, computability is extended to abstract settings. Very loosely, one may say that this approach is similar to the technique used for "smooth manifolds". Given the good old notion of differentiability over R^n, by a system of local coordinates one defines differentiability in abstract spaces.

§.3 The generalized *'-computability

Independently of any "limit" or topological technique, in 1955 Myhill and Shepherdson gave a simple caractherization of type 2 computable operator, by turning $P\omega$ into a monoid $(P\omega,\cdot)$ (see Rogers (1967), Scott (1976)). Define then $\Phi:P\omega \to P\omega$ is an <u>enumeration operator</u> iff, for some r.e. set A, $\forall B \in P\omega\ f(B) = A\cdot B$.

(It is easy to see that $f:P\omega \to P\omega$ is an enumeration operator iff $f \in D(P\omega,P\omega)_c$, i.e. f is continuous and computable in the category of domains).

Consider now a T_0-topological space X, with a countable base $\{U_n\}_{n\in\omega}$, and define the embedding $I_X:X \to P\omega$ by $I_X(x) = \{n/x \in U_n\}$. Notice that the Scott topology top_S on $P\omega$ may be equivalently defined as the finest topology such that, for any countably based T_0 space X, I_X is continuous.

Let (X,top) and (Y,top) be separable T_0 spaces; define then $g:X \rightharpoonup Y$ is a <u>partial computable operator</u> iff for some enumeration operator f, the following diagram commutes:

(4)
$$\begin{array}{ccc} X & \xrightarrow{\ g\ } & Y \\ {\scriptstyle I_X}\downarrow & & \downarrow{\scriptstyle I_Y} \\ P\omega & \xrightarrow[\ f\ ]{} & P\omega \end{array} \quad .$$

Note that range $(f\restriction I_X(X))$ does not need to be included in $I_Y(Y)$. Thus g, in general, is partial. This method generalizes the idea used in Barendregt & Longo (1982) to characterize classes of partial recursive operators (some of which extendible to total ones), by various combinations of embeddings called * and'. In particular, consider $P\omega$ with the Cantor set topology top_C, i.e. endow 2^ω with the induced topology by the Baire space ω^ω and take $P\omega \cong 2^\omega$. Then $(P\omega,top_C)$ is a countably based compact metric space. By taking $X = Y = (P\omega,top_C)$, and a variant of (4), one obtains the partial Turing operators, i.e. as defined by oracle Turing machines. Some of these partial operators are not extendible to total ones (Barendregt & Longo (1982): this is shown by observing that total continuous functions on compact metric spaces are uniformely continuous and that a uniform modulus of continuity gives truth-table reducibility, which is strictly stronger than Turing reducibility. Ruggeri (1984) proves that the total Turing operators are exactly the computable elements in the limit structure $L(P\omega,P\omega)$, over $(P\omega,top_C)$.

It is probably worth looking at the technique in (4) as a simple tool for giving a model of partial (and total) computability over abstract data types which can't be naturally structured as domains and where partial operations are essential.

§.3½ From typed to type-free

The relevance of the notion of type in Logic and Computer science is well kown. Types are constraints on expressions in a language. Thus, in Logic, they allow full expressiveness by avoiding paradoxes. In Computer Science, consistent type disciplines ensure that any expression satisfying the constraints will not produce run-time errors. Type checking in Computer Science may be compared to a very usual procedure in Phisics, when one checks that, in an equation, a force faces a force etc... introducing by this a preliminary and basic syntactic verification of correcteness.

Type-free languages, though, are commonly used. In Logic they will never take us into the fregean paradise, but in Computer Science they are handy and enough powerful.

There are several ways to go from models of typed languages to models of type-free calculi (Combinatory Logic and λ-calculus). In the next section some are just listed, because of the space limitations imposed by the Publisher.

§.4 Types and models for type-free calculi (summary)

1 - Consider a non trivial domain $X^{(0)} = X$ and set $X^{(n+1)} = D(X^{(n)},X^{(n)})$, the "next"higher type. At the limit (an "inverse" limit in the sense of category theory) one obtains $X^{\infty} = D(X^{\infty},X^{\infty})$ and, hence, a (strongly) extensional model of λ-calculus. This was done in Scott (1972) and, up to now, it seems to be the strongest method also for solving more general recursive domain equations in Computer Science (see Plotkin & Smith (1982)).

2 - Let $X^{(1)} = P^{(1)}$, the domain of partial number theoretic functions, then any $P^{(n)}$ (= $X^{(n)}$ above), for $n > 0$, yields a model of λ-calculus. In Longo & Moggi (1984 pr) this is sketched after proving that, any time

one can "relatively enumerate" a type by the type below and two further simple conditions are satisfied, the former yields a model of Combinatory Logic (i.e. of the theory of Combinators, as functional language). Thus the semantics of λ-calculus may be given also at any finite higher type. Moreover, this is done over an interesting type structure.

3 - Barendregt et al. (1983) give a model of λ-calculus whose elements are collections of type-simbols. Any (type-free) term is interpretd by the filter of types assigned to it by the extended type discipline in Coppo et al. (1981). As a consequence, they obtain an elegant completeness result for type-assignement (type checking). Following Coppo (1983), a completeness theorem with more semantic flavour is given in Longo & Martini (1984), by looking at higher type functionals within $P\omega$.

It is worth noticing that all the constructions in 1,2 and 3 may be carried on by just taking the computable elements in the intended domains (Kanda (1980)). Observe also that the idea in 2 of giving a relative enumeration of type $n+2$ by type $n+1$, say, which turns $P^{(n+1)}$ into a model for type-free calculi, is the key fact in the alternative approach (with no continuity involved) to the higher type partial functionals in Longo & Moggi (1983). Limit notions, though, are essential in the proof of the basic properties of that direct approach, see Longo (1984) for a discussion.

§.4½ Operational versus denotational semantics

There are two main aspects of the relation between formal operations and meanings in mathematical structures:

1 - Formal definability vs. computability of objects.

2 - Operationally convertible vs. semantically equivalent programs (as a special case of provable vs. true properties in interesting models).

As usual, we look at programming languages (or at languages for computations) just as at formal languages of Logic. The main difference is that the former and their semantics are much more difficult to handle, for they have not been originally invented for the formalization of semantic notions, but for their operational use, possibly over a running computer. This is why the Model Theory of progamming languages has started only some 15 years ago and the interesting results form a fast growing, deep, difficult topic.

§.5 Computations and values

1 - In the previous sections the notion of computable element in general semantic domains has been briefly surveyed. It is a basic completeness question to ask which computable objects are definable by the intended formal languages. (As it may be easely understood, the interpretation of closed terms of finitary languages can be simply shown, in general, to be computable).

As for the simplest model of λ-calculus, $P\omega$, a natural extension of the pure language may be given, the language LAMBDA, such that all computable elements (i.e. all r.e. sets) are definable (Scott (1976)). Moreover, there is a uniform effective way to go from terms possessing normal form to their interpretation as recursive sets; plenty of interesting terms of the pure language, though, have an r.e. non recursive interpretation (Giannini & Longo (1982)).

More work may be found for typed λ-calculus and some pratically usful extensions of it. In this case, the full generality of the previously skeched theory of computability in (higher type) domains is required. (As for λ-definability in the full type hierarchy, i.e. in the category of sets, one should consult Plotkin (1980) and Statman (1980)).

In Plotkin (1977) some basic results are given by which a simple extension of typed λ-calculus, with parallel facilities, defines all computable elements, in each semantic type. Moreover, Streicher (1983) discusses a metalanguage, with recursion and if-then-else, for the purpose of definability of application and functional abstraction. The interesting goal of the latter work would be the formal definition, in the metalanguage, of an interpreter for the metalanguage and, then, a completeness result for compiler definability over domains specified by recursive domain equations (see Kanda (1980), Plotkin & Smith (1982).

2 - Everybody should agree on the basic role that λ-calculus had in the origin and developements, in Logic, of Computability Theory, and, in Computer Science, of operational (reductions) and denotational semantics

of functional programming languages. The point is that λ-calculus, or Combinatory Logic, may be regarded as a "minimal" theory of functions. As a matter of fact, functions are based an application (apply function f to argument x,f(x)) and abstraction (turn expression (....) into a function of x, λx.(....)). This is exactly how λ-terms are constructed and, clearly, any functional language must possess these facilities. Moreover, by Kleene's theorem on λ-definability, application and abstraction are sufficient for computing all partial recursive functions.

There is an other classical relation between computions and λ-calculus as axiomatizable theory. This is expressed by Böhm's 1968 theorem (see Barendregt (1984)). Namely, terms possessing normal form (i.e. strongly convergent) may be consistently equated iff they are provably equal. Thus, in no model "terminating" programs may turn out to be equivalent, unless they are equivalent in all models. This is a strong, relevant information on a relative Hilbert-Post completeness of deductions and effective computations of functions. Thus, given a model one is interested in, what remains to be studied are the properties which hold in it, besides the provable ones, for there are plenty of interesting terms with no normal form. Syntactic characterizations of true equalities are usually given by combining a technique due to Scott, Hyland and Wadsworth (approximation theorem) and an extension of Böhm's work (Barendregt's "Böhm-out"). In this way, Wadsworth (1976) and Hyland (see Barendregt (1984)) characterized the semantic equivalences in Scott's D^{∞}(X^{∞} in 4.1) and Pω. In Barendregt & Longo (1980) the theory of Plotkin's Tω model, the partial 01-valued maps, is given. In all of these models exactly the terms with the same Böhm-tree coincide (up to possibly infinite η-expansion, as for D^{∞}). The Böhm-tree of a λ-term is constructed, roughly, by taking the leftmost reductions until one gets a term of the form $\lambda x_1 \ldots x_n . yM_1 \ldots M_p$ and then proceeding similarly with M_j; this may give an infinite tree, which displays the possibly infinite operational behaviour of a term. Then the characterization results are proved as follows. The approximation theorem says that the value of a term is the l.u.b. (and, hence, a limit point, see §.2) of the interpretation of the "cuts", at finite depth, of its Böhm-tree; hence, Böhm-tree equal terms coincide. Conversely, if two terms differ in some node of their (possibly infinite) Böhm-trees, then by the Böhm-out technique one may take out that node and exhibit the difference in the model. Thus the pure λ-calculus, as an r.e. theory, is not complete w.r.t. these models, for Böhm-tree equal terms do not need to be provably equal. By the discussion carried on in this lecture, though, we may say that there is a precise limiting sense in which reduction rules are complete for the purposes of semantic evaluation; moreover, by the operational relevance of Böhm-trees, terms are given the right meanings from a computational point of view. (See Wand (1983/4) for recent applications of Böhm-trees).

In order to avoid to prove again and again similar results on single different structures, more work should be done in the right general setting. A preliminary attempt to look at once at a large class of models is done in Longo (1982), by the notion of "approximable application" satisfied by models constructed over arbitrary sets by a simple technique. In view of the deep and elegant connections of λ-calculus with category theory, a category-theoretic analysis of theories of models should be also worked out. Apparently, though, categorical notions seem too general to look inside valid equalities and program equivalences, which often depend on minor differences in the hardware of models.

A few specific and interesting applications have been given by relating termination of programs and their semantics; by this, semantic equivalence guarantees similar behaviours in all syntactic contexts (see Plotkin (1977) or Gordon (1973), for LISP, and Milne (1974), for an extension of ALGOL 60). However, in this author's opinion, the deep insight given by the Böhm-out and approximation theorems hasn't yet been fully exploited for the investigation of semantic properties of (functional) programming languages. One should consider, say, that even from the point of view of Model Theory, an highly developed topic in Logic, such a surprising connection, as established in λ-calculus between theories and mathematical models, may be regarded as an original and technically deep achievement, which is surely worth applying more extensively.

Acknowledgements. I am endebted to my students Francesco Ruggeri, Eugenio Moggi and Simone Martini for the joint work, the discussions we had and for their help in revising this lecture.

REFERENCES

Albano A., Cardelli L., Orsini R. (1983) "Galileo: a strongly typed, Interactive Conceptual Language" Tech. Note 30 C.S. Dept. Toronto (revised: Tech. rep. 83-11271-2, Bell Lab.).

Barendregt H. (1978) "The type-free lambda-calculus" in Handbook of Math. Logic (Barwise ed.), North-Holland.

Barendregt H. (1984) The lambda-calculus: its syntax and semantics (revised and expanded edition), North-Holland.

Barendregt H., Coppo M., Dezani M. (1983) "A filter lambda model and the completeness of type assignement" J. Symb. Logic 48, 4, 931-940.

Barendregt H., Longo G. (1980) "Equality of λ-terms in the model Tω" in To H.B. Curry: essays in Combinatory Logic, λ-calculus and formalism (Hindley, Seldin eds) Academic Press.

Barendregt H., Longo G. (1982) "Recursion theoretic operators and morphisms of numbered sets" Fund. Math., CXIX, 49-62.

Coppo M. (1983) "Completeness of Type Assignement in continuous lambda-models" Teor. Comp. Sci. (to appear).

Coppo M., Dezani M., Venneri B. (1981) "Functional character of solvable terms" ZML, 27, 45-58.

Egli H., Constable R. (1976) "Computability concepts for programming languages semantics" Theor. Comp. Sci. 2, 133-145.

Ershov Yu.L. (1972) "Computable functionals of finite type" Algebra and Logic 11, 4, 367-437.

Ershov Yu.L. (1974) "Maximal and everywhere defined functionals" Algebra and Logic 13, 4, 374-397.

Ershov Yu.L. (1976) "Model C of the continuous functionals" Logic Coll. 76 (Gandy, Hyland eds.), North-Holland.

Gandy R., Hyland M. (1976) "Computable and recursively countable functions of higher type" Logic Coll. '76 (Gandy, Hyland eds), North-Holland.

Gordon M. (1973) "Evaluation and denotation of pure LISP programs" Ph. D. Thesis, Edinburgh.

Gordon M., Milner R., Wadsworth C. (1979) Edinburgh LCF LNCS -78, Springer-Verlag.

Heller A. (1983) "Dominical Categories and Recursion Theory" Underground, New York.

Hyland M. (1976) "The countable functionals" Underground, Oxford.

Hyland M. (1979) "Filter spaces and the continuous functionals" Ann. Math. Logic 16, 101-143.

Kanda A. (1980) "Fully effective solution of recursive domain equations" Ph. D. Thesis, Warwick.

Kelley J. (1955) General topology Springer- Verlag.

Kleene S.C. (1959) "Recursive functionals and quantifiers of finite type I" Trans. A.M.S. 91, 1-52.

Kleene S.C. (1959a) "Countable functionals" in Constructivity in Mathematics (Heyting ed.), North-Holland.

Kreisel G. (1959) "Interpretation of Analysis by means of functionals of finite type" in Constructivity in Mathematics (Heyting ed.), North-Holland.

Kuratowski C. (1952) Topologie, vol. 1, Warsaw.

Longo G. (1983) "Set-theoretical models of lambda-calculus: theories, expansions, isomorphisms" Ann. Pure Applied Logic (formerly: Ann. Math. Logic), 24, 153-188.

Longo G. (1984) "Continuous structures and analytic methods in Computer Science" Proceedings CAAP (Courcelle ed.), Cambridge University Press.

Longo G., Moggi E. (1983) "The Hereditary Effective Partial Functionals and Recursion Theory in Higher Types" J. Symb. Logic (to appear).

Longo G., Moggi E. (1984 an) "Cartesian Closed Categories and partial morphisms for effective type structures I" Symp. Semantics Data Types LNCS, Springer-Verlag (to appear).

Longo G., Moggi E. (1984 pr) "Gödel-numberings, principal morphisms, combinatory algebras" MFCS '84 (Chytil ed.) LNCS, Springer-Verlag.

Mac Lane S. (1971) Categories for the Working Mathematician, Springer-Verlag.

Milne R. (1974) "The formal semantics of computer languages and their implementations" Ph. D. Thesis, Cambridge.

Moschavakis Y. (1963) "Recursive analysis" Ph. D. Dissertation, Madison.

Normann D. (1980) Recursion on the countable functionals LNM 811, Springer-Verlag.

Plotkin G. (1977) "LCF considered as a programming language" Theor. Comp. Sci. 5, 223-255.

Plotkin G. (1980) "Lambda-definability in the full type hierarchy" in To H.B. Curry: essays... (Hindley, Seldin eds), Academic Press.

Plotkin G., Smyth M. (1982) "The category-theoretic solution of recursive domain equations" C.S. Dept., Edinburgh.

Rogers H. (1967) Theory of recursive functions and effective computability, McGraw-Hill.

Ruggeri F. (1984) "Spazi di filtri ed operatori di Turing" Tesi di Laurea in Matematica, Pisa.

Scarpellini B. (1971) "A model for bar recursion of higher types" Comp. Math. 23, 113-153.

Scott D.S. (1969) "A theory of computable functions in higher types" Seminar Notes, University of Oxford.

Scott D.S. (1970) "Outline of a mathematical theory of computation" IV Annual Princeton Conf. on Info. Syst. and Sci., 169-176.

Scott D.S. (1976) "Data types as lattices" SIAM J. Comp., 5, 522-587.

Scott D.S. (1982) "Some ordered sets in Computer Science" in Ordered sets (Rival ed.), Reidel.

Spreen D. (1983) "Effective operators in a topological setting" Logic Coll. 83, LNM, Springer-Verlag (to appear).

Statman R. (1980) "Completeness, invariance and lambda-definability" J. Symb. Logic (to appear).

Steicher T. (1984) "A solution for the definability problem for deterministic domains" Inst. Math. J. Kepler Univ., Linz.

Wand M. (1983) "Loops in Combinator-Based Compilers" Info. Contr. (to appear).

Wand M. (1984) "A Types-as-Sets Semantics for Milner-Style Polymorphism" 11th ACM Symp. Principle of Progr. Lang..

Weihrauch K., Schäfer G. (1983) "Admissible representation of effective cpo's" Theor. Comp. Sci. (to appear).

Giannini P., Longo G. (1982) "Effectively given domains and lambda-calculus semantics" Info. Contr. (to appear).

Longo G., Martini S. (1984) "Higher type computability, $P\omega$ and the completeness of type assignement" (Prelim. incomplete version: Proc. STACS, LNCS, Springer-Verlag).

TRACES, HISTORIES, GRAPHS: INSTANCES OF A PROCESS MONOID

Antoni Mazurkiewicz
IPI PAN, PKiN, S.P.22
PL 00-901 Warszawa

INTRODUCTION

The algebra of strings and of sets of strings has been proved to be of great importance for the theory of sequential systems. Many various models of computations have been investigated by means of sets of strings they can generate (or accept). Such notions, as finite automata, sequential machines, regular expressions, are strongly connected with strings of symbols representing actions occurring in real systems, since the behaviour of such systems can be described by sets of interpreted strings. The algebra of strings has established a basis for all subsequent theoretical investigations.

In case of concurrent systems, strings of symbols - perfectly suited for modelling sequential processes - are not so generally available. Since some events occurring in concurrent systems are independent of each other, there is no causal connection putting them in one linearly ordered sequence; instead, their occurrences form a partially ordered set. There is no possibility to impose any linear order in which events must, or should, occur, without destroying their independence and reducing the system to a sequential one. Hence, there is a need to establish a suitable algebra of "partial strings" and to find some basic properties of it.

Some notions of processes with partially ordered set of event occurrences have been already elaborated by different authors at different time, to enable analysis of concurrent systems and to create a solid basis for developing their theory. The first such a notion is due to Petri [P], [GLT], [R], and other notions - either directly, or indirectly - have been derived from it. In this approach processes are represented by occurrence nets; their finite version is close to occurrence graphs considered below and the infinite can be obtained via the limit construction. The second was the notion of traces [M] introduced to facilitate formal reasonong about concurrent program schemata; dependence graphs, also described in the present paper, have been thought as graphical representations of traces. Almost at the same time the notion of string vectors [S1], called here histories, was introduced for a formal description of the path expressions semantics, converted next to the COSY system [LSB].

The aim of this paper is to compare the above notions of processes, to find some common features of them, and if possible, to define an abstract algebra comprising features of all of them. We shall not follow the original definitions in details; we hope however, that the essential properties of the authors' definitions as well as their intentions will be preserved. It will be shown that all these algebras are instances of one algebra: a monoid with some characteristic axioms. Since in this paper we are not interested in any concrete concurrent system, no restrictions on the sets of processes will be imposed (as conditions concerning a mutual precedence of events in processes). The intention of the author is only to characterize any possible processes in a given event domain and to leave the question of properties of their sets to more specific theories of concurrent systems (similar to the theory of formal languages in sequential systems).

PRELIMINARY NOTIONS

The standard mathematical notation is used through the paper, with few exceptions. If c', c" are conditions, then (c',c") will denote their conjunction. The set of all subsets of a set A will be denoted by $\mathbb{B}(A)$. The cartesian product of a family Z of sets will be denoted by $\mathbb{P}(Z)$. The braces around singletons will be omitted, if it causes no ambiguity.

We assume the reader to be familiar with elements of the theory of abstract algebras and of relations; only some notions from these theories will be recalled here. For any relations Q, R, the composition QR is a relation such that $xQRy = \exists z: xQzRy$. A relation R is an ordering, if R is reflexive, transitive, and antisymmetric (R is antisymmetric, if $(xRy, yRx) \Rightarrow x = y$); R is a linear ordering, if R is an ordering and $\forall x,y \in dom(R)$: (xRy or yRx). For any relation R, R^* is the transitive and reflexive closure of R. An ordering R is discrete, if $R = (R - RR)^*$. The Hase diagram of a discrete ordering R is a graph with dom(R) as the set of nodes and (R - RR) as the set of arcs. If f is a function and R is a relation, then fR, Rf, are relations such that $xfRy \Leftrightarrow f(x)Ry$, $xRfy \Leftrightarrow \exists z: (xRz, f(z) = y)$.

Let $w = (a_1, \ldots, a_m)$ be a finite sequence; each pair (a_i, n_i) with $n_i = \text{card}\{j \mid j \leqslant i,\ a_i = a_j\}$, $1 \leqslant i \leqslant m$, is called an occurrence of a_i in w. The relation

$$\text{Ord}(w) = \{((a_i,n_i),(a_j,n_j)) \mid i \leqslant j,\ 1 \leqslant i,j \leqslant m\}$$

is called the natural ordering in w.

A monoid is an algebra with one associative operation $\cdot$ (the composition operation) and one constant e (the neutral element of the composition). An element a of a monoid is an atom, if it is not neutral and $a = a' \cdot a''$ implies either a' or a" to be neutral. A monoid A is generated by a set G, if each element of A is composed of a finite number of elements of G (by convention, the neutral element is composed of zero elements of G). A monoid is atomic, if it is generated by the set of its atoms. Let E be a set; then E^* denotes the set of all finite sequences of elements of E. A monoid with E^* as the set of its elements, concatenation of sequences as the composition operation, and the empty sequence as the neutral element, is called a monoid of strings over E. In expressions denoting compositions in a monoid, the composition sign will be systematically omitted.

EVENT DOMAINS

By an <u>event domain</u> we shall understood here any ordered triple $\underline{\underline{E}} = (X, E, R)$, where

X is a set (of <u>objects</u> of $\underline{\underline{E}}$),
E is a set (of <u>events</u> of $\underline{\underline{E}}$),
$R \subseteq X \times E$, cod(R) = E.

Intuitively, events of an event domain can be regarded as changes of states of its objects. Let $\underline{\underline{E}} = (X, E, R)$ be an event domain. If $(x,e) \in R$, we say that e <u>concernes</u> x, or e <u>can happen</u> to x, or x is <u>subjected</u> to e. Since in concurrent systems events need not concerne all system objects, to specify a system it is necessary to say which events can happen to an object (the scope of an object), or equivalently, which objects are subjected to an event (the range of an event). Is a relation R given, both kinds of information are available. Objects are treated as indivisible items; consequently, all events concerning the same object can happen only sequentially, i.e. they are mutually dependent. On the contrary, events which do not concerne any common object can happen independently of each other.

Let $\underline{\underline{E}} = (X, E, R)$ be an event domain fixed from now on (except examples). All subsequent notions will be related to this domain. Put

$$D = R^{-1}R,\ I = E \times E - D,\ X_e = \{x \mid (x,e) \in R\},\ E_x = \{e \mid (x,e) \in R\}.$$

The relations D, I, are called the *dependency* in $\underline{\underline{E}}$, *independency* in $\underline{\underline{E}}$, respectively, the sets X_e, E_x, the *range* of e, the *scope* of x, resp.

Example 1. The triple (X, E, R) with

$$X = \{1,2\},\ E = \{a,b,c\},\ R = \{(1,a),(2,a),(1,b),(2,c)\},$$

is an event domain, in which

$D = \{(a,a),(b,b),(c,c),(a,b),(b,a),(a,c),(c,a)\}$, (dependency)
$I = \{(b,c),(c,b)\}$, (independency)
$X_a = \{1,2\}$, $X_b = \{1\}$, $X_c = \{2\}$, (ranges of a, b, c)
$E_1 = \{a,b\}$, $E_2 = \{a,c\}$, (scopes of 1, 2).□

Clearly, dependency and independency are symmetric relations for each event domain, dependency is reflexive, independency is irreflexive.

Let $\cdot$ denote the concatenation of sequences, ε the empty sequence. The monoid

$$\underline{\underline{S}} = (E\ ,\ \cdot,\ \varepsilon)$$

will be referred to as the monoid of strings over $\underline{\underline{E}}$. Properties of such monoids are commonly known and will not be quoted here. Recall only that the ordering of event occurrences in strings (the natural ordering) is given by Ord.

Example 2. The Hase diagram of the ordering Ord(abbca) gives Figure 1.□

$$(a,1) \longrightarrow (b,1) \longrightarrow (b,2) \longrightarrow (c,1) \longrightarrow (a,2)$$

Fig.1

TRACES

Traces have been thought as representants of processes generated by concurrent schemata on the same level of abstraction as strings being representants of sequential processes. To avoid a combinatorial explosion of cases and to preserve concurrency existing in parallel systems, some execution sequences (firing sequences) have been identified into traces; strings identified in the trace construction are the same as those identified by an anticipated interpretation of action symbols by concrete actions of analysed system. Thus, identification in trace approach is made *prior* to an interpretation; under any reasonable interpretation, i.e. any interpretation giving the same values to strings which differ from each other only in the ordering of concurrently executed actions, the final effect of analysis is the same, independently of which objects have been used: traces, or strings.

Let $\equiv$ be the least congruence in $\underline{\underline{S}}$ such that

$$(a',a'') \in I \Rightarrow a'a'' \equiv a''a'.$$

The quotient algebra $\underline{\underline{T}} = \underline{\underline{S}}/\equiv$ is called the *trace algebra* over $\underline{\underline{E}}$; its elements, i.e. equivalence classes of the congruence $\equiv$, are called *traces* over $\underline{\underline{E}}$. Clearly, $\underline{\underline{T}}$ as an algebra arising from a monoid by factorization, is a monoid too. For each $w \in E$, let $[w]$ be the trace represented by w. Let

$$\hat{\underline{\underline{T}}}(t) = \bigcap \{ Ord(w) \mid t = [w] \}.$$

Theorem 1. $\hat{\underline{\underline{T}}}(t)$ is an ordering for each trace t over $\underline{\underline{E}}$. □

$\hat{\underline{\underline{T}}}$ treated as a function on traces will be called the *trace ordering* function.

Interpretation of the ordering $\hat{\underline{\underline{T}}}(t)$ is the following. Let t represent a process and let each w with $[w] = t$ be regarded as an observation of t; then $w \equiv v$ means that w and v are observations of the same process. In all observations the ordering of event occurrences is linear; if some of them discover an opposite ordering of the same event occurrences, it means that these occurrences are actually not ordered in the observed process, and that the difference noticed by observers results only because their specific point of view. Thus, such an ordering should not be taken into account, and consequently two event occurrences can be considered as ordered in a process if and only if they occur in the same order in all possible observations of the process (all observers agree with each other on the same order).

Example 3. Let $\underline{\underline{E}}$ be the same event domain as in the first example; then

$$[abbca] = \{abbca,\ abcba,\ acbba\}.$$

Thus, there is no ordering between occurrences of b and c in abbca . □

HISTORIES

The notion of histories has been introduced for a formal semantic description of systems composed of cooperating sequential components. The main idea is to represent concurrent processes by a collection of individual histories of objects involved in these processes; such an individual history is a string of events changing states of only one object. This approach, appealing directly to the intuitive meaning of parallel processing, is particularly well suited to CSP-like systems, where individual components generate their processes independently of each other with one exception: an event concerning a number of components can occur only coincidently in all these components ("shake hands" or "rendez vous" synchronization principle). Such a rule, in more or less explicit form, is present in majority of concurrent systems; it is also reflected in the definitions given below. The presentation and the terminology used here are adjusted to the present purposes and differ from those of the authors.

Let $\bar{M} = \mathbb{P}(\{E_x^* \mid x \in X\})$; thus, each $\bar{w} \in \bar{M}$ is a function from X to E^* such that $\bar{w}(x) \in E_x^*$ for each $x \in X$. The concatenation operation in $\bar{M}$ is defined componentwise:

$$\bar{w}'\bar{w}''(x) = \bar{w}'(x)\bar{w}''(x), \text{ for all } x \in X;$$

denote by $\cdot$ this operation and put $\bar{\varepsilon}(x) = \varepsilon$ for each $x \in X$; then $\underline{\underline{M}} = (\bar{M}, \cdot, \bar{\varepsilon})$ is a monoid, as a product of monoids. Define $\bar{e}$, for each $e \in E$, as an element of $\bar{M}$ such that

$$\bar{e}(x) = \begin{cases} e, & \text{if } x \in X_e, \\ \varepsilon, & \text{if } x \notin X_e, \end{cases}$$

and put $\bar{H}_0 = \{\bar{e} \mid e \in E\}$. The submonoid $\underline{\underline{H}}$ of $\underline{\underline{M}}$ generated by $\bar{H}_0$ will be called the monoid of <u>global</u> <u>histories</u>, or simply <u>histories</u>, over $\underline{\underline{E}}$. If $\bar{w}$ is a history, then $\bar{w}(x)$, for each $x \in X$, is said to be the <u>individual</u> <u>history</u> of x in $\bar{w}$.

<u>Example</u> 4. For the event domain as in the previous examples, $\bar{w} = (abba, aca)$ is a history, since $\bar{w} = (a,a)(b,\varepsilon)(b,\varepsilon)(\varepsilon,c)(a,a)$. □

Define

$$\hat{\underline{\underline{H}}}(\bar{w}) = (\bigcup \{Ord(\bar{w}(x)) \mid x \in X\})^*.$$

<u>Theorem</u> 2. For each history $\bar{w}$, $\hat{\underline{\underline{H}}}(\bar{w})$ is an ordering. □

We shall refer to $\hat{\underline{\underline{H}}}$ as the <u>history</u> <u>ordering</u> function.

Let interprete the ordering $\hat{\underline{\underline{H}}}(\bar{w})$ in terms of observations. Each individual history in $\bar{w}$ can be treated as the result of a local observation of $\bar{w}$ limited to the scope of a single object only; from such a point of view, an observer can see only events concerning the object he is localized at; remaining events are not visible for him. Thus, being localized at x he notices the ordering $Ord(\bar{w}(x))$ only. In order to discover the ordering of event occurrences in the global history $\bar{w}$, all such observations have to be put together; then events observed coincidently from different objects form a sort of links between individual observations and make possible to discover an ordering between event occurrences concerning separate objects. The rule is: an event occurrence c" follows another event occurrence c', if there is a sequence of event occurrences beginning with c', ending with c", in which every element follows its predecessor according to the individual history containing both of them (Figure 2).

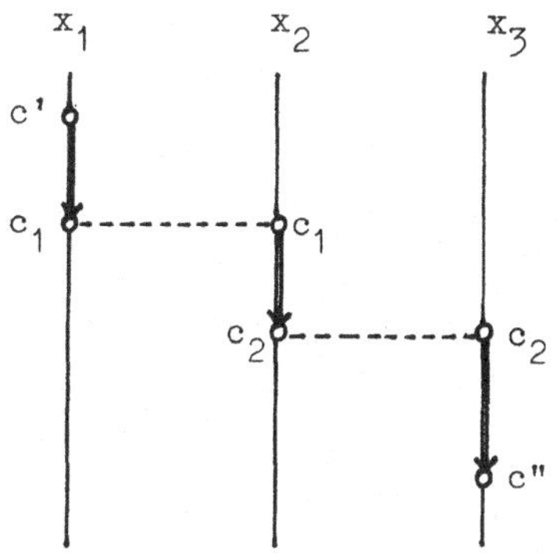

Fig.2

Such a principle of ordering is used by historians to establish a chronology of events concerning remote and separate historical objects. Observe that the greater dependency is contained in the considered event domain, the shorter the above mentioned sequence can be; that is why historians try to discover events with possibly wide range, like eclipses, as reference points in their chronologies. Comparing the trace ordering with the history ordering we can see that while in the former all event occurrences are visible to each observer and observations made by them are compared for rejecting a subjective and irrelevant information, in the history ordering only a part of all event occurrences is visible by a single observer an a comparison of all observations is made to gain an additional and relevant information If the latter method is compared to a historians method, the former can be similarly compared to that of physicists.

Let $P: E^* \longrightarrow \bar{M}$ be defined as follows:

$$P(\varepsilon) = \bar{\varepsilon},$$
$$P(we) = P(w)\bar{e}, \quad \text{for } e \in E,\ w \in E .$$

The mapping P is called the distribution function, and each w such that $P(w) = \bar{w}$ is called a sequentialization of $\bar{w}$. By the definition of $\underline{\underline{H}}$, each history has (a number of) sequentializations.

Theorem 3. $\hat{\underline{\underline{H}}}(\bar{w}) = \bigcap \{Ord(w) \mid P(w) = \bar{w}\}$.□

Thus, the history ordering is the common ordering observed in all sequentializations of this history.

DEPENDENCE GRAPHS

Dependence graphs have been thought as graphical representations of traces, in which the ordering of event occurrences is made more explicit. This notion refers directly to understanding processes as partially ordered sets of events occurrences; such an approach seems to be promising, but there are some drawbacks in it: dealing with graphs as objects of an algebra we have to do with a relatively complex composition mechanism. Instead, graph representation of processes offers a considerable advantage in self-explaining facilities.

Dependence graphs are finite, directed graphs with nodes labelled with events. Call two nodes dependent, if they are labelled with dependent events. Two dependence graphs are isomorphic, if there is a bijection between their sets of nodes preserving arc connections and labelling. All properties of dependence graphs are given up to such an isomorphism. The definition of dependence graphs is the following. A triple

$$g = (V,\ A,\ f)$$

is a dependence graph, or briefly, a d-graph over $\underline{\underline{E}}$, if

V is a finite set (of nodes of g),

$A \subseteq V \times V$ (the set of arcs of g),
$f: V \longrightarrow E$ (the labelling of g),

such that

$$AA^{*} \cap id_V = \emptyset,$$
$$A \cup A^{-1} \cup id_V = fDf^{-1};$$

the first condition ensures cycle freeness of g, the second means that two different nodes of g are connected by an arc if and only if they are dependent (id_V denotes the identity relation in V).

Example 5. for the same event domain as before, the graph in Figure 3 is a d-graph. □

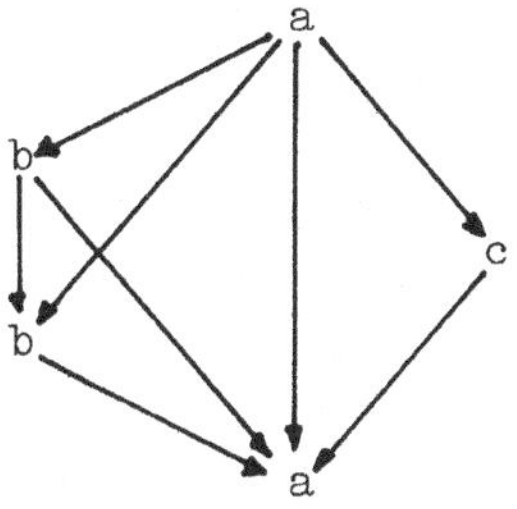

Fig.3

A directed labelled graph $g = (V, A, f)$ is said to be the composition of d-graphs g_1, g_2, if there are graphs (V_1, A_1, f_1), (V_2, A_2, f_2), isomorphic to g_1, g_2, respectively, such that:

$$V = V_1 \cup V_2, \quad V_1 \cap V_2 = \emptyset,$$
$$A = A_1 \cup A_2 \cup f_1 D f_2^{-1},$$
$$f = f_1 \cup f_2.$$

It means that g arises from the disjoint union of g_1, g_2, by adding to it new arcs leading from each node of g_1 to each node of g_2 provided they are dependent.

Proposition 1. For each d-graphs g_1, g_2, over $\underline{\underline{E}}$, the composition of g_1, g_2 exists, is unique (up to isomorphism), and is a d-graph over $\underline{\underline{E}}$. □

Denote the composition of d-graphs g_1, g_2 by $g_1 \oplus g_2$. Figure 4 shows an example of the composition; the dotted arrows represent arrows added by the composition.

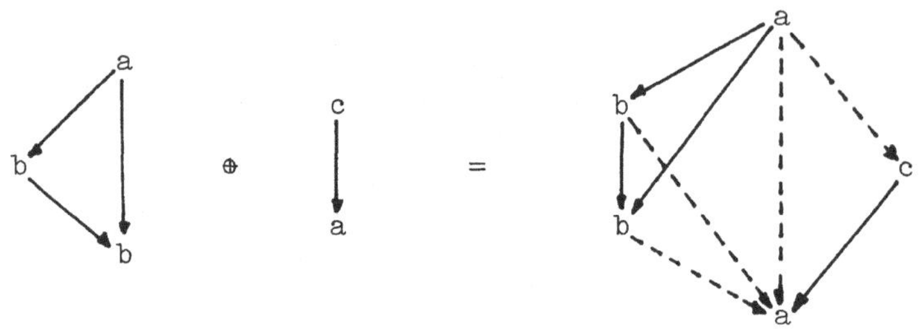

Fig.4

Put $\lambda = (\emptyset, \emptyset, \emptyset)$; clearly, λ is a d-graph. Let G be the set of all d-graphs over $\underline{\underline{E}}$.

Theorem 4. $\underline{\underline{G}} = (G, \oplus, \lambda)$ is a monoid. □

Let $g = (V, A, f)$ be an arbitrary d-graph over $\underline{\underline{E}}$; let

$$h(v) = (f(v), n),$$

where $n = \mathrm{card}\,\{u \in V \mid uA^{*}v,\ f(u) = f(v)\}$, and put

$$\hat{\underline{\underline{G}}}(g) = h^{-1}A^{*}h.$$

Theorem 5. For each d-graph g, $\hat{\underline{\underline{G}}}(g)$ is an ordering. □

Observe that h assigns to each node of g an occurrence of its label; the ordinal number of this occurrence is the number of all nodes of g with the same label, from which the considered node is reachable. Thus, $\hat{\underline{\underline{G}}}(g)$ is a natural path ordering of nodes of g, representing event occurrences in g.

Let $\langle\,\rangle : E^{*} \longrightarrow G$ be a mapping defined as follows:

$$\langle \varepsilon \rangle = \lambda,$$
$$\langle we \rangle = \langle w \rangle \oplus (e,(e,e),(e,e)), \text{ for } e \in E,\ w \in E^{*}.$$

Clearly, w is a d-graph for each $w \in E^{*}$.

Proposition 2. For each $g \in G$ there is $w \in E^{*}$ such that $g = \langle w \rangle$. □

Theorem 6. $\hat{\underline{\underline{G}}}(g) = \bigcap \{\mathrm{Ord}(w) \mid \langle w \rangle = g\}$. □

OCCURRENCE GRAPHS

Occurrence graphs are constructs similar to dependence graphs, but defined in another way. While in dependence graphs the ordering of event occurrences is (almost) explicit and occurrences of events are implied by construction, in occurrence graphs both of them are explicit, since these graphs are defined as orderings of event occurrences. Dependence graphs are abstract ones; the choice of their nodes is arbitrary; occurrence graphs are concrete, with prescribed sets of nodes as event occurrences sets.

Let $Z = E \times \{1,2, \ldots\}$; define $B \subseteq Z \times Z$ as follows:

$$B = \{((a,n),(b,m)) \mid aDb,\ n \geqslant 1,\ m \geqslant 1\},$$

and for each $w \in E^*$ put

$$s(w) = (Ord(w) \cap B)^*.$$

Thus, s is a mapping from E^* to the power set of $Z \times Z$. By the above definition, s(w) arises from the natural ordering Ord(w) by first, deleting from it all pairs of occurrences of independent events, and next, by reflexive and transitive closure of the obtained relation. It is clear that s(w) is an ordering for each $w \in E^*$.

Example 6. For the same event domain as before, Figure 5 gives the Hase diagram of the ordering s(abbca).□

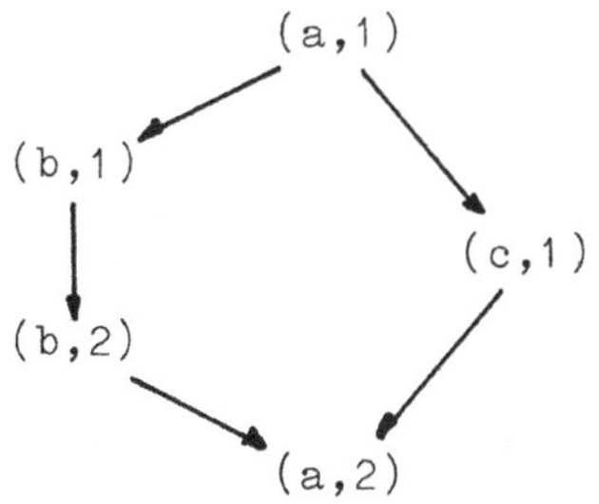

Fig.5

For each $w \in E^*$, s(w) will be called an occurrence graph generated by w. Let $Q = s(E^*)$; it means that Q is the set of all occurrence graphs generated by elements of E^*.

Proposition 3. The mapping s is congruent in $\underline{\underline{S}}$, i.e. $s(v') = s(w')$, $s(v'') = s(w'')$ implies $s(v'v'') = s(w'w'')$, for each $v',v'',w',w'' \in E^*$.

By the above proposition the following definition of the composition of occurence graphs is correct: for each $q', q'' \in Q$, the composition of q' with q'' is an occurrence graph $q' \cdot q''$ such that

$$q' = s(v'),\ q'' = s(v'') \Rightarrow q' \cdot q'' = s(v'v''),$$

for all $v',v'' \in E^*$. Put $\underline{\underline{Q}} = (Q, \cdot, \emptyset)$.

Theorem 7. $\underline{\underline{Q}}$ is a monoid.

The monoid $\underline{\underline{Q}}$ will be referred to as the monoid of occurrence graphs over $\underline{\underline{E}}$. Let, for each $q \in Q$, $\hat{\underline{\underline{Q}}}(q) = q$; $\hat{\underline{\underline{Q}}}$ will be said to be the occurrence graph ordering function.

Theorem 8. $\hat{\underline{\underline{Q}}}(q) = \bigcap \{\mathrm{Ord}(w) \mid s(w) = q\}$.

AN AXIOMATIC PROCESS MONOID

As it could be observed, all monoids of processes described above have many common features although they differ considerably in the nature of objects their elements are constructed from. A natural question arises, whether it is possible to abstract from these differences and to define axiomatically a class of algebras retaining common basic properties of processes, as ordering of event occurrences, laws of composition, etc., and such that the concrete algebras considered above could serve as particular instances, or even representations. The answer is positive; algebras defined below meet such requirements by a suitable choice of their specific axioms.

Let in what follows p,q,r, ... , with possible primes and subscripts, always denote elements of the considered monoid, while symbols a,b,c, ... , similarly decorated, denote its atoms. As usual, the composition sign will be omitted in expressions.

By a process monoid we shall mean any atomic monoid

$$\underline{\underline{P}} = (P, \cdot, \theta)$$

satisfying the following axioms:

$$pq = \theta \Rightarrow (p = \theta,\ q = \theta), \tag{A1}$$

$$pq = rs \Rightarrow \exists\, p_1, p_2, q_1, q_2 : (p = p_1 p_2,\ q = q_1 q_2,\ r = p_1 q_1,\ s = p_2 q_2). \qquad (A2)$$

Elements of P will be called processes, atoms of $\underline{\underline{P}}$ will be referred to as actions. Figure 6a explains the meaning of (A2); this axiom is a generalization of Lévi's Lemma [L] for monoids of strings (Figure 6b) which states that for each strings p,q,r,s:

$$pq = rs \Rightarrow \exists\, t: (p = rt,\ s = tq \text{ or } r = pt,\ q = ts).$$

Axioms (A1), (A2) are independent: let $P_1 = (\{\theta, a\}, \cdot, \theta)$ be a monoid with $aa = \theta$; then P_1 is an atomic monoid, (A2) holds, but (A1) does not; let $P_2 = (\{\theta, a, b\}, \cdot, \theta)$ be a monoid with $pq = b$ for each $p, q \neq \theta$; then P_2 is an atomic monoid, (A1) holds, but (A1) does not.

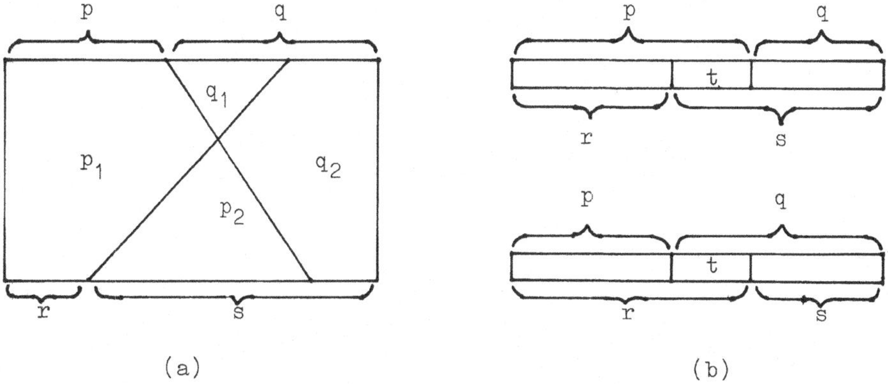

Fig.6

Proposition 4. (A1, A2) <=> (B1, B2, B3), where B1, B2, B3 are the following:

$$pa \neq \theta, \qquad (B1)$$
$$pa = qa \Rightarrow p = q, \qquad (B2)$$
$$(pa = qb,\ a \neq b) \Rightarrow (ab = ba,\ \exists\, r: (p = rb,\ q = ra)); \qquad (B3)$$

similarly, (A1, A2) <=> (C1, C2, C3), where C1, C2, C3 are the following:

$$ap \neq \theta, \qquad (C1)$$
$$ap = aq \Rightarrow p = q, \qquad (C2)$$
$$(ap = bq,\ a \neq b) \Rightarrow (ab = ba,\ \exists\, r: (p = br,\ q = ar)).\ \square \qquad (C3)$$

Theorem 9. Monoids $\underline{\underline{T}}$, $\underline{\underline{H}}$, $\underline{\underline{G}}$, and $\underline{\underline{Q}}$ are process monoids. $\square$

Let us quote some general properties of process monoids:

$$pa = b_1 \dots b_m \Longleftrightarrow \exists\, i: (1 \leq i \leq m,\ b_i = a,\ b_1 \dots b_{i-1} b_{i+1} \dots b_m = p,\ \forall\, j: (1 \leq j < i \Rightarrow b_j a = a b_j));$$

$$a_1 \dots a_n = b_1 \dots b_m \Rightarrow (n = m,\ (a_1, \dots, a_n) \text{ is a permutation of } (b_1, \dots, b_m));$$

$$pq'r = pq''r \Longleftrightarrow q' = q''.$$

Let $\underline{\underline{P}} = (P, \cdot, \theta)$ be a process monoid and let P_0 be the set of its atoms. Define a mapping $d: P_0^* \longrightarrow P$ inductively:

$$d(\varepsilon) = \theta,$$
$$d(wa) = d(w)a, \text{ for } w \in P_0^*,\ a \in P_0,$$

and put

$$\hat{\underline{\underline{P}}}(p) = \bigcap \{\mathrm{Ord}(w) \mid d(w) = p\}.$$

<u>Theorem</u> 10. $\hat{\underline{\underline{P}}}(p)$ is an ordering for each $p \in P$.□

$\hat{\underline{\underline{P}}}$ will be called the <u>process</u> <u>ordering</u> function. Thus, $\hat{\underline{\underline{P}}}(p)$ is an ordering of action occurrences of p; if $((a,n),(b,m)) \in \hat{\underline{\underline{P}}}(p)$ and $(a,n) \neq (b,m)$, we say that (a,n) occurs earlier than (b,m) in p, or equivalently that (b,m) occurs later than (a,n) in p. From (A2) it follows that in $\hat{\underline{\underline{P}}}(pq)$ each action in p occurs not later than any action in q, and each action in q occurs not earlier than any action in p (i.e. (p,q) is a Dedekind decomposition of $r = pq$, provided $p \neq \theta \neq q$).

Properties of process monoids depend upon specific axioms added to A1 and A2, in particular on axioms concerning commutativity of atom compositions. Let $\underline{\underline{P}}$ be a process monoid with P_0 as the set of its atoms we shall consider three main cases.

<u>Theorem</u> 11. If for all $a,b \in P_0$ $ab = ba \Longleftrightarrow a = b$, then $\underline{\underline{P}}$ is isomorphic to the monoid of strings over P_0.□

We say that $\underline{\underline{P}}$ is a process monoid <u>over</u> an event domain $\underline{\underline{E}}$, $\underline{\underline{E}} = (X, E, R)$ I is the independency of $\underline{\underline{E}}$, if there is a bijection $h: E \longrightarrow P_0$ such that for each $a,b \in P_0$

$$ab = ba \Longleftrightarrow a = b \text{ or } a(h^{-1} I h)b.$$

<u>Theorem</u> 12. There exist exactly one (up to isomorphism) process monoid over a given event domain.□

<u>Corollary</u>. Monoids $\underline{\underline{T}}$, $\underline{\underline{H}}$, $\underline{\underline{G}}$, $\underline{\underline{Q}}$ are mutually isomorphic.□

Since ordering functions of $\underline{\underline{T}}$, $\underline{\underline{H}}$, $\underline{\underline{G}}$, $\underline{\underline{Q}}$ are process ordering functions, and isomorphic process monoids clearly have isomorphic ordering functions, $\hat{\underline{\underline{T}}}$, $\hat{\underline{\underline{H}}}$, $\hat{\underline{\underline{G}}}$, $\hat{\underline{\underline{Q}}}$ are isomorphic to each other.

Let A be a set; each function $z: A \longrightarrow \{1,2, \ldots\}$ such that

$$\sum \{z(a) \mid a \in A\} \text{ is finite,}$$

will be called a (finite) <u>multiset</u> over A. If z', z" are multisets over A, then their sum defined as a function (z'+ z") by

$$(z'+ z'')(a) = z'(a) + z''(a), \text{ for all } a \in A,$$

is a multiset over A. Let Z be the set of all finite multisets over A; then $\underline{\underline{Z}} = (Z, +, \emptyset)$ is a monoid referred to as the <u>monoid of multisets</u> over A.

<u>Theorem</u> 13. If for all $a,b \in P_0$ $ab = ba$, then $\underline{\underline{P}}$ is isomorphic to the monoid of multisets over P_0.

Thus, depending on commutativity properties of $\underline{\underline{P}}$, it can represent monoids with a "maximal" ordering, as monoids of strings, as well as those with "minimal" ordering, as monoids of multisets. Between these two extremities monoids with partially ordered sets of element occurrences as elements are placed.

SOME PERSPECTIVES

Process monoids form a first step for developing algebras of <u>activities</u>, i.e. sets of processes. Let $\underline{\underline{P}}$ be a process monoid; then similarly as in case of string monoids, the composition in $\underline{\underline{P}}$ can be lifted to the composition of subsets in $\underline{\underline{P}}$:

$$XY = \{pq \mid p \in X, q \in Y\},$$

for each sets X, Y, of elements of $\underline{\underline{P}}$. Having defined composition of activities, some other operations (as e.g. iteration) can be introduced. Finding properties of activities defined by means of such operations, classifying their families, describing their hierarchies - these are only some issues connected with and promoted by process monoids. As an example of properties of such lifted algebra let us quote a fact already applied in the concurrent system analysis.

Let $\underline{\underline{P}}$ be a process monoid over an event domain $\underline{\underline{E}}$. Let P_0 be the set of atoms of $\underline{\underline{P}}$, E the set of events of $\underline{\underline{E}}$, and $h: E \longrightarrow P_0$ a bijection. Extend h to a homomorphism from $\underline{\underline{S}}$ over $\underline{\underline{E}}$ to $\underline{\underline{P}}$ by:

$$h(\varepsilon) = \theta,$$
$$h(wa) = h(w)h(a), \text{ for } w \in E^*, a \in E,$$

and in turn extend this homomorphism to subsets of E^*:

$h(X) = \{h(w) \mid w \in X\}$, for $X \subseteq E^*$.

Let $f: \mathbb{B}(E^*) \longrightarrow \mathbb{B}(E^*)$ be an arbitrary mapping; we say that f is <u>congruent</u>, if for each $X,Y \subseteq E^*$:

$$h(X) = h(Y) \Rightarrow h(f(X)) = h(f(Y)),$$

and that f is <u>monotonic</u>, if

$$X \subseteq Y \Rightarrow f(X) \subseteq f(Y).$$

If f is congruent, P is the set of processes of $\underline{\underline{P}}$, then there is a mapping $g: \mathbb{B}(P) \longrightarrow \mathbb{B}(P)$ such that $h(f(X)) = g(h(X))$, for each $X \subseteq E$. Denote this mapping by $h(f)$. Then we have

<u>Theorem</u> 14. If $f: \mathbb{B}(E^*) \longrightarrow \mathbb{B}(E^*)$ is monotonic and congruent, and X_0 is the least fixed point of f, then $h(X_0)$ is the least fixed point of $h(f)$.□

Thus, knowing how to solve fixed point equations for sets of strings, we know how to solve them for sets of processes of an arbitrary process monoid.

Other issues are connected with infinite processes. Although elements of process monoid are finite objects, there is a possibility to deal with infinite processes within the framework of these monoids; namely, it is possible to use the direct limit construction. Let $\underline{\underline{P}} = (P, \cdot, \theta)$ be a process monoid. For each $p \in P$ let the set

$$\text{pref}(p) = \{q \mid \exists\, r: p = qr\}$$

be called the set of <u>prefixes</u> of p, and for each $R \subseteq P$ let

$$\text{pref}(R) = \bigcup \{\text{pref}(r) \mid r \in R\}.$$

A subset R of P is directed, if for each $p,q \in R$ there is $r \in R$ such that $p,q \in \text{pref}(r)$. Each directed subset R of P uniquely determines an object, called the limit of R and denoted by lim(R), together with its set of prefixes, which is defined as

$$\text{pref}(\lim(R)) = \text{pref}(R).$$

All elements of P are limits of some finite directed subsets of P, and limits of all finite directed subsets of P are elements of P. Limits of infinite directed subsets of P can be then considered as infinite processes with properties determined by their sets of prefixes.

<u>Example</u> 7. Let $\mathbb{N} = \{1,2, \dots \}$, $\underline{\underline{E}} = (X, E, R)$ with $X = \mathbb{N}$, $E = \{a_i \mid i \in \mathbb{N}\}$, $R = \{(i,a_j) \mid i,j \in \mathbb{N},\ i \leqslant j \leqslant i+1\}$. Let $\underline{\underline{P}}$ be a process monoid over $\underline{\underline{E}}$. Define, for each $i \in \mathbb{N}$, processes p_i, q_i in $\underline{\underline{P}}$ as follows:

$q_0 = \theta$, $p_0 = \theta$,

$q_i = a_i q_{i-1}$, $p_i = p_{i-1} q_i$, for $i \in \mathbb{N}$,

and put $R = \{p_i | i \in \mathbb{N}\}$. R is a directed set; the dependence graph of lim(R) (without arcs following from others by transitivity) is given in Figure 7.

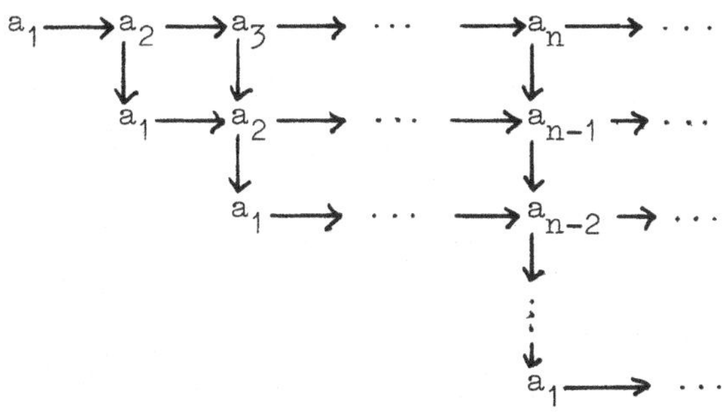

Fig.7

Each individual history in lim(R) is infinite, e.g. for object i it is the following: $a_i a_{i+1} a_i a_{i+1} \cdots$. Now, for the same process monoid, let define processes t_i, for all $i \in \mathbb{N}$, in the following way:

$t_0 = \theta$, $t_1 = a_1$,

$t_{i+1} = t_{i-1} a_{i+1} a_i$, for each $i \in \mathbb{N}$,

and put $Q = \{t_i | i \in \mathbb{N}\}$. Q is a directed set; its limit lim(Q), in the same convention as the previous one, is presented in Figure 8.

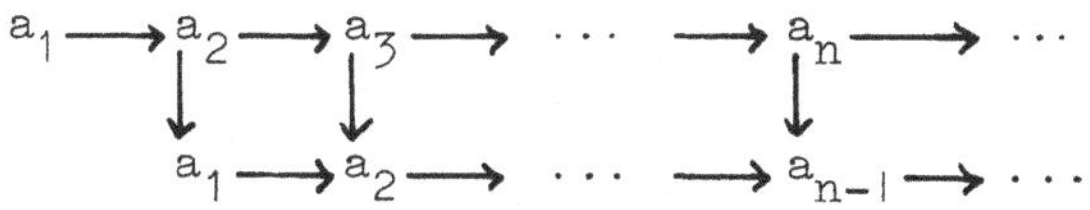

Fig.8

Here each individual history is finite, e.g. i-th one is: $a_i a_{i+1} a_i$, although the whole process is infinite.

CONCLUDING REMARKS

Some monoids with partially ordered sets of event occurrences (thought as representations of discrete concurrent processes) has been presented. It turned out that all these monoids are mutually isomorphic and form instances of a monoid, called here a process monoid, which could be called a monoid of (finite) partial orderings. Two axioms on decomposability of elements define this monoid; adding to them some specific axioms one can convert the process monoid into a monoid of some special ordering.

We hope that the presented approach can be helpful in developing an algebra of concurrent behaviours and that the comparison made above put some light on mutual relationships between different approaches to a formal description of concurrent systems.

There is a number of results which have influenced on the presented approach and it is impossible quote all of them; let mention only the main ones: the general notion of nonsequentiall processes P , R , P1 ; the notion of history S1 , S , and also K ; the history ordering L ; the notion of occurrence graphs GLT ; the notion of traces M . There are also researches on topics mentioned in the paper or close to them: on theory of traces AR , BMS , BMS1 , BMS2 , J , K2 ; on graph rewriting systems KR , JKRE ; R2 , ST on processes in Petri nets.

Acknowledgements

The idea of the present paper arose in effect of many discussions with Prof.Dr. J. Winkowski from Institute of Computer Science of PAS in Warsaw; the author wish to acknowledge here his influence on ideas presented in this paper. The helpful atmosphere and excellent conditions created for the author in the institute F1 of GMD, Bonn, during preparation of the final version of this paper, are also gratefully acknowledged.

References

AR Aalsbersberg,Y.J., Rozenberg,G.: Trace Theory - a Survey, Technical Report, Inst. of Appl.Math. and Comp.Sci., University of Leiden, 1984

BMS Bertoni,A., Mauri,G.,Sabadini,N.: Equivalence and Membership Problems for Regular Trace Languages, LNCS 88, Springer Verlag, 1980

BMS1 -: Behaviour of CE Petri Nets an Recognizable Trace Languages, Internal Report Istituto di Cibernetica Milano, 1982

BMS2 -: A Hierarchy of Regular Trace Languages and Some Combinatorial Applications, Proc. II World Conf. on Mathematics at service of the man, 1982

GLT Genrich,H. J., Lautenbach,K., Thiagarajan,P.S.: Elements of General Net Theory, in Net Theory and Application (Brauer, W. editor)LNCS 84, Springer Verlag, 1980

J Janicki,R.: Mazurkiewicz Traces Semantics for Communication Sequential systems, 5th European Workshop on Applications and Theory of Petri Nets, 1984

JKRE Janssen,D.,Kreowski,H.J.,Rozenberg,G.,Ehrig,H.: Concurrency of Node-Label-Controlled Graph Transformations, Proc. of 8th Conf. on Graph-Theoretic Concepts in Computer Science, Hanser Verlag, 1982

K Knuth,E.&al: A Study of the Projection Operation, Springer Verlag, Inf. Fachb. 52, 1982

K1 Knuth,F.: Petri Nets and Trace Languages, Proc. Ist European Conf. on Parallel and Distributed Processing, Toulouse,1979

KR Kreowski,H.J.: Graph Grammar Derivation Processes, manuscript, 1984

L Lamport,L.: Time, Clocks, and the Ordering of Events in Distributive Systems, CACM,7, 1978

LSB Lauer,P.E., Shields,M.W., Best,E.: Formal Theory of the Basic COSY Notation, Univ. of Newcastle upon Tyne, Comp.Lab. TR 143, 1979

M Mazurkiewicz,A.: Concurrent Program Schemes and Their Interpretations, Aarhus University, DAIMI Rep. PB 78, 1977

M1 -: The Semantics of Concurrent Systems: A Modular Fixed Point Trace Approach,5th European Workshop on Applications and Theory of Petri Nets, 1984

P Petri,C.A.: Non-Sequential Processes. ISF Report 77.05, Gesellschaft für Mathematik und Datenverarbeitung mbH Bonn, 1977

P1 -: Fundamentals of the Representation of Discrete Processes, ISF Report 82.04, Gesellschaft für Mathematik und Datenverarbeitung mbH Bonn, 1982

R Reisig,W.: Petrinetze, Springer Verlag, Berlin-Heidelberg-New York, 1982

S Shields,M.W.: NonSequential Behaviour:1, University of Edinburgh, Dept. of Comp.Sci., Internal Report CSR-120-82, 1982

S1 -: Adequate Path Expressions, Proc. of Symposium on the Semantics of Concurrent Computation, LNCS 70, Springer Verlag, 1979

ST Starke,P.: Processes in Petri Nets, Elektron. Informationsverarbeitung und Kybernetik, 17, 1981

RECENT RESULTS ON AUTOMATA AND INFINITE WORDS

Dominique Perrin
L.I.T.P., Université Paris 7
2, Place Jussieu - 75251 Paris Cedex 05 - France

0. INTRODUCTION

The theory of automata on infinite words was initiated by Buchi and McNaughton at the beginning of the sixties, with a flavour of mathematical logic. As for the rest of automata theory, it uses concepts and methods which are familiar to computer scientists. The consideration of infinite words, which could in fact seem at first glance to be highly philosophical, has indeed a natural interpretation when one is wanting to study properties of systems that are not supposed to eventually stop, such as operating systems for instance. The use of automata on ordinary words has now become standard in several areas of algorithm design and I think that the same will happen in the next future with automata on infinite words.

From the mathematical point of view, the study of automata on infinite words is substantially more difficult than the ordinary one. The main problem is that, contrary to the ordinary case, it is not possible to restrict one's attention to deterministic automata. However, a number of intersting results have been obtained during the last years and the aim of this paper is to describe some of them.

The paper is divided into four sections. In the first one, I recall some of the basic definitions and results. Part of them can be found in Chapter XIV of Eilenberg's book [7], whose notation I follow. In the second section, I discuss the matter of star free sets of infinite words. A new result is proved there, according to which one can decide wether a given recognizable set of infinite words is star free. In the third section, an extension of the previous notions to two-sided infinite words is presented. This extension to a higher ordinal has two main features : (1) most results holding in the one sided case still hold in this case, though it is by no means a trivial extension ; (2) extensions to other ordinals such as the theory of infinite trees present quite different features (see [19] for instance, where it is proved that no equivalent of McNaughton's theorem holds on infinite trees). In the last section I present some results on unambiguity.

1. RECOGNIZABILITY

Let A be a (possibly infinite) set called the alphabet. We denote by $A^{\mathbb{N}}$ the set of all one-sided infinite words

$$\alpha = \alpha_0\alpha_1\alpha_2\ldots \qquad (\alpha_i \in A)$$

The set of ordinary (finite) words is denoted as usual by A^*. A morphism

$$\phi : A^* \to M$$

from A^* onto a finite monoid M is said to recognize a subset U of A^N if there exists a set $P \subset M \times M$ such that

$$U = \bigcup_{(m,n)\in P} \phi^{-1}(m)\,[\phi^{-1}(n)]^\omega$$

A set $U \subset A^N$ is said to be recognizable if there exists a morphism ϕ from A^* onto a finite monoid that recognizes U.

An equivalent definition of the family $Rec(A^N)$ of recognizable subsets of A^N is the family of sets of the form

$$U = \bigcup_{i=1}^{n} U_i V_i^\omega$$

for U_i, $V_i \in Rec(A^*)$.

A morphism $\phi : A^* \to M$ from A^* onto a finite monoid M is said to saturate a subset U of A^N if for any $m,n \in M$ one has either

$$\phi^{-1}(m)\,[\phi^{-1}(n)]^\omega \cap U = \emptyset$$

or

$$\phi^{-1}(m)\,[\phi^{-1}(n)]^\omega \subset U$$

Obviously, if ϕ saturates U, it also saturates its complement. More precisely, the family of sets saturated by ϕ is the boolean algebra of sets having trivial intersection with the sets $\phi^{-1}(m)\,[\phi^{-1}(n)]^\omega$.

In all above definitions, one may replace the pairs $(m,n) \in M \times M$ by a pair of an element m of M and an idempotent e of M since one has $[\phi^{-1}(n)]^\omega = [\phi^{-1}(e)]^\omega$ when e is the idempotent which is a power of n. The following two propositions relate the notions of recognizability and saturation.

PROPOSITION 1.1 Let ϕ be a morphism from A^* onto a finite monoid which saturates a set $U \subset A^N$. Then ϕ recognizes U and U is a union of sets of the form

$$\phi^{-1}(m)\,[\phi^{-1}(e)]^\omega$$

for $m \in M$ and e an idempotent of M.

The above proposition is an easy consequence of the following lemma that we prove directly, although it can be viewed as a particular case of Ramsey's theorem.

LEMMA 1.2 Let ϕ be a morphism from A^* onto a finite monoid M. For each $\alpha \in A^N$ there exists an $m \in M$ and an idempotent e in M such that

$$\alpha \in \phi^{-1}(m)\,[\phi^{-1}(e)]^\omega$$

Proof : We proceed by induction on Card(M). Let

$$R = \{r \in M \mid rM = r\}$$

be the set of right zeroes of M. We first consider the case $\alpha \in [\phi^{-1}(R)]^{\omega}$. In this case, we select on $r \in R$ which occurs infinitely often, i.e. such that

$$\alpha = u_0 v_0 u_1 v_1 \ldots$$

with $\phi(v_i) = r$. Then $\phi(v_i u_i) = r$ and therefore $\alpha \in u_0 [\phi^{-1}(r)]^{\omega}$. Since r is an idempotent, the property is proved. Otherwise, one has $\alpha = u\beta$ where β has no left factor in $\phi^{-1}(R)$. Let $n \in M$ be such that β has an infinity of left factors in $\phi^{-1}(n)$, and let N be the monoid $N = \{t \in M \mid nt = n\}$. One has

$$\beta = v_0 v_1 v_2 \ldots$$

with $\phi(v_0 v_1 \ldots v_i) = n$ and therefore $\phi(v_i) \in N$ for all $i \geq 1$. We apply the induction hypothesis to the morphism ψ from B^* (where B is in bijection with the v_i) into N. Since $n \notin R$, one has Card(N) < Card(M). Therefore, by the induction hypothesis, there exist $s \in N$ and an idempotent e in N such that $\beta \in \phi^{-1}(s)\, [\phi^{-1}(e)]^{\omega}$ which proves the property.□

The converse of Proposition 1.1 is not true in general as shown in the following example :

EXAMPLE 1.1 Let $\phi : \{a,b\}^* \to Z/2$ be the morphism defined by

$$\phi(a) = 0, \; \phi(b) = 1$$

then $[\phi^{-1}(0)]^{\omega} = \{\alpha \in \{a,b\}^N \mid |\alpha|_b \text{ is even or infinite}\}$

$$\phi^{-1}(1)\, [\phi^{-1}(0)]^{\omega} = \{\alpha \in \{a,b\}^N \mid |\alpha|_b \text{ is odd infinite}\}$$

Thus ϕ recognizes the first set but it does not saturate it since the first set has a non trivial intersection with the second one. □

A set of the form $\phi^{-1}(m)\, [\phi^{-1}(e)]^{\omega}$ as in Proposition 1.1 is called a simple recognizable set. Accordingly, any recognizable set is a finite union of simple ones.

To prove a partial converse to Proposition 1.1, we recall the definition of the Schützenberger's product of two monoids M and N. It is the set

$$M \diamond N = \{(m,\rho,n) \mid m \in M,\ \rho \subset M \times N,\ n \in N\}$$

equipped with the product

$$(m,\rho,n)(m',\rho',n') = (mm',\ m\rho' \cup \rho n',\ n')$$

where $m\rho' = \{(mr',s') \mid (r',s') \in \rho'\}$ and $\rho n' = \{(r,sn') \mid (r,s) \in \rho\}$.

Let $\phi : A^* \to M$ and $\psi : A^* \to N$ be two morphisms. Then one denotes by

$$\phi \diamond \psi : A^* \to M \diamond N$$

the morphism defined for each $w \in A^*$ by

$$\phi \diamond \psi(w) = (\phi(w),\ \rho(w),\ \psi(w))$$

with

$$\rho(w) = \{(\phi(u),\ \psi(v)) \mid w = uv\}$$

The following result is implicitly contained in [16].

PROPOSITION 1.3 <u>Let</u> $\phi : A^* \to M$ <u>be a morphism from</u> A^* <u>into a finite monoid</u> M <u>and let</u> U <u>be a subset of</u> A^N. <u>If</u> ϕ <u>recognizes</u> U, <u>then</u> $\phi \diamond \phi$ <u>saturates</u> U.

Proof : Let $M' = M \diamond M$ and $\phi' = \phi \diamond \phi$. It is enough to prove that ϕ' saturates each set U of the from $U = \phi^{-1}(m)\,[\phi^{-1}(e)]^\omega$ with $m \in M$ and $e = e^2 \in M$. Suppose that for $m' \in M'$ and $e' = e'^2 \in M'$ there exists a word α such that

$$\alpha \in \phi^{-1}(m)\,[\phi^{-1}(e)]^\omega \cap \phi'^{-1}(m')\,[\phi^{-1}(e')]^\omega$$

Let $m' = (m_1, \rho, m_1)$ and $e' = (e_1, \epsilon, e_1)$. Taking advantage of the fact that e and e' are idempotents, one may write

$$\alpha = u_o v_o w_o v_1 w_1 \ldots$$

with $\phi(u_o) = m$, $\phi(v_i w_i) = e$, $\phi'(u_o v_o) = m'e'$, $\phi'(w_i v_{i+1}) = e'$ (see figure below)

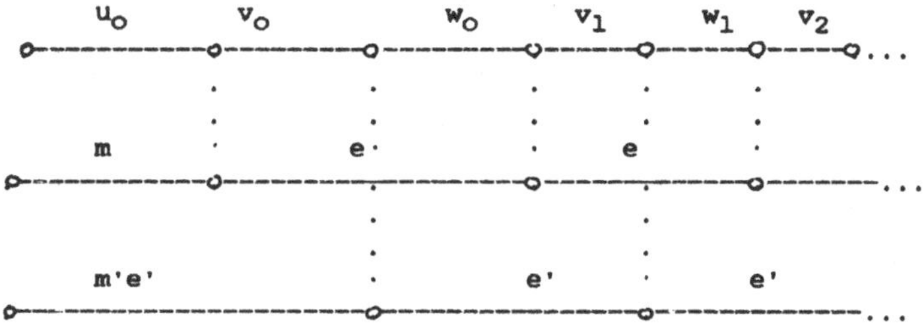

Let now β be any element of the form

$$\beta = r_o s_o s_1 s_2 \ldots$$

with $\phi'(r_o) = m'$, $\phi'(s_i) = e'$. By the definition of ϕ', one has

$$r_o s_o = x_o y_o,\quad s_i = z_i y_{i+1} \qquad (i \geq 0)$$

with $\phi(x_o) = \phi(u_o)$, $\phi(y_o) = \phi(v_o)$ and for all $i \geq 0$ $\phi(z_i) = \phi(w_i)$, $\phi(y_{i+1}) = \phi(v_{i+1})$. This implies

$$\phi(x_o) = m,\quad \phi(y_i z_i) = e$$

whence $\beta \in \phi^{-1}(m)\,[\phi^{-1}(e)]^\omega$. This proves that ϕ' saturates U. □

EXAMPLE 1.2 Let us come back to the morphism $\phi : \{a,b\}^* \to Z/2$ of Example 1.1 The morphism $\phi' = \phi \diamond \phi$ sends A^* onto a monoid M' with three elements made up with an identity equal to $\phi'(a)$ and a two elements group $\{\phi'(b), \phi'(bb)\}$. In fact, one has

$$\phi'(a) = (0, (0,0),0),\ \phi'(b) = (1, (1,0)+(0,1),1), \phi'(bb) = \{0,(0,0)+(1,1),0\}.$$

The sets recognized by ϕ' are

$\phi'^{-1}(b)\ [\phi'^{-1}(a)]^\omega$: odd number of b's

$\phi'^{-1}(bb)\ [\phi'^{-1}(a)]^\omega$: even number of b's

$[\phi'^{-1}(bb)]^\omega$: infinite number of b's

and ϕ' therefore saturates all sets recognized by ϕ. □

We say that a congruence θ on A^* saturates a set $U \subset A^N$ if the canonical morphism on the quotient saturates U. If θ,θ' are two congruences that saturate U, their upper bound $\theta \vee \theta'$ also saturates U. Therefore, for any set $U \subset A^N$ there exists a maximal congruence that saturates U. As for subsets of A^*, it is called the syntactic congruence of U. It has been introduced by A. Arnold in [2] who has proved that for a recognizable subset U of A^N it is characterized in the following way. Define for each word v two sets $\Gamma(v)$ and $\Delta(v)$ by

$$\Gamma(v) = \{(u,w,x) \in A^* x A^* x A^* \mid uvwx^\omega \in U\}$$

$$\Delta(v) = \{(u,x,y) \in A^* x A^* x A^* \mid u(xvy)^\omega \in U\}$$

Then v,v' are equivalent modulo the syntactic congruence iff $\Gamma(v) = \Gamma(v')$ and $\Delta(v) = \Delta(v')$.

The quotient of A^* by the syntactic congruence of U is of course called the syntactic monoid of U.

Instead of starting with a morphism recognizing a set $U \subset A^N$, one may start with a finite automaton

$$\mathcal{A} = (Q,I,T)$$

and consider the set $U \subset A^N$ of all the infinite words α which are the label of a path starting in I and going infinitely often through the set T of terminal states. This is often called a Büchi automaton and one says that it recognizes U. This is the usual definition of recognizability for subsets of A^N. It is of course equivalent to the definition given previously. One may indeed consider the morphism

$$\phi : A^* \to M$$

where M is the monoid of transition of the automaton $\mathcal{A}$. The image under ϕ of a word $w \in A^*$ is the set of all pairs $(p,q) \in QxQ$ such that there exists a path from p to q with label w. The classical construction performed to recognize the complement of U is the following : one considers the relation $\tau(w)$ defined as the set of pairs $(p,q) \in Q \times Q$ such that there exists a path from p to q with label w and which moreover use at least one element of T. One then considers for a word w the relation $\psi(w)$ over two copies of Q with matrix representation

$$\psi(w) = \begin{bmatrix} \phi(w) & \tau(w) \\ 0 & \phi(w) \end{bmatrix}$$

Then ψ is obviously a morphism and it is easy to verify that ψ saturates the set U recognized by the automaton A. In particular, ψ recognizes the complement of U.

It is worthwhile to observe that the morphism ψ can be obtained by representing the monoid $M' = M \diamond M$ by binary relations. In fact, let M be any monoid of relations over a set Q and let T be a subset of Q. For each element $m' = (m,\rho,n)$ of M' define a relation $\Theta(m')$ over two copies of Q defined by its matrix representation as

$$\Theta(m') = \begin{bmatrix} m & r \\ 0 & n \end{bmatrix}$$

where r is the relation over Q defined by $(p,q) \in r$ iff there exists a $t \in T$ and $(u,v) \in \rho$ such that $(p,t) \in u$ and $(t,q) \in v$. It is not difficult to verify that Θ is in fact a morphism.

EXAMPLE 1.3 Coming back again to the example concerning the parity of the number of occurrences of b, we now use the automaton shown below :

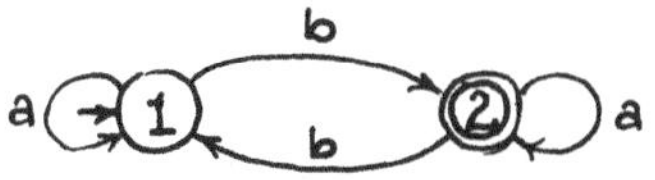

with $Q = \{1,2\}$, $I = \{1\}$, $T = \{2\}$. The above construction leads to the representation by binary relations defined by

$$\psi(a) = \begin{bmatrix} 1 & 0 & 0 & 0 \\ 0 & 1 & 0 & 1 \\ 0 & 0 & 1 & 0 \\ 0 & 0 & 0 & 1 \end{bmatrix} \qquad (b) = \begin{bmatrix} 0 & 1 & 0 & 1 \\ 1 & 0 & 1 & 0 \\ 0 & 0 & 0 & 1 \\ 0 & 0 & 1 & 0 \end{bmatrix}$$

which is a faithful representation by relations of the monoid defined in Example 1.2. □

An other way of using automata on infinite words involves the notion of a Muller automaton. This is an automaton

$$A = (Q,I,\mathcal{T})$$

with $\mathcal{T}$ a family of subsets of Q. An infinite word $\alpha \in A^{\mathbb{N}}$ is recognized by the automaton if there is a path with label α such that the set of states appearing infinitely often is an element of the family $\mathcal{T}$. It is not difficult to show that this new notion of recognizability coincides with the previous ones.

A fundamental theorem, due to R. McNaughton, says that for each recognizable set $U \subset A^N$ there exists a deterministic Muller automaton recognizing U. This can be rephrased in the following way : for a set $X \subset A^*$, denote by

$$U = \vec{X}$$

the limit of X which is the set of infinite words $\alpha \in A^N$ having an infinity of left factors in X. Then, in an equivalent way, McNaughton's theorem says that the family $Rec(A^N)$ is equal to the boolean closure of the family of limits of sets in $Rec(A^*)$:

$$Rec(A^N) = (\overrightarrow{Rec(A^*)})^B$$

The link between the two formulations of McNaughton's theorem is the fact that U can be recognized by a deterministic Buchi automaton if and only if it is in $\overrightarrow{Rec(A^*)}$.

Several proofs of McNaughton's theorem have been given (see [10], [7], [18]), including one by Schutzenberger [15] which uses the following lemma. One starts with a simple recognizable set $U = YZ^\omega$ with

$$Y = \phi^{-1}(m), \quad Z = \phi^{-1}(e)$$

Let $W = \{w \in A^* \mid Z \cap A^* w A^* = \emptyset\}$, $V = W - WA^+$ and $R = A^*V$. Let also $D = Z - ZA^+$.

PROPOSITION 1.4 One has

(1) $Z^\omega = \overrightarrow{ZD}$

(2) $YZ^\omega = \overrightarrow{YZD} - \vec{R}$.

A natural question left open by McNaughton's theorem is how one can decide whether a recognizable set U is a limit. The answer is given by the following result, due to H. Landweber [8].

THEOREM 1.5 Let $\mathcal{A} = (Q,i,\mathcal{T})$ be a deterministic Muller automaton recognizing a set $U \subset A^N$. Then

$$U \in \overrightarrow{Rec(A^*)}$$

iff U is still recognized by the Muller automaton $(Q,i,\hat{\mathcal{T}})$ where $\hat{\mathcal{T}}$ is the family of subsets of Q containing an element of $\mathcal{T}$.

It is not difficult to prove the "only if" part of the theorem. To prove the converse, one considers a deterministic Muller automaton $\mathcal{A} = (Q,i,\mathcal{T})$ with $\mathcal{T} = \hat{\mathcal{T}}$ recognizing a set U. One then builds a new automaton

$$\mathcal{B} = (Q \times \mathcal{P}(Q), (i,\emptyset), R)$$

where $R = \{(q,\emptyset) \mid q \in Q\}$ and with the transitions

$$(q,S).a = \begin{cases} (q.a,\ \emptyset) & \text{if } S+q.a \in T \\ (q.a,\ S+q.a) & \text{otherwise} \end{cases}$$

This automaton is deterministic and obviously recognizes (in Büchi's sense) the same set U as the Muller automaton A.

EXAMPLE 1.4 Consider again the automaton of Example 1.3

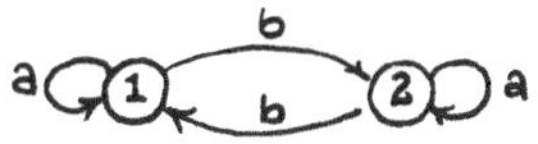

but this time used as a Muller automaton with $T = \{\{1,2\}\}$. Since $T = \hat{T}$ it recognizes a set which is a limit and it is in fact the set

$$U = (a^*b)^\omega = \overrightarrow{(a+b)^*b}$$

of words having an infinite number of occurrences of b. The previous construction leads to the automaton

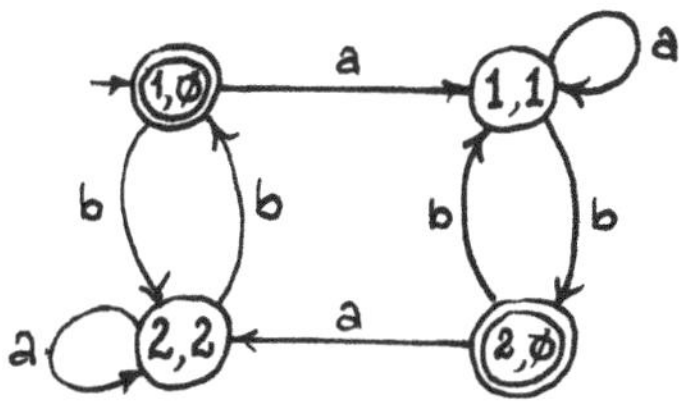

giving a strange way of testing an infinite number of b's. □

Two-way automata on infinite words have been considered by J.P. Pécuchet [13]. He has proved that they recognize the same sets as one-way automata, therefore generalizing Shepherdson's result for two-way automata on ordinary words.

2. APERIODIC SETS

The family of <u>star free</u> subsets of A^* denoted by $Sf(A^*)$ is defined as the closure under set-product and boolean operations of the powerset of the alphabet A. Further, a monoid M is said to be <u>aperiodic</u> if there exists an integer $n \geq 0$ such that the identity

$$x^{n+1} = x^n$$

holds for all x in M. According to a classical theorem of Schützenberger (see [7] or [8] or [15]) a recognizable set $X \subset A^*$ is star free if and only if its syntactic monoid is aperiodic. This proves in particular that it is decidable wether a recognizable set is star free. For sets of infinite words, one has the following result, conjectured by R. Ladner and proved by W. Thomas in [17].

THEOREM 2.1 Let U be a recognizable subset of A^N. The following conditions are equivalent :

(i) $U = \bigcup_{i=1}^{n} X_i Y_i^{\omega}$ with $X_i, Y_i^* \in Sf(A^*)$

(ii) U is a boolean combination of limits of star free sets :

$$U \in (\overrightarrow{Sf(A^*)})^B$$

(iii) U can be obtained from $\emptyset$ by a finite number of boolean operations and product on the left by an element of $Sf(A^*)$.

We shall denote by $Sf(A^N)$ the family of star free subsets of A^N which is defined by the above equivalent conditions.

There is a striking analogy between McNaughton's theorem and the above one. This is emphazized in [14] where a result having both ones as particular cases is proved. The framework used to state the result is that of varieties of semigroups.

The proof of Theorem 2.1 can be essentially handled using Proposition 1.4. Indeed, let F_n be the family of sets satisfying condition (ni). One has trivially

$$F_2 \subset F_1^B$$

Now Proposition 1.4 shows that $F_1 \subset F_2$, whence $F_1 = F_2$. Finally, $F_3 \subset F_1$ is clear since F_1 is closed under boolean operations and $F_2 \subset F_3$ comes from the fact that the complement of a limit can be written as a finite union of sets of the form $Y(A^N - ZA^N)$.

We shall now prove the following result which implies in particular that the property of being star free is decidable within $Rec(A^N)$.

THEOREM 2.2 A recognizable set $U \subset A^N$ is star free iff its syntactic monoid is aperiodic.

Proof : Suppose first that U is star free. By condition (i) there exists a morphism ϕ from A^* onto an aperiodic monoid M recognizing U. Now, by Proposition 1.3, the morphism $\phi \diamond \phi$ saturates U. But it is well known (see for instance [7]) that $M \diamond M$ is aperiodic if M is. Therefore the syntactic monoid of U is aperiodic.

Conversely, if U is recognized by a morphism onto an aperiodic monoid, it satisfies certainely condition (i). □

Star free subsets of A^N have an interesting characterization in terms of logic. Let indeed $L(A)$ be the set of first order formulas on N with a relation $<$ and a set of predicates $\Pi_a(i)$ for $a \in A$. An infinite word $\alpha \in A^N$ satisfies the formula $\phi \in L(A)$ written $\alpha \models \phi$, if the formula is true when the variables are interpreted in N, the relation $<$ as the usual ordering on the integers and the predicate $\Pi_a(i)$ as : the i-th letter of α is an a.

For a formula $\phi \in L(A)$, we denote by

$$M(\phi) = \{\alpha \in A^N \mid \alpha \models \phi\}$$

the set of words satisfying the formula. We say that the set $M(\phi)$ is <u>defined</u> by formula ϕ. The following result has been proved by W. Thomas [17].

THEOREM 2.3 <u>A set</u> $U \subset A^N$ <u>is star free iff it can be defined by a first order formula</u> $\phi \in L(A)$.

An analogous result for finite words also holds and it was one of the motivations that lead R. Mc Naughton to consider star free sets.

EXAMPLE 2.1 Let $A = \{a,b\}$ and

$$U = (a+b)^* a^\omega$$

then U is star free according to condition (i) of Theorem 2.1. An expression in the form of condition (ii) is

$$U = (a+b)^\omega - a^* b (a+b)^\omega$$

The syntactic monoid of U is composed of three elements $1, \alpha = \phi(a^+)$, $\beta = \phi(A^* b A^*)$. This gives the following decomposition of U into simple recognizable sets :

$$U = (a+b)^* b a^\omega + a^\omega$$

A simple formula that defines U is

$$\exists x \; (\forall y \; ((x<y) \rightarrow \Pi_a(y))). \qquad \square$$

3. TWO-SIDED INFINITE WORDS

We consider now two-sided infinite words, that is to say elements $\alpha \in A^Z$ written

$$\alpha = \ldots \alpha_{-1} \alpha_0 \alpha_1 \alpha_2 \ldots$$

with $\alpha_i \in A$. Let $\mathcal{A} = (Q,I,T)$ be a finite automaton. THe set of words $\alpha \in A^Z$ <u>recognized</u> by $\mathcal{A}$ is, by definition, formed by the labels of two-sided infinite paths in $\mathcal{A}$ going infinitely often through I on the left and infinitely often through T on the right. A set $W \subset A^Z$ is said to be recognizable, written $W \in Rec(A^Z)$, if it can be recognized by a finite automaton.

Observe an essential feature of this definition : a recognizable set is invariant under the shift. This is the point which makes both the interest of the objects and also the difficulty in their study since it adds a new phenomenon which was not present (nor possible to define) with one-sided infinite words. To see the relationship with the one sided case, we introduce the notation $[U,V]$ for $U,V \subset A^N$ to represent the set of all $\alpha \in A^Z$ such that

$$\alpha_0 \alpha_{-1} \alpha_{-2} \ldots \in U, \; \alpha_0 \alpha_1 \alpha_2 \ldots \in V$$

whereas $\widetilde{UV}$ is the closure of $[U,V]$ under the shift. We also note ${}^{\omega}X$ for the reversal of X^{ω}. Then it is not difficult to prove that the three following conditions are equivalent :

(i) $W \in \mathrm{Rec}(A^{\mathbb{Z}})$

(ii) W is shift invariant and is a finite union of sets $[U,V]$ for $U,V \in \mathrm{Rec}(A^{\mathbb{N}})$

(iii) W is a finite union of sets ${}^{\omega}XYZ^{\omega}$ for $X,Y,Z \in \mathrm{Rec}(A^{*})$.

Condition (ii) can be used to prove easily that the family $\mathrm{Rec}(A^{\mathbb{Z}})$ is closed under complementation.

The notion of a morphism $\phi : A^{*} \to M$ that recognizes or saturates a set $U \subset A^{\mathbb{Z}}$ extends without difficulty to the two-sided case. The main problem is the extension of McNaughton's theorem.

For this, we first introduce the notion of <u>two-sided limit</u> : for $X \subset A^{+}$, the two-sided limit of X, denoted by $\overleftrightarrow{X}$ is the set of $\gamma \in A^{\mathbb{Z}}$ such that

$$\gamma_{-n}\gamma_{n+1}\cdots\gamma_{m-1}\gamma_{m} \in X$$

for infinitely many $n \geq 0$ and infinitely many $m \geq 0$. The following result, proved in [12] by M. Nivat and myself, generalizes McNaughton's theorem to the two-sided case.

THEOREM 3.1 <u>The family</u> $\mathrm{Rec}(A^{\mathbb{Z}})$ <u>is the boolean closure of the two-sided limits of elements of</u> $\mathrm{Rec}(A^{*})$:

$$\mathrm{Rec}(A^{\mathbb{Z}}) = (\overleftrightarrow{\mathrm{Rec}(A^{*})})^{B}$$

The proof of this result relies on a generalization to the two-sided case of Proposition 1.4. One starts with a simple recognizable set

$$W = {}^{\omega}XYZ^{\omega}$$

with $Y = \phi^{-1}(m)$ and Y,Z the minimal generating sets of $\phi^{-1}(e)$, $\phi^{-1}(f)$ for two idempotents $e,f \in \phi(A^{*}) = M$. One then considers

$$G = X - A^{+}X, \quad D = Z - ZA^{+}$$

and two sets $L = KA^{*}$, $R = A^{*}J$ where K,J are the basis of the two sided ideals which are the complements of the set of factors of X^{*},Z^{*} respectively.

PROPOSITION 3.2 <u>One has the equality</u>

$${}^{\omega}XYZ^{\omega} = \overleftrightarrow{GXYZD} - \overleftrightarrow{L} - \overleftrightarrow{R}.$$

The relationship with deterministic automata goes through the definition of a new type of two-sided automaton or, equivalently, through the following notion : define the family $Det(A^Z)$ of *deterministic* sets as that of shift-invariant subsets of A^Z which are finite unions of sets of the form :

$$[U,V]$$

with $U,V \in Det(A^N) = Rec(A^*)$. One then has the following result proved by D. Beauquier [3].

PROPOSITION 3.3 *The family* $Det(A^Z)$ *is equal to that of two sided limits of recognizable sets of* A^* :

$$Det(A^Z) = \overleftrightarrow{Rec(A^*)}$$

EXAMPLE 3.1 Let $A = \{a,b\}$ and consider

$$W = {}^{\omega}a(a+b)^*b^{\omega}$$

which can be recognized by the automaton

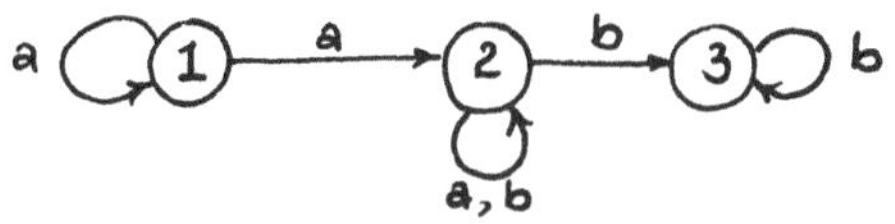

An expression according to Theorem 3.1 is

$$W = \overleftarrow{A}^* - \overleftrightarrow{A^*a} - \overleftrightarrow{bA^*}. \qquad \square$$

An interesting result on two-sided infinite words has been obtained by L. Compton [5]. To state it, say that $\alpha \in A^Z$ is *generic* if each word $w \in A^*$ appears an infinity of times on the right and on the left in α. Then, one has the following result [5] :

THEOREM 3.4 *A recognizable subset* W *of* A^Z *either contains all generic words or none of them.*

This result says, in other terms, that it is not possible to distinguish two generic words using a finite automaton.

4. UNAMBIGUITY

Since it is not always possible to restrict one's attention to deterministic Büchi automata, the problem of unambiguity in the recognition of infinite words is more difficult than in the case of finite ones. However deterministic Muller automata can be used to obtain unambiguous representations as follows.

We say that a family (X_i,Y_i), $1 \leq i \leq n$ of sets in A^* is *unambiguous* if for each infinite word $\alpha \in A^N$ there is at most one decomposition in the form

$$\alpha = xy_1y_2\ldots$$

with $x \in X_i$, $y_j \in Y_i$ and $1 \leq i \leq n$. This means that the set

$$U = \bigcup_{i=1}^{n} X_i Y_i^{\omega}$$

is described in this way by an unambiguous expression. The following result is due to A. Arnold [1] :

THEOREM 4.1 For each recognizable set $U \subset A^N$ there exists an unambiguous family (X_i, Y_i) defining U, that is such that $U = \bigcup_{i=1}^{n} X_i Y_i^{\omega}$.

An other kind of unambiguity result was suggested to me by considering two-sided infinite words. The basic idea is to use automata on infinite words that start reading "at infinity". More precisely, let us say that an automaton

$$\mathcal{A} = (Q, I, T)$$

is codeterministic if for any two arrows $(p,a,q),(p',a,q')$ labeled by the same letter, $q = q'$ implies $p = p'$. The following result was obtained by D. Beauquier and myself [4]. It was in fact, previous to our publication, announced by A. Mostowski in [11] but we have not been able to understand his proof.

THEOREM 4.2 Any recognizable set in A^N can be recognized by a codeterministic automaton.

The construction used to build a codeterministic automaton is the following : let $U = YZ^{\omega}$ be a simple recognizable set with $Y = \phi^{-1}(m)$, $Z = \phi^{-1}(e)$ and e an idempotent. Consider a codeterministic automaton $\mathcal{A}$ that simultane ously recognizes YZ and GZ with $G = Z - A^+Z$. Let I be the set of initial states for YZ and J that of GZ. Then one can prove that the codeterministic automaton $\mathcal{A}$ having I as initial states and J as terminal states recognizes YZ^{ω}.

EXAMPLE 4.1 Let $A = \{a,b\}$ and consider $U = (a^*b)^{\omega}$. Then the following codeterministic automaton recognizes U :

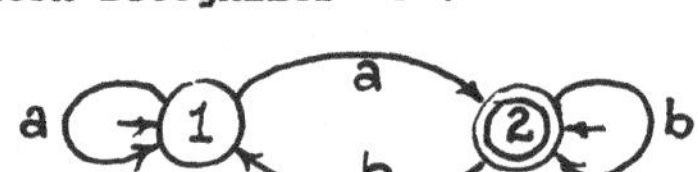

The behaviour of this automaton can be understood as follows : if, in state 1, the input is an a, there it has to make a transition to state 2 if the next input is going to be a b ; otherwise it would stop on the next b. In terms of the above construction, one has

$$Y = 1, \quad Z = A^*b, \quad G = b$$

A codeterministic automaton recognizing YZ and GZ is shown below

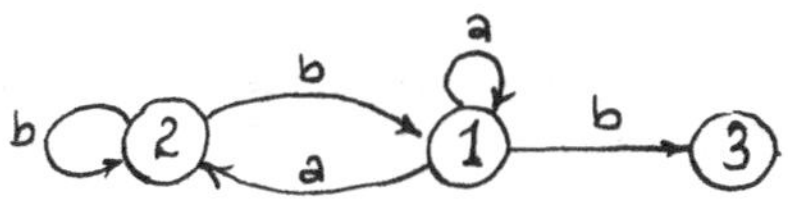

With the above notations, one has I = {1,2}, J = {1}. The automaton obtained by using I as initial states and J as terminal states is almost identical to the above one, up to removal of the useless state 3. □

BIBLIOGRAPHY

[1] Arnold, A., 1983, Rational ω-languages are non-ambiguous, Theoret. Comput. Sci., 26, 221-224.

[2] Arnold, A., 1984, A syntactic congruence for rational ω-languages, to appear in Theoret. Comput. Sci.

[3] Beauquier, D., 1984, Bilimites de langages reconnaissables, to appear in Theoret. Comput. Sci.

[4] Beauquier, D., Perrin, D., 1984, Automates codeterministes sur les mots infinis, to appear.

[5] Compton, L., 1984, in Progress in Combinatorics on Words, Academic Press.

[6] Büchi, J.R., 1962, On a decision method in restricted second order arithmetic, in Logic, Methodology and Philosophy of Science, (Proc. 1960 Internat. Congr.), Stanford University Press, Stanford, Calif., 1-11.

[7] Eilenberg, S., 1974, Automata, Languages and Machines, Vol. A, Academic Press, New York, Vol. B, 1976.

[8] Lallement, G., 1979, Semigroups and Combinatorial Applications, Wiley.

[9] Landweber, L.H., 1969, Decision problems for ω-automata, Math. Syst. Theory, 3, 376-384.

[10] McNaughton, R., 1966, Testing and generating infinite sequences by a finite automaton, Information and Control, 9, 521-530.

[11] Mostowski, A., 1982, Determinancy of sinking automata on infinite trees and inequalities between various Rabin's pair indices, Information Processing Letters, 15, 159-163.

[12] Nivat, M., Perrin D., 1982, Ensembles reconnaissables de mots biinfinis, Proc. 14th ACM Symp. on Theory of Computing, 47-59.

[13] Pécuchet, J.P., 1983, Automates boustrophédons et mots infinis, à paraitre dans Theoret. Comput. Sci.

[14] Perrin D., Variétés de langages et mots infinis, C.R. Acad. Sci. Paris, 295, 595-598.

[15] Pin, J.E., Variétés de langages formels, Masson, 1984.

[16] Schützenberger, M.P., 1972, Sur les relations rationnelles fonctionnelles, in Automata, Languages and Programming (M. Nivat ed.) North Holland, 103-114.

[17] Thomas, W., 1979, Star free regular sets of ω-sequences, Inform. Control, 42, 148-156.

[18] Thomas, W., 1981, A combinatorial approach to the theory of ω-automata, Inform. Control, 48, 261-283.

[19] Thomas, W., 1982, A hierarchy of sets of infinite trees, in Theoretical Computer Science, Springer Lecture Notes on Comput. Sci., 145.

VLSI ALGORITHMS AND ARCHITECTURES*

Franco P. Preparata
Coordinated Science Laboratory
University of Illinois at Urbana-Champaign
Urbana, IL 61801/USA

ABSTRACT

In this paper we consider the relationship between algorithms and parallel VLSI architectures. Starting from the premise that VLSI is the natural habitat for parallel algorithmics, we outline the desirable features of VLSI architectures. Next we discuss the notion of algorithmic paradigm, as the data transfer pattern of a class of specific algorithms. The recursive combination paradigm (applicable to merging, sorting, FFT, permutation, etc.), is naturally mapped to the awkward binary cube architecture. Thus we analyze four viable architectures - the shuffle-exchange, the linear array, the mesh, and the cube-connected-cycles - which emulate the binary cube. Finally, we illustrate the mechanisms of pipelining, pleating, and mixing, which play a significant role in matching algorithms and architectures.

1. Introduction

Algorithm design, concurrency of operation, and device technology are the major avenues along which computer science and engineering have traditionally pursued their central goal of increasing computational throughput.

Algorithm design seeks to develop better procedures and data structures that will reduce the time to solve given problems on a given computing system.

Concurrency of operation seeks to achieve a better utilization of available hardware by overlapping activities which use disjoint parts of the computing system. This principle has been adopted since the early days, typically in overlapping the activities of the CPU and of peripheral channels. Within the framework of the von Neumann machine, concurrency is essentially the exposure of parallelism in sequential programs. Although this endeavor is significant and worth pursuing, the fullest exploitation of concurrency is the design of parallel algorithms, that is, algorithms to be executed by a system of cooperating processors. This activity could be aptly

*This work has been supported in part by the National Science Foundation under Grant MCS-81-05552.

called parallel algorithmics.[1] Of course, this approach postulates the deployment of a very large number of processors, and until recently has been by and large the domain of academic speculation. What has drastically changed the landscape, giving concreteness and impetus to parallel algorithmics, has been the far reaching technological revolution represented by Very-Large-Scale-Integration (VLSI).

Indeed, since the invention of the transistor, the development of device technology has been a remarkable success story. Since its very infancy, the transistor has been "integrated", i.e., a device whose portions are obtained by modifying the physical properties of a monolithic piece of material. Thus, transistor technology has evolved on two fronts: amelioration of switching speed and increase of integration. The first line has been characterized by steady progress, but the revolution has occurred in the domain of integration, which after a rather unimpressive start, has now reached mind-boggling capabilities. These capabilities far exceed the immediate needs of system architects. While normally the development of engineering disciplines is technology-bound, today the bottleneck in the computing field is of a system-theoretical nature.

VLSI makes a reality of the long-wished possibility to deploy a very large number of cooperating processors. On the other hand, the cost of VLSI - both of design and of fabrication - is roughly proportional to the chip area. Thus, best utilization of the technology occurs when each portion of the chip uninterruptedly participates to computation. This is a paraphrase of concurrency, that is, VLSI offers the natural habitat for parallel algorithmics.

This paper is organized as follows. In Section 2 we contrast the models of sequential and parallel processing, and discuss the emerging difficulties. We also review the VLSI model of computation and introduce the notion of broad-purpose architectures. In Section 3 we explore the relationship between algorithmic paradigms and their architectures, with a case study of the recursive combination paradigm. Finally, in Section 4 we illustrate the interesting design mechanisms called pipelining, pleating, and mixing.

2. Models of Sequential and Parallel Computation

The characteristic feature of the traditional sequential processor (the von Neumann paradigm, or its abstraction called Random-Access-Machine (RAM) [1]) is a single CPU-memory channel, which allows the transmission of one computer word per use. This feature - commonly referred to as von Neumann's bottleneck - is at the same time a severe constraint on data communication and a powerfully simplifying element in algorithm design and analysis. The usefulness of the RAM model of computation stems from its being applicable to any problem, i.e., from its being general-purpose.

Parallel computation aims at overcoming the von Neumann's bottleneck. Ideally,

---

(1) We do not consider in this paper a third and very important "brand" of concurrency, i.e., distributed computing (computer networks).

one wishes to obtain a model of parallel computation with a role analogous to that of the RAM in sequential computation. We begin with some definitions:

Definition 1. A (computing) module is a combinational or sequential boolean machine. [A module is an atomic system constituent not analyzable in its internal structure.]

Definition 2. A parallel architecture is a directed graph $G = (V,E)$. [V is a set of computing modules and E is a set of arcs, each of which provides the necessary data path between the modules it connects. A parallel architecture is also called a "computation graph".]

Definition 3. Given a set of variables $X = \{x_1,x_2,\ldots,x_m\}$ (each stored in a module) and a set of constants K, a parallel step σ is a set of assignments of the form $x_i := f_i(X \cup K)$, $i = 1,\ldots,m$. A parallel computation is a sequence of parallel steps $\sigma_1,\sigma_2,\ldots,\sigma_p$. A parallel algorithm is a procedure executed as a parallel computation.

Definition 4. A parallel architecture supports a given parallel algorithm $\mathcal{A}$ if it provides an arc for each data transfer occurring in any execution of $\mathcal{A}$. [For example, if steps σ_ℓ of $\mathcal{A}$ contains the assignment $x_i := x_j + x_k$, there is an arc to the module storing x_i from each of the modules storing x_j and x_k, respectively.]

It follows that a truly general-purpose parallel computer is one that supports any parallel algorithm, that is, a maximally connected computation graph. Algorithms for significant problems, however, may not require arbitrary parallel steps, but only a special subset thereof. In addition, one may expect that the cost of a parallel architecture on a given set of modules be an increasing function of the intricacy of its interconnection. These costs are to be quantified in a technological framework. We shall consider as a framework the VLSI model of computation [2,3,4], which hinges crucially on the fact that processors and their interconnection are realized in the same physical medium. So we have:

Definition 5. In the synchronous VLSI model of computation, a chip is a planar embedding (layout) of a computation graph so that:

1. Each constituent of the graph (module or wire) is assigned a region of the plane;
2. Any point of the plane is assigned to at most $\nu \geq 2$ constituents;
3. Each constituent has minimum size requirements;
4. The computation graph is a synchronous boolean machine [i.e., transmission of a signal on a wire takes unit time τ].

In the above definition, the size requirements of the constituents, the values of ν (number of "layers") and τ are fixed within any given technology. A basic cost parameter is the area A of the chip, which can be measured, for example, as the area of the smallest rectangle enclosing the layout of the computation graph. The notion of chip area, which reflects the cost of both processors and wires, captures the two fundamental aspects of computing - processing and communication - and supersedes the traditional component count.

Given a computation graph G and a problem $\mathcal{P}$, one may design an algorithm $\mathcal{A}$ to solve $\mathcal{P}$ on G. The number of time units used by the execution of $\mathcal{A}$ gives the *computation time* T of $\mathcal{A}$ on G: T is the second basic cost parameter, relating an algorithm to a computation graph. The two parameters A and T embody a natural resource trade-off: indeed, on the basis of arguments on information transfer and accessibility, two measures of complexity, intrinsic to the synchronous VLSI model, are the products AT^2 and AT. In the sequel, we shall refer to the AT^2 measure alone. This leads to the following definition:

Definition 6. A computation graph G laid out in area A is *optimal* for a given problem $\mathcal{P}$ if it supports an algorithm $\mathcal{A}$ for $\mathcal{P}$ running in time T, so that AT^2 is of the same order as the theoretical minimum.

This definition provides a focus to VLSI computation research. The design objective is clearly the discovery of "versatile" architectures, i.e., architectures that are optimal for several problems. However, the pair (architecture, algorithm) exhibits an undesirable circularity (the nature of the algorithm suggests the architecture, but the architecture specification precedes the algorithm specification): this is a basic difficulty of parallel computation. Fortunately, today the subtle relation between algorithms and architectures is less inscrutable. While a few years ago it was not uncommon to suspect that each problem may have demanded its own specific (special-purpose) architecture, today there is growing consensus that a few simple and highly structured architectures are suitable for the solution of large classes of problems. In other words, there appears to be a small set of very versatile interconnections, which can be aptly called *broad-purpose architectures*. A viable system concept, accommodating broad-purpose architectures, is a direct evolution of the von Neumann paradigm (Figure 1). A host processor executes a sequential program by controlling the data flow between memory and processing units. The latter, rather than familiar ALUs, are now large parallel computing arrays, each capable of executing extremely complex commands, such as SORT, INVERT-MATRIX, FFT, etc.; correspondingly, the data paths must have a bandwidth adequate to sustain the data rate required by the arrays. It should be noted that the concept is not new,

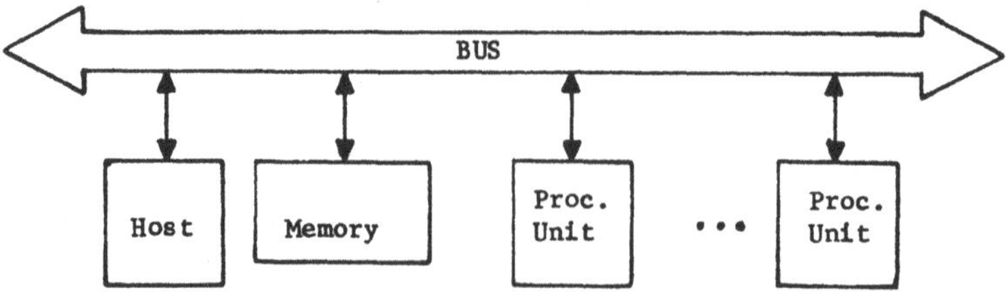

Figure 1. System concept. "Processing units" are, in turn, parallel architectures.

since conventional ALU's could be viewed as small-scale parallel systems for the execution of arithmetic algorithms. In order to enrich the instruction repertoire of the system, each of these computing arrays should be a broad-purpose architecture. The corresponding computation graphs should satisfy the following qualitative requirements, suggested by the layout constraints and by general design principles:

Requirement 1. (Degree-boundedness). The degree of each node should be bounded by a small number (4 or 6, for example);

Requirement 2. (Size-independence). The structure of the deployed modules should be independent of the system size (i.e., the total number N of modules);

Requirement 3. (Modularity). The layout of the graph should be modular, i.e., obtainable by simply combining layouts of analogous graphs of smaller size.

These features are exhibited, with minor exceptions, by the broad-purpose architectures to be described in the next section. We shall also make use of a restricted notion of "emulation" of an architecture by another architecture, as expressed by the following definition:

Definition 7. Given two N-node architecture graphs $G_1 = (V_1, E_1)$ and $G_2 = (V_2, E_2)$, we say that G_2 emulates G_1 (denoted by $G_2 \propto G_1$) for a given algorithm $\mathcal{A}$ if the following holds:

(i) At each step in the execution of $\mathcal{A}$ on both G_1 and G_2, there is a one-to-one correspondence between sets V_1 and V_2 (i.e. $v_1 \in V_1$ and $v_2 \in V_2$ correspond to each other if they contain the same data);

(ii) for each one-step global data transfer on G_1, there is an algorithm supported by G_2 accomplishing the corresponding data transfer on G_2.

Emulation is, obviously, a reflexive and transitive relation.

3. The Recursive Combination Paradigm and Its Architectures

Unfortunately, we are still far from elucidating the relationship between VLSI architectures and algorithm design. What we know today is a collection of several originally isolated results, from which, however, some common patterns and principles begin to emerge; the seeds of a future theory.

The ideal objective is to be able to classify algorithms according to paradigms, determined exclusively by their communication pattern. It would then be possible to match paradigms and architectures in a natural way. Some progress has been made in this direction, and we shall illustrate it by the case study of a significant paradigm, recursive combination, described below.

In the recursive combination paradigm, both input and output are each N-component vectors of data items (operands), and N is assumed to be of the form $N = 2^d$ for simplicity. The problems to be considered are those solved by a call (ALGORITHM(0, N-1)) of the recursive procedure ALGORITHM($\ell, \ell+m-1$) given below (the input is initially stored in an array T[0:N-1], which after the execution of ALGORITHM will contain the output):[2]

```
proc    ALGORITHM(ℓ,ℓ+m-1)
begin   if (m > 2) then ALGORITHM(ℓ,ℓ+m/2-1), ALGORITHM(ℓ+m/2,m-1);
        foreach i : ℓ ≤ i < ℓ + m/2 par do (T[i],T[i+m/2]) ← OPER (T[i],T[i+m/2])
end
```

Here OPER is an operation (frequently referred to as "butterfly") which updates the values of T[i] and T[i+m/2] and may depend upon both m and i. Although extremely simple, ALGORITHM is the paradigm of several extremely significant computations. Specifically, as it has been shown in [5,6] that there are algorithms for such fundamental problems as merging, sorting, Fast-Fourier Transform, convolution, integer multiplication, permutations, etc., which are either instances of the above scheme or simple combinations of such instances.

To determine an architecture which naturally supports ALGORITHM, it is first convenient to transform the previously given recursive version of it into its iterative counterpart. ($Bit_i(j)$ denotes the coefficient of 2^i in the binary expansion of integer j.)

```
proc  ALGORITHM(0,N-1)
begin for i ← 0 until d-1 do
      foreach j: 0 ≤ j < N
        par do if (bit_i(j)=0) then (T[j],T[j+2^i]) ← OPER_i([T[j],T[j+2^i])
end
```

We now identify each cell of the array T[0:N-1] with a node of a graph and draw an edge between any two nodes involved in a data transfer specified by the procedure. Thus any two nodes corresponding to the pairs $\{(T[j],T[j+2^i]) : 0 \le j < N,\ 0 \le i < d\}$ are connected by an edge. It is immediately recognized that the resulting graph (Figure 2) has the topology of the multi-dimensional binary cube (briefly cube).

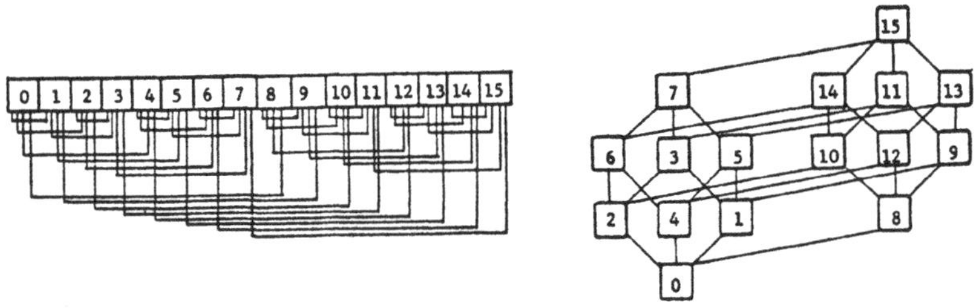

Figure 2. The binary cube naturally supports ALGORITHM.

---

(2) Here and hereafter, algorithms will be presented in modified Pidgin Algol, which contains the additional construct: foreach m: <cond(m)> par do <action>, which specifies an action to be executed in parallel on all positions (cells or modules) m for which a certain condition holds.

If we examine the cube from the point of view of the reasonable requirements earlier set forth for architecture graphs, we see that none of them is satisfied for sufficiently large number of dimensions. Thus the binary cube, while a computationally fundamental communications structure, is not per se a viable architecture graph.

Upon closer examination, however, we note that ALGORITHM does not strictly make full use of the communication capabilities of the cube. Indeed, referring to the above iterative version of ALGORITHM, we realize that cube edges are not used in a capricious sequence. Rather, they are grouped by "dimension" (there are d dimensions), and dimensions are used in succession according to a fixed schedule, as specified by the for-loop. This succession, called ASCEND, is denoted $E_0, E_1, \ldots, E_{d-1}$. The observation about the use of edges - one dimension at a time, in a rigid sequence - is of crucial importance in seeking viable substitutes for the cube graph, i.e. emulators of the cube according to Definition 7.

In a graph to be used for ALGORITHM, we call an exchange edge an edge used to execute operation OPER. Similarly, we call a transfer edge a (possibly unidirectional) edge used to transfer data between a pair of vertices. Clearly, in the d-dimensional cube, all $d2^{d-1}$ edges are exchange edges and there is no transfer edge.

We note, however, that at any step only 2^{d-1} exchange edges are in use, the others being idle. This observation is at the basis of all graphs so far proposed to emulate the cube; indeed at most 2^d exchange edges are provided, and the graph is completed with transfer edges, so that all data pairs interacting in OPER (i.e., pairs of the type $(T[j],T[j+2^i])$ for $0 \le j < N$, $0 \le i < d$, $bit_i(j) = 0$) are correctly brought to reside for the duration of one step in a pair of nodes connected by an exchange edge. We shall now discuss these graphs in detail. We mention that ALGORITHM on an N-component vector satisfies the lower bound $AT^2 = \Omega(N^2)$ [2].

3.1. The Shuffle-Exchange Graph

Let V[j] denote the node initially containing T[j], $0 \le j < 2^d$. The edge set of the shuffle-exchange graph is as follows

exchange edges (undirected): $\{<V[2j],V[2j+1]>: 0 \le j < 2^{d-1}\}$

transfer arcs (directed): $\{(V[j],V[j2^{(d-1)} \bmod (2^d-1)]): 0 < j < 2^d-1\}$.

Note that, if we reverse the direction of all transfer arcs, we obtain the set $\{(V[j],V[2j \bmod (2^d-1)]): 0 < j < 2^d-1\}$. Then, it is readily realized from elementary number theory that transfer arcs form cycles, whose lengths divide d. After loading V[j] with T[j], $0 \le j < N$, execution of ALGORITHM proceeds as follows:

```
proc  SE-ALGORITHM
begin for i ← 0 until d-1 do
      foreach (exchange edge) par do OPER_i;(*exchange step*)
      foreach (transfer edge) par do transfer; (*shuffle step*)
end
```

To convince ourselves that this is a correct emulation, we notice that at the (i+1)-st step nodes V[2j] and V[2j+1], which are exchange-connected, respectively contain data items $T[2j.2^i \bmod (2^d-1)]$ and $T[(2j+1)2^i \bmod (2^d-1)] = T[2j.2^i \bmod (2^d-1)+2^i]$, as specified by ALGORITHM.

As to the time performance of the shuffle-exchange graph,(3) we note that each shuffle step and exchange step use a fixed amount of time. Hence the running time of ALGORITHM is still O(logN), with an insignificant loss with respect to the execution on the cube.

The shuffle-exchange graph complies with Requirements 1 (the degree of each node is at most 3) and 2. However, since each value of d yields a specific cycle structure, the layout is not modular (Requirement 3), but can be realized with an optimal area $A = \theta((N/\log N)^2)$ [8].

3.1. The Linear Array

With the usual notation, we have the following edge set:

exchange edges (undirected): $\{<V[2j],V[2j+1]>: 0 \leq j < 2^{d-1}\}$

transfer edges (undirected): $\{<V[j],V[j+1]>: 0 \leq j < 2^d\}$.

Thus the exchange edges are a subset of the transfer edges, and the topology is just that of an N-node chain. The transfer edges must be used to achieve the correct arrangements of data postulated by the various exchange steps of ALGORITHM. Specifically, referring to the shuffle-exchange graph, all we need to do is to emulate the shuffle step on the linear array. In other words, we establish the relation LINEAR ARRAY ∝ CUBE by establishing LINEAR ARRAY ∝ SHUFFLE EXCHANGE and using the already established fact SHUFFLE EXCHANGE ∝ CUBE.

Letting U[j] denote the operand currently stored in V[j], we define the following procedure acting on the subarray $V[\ell 2^r:(\ell+1)2^r-1]$ (for some $1 \leq r \leq d$ and $0 \leq \ell < 2^{d-r}$):

<u>proc</u> SHUFFLE(ℓ,r)
<u>begin</u> <u>for</u> b ← $2^{r-1}-1$ <u>down</u> <u>to</u> 1 <u>do</u>
<u>foreach</u> m:m = $\ell 2^r+s$, s = 2p+b, $0 \leq p < 2^{r-1}-b$ <u>par</u> <u>do</u> U[m] ↔ U[m+1]
<u>end</u>

A simple analysis shows that SHUFFLE(0,d) runs in time $O(2^d)$. It is also easily realized that the required adjacencies are obtained by letting SHUFFLE successively act on each of 2^{d-i} contiguous subarrays of length 2^i. Thus we have the following linear-array emulation of the binary cube [5]:

(3) This graph owes its name to the fact that the permutation represented by the inverse of the transfer acting on an array is a perfect riddle shuffle of the two halves of the array [7].

```
   proc LA-ALGORITHM(0,d)
1. begin for i ← 0 until d-1 do
2.         begin foreach ℓ: 0 ≤ ℓ < 2^(d-i) par do SHUFFLE(ℓ,i+1);
3.               foreach (exchange edge) par do OPER_i
           end;
4.       for i ← 1 until d-1 do
5.               foreach ℓ: 0 ≤ ℓ < 2^(d-i) par do SHUFFLE(ℓ,i+1)
   end
```

We remark that the step in line 2 is dummy for $i = 0$, and takes time proportional to 2^i. Thus the time globally used by line 2 is proportional to 2^d. The for-loop in lines 4 and 5 is necessary to restore the original order of the operands. In conclusion the total running time of the emulation is $O(2^d) = O(N)$.

Obviously, the linear array graph complies with Requirements 1, 2, and 3. Its simplicity is offset by the drawback that the running time grows from $O(\log N)$ to $O(N)$.

3.3. The Rectangular Mesh

The linear array can now be used as a building block for more complex graphs. Recall that the d cube dimensions are processed in the order $E_0, E_1, \ldots, E_{d-1}$. Imagine now to remove from the d-dimensional cube all edges corresponding to the higher $k < d$ dimensions. Thus we obtain a collection of 2^k $(d-k)$-dimensional cubes. Each of them can be emulated in parallel by a linear-array of length 2^{d-k}. Once processing on these arrays is completed, what remains to be done is the processing pertaining to the higher k dimensions, which is done by arranging the just described arrays as the rows of a $2^k \times 2^{d-k}$ grid. We then introduce edges on the columns so that the column appears as linear arrays and can be used to perform the computation pertaining to the higher k dimensions. In this manner a new graph - the rectangular mesh - is generated, which is a valid emulator of the cube. The mesh clearly satisfies Requirements 1, 2, and 3. While the layout area is still $O(2^d) = O(N)$ (as for the linear array), the computation time is now $T = O(\max(2^k, 2^{d-k}))$. Thus by choosing $k = \lfloor d/2 \rfloor$ (square mesh), we obtain $T = O(\sqrt{N})$. From $AT^2 = \Omega(N^2)$, we recognize that the square mesh is an optimal emulator of the cube.

3.4. The Cube-Connected-Cycles

We start from a $2^r \times 2^{d-r}$ rectangular mesh (r an integer, with $\lceil \log(d-r) \rceil \leq r \leq d/2$), where cube dimensions $E_0, \ldots, E_{r-1}$ are assigned to the columns and $E_r, \ldots, E_{d-r-1}$ to the rows. We begin by removing all row connections (see Figure 3a) and by extending each of the resulting column arrays with additional $(d-r)$ modules. The added modules (arranged as a rectangular $(d-r) \times 2^{d-r}$ mesh) are denoted $P(i,j)$, where $0 \leq i < 2^{d-r}$ identifies the column, and $0 \leq j < (d-r)$ identifies the row (from bottom to top). An exchange edge is added between $P(h,j)$ and $P(k,j)$ if and only if the binary spellings of h and k differ exactly in the coefficient of 2^j. The resulting interconnection in the upper mesh - sometimes called the "unfolded binary cube" - can be used to emulate the edges removed in the lower mesh. Indeed, the lower mesh processors are denoted $A(h,k)$, with the same

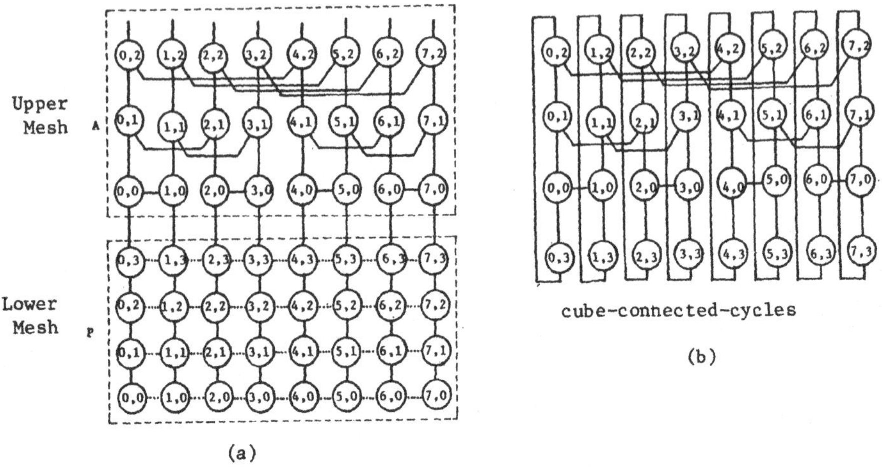

Figure 3. Transformation of a mesh to the cube-connected-cycles.

convention. Suppose that, for $j = 2^r h+k$, operand T[j] is initially stored in A(h,k). While dimensions $E_0,\ldots,E_{r-1}$ are processed by the existing column connection in the lower "mesh", dimensions $E_r,\ldots,E_{d-r-1}$ are processed by pipelining the data in the lower arrays through the upper arrays and outputting the results. Clearly the network of Figure 3a is an emulator of the cube with $T = O((d-r) + 2^r)$. During the pipelining operation, there are at most 2^r modules containing data in each $(d-r+2^r)$-module column array. Specifically A(h,j) and P(h,j) are never simultaneously in use, for any h and for $0 \le j < d-r$. This shows that they can be identified, and this yields the cube-connected-cycles (CCC) architecture shown in Figure 3b [5]. It can be shown that the CCC can be laid out in optimal area $A = O(N^2/T^2)$; in addition, by selecting $r = \lceil \log(d-r) \rceil$, we obtain $T = O(\log N)$, which shows that the CCC can emulate the cube with no significant loss of speed. Besides Requirements 1 and 2, the CCC also satisfies Requirement 3, since a CCC with $k.2^k$ modules can be easily obtained by combining two CCCs with $(k-1)2^{k-1}$ modules; the construction is vaguely suggestive of a tree structure.

4. Pipelining, Pleating, and Mixing

There are some simple mechanisms, each with distinct features and effects, which are, in varying degrees, recognizable in a variety of applications and could be labelled "design principles". At this stage it is wise to refrain from such ambitious denotations, and restrict our attention at how these mechanisms work in connection with the recursive combination paradigm. These mechanisms are "pipelining", "pleating", and "mixing".

Pipelining. Divide-and-conquer is a general algorithmic strategy applicable to numerous problems. This entails subdividing a given problem into a collection of subproblems, solving these, and then suitably combining the solutions of the latter. The recursive combination paradigm naturally lends itself to this strategy, for example, by assigning dimensions $E_0, E_1, \ldots, E_{d-r-1}$ to the subproblems, and dimensions $E_{d-r}, \ldots, E_{r-1}$ to the combination phase. This assignment entails decomposing the original 2^d-component data vector into 2^r contiguous segments (each a 2^{d-r}-component vector). Solving the subproblems consists of feeding in succession these segments to a cascade of networks $\mathcal{N}_0, \ldots, \mathcal{N}_{d-r-1}$ (the "pipeline") where $\mathcal{N}_i$ executes the following action:

for each $j: 0 \leq j < 2^{d-r}$ pardo if $\mathrm{bit}_i(j) = 0$ then $(T[j], T[j+2^i] \leftarrow \mathrm{OPER}_i(T[j], T[j+2^i]$.

This mechanism we call pipelining. The combination phase is done by applying ALGORITHM to homologous components of the processed vectors. This technique finds its application in the CCC, where the combination is done by the LA-ALGORITHM on a collection of arrays (Section 3.2).(4)

Pleating. Up to this point we have implicitly considered problems whose paradigm is a fixed sequence of applications of ALGORITHM (such as merging, DFT, cyclic shifts, etc.). Other problems, instead, use (a specialization of) ALGORITHM as a recursive call: typical in this category are merge-sort (both bitonic and odd-even) and the calculation of symmetric functions. It has been recognized that the "fast" emulators of the cube (the Shuffle-Exchange and the CCC) are not optimal architectures for such problems. The reason for this inadequacy is readily discovered. Referring, for concreteness, to bitonic sort, the cube dimensions are used in the following d-step schedule:

step\dim

E_0

E_0, E_1

$\cdot$

$\cdot$

$\cdot$

$E_0, E_1, \ldots, E_{d-1}$.

This schedule suggests to tailor at each step the size of the machines used to the number of processed dimensions, and, possibly, to reconfigure these machines into larger machines as the number of dimensions grows. Specifically, let a c-dimensional machine denote an architecture supporting the above schedule for d = c, and let

(4) The same technique is to be found in the systolic matrix multiplier [9].

$\beta \geq 1$ be a small integer. Pleating consists of a reconfiguring 2^β c-dimensional machines into a $(\beta + c)$-dimensional machine.

Note that both the linear array and (consequently) the rectangular mesh realize the pleating mechanism naturally, i.e., with no structural addition. The pleating mechanism has been successfully applied to the CCC [10] with $\beta \geq 2$, thereby obtaining an AT^2-optimal fastest bitonic sorter (called pleated-CCC).

Mixing. Suppose we have two AT^2-optimal architectures, $\mathcal{N}_1$ and $\mathcal{N}_2$, for a given problem, with significantly different computation times, T_1 and T_2 ($T_1 < T_2$), respectively. Our objective is to obtain an AT^2-optimal architecture $\mathcal{N}$ for the problem, whose computation time T is in the range $[T_1, T_2]$. For problems fitting the recursive combination paradigm - and, presumably, for other problems with analogous decomposability - the following strategy achieves the objective. For an integer s that divides N, construct one of the two architectures, say $\mathcal{N}_i$, for N/s operands: let $A_i(N/s)$ and $T_i(N/s)$ be the corresponding area and time. Next replace each module of $\mathcal{N}_i$ with an architecture $\mathcal{N}_{3-i}$ for s operands, with $A_{3-i}(s)$ and $T_{3-i}(s)$ (correspondingly the data paths in $\mathcal{N}_i$ are expanded by a factor $\sqrt{A_{3-i}(s)}$). $\mathcal{N}_i$ and $\mathcal{N}_{3-i}$ are called the outer and inner structures, and the mechanism just described is called mixing. If $AT^2 = \Omega(N^\alpha)$, for some real α, then the new architecture $\mathcal{N}$ is trivially AT^2-optimal. Indeed, we have $A = A_i(N/s).A_{3-i}(s)$ and $T = T_i(N/s).T_{3-i}(s)$, whence

$$AT^2 = A_i(N/s).A_{3-i}(s).T_i^2(N/s).T_{3-i}^2(s) = \Omega((N/s)^2).\Omega(s^2) = \Omega(N^2).$$

This mechanism can be applied, for example, to obtain AT^2-optimal architectures with $T \in [\Omega(\log N), O(\sqrt{N})]$ with the following (inner,outer) pairs: (CCC,mesh), (mesh,CCC), (shuffle-exchange, mesh).(5)

5. Conclusions

In this paper we have attempted to shed some light on the relationship between parallel algorithms and VLSI architectures. Although much research is needed to understand this complex interaction, the patterns emerging from recent research are very useful guidelines.

Several important problems can be solved by algorithms belonging to a few highly structured classes, called paradigms. Once a paradigm is found, the next step is the development of VLSI architectures supporting that paradigm. We have illustrated this approach with the case study of recursive combination. In addition, we have illustrated the general mechanisms of pipelining, pleating and mixing, which play a significant role in matching algorithms and architectures.

It appears that a few powerful and highly regular architectures can be used to satisfy a majority of computational requirements. It is refreshing to note that architectures which are appealing from their purely geometric properties (some sort of aesthetic category) turn out to possess remarkable computational significance.

---

(5) The same mechanism has been applied to matrix multiplication [9].

References

1. A. V. Aho, J. E. Hopcroft, and J. D. Ullman, The Design and Analysis of Computer Algorithms, Addison-Wesley, Reading, MA, 1974.

2. R. P. Brent and H. T. Kung, "The chip complexity of binary arithmetic," Journal of the ACM, vol. 28, n. 3, pp. 521-534, July 1981.

3. C. D. Thompson, "Area-time complexity for VLSI," Proc. of the 11th Annual ACM Symposium on the Theory of Computing (SIGACT), pp. 81-88, May 1979.

4. J. E. Vuillemin, "A combinatorial limit to the computing power of VLSI circuits," IEEE Transactions on Computers, C-32, n. 3, pp. 294-300, March 1983.

5. F. P. Preparata and J. Vuillemin, "The cube-connected-cycles: a versatile network for parallel computation," Communications of the ACM, vol. 24, n. 5, pp. 300-309, May 1981.

6. F. P. Preparata and J. Vuillemin, "Area-time optimal VLSI networks for computing integer multiplication and Discrete Fourier Transform," Proc. of I.C.A.L.P., Acre, Israel, pp. 29-40, June 1981.

7. H. S. Stone, "Parallel processing with the perfect shuffle," IEEE Transactions on Computers, C-20, n. 2, pp. 153-161, February 1971.

8. D. Kleitman, et al., "An asymptotically optimal layout for the shuffle-exchange graph," J. of Comp. and Syst. Sci., vol. 26, n. 3, pp. 339-361, June 1983.

9. F. P. Preparata and J. Vuillemin, "Area-time optimal VLSI network for multiplying matrices," Information Processing Letters, vol. 11, n. 2, pp. 77-80, October 1980.

10. G. Bilardi and F. P. Preparata, "An architecture for bitonic sorting with optimal VLSI performance," IEEE Transactions on Computers, to appear (May 1984).

DECIDABILITY OF MONADIC THEORIES

A. L. Semenov
Academy of Sciences
Moscow

On the geographical map of logical theories we now recognize an immense ocean of undecidable and sparse islands of decidability. Monadic decidable theories constitute an important archipelago on the map. Büchi's introduction of automata technique opened a new epoch in the exploration of monadic theories. Rabin's discovery of decidability for monadic theory of the tree can be compared by its importance for mathematical logic only with Tarski theorem on decidability of the real field.

The purpose of this paper is to survay some results and methods obtained in Moscow during last five years as a further developemnt of Büchi's and Rabin's achievements. In the first part we consider some extensions of Büchi's result - namely decidability of monadic theories of structures $\langle N; \leqslant, f \rangle$, where f is a unary function. In the second part we discuss some ideas of a new proof for Rabin's theorem and extensions of this theorem. In the third part we mention some results concerning weak monadic theories. Monadic theory of a structure $\mathcal{S}$ will be denoted by $\mathcal{MTS}$. N is the set of all non-negative integers, ω-word is an infinite sequence of symbols from a finite alphabet; its limit is the set of symbols which have an infinite number of occurences in it.

I. MONADIC THEORIES OF UNARY FUNCTIONS ON N .

Let f be a fixed $N \longrightarrow N$ function with a finite range. We can try to express different properties of f using the monadic language of the structure $\langle N; \leqslant \rangle$ and ask if there is an algorithm to decide the resulting theory $\mathcal{MT}f \rightleftharpoons \mathcal{MT}\langle N; \leqslant, f \rangle$. In <u>this</u> part we will use abbreviation $\mathcal{MT}$ for $\mathcal{MT}\langle N; \leqslant \rangle$ (another notation is S1S). The first results on these theories were obtained by C. Elgot and M. Rabin in their paper [ER]. Their idea was very simple and powerful. Any sentence Φ in the language of $\mathcal{MT}f$ can be treated as a formula in the lantuage of $\mathcal{MT}$ with a single functional variable f . Using Büchi's method we can construct an automaton which accepts ω-word representing f if and only if the sentence Φ is true. As was proved by Elgot and Rabin the acceptance condition can be effectively verified for f being characteristic functions of sets of squares, factorials and so on. Siefkes in [S1] studied the class of functions, for which the decidability can be proved by Elgot - Rabin method. In [S2] he settled a question on the decidability of $\mathcal{MT}f$, where

$$f(x) = \text{sign}\,\sin x\ .$$

For this function nor positive answer can be given by Elgot - Rabin - Siefkes method, neither negative can be given by theorems by Siefkes and Thomas (see below). The study of the function sign sin x became a starting point of our investigation. This function is an example of an almost periodic function of symbolic dynamics. Symbolic dynamics studies the behaviour of dynamical systems by using corresponding discrete systems, elements of which are ω-words (or symbolic sequences) (see [H]). We will illustrate the idea by one example of connection between continuous and symbolic dynamic systems. Let us take as our dynamic system a torus and a transformation of it, say translation $T : x \longmapsto x + v$.

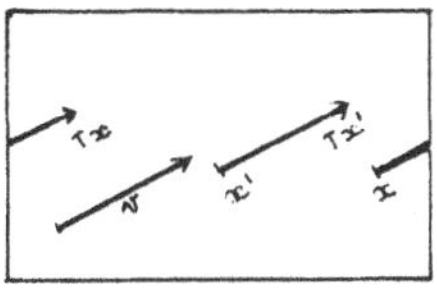

The trajectory of x will be x, Tx, TTx, ... Then we cut the torus into four parts and consider symbolic trajectory - the sequence

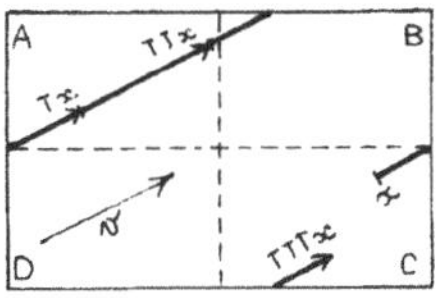

of parts in which x, Tx, ... lie. For x it will be CAAC... . The key-point is that the symbolic trajectory describes the trajectory well enough. It is clear now how such functions as sign sin x appear in symbolic dynamics. Different other examples of sequences, such as Thue, Morse or Keane examples are almost periodic (see [J1]). Now we recall this notion. We formulate it in terms of ω-words. A function f (an ω-word) is almost periodic (a.p.) if for any word u there exists an integer Δ such that either

1) all occurences of u in f are contained in the initial segment of f of length Δ, or

2) the word u occurs in every segment of f having length Δ.

Actually particular examples of a.p. functions constructed in symbolic dynamics are effectively almost periodic, that is an integer Δ can be found by u algorithmically and we can decide, which alternative 1) or 2) is true. The function sign sin x is effectively a.p. as well as other sequences mentioned above.

The following question was the origin of the general study of decidability criterion for $\mathcal{M}\mathcal{T}$f : Do the finite-automata transformations preserve almost periodicity? The answer is - "Yes" and in fact leads to the characterisation of decidability of $\mathcal{M}\mathcal{T}$f for almost periodic f.

<u>Decidability theorem (almost periodic case)</u>. For any a.p. f $\mathcal{M}\mathcal{T}$ f is dècidable $\Longleftrightarrow$ f is computable and effectively almost periodic.

So $\mathcal{M}\mathcal{T}$ sign sin x as well as monadic theories of many other particular functions are decidable. The theorem is a special case of the general decidability theorem we shall consider now.

The notion of congruence $\mathcal{C}$ (with finite set of $\mathcal{C}$-classes, we call it simply congruence) on a free monoid Σ^* was used in many works on automata on ω-words (see [B1], [T2]). Let $\mathcal{C}$ be such a congruence. A <u>$\mathcal{C}$-index</u> is a nonempty word over the alphabet of $\mathcal{C}$-classes. The length of it does not exceed $|\mathcal{C}|$ - the number of $\mathcal{C}$-classes. For every $\mathcal{C}$-index $\varkappa$ we define its value $\langle\langle\varkappa\rangle\rangle$ by induction. If K is a $\mathcal{C}$-class, $\langle\langle K\rangle\rangle \rightleftharpoons \Sigma \cap K$; $w \in \langle\langle\varkappa K\rangle\rangle \Longleftrightarrow w \in K$ & there exist prefix and suffix of w belonging to $\langle\langle\varkappa\rangle\rangle$ & each subword of w, which belongs to $\langle\langle\varkappa\rangle\rangle$, is its prefix or suffix.

The idea here is to find a repetition structure in an ω-word modulo $\mathcal{C}$. Firstly, we recognize neighborous letters which are $\mathcal{C}$-congruent, then - $\mathcal{C}$-congruent segments having appeared in our search for letters, and so on.

<u>Decidability theorem</u>. The decision problem for $\mathcal{M}\mathcal{T}\langle N; \leqslant, W\rangle$ is equivalent to the problem: to compute f and

for every congruence $\mathcal{C}$ to find the set of all $\mathcal{C}$-classes for which no element of them occurs in some end of W &

to find such an end.

The condition of the decidability given by the theorem is easy to be verified in the cases in which decidability was proved by the use of Elgot - Rabin - Siefkes theorem. We have mentioned yet one consequence of the decidability theorem - a.p. case. Another consequence can be trivially obtained from the theorem by substituting all regular sets to indices.

Corollary. For any W the decision problem for $\mathcal{MT}$W is equivalent to the problem of both :

1) to calculate W

2) to decide for every regular set A if there is an end of W which does not contain elements of A as subwords and if so - to find such an end.

Z-case. In the symbolic dynamics and other applications it is important to consider Z-word - a bisequence of letters from a finite alphabeth Σ . The set of all such words is denoted by Σ^Z. Let $w \in \Sigma^Z$. If we are interested in the decidability of $\mathcal{MT}W = \mathcal{MT}\langle Z; \leqslant, W\rangle$, two main cases arise. In accordance with symbolic dynamics we call a Z-word recurrent if no word has the first or the last occurence in it.

Theorem on zero undefinability. For any $W \in \Sigma^Z$ the following conditions are equivalent :

1) 0 is not definable in $\mathcal{MT}$W

2) W is recurrent.

It is clear that in the nonrecurrent case the decidability of $\mathcal{MT}$W and the decidability of the pair of theories obtained by cutting W into two ω-words are mutually reducible. In the recurrent case the situation is more simple and in some sense more exotic.

Decidability theorem (recurrent Z-case). For any recurrent Z-word W the decision problem for $\mathcal{MT}$W is algorithmically equivalent to the problem: for every congruence $\mathcal{C}$ and $\mathcal{C}$-index $\varkappa$ recognize if there is an occurence of some element from $\langle\!\langle \varkappa \rangle\!\rangle$ in W .

We see that for recurrent W monadic theory $\mathcal{MT}$W is completely defined by the set of all subwords of W . In particular, all W for which this set is Σ^* have the same $\mathcal{MT}$W. From the theorem we get the decidability of it. In symbolic dynamic such Z-words are called transitive; they constitute a subset in Σ^Z of measure 1 . This leads us to some consequencesfor symbolic dynamics. One of them is the following :

Corollary (Gleason - Welch, see [H], sect. 11). For a given endomorphism f of the shift dynamical system, each recurrent transitive point x has the same cardinality of $f^{-1}(x)$.

We formulate the uniformisation problem for the theory in the following way. For any formula $\phi(X, Y)$ in the language of $\mathcal{MT}$ construct a transducer T, which for given input X produces some output Y, for which $\phi(X, Y)$ is true if such Y exists, and stops if there is no such Y.

It is well known that if transducer is understood as finite-state transducer, the uniformisation is not possible in all cases. It can be effectively constructed if it does exist. (See [R2] for historical references.) Transducer which perform uniformisation in all cases must have an unbounded memory and must get some information about infinite behaviour of ω-word.

We introduce the notion of minimal transducers which form (in some informal sense) minimal extension of the class of finite state transducers.

Let us fix a family of finite sets: input alphabet Σ, output alphabet Ω, two sets of marks - Π-marks (they form the set Π) and Γ-marks (they form the set Γ). Minimal transducers have input tape, output tape, control unit (a finite deterministic automaton) and memory. A transducer activity at any moment is determined by the part of input tape which it has not read, control state and memory state. State transition is determined (as in finite automata) by the previous state and input symbol. Memory state is defined by two marking maps which assign values to marks. The first is a total map $p : \Pi \longrightarrow \Omega^*$, the second is partial, $g : \Gamma \longrightarrow \Omega^*$. All elements from the image of g have equal lengths. Every control state has its printing mark - an element of Π and marking move - a pair of partial maps P, G :

$$P : \Pi \longrightarrow \Gamma, \quad G : \Gamma \longrightarrow \Gamma \times \Omega.$$

Each step begins with reading an input symbol, the control state changes, and then three actions are peformed: Γ-transition, printing, Π-transition. The marking changes in all these actions and each action uses the marking which appears after the previous action.

Γ-transition. For every $\gamma \in \Gamma$, if $G(\gamma)$ is undefined, then $g(\gamma)$ becomes also undefined. If $G(\gamma) = \langle \delta, \sigma \rangle$, then $g(\gamma) \longleftarrow g(\delta)$ ($\longleftarrow$ is the assignement). So, the value of γ becomes equal to the previous value of some Γ-mark extended by one symbol.

Printing. The value of printing mark is printed. Let it be x. Then we divide (if possible) the value of each mark by x from left. If it is impossible for a Π-mark, its value becomes Λ, if it is impossible for some Γ-mark, its value becomes undefined.

Π-transition. For every Π-mark ρ, if $P(\rho)$ is undefined, then the value of ρ remains unchanged, if it is defined, then $P(\rho) \leftarrow g(P(\rho))$.

To end the description we note that the work of a transducer begins in some initial control state with some initial marking. It can be proved that if a minimal transducer is total, then it has a finite delay.

Theorem (definability of minimal transducers). Given a minimal transducer one can construct a formula $\phi(X, Y)$ in the monadic language of $\mathcal{MT}$ which defines the same function on ω-words as the transducer.

The technique developed for proving the decidability theorem gives us tools for the uniformisation. We state a uniformisation theorem in a simplified style. Given a family of regular sets R and an ω-word W we define W_R as any ω-word, obtained from W by marking two symbols of W by benchmarks. The part of W following the first benchmark contains the occurences of the elements of only those members from R the elements of which occur in each end of W; between the first and the second marks you can meet elements of all those members.

A uniformisation theorem. Given a formula $\phi(X, Y)$ in the language of $\mathcal{MT}$ one can construct a minimal transducer T and a finite family of regular sets R such that T transforms any W_R into some U, for which $\phi(W, U)$ holds and it falls into some rejecting state if no such U exists.

For every W all images of W obtained by minimal transducers constitute a model of $\mathcal{MT}W$. For almost periodic W these images are almost periodic. All a.p. ω-words form a model for $\mathcal{MT}$.

Unbounded functions. Till now we were interested in functions with finite range only. The major reason for it is that usually the theory $\mathcal{MT}f$ is undecidable for f with infinite range. Actually, as follows from a theorem of Thomas [T1], for every strictly increasing f which is not definable in $\mathcal{MT}$ the theory $\mathcal{MT}f$ is undecidable. It is easy to prove that every strictly increasing function definable in $\mathcal{MT}$ coincides, for a sufficiently large x, with x + c . As it was mentioned in [ER], the theory $\mathcal{MT}f$ is decidable if f is definable in some decidable theory $\mathcal{MT}g$ where g has a finite range. A question naturally arises : is there any decidable $\mathcal{MT}f$ where f is definable in no theory of the form $\mathcal{MT}g$, g having a finite range ? Such f does exist (see [Se2]).

II. RABIN THEOREM. A NEW PROOF AND GENERALIZATION.

In his report on the International congress of Mathematicians in Nice, 1970, Rabin stated as number 1 the following problem :

"Find a simpler proof for theorem 2(ii), possibly avoiding the transfinite induction used in [2]".(Theorem 2(ii) is the so called "Complementation lemma" and [2] is Rabin's publication in Transaction AMS - our reference [R1].)

In 1979 a new proof of Rabin's theorem was given by An. A. Muchnik - a student at Moscow University. In May, 1980 the (unpublished) paper of Muchnik was approved as his magister thesis (see [M1]). For technical reasons it is not published yet, a very meagre compensation was lectures on it in Moscow University (A. L. Semenov 1979/80, A. G. Dragalin, A. Shen' 1980/81). In 1981 (see [GH]) Rabin's problem was solved also by Yu. Gurevich and L. Harrington. Their solution is very close to Muchnik's. There are some differences in general idea, the Gurevich - Harrington proof is more "game-oriented". Other differences will be briefly discussed later. Gurevich - Harrington method also gives a new proof for the Stupp's generalization of Rabin's theorem (see [Sh]). The same is true for Muchnik's proof, theorem 2 below is a further generalization of Stupp's theorem. This and some other results were obtained thanks to stimulating discussions with P. Schupp.

Muchnik used finite automata in his proof. His automata are a generalization of Rabin's. The most important distinction is the introduction of deadends, it will be discussed later. In the simplest case of automata without input the notion of automaton in fact coincides with the notion of game of a special kind. This case corresponds to the problem of emptiness part in Rabin's proof. So we begin with games.

<u>Games</u>. Our games have two players : White and Black. White positions (W-positions) are elements of some set Π , Black positions (B-positions) are subsets of Π . The <u>rule</u> of a game is a set R of pairs of the form $\langle p, P\rangle$, where p is a W-position, P is a B-position. <u>Party</u> is a sequence

$$p_0,\ P_0,\ p_1,\ P_1,\ \dots$$

where $p_0, p_1, \dots$ are W-positions, $P_0, P_1, \dots$ are B-positions and for every i

$$\langle p_i, P_i\rangle \in R$$

$$p_{i+1} \in P_i \ .$$

So, White plays by choosing an element from R and giving a position to Black. Black plays by choosing an element of his position and giving it to White. Each party is either white party or black party, not both. Finite party is a finite sequence defined in a similar way; it is not a party.

White strategy (W-strategy) is a map φ from finite parties $p_0, P_0, \dots, p_n$ into B-positions P_n for which $\langle p_n, P_n\rangle \in R$. White wins in the game with an initial position p_0 if every party

$$p_0,\ P_0,\ \dots$$

where

$$P_i = \varphi(p_0, P_0, \dots, p_i), \quad i = 0, 1, \dots$$

is a white party; similarly for black.

Now we have to introduce a particular sort of games for which the division of parties into white and black will be of a special kind. Let Π have the form $\Pi = F \times S$, where F is a nonempty field of the game, S is a nonempty finite set of states of the game; let σ be a projection : $\sigma\langle x, s\rangle = s$. The notation of a party $p_0, P_0, \dots$ is an ω-word

$$\sigma(p_0)\ \ \sigma(p_1)\ \ \sigma(p_2)\ \dots\ .$$

All subsets of S are divided into white and black. A party is white if the limit of its notation is white, otherwise it is black. Now we will consider the monadic language of the monadic structure $\langle \Pi\, ;\, R, \sigma\rangle$. The atomic formulas of this language are of the form $x \in X$, $R(x, X)$, $\sigma(x) = s$, where x is an individual variable, X is a set variable, $s \in S$. So the language is determined by S.

Theorem 1. (Winner definability). There is an algorithm which for every set of states S and a family of white subsets of it constructs a monadic formula $\phi(x)$ such that for every game with this set of states and an arbitrary set of white states and an arbitrary field and rule it holds :

White wins in the game with initial position $p_0 \Longleftrightarrow \phi(p_0)$ is true.

An important and obvious consequence of this theorem is the following. Let us have a monadic structure $\langle F, \mathfrak{N}\rangle$. Sometimes we can define a game on it by introducing a set S , a family of formulas $R_{s,t}(x, X)$ of the monadic language of signature π , where $R_{s,t}(x, X) \Longleftrightarrow R(\langle x, s\rangle , \langle X, t\rangle)$, a family of white subsets of S and an initial state $s_0 \in S$. We shall suppose that each of $R_{s,t}$ is defined by a fixed formula in the monadic language of $\langle F, \pi\rangle$.

<u>Corollary</u>. There is an algorithm which for every set of states S , family of formulas $R_{s,t}$, family $L \subseteq 2^S$ and a state s_0 constructs a formula ϕ in the monadic language of signature π such that for arbitrary structure $\langle F, \pi\rangle$ we have :

White wins in the game with the field F , the set of states S , the rule defined by $R_{s,t}$, the family of white subsets L , and the initial position $\langle x, s_0\rangle \Longleftrightarrow \phi(x)$ is true.

So, if a game is definable in the monadic language of some monadic structure, then the set of winning positions for it is definable in the same language.

Now we pass to the Muchnik's generalization of Rabin's theorem. It is also a generalization of a Stupp's theorem. The method of the proof for this theorem and Rabin theorem is the same and it uses no transfinite induction.

Let $\mathcal{S} = \langle A; \ldots\rangle$ be a monadic structure. We construct some new monadic structure <u>Tree</u>($\mathcal{S}$) - the tree structure over $\mathcal{S}$. Its elements are finite sequences of elements of $\mathcal{S}$. The signature of it contains the relation $x \prec y$ - " x is a proper initial segment of y " . Then, for each formula $(\overline{x}, \overline{X})$ of the monadic language of $\mathcal{S}$ the signature contains the symbol $\hat{\varphi}$ with the interpretation

$$\hat{\varphi}(\overline{y}, \overline{Y}) \Longleftrightarrow \exists z \in A^*, \overline{x}, \overline{X} \; (\; \overline{y} = z\overline{x} \quad \overline{Y} = z\overline{X} \quad \varphi(\overline{x}, \overline{X}) \;).$$

Here, $\overline{y}$, $\overline{Y}$ ... are vectors over the elements of A^*, 2^{A^*}, etc. Finally, for each formula $\varphi(t, \overline{x}, \overline{X})$ we place to the signature of <u>Tree</u>($\mathcal{S}$) the symbol $\hat{\hat{\varphi}}$ with the interpretation

$$\hat{\hat{\varphi}}(\overline{y}, \overline{Y}) \Longleftrightarrow \exists z \in A^*, t \in A, \overline{x}, \overline{X} \; (\; \overline{y} = zt\overline{x} \;\&\; \overline{Y} = zt\overline{X} \;\&\; \varphi(t, \overline{x}, \overline{X}) \;).$$

<u>Reduction theorem for tree structures</u>. There is an algorithm to construct for every sentence in the monadic language of <u>Tree</u>($\mathcal{S}$) an equivalent sentence in the monadic language of $\mathcal{S}$.

If we do not use $\hat{\hat{\varphi}}$ in the construction of <u>Tree</u>($\mathcal{S}$) and suppose that $\mathcal{S}$ is an elementary structure, we obtain the theorem of Stupp (see [Sh], [GH]). As a special case for $\mathcal{S}$ of cardinality two we get Rabin's theorem.

About the proof. Firstly the proper notion of automaton is introduced and the theorem on automata definability for definable subsets of Tree($\mathcal{S}$) is proved. Secondly, it is noted that the automaton constructed for a closed formula is a game in essence. So we can use theorem 1 to obtain a sentence in the monadic language of $\mathcal{S}$ which is true iff the automaton accepts the tree, that is equivalent to the truthness of the given sentence in the monadic language of Tree($\mathcal{S}$).

The proof uses the same construction as in theorem 1. Naturally the most difficult part is the construction of an automaton, especially for the negation of a formula.

The notion of automaton. An automaton has a finite set of states S, and an input alphabet Σ (usually $\Sigma = \{0, 1\}^n$ for some n). The set of positions is $\Pi \rightleftharpoons A \times S$. Transition rule is a family of definable in $\mathcal{MTS}$ relations $R_\alpha \subseteq \Pi \times 2^\Pi$. Initial condition is a family of $\mathcal{MTS}$-definable predicates $I_\alpha \subseteq \Pi$. Some subsets of S are white. A run ϑ of the automaton on a marked tree (that is a map $\mu : A^* \longrightarrow \Sigma$) is a set of pairs $\langle x, \langle s_0, s_1\rangle\rangle$, where $x \in A^+$; for $x \in A$ $\langle x, s_1\rangle \in I_{\mu(x)}$ and, for every $x \in A^+$, $a \in A$ if $\langle xa, \langle s_0, s_1\rangle\rangle \in \vartheta$ then

$$\langle\langle a, s_1\rangle , \{\langle b, s_1'\rangle \mid b \in A \ \&\ \langle xab, \langle s_1, s_1'\rangle\rangle \in \vartheta\}\rangle \in R_{\mu(x)} .$$

Now we shall formulate the acceptance condition for the automaton. Let μ be a marked tree. Then it is accepted iff there is an accepting run ϑ of the automaton on it. The run is accepting if for any path $x_0 = a_0, x_1 = x_0a_1, \ldots, x_{i+1} = x_ia_i, a_i \in A$ and any sequence of states $s_0, s_1, \ldots$, where $\langle s_i, s_{i+1}\rangle \in \vartheta(x_i)$, the limit of the sequence is white. So the "actual state" of automaton in the vertex x_i is s_{i+1} and we use pairs for the connection of times.

As we see, for a fixed run an element of tree does not determine a pair of the automaton states in it. But actually every automaton can be effectively transformed into one for which this kind of determinism takes place. Let us call these automata standard. (This is a straightforward generalization of Rabin's automata.)

The proof of automata definability now proceeds in the usual way. We construct automata defining signature relations and then prove closeness under $\vee$, $\exists$, $\neg$ for standard automata. The main point again is a complementation lemma. For its proof the notion of automata with deadends is introduced. A set of deadends plays in the description of automata the role similar to that of states, but no transition from deadends is permitted. Deadends are also used as marks of the vertices in the tree. Then, if the automaton with deadends arrives to some vertex marked by a deadend then the acceptance condition demands that the automaton must be in a deadend (not in state), and the deadend must coïncide with the mark in the vertex. The complementation lemma is now proved using the induction by number of states (the number of deadends is irrelevant). The steps are similar to the proof of the first theorem. No transfinite induction is used! The states of automaton for the complement are the permutations of the given automaton $\mathfrak{A}$. (This method adds three levels of exponentation at each quantifier change in the decision procedure.) In constructing its rule we use the following operation acting on permutations. We suppose that all states of $\mathfrak{A}$ are cyclically ordered. Let w be a permutation and s - a state. If w = usv and v's is the reordering of sv in the cyclical order, then s transforms w into uv's. This method of storing the information on the history is used in constructing accepting runs for automaton accepting the complement. The technical improvements we have discussed very briefly lead us to the short and transparent proof of Rabin's theorem using only a few simple ideas. Gurevich and Harrington's proof is, as we have said, very close to Muchnik's as can be understood from [GH], but there are no mentions about such constructions as deadends and methods for storing a history.

<u>Uniformisation</u>. As it was conjectured by Siefkes in [S3], the definable uniformisation for S2S (the tree structure over the structure with two elements) is impossible. That conjecture was proved by Gurevich & Harrington in a strong form of nonuniformisability of some relation in $2^{\{0,1\}^*} \times \{0,1\}^*$. The relation is the natural $\in$-relation. So the choice function is indefinable in S2S. On the other hand, any relation in $\{0,1\}^* \times 2^{\{0,1\}^*}$ (and, consequently in $\{0,1\}^* \times \{0,1\}^*$) has a definable uniformisation. It was also proved by Muchnik.

III. WEAK MONADIC AND ELEMENTARY THEORIES.

There is a natural correspondence between finite subsets of N and N based on finary coding (see [B2]). This correspondence was used in [B], [ER] for treating the weak monadic theory of $\langle N; \leq \rangle$ as elementary. The addition of natural numbers and predicate Pw_2 "to be a power of two" are definable in this theory. Actually the set of relations definable in this theory coincides with the set of relations definable by finite automata using binary encoding of integers. As it was conjectured in [McN] and proved in [Se1] (and also can easily be obtained from results of [Se2]), there are some relations definable by finite automata but not definable in elementary theory of $\langle N; +, Pw_2 \rangle$. But we can use another unary predicate instead of Pw_2 to characterize the definability.

<u>Theorem</u>. Let P_0 be a set of all natural numbers with binary codings from the set

$$\{000, 001\}^* \ 010 \ \{000,001\}^* \cup \{000,001\}^*.$$

Then a relation is definable by a finite automaton iff it is definable in elementary theory of $\langle N; +, P_0 \rangle$.

The decidability theorem for elementary theories of structures $\langle N; \leq, W \rangle$, where W is an ω-word, and other results from the first part of this paper can be proved in a modified form. The modification is made by restricting the set of congruences to special congruences only. We call a congruence special if it corresponds to the homomorphism on a finite monoid with no subgroups. The connection of this notion with elementary theories was found by Schützenberger (see also [M2] for a simple proof of Schützenberger's theorem). Some other questions on decidability of the elementary theories of $\langle N; +, f \rangle$, $\langle N; \leq, f \rangle$ were discussed in [Se1], [Se2]. In particular, in [Se2] a predicate P was constructed for which elementary theory of $\langle N; +, P \rangle$ is undecidable but all relations definable in it are decidable. (This answers a question of Büchi from [B1].)

Elgot and Rabin asked whether there is a maximal decidable theory, e.g. a decidable theory which becomes undecidable if we add to it some predicate which is not definable in it. As Soprunov proved, there is no maximal decidable weak monadic theory.

REFERENCES

[B1] Büchi J.R.: On a decision method in restricted second order arithmetic. Proc. 1960 Intern. Congr. for Logic, Methodol. and Philos. of Sci., Stanford, 1962, pp. 1 - 11.

[B2] Büchi J.R.: Weak second-order arithmetic and finite automata. Z.Math. Logik und Grundl. Math. 6(1960), N° 1, pp. 66 - 92.

[ER] Elgot C.C., Rabin M.O.: Decidability and undecidability of extentions of second (first) order theory of (generalized) successor. J. Symbolic Logic 31(1966), N° 2, pp. 169 - 181.

[GH] Gurevich Yu., Harrington L.: Trees, automata and games. Proc. ACM STOC , May 1982, pp. 60 - 65.

[H] Hedlund G.A.: Endomorphisms and automorphisms of the shift dynamical systems. Math. syst. theory 3(1971), N° 4, pp. 320 - 375.

[J] Jacobs K.: Maschinenerzeugte 0-1-Folgen, Selecta Mathematica, II, Springer-Verlag, 1970.

[M1] Мучник Ан. А. [Muchnik An.A.] : Игры на бесконечных деревьях и автоматы с тупиками. Новое доказательство разрешимости монадической теории нескольких следований. [Games on infinite trees and automata with deadends. A new proof of the decidability for the monadic theory of several successors.] Semiotica i Informatica 1984 (to appear) .

[M2] Мучник Ан.А. [Muchnik An.A.] : Новое доказательство теоремы Шютценберже о моноидах без нетривиальных подгрупп. [A new proof of Schützenberger's theorem on monoids without nontrivial subgroups.] Matematicheskaia logika, mat. lingvistika i teoria algoritmov. Kalinin 1983, pp. 65 - 68 .

[McN] McNaughton R.: Rev. of [B2]. J. Symb. Logic 28(1983) N°2, pp. 100 - 102.

[R1] Rabin M.O.: Decidability of second-order theories and automata on infinite trees. Transactions AMS 141(1969), pp. 1 - 35.

[R2] Rabin M.O.: Automata on infinite objects and Church's problem. AMS, Providence, 1972.

[S1] Siefkes D.: Decidable extentions of monadic second order successor arithmetic. Automatentheorie un Formale Sprachen, Mannheim, 1969, pp. 441 - 472.

[S2] Siefkes D.: Undecidable extentions of monadic second order successor arithmetic. Z. Math. Logik und Grundlagen der Mathematik 17(1971), N° 5, pp. 383 - 394.

[S3] Siefkes D.: The recursive sets in certain monadic second order fragments of arithmetic. Arch. Math. Logik, 17(1975), N° 1-2, pp. 71 - 80.

[Se1] Семенов А.Л. [Semenov A.L.] : О некоторых расширениях арифметики сложения натуральных чисел. [On certain extensions of the arithmetic of addition of natural numbers.] Iz. Akad. Nauk SSSR 43(1979), N° 5, pp. 1175 - 1195; Math. USSR - Izv. 15(1980), N° 2, pp. 401 - 418.

[Se2] Семенов А.Л. [Semenov A.L.]: Логические теории одноместных функций на натуральном ряде. [Logical theories of unary functions on natural numbers.] Izv. Akad. Nauk SSSR 47(1983),

N° 3, pp. 623 - 658.

[Sh] Shelah S.: The monadic theory of order. Annals of Math. 102 (1975), pp. 379 - 419.

[T1] Thomas W.: A note on undecidable extentions of monadic second order arithmetic. Z. Math. Logik und Grundlagen Math. 17 (1975), N° 1-2, pp. 43 - 44.

[T2] Thomas W.: A combinatorial approach to the theory of ω-automata. Information and Control 48(1981), pp. 261 - 283.

ON THE EHRENFEUCHT CONJECTURE ON TEST SETS AND ITS DUAL VERSION

J. Albert
Dept. of Mathematics
University of Würzburg
F.R.G.

Abstract: The Ehrenfeucht Conjecture on test sets states the following: Each language L over some finite alphabet contains a finite subset F (a "test set") such that for each pair (g,h) of homomorphisms it holds g(x) = h(x) for all x in F if and only if g(x) = h(x) for all x in L. In this paper we investigate the connections of this conjecture to its dual form where finite representation for any set of pairs of homomorphisms is stated. We also point out similarities and differences of these conjectures to well-known constructions of bases of linear vector-spaces.

1. INTRODUCTION

One of the important parts of traditional formal language theory is the study of homomorphisms applied to sets of words. The power of homomorphisms as language generating devices in the form of equality sets has been demonstrated by a number of deep results; e.g. in [7], [15] a homomorphic generation of any recursively enumerable language was presented and [14] showed that even the complements of recursively enumerable languages are obtained by a homomorphic device. There are also many connections between homomorphic equivalence on languages and equivalences of transductions as shown in [2], [3], [5], [7], [11]. Solving the question of homomorphic equivalence on given languages is closely related to the existence of so-called test sets. A test set for a specific language is nothing but a finite subset of this language, such that all pairs of homomorphisms which agree on this finite subset also agree on the whole language. The well-known Ehrenfeucht-Conjecture, which is several years older than this notion of a test set can therefore be stated as: "For every language there exists a test set".

For an excellent survey of the results obtained on test sets and equality sets before 1980 see [8]. By now, the Ehrenfeucht Conjecture is known to hold for several classes of languages, e.g. languages over any two-element alphabet [12], simple DOL-languages [10], DOL-languages with fair distribution of letters [17], context-free languages [2], rich languages, commutatively closed languages and (1,2)-complete languages [3], level languages, 2-bounded languages, 3-bounded languages of Parikh-rank 1 [4].

In the following chapter we will first discuss some basic concepts arising naturally when studying the language-theoretic background of the Ehrenfeucht Conjecture. In chapter 3 we investigate several variations of this conjecture and their dependencies.

2. NOTATION AND BASIC RESULTS

To establish notation we will give here a short survey of some of the notions used in the sequel; all others can be found in most standard text books on formal language theory, e.g. [1], [5], [16], [18], [19]. For the following we will always assume that an alphabet is a finite set with a fixed order of its letters. By $\#_a(x)$ we denote the total number of occurrences of the letter a in a word x. For an alphabet $\Sigma = \{a_1,\ldots,a_n\}$ and a word x over Σ, $p(x) = (\#_{a_1}(x),\ldots,\#_{a_n}(x))^T$ denotes the Parikh-vector of x. We extend this definition to sets of words by $p(L) = \{p(x) \mid x \in L\}$ for each language $L \subseteq \Sigma^*$.

Let $L \subseteq \Sigma^*$ and g, h be two homomorphisms $g : \Sigma^* \to \Delta^*$, $h : \Sigma^* \to \Delta^*$. Then we say that g and h agree on L if $g(x) = h(x)$ for all $x \in L$. The equality set for g and h is defined as $E(g,h) = \{x \in \Sigma^* \mid g(x) = h(x)\}$ Thus, two homomorphisms g, h agree on L iff $L \subseteq E(g,h)$. For the following we can always assume that the considered homomorphisms have the same target alphabet as given in their domain. By associating now a set of pairs of homomorphisms to a fixed language we are giving a definition which can be considered dual to the introduction of equality sets.

Definition 2.1: For a finite alphabet Σ let $H_\Sigma = \{(g,h) \mid g, h$ homomorphisms $g, h : \Sigma^* \to \Sigma^*\}$. Then for $L \subseteq \Sigma^*$, $H(L) \subseteq H_\Sigma$ is the set of all pairs of homomorphisms, which agree on L.
$H(L) = \{(g,h) \in H_\Sigma \mid E(g,h) \supseteq L\}$.

<u>Definition 2.2:</u> For a language $L \subseteq \Sigma^*$ a finite set $C \subseteq \Sigma^*$ is called checking set for L if $H(C) = H(L)$.
C is called test set for L if C is a checking set for L and $C \subseteq L$.

The cardinality of each checking set C for a language L is bounded from below by the rank of $p(L)$, as shown in [3].

In a very natural way for a language $L \subseteq \Sigma^*$ and the set $H(L)$ there is a corresponding counterpart again contained in Σ^* which includes all words from Σ^* on which all pairs of homomorphisms from $H(L)$ agree.

<u>Definition 2.3:</u> Let $L \subseteq \Sigma^*$ then the maximal extension $M(L)$ is defined as $M(L) = \{x \in \Sigma^* \mid \text{for all } (g,h) \in H(L) \text{ it holds } g(x) = h(x)\}$

We will list now several simple facts about maximal extensions which show the nice algebraic and set-theoretic properties of this operation on languages.

<u>Lemma 2.4:</u> Let $L, L' \subseteq \Sigma^*$.
a) If $L \subseteq L'$ then $H(L) \supseteq H(L')$ and $M(L) \subseteq M(L')$.
b) $H(L) = H(M(L))$
c) $H(L) \supseteq H(L')$ iff $M(L) \subseteq M(L')$
d) $M(L) \supseteq L^*$
e) $M(L) = M(L^*) = (M(L))^* = M(M(L))$

This follows immediately from the definitions of the corresponding sets. Furthermore, we get some straightforward relations between maximal extensions, checking sets and test sets.

<u>Lemma 2.5:</u> Let $L \subseteq \Sigma^*$ and $C \subseteq \Sigma^*$ finite.
a) C is checking set for L iff $M(C) = M(L)$
b) C is test set for L iff $C \subseteq L$ and $L \subseteq M(C)$
c) $$M(L) = \bigcup_{\substack{F \text{ checking} \\ \text{set for } L}} F = \bigcap_{L \subseteq E(g,h)} E(g,h)$$

For the investigations of checking sets (test sets) of minimal cardinality the notion of "homomorphic independence" turns out to be crucial.

<u>Definition 2.6:</u> A language $L \subseteq \Sigma^*$ is called homomorphically independent (h-independent) if for each $x \in L$ it holds $x \notin M(L - \{x\})$.

Let us shortly mention some of the relations between homomorphically independent sets of words and associated linearly independent sets of Parikh-vectors and also point out the differences.

<u>Lemma 2.7:</u>

a) If $C \subseteq \Sigma^*$ such that $p(C)$ is a linearly independent set of vectors then C is homomorphically independent.

b) There exist finite homomorphically independent sets $C \subseteq \Sigma^*$ such that $p(C)$ is not linearly independent.
There exist finite homomorphically independent sets $C \subseteq \Sigma^*$ and $x \in \Sigma^*$ such that $x \notin M(C)$ and $C \cup \{x\}$ is not homomorphically independent.

These statements are not hard to verify; for details see [4].

3. ON SOME VARIATIONS OF THE EHRENFEUCHT CONJECTURE

Let us recall the aforementioned Ehrenfeucht Conjecture on the existence of test sets in its detailed form:

<u>Conjecture 3.1:</u> For each finite alphabet Σ and each language $L \subseteq \Sigma^*$ there exists a finite subset $F \subseteq L$ such that for each pair g, h of homomorphisms defined on Σ^* it holds: $g(x) = h(x)$ for all $x \in L$ if and only if $g(x) = h(x)$ for all $x \in F$.

It can be reformulated equivalently as: For each language $L \subseteq \Sigma^*$ there exists a finite subset $F \subseteq L$ such that $M(F) \supseteq L$.
Within the last five years this conjecture has been related to several hard open problems, e.g. it was shown in [9] that the validity of the Ehrenfeucht Conjecture would imply the decidability of the HDOL sequence equivalence problem as well as the decidability of the DTOL sequence equivalence problem. Besides these relations there are strong connections of the Ehrenfeucht Conjecture to the theory of binary relations on free monoids and especially to rational and pushdown transductions (c.f. [9]).

Let us continue now stating a somewhat stronger version of Conjecture 3.1. Here a uniform bound on the cardinality of test sets, depending only on the cardinality of the underlying alphabet, is assumed.

Conjecture 3.2: For each finite alphabet Σ there exists a number $k \geq 1$ such that for each $L \subseteq \Sigma^*$ there is an $F \subseteq L$, $\text{card}(F) \leq k$, such that $M(F) \supseteq L$.

Partial confirmations of this conjecture were obtained in [3] for the rich, the commutatively closed and the (1,2)-complete languages. In [13] one can find the result about languages over two-letter alphabets saying that test sets of cardinality 3 are sufficient and it is even conjectured for this two-letter-case, that test sets of cardinality 2 are always available.

We proceed now by examining the negation of the Ehrenfeucht Conjecture.

Lemma 3.3: The Ehrenfeucht Conjecture (Conjecture 3.1) does not hold iff there exists a finite alphabet Σ and for each $i \geq 1$ there exist $x_i \in \Sigma^*$, homomorphisms $g_i, h_i : \Sigma^* \to \Sigma^*$ such that for each $j \in \{1,\ldots,i-1\}$ $g_i(x_j) = h_i(x_j)$ and $g_i(x_i) \neq h_i(x_i)$.

Proof: If the Ehrenfeucht Conjecture is not valid, there is an $L \subseteq \Sigma^*$ such that for each finite subset $F \subseteq L$ $M(F) \subsetneq M(L)$.
Let $F_1 = \{x_1\}$ where x_1 is any element of L. Then by the above assumptions there exists an $x_2 \in L$ and homomorphisms g_2, h_2 such that $g_2(x_1) = h_2(x_1)$, $g_2(x_2) \neq h_2(x_2)$. Then define $F_2 = \{x_1,x_2\}$, etc.

For the converse, let $L = \{x_i \mid i \geq 1\} \subseteq \Sigma^*$. Then for each finite subset $F \subseteq L$ there is a number $t \geq 1$ such that $F \subseteq \{x_i \mid 1 \leq i \leq t\}$. Thus, there is an $x_{t+1} \in L$ and there exist homomorphisms g_{t+1}, h_{t+1} such that $F \subseteq E(g_{t+1}, h_{t+1})$, but $g_{t+1}(x_{t+1}) \neq h_{t+1}(x_{t+1})$.

This set of equalities and inequalities for the words x_i and the homomorphisms g_i, h_i can be depicted in an infinite grid in an intuitively appealing way:

	x_1	x_2	x_3	x_4	. . .
(g_1,h_1)	$\neq$				
(g_2,h_2)	=	$\neq$			
(g_3,h_3)	=	=	$\neq$		
(g_4,h_4)	=	=	=	$\neq$	
⋮					

With this representation in mind it is straightforward that the following lemma holds:

Lemma 3.4: For each $i \geq 1$ let $x_i \in \Sigma^*$ and g_i, h_i be homomorphisms, g_i, $h_i : \Sigma^* \to \Sigma^*$ such that for each $j \in \{1,\dots,i-1\}$ $g_i(x_j) = h_i(x_j)$ and $g_i(x_i) \neq h_i(x_i)$. Then for each infinite subset I of the natural numbers, $I \subseteq \mathbb{N}$, the language $L_I = \{x_t \mid t \in I\}$ disproves the Ehrenfeucht Conjecture.

The restriction to an infinite subset I of $\mathbb{N}$ corresponds to the elimination of rows and columns in the above grid. Since only rows and columns which bear the same indices are taken out, we are left with a structure of exactly the same form.

The conjecture dual to the Ehrenfeucht Conjecture now states a property of the subsets of H_Σ.

Conjecture 3.5: For each finite alphabet Σ and each subset $S \subseteq H_\Sigma$ there exists a finite subset $T \subseteq S$ such that

$$\bigcap_{(g,h) \in T} E(g,h) = \bigcap_{(g,h) \in S} E(g,h) .$$

The connections of this conjecture to the Ehrenfeucht Conjecture may not be apparent yet, so we first remind of the representation of M(L) as the intersection of all equality set including L (Lemma 2.5c). Thus, the validity of Conjecture 3.5 would imply that for each $L \subseteq \Sigma^*$ there exists a finite subset $K \subseteq H(L)$ such that

$$M(L) = \bigcap_{(g,h) \in K} E(g,h)$$

This duality of Conjectures 3.1 and 3.5 becomes even more obvious, if the negation of Conjecture 3.5 is examined.

Lemma 3.6: The Conjecture 3.5 does not hold iff there exists a finite alphabet Σ and for each $i \geq 1$ there exist $x_i \in \Sigma^*$, homomorphisms $g_i, h_i : \Sigma^* \to \Sigma^*$ such that for each $j > i$ $g_i(x_j) = h_i(x_j)$ and $g_i(x_i) \neq h_i(x_i)$.

For the details see [4]. If we depict the relations of $g_i(x_j)$, $h_i(x_j)$ in a grid again, we get:

	x_1	x_2	x_3	x_4	. . .
(g_1,h_1)	$\neq$	=	=	=	
(g_2,h_2)		$\neq$	=	=	
(g_3,h_3)			$\neq$	=	
(g_4,h_4)				$\neq$	
⋮					

For the same reasons as given in Lemma 3.4 we can choose again any infinite subset of indices of $\mathbb{N}$ to obtain a subset of H_Σ with the same properties as S.

The validity of Conjecture 3.5 can be shown e.g. for pairs of homomorphisms with k-bounded balance on specific languages and also pairs of periodic homomorphisms as demonstrated in [4].

The Ehrenfeucht Conjecture and Conjecture 3.5 are now related both to a weaker statement about homomorphically independent sets.

Conjecture 3.7: For each finite alphabet Σ and each language $L \subseteq \Sigma^*$ it holds: L can be homomorphically independent only if L is finite.

Lemma 3.8: The Conjecture 3.7 does not hold iff there exists a finite alphabet Σ and for each $i \geq 1$ there exist $x_i \in \Sigma^*$, homomorphisms $g_i, h_i : \Sigma^* \to \Sigma^*$ such that for each $j \geq 1$ $g_i(x_j) = h_i(x_j)$ if $i \neq j$ and $g_i(x_i) \neq h_i(x_i)$.

Corollary 3.9: If at least one of the conjectures 3.1, 3.5 is valid, then Conjecture 3.7 holds too.

Similarly, we can establish the connections of Conjecture 3.2 and its dual form, where the cardinality of representative subsets is uni-

formly bounded, to the corresponding statement about bounded cardinality of homomorphically independent sets.
The relations of those conjectures can be made clear easily considering the negations of these statements in terms of grids. And by similar arguments one concludes that both, Conjecture 3.2 and its dual form, are derivable from a conjecture on strictly increasing chains of maximal extensions.

Conjecture 3.10: For each finite alphabet Σ there exists a number $k \geq 1$ such that for all $t \geq 1$ and $x_1,\ldots x_t \in \Sigma^*$ it holds:
$M(\{x_1\}) \subsetneq \ldots \subsetneq M(\{x_1,\ldots,x_t\})$ implies that $t \leq k$.

Lemma 3.11: The validity of Conjecture 3.13 implies the validity of Conjecture 3.2 and its dual form.

To us it seems, that thinking of the mentioned conjectures in terms of grids might be a valuable basis for proving (or disproving) them finally. Some first steps were taken in [4]; assuming the existence of languages which disprove the Ehrenfeucht Conjecture, strong structural restrictions on those languages can be enforced by the application of Lemma 3.4.

REFERENCES

[1] Aho, A.V., Ullman, J. D.: "The Theory of Parsing, Translation and Compiling", vol. 1, Prentice-Hall (1972)

[2] Albert, J., Culik II, K., Karhumäki, J.: Test sets for contextfree languages and systems of equations over a free monoid, Inf. and Control 52, 2, 172-186 (1982)

[3] Albert, J., Wood, D.: Checking sets, test sets, rich languages and commutatively closed languages, J. Comp. System Sci. 26, 1, 82-91 (1983)

[4] Albert, J.: On test sets, checking sets, maximal extensions and their effective constructions, Forschungsbericht 129, Universität Karlsruhe (1983)

[5] Berstel, J.: "Transductions and Context-Free Languages", B.G. Teubner (1979)

[6] Culik II, K.: A purely homomorphic characterization of recursively enumerable sets, J. ACM 26,2, 345-350 (1979)

[7] Culik II, K.: Some decidability results about regular and pushdown translations, Inform. Proc. Letters 8, 5 - 8 (1979)

[8] Culik II, K.: Homomorphisms: decidability, equality and test sets, in "Formal Language Theory, Perspectives and Open Problems",

R. Book ed., Academic Press (1980)

[9] Culik II, K., Karhumäki, J.: Systems of equations over a free monoid and Ehrenfeucht conjecture, 9. ICALP, Aarhus, Lect. Notes in Comp. Sci. 140, Springer Verlag, 128-140 (1982)

[10] Culik II, K., Karhumäki, J.: On the Ehrenfeucht Conjecture for DOL languages, RAIRO, to appear

[11] Culik II, K., Salomaa, A.: On the decidability of homomorphism equivalence for languages, J. Comp. System Sci. 17, 20-39 (1978)

[12] Culik II, K., Salomaa, A.: Test sets and checking words for homomorphism equivalence, J. Comp. System Sci. 20, 379-395 (1980)

[13] Ehrenfeucht, A., Karhumäki, J., Rozenberg, G.: On binary equality languages and a solution to the Ehrenfeucht Conjecture in the binary case, J. of Algebra, to appear

[14] Ehrenfeucht, A., Rozenberg, G., Ruohonen, K.: A morphic representation of complements of recursively enumerable sets, J. ACM 28, 4, 706-714 (1981)

[15] Engelfriet, J., Rozenberg, G.: Fixed point languages, equality languages, and representations of recursively enumerable languages, J. ACM 27, 3, 499-518 (1980)

[16] Harrison, M.A.:"Introduction to Formal Language Theory", Addison-Wesley (1978)

[17] Maon, Y., Yehudai, A.: Test sets for checking morphism equivalence on languages with fair distribution of letters, Technical report, Tel-Aviv University, (1983)

[18] Rozenberg, G., Salomaa, A.: "The Mathematical Theory of L Systems", Academic Press (1980)

[19] Salomaa, A.:"Formal Languages", Academic Press (1973)

SPARSE ORACLES, LOWNESS, AND HIGHNESS[†]

José L. Balcázar
Facultat d'Informàtica
Universitat Politècnica de Barcelona
Barcelona, 34, Spain

Ronald V. Book
Department of Mathematics
University of California at Santa Barbara
Santa Barbara, Ca. 93106, U.S.A.

Uwe Schöning
Institut für Informatik
Universität Stuttgart
D-7000 Stuttgart 1, West Germany

1. INTRODUCTION

The polynomial-time hierarchy has been studied extensively since it lies between the class P of languages accepted deterministically in polynomial time and the class PSPACE of languages accepted (deterministically or nondeterministically) in polynomial space. Since the P =? NP problem is still unsolved, it is not known whether the hierarchy is nontrivial, although it is known that there is a relativization which allows the hierarchy to exist to at least three levels [3]. The purpose of the present paper is to study the role that structural notions such as being sparse, having polynomial-size circuits, etc., have in determining the underlying structure of complexity classes and in particular the underlying structure of the polynomial-time hierarchy.

Schöning [15] has considered a decomposition of the class NP which depends on the number of distinct levels in the polynomial-time hierarchy. Call a set A in NP "low" if for some n, $\Sigma_n^P(A) \subseteq \Sigma_n^P$ and call a set B in NP "high" if for some n, $\Sigma_{n+1}^P \subseteq \Sigma_n^P(B)$. Thus, if A is low, then with respect to the operator $\Sigma_n^P(\ )$, A does not encode the power of a quantifier, but if A is high, then with respect to the operator $\Sigma_n^P(\ )$, A does encode the power of a quantifier. It is easy to see that if a set in NP is both high and low, then the polynomial-

[†]This research was supported in part by the U.S.A.-Spanish Joint Committee for Educational and Cultural Affairs, by the Deutsche Forschungsgemeinschaft, and by the National Science Foundation under Grant No. MCS83-12472.

time hierarchy collapses. On the other hand, if the polynomial-time hierarchy extends to infinitely many levels, then the collection of low sets and the collection of high sets are disjoint (and there are sets in NP that are neither high nor low). Ko and Schöning [9] have shown that the sets in NP with polynomial-size circuits (hence, also the sparse sets in NP) are low. Thus, if a sparse set in NP is high, then the polynomial-time hierarchy extends only finitely many levels.

Here we extend the arguments of Ko and Schöning regarding lowness to sets that are not necessarily in NP. Let PH be the union of the classes in the polynomial-time hierarchy. We define the notions of "generalized low" and "generalized high" and show that there is a set in PH that is simultaneously generalized low and generalized high if and only if the hierarchy collapses. We also consider the notions of "extended low" and "extended high" that apply to sets that are not necessarily in PH. There are two principal results. First, either every sparse set in PH is generalized high or no sparse set in PH is generalized low. Second, either every sparse set is extended high or no sparse set is extended high.

2. PRELIMINARIES

It is assumed that the reader is familiar with the basic concepts from the theories of automata, computability, and formal languages. Some of the concepts that are most important for this paper are reviewed here, and notation is established.

For a string w, $|w|$ denotes the length of w. The empty string is denoted by e, $|e| = 0$. For a set S, $\|S\|$ denotes the cardinality of S. It is assumed that all sets of strings are taken over some fixed alphabet Σ that includes $\{0,1\}$. If $A \subseteq \Sigma^*$, then $\bar{A} = \Sigma^* - A$. For a set S and an integer $n \geq 0$, let $S_{\leq n}$ denote $\{x \in S \mid |x| \leq n\}$.

Let $<$ denote any standard polynomial-time computable total order defined on Σ^*. For a finite set $S \subset \Sigma^*$, say $S = \{y_1, \ldots, y_n\}$ where $i < j$ implies $y_i < y_j$, let $\langle S \rangle = \% y_1 \% \ldots \% y_n \%$ where $\%$ is a symbol not in Σ. Let $\langle \phi \rangle = \%$. We consider $\langle , \rangle$ to be an encoding function. Notice that if $S \in \Sigma^*$ is a finite set and $y \in \Sigma^*$, then the predicate "y is in S" can be computed in polynomial time from the inputs y and $\langle S \rangle$. We use this same notation for pairing functions so that $\langle x_1, \ldots, x_n \rangle$ denotes $\langle \{x_1, \ldots, x_n\} \rangle$ when each x_i is in Σ^*, and $\langle x,S \rangle$ denotes $\langle \{x\} \cup S \rangle$ when $x \in \Sigma^*$ and $S \subset \Sigma^*$ with S being finite.

For an integer $k > 1$, a polynomial p, and a set L, we may define a set A as follows: $x \in A$ if and only if $(\exists y_1)_p \ldots$

$(Q_k y_k)_p(\langle x, y_1, \ldots, y_k\rangle \in L)$. The quantifiers are intended to alternate between existential and universal so that if k is even, then Q_k is universal, and if k is odd, then Q_k is existential. For each i, the domain of y is bounded: $|y_i| \le p(|x|)$. Similarly, we may define a set B as follows: $x \in B$ if and only if $(\forall y_1)_p \ldots (Q_k y_k)_p(\langle x, y_1, \ldots, y_k\rangle \in L)$. Now Q_k is universal if k is odd and existential if k is even.

For set A, $B \subseteq \Sigma^*$, the _join_ of A and B is defined as $A \oplus B = \{0x \mid x \in A\} \cup \{1y \mid y \in B\}$.

We assume standard definitions of oracle machines and relativized complexity classes specified by deterministic or nondeterministic oracle machines that are time-bounded or space-bounded. See [8], [11], [15], or [18] for details.

Now we describe some well-known properties of the polynomial-time hierarchy. We refer the reader to the papers of Stockmeyer [17] and Wrathall [18] where the properties of this hierarchy were first described.

Definition 2.1. (a) Let A be a set. Define $\Sigma_0^P(A) = \Pi_0^P(A) = \Delta_0^P(A) = P(A)$, and for each integer $i \ge 0$, define $\Delta_{i+1}^P(A) = P(\Sigma_i^P(A))$, $\Sigma_{i+1}^P(A) = NP(\Pi_i^P(A))$, and $\Pi_{i+1}^P(A) = \text{co-}\Sigma_{i+1}^P(A)$. The structure $\{(\Delta_i^P(A), \Sigma_i^P(A), \Pi_i^P(A))\}_{i\ge 0}$ is the _polynomial-time hierarchy relative to A_. Define $PH(A) = \cup_{i\ge 0}\, \Sigma_i^P(A)$. (b) For each integer $i \ge 0$, define $\Delta_i^P = \Delta_i^P(\phi)$, $\Sigma_i^P = \Sigma_i^P(\phi)$, and $\Pi_i^P = \Pi_i^P(\phi)$. The structure $\{(\Delta_i^P, \Sigma_i^P, \Delta_i^P)\}_{i\ge 0}$ is the _polynomial-time hierarchy_. Define $PH = \cup_{i\ge 0}\, \Sigma_i^P$.

Recall that for each $k \ge 1$, a set A is in Σ_k^P if and only if there is a set $B \in P$ and a polynomial p such that for all x, $x \in A$ if and only if $(\exists y_1)_p \ldots (Q_k y_k)_p(\langle x, y_1, \ldots, y_k\rangle \in B)$. A similar characterization of the classes Π_k^P, $k \ge 1$, also holds.

It is known [3] that there is a set B such that $\Sigma_1^P(B) \subsetneq \Sigma_2^P(B) \subsetneq \Sigma_3^P(B)$ but it is not known whether there is any set C such that the polynomial-time hierarchy relative to C is infinite.

Clearly, for every set A, $PH(A) = \cup_{i\ge 0}\, \Pi_i^P(A)$ and $PH(A) = \cup_{i\ge 0}\, \Delta_i^P(A)$. For the most part the definition of $PH(A)$ as $\cup_{i\ge 0}\, \Sigma_i^P(A)$ is the most useful here.

3. THE DECOMPOSITION OF PH

There are a number of results that give conditions on NP under which the polynomial-time hierarchy "collapses," that is, it has only finitely many distinct levels. For example, Mahaney [13] has shown that $P = NP$ if and only if there is a sparse set S that is $\leq_m^P$-complete for NP, and Long [10] has shown that if there is a sparse set S such that $NP \subseteq P(S)$ and S is in Δ_2^P, then $PH = \Delta_2^P$. Schöning [15] developed a decomposition theory for NP that generalizes many of these results.

<u>Definition 3.1</u> [15]. For each integer $n > 0$, let $L_n = \{A \in NP \mid \Sigma_n^P(A) \subseteq \Sigma_n^P\}$ and let $H_n = \{A \in NP \mid \Sigma_{n+1}^P \subseteq \Sigma_n^P(A)\}$. Let $LH = \cup_{n\geq 0} L_n$ and $HH = \cup_{n\geq 0} H_n$.

<u>Proposition 3.2</u> [15]. (a) The polynomial-time hierarchy has only finitely many levels if and only if $LH \cap HH \neq \phi$ if and only if for some $n > 0$, $L_n = H_n = LH = HH = NP$. (b) The polynomial-time hierarchy has infinitely many levels if and only if $NP - (LH \cup HH) \neq \phi$.

One of the contributions of the present paper is to extend the techniques of [15] to develop a decomposition of the class PH that is similar to the decomposition of NP. In this section we describe this decomposition. In Section 4 we study some of the conditions that cause the polynomial-time hierarchy to collapse, conditions that can be defined on classes in this decomposition.

<u>Definition 3.3</u>. For every $i, j > 0$, define $L(i,j) = \{A \mid \Sigma_i^P(A) \subseteq \Sigma_j^P\}$ and define $H(k,\ell) = \{A \mid \Sigma_k^P \subseteq \Sigma_\ell^P(A)\}$.

Notice that for each $n > 0$, $L_n = L(n,n)$ NP and $H_n = H(n+1, n) \cap NP$. We call sets "low" if they lie in $LH = \cup_{n>0} L_n$ and "high" if they lie in $HH = \cup_{n>0} H_n$, and we use the term "generalized lowness" to refer to properties of sets in $\cup_{i,j} L(i,j)$ and the term "generalized highness" to refer to properties of sets in $\cup_{k,\ell} H(k,\ell)$.

Consider generalized lowness. Clearly, if $A \in L(i,j)$, then $A \in \Sigma_j^P$ so that $\cup_{i,j} L(i,j) \subseteq PH$. On the other hand, if the polynomial time hierarchy is infinite, then it makes no sense to take $i > j$ when considering $L(i,j)$ since for $i > j$, $L(i,j) = \phi$ if $\Sigma_j^P \neq \Sigma_{j+1}^P$ and $L(i,j) = PH$ if $\Sigma_j^P = \Sigma_{j+1}^P$. Thus, we consider classes $L(i,j)$ only when $i \leq j$.

Let $A \in \Sigma_n^P$. Which lowness properties are implied? That is, which pairs (i,j) are candidates for $A \in L(i,j)$? Since $A \in \Sigma_n^P$, for any i, $\Sigma_i^P(A) \subseteq \Sigma_i^P(\Sigma_n^P) = \Sigma_{i+n}^P$ so that if $j \geq i+n$, then we have $A \in L(i,$ trivially. Thus, if $A \in \Sigma_n^P$, then the lowness property $A \in L(i,j)$ ha significance only if $i+n > j$.

Consider generalized highness. Clearly, if $k \leq \ell$, then every set A satisfies $\Sigma_k^P \subseteq \Sigma_\ell^P(A)$ so that $H(k,\ell) = 2^{\Sigma^*}$. Thus, we consider classes $H(k,\ell)$ only when $k > \ell$.

Let $A \in \Sigma_n^P$. Which highness properties are implied? That is, which pairs (k,ℓ) are candidates for $A \in H(k,\ell)$? If A is $\leq_T^P$-complete for Σ_n^P, then $\Sigma_\ell^P(A) = \Sigma_{\ell+n}^P$ so that for $k > \ell+n$, $A \in H(k,\ell)$ implies that the polynomial-time hierarchy collapses. More precisely, if $k > \ell+n$, then $\Sigma_n^P \cap H(k,\ell) = \phi$ if $\Sigma_{\ell+n}^P \neq \Sigma_{\ell+n+1}^P$ and $\Sigma_n^P \cap H(k,\ell) = PH$ if $\Sigma_{\ell+n}^P = \Sigma_{\ell+n+1}^P$. Thus, if $A \in \Sigma_n^P$, then the highness property $A \in H(k,\ell)$ has significance only if $k \leq \ell+n$.

Schöning [15] showed that the polynomial-time hierarchy collapses if and only if there exist m and n such that $L_m \cap H_n \neq \phi$. We generalize this result as follows.

Theorem 3.4. If $L(i,j) \cap H(k,\ell) \neq \phi$ and $j+\ell < i+k$, then the polynomial-time hierarchy collapses to $\Sigma_{\max\{i,\ell\}+j-i}^P$.

Proof. Let $A \in L(i,j) \cap H(k,\ell)$. Since $A \in L(i,j)$, $\Sigma_i^P(A) \subseteq \Sigma_j^P$ so that $\Sigma_{\max\{i,\ell\}}^P(A) \subseteq \Sigma_{\max\{i,\ell\}+j-i}^P$. Since $A \in H(k,\ell)$, $\Sigma_k^P \subseteq \Sigma_\ell^P(A)$ so that $\Sigma_{k+\max\{i,\ell\}-\ell}^P \subseteq \Sigma_{\max\{i,\ell\}}^P(A)$. Thus, $\Sigma_{k+\max\{i,\ell\}-\ell}^P \subseteq \Sigma_{\max\{i,\ell\}+j-i}^P$ which implies that $PH = \Sigma_{\max\{i,\ell\}+j-i}^P$ provided that $k-\ell > j-i$ or, equivalently, $j+\ell > i+k$. □

Now suppose that $A \in \Sigma_n^P \cap L(i,j) \cap H(k,\ell)$. As noted above, this has significance only if $i \leq j < i+n$ and $\ell < k \leq \ell+n$. These inequalities do not necessarily imply $j+\ell < i+k$. However, if $k-\ell \geq n$ so that $k-\ell = n$, then the hypotheses of Theorem 3.4 hold.

4. GENERALIZED AND EXTENDED LOWNESS AND SPARSE ORACLES

The property of being generalized low asserts that the usefulness of a set as an oracle set is quite restricted. If $A \in L(i,j)$ where $i \leq j$, then A does not have the power of more than $j-i$ alternating quantifiers since $\Sigma_i^P(A) \subseteq \Sigma_j^P$. What type of sets have this property? Ko and Schöning [9] showed that sparse sets in NP have this property. As noted in Section 3, any generalized low set is in PH, so we may ask if every sparse set in PH has this property. Here we prove a somewhat stronger result.

Theorem 4.1. For any set A, if there is a sparse set S and integers n, k with $n > k$ such that $A \in \Sigma_k^P(S)$ and $A \in \Sigma_n^P$, then A is in $L(3, n+2)$.

Proof. Since $A \in \Sigma_k^P(S)$, there is a polynomial p and a deterministic polynomial time-bounded oracle machine M such that for all inputs x, $x \in A$ if and only if $(\exists y_1)_p \ldots (Q_k y_k)_p(\langle x, y_1, \ldots, y_k\rangle \in L(M,S))$.

Let $B = \{\langle x,T\rangle \mid T$ is a finite table and $(\exists y_1)_p \cdots (Q_k y_k)_p \langle x, y_1, \ldots, y_k\rangle \in L(M,T)\}$. Clearly, B is in Σ_k^P. Since S is sparse, there is a polynomial q such that for all n, $\|S_{\le n}\| \le q(n)$.

Consider any $C \in \Sigma_3^P(A)$. There is a polynomial r such that for all u, $u \in C$ if and only if there exists a table T of size at most $r(|u|)$ such that (i) for all x with $|x| \le r(|u|)$, $\langle x,T\rangle \in B$ if and only if $x \in A$, and (i) $(\exists z_1)_r (\forall z_2)_r (\exists z_3)_r \langle\langle u, z_1, z_2, z_3\rangle, T\rangle \in B$.

The predicate described by (i) is in Π_{n+1}^P since $B \in \Sigma_k^P$, $A \in \Sigma_n^P$, and $k < n$. The predicate described by (ii) is in Σ_{k+3}^P since $B \in \Sigma_k^P$. Thus, membership in C is described by a predicate of the form $\exists\left(\Pi_{n+1}^P \text{ and } \Sigma_{k+3}^P\right)$ so that C is in $\Sigma_{\max(n+2,k+3)}^P = \Sigma_{n+2}^P$, since $k < n$. Hence, $\Sigma_3^P(A) \subseteq \Sigma_{n+2}^P$ so that $A \in L(3, n+2)$. □

Observe that $A \in L(3, n+2)$ for a set $A \in \Sigma_n^P$ is a "meaningful" lowness condition as described in Section 3 since $3 \le n+2 < 3+n$. Also, notice that in the case $k = 0$ and $n = 1$, we have $A \in NP$ and $A \in P(S)$ so that $A \in L(3,3)$. Since $NP \cap L(3,3) = L_3$, we see that sets in NP with polynomial size circuits are in L_3. This was first proved by Ko and Schöning [9].

Notice that in Theorem 4.1, the condition that S is sparse could be replaced by the condition that S has polynomial size circuits and the conclusions will still hold. The idea that a set is in a certain level of the polynomial-time hierarchy but is also in a lower level of that hierarchy relativized to a sparse set might be called "pseudo-sparse." This notion encompasses such properties as a set being in APT [14], a set having Σ_k^P-circuits [9], a set having small generators [16,19], and other properties that can be translated to being easily recognized relative to a sparse set.

From one viewpoint the notion of generalized lowness should not force sets to be in PH but rather should apply to all sets with some particular property, e.g., to all sparse sets. This requires a change in the definitions. One way to accomplish this is as follows.

Definition 4.2. For each $n \ge 0$, let $EL_n = \{A \mid \Sigma_n^P(A) \subseteq \Sigma_{n-1}^P(A \oplus SAT)\}$ and let $EH_n = \{A \mid \Sigma_n^P(A \oplus SAT) \subseteq \Sigma_n^P(A)\}$. Let $EL = \cup_n EL_n$ and $EH = \cup_n EH_n$. A set is extended low if it is in EL and is extended high if it is in EH.

Let K_S be any set that is $\le_m^P$-complete for PSPACE. Clearly, for all n, $\Sigma_n^P(K_S) = \Sigma_{n-1}^P(K_S \oplus SAT) = \Sigma_n^P(K_S \oplus SAT) = PSPACE$. Thus, K_S is in $EL \cap EH$, so EL and EH are not disjoint. However, it is easy to see that $EL \cap EH \cap PH$ is empty if and only if the polynomial-time hierarchy is infinite.

Theorem 4.3. If S is sparse or if S has polynomial size circuits then $S \in EL_3$.

The proof of Theorem 4.3 is similar to that of Theorem 4.1.

Recall that a set A has polynomial-size circuits if and only if there is a sparse set S such that $A \in P(S)$.

The notions of extended lowness and extended highness appear to be natural extensions of the basic idea of Schöning that was developed in [15]. It is easy to see that for every n, $EL_n \cap NP = L_n$ and $EH_n \cap NP = H_n$.

5. GENERALIZED HIGHNESS AND SPARSE SETS

The property of being generalized high asserts that using such a set as an oracle set is an advantage. If $A \in H(k,\ell)$ where $k > \ell$, then A has the power of at least $k-\ell$ alternating quantifiers since $\Sigma_k^P \subseteq \Sigma_\ell^P(A)$. What type of sets have this property? Schöning [15] showed that certain types of complete sets for NP have this property, and it is easy to see that similar sets in each of the classes Σ_i^P, $i > 0$, have this property.

It is known [10,13] that if a sparse set is hard for NP with respect to $\leq_T^P$ or even weaker reducibilities such as $\leq_T^{SN}$, then the polynomial-time hierarchy collapses. Being hard for any of the classes Σ_i^P, $i > 0$, represents a certain "highness." Thus, we consider the situation in which a sparse set is generalized high. We need the following notion.

Definition 5.1. A set A is self-reducible if there exists a deterministic polynomial time-bounded oracle machine M such that (i) on input of size n, M queries the oracle only about strings of length at most $n-1$ and (ii) $L(M,A) = A$.

This definition captures the essential idea of the seemingly more general notions due to Ko [8] and Meyer and Paterson [14]. Notice that there are $\leq_m^P$-complete sets for NP that are self-reducible, and that at each stage Σ_k^P of the polynomial-time hierarchy, the $\leq_m^P$-complete set for Σ_k^P described by Wrathall [18] is self-reducible.

Lemma 5.2. If A is a self-reducible set and there is a $k > 0$ and a sparse set S such that $A \in \Sigma_k^P(S)$, then $A \in L(2, k+2)$.

Now we have our result on sparse sets that are generalized high.

Theorem 5.3. If there exists a sparse set that is generalized high, then the polynomial-time hierarchy collapses. That is, if there is a sparse set S such that for some k, ℓ with k > ℓ, $S \in H(k,\ell)$, then $\Sigma^P_{\ell+2} = \Pi^P_{\ell+2}$ so $PH = \Sigma^P_{\ell+2}$.

Proof. Suppose that $\Sigma^P_k \subseteq \Sigma^P_\ell(S)$ where S is sparse and k > ℓ. Let A be a set that is both self-reducible and also $\leq^P_m$-complete for Σ^P_k. Since A is self-reducible and $A \in \Sigma^P_k \subseteq \Sigma^P_\ell(S)$, $\Sigma^P_2(S) \subseteq \Sigma^P_{\ell+2}$ by Lemma 5.2. Since A is $\leq^P_m$-complete for Σ^P_k, $\Sigma^P_2(A) \subseteq \Sigma^P_{\ell+2}$ implies that $\Sigma^P_{k+2} \subseteq \Sigma^P_{\ell+2}$. Since $k+2 > \ell+2$, $\Pi^P_{\ell+2} \subseteq \Sigma^P_{k+2}$ so that $\Sigma^P_{k+2} \subseteq \Sigma^P_{\ell+2}$ implies that $\Pi^P_{\ell+2} \subseteq \Sigma^P_{\ell+2}$. Hence, $PH = \Sigma^P_{\ell+2} = \Pi^P_{\ell+2}$. □

Corollary 5.4. If A is sparse and A is extended high, then the polynomial-time hierarchy collapses.

Theorem 5.3 generalizes a number of results that assert conditions that force the polynomial-time hierarchy to collapse. Some of these conditions are as follows: (a) there is a sparse set S such that SAT $\leq^P_m$ S [13]; (b) there is a sparse or co-sparse set S that is $\leq^P_T$-complete for NP [10,13]; (c) NP has polynomial-size circuits [7]; (d) NP $\subseteq$ APT [14]; (e) NP = R [1]; (f) there is a sparse set S and an integer k such that $PH \subseteq \Sigma^P_k(S)$; (g) NP is p-selective [8]; (h) NP $\subseteq$ BPP [1].

6. MAIN RESULTS

In Theorem 5.3 we showed that the existence of a sparse set that is generalized high causes the polynomial-time hierarchy to collapse. Clearly, if the polynomial-time hierarchy does collapse, then every sparse set in the polynomial-time hierarchy is generalized high. Thus, we have the first of our two main theorems.

Theorem 6.1. Either *every* sparse set in PH is generalized high or *no* sparse set in PH is generalized high.

In Corollary 5.4 we showed that the existence of a sparse set that is extended high causes the polynomial-time hierarchy to collapse. But it is conceivable that one sparse set might be extended high while another sparse set is not extended high. However, this cannot be the case

Proposition 6.2 [12]. If the polynomial-time hierarchy collapses, then for every sparse set S, the polynomial-time hierarchy relative to S collapses.

Thus, if the polynomial-time hierarchy collapses, then for every sparse set S there is an integer k > 0 such that $PH(S) = \Sigma^P_k(S)$.

Since NP $\subseteq$ PH $\subseteq$ PH(S), this means that $\Sigma_k^P(S \oplus SAT) = \Sigma_k^P(S)$ so that S is extended high. Thus, we have our other result.

Theorem 6.3. Either every sparse set is extended high or no sparse set is extended high.

REFERENCES

1. L. Adleman, Two theorems on random polynomial time. Proc. 19th IEEE Symp. Foundations of Computer Science (1978), 75-83.
2. T. Baker, J. Gill, and R. Solovay, Relativizations of the P =? NP question. SIAM J. Computing 4 (1975), 161-173.
3. T. Baker and A. Selman, A second step towards the polynomial-time hierarchy. Theoret. Comput. Sci. 8 (1979), 177-187.
4. P. Berman, Relationships between density and deterministic complexity of NP-complete languages. Proc. 5th ICALP, Lecture Notes in Computer Science 67 (1978), 63-71.
5. L. Berman and J. Hartmanis, On isomorphisms and density of NP and other complete sets. SIAM J. Computing 6 (1977), 305-322.
6. S. Fortune, A note on sparse complete sets. SIAM J. Computing 8 (1979), 431-433.
7. R. Karp and R. Lipton, Some connections between nonuniform and uniform complexity classes. Proc. 12th ACM Symp. Theory of Computing (1980), 302-309.
8. K. Ko, On self-reducibility and weak P-selectivity. J. Comput. Syst. Sci. 26 (1982), 209-221.
9. K. Ko and U. Schöning, On circuit-size complexity and the low hierarchy in NP. SIAM J. Computing 13 (1984), to appear.
10. T. Long, A note on sparse oracles for NP. J. Comput. Syst. Sci. 24 (1982), 224-232.
11. T. Long, Strong nondeterministic polynomial-time reducibilities. Theoret. Comput. Sci. 21 (1982), 1-25.
12. T. Long and A. Selman, Relativizing complexity classes with sparse oracles. Unpublished abstract, June 1983.
13. S. Mahaney, Sparse complete sets for NP: solution to a conjecture of Berman and Hartmanis. J. Comput. Syst. Sci. 25 (1982), 130-143.
14. A. Meyer and M. Paterson, With what frequency are apparently intractable problems difficult? M.I.T. Technical Report, Feb. 1979.
15. U. Schöning, A low- and a high-hierarchy in NP. J. Comput. Syst. Sci. 27 (1983), 14-28.
16. U. Schöning, A note on small generators. Submitted for publication.
17. L. Stockmeyer, The polynomial-time hierarchy. Theoret. Comput. Sci. 3 (1976), 1-22.
18. C. Wrathall, Complete sets and the polynomial-time hierarchy. Theoret. Comput. Sci. 3 (1976), 23-33.
19. C. Yap, Some consequences of non-uniform conditions on uniform classes. Theoret. Comput. Sci. 27 (1984), 287-300.

COMPUTABILITY OF PROBABILISTIC PARAMETERS FOR SOME CLASSES OF FORMAL LANGUAGES

Joffroy BEAUQUIER (x)(xx)
Loÿs THIMONIER (xx)

(x) L.I.T.P. -U.E.R. de Mathématiques - Université Paris 7- 2 place Jussieu - 75221 PARIS Cédex 05 - FRANCE

(xx) Institut d'Informatique - U.E.R. de Mathématiques - Université de Picardie - 33, rue St Leu - 80039 AMIENS Cédex FRANCE.

ABSTRACT. :

In a previous paper [BT 83], some probabilistic notions of density and waiting time for a formal language have been studied. We prove here that these probabilistic parameters are computable with an arbitrary precision for some families of languages : the languages with an end marker ; the prefix-free regular sets, with matricial algorithms on Markov chains related to deterministic finite-state automata ; at the end, the prefix-free languages of palindrom words, for which the use of counting generating series yields new results in the equally likely case, already studied in [BT 83], and allows to give partial answers in the general case.

I. - INTRODUCTION.

During the past years, several studies have been made in order to develop a notion of density for formal languages (see, for instance [Ber 73] or [HP 83]). Most of them are relevant from the combinatorics of the free monoïd.

Nevertheless, in a previous paper [BT 83], we have shown how probabilistic notions as Bernouillian density or Bernouillan conditional waiting time could be used for this type of problems. The meaning of the Bernouillian parameters of a language is quite simple. Given a formal language L over an alphabet A and a probability distribution p over A, we draw (with replacement) letters of A according to p until the sequence of drawn letters constitues a word of L. The Bernouillian density of L is simply the probability, acting this way, to obtain a word of L whereas the Bernouillian waiting time of L is the average number of drawings necessary to obtain this word. In the present paper, we are just interested in the (apparently) simple question : L and p being given, how to compute the Bernouillian parameters of L. To compute means here : to be able, a real number ε beinggiven, to obtain a value of the density or of the waiting time with a precision ε. This problem appears as being, in fact, not simple at all, for arbitrary languages.

We prove that a computation is feasible for two large classes of languages : the languages with an end marker and the prefix-free regular sets. The last part can be viewed as an example of difficulties which can appear in the general case. We consider the prefix-free languages of palindrom words. Their treatment pinpoints the combinatorial difficulties of the problem and the important role of the counting generating series of these languages, especially of their analytic properties. So that likewise [Fla 83], this paper can be viewed as an example of the use (elementary here) of analytic techniques for proving combinatorial results.

In a 1st part (preliminaries) we give a very short proof, thanks to a Bernouilli process and our notion of density, of the following theorem : the Bernouilli measure of a prefix-free language is inferior to 1 (proposition II.1.5.). Then comparing with similar notions of other authors, we remark that, unlike previously, our density is not 0 for many usual languages.

In the 2nd part, we show, for the languages with an end marker, how the computation of the Bernouillian parameters is possible with an arbitrary precision (theorems III.2. and III.4.).

In a 3st part we consider the regular prefix-free languages : a similar result of computability is obtained with matricial algorithms resulting from the association on an absorbing Markov chain to a deterministic finite-state automaton recognizing the language (theorems IV.2.1 and IV.2.2.). Moreover we characterize the regular languages whose density equals the maximum 1 : if this result holds for one probability distribution, it is still true for any other distribution (theorems IV. 2.3. and IV.2.4.). We can also show a regular language with a given density δ belonging to $]0,1[$ (theorem IV.2.5.).

Finally in a 4th part we study the prefix-free languages of palindrom words, for which the use of counting generating series yields new results in the equally likely case, already studied in [BT 83], and allows to give partial answers in the general case (theorems V.2,V.3, V.4 V.6.).

II. - PRELIMINARIES.

II. 1. STOCHASTIC PROCESSES AND PREFIX-FREE LANGUAGES.

In this section, we will show how a particular stochastic process, namely the Bernouilli process, yields a new, very short proof of the well known fact, that the Bernouilli measure of a prefix-free language is inferior to 1. The reader is assumed to be familiar with classical notions of formal languages theory (as in [Har 78]), of discrete probabilities theory and of stochastic processes theory (as in [Chu 79] or [HPS 71]).

Let A be a finite alphabet. A* is the free monoïd generated by A with ε the empty word. A^+ is A* $-\{\varepsilon\}$ and $\mathcal{P}(A^*)$ is the set of the subsets (languages) of A*.

Definition II.1.1. : Let L be a language in A*. L is a code iff $x_1 x_2 \dots x_n = y_1 y_2 \dots y_m$, where $x_i, y_j \in L$ for $i = 1,2,\dots,n$, $j = 1,2,\dots m$, implies i) $n = m$ ii) $x_i = y_i$ for $i = 1,2,\dots,n$.

Definitions II.1.2. : Let L be a language in A* and let w be a word in L. w is said prefix-free for L iff $w = w_1 w_2$ and $w_1 \in L$ involves $w_2 = \varepsilon$. A language L in A* is said to be prefix-free if any w in L is prefix-free for L.

Clearly, a prefix-free language is a code. We note Pref(L), for any language L, the set of words w in L that are prefix-free for L. So that L is prefix-free iff L = Pref(L).

Definition II.1.3. : A Bernouilli measure on A* is a mapping $p : \mathcal{P}(A^*) \to (0, \infty)$ such that :
(i) $\forall a \in A$ $p(a) > 0$ and $p(A) = 1$ (ii) $p(\varepsilon) = 1$; $\forall w_1, w_2 \in A^*$ $p(w_1 w_2) = p(w_1)\, p(w_2)$ (iii) for $L \subset A^*$, $p(L) = \sum_{w \in L} p(w)$ (countable sum equal to $\lim_{n\to\infty} p(L \cap (\bigcup_{i=1}^{n} A^i))$)

There is a not straightforward proof of the fact that, if $L \subset A^*$ is a code and p is a Bernouilli measure on A*, then $p(L) \leqslant 1$ [HP 83].

Definition II.1.4. : Let L be a language over $A = \{a_1, \dots a_m\}$. Suppose A provided with the probability distribution $\{p_1, \dots p_m\}$ ($\sum_{i=1}^{m} p_i = 1$). A Bernouilli process related to L is a sequence of drawings with replacement of letters in A. This sequence is finite or finite depending on the following conditions : (1) As soon as the sequence of drawn letters constitues a word in L, the process stops (then this word is prefix-free for L). (2) The process can never stop : it produces then an infinite word, whose no left factor (prefix) is in L. Then we can prove very quickly that $p(L) \leqslant 1$, with L prefix-free, thanks to the:

Proposition II.1.5. : Let $L \subset A^*$ be a prefix-free language and let p be a Bernouilli measure over A^*. Then p(L) is the probability $\delta_{L,p}$ of obtaining a finite Bernouilli process related to L (with the restriction of p to A as a probability distribution).

II.2. BERNOUILLIAN PARAMETERS OF A FORMAL LANGUAGE.

Recall (cf. [BT 83]) that two notions naturally appear when one draws letters and considers the obtained words : first the outcome frequency of a prefix-free word in L during a large number of Bernouilli processes related to L (corresponding to the same probability distribution p). Secondly, when considering finite processes, the waiting time, that is the average number of drawings until obtaining a word in L.

According to the law of large numbers, the first notion corresponds to the probability $\delta_{L,p}$ of obtaining a finite process, that is producing a prefix-free word in L.

Let X_L be the random variable with values in $(\mathbb{N} - \{0\}) \cup \{\infty\}$ that corresponds to the number of drawings until obtaining a word in L (the value is ∞ if the process is infinite). We set :

$\pi_{L,n} = \text{proba}\ (X_L = n) = \sum_{w \in \text{Pref}(L) \cap A^n} p(w)$; then $\delta_{L,p} = \sum_{n=1}^{\infty} \pi_{L,n}$.

Definition II.2.1. : The counting generating series of an infinite language L over a finite alphabet A is

$\gamma(Z) = \sum_{n=1}^{\infty} \gamma_n Z^n$, with $\gamma_n = \text{card}\ (L \cap A^n)$ [Ber 73] [Ber 77] [Fla 83]

When the distribution p is equally likely (for i=1,2,...m, $p_i = \frac{1}{m}$), we note $\delta_{L,eq}$ for $\delta_{L,p}$; here we obtain $\delta_{L,eq} = \gamma(1/m)$ (γ series related to Pref(L)).

According to the law of large numbers, the second notion corresponds to the conditional expectation $\alpha_{L,p}$ (the condition being that the process is finite) of the random variable X_L.

Then $\alpha_{L,p} = \frac{1}{\delta_{L,p}} \sum_{n=1}^{\infty} n\, \pi_{L,n}$; in particular $\alpha_{L,eq} = \frac{1}{m} \frac{\gamma'(1/m)}{\gamma(1/m)}$ (with γ' for the derivative of γ).

We consider now the differences between these two notions and the similar notions of other authors ([Ber 73] , [BPS 84] , [HP 83]). Contrary to the already existing notions of density, $\delta_{L,p}$ always exists for any language L, and its values are different from 0 for usual languages (cf. [BT 83]). E.g. the density $\xi_{L,p}$ introduced in [BPS 84] and HP 83 is the limit (provided it exists) in the Cesaro sense of the sequence $\tilde{\pi}_{L,n}$, with $\tilde{\pi}_{L,n} = \sum_{w \in L \cap A^n} p(w)$

(i.e. $\tilde{\delta}_{L,p} = \lim_{n \to \infty} \frac{\tilde{\pi}_{L,1} + \dots + \tilde{\pi}_{L,n}}{n}$). If L is a prefix-free language, it is easy to see that $\tilde{\delta}_{L,p} = 0$. $\alpha_{L,p}$ $\delta_{L,p} = \sum_{n=1}^{\infty} n\, \pi_{L,n}$ corresponds to the average length $\lambda(L)$ introduced in [BP 84] , [HP 83], with the difference that our notion $\alpha_{L,p}$ can be observed during a stochastic process ; $\alpha_{L,p}$ can obviously be infinite (cf. the language $D_1'^*$ in the equally likely case [BT 83]).

III. THE LANGUAGES WITH AN END MARKER

Let $A = \{a_1, a_2, \dots a_m\}$ be the alphabet provided with the probability distribution $\{p_1, p_2, \dots p_m\}$ and let L be a language in A*. If a marker # is introduced, the alphabet becomes $A' = A \cup \{\#\}$ which is equipped with a new probability distribution $\tilde{p}$ $\{\lambda p_1, \lambda p_2, \dots, \lambda p_m, p_\# = 1 - \lambda\}$. Clearly, $L\# = \{w\# \ / w \in L\}$ is a prefix-free language.

We prove the following :

. Proposition III.1. : (i) $\delta_{A^*_N, \tilde{p}} = 1$ (ii) the remainder of order N $R_N = \sum_{n>N} \pi_{A^*\#,n}$ equals $(1-p_\#)^N$. It yields, from $\pi_{L\#,n} \leq \pi_{A^*\#,n}$:

Theorem III.2. : Let L be a language over the alphabet A and let # be a symbol not in A. Then $\delta_{L\#,\tilde{p}}$ is computable with any given precision

. Proposition III.3.

(i) $\alpha_{A^*\#,\tilde{p}} = \frac{1}{p_\#}$ (ii) the remainder of order N $S_N = \sum_{n>N} n\, \pi_{A^*\#,n}$ equals $(1-p_\#)^N (N + \frac{1}{p_\#})$. Likewise :

Theorem III.4. : Let L be a language over the alphabet A and let # be a symbol not in A. Then $\alpha_{L\#,\tilde{p}}$ is computable with any given precision.

IV. - THE REGULAR PREFIX-FREE LANGUAGES : ASSOCIATION OF AN ABSORBING MARKOV CHAIN TO A FINITE-STATE AUTOMATON.

IV. 1. PRELIMINARIES : ABSORBING HOMOGENEOUS MARKOV CHAINS WITH A FINITE STATE SPACE.

The classical notions about Markov chains with a countable state space can be found in [Chu 79] and [HPS 71] .

Consider a system that can be in a finite number s of states ; let $\mathcal{S}$ denote the set of states (state space) : we can assume that $\mathcal{S}$ is the subset $\{1, 2, \dots, s\}$. Let the system be observed at discrete moments of time n = 0,1,2,... and let X_n denote the random variable whose value is the state of the system at time n.

The Markov property is : $\forall n \geq 0$, $\forall i_0, i_1, \ldots, i_{n+1} \in \{1,2,\ldots,s\} = \mathcal{S}$ Proba $(X_{n+1}=i_{n+1} / X_0=i_0, X_1=i_1, \ldots X_n=i_n)$ = Proba$(X_{n+1}=i_{n+1} / X_n=i_n)$ (Proba (A/B) denotes the conditional probability of the event A given the event B).

The conditional probabilities Proba$(X_{n+1}=j / X_n=i)$ are the transition probabilities ; then are stationary when they are independent of n.

Definitions IV.1.1. : A Markov chain with m states is a sequence $(X_n)_{n\geq 0}$ of random variables whose values i_n belong to $\{1,2,\ldots,s\}$, verifying the Markov property. It is homogeneous if the transition probabilities are stationary : we assume it from now on.

So we can consider p_{ij}=Proba$(X_{n+1}=j / X_n=i)$, which defines a(s,s) transition matrix P such that (1) $\forall i,j$, $p_{ij} \in [0,1]$. The system beeing in the state i at the time n will be in any state j belonging to $\{1,2,\ldots,s\}$ at the time n+1, whence (2) $\forall i$, $\sum_{j=1}^{s} p_{ij}=1$.

To an homogeneous Markov chain with s states, we can associate a valued graph $G(\mathcal{S},\gamma)$, with $\mathcal{S}=\{1,\ldots,s\}$ the set of the vertices of the graph and γ the multivalued mapping : $i \in \mathcal{S} \to \{(j,p_{ij}) / j = 1 \text{ à } s\}$. It is easily be seen that Proba$(X_1=i_1,\ldots,X_n=i_n/X_0=i_0) = p_{i_0 i_1} p_{i_1 i_2} \cdots p_{i_{n-1} i_n}$

Definitions IV.1.2. : -) The state i is absorbing if $p_{ii}=1$ (then $j \neq i \Rightarrow p_{ij}=0$: once the system is arrived at the state i, it will remain in this state) ; let $\mathcal{S}_A$ be the set of the absorbing states
-) A Markov chain is absorbing if : i) $\mathcal{S}_A \neq \emptyset$; (ii) For every non absorbing state i, there is a finite sequence of transitions leading at last to one absorbing state.

From now on we consider only absorbing Markov chains.
Let i be a non absorbing state : there exists a probability, different from 0, not to come back in i (once the system is arrived at an absorbing state, it will remain in it) ; i is said to be transient ; we denote by $\mathcal{S}_T$ the set $\mathcal{S} - \mathcal{S}_A$.

We are concerned with the possible absorption, by starting from i belonging to $\mathcal{S}_T$, in a part $\mathcal{S}_F$ (final states) of $\mathcal{S}_A$. It is convenient to number the states from 1 to s, so as to the first states correspond to $\mathcal{S}_T$, the (s-r) last states correspond to $\mathcal{S}_F$ followed by $\mathcal{S}_A$; then P splits in blocks :

$$p = \begin{array}{ll} i = 1 \text{ to } r & \\ i = r+1 \text{ to } s & \begin{array}{l} \mathcal{S}_F \\ \mathcal{S}_A \\ \mathcal{S}_F \end{array} \end{array} \left[\begin{array}{c:c:c} Q & S_1 & S_2 \\ \hdashline (0) & \begin{matrix} 1 & & (0) \\ & \ddots & \\ (0) & & 1 \end{matrix} & (0) \\ \hdashline (0) & (0) & \begin{matrix} 1 & & (0) \\ & \ddots & \\ (0) & & 1 \end{matrix} \end{array} \right]$$

Let $\pi_{i, \mathcal{S}_F}$ be the probability of absorption from a transient state i in $\mathcal{S}_F$.

Proposition IV.1.3. (First mean value rule) : $\pi_{i, \mathcal{S}_F} = \sum_{j=1}^{s} p_{ij} \pi_{j, \mathcal{S}_F}$

Proposition IV.1.4. : $\forall i$, $\pi_{i, \mathcal{S}_F}$ is computable by using the fundamental matrix $(I_r - Q)^{-1}$ (I_r is the (r,r) identity matrix) of the absorbing Markov chain.

Proposition IV.1.5. : $\forall i \in \mathcal{S}_T = \{1, \ldots, r\}$, $\pi_{i, \mathcal{S}_A} = 1$.

Corollary : if (i) from the state 1 it is possible to reach all the absorbing states (it is the case if the Markov graph is geometrically a tree whose root is the state 1, where considering always 1 single link between 2 different vertices instead of 1 et 2 transition arrows (and deleting every transition arrow from an absorbing state to itself)) (ii) $\mathcal{S}_A \supsetneqq \mathcal{S}_F$, then $\pi_{1, \mathcal{S}_F} < 1$.

Let Y_i be the random variable whose value is the number of transitions next to the absorption in $\mathcal{S}_F$ (if there is an absorption), ∞ otherwise. We consider now the conditional expectation $\alpha_{i, \mathcal{S}_F} = E(Y_i$ / the condition beeing the absorption in $\mathcal{S}_F$). The following results are less known : the propositions IV.1.6. and IV.1.7. can be found in [Rad 70], the resulting proposition IV.1.8. is new.

Proposition IV.1.6. (Second mean value rule) :

If $\mathcal{S}_F = \mathcal{S}_A$, $\alpha_{i, \mathcal{S}_A} = 1 + \sum_{j=1}^{r} p_{ij} \; \alpha_{j, \mathcal{S}_A}$

Proposition IV.1.7. : $\forall i \in \mathcal{S}_T$, $\alpha_{i, \mathcal{S}_A}$ is computable (it is the sum of the i th row of the fundamental matrix $(I_r - Q)^{-1}$ of the absorbing Markov chain).

Proposition IV.1.8. : If $\mathcal{S}_F \subsetneqq \mathcal{S}_A$, $\forall i$, $\alpha_{i, \mathcal{S}_F}$ is computable by using the second mean value rule, thanks to a reduction and a probabilistic renormalization of the graph and of the transition matrix. More precisely : i) $\forall i$ transient state, $\forall j \in \mathcal{S}_T \cup \mathcal{S}_F$,

p_{ij} becomes $p_{ij} \times \dfrac{1}{1 - \sum_{k \in \mathcal{S}_A - \mathcal{S}_F} p_{ik}} = p'_{ij}$ ii) P becomes P' =

$$P' = \begin{array}{c} i=1 \text{ to } r \\ \mathcal{S}_F \end{array} \left(\begin{array}{c|c} Q' & S'_1 \\ \hline (0) & \begin{matrix} 1 & & (0) \\ & \ddots & \\ (0) & & 1 \end{matrix} \end{array} \right)$$

IV.2. ASSOCIATION OF AN ABSORBING MARKOV, DESCRIBING A BERNOUILLI PROCESS, TO A FINITE STATE AUTOMATON :

We consider now a regular prefix-free language L and a finite-state automaton recognizing L. In the frame of a Bernouilli process related to L, over the alphabet $A = \{a_1,..,a_m\}$ with the probability distribution $\{p_1,...,p_m\}$, this automaton is, by a natural manner, in correspondance with an absorbing Markov chain :

-) The states of the automaton become the states of the chain (the initial state becoming the first state 1)

-) The production of a letter a_i by the automaton corresponds to a transition of probability p_i;

-) The final states, producting words of L, become absorbing states ;

-) A set of states is closed if it is impossible to go out of it after the entering into it : any closed set of states containing no final state becomes a single absorbing state.

-) The other states become transient states : finally any transient state leads to a word of L (final absorbing state) or not (non final absorbing state), therefore in all the cases to an absorbing state.

The propositions IV.1.4. and IV.1.7. immediately yield :

Theorem IV.2.1. : $\delta_{L,p}$ is computable (with a matrix) ; it is $\pi_{1,\mathcal{S}_F}$

Theorem IV.2.2. $\alpha_{L,p}$ is computable (with a matrix), independently from $\delta_{L,p}$ (the definition with a series leads usually to compute first the product $\alpha_{L,p}\ \delta_{L,p}$) : it is $\alpha_{1,\mathcal{S}_F}$.

The proposition IV.1.5. and its corollary yield also immediately, since here the Markov graph, without the transition arrows but with only links between the vertices, is geometrically a tree whose root is the initial state 1 :

Theorem IV.2.3. If the associated absorbing Markov chain is verifying $\mathcal{S}_A = \mathcal{S}_F$ (all the absorbing states are final states), then $\delta_{L,p} = 1$. This result is independent from the probability distribution p over the alphabet A (it is connected with the "geometry" of the graph).

Theorem IV.2.4. : If $\mathcal{S}_A \supsetneq \mathcal{S}_F$ (i.e; there exists a non final absorbing state), then $\delta_{L,p} < 1$ (this result is also independent from the probability distribution p).

At the end, we consider $L = \{x\}^+$, the language of the words over $A = \{x,y\}$

with only letters equal to x ; it is a regular language, with $Pref(L) = \{x\}$ and verifies $\delta_{L,p} = \delta_{Pref(L),p} = p_x$, whence :

Theorem IV.2.5. : $\forall \delta \in]0,1[$, there exists a probability distribution p over $A = \{x,y\}$ and a regular language over A with δ for density $\delta_{L,p}$ (it can be immediately generalized to an alphabet of more than 2 Letters)

V. - THE CASE OF PALINDROM LANGUAGES.

PAL_m is the language of palindrom words (w is equal to its mirror-image) over $\{a_1, a_2, .. a_m\}$ provided with the probability distribution $p = \{p_1, p_2, .., p_m\}$. For $m \geqslant 3$ (the case m=2 is well known in probability theory) it has been obtained [BT 83] approximates formulas and asymptotic results, in the equally likely case, by means of a combinatorial lemma [BT 83] about γ_n, the number of prefix-free words of length n in PAL_m : $\gamma_{2n-1} = \gamma_{2n} = m^n \sum_{i=2}^{n} \gamma_i \, m^{n-i}$.

We present here a method, in the equally likely case and in a part of the general case, for computing the parameters with an arbitrary precision, and giving upper and lower bounds, useful when only a comparison of parameters is needed. The main idea is the use of the counting generating series [II.2] of $Pref(PAL_m)$: $\gamma(z) = \sum_{n=1}^{\infty} \gamma_n z^n$, whith a convergence radius ρ verifying $\rho \in [1/m, 1[$, which represents an analytic function for $|z| < \rho$. More generally, the analytic properties of $\gamma(z)$, for a given language L, are related to asymptotic properties of the sequence (γ_n), possibly giving results about the structure of L [Fla 83].

We prove the following :

.Proposition V.1. $\rho = 1/\sqrt{m}$ (the combinatorial lemma allows to obtain a polynomial lower bound of γ_n).

. If $p_+ = \max(p_1, p_2, \ldots p_m)$, then $p_+ \geqslant 1/m$ and $\delta_{PAL_{m,p}} \leqslant \gamma(p_+)$. It yields :

. Theorem V.2. : if $p_+ \in [1/m, 1/\sqrt{m}[$, $\delta_{PAL_{m,p}} \leqslant m\, p_+^2 \,(1 + mp_+)/(1 - mp_+^2)$

. Theorem V.3. : for $m \geqslant 2$, if $p_+ \in [1/m, 1/\sqrt{m}[$, then the remainder R_{2N} of order 2N in the series $\delta_{PAL_{m,p}}$ verifies $R_{2N} \leqslant (p_+ + 1)(mp_+^2)^{N+1} / p_+ (1 - mp_+^2)$.

If $p_- = \min(p_1, p_2, .. p_m)$ then $p_- \leqslant 1/m$ and $\delta_{PAL_{m,p}} \geqslant \gamma(p_-)$. It yields :

. Theorem V.4. : $\delta_{PAL_{m,p}} \geqslant mp_-^2 + \frac{m}{m-1} p_-^3 (p_- + 1) \left[\frac{m(m-3)}{1 - p_-^2} + \frac{m+1+2mp_-^2}{1 - p_-^4 m} \right]$

. Theorem V.5. : if $p_+ \in \left[\frac{1}{m}, \frac{1}{\sqrt{m}}\right[$ $\alpha_{PAL_{m,p}} \, \delta_{PAL_{m,p}} \leqslant mp_+^2 \left[-m^2 p_+^3 + 3mp_+ + 2\right] / (1 - mp_+^2)^2$

. Theorem V.6. : if $p_+ \in \left[\frac{1}{m}, \frac{1}{\sqrt{m}}\right[$, then the remainder S_{2N} of order 2N in the series $\alpha_{PAL_{m,p}} \, \delta_{PAL_{m,p}}$ verifies

$$S_{2N} \leqslant \frac{(mp_+^2)^{N+1} \left[-2Nmp_+^3 - m(2N-1)\, p_+^2 + (2N+2)\, p_+ + (2N+1)\right]}{p_+ (1 - mp_+^2)^2}$$

VI. - CONCLUSION.

It has been shown that density and conditional waiting time are effectively computable for large classes of formal languages.

From a theoretical point of view, one can think to extend these results to other classes. A good tool to achieve that could be the counting generating series of a language, whose analytic properties are certainly of the highest interest ([Fla 83], [Sed 83], [Thi 84]).

From a more practical point of view, the effective computability of probabilistic parameters gives new insights into the domain of probabilistic algorithms. Indeed, each time that the sequences of actions of a given process are, a priori, known to be in a given language L, the ability to compute the density of L allows, at each time, to predicate the most probable sequence of actions to come in the future. That idea is at the basis of actual works of the authors especially about probabilitic paging algorithms and probabilistic algorithms of pattern recognition after transmission through a channel [BT 84].

ACKNOWLEDGEMENT :

The authors want particularly to thank Philippe FLAJOLET for stimulating remarks.

REFERENCES :

[Ber 73] J. BERSTEL. -"Sur la densité asymptotique des langages formels", in M. NIVAT Ed., Automata, languages and Programming, North Holland P.C., Amsterdam (1973), pp. 345-358.

[Ber 77] J. BERSTEL, editor. - "Séries formelles en variables non commutatives et applications", Actes de la 5° Ecole de Printemps d'Informatique Théorique, Vieux Boucau (1977), published by L.I.T.P. and E.N.S.T.A, Paris.

[BPS 84] J. BERSTEL, D.PERRIN, M.P. SCHUTZENBERGER, -"Théorie des codes"- chap. VI (densités) L.I.T.P. report 84-5 and "The theory of codes" - Academic Press - To appear-

[BT 83] J. BEAUQUIER, L. THIMONIER. - "Formal languages and Bernouilli processes". L.I.T.P. report 83-30 - To appear in Colloquia Mathematica Societatis Janos Bolyai - (Hungary-September 1983).

[BT 84] J. BEAUQUIER. L. THIMONIER - "Formal languages, probabilities, paging and decoding algorithms"- L.I.T.P. report (to appear in 1984) - Submitted at the 25 th F.O.C.S. -I.E.E.E. (USA 1984)

[Chu 79] Kai Lai CHUNG. - "Elementary probability theory with stochastic processes" - Springer Verlag, (1979)

[Fla 83] P. FLAJOLET. - "Ambiguity and transcendence" - Proceedings of the 16th Annual Symposium on Theory of Computing (USA 1984)

[Har 78] M.A. HARRISON. - "Introduction to Formal language Theory"- Addison Wesley P.C. (1978)

[HP 83] G. HANSEL, D. PERRIN. - "Codes and Bernouilli Partitions" - Math. Systems Theory 16, 133-157 (1983).

[HPS 71] P. HOEL, S. PORT, G. STONE. -"Introduction to stochastic processes" Houghton Mifflin Company ed. (1971).

[Rad 70] L. RADE. - "The teaching of probability and statistics" proceedings of the 1st C.S.M.P. International Conference - Edited by L. Rade, Almqvist & Wiksell Forlag AB, Stockholm (1970).

[Sed 83] R. SEDGEWICK. - "Mathematical Analysis of Combinatorial Algorithms" in G. LOUCHARD, G. LATOUCHE editors, Probability theory and computer science, Academic press 1983.

[Thi 84] L. THIMONIER. - "A methodology for computing probabilistic parameters of formal languages : generating functions, non ambiguity and Dyck languages".
-L.I.T.P. report (to appear in 1984) -
Submitted at the 2nd Symposium on Theoretical Aspects of Computer Science (West Germany January 1985)

A TRUELY MORPHIC CHARACTERIZATION OF RECURSIVELY ENUMERABLE SETS

Franz Josef Brandenburg

Lehrstuhl für Informatik, Universität Passau
Postfach 2540, 8390 Passau, Federal Republic of Germany

ABSTRACT

Continuing recent research of [3-9] et al. we study the composition of homomorphisms, inverse homomorphisms and twin-morphisms $<g,h>$ and $<g,h>^{-1}$, where $<g,h>(w) = g(w) \cap h(w)$ and $<g,h>^{-1}(w) = g^{-1}(w) \cap h^{-1}(w)$. We investigate some properties of these morphic mappings and concentrate on a characterization of the recursively enumerable sets, which says: For every recursively enumerable set L there exist four homomorphisms such that

$$L = f_1 \circ <f_2,f_3> \circ f_4^{-1}(\{\$\})$$

and $$L = h_1 \circ <h_2,h_3>^{-1} \circ h_4^{-1}(\{\$\}),$$

and four homomorphisms are minimal for such representations.

INTRODUCTION

Homomorphisms over free monoids are the simplest and most natural operations on formal languages. They have been used in many classical theorems, such as the Chomsky-Schützenberger Theorem and Greibach's theorem on the representation of the context-free languages. These representations hold accordingly for other classes of languages, such as the recursively enumerable sets. See [10]. From an algebraic point of view they provide a characterization of a particular class of languages in terms of primitive operations and a basis consisting of a single language from the class. This language reflects the structure inherent to all languages of the class. Here we restrict ourselves to operations built from homomorphisms and aim at a simple language in the basis. Since homomorphisms are closed under composition and, e.g., the regular sets are closed under homomorphism and inverse homomorphism we must use more complex operations to jump out of this class. This is done by pairing homomorphisms and inverse homomorphisms. The morphic operations so defined are very powerful. They are based on the notion of equality sets, which together with homomorphisms, intersection with regular sets and the MIN-operation have been used recently for simple representations of the recursively enumerable sets. See [2-5,7]. Another direction of recent research aims at purely homomorphic characterizations of the regular sets. See [6,8,9]. In this paper we combine these approaches. Immediately, we obtain a truely morphic characterization of the recursively enumerable sets from the trivial set {\$}, which is then tuned to four homomorphisms, which is least possible.

TWIN-MORPHISMS AND INVERSE TWIN-MORPHISMS

We assume familiarity with the basic notions from formal language theory. See, e.g., [10].

DEFINITION. A homomorphism between free monoids is a mapping $h : X^* \to Y^*$ with $h(xy) = h(x)h(y)$ for all $x,y \in X^*$. h is nonerasing, if $h(a) \neq \lambda$ and alphabetic, if $h(a) \in Y \cup \{\lambda\}$ for every $a \in X$, where λ denotes the empty string. h is uniform, if there is a string $u \in Y^*$ and $h(a) = au$ for every $a \in X$.

DEFINITION. For homomorphisms g and h and a string w define $\langle g,h\rangle(w) = g(w) \cap h(w)$ and $\langle g,h\rangle^{-1}(w) = g^{-1}(w) \cap h^{-1}(w)$.
$\langle g,h\rangle$ is called the twin-morphism or homomorphic equality of g and h and $\langle g,h\rangle^{-1}$ is called inverse twin-morphism or inverse homomorphic equality.
Define the equality set of g and h by $Eq(g,h) = \{w \in X^* \mid g(w) = h(w)\}$.

Homomorphic equality and inverse homomorphic equality operations have been studied in [3,4]. There the generalization to n-tuples of homomorphisms has been discussed, too. For equality sets see, e.g. [2,4,5,7,10].

The following properties of twin-morphisms and inverse twin-morphisms are obvious from the definitions.

LEMMA 1. For homomorphisms g and h and a string w,

$$\langle g,h\rangle(w) = \begin{cases} g(w), & \text{if } g(w) = h(w) \text{ (if } w \in Eq(g,h)) \\ \text{undefined, otherwise} \end{cases}$$

$$\langle g,h\rangle^{-1}(w) = \{v \mid g(v) = h(v) = w\} = Eq(g,h) \cap g^{-1}(w).$$

LEMMA 2. For homomorphisms g and h and strings u and v,

$$\langle g,h\rangle(u)\langle g,h\rangle(v) \subseteq \langle g,h\rangle(uv), \text{ and}$$

$$\langle g,h\rangle^{-1}(u)\langle g,h\rangle^{-1}(v) \subseteq \langle g,h\rangle^{-1}(uv),$$

where in the first case either equality holds or $\langle g,h\rangle(u)$ or $\langle g,h\rangle(v)$ is undefined (and thus empty).

Clearly, for every homomorphism h, $\langle h,h\rangle = h$ and $\langle h,h\rangle^{-1} = h^{-1}$. Hence, twin-morphisms are a generalization of homomorphisms and inverse twin-morphisms are a generalization of inverse homomorphisms. In general, a twin-morphism is a partial mapping, which preserves the concatenation of strings, and thus the morphism property.

LEMMA 3. For homomorphisms g and h and a language L,

$$\langle g,h\rangle(L) = g(Eq(g,h) \cap L), \text{ and}$$

$$\langle g,h\rangle^{-1}(L) = Eq(g,h) \cap g^{-1}(L) = Eq(g,h) \cap g^{-1}(L) \cap h^{-1}(L).$$

Lemma 3 and the set theoretic identities $f(X) \cap Y = f(X \cap f^{-1}(Y))$ and $f^{-1}(X \cap Y) = f^{-1}(X) \cap f^{-1}(Y)$ imply that twin-morphisms are closed under right-composition with homomorphisms and inverse twin-morphisms are closed under left-composition with inverse homomorphisms.

The reversed compositions do not hold, in general. Hence, (inverse) twin-morphisms are not closed under composition, contradicting the composition property of the building homomorphisms.

LEMMA 4. For homomorphisms f,g and h

$$\langle f,g\rangle \circ h = \langle fh,gh\rangle \quad \text{and} \quad f^{-1} \circ \langle g,h\rangle^{-1} = \langle gf,hf\rangle^{-1}$$

If f is an injective homomorphism, then

$$f \circ \langle g,h\rangle = \langle fg,fh\rangle \quad \text{and} \quad \langle g,h\rangle^{-1} \circ f^{-1} = \langle fg,fh\rangle^{-1}$$

LEMMA 5. There exist homomorphisms $g,h_1,\ldots,h_4$ such that for all homomorphisms α,β

$$g \circ \langle h_1,h_2\rangle \neq \langle \alpha,\beta\rangle \quad \text{and} \quad \langle h_3,h_4\rangle^{-1} \circ g \neq \langle \alpha,\beta\rangle^{-1}.$$

Proof. Let g map into $\{a\}^*$ where a is a letter. Define $h_1(b) = b$, $h_1(a) = h_1(d) = a$, $h_1(c) = c$, $h_2(a) = b^2$, $h_2(b) = a^2$, $h_2(c) = c\,b^2$ and $h_2(d) = \lambda$. Then $\langle h_1,h_2\rangle(R) = \{c\,b^2a^4\ldots b^{2^{n-1}}a^{2^n} \mid n \geq 2, n \text{ even}\}$ and $L_1 = g \circ \langle h_1,h_2\rangle(R) = \{a^{2^n-1} \mid n \geq 3, n \text{ odd}\}$, where $R = \{c\}\{a,b\}^*\{d\}^*$. However, $L_1 \neq \langle\alpha,\beta\rangle(Q)$ for all α,β and all regular sets Q. To see this observe that $L_1 \subseteq \{a\}^*$ and $L_1 = \langle\alpha,\beta\rangle(Q)$ implies that α and β map into $\{a\}^*$ (or can be restricted in this way). Then $Eq(\alpha,\beta)$ and $<\alpha,\beta>(Q)$ can be accepted by a one-counter machine and thus are context-free languages, whereas L_1 is not.

Define $h_3(a) = h_4(a') = a$, $h_3(b) = h_4(b') = b$, and $h_3(a') = h_3(b') = h_4(a) = h_4(b) = \lambda$. Then $L_2 = <h_3,h_4>^{-1} \circ g^{-1}(\{a\}^*)$ is the hardest equality set and non-context-free. See [2,7]. As above, if $<\alpha,\beta>^{-1}(\{a\}^*)$, then α and β map into $\{a\}^*$ and $<\alpha,\beta>^{-1}(\{a\}^*)$ is a context-free language. □

For more results on twin-morphisms and inverse twin-morphisms such as the closure of particular classes of languages under these operations we refer the interested reader to [3,4].

REPRESENTATION THEOREM

A starting point for our main result is the following representation theorem from [3,4]. It shall be reproved in Theorem 1 in a slightly different form.

PROPOSITION 1. For every recursively enumerable set L there exist alphabetic homomorphisms f_1,f_2,f_3 and a regular set R such that

$$L = f_1 \circ <f_2,f_3> (R)$$

and

$$L = f_1 \circ <f_2,f_3>^{-1}(R).$$

Another starting point is a characterization of the regular sets from [9].

PROPOSITION 2. For every regular set R there exist a uniform homomorphism g_1, a homomorphism g_2 and an alphabetic homomorphism g_3 such that

$$R = g_1^{-1} \circ g_2 \circ g_3^{-1}(\{\$\}).$$

Combining these approaches and using Lemma 4 and the fact that every uniform homomorphism is injective we immediately obtain truely morphic characterizations of the recursively enumerable sets from $\{\$\}$ of the form

$$L = f_1 \circ <f_2,f_3> \circ g_1^{-1} \circ g_2 \circ g_3^{-1}(\{\$\}) \quad \text{and} \quad L = f_1 \circ <g_1f_2,g_1f_3>^{-1} \circ g_2 \circ g_3^{-1}(\{\$\}).$$

Exploiting common properties and goals of the homomorphisms these representations can be tuned to four homomorphisms, and the homomorphisms can be restricted to nonerasing and alphabetic, respectively.

At their first appearance Proposition 1 and Proposition 2 look quite different. They start from different basic languages, Proposition 1 additionally glues a pair of homomorphisms into a new mapping, and most importantly, they reach extremely different families of languages, namely the recursively enumerable sets, and the regular sets. However, both characterizations are built on the same phenomenon: the equality of homomorphic images of individual strings.

This is directly and consequently modelled by our new morphic mappings $\langle g,h\rangle$ and $\langle g,h\rangle^{-1}$. On the other hand it is the key to the characterization of Proposition 2, where it is modelled by a sequence of homomorphisms and inverse homomorphisms in the following way:

Let $A = (Q,\Sigma,\delta,q_0,F)$ be a nondeterministic finite automaton with m states $\{q_0,\ldots,q_{m-1}\}$. Suppose that no transition reaches q_0 and no transition leaves any final state, and that A makes no λ-moves with the exception of $\delta(q_0,\lambda) \in F$ iff $\lambda \in L(A)$. Determined by δ define an alphabet $\Delta = \{(q,a,q') \mid q' \in \delta(q',a)\}$, where $q,q' \in Q$, $a \in \Sigma \cup \{\lambda\}$ and $a = \lambda$ iff $q = q_0$, $q' \in F$ and $\lambda \in L(A)$. Let z be a new symbol. Define homomorphisms $g_3 : \Delta^* \to \{\$\}^*$ with $g_3(q,a,q') = \lambda$, if $q' \notin F$ and $g_3(q,a,q') = \$$, if $q' \in F$, $g_2 : \Delta^* \to (\Sigma \cup \{z\})^*$ with $g_2(q_i,a,q_j) = z^i a z^{m-j}$, if $q_j \notin F$ and $g_2(q_i,a,q_j) = z^i a a^m$, if $q_j \in F$, $\pi : \Delta^* \to \Sigma^*$ with $\pi(q,a,q') = a$, and $g_1 : \Sigma^* \to (\Sigma \cup \{z\})^*$ by $g_1(a) = a z^m$.

Encode each computation of A on an input $w = a_1 a_2 \ldots a_n$ into a string $\pi^{-1}(w) = (q_{i_1},a_1,q'_{i_1})(q_{i_2},a_2,q'_{i_2})\ldots\ldots(q_{i_n},a_m,q'_{i_n})$. $\pi^{-1}(w)$ is a <u>valid sequence</u> iff $q_{i_1} = q_0$, $q_{i_m} \in F$ and most importantly $q'_{i_j} = q_{i_{j+1}}$ for $j = 1,\ldots,n-1$. The validity can be checked by homomorphisms doing a local equality test. Observe that $Eq(g_2, g_1 \circ \pi) = \{v_1 v_2 \ldots v_k \mid \text{each } v_i \in \Delta^* \text{ is a valid sequence, } k \geq 0\}$ and $Eq(g_2, g_1 \circ \pi) \cap g_3^{-1}(\{\$\}) = \{v \in \Delta^* \mid v \text{ is a valid sequence}\}$.

Hence, $w \in L(A)$ if and only if $w \in g_1^{-1} g_2 g_3^{-1}(\{\$\})$.

Note that the above constructions are a simple and direct proof of Proposition 2.

<u>THEOREM 1.</u> For every recursively enumerable set L there exist homomorphisms $f_1,\ldots,f_4$ and h_1,h_2,h_3 with f_2,f_3,h_2 and h_3 nonerasing and f_1,f_4 and h_1 alphabetic, and a fixed alphabetic homomorphism $h_4 : \{0,1,\$\}^* \to \{\$\}^*$ with $h_4(0) = h_4(1) = \lambda$ and $h_4(\$) = \$$, such that

$$L = f_1 \circ \langle f_2,f_3\rangle \circ f_4^{-1}(\{\$\})$$

and

$$L = h_1 \circ \langle h_2,h_3\rangle^{-1} \circ h_4^{-1}(\{\$\}).$$

<u>Proof.</u> Let $L = L(M')$ for a deterministic, single tape Turing machine M'. Let X be the input alphabet of M'. In general, M' enlarges its work space during a computation, which will imply erasing homomorphisms. Hence, we consider the deterministic linear-bounded automaton $M = (K, X \cup \{d\}, Y, \delta, k_0, F)$ associated with M'. If M' operates on an input w', then M simulates M' on an input $w = w'd\ldots d$, where d is a new symbol.

The homomorphisms $f_1,\ldots,f_4$ and h_1,h_2,h_3 depend on the Turing machine M. Their domain and their range is determined by K, Y and δ. f_4 however, is independent of L and of M.

For every input string $w \in L(M)$ accepted by M there is a sequence of instantaneous descriptions $ID_0, ID_1, \ldots, ID_t$ making up an accepting computation of M on w. Each ID_i is a string in $Y^* K Y^*$ of length n for some $n \geq 1$, $ID_0 = q_0 w$ and $ID_t \in Y^* F Y^*$. Suppose that $t \geq 3$ so that M makes at least three steps before it may come to acceptance. Encode each such sequence of instantaneous descriptions by tripling into a string of the format of $C(w)$.

$$C(w) = [\overline{\overline{ID_0}}\overline{\overline{\#}}, \overline{ID_0}\overline{\#}, ID_1\#][\overline{ID_0}\overline{\#}, ID_1\#, ID_2\#]\ldots\ldots[ID_{i-1}\#, ID_i\#, ID_{i+1}\#]\ldots$$

$$\ldots[ID_{t-2}\#, ID_{t-1}\#, \overline{ID_t}\overline{\#}][ID_{t-1}\#, \overline{ID_t}\overline{\#}, \overline{\overline{ID_t}}\overline{\overline{\#}}].$$

Here the tripling of strings $x = a_1 a_2 \ldots a_n$, $y = b_1 b_2 \ldots b_n$ and $z = c_1 c_2 \ldots c_n$ of equal length n means a string $[x,y,z] = (a_1,b_1,c_1)(a_2,b_2,c_2)\ldots.(a_n,b_n,c_n)$ of length n over the triple alphabet. Formally, $C(w)$ is a string from a subset of $(\Delta \times \Delta \times \Delta)^*$ as indicated by the barring and the endmarker $\#$, where $\Delta = \Gamma \cup \overline{\Gamma} \cup \overline{\overline{\Gamma}}$ and $\Gamma = K \cup Y \cup \{\#\}$. Here barring means a new copy of the alphabet.

We wish nonerasing homomorphisms f_2, f_3, h_2, h_3. Therefore attach a head $H(w)$ and a tail $T(w)$ to each $C(w)$. $H(w)$ and $T(w)$ are strings over new symbols of the following form: For every symbol α and its barred copy $\overline{\alpha}$ and every $n \geq 1$ let $\underline{\alpha}^n$ denote any string from $\{\alpha^i \overline{\alpha}\, \alpha^{n-i-1}\}$. Let $\alpha, \beta, \tau, \sigma, \not c, \$$ be new symbols and define $\Sigma = (\Delta \times \Delta \times \Delta) \cup \{\not c, \alpha, \beta, \tau, \sigma, \$\} \cup \{\not c, \alpha, \beta, \tau, \sigma, \$\}$. For a string of the form of $C(w)$ let $n = |[\overline{\overline{ID_0}}\overline{\overline{\#}}, \overline{ID_0}\overline{\#}, ID_1\#]|$. Thus n is the length of each block of $C(w)$ endmarked by $\#$'s and $n = |w| + 2$. If n is even, then $H(w) = \not c\ \underline{\alpha}\ \underline{\beta}^2 \ldots.\underline{\beta}^{n-2}\ \underline{\alpha}^{n-1}$ and $T(w) = \underline{\tau}^{n-1}\ \underline{\sigma}^{n-2} \ldots.\underline{\sigma}^2 \underline{\tau}\$$, and if n is odd, then $H(w) = \overline{\not c}\ \underline{\beta}\ \underline{\alpha}^2 \ldots.\underline{\beta}^{n-2}\ \underline{\alpha}^{n-1}$ and $T(w) = \underline{\tau}^{n-1}\ \underline{\sigma}^{n-2} \ldots.\underline{\tau}^2 \underline{\sigma}\overline{\$}$.

Define homomorphisms g_2 and g_3 over Σ^* by the table

	$\not c$	$\overline{\not c}$	α	$\overline{\alpha}$	β	$\overline{\beta}$	τ	$\overline{\tau}$	σ	$\overline{\sigma}$	$\$$	$\overline{\$}$
g_2	$\not c$	$\not c$	α	α	β	β	σ	$\sigma\sigma$	τ	$\tau\tau$	$\tau\$$	$\sigma\$$
g_3	$\not c\alpha$	$\not c\beta$	β	$\beta\beta$	α	$\alpha\alpha$	τ	τ	σ	σ	$\$$	$\$$

and extend them to $(\Delta \times \Delta \times \Delta)^*$ by $g_2((a,b,c)) = (a,b)$, if $a \notin \overline{\overline{\Gamma}}$, $g_2((a,b,c)) = \alpha$, if $a \in \overline{\overline{\Gamma}}$, $g_3((a,b,c)) = (b,c)$, if $c \notin \overline{\Gamma}$ and $g_3((a,b,c)) = \tau$, if $c \in \overline{\Gamma}$.

Notice that $g_2(H(w)\,C(w)\,T(w)) = g_3(H(w)\,C(w)\,T(w)) = \not c\,\underline{\alpha}\,\underline{\beta}^2 \ldots \underline{\alpha}^n\ [\overline{ID_0}\overline{\#}, ID_1\#] \ldots \ldots [ID_{i-1}\#, ID_i\#]\ldots\ldots[ID_{t-1}\#, \overline{ID_t}\overline{\#}]\ \underline{\tau}^n \ldots \underline{\sigma}^2 \underline{\tau}\$$, if n is even, and that $g_2(H(w)\,C(w)\,T(w)) = g_3(H(w)\,C(w)\,T(w)) = \overline{\not c}\,\underline{\beta}\,\underline{\alpha}^2 \ldots \underline{\alpha}^n\ [\overline{ID_o}\overline{\#}, ID_1\ ]\ldots\ [ID_{i-1}\#, ID_i\#] \ldots \ldots[ID_{t-1}\#, \overline{ID_t}\overline{\#}]\ \underline{\tau}^n \ldots \underline{\tau}^2\ \underline{\sigma}\overline{\$}$, if n is odd, provided that $H(w)$, $C(w)$ and $T(w)$ are strings of the proper format. Because of the triple encoding, this can be specified by a regular set R. See [3,4]. Then $\{H(w)\,C(w)\,T(w) \mid w \in L(M)\} = Eq(g_2, g_3) \cap R = Eq(g_2, g_3) \cap g_2^{-1}(g_2(R))$.

Let $R = L(A)$ for a deterministic finite automaton $A = (Q, \Sigma, \delta, q_0, F)$. Suppose that A has m states so that $Q = \{q_0, \ldots, q_{m-1}\}$. Because of the particular format of the strings $H(w)\,C(w)\,T(w)$, the initial state q_0 is never reached by a move and there is a unique final state q_f, which can only be reached upon reading $\$$ or $\overline{\$}$.

According to the transitions of A define a new alphabet $\Omega \subseteq Q \times \Sigma \times Q$ by $(q,a,q') \in \Omega$ if and only if $\delta(q,a) = q'$. Hence, Ω depends on A, which depends on the fixed form of the strings $H(w)C(w)T(w)$, and on the transitions of the Turing machine M. Thus Ω depends on M. Ω is the domain of the homomorphisms f_2, f_3, f_4, h_1, h_2 and h_3.

Let z be a new symbol, not in Ω, and let φ be a uniform homomorphism with $\varphi(a) = a\, z^m$ for every a. Recall that A has m states $\{q_0,\dots,q_{m-1}\}$. Define nonerasing homomorphisms f_2 and f_3 over Ω. f_2 and f_3 can be seen as extensious of g_2 and g_3 modelling both $Eq(g_2,g_3)$ and the regular set R. For every $(q_i,a,q_j) \in \Omega$ with $a \in \Sigma$ let $f_2((q_i,a,q_j)) = \varphi(g_2(a))$, $f_3((q_i,a,q_j)) = z^i g_3(a) z^{m-j}$, if $a \notin \{\$,\bar{\$}\}$ and $|g_3(a)| = 1$, $f_3((q_i,a,q_j)) = z^i\, \xi\, z^m\, \eta\, z^{m-j}$, if $g_3(a) = \xi\,\eta$ with symbols ξ and η, and $f_3((q_i,\$,q_j)) = f_3((q_i,\bar{\$},q_j)) = z^i\ \$\ z^m$. Let f_4: $\Omega^* \to \{\$\}^*$ be an alphabetical homomorphism with $f_4((q_i,\$,q_f)) = f_4((q_i,\bar{\$},q_f)) = \$$, and $f_4(a) = \lambda$, otherwise.

Then $Eq(f_2,f_3) \cap f_4^{-1}(\{\$\}) = \{c_1c_2\dots c_r \mid c_i = (p_i,a_i,p_i') \in \Omega,\ a_1a_2\dots a_r = H(w)C(w)T(w)$ for some $w \in L(M)$, $p_1p_2\dots p_r$ is the sequence of states of the finite automaton A controlling $a_1\dots a_r$ with $p_1 = q_0$, $p_i' = \delta(p_i,a_i)$, $p_r' = q_f\}$. To see this, observe that $h_4^{-1}(\{\$\}) = \{c_1c_2\dots c_r\dots c_s \mid c_i = (p_i,a_i,p_i') \in \Omega$, $p_r' = q_f$ and $p_i' \neq q_f$ for all $i \neq r\}$. For each such $v = c_1\dots c_r\dots c_s$, $f_2(v) = \varphi(g_2(a_1\dots a_s))$. $f_2(v) = f_3(v)$ implies $p_1 = q_0$, otherwise $f_3(v)$ begins with z. Because of the choice of A, $a_1 = ¢$ or $a_1 = \bar{¢}$ and $p_1' = \delta(q_0,¢)$ or $p_1' = \delta(q_0,\bar{¢})$. Consider $a_1 = ¢$ and let $p_1' = q_j$. Then $f_2((q_0,¢,q_j)) = ¢\, z^m$ and $f_3((q_0,¢,q_j)) = ¢\, z^m\, \alpha\, z^{m-j}$. Now, $c_2 = (p_2,a_2,p_2')$ implies $p_2 = q_j$ and $a_2 = \bar{\alpha}$, otherwise, if $p_2 \neq q_j$ then $f_3(c_1c_2) = ¢\, z^m\, \alpha\, z^{\ell}\dots$ with $\ell \neq m$ whereas $f_2(c_1c_2) = ¢\, z^m\, \alpha\, z^m$. If $a_2 \neq \bar{\alpha}$, then $a_2 = \alpha$, however there is no transition $\delta(q_j,\alpha)$ in A and no symbol (q_j,α,q_j') in Ω. Continuing in this way we see that f_2 "chases" f_3 on the prefixes of $c_1c_2\dots c_r$. The blocks of exactly m z's under f_2 imply $p_i' = p_{i+1}$ by f_3. For the same reason $p_r' = q_f$, and no other p_i' or p_i equals q_f. So f_2 and f_3 simulate the behaviour of the finite automaton A. Moreover, $g_2(a_1a_2\dots a_r) = g_3(a_1a_2\dots a_r)$, so that $a_1a_2\dots a_r = H(w)C(w)T(w)$ for some $w \in L(M)$.

Notice that $R = f_2^{-1}(f_3(f_4^{-1}(\{\$\}))) = f_3^{-1}(f_2(f_4^{-1}(\{\$\})))$.

Finally, define a homomorphism $f_1 : (\Delta \times \Delta \cup \{¢,\alpha,\beta,\tau,\sigma,\$\} \cup \{z\})^* \longrightarrow X^*$, which erases all symbols but those from $\Delta \times \Delta$ of the form $(\bar{a},b)$ with $a \in X$. Let $f_1((\bar{a},b)) = a$. Thus f_1 is alphabetic and retrieves the very input string $w \in L$ from $[\overline{ID}_0\bar{\#}, ID_1\#]$ erasing $\bar{q}_0$ and the dummy symbols $\bar{d}$ from $\overline{ID}_0$.

Hence, $L = f_1 \circ \langle f_2,f_3\rangle \circ f_4^{-1}(\{\$\})$, which proves the first representation.

For the second we extend the homomorphisms f_2 and f_3. Let $\{b_0,b_1,\dots,b_r\} = \Delta \times \Delta \cup \{\not c,\alpha,\beta,\tau,\sigma,\$\} \cup \{z\}$ be the range alphabet of h_2,h_3 with $\$ = b_0$. Define a homomorphism $\psi : \{b_0,\dots,b_r\} \to \{0,1,\$\}^*$ by $\psi(b_i) = 01^i0$, if $i \geq 1$ and $\psi(\$) = \$$. Let $h_2 = \psi f_2$ and $h_3 = \psi f_3$. Then $Eq(h_2,h_3) = Eq(f_2,f_3)$, $h_2^{-1}(h_4^{-1}(\{\$\})) = h_2^{-1}(\{0,1\}^*\{\$\}\{0,1\}^*) =$ $\{(q_1,a_1,q_1')(q_2,a_2,q_2')\dots(q_\nu,a_\nu,q_\nu')\dots(q_\mu,a_\mu,q_\mu') \mid (q_i,a_1,q_i') \in \Omega,\ q_\nu' = q_f,$ $a_\nu \in \{\$,\bar{\$}\}$ and $q_i' \neq q_f$ for $i \neq \nu$, and $Eq(h_2,h_3) \cap h_2^{-1}(h_4^{-1}(\{\$\})) = \{H(w)C(w)T(w) \mid w \in L(M)\}$. Let $h_1 = f_1f_2$. Then $L = h_1 \circ \langle h_2,h_3\rangle^{-1} \circ h_4^{-1}(\{\$\})$, and the proof is complete. □

THEOREM 2. The characterizations from Theorem 1 are optimal. I.e., starting from the basis $\{\$\}$ one cannot characterize the recursively enumerable sets by our morphic mappings using three homomorphisms, or four homomorphisms in another sequence, or four homomorphisms three of which are nonerasing.

Proof. Clearly, a twin-morphism or an inverse twin-morphism must be used to jump beyond the regular sets. If L is a finite language, then $h(L)$ and $\langle g,h\rangle(L)$ are finite, and $h^{-1}(L)$ and $\langle g,h\rangle^{-1}(L)$ are regular sets. By results from [3,4], a twin-morphism or an inverse twin-morphism maps a regular set into (a subclass of) the class MULTI-RESET, which, e.g., is a subclass of the class of languages accepted in realtime by nondeterministic Turing machines, which is a proper subclass of the recursively enumerable sets. Hence, three homomorphisms or four homomorphisms in another sequence cannot yield the representations of Theorem 1. Accordingly, f_4 or h_4 cannot be chosen to be nonerasing, and with f_1 and h_1 nonerasing one ends up in MULTI-RESET. □

CONCLUSION

Our characterizations from Theorem 1, $L = f_1 \circ \langle f_2,f_3\rangle \circ f_4^{-1}(\{\$\})$ and $L = h_1 \circ \langle h_2,h_3\rangle^{-1}(\{0,1\}^*\{\$\}\{0,1\}^*)$ impressively demonstrate the power of representation of homomorphisms using their strength of comparing or matching equal strings. Vice-versa they show that recursively enumerable sets can be represented with weak control mechanisms, so weak that they can be captured by four homomorphisms. That's surprising, at least to me.

There are some open problems when the leading homomorphisms are nonerasing. Can every language of the form $L = g_1(Eq(g_2,g_3) \cap R)$ be represented by $L = f_1 \circ \langle f_2,f_3\rangle \circ f_4^{-1}(\{\$\})$ or $L = h_1 \circ \langle h_2,h_3\rangle^{-1} \circ h_4^{-1}(\{\$\})$, if f_1,g_1 and h_1 are nonerasing homomorphisms and R is a regular set? Such representations hold for extensions to n-tuples $\langle h_1,\dots,h_n\rangle$, $\langle h_1,\dots,h_n\rangle^{-1}$ and $Eq(h_1,\dots,h_n)$. This follows by applying the techniques developed in this paper to results in [3,4].

On the Herbrand Kleene Universe for Nondeterministic Computations

Manfred Broy

Universität Passau
Fakultät für Informatik
Postfach 25 40
8390 Passau

Abstract

For nondeterministic recursive equations over an arbitrary signature of function symbols including the nondeterministic choice operator "or" the interpretation is factorized. It is shown that one can either associate an infinite tree with the equations, then interprete the function symbol "or" as a nondeterministic choice operator and so mapping the tree onto a set of infinite trees and then interprete these trees. Or one can interprete the recursive equation directly yielding a set-valued function. Both possibilities lead to the same result, i.e. we obtain a commuting diagram. This explains and solves a problem posed in [Nivat 80]. Basically the construction gives a generalisation of the powerdomain approach applicable to arbitrary nonflat (nondiscrete) algebraic domains.

1. Introduction

In the algebraic semantics as described in [Nivat 75] it is shown that a recursive equation can either be interpreted directly in an algebraic domain D or first transformed into an infinite tree (the Herbrand Kleene interpretation) and then interpreted in D with identical results. As pointed out in [Nivat 80] for nondeterministic equations containing the nondeterministic choice operator "or" one would like to have analogous techniques. One would like to transform a recursion equation with the nondeterministic choice operation "or" either at first into an infinite tree (without interpreting "or", just considering it as a binary function symbol), then interprete "or" yielding a set of infinite (deterministic) trees, that can be interpreted then. Or one could directly interprete the recursive equation over set-valued functions. Like in the deterministic case one is interested that all these interpretations form a commuting diagram. The function mapping applications of recursively defined functions onto sets of possibly infinite trees is what we call nondeterministic Herbrand Kleene interpretation.

As pointed out in [Nivat 80], it seems difficult or even impossible to define the nondeterministic Herbrand Kleene interpretation. In fact it is impossible in the classical way as long as one wants to consider a domain (a "powerdomain") of infinite trees, and just monotonic and continuous functions for it. However, applying the techniques of [Broy 82], that have originally been developed to give a

REFERENCES

[1] J. Berstel, "Transductions and Context-Free Languages", Teubner 1979.

[2] R. Book and F.J. Brandenburg, "Equality Sets and Complexity Classes" SIAM J. Comput. 9, 729-743, (1980).

[3] F.J. Brandenburg, "Extended Chomsky-Schützenberger Theorems", Lecture Notes in Computer Science 140, 83-93, (1982).

[4] F.J. Brandenburg, "Representations of language families by homomorphic equality operations and generalized equality sets", to appear in Theoret. Comput. Sci.

[5] K. Culik II, "A purely homomorphic characterization of recursively emunerable sets", J. Assoc. Comput. Mach. 26, 345-350, (1979).

[6] K. Culik II, F. Fich, and A. Salomaa, "A homomorphic characterization of regular languages", Discrete Applied Math. 4, 149-152, (1982).

[7] J. Engelfriet and G. Rozenberg, "Fixed point languages, equality languages and representations of recursively enumerable languages", J. Assoc. Comput. Mach. 27, 499-513, (1980).

[8] J. Karhumäki and M. Linna, "A note on morphic characterization of languages", Discrete Applied Math. 5, 243-246, (1983).

[9] M. Latteux and J. Lequy, "On the composition of morphisms and inverse morphisms", Lecture Notes in Computer Science 154, 420-432, (1983).

[10] A. Salomaa, "Juwels of Formal Language Theory", Computer Science Press 1981.

denotational semantics to concurrent, communicating programs, the reason for the problem can be explained and a "nonclassical" solution can be envisaged. Combining several orderings nondeterministic Herbrand Kleene interpretations can be defined, such that the resp. constructions commute and the techniques of "algebraic" semantics by interpretations of recursive equations as infinite trees can be carried over to nondeterministic computations. In particular, such techniques can be applied for arbitrary nonflat (nondiscrete) domains. Thus a generalized construction for a nondeterministic Herbrand-Kleene interpretation is obtained where the problems of interpreting nondeterministic choice are separated from the problems of classical determinate interpretations. In this version all proofs are omitted. An extended version can be obtained by the author containing all proofs.

2. Basic Definitions

A signature $\Sigma = (S, F)$ is a pair consisting of a set S of sorts and a set F of function symbols with some fixed functionality $s_1 \times \ldots \times s_n \rightarrow s_{n+1}$ for each $f \in F$ with $s_i \in S$. For our purpose it is sufficient to consider just a one-element set of sorts and an arbitrary set F of function symbols not containing $\bot$ and "or". In particular we consider

$$\begin{array}{lll} \Sigma = (\{s\}, F) & \text{where} & \bot \notin F, \text{ or} \notin F, \\ \Sigma^+ = (\{s\}, F^+) & \text{where} & F^+ = F \cup \{\bot, \text{or}\}, \\ \Sigma^\bot = (\{s\}, F^\bot) & \text{where} & F^\bot = F \cup \{\bot\}, \end{array}$$

where the functionality of $\bot$ is $\rightarrow s$ and that of "or" is $s \times s \rightarrow s$.

A (total) <u>Σ-algebra</u> $A = \langle \{s^A\}_{s \in S}, \{f^A\}_{f \in F} \rangle$ consists of a family of carriersets s^A for each sort $s \in S$ and a family of functions f^A for each $f \in F$ with a functionality according to the functionality of f.

A is called an <u>algebraic Σ-algebra</u> iff all s^A form consistently complete countably algebraic domains and all f^A denote countinuous functions.

The <u>termalgebra</u> W_Σ is the Σ-algebra, the sorts of which consist of all well-formed terms (of the resp. sorts) formed from the function symbols in F and the functions consisting of the corresponding "term-construction" operations. By $I_A(t)$ we denote the interpretation of a term t in the algebra A.

W_{Σ^+} (and also $W_{\Sigma^\bot} \subseteq W_{\Sigma^+}$) can be partially ordered by (cf. [Nivat 75]):

$$t \sqsubseteq r \text{ iff } t = \bot \vee (t = f(t_1,\ldots,t_n) \wedge r = f(r_1,\ldots,r_n) \wedge t_1 \sqsubseteq r_1 \wedge \ldots \wedge t_n \sqsubseteq r_n)$$

Since "$\sqsubseteq$" defines a partial ordering with least element $\bot$ we can form the ideal completion leading to the algebra of finite and infinite terms or trees (also called "magmas"). By $W^\infty_{\Sigma^+}$ and $W^\infty_{\Sigma^\bot}$ the ideal completions of W_{Σ^+} and $W_{\Sigma^\bot}$ resp. are denoted. As demonstrated in [Nivat 75] the interpretation I_A can be continuously extended in a unique way to "infinite" terms from $W^\infty_{\Sigma^\bot}$. This extension will be denoted by I^∞_A in the following.

3. On Nondeterministic Herbrand Kleene Interpretations

Let A be an algebraic Σ-algebra. By HK we abreviate the Herbrand-Kleene Universe, i.e. the algebra $W^{\infty}_{\Sigma}\bot$. It is our goal to define mappings (the application of I^{∞}_{A} to sets of terms is assumed to be a shorthand for the elementwise application):

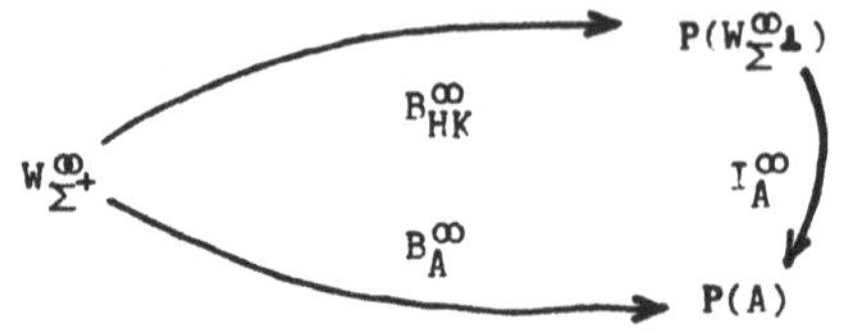

where the following equations are to hold:

$B^{\infty}_{A}[t] \;=\; \{\, I^{\infty}_{A}(t) \,\}$, for $t \in W^{\infty}_{\Sigma}\bot$;

$B^{\infty}_{A}[or(t1,\, t2)] = B^{\infty}_{A}[t1] \cup B^{\infty}_{A}[t2]$,

$B^{\infty}_{A}[f(t_1,\ldots,t_n)] = \{\, f^{A}(a_1,\ldots,a_n) \colon \forall\, i,\; 1 \leq i \leq n \colon a_i \in B^{\infty}_{A}[t_i] \,\}$ for $f \in F$

Note, that variables can be seen as nullary function symbols. The intended function B^{∞}_{HK} is called <u>nondeterministic Herbrand Kleene interpretation</u>. For finite terms t the sets $B^{\infty}_{HK}[t]$ and $B^{\infty}_{A}[t]$ are uniquely defined, if we interprete the equations above as an inductive definition. However, this does not work for infinite terms t. Actually, at a first glance it is not clear whether such a function B^{∞}_{HK} actually exists and is uniquely determined. In spite of this we can prove that B^{∞}_{HK} cannot be monotonic.

<u>Lemma 1</u>: There does not exist an ordering on $P(W^{\infty}_{\Sigma}\bot)$ such that B^{∞}_{HK} is monotonic.

<u>Proof</u>: Consider the following terms:

$t1 = or(s(s(\bot)),\, \bot)$, $\qquad B^{\infty}_{HK}[t1] = \{s(s(\bot)),\, \bot\}$

$t2 = or(s(s(\bot)),\, or(s(\bot),\, \bot))$, $\qquad B^{\infty}_{HK}[t2] = \{s(s(\bot)),\, s(\bot),\, \bot\}$

$t3 = or(s(s(\bot)),\, or(s(s(\bot)),\, \bot))$, $\qquad B^{\infty}_{HK}[t3] = \{s(s(\bot)),\, \bot\}$

We have $t1 \sqsubseteq t2 \sqsubseteq t3$; so the monotonicity of B^{∞}_{HK} would immediately imply

$$B^{\infty}_{HK}[t1] \sqsubseteq B^{\infty}_{HK}[t2] \sqsubseteq B^{\infty}_{HK}[t3];$$

Since $B^{\infty}_{HK}[t1] = B^{\infty}_{HK}[t3]$ one immediately obtains by the antisymmetry of a partial ordering $B^{\infty}_{HK}[t1] = B^{\infty}_{HK}[t2]$ which obviously gives a contradiction.

□

The proof shows a problem arising with nondeterministic computations. Scott's theory of computation is based on the principle of approximation: every object is determined uniquely by the set of its finite approximations (forming an ideal). For nondeterministic computations one has to consider sets of objects. A set of objects is approximated by a set of approximations for these objects. However, unfortunately, if such objects $x \neq y \neq z$ are in the partial order, i.e. if $x \sqsubseteq y \sqsubseteq z$ holds, then every approximation of y is also an approximation of z; every set approximating $\{x, z\}$ contains approximations of x and thus of y. So an approximation of $\{x, z\}$ is always an approximation of $\{x, y, z\}$ and vice versa. Both are not distinguishable by the classical approximation order, the Egli-Milner order. This explains one of the basic anomalies in the generalized (to nonflat domains) powerdomain approach (cf. [Plotkin 76], [Smyth 78]).

Nondeterministic interpretations do not only describe one computation, but a set of feasible computations. So one has two independent notions of approximations: the classical approxiation "is less defined" of computations in the sense of Scott and the approximation "has a wider spectrum of choices" on the level of nondeterminism. It is not surprising that it is not possible to combine these two distinct notions of approximation into one single partial ordering without running into problems.

4. Some definitions on powersets and powerdomains

Let DOM be a consistently complete, countably algebraic domain; for $S, S1, S2 \subseteq DOM$ we define:

$MIN(S) = \{x \in S : \forall y \in S : y \sqsubseteq x \Rightarrow x = y\}$,

$$CLOSE(S) = \{x \in DOM : \exists S0 \subseteq S: ((\forall a, b \in S0 \, \exists z \in S0 : a \sqsubseteq z \wedge b \sqsubseteq z) \wedge x = \text{lub } S0) \vee ((\forall a, b \in S0 \, \exists z \in S0 : z \sqsubseteq a \wedge z \sqsubseteq b) \wedge x = \text{glb } S0)\},$$

$CONE(S) = \{x \in DOM : \exists y, z \in S : y \sqsubseteq x \sqsubseteq z\}$.

S is called <u>convex</u> iff $CONE(S) = S$; S is called <u>closed</u> iff $CLOSE(S) = S$.

The following preorderings are used (cf. [Plotkin 76], [Smyth 78]):

$S1 \sqsubseteq_M S2$ iff $\forall y \in S2 \, \exists x \in S1 : x \sqsubseteq y$,
$S1 \sqsubseteq_{EM} S2$ iff $S1 \sqsubseteq_E S2 \wedge S1 \sqsubseteq_M S2$.

Over nonflat (nondiscrete) domains these relations just define preorderings. What sets are identified if we try to make these relations into orderings can be seen from the following lemma:

<u>Lemma 2</u>: For closed sets $S1, S2$ we have:

$S1 \sqsubseteq_M S2$ iff $MIN(S1) \sqsubseteq_M MIN(S2)$,
$S1 \sqsubseteq_{EM} S2$ iff $CONE(S1) \sqsubseteq_{EM} CONE(S2)$.

This lemma shows one pathological property of the powerdomains based on these "orderings": In a powerdomain particular distinct sets are considered as being equivalent, i.e. the powerdomain constructions actually consider classes of equivalent sets. Sets may not only be equivalent, because they cannot be distinguished by the orderings above. Due to the principle of finite approximability and continuity two sets are considered to be equivalent in a countably algebraic powerdomain based on some ordering, iff the classes of finite sets of finite elements that approximate these sets in the sense of these orderings are identical. Let FDOM denote the set of finite elements from DOM.

We take here a very concrete set-theoretic view of powerdomains. Their elements are just represented by subsets of P(DOM), i.e. by particular elements of the powerset over DOM. These representations are chosen in a very particular way which is most convenient for our construction.

The power domain PD(DOM) of erratic nondeterminism (also called Plotkin power domain or Egli-Milner power domain) is defined as follows. The set of finite elements is represented by the convex hull of finite sets of finite elements:

$$GD = \{CONE(S) : S \subseteq FDOM \wedge |S| < \infty\}.$$

We immediately can prove that $(GD, \sqsubseteq_{EM})$ is po-set. PD(DOM) is defined as the ideal-completion of $(GD, \sqsubseteq_{EM})$. We choose as representation for PD(DOM) a subset of P(DOM), such that every ideal $I \subseteq GD$ is represented by

$$\{x \in DOM : \forall S1 \in I, y \in FDOM : y \sqsubseteq x \Rightarrow \exists S2 \in I, S1 \sqsubseteq_{EM} S2, z \in S2 : y \sqsubseteq z \sqsubseteq x\}.$$

By

$$PDOM : P(DOM) \rightarrow PD(DOM)$$

we denote the function mapping every set $S \subseteq DOM$ on its power domain representation. It is defined by

$$PDOM(S) = \{x \in DOM : \forall S1 \in GD, y \in FDOM : y \sqsubseteq x \wedge S1 \sqsubseteq_{EM} S \Rightarrow \exists S2 \in GD, S1 \sqsubseteq_{EM} S2 \sqsubseteq_{EM} S, z \in S2 : y \sqsubseteq z \sqsubseteq x\}.$$

Note, that we have chosen a $\subseteq$-maximal representation for PDOM(S), i.e. the $\subseteq$-maximal set in the class of sets that are $\sqsubseteq_{EM}$-equivalent w.r.t. $\sqsubseteq_{EM}$-approximations by finite sets of finite elements.

The power domain of demonic nondeterminism (also called Smyth power domain) is defined as follows:

$$PM(DOM) = \{ MIN(S) : S \in PD(DOM) \}$$
$$PMDOM : P(DOM) \rightarrow PM(DOM)$$
$$PMDOM(S) = MIN(PDOM(S))$$

Note, that we have chosen a $\subseteq$-minimal representation for PMDOM(S), i.e. the $\subseteq$-minimal set in the class of closed, finitely approximable sets that are $\sqsubseteq_M$-equivalent w.r.t. $\sqsubseteq_M$-approximations by finite sets of finite elements.

On function domains we use the classical ordering $\sqsubseteq^*$ obtained bt applying $\sqsubseteq$ pointwise. Analogously we write lub^* for the lub on the function domain ordered by $\sqsubseteq^*$.

Unfortunately the simple powerset without the empty set ordered by inclusion ordering does not form a domain. For very obvious reasons we do not accept the empty set as element, since the set of possible computations of a nondeterministic program can never be empty. However $(P(DOM)\setminus\{\emptyset\}, \subseteq)$ forms a predomain, i.e. it has all properties of a domain besides the existence of a least element. We restrict ourselves to closed sets, i.e. to sets S where with every directed set in S its least upper bound is also in S. This is motivated by the concept of finite observability. Every object should be determined by its finite approximations. Accordingly the power predomain of closed sets is defined as follows:

$$PC(DOM) = \{S \subseteq DOM : S = CLOSE(S)\}.$$

A function $f : P(DOM) \to P(DOM)$ is called closely union continuous iff

$$f(CLOSE(\bigcup x_i)) = CLOSE(\bigcup f(x_i))$$

for every $\subseteq$-chain $\{x_i\}_{i \in N}$, $x_i \in P(DOM)$. Similarly a functional

$$T : (DOM \to P(DOM)) \to (DOM \to P(DOM))$$

is called closely union continuous iff

$$T[CLOSE\cdot(\bigcup^* f_i)] = CLOSE\cdot(\bigcup^* T[f_i])$$

for every $\subseteq$-chain $(f_i)_{i \in N}$, f_i: DOM -> P(DOM). Here $\cdot$ denotes the composition of functions and $\bigcup^* f_i$ denotes the elementwise union of the set-valued function f_i.

Based on these general definitions we now are going to give a general construction for a nondeterministic Herbrand-Kleene interpretation.

5. Nondeterministic Interpretations

Let A be an algebraic $\Sigma^\perp$-algebra with least element $\perp_A$. The nondeterministic interpretation of terms $t \in W^\infty_{\Sigma^+}$ in A is to be defined by the function:

$$B^\infty_A : W^\infty_{\Sigma^+} \to P(A)$$

fullfilling the equations:

$$B^\infty_A[or(t1, t2)] = B^\infty_A[t1] \cup B^\infty_A[t2],$$
$$B^\infty_A[f(t_1,\ldots,t_n)] = \{ f^A(a_1,\ldots,a_n) : \forall i, 1 \leq i \leq n: a_i \in B^\infty_A[t_i] \}.$$

Note, that we do not require that the images of B^∞_A are in PD(A) but just in P(A), the powerset over A. It is not difficult to show, that there are many distinct functions that fullfill these equations, since these equations do not characterize B^∞_A uniquely. Such a situation is well-known from classical fixed point theory. Besides the least fixed point generally there are many other fixedpoints (optimal, maximal fixed points etc.). What we would like to do is to proceed like in classical fixed point theory: we would like to find an appropriate ordering on P(A) such that B^∞_A can be specified as the least fixed point. But according to our Lemma 1 such an ordering does not exist. So we have to use more refined techniques instead. In particular now we have to prove that the class of functions fullfilling the equations above is nonempty; to define uniquely which function from this class is to be taken. We are going to define the function B^∞_A in three steps:

1. Step: We define an approximation BP_A^∞ for the intended function B_A^∞ in the powerdomain of erratic nondeterminism, such that the following relation should hold

$$BP_A^\infty[t] = PDOM(\ B_A^\infty[t]\ ).$$

Define

$$BP_A^\infty : W_{\Sigma+}^{\infty} \to PD(A)$$

by the least fixed point in the set of functions $W_{\Sigma+}^{\infty} \to PD(A)$ fullfilling the equations:

$$BP_A^\infty[or(t1,\ t2)] = CONE(BP_A^\infty[t1]\ \cup\ BP_A^\infty[t2])$$
$$BP_A^\infty[f(t_1,\dots,t_n)] = CONE(\{f^A(a_1,\dots,a_n):\ \forall\ i,\ 1 \le i \le n:\ a_i \in BP_A^\infty[t_i]\ \})$$

Lemma 3: BP_A^∞ is consistently defined, $\sqsubseteq_{EM}$-monotonic, and $\sqsubseteq_{EM}$-continuous.

2. STEP: We define an approximation for B_A^∞ in the powerdomain of demonic nondeterminism. For $t \in W_{\Sigma\bot}^{\infty}$ we have according to the results of [Nivat 75]:

$$BP_A^\infty[t] = \{I_A^\infty(t)\}.$$

According to the lemma in section 4

$$MIN \cdot BP_A^\infty : W_{\Sigma+}^{\infty} \to PM(W_{\Sigma\bot}^{\infty})$$

defines a monotonic and continuous mapping, too.

Lemma 4: $MIN \cdot BP_A^\infty$ is consistently defined, $\sqsubseteq_M$-monotonic, and $\sqsubseteq_M$-continuous. It is the least fixed point in the set of functions $W_{\Sigma+}^{\infty} \to PM(A)$ fullfilling the equations:

$$MIN(BP_A^\infty[or(t1,\ t2)]) = MIN(BP_A^\infty[t1]\ \cup\ BP_A^\infty[t2]),$$
$$MIN(BP_A^\infty[f(t_1,\dots,t_n)]) = MIN(\{f^A(a_1,\dots,a_n)\ :\ \forall\ i,\ 1 \le i \le n:\ a_i \in BP_A^\infty[t_i]\ \}).$$

3. Step: We define B_A^∞ as the $\le$-least closed function with

$$MIN \cdot BP_A^\infty \le^* B_A^\infty \le^* BP_A^\infty,$$

that fullfills the specifying equations.

Proof of the existence of B_A^∞ (constructive):

(1) Define $B_A^i : W_{\Sigma+}^{\infty} \to PC(A)$ by

$$B_A^0 = MIN \cdot BP_A^\infty$$
$$B_A^{i+1}[or(t1,\ t2)] = B_A^i[t1]\ \cup\ B_A^i[t2],$$
$$B_A^{i+1}[f(t_1,\dots,t_n)] = \{f^A(a_1,\dots,a_n):\ \forall\ i,\ 1 \le i \le n:\ a_i \in BP_A^i[t_i]\ \}.$$

(2) We have $B_A^i \le^* B_A^{i+1}$ because the definitions above of B_i are $\le^*$-monotonic in B and

$$B_A^0[t] = MIN(BP_A^\infty[t]) = \begin{cases} MIN(BP_A^\infty[t1] \cup BP_A^\infty[t2]) \\ MIN(\{f^A(a_1,\dots,a_n) : a_i \in BP_A^\infty[t_i]\}) \end{cases}$$

$$\begin{rcases} \le MIN(BP_A^\infty[t1]) \cup MIN(BP_A^\infty[t2]) \\ \le \{f^A(a_1,\dots a_n) : a_i \in MIN(BP_A^\infty[t_i])\} \end{rcases} = B_A^1[t].$$

(3) Define $B_A^\infty = CLOSE \cdot \cup^* B_A^i$; since all language constructs are closely union-continuous, B_A^∞ is a fixed point of the defining equations.

<u>Corollary</u>: With the definitions of the previous lemma we have for all terms t:

$$PMDOM(B_A^\infty[t]) = MIN(PDOM(B_A^\infty[t])) = MIN(BP_A^\infty[t]) = MIN(B_A^\infty[t]) \le$$
$$B_A^\infty[t] \le CONE(B_A^\infty[t]) \le PDOM(B_A^\infty[t]) = BP_A^\infty[t].$$

For $A = W_\Sigma^{\infty}\bot$ the function B_A^∞ will be denoted by

$$B_{HK}^\infty : W_\Sigma^{\infty}{+} \to P(W_\Sigma^{\infty}\bot)$$

and called the nondeterministic <u>Herbrand-Kleene Interpretation</u>. B_A^∞ has the following properties:

- $PDOM \cdot B_A^\infty$ is $\sqsubseteq_{EM}$-continuous and $\sqsubseteq_{EM}$-least fixed point of the defining equation in $PD(A)$,
- $B_A^\infty[t] = \{I_A^\infty(t)\}$ for $t \in W_\Sigma\bot$,
- B_A^∞ is the $\le^*$-least function that is $\sqsubseteq_{EM}$-equivalent to $BP(A,\infty)$.

The basic results of the definitions above can be condensed into following theorem:

Theorem: For every algebraic Σ-algebra A the following diagram commutes

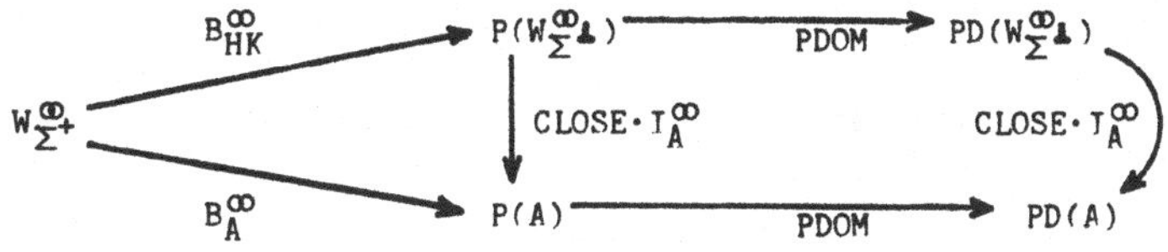

This means that it is possible either to interprete a term $t \in W_\Sigma^{\infty}{+}$ immediately in P(A) as representing a set of elements of A or at first interprete these term elementwise to obtain a set over A. The results of both ways of proceeding coincides, the diagram commutes.

6. Concluding Remarks

The basic idea presented above comes from a particular view of nondeterminism: a nondeterministic program does not specify just one computation (for every input), but rather a class of computations. The technique applied in this paper was used the first time in [Broy 82] to define a denotational semantics for concurrent, communicating, nondeterministic programs. In the preceding section it is demonstrated, that this technique is general in the sense, that is can applied to nondeterministic computations over arbitrary domains.

Note that one problem remains: All sets $B^{\infty}_{HK}[t]$ and $B^{\infty}_{A}[t]$ are closed. So if we consider the infinite term t defined by $\Phi(\bot)$ where $\Phi(x) = or(s(x), \Phi(s(x)))$ is defined like in the example above, one may expect that B^{∞}_{HK} does not contain the infinite tree s^{∞} defined by $s^{\infty} = s(s^{\infty})$. However this tree s^{∞} is an element of B^{∞}_{HK} since $B^{\infty}_{HK}[t]$ is always closed. This seems only at a first sight as a drawback, but is a consequence of the notion of finite observability and a particular interpretation of $\bot$. If all finite approximations of an element x are members of $B^{\infty}_{HK}[t]$ so has to be x. Note, that this has nothing to do with fairness, since fairness rather corresponds to the tree T associated with Φ of the problem above. Note that $B^{\infty}_{HK}[T]$ does not contain s^{∞}.

Acknowledgement
I gratefully acknowledge a number of valuable comments by Andre Arnold and Frederike Nickl.

References

[Broy 82]
M. Broy: Fixed point theory for communication and concurrency. In: D. Bjørner (ed): IFIP TC2 Working Conference on "Formal Description of Programming Concepts II". Garmisch, June 1982, Amsterdam-New York-Oxford: North Holland Publ. Company 1983, 125-147

[Nivat 75]
M. Nivat: On the interpretation of recursive polyadic program schemes. Symposia Mathematica 15, 1975, 255-281

[Nivat 80]
M. Nivat: Nondeterministic programs: an algebraic overview. S.H. Lavington (ed.): Information Processing 80, Proc. of the IFIP Congress 80, Amsterdam - New York - Oxford: North-Holland Publ. Comp. 1980, 17-28.

[Plotkin 76]
G. Plotkin: A powerdomain construction. SIAM J. on Computing 5, 1976, 452-486

[Smyth 78]
M. B. Smyth: Power domains. J. CSS 16, 1978, 23-36

AN INVESTIGATION OF CONTROLS FOR CONCURRENT SYSTEMS BY ABSTRACT CONTROL LANGUAGES

H.D.Burkhard
Sektion Mathematik
Humboldt-Universität
DDR-1086 Berlin, PSF 1297

Abstract: The behaviour of the controlled system determines the control. This concise statement summarizes our approach to the investigation of controls. Using abstract languages to define behaviour and subbehaviour we are able to describe and to study different types of control rules and properties to be realized by control like deadlock avoidance, liveness and fairness.

Introduction

Control is one of the central problems in studies of concurrent systems and programs. It is inherent in conflict resolution, scheduling, synchronization, program semantics, wherever decisions on choices are made. There is a common understanding of control, but general definitions and considerations are missing.

From its use in the literature, two aspects appear: Application of certain control rules (queues, priorities, choice of maximal sets of concurrently performable actions etc.) and controls which are defined as restrictions of the behaviour in order to enforce properties like deadlock avoidance, termination and fairness. The second approach may be misleading, for example: "The choice of all fair executions" appears as an obscure notion from the view-point of control (cf. section 3 of this paper). Both aspects should be considered on a common base since realizations of properties by control rules (e.g. fairness by queues) are an important subject.

Two observations are essential to come to general considerations:

1. Control is considered as a restriction with respect to the behaviour of the system to be controlled.
2. Each restriction of the behaviour of the uncontrolled system determines a control since all decisions of the control are well-defined.

This correspondence between control and behaviour is employed for our purposes. Now it depends on the descriptions of the behaviour which problems of control can be examined. As we shall show in this paper, abstract languages describing the external behaviour are a convenient tool for many such problems.

Special systems can be examined by the corresponding families of languages, and various types of controls can be specified in terms of languages using the notion of control principles (section 1). Special emphasis is put on the regular languages and to the languages of firing sequences of finite Petri nets as well as to the problems of deadlock avoidance, liveness and fairness. While many considerations can be performed using only the languages without reference to the structure of a system, the languages are supposed to be given in a suitable form with respect to the decidability results etc.

Having in this way a suitable framework to speak about controls, we can compare controls of different types. We are able to investigate stepwise refinements of controls in order to realize different properties. It will be shown that the order of such refinements plays an important role (section 2). In general one has to take into account that properties are not preserved under controls, and verification results for an uncontrolled system need not be relevant for the controlled system.

Section 3 investigates the problem whether all executions corresponding to a special property can be realized by a uniform control. Such controls do, in general, not exist for fairness properties, but they do exist and are characterized for deadlock avoidance and liveness.

The last section deals with decidability results (existence of controls, realization of controls by finite automata as control devices, properties of automata controlled systems). It happens that generation of controls is preferable to analysis of controlled systems.

Examples and proofs are given in /Bu84.1/ and /Bu84.2/. The following notations are used:

T^* (T^ω) is the set of all finite (infinite) sequences over the alphabet T, e denotes the empty word. By $\pi(v,t)$ we denote the number of occurences of t in $v \in T^* \cup T^\omega$. A sequence $u \in T^*$ is a prefix of $v \in T^* \cup T^\omega$ ($u \sqsubseteq v$), if there exists a sequence v' with $v = uv'$. The closure of a language $L \subseteq T^*$ with respect to prefixes is denoted by $\overleftarrow{L} := \{ u \;/\; \exists v \in L: u \sqsubseteq v \}$. The adherence of L is defined by $Adh(L) := \{ w \;/\; w \in T^\omega \;\&\; \overleftarrow{w} \subseteq L \}$, whereby $\overleftarrow{w} := \{u \;/\; u \in T^* \;\&\; u \sqsubseteq w\}$. The powerset of a set A is denoted by P(A).
$\exists^\infty$ denotes "infinitely many", $\forall^\infty$ denotes "almost all".

1. Definition of control principles

We consider (controlled or uncontrolled) systems by means of their behaviour, given by languages L over a finite fixed alphabet T with at least two elements. We suppose these languages to be not empty and closed with respect to prefixes since control may influence the behaviour at any time. The control of a system is regarded as a restriction of its possibilities, thus the language L' of a controlled system is always a subset of the language L of the original (uncontrolled) system.

(1) Definition

(1) $CONT := \{ L \;/\; L \subseteq T^* \;\&\; \emptyset \neq L = \overleftarrow{L} \}$
is the family of all control languages over T.

(2) $cont(L) := P(L) \cap CONT$
is the family of all control languages for $L \in CONT$.

Since the behaviour of a control (the decisions to be made with respect to L) is defined by a language $L' \in cont(L)$, the family cont(L) describes all the possible controls of the system with the behaviour given by L.

Having a special way to perform controls (like scheduling disciplines) we obtain a special subset of cont(L). Having also in mind special conditions to be satisfied (like fairness, conflict resolution etc.) we are going to study subsets of cont(L):

(2) Definition

(1) A control principle is a mapping $c: \mathrm{CONT} \longrightarrow P(\mathrm{CONT})$ with $c(L) \subseteq \mathrm{cont}(L)$ for all $L \in \mathrm{CONT}$.

(2) A control principle c is said to be extensional if there exists a subset $C \subseteq \mathrm{CONT}$ such that $c(L) = P(L) \cap C = \mathrm{cont}(L) \cap C$.

If c is extensional, then c is monotonous: $L \subseteq L'$ implies $c(L) \subseteq c(L')$, and the set C is uniquely determined by $C = c(T^*) = c(\mathrm{CONT})$.

We write $c(C') := \bigcup_{L \in C'} c(L)$ for sets $C' \subseteq \mathrm{CONT}$.

(3) Definition

The extensional control principles c = dfr, live, imp, fair, just, pfin, preg, prec, pren and fnl are defined by the corresponding sets $C \subseteq \mathrm{CONT}$ which for $L \in \mathrm{CONT}$ fulfil:

(1) $L \in \mathrm{DFR} \iff \forall u \in L\ \exists t \in T: ut \in L$

(2) $L \in \mathrm{LIVE} \iff \forall u \in L\ \ \forall t \in T\ \exists u' \in T^*: uu't \in L$

(3) $L \in \mathrm{IMP} \iff \forall w \in \mathrm{Adh}(L)\ \forall t \in T: \pi(w,t) = \omega$

(4) $L \in \mathrm{FAIR} \iff \forall w \in \mathrm{Adh}(L)\ \forall t \in T: (\exists^{\infty} u \subseteq w: ut \in L) \longrightarrow \pi(w,t) = \omega$

(5) $L \in \mathrm{JUST} \iff \forall w \in \mathrm{Adh}(L)\ \forall t \in T: (\forall^{\infty} u \subseteq w: ut \in L) \longrightarrow \pi(w,t) = \omega$

(6)-(10) PFIN (PREG, PREC, PREN) denotes the family of prefix closed finite (regular, recursive, recursively enumerable) languages, and FNL is the family of firing languages of finite Petri nets.

By dfr(L) and live(L) we can consider the deadlockfree and live controls, respectively, for L; preg(L), e.g., describes all those controls of L where the controlled systems can be modelled by finite automata. By imp(L), fair(L), just(L) we have specified the controlled systems satisfying the fairness notions in the paper /LPS/ (impartiality, fairness, justice). Due to their extensionality, these fairness properties are defined without reference to the original uncontrolled systems. We have, e.g., $\{a\}^* \in \mathrm{fair}(\{a,b\}^*)$. Thus, there may be a need to consider the related non-extensional control principles rfair and rjust (relatively fair, relatively justice - with respect to the corresponding uncontrolled system). The control principles "conflict resolution" and "nonblocking" are also not extensional:

(4) Definition

(1) $L' \in \mathrm{rfair}(L) \iff \forall w \in \mathrm{Adh}(L')\ \forall t \in T: (\exists^{\infty} u \subseteq w: ut \in L) \longrightarrow \pi(w,t) = \omega$

(2) $L' \in \mathrm{rjust}(L) \iff \forall w \in \mathrm{Adh}(L')\ \forall t \in T: (\forall^{\infty} u \subseteq w: ut \in L) \longrightarrow \pi(w,t) = \omega$

(3) $L' \in \mathrm{crs}(L) \iff \forall ut, ut' \in L': t \neq t'\ \&\ t,t' \in T \longrightarrow utt' \in L$

(4) $L' \in \mathrm{nbl}(L) \iff \forall u \in L': (\exists t \in T: ut \in L) \longrightarrow (\exists t' \in T: ut' \in L')$

(5) Theorem

The problems "$L' \in c(L)$?" are decidable for the control principles c of the Definitions (3),(4) and languages $L,L' \in \mathrm{PREG}$ or $L,L' \in \mathrm{FNL}$, except for the open problem "$L' \in \mathrm{fnl}(L)$?" in the case $L' \in \mathrm{PREG}$. This includes the decidability of the problems "$L \in C$?" for the sets C of the Definitions (3.1-9) (FNL excluded) and $L \in \mathrm{PREG} \cup \mathrm{FNL}$.

The proof for the FNL-languages uses the decidability of the reachability problem /K/ (for dfr, live, fnl, crs, nbl), the coverability tree construction (for pfin and, generalizing methods from /VJ/ and /Bu83.2/, for imp, fair, just, rfair, rjust) and a result from /VV/ (for preg).

(6) Definition

The control principle c is covered by c', $c \leqq c'$,
iff $c(L) \subseteq c'(L)$ holds for all $L \in CONT$.

If c is covered by c' then the properties of c' are preserved by c, for example, all live controls are deadlockfree since live $\leqq$ dfr.

(7) Theorem

For $card(T) > 2$ we have the relations between control principles as represented by the following figure:

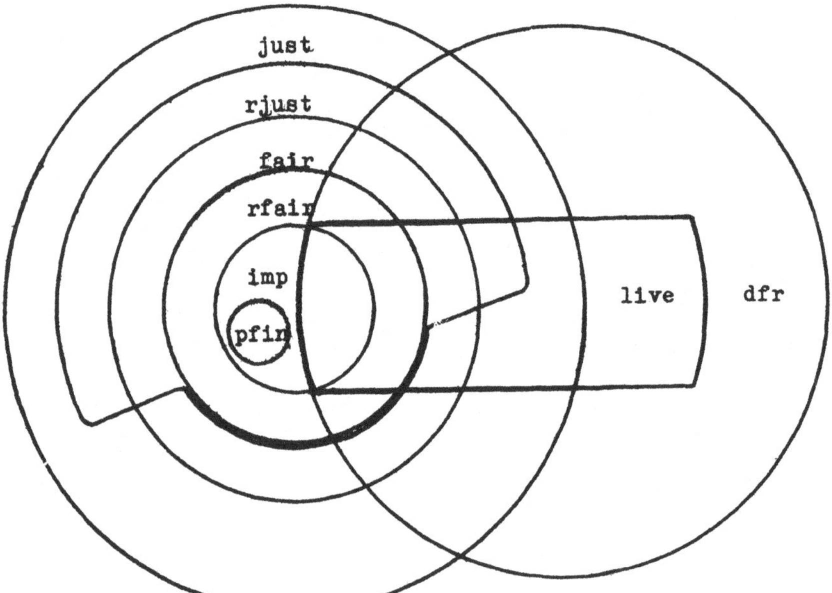

Since control devices often work as finite automata we consider finite nondeterministic automata $\underline{A} = (P(T), T, Z, h, z_0)$ as "control automata" where P(T), T, Z are the finite sets of inputs, outputs and states, respectively, $z_0 \in Z$ is the initial state, and $h: Z \times P(T) \longrightarrow P(T \times Z)$ is the output/next-state function with

$$\emptyset \neq \{ t \ / \exists z' : (t,z') \in h(z,U) \} \subseteq U \quad \text{for all } z \in Z, U \in P(T) \setminus \{\emptyset\}.$$

(The last conditions assures that automata work nonblocking.)

A language L under the control of a control automaton $\underline{A}$ results in $L/\underline{A}$ as the language of the controlled system with $e \in L/\underline{A}$ and

$$t_1 \ldots t_n \in L/\underline{A} \iff \begin{array}{ll} (1) & t_1 \ldots t_n \in L \\ (2) & \exists z_1 \ldots \exists z_n \in Z \ \forall i=0,\ldots,n-1: \\ & (t_{i+1}, z_{i+1}) \in h(z_i, \{t \,/\, t_1 \ldots t_i t \in L\}). \end{array}$$

A control automaton $\underline{A}$ can be considered as a R-robot working in the R-environment L in the sense of /B/, and a related approach for system controls was given in /AN/. Controls of Petri nets by such control automata have been studied in /Bu82/, /Bu83.1/, /BS/. Obviously, priority rules or fifo-queues can be realized by finite control automata.

(8) Definition

For a non-empty class aut of control automata we define the control principle aut by $\text{aut}(L) := \{ L/\underline{A} \ / \ \underline{A} \in \underline{\text{aut}} \}$.

The control principles aut are not extensional since they are not monotonous. To characterize the results of automata controls we have:

(9) Theorem

It holds for the class aut1 of all control automata and $\text{card}(T) > 2$:
$\text{FNL} \subsetneq \text{aut1}(\text{FNL}) \subsetneq \text{PREC} = \text{aut1}(\text{PREC}) \subsetneq \text{PREN} \subsetneq \text{aut1}(\text{PREN})$,
$\text{PREG} = \text{aut1}(\text{PREG})$, PREG and aut1(FNL) are incomparable.

2. Composition of control principles

(10) Definition

The conjunction $c \,\&\, c'$ and the superposition $c * c'$ of two control principles c, c' are defined by

$c \,\&\, c'(L) := c(L) \cap c'(L)$ and

$c * c'(L) := c(c'(L))$, respectively, for all $L \in \text{CONT}$.

The superposition reflects the stepwise construction of controls, e.g., aut * crs * live means first construction of a live system, then resolution of conflicts and finally realization by an automaton from aut. The goal is to obtain a control from aut & crs & live. But it turns out that such stepwise constructions need not result in controls satisfying the desired properties, for example: Since $\text{crs} * \text{live} \leq \text{crs} \,\&\, \text{live}$ is not true, conflict resolution for a live system need not result in a live system (consider $\{a\}^* \in \text{crs}(\{a,b\}^*)$). This observation is important for some analyzing methods, too: The verification of some properties for an uncontrolled system is in general not relevant concerning the controlled system (cf. /BS/ for the study of liveness and deadlock avoidance in Petri nets working under different firing rules). The positive aspect of this observation is the possibility to control systems in order to satisfy properties which do not hold for the uncontrolled system.

(11) Definition

(1) A control principle c is left-invariant
iff $c' * c \leq c$ for all c'
(or equivalently: iff $\text{cont} * c \leq c$, iff $\text{cont} * c = c$).

(2) A control principle c is right-invariant
iff $c * c' \leqq c$ for all c'
(or equivalently: iff $c * \text{cont} \leqq c$, iff $c * \text{cont} = c$,
iff c is monotonous).

(12) Proposition
Extensional control principles are right-invariant.

The converse is not true, e.g., crs is not extensional but it holds:

(13) Proposition
crs is right-invariant and left-invariant.

The next theorem points to some possibilities to realize a c & c'-control by stepwise constructions:

(14) Theorem
(1) c right-invariant, c' left-invariant $\Longrightarrow$ $c * c' \leqq c \& c'$.
(2) c extensional, c' left-invariant $\Longrightarrow$ $c * c' = c \& c'$.

It follows, for example, live $*$ crs = live & crs (but crs $*$ live $\nleqq$ live) and crs $*$ rfair $\lneqq$ crs & rfair (but rfair $*$ crs $\nleqq$ rfair). Hence, the order of stepwise refinements may be important. Since the control principles aut are neither right-invariant nor left-invariant, properties may change (and may be changed) by controlling systems by automata.
Another approach to preservation of properties is due to fixed point considerations:

(15) Theorem
For extensional c and arbitrary c' we have:
$c' * c \leqq c$ iff $c'(C) \subseteqq C$.

Applications (like preg $*$ live $\nleqq$ live) are left to the reader.

3. Unitarity

(16) Definition
A control principle c is called unitary iff
$\forall L \in \text{CONT}: c(L) \neq \emptyset \implies L(c) := \bigcup c(L) \in c(L)$
(iff a maximum element L(c) exists in the nonempty sets c(L)).

If L(c) exists, it gives the least restrictive control of L with respect to the c-properties. Since live is unitary, one may consider L(live) as "the" live control of L. But in general it is meaningless to speak of "the fair control" because fair is not unitary.

(17) Proposition
If c is extensional and C is closed under arbitrary unions, then c is unitary.

It follows, that dfr and live are unitary. All other control principles from the Definitions (3) and (4), except nbl, are not unitary. As concerns the control principles c = imp, fair, just, rfair, rjust consider for example:

$L = \{a,b\}^*$, $L_u := \overleftarrow{u\cdot(ba)^\omega} \in c(L)$ for all $u \in T^*$,

such that $\bigcup c(L) = L \notin c(L)$.

(18) Theorem (Suppose $live(L) \neq \emptyset$.)

(1) $L(live) = \{u \;/\; \exists w \in Adh(L){:}\; u \leq w \;\&\; \forall t \in T{:}\; \pi(w,t) = \omega\}$.

(2) If $L \in PREG \cup FNL$:
$L(live) = \{u \;/\; \exists u',v \in T^*{:}\; uu'v^\omega \in Adh(L) \;\&\; \forall t \in T{:}\; \pi(v,t) > 0\}$.

(3) $L \in PREG \implies L(live) \in PREG$.

(4) $L \in FNL \implies L(live) \in PREC$ (but in general $\notin FNL$).

(5) $L \in PREC$ does not imply $L(live) \in PREN$.

(6) There exists $L \in FNL$ with $live(L) \neq \emptyset$ and $live \,\&\, fnl(L) = \emptyset$.

(7) For $L \in PREG \cup FNL$: $live(L) \neq \emptyset \implies live \,\&\, preg(L) \neq \emptyset$.

The Theorem holds for dfr if the statements about the occurences of t are omitted.

Remark: Following some ideas of /C/ we can define "impartiality with respect to a delay function $d{:}\; T^* \times T \to \mathbb{N}$" by a family d-IMP where

$$L \in d\text{-}IMP \iff \forall uv \in L \;\forall t \in T{:}\; |v| > d(u,t) \longrightarrow \pi(v,t) > 0 .$$

Then the extensional control principle d-imp ($\lneqq$ imp) is unitary, and IMP is the union of the families d-IMP for all delay functions d. A related result holds for just/rjust, while the union of the families d-FAIR considered in /Bu83.2/ is in general only a subset of FAIR (thus, L.Czaja's problem to characterize fairness without using infinite behaviour is still open in this case).

4. Existence of controls and their realization by finite automata

By the problem "$c(L) \neq \emptyset$?" we ask after the existence of controls (specified by c) for given systems (specified by L).

(19) Theorem

(1) $c(L) \neq \emptyset$ holds for c = imp, fair, just, rfair, rjust (by pfin $\leq$ c).

(2) $c \,\&\, nbl(L) \neq \emptyset$ holds for c = fair, just, rfair, rjust,
$c \,\&\, dfr(L) \neq \emptyset$ holds for c = fair, just
(by application of appropriate queue regimes as in /Bu81/).

(3) $imp \,\&\, dfr(L) \neq \emptyset \iff live(L) \neq \emptyset$,
$imp \,\&\, nbl(L) \neq \emptyset \iff imp \,\&\, dfr(L) \neq \emptyset \;\vee\; pfin \,\&\, nbl(L) \neq \emptyset$.

(20) Theorem

The problems "$c(L) \neq \emptyset$?" are decidable for $L \in PREG \cup FNL$ and c = dfr, live, imp & dfr, imp & nbl, pfin & nbl.

The proof uses the decidability of the reachability problem /K/ and the decidability of the condition in the Theorem (18.2), especially in the case u = e. Proofs for dfr and live are given in /VJ/, /Bu83.1/.

(21) Theorem

(1) Suppose c ∈{fair, just, rfair, rjust}. There exists a control automaton $\underline{A}$ such that L/$\underline{A}$ ∈ c & nbl(L) holds for all L ∈ CONT.

(2) Suppose c ∈{dfr, live, imp & nbl, imp & dfr, fair & dfr, just & dfr}. For each L ∈ PREG ∪ FNL with c(L) ≠ ∅ there exists a control automaton $\underline{A}$ such that we have L/$\underline{A}$ ∈ c(L). The assertion is not true for c ∈{rfair&dfr, rjust&dfr}, L ∈ FNL.

The difference between (21.2) and (21.1) cannot be overcome: The automata referred to in (21.2) must in fact be constructed individually (cf. /BS/). The proof uses (18)-(20) and the following proposition:

(22) Proposition

preg & nbl ≤ aut1   (aut1 denotes the class of all control automata).

Using methods from /Bu82/ (simulation of counter machines by automata controlled Petri nets) and a construction from /BS/, it can be shown:

(23) Theorem

Suppose T to be sufficiently large (e.g. card(T) > 6n + 53 where n is the number of states of a counter machine computing a function with a nonrecursive domain).

(1) Let $\underline{A}$ be a control automaton with L/$\underline{A}$ ∈ crs(L) for all L ∈ FNL (i.e., $\underline{A}$ represents a conflict resolution rule for Petri nets). Then all of the problems "L/$\underline{A}$ ∈ c(L)?" , c ∈{dfr, live, imp & dfr, imp & nbl} , are undecidable for the languages L ∈ FNL.

(2) All of the problems "L/$\underline{A}$ ∈ c(L)?", c ∈{dfr,live,pfin,preg,crs, imp,fair,just,rfair,rjust, imp & nbl,...,rfair & nbl, imp & dfr, ...,rfair & dfr} , are undecidable with respect to arbitrary control automata $\underline{A}$ and arbitrary languages L ∈ FNL.

The Theorem points out that in the case of Petri nets the construction of an appropriate control is in a sense easier than the verification of a property. It follows from the Theorems (5) and (9):

(24) Theorem

All the problems "L/$\underline{A}$ ∈ c(L)?" , where c is a control principle as in Theorem (23.2), are decidable with respect to arbitrary control automata $\underline{A}$ and arbitrary languages L ∈ PREG.

Conclusions

It was shown that many problems of control can be studied on a very high level of abstraction. This framework should be useful to study and to compare different approaches to systems, controls and their properties, as for example given in /AN/, /APS/, /Be/, /Bu81/, /BS/, /C/, /CV/, /LPS/, /SM/.

The use of our approach is restricted by considering systems to be equivalent if they have the same behaviour regarding abstract languages (but there may be possibilities for different controls based on different internal structures). In some cases this problem can be overcome even by other notions of external behaviour. To be closer to concurrency, the alphabet T may be chosen as the powerset of another set of actions. In such models the MAX-semantics /SM/, /Bu81/ can be considered (with appropriate definitions reflecting liveness and fairness properties). Another approach to concurrency by partially ordered sequences may lead to the study of the corresponding behaviour and controls.

The general notion of control principles opens the way to further examinations of properties like invariance and unitarity. Stochastic controls may be studied by measuring the sets c(L).

Acknowledgement: I want to express my thanks to P.H. Starke and H.W. Pohl for various helpful discussions.

References:

/AN/ Arnold,A., Nivat,M.: Controlling behaviours of systems, some basic concepts and some applications. LNCS 88(1980),113-122.

/APS/ Apt,K.R., Pnueli,A., Stavi,J.: Fair termination revisited - with delay. Publ.du L.I.T.P.,Univ.Paris VII, 82-51(1982).

/BE/ Best,E.: Relational semantics of concurrent programs (with some applications). Proc. IFIP TC-2 Conf. on Formal Descriptions of Programming Concepts, Garmisch-Partenkirchen 1982.

/B/ Budach,L.: Environments, Labyrinths and Automata. LNCS 56 (1977), 54-64.

/Bu81/ Burkhard,H.D.: Ordered firing in Petri nets. EIK 17 (1981)2/3,71-86

/Bu82/ Burkhard,H.D.: What gives Petri nets more computational power. Preprint 45, Sekt.Mathematik d.Humboldt-Univ.Berlin 1982.

/Bu83.1/ Burkhard,H.D.: Control of Petri nets by finite automata. Fundamenta Informaticae VI.2 (1983), 185-215.

/Bu83.2/ Burkhard,H.D.: On the control of concurrent systems with respect to fairness and liveness conditions. Intern. Summer School of the Progr.Language LOGLAN-82, Zaborów 1983.

/Bu84.1/ Burkhard,H.D.: An investigation of controls for concurrent systems based on abstract control languages. Sekt.Mathe. d.Humboldt-Univ.Berlin 1984 (to appear as Preprint).

/Bu84.2/ Burkhard,H.D.: Untersuchung von Steuerproblemen nebenläufiger Systeme auf der Basis abstrakter Steuersprachen. Sekt.Mathe. d.Humboldt-Univ.Berlin 1984 (to appear as Seminarbericht).

/BS/ Burkhard,H.D., Starke,P.H.: A note on the impact of conflict resolution to liveness and deadlock in Petri nets. To appear in Fundamenta Informaticae.

/C/ Czaja,L.: Are infinite behaviours of parallel system schemata necessary? LNCS 148 (1983), 108-117.

/CV/ Carstensen,H., Valk,R.: Infinite behaviour and fairness in Petri nets. Proc. 4th European Workshop on Application and Theory of Petri Nets, Toulouse 1983, 104-123.

/K/ Kosaraju,S.R.: Decidability of reachability in vector addition systems. Proc. 14th Ann. ACM Symp. on Theory of Computing 1982, 267-281.

/LPS/ Lehmann,D., Pnueli,A., Stavi,J.: Impartiality, justice, fairness: The ethics of concurrent termination. LNCS 115 (1981), 264-277.

/SM/ Salwicki,A., Müldner,T.: On the algorithmic properties of concurrent programs. LNCS 125 (1981), 169-197.

/VJ/ Valk,R., Jantzen,M.: The residue of vector sets with applications to decidability problems in Petri nets. Bericht IFI-HH-B-101/84, Fachbereich Informatik, Univ. Hamburg.

/VV/ Valk,R., Vidal-Naquet,G.: Petri nets and regular languages. J.Comp.Syst.Sci. 23 (1981) 3, 299-325.

ON GENERALIZED WORDS OF THUE-MORSE

A. Černý
Department of Theoretical Cybernetics
Comenius University
Mlynská dolina, 842 15 Bratislava
Czechoslovakia

INTRODUCTION

In the theory of formal languages a great number of deep results concerning the structure of a language or even of a family of languages have been achieved. On the other hand one knows less about the internal structure of words. In structural investigations, infinite words serve often as a useful tool. For example, if some hereditary property P of words is considered (P is hereditary if from the validity of P for some word follows the validity of P for all its factors) an infinite word with the property P gives us an infinite set of finite words with the same property.

Axel Thue in his remarkable works [7,8] on infinite sequences of symbols has shown the existence of infinite words over three-letter alphabets without squares of nonempty words as factors. The construction of such a word in [7] is based on an infinite sequence $\underline{t}$ over the two-letter alphabet $\{0,1\}$, not containing a factor of the form xvxvx, x being a letter and v a nonempty word. The i-th symbol of $\underline{t}$ can be described as the parity of occurrences of the symbol 1 in binary notation of the natural number i. The same sequence $\underline{t}$ appears in a paper of Morse [5] on symbolic dynamics. Therefore we will call $\underline{t}$ the sequence of Thue-Morse.

In the sense of Cobham [3], $\underline{t}$ is an example of a simple uniform tag sequence. In [2] where some algebraic properties of uniform tag sequences have been investigated, generalized sequences of Thue-Morse have been introduced. In such a generalized sequence the i-th symbol denotes the parity of occurrences of some fixed factor w over $\{0,1\}$

in the binary notation of i. In the presented paper we show that in such generalized words of Thue-Morse all the factors are of a bounded power. More precisely, there are no factors of the form $(xv)^k x$ where $k = 2^{|w|}$, x is a letter and v is a word.

NOTATIONS AND DEFINITIONS

Let A^* be the free monoid generated by a finite alphabet A, with the neutral element ε. Let $A^+ = A^* - \{\varepsilon\}$. Let A^ω be the set of all infinite (to the right) sequences of elements of A. Let $A^\infty = A^* \cup A^\omega$. The elements of A^∞ will be called <u>words</u> (finite or infinite). A word $x \in A^*$ is a <u>factor</u> of a word $y \in A^\infty$, iff $y = zxt$ for some $z \in A^*$, $t \in A^\infty$. x is called <u>initial</u> /<u>terminal</u>/ {<u>proper</u>} <u>factor</u> of y iff $z = \varepsilon$ /$t = \varepsilon$/ $\{zt \neq \varepsilon\}$. The length $|x|$ of a finite word x is the number of its symbols, $|\varepsilon| = 0$.

Let $\varphi : A^* \rightarrow B^*$ be a morphism of monoids. φ can be extended to the mapping $\varphi : A^\infty \rightarrow B^\infty$ satisfying $\varphi(xy) = \varphi(x)\varphi(y)$ for all $x \in A^*$, $y \in B^\infty$. φ is called <u>prolongable in</u> $a \in A$ iff $\varphi(a) = ax$ for some $x \in A^+$. In this case for each $n \geqslant 0$ $\varphi^n(a)$ is a propper initial factor of $\varphi^{n+1}(a)$. There exists a limit $\underline{z} = \lim_{n\to\infty} \varphi^n(a) \in A^\infty$ such that each $\varphi^n(a)$ is an initial factor of $\underline{z}$. Moreover, $\underline{z}$ is a fixpoint of φ, i.e. $\varphi(\underline{z}) = \underline{z}$. The morphism φ is called <u>m-uniform</u> for some $m \geqslant 0$ iff $|\varphi(b)| = m$ for all $b \in A$.

Let $\mu : A^i \rightarrow A^j$ be a mapping, $i, j \geqslant 1$. μ can be extended to the mapping $\mu : A^\omega \rightarrow A^\omega$ defined by $\mu(x_0x_1x_2\ldots) = y_0y_1y_2\ldots$ where $y_{k.j}y_{k.j+1}\ldots y_{k.j+j-1} = \mu(x_{k.i}x_{k.i+1}\ldots x_{k.i+i-1})$ for all $k \geqslant 0$. This extension is called <u>(i,j)-substitution.</u>

We will use two devices for a formal description of infinite words - uniform tag systems and sorting automata. A <u>tag system</u> is a quintuple $T = (\Sigma, a, \sigma, \Gamma, \tau)$ where Σ and Γ are alphabets, $\sigma : \Sigma^* \rightarrow \Sigma^*$ is a morphism prolongable in $a \in \Sigma$, $\tau : \Sigma^* \rightarrow \Gamma^*$ is a morphism such that $\tau(\Sigma) \subseteq \Gamma$. The <u>internal</u> /<u>external</u>/ <u>tag sequence</u> generated by T is $intseq_T = \lim_{n\to\infty} \sigma^n(a)$ /$seq_T = \lim_{n\to\infty} \tau(\sigma^n(a)) = \tau(intseq_T)$ /. The tag system and the corresponding sequences are called <u>m-uniform</u> iff σ is m-uniform.

Let $m > 0$. Denote $[m] = \{0,1,\dots,m-1\}$. A <u>sorting automaton</u> over $[m]$ is a quintuple $A = (S, \delta, s_0, F, G)$ where S is a finite set (of states), $s_0 \in S$ is the initial state, $\delta : S \times [m] \to S$ is the transition function satisfying $\delta(s_0,0) = s_0$, G is an alphabet and $F = \{F_g\}_{g \in G}$ is a dijoint partition of S. δ can be extended to the domain $S \times [m]^*$ as usual by finite automata. The <u>state /sorting/ sequence</u> of the automaton A is defined by $state_A = y_0y_1\dots \in S^\omega$ / $sort_A = x_0x_1\dots \in G^\omega$/ where $y_i = s \in S$ iff $\delta(s_0, i_{[m]}) = s$, $i_{[m]}$ being the binary notation of i (since $\delta(s_0,0) = s_0$ there are no problems with leading zeros), and $x_i = g$ iff $y_i \in F_g$. Thus the notion of sorting automaton is a slight generalization of the notion of finite automaton which in fact sorts into two classes of objects (accepted - rejected). The relation between tag systems and sorting automata is expressed in the following proposition.

<u>Proposition 1</u> [3]. Let $T = (\Sigma, a, \sigma, \Gamma, \tau)$ be an m-uniform tag system and let $A = (\Sigma, \delta, s_0, F, \Gamma)$ be a sorting automaton over $[m]$ such that $\delta(s,i)$ = the i-th symbol of $\sigma(s)$, $s_0 = a$ and $s \in F_g$ iff $\tau(s) = g$, where $s \in S$, $i \in [m]$, $g \in \Gamma$. Then $intseq_T = state_A$ and $seq_T = sort_A$.

Finally, let us define the generalized words of Thue-Morse. Let $w \in \{0,1\}^* - 0^*$. Denote $\underline{a}_w = a(0)a(1)a(2)\dots$ the infinite word with the i-th symbol $a(i) = \#_w(i_{[2]}) \bmod 2$, where $\#_w(x)$ denotes the number of occurrences of the factor w in the word x and $i_{[2]}$ is the binary notation of i with at least $|w|$ leading zeros. For example, 000010101010101 contains five occurrences of the factor 0101. In the case $w = 1$ we obtain $\underline{a}_1 = \underline{t} = 0110100110010110\dots$ - the word of Thue-Morse. From [2] we know the following important property of the words $\underline{a}_w$.

<u>Proposition 2</u> [2]. Let $w \in \{0,1\}^* - 0^*$, $k = 2^{|w|-1}$, let μ be a (k,2k)-substitution on $\{0,1\}^\omega$ defined by $\mu(x_0x_1\dots x_{k-1}) = y_0y_1\dots y_{2k-1}$, $y_i = x_{i/2} + \chi_w(i) \pmod 2$, $i \in \{0,1,\dots,2k-1\}$ where $\chi_w(i) =$ <u>if</u> w is a terminal factor of $i_{[2]}$ <u>then</u> 1 <u>else</u> 0.
Then $\mu(\underline{a}_w) = \underline{a}_w$.

As one can easily see, there is exactly one $j \in \{0,1,\dots,2k-1\}$ such that $\chi_w(j) = 1$.

PROOF OF THE RESULT

Our goal is to prove that there are no factors of the form $(xv)^p x$, $p = 2^{|w|}$, x being a symbol and v a word, in $\underline{a}_w$. It is well known to be true [4,6] for $w = 1$, thus in the following we consider $w \in \{0,1\}^*$ to be a fixed word of length at least 2. The result will be reached using a series of lemmas. The proofs will be sketched or omitted. For the detailed proofs see [1]. In the first lemma, the minimal sorting automaton for $\underline{a}_w$ is described. Since the notion of sorting automaton is derived directly from the notion of finite automaton, the results from the theory of finite automata concerning minimality can be applied to sorting automata, too.

Lemma 1. Let $A_w = (S, \delta, s, \{F_0,F_1\}, \{0,1\})$ be a sorting automaton over [2] , where

$S = \{ \langle\alpha\rangle_0 , \langle\alpha\rangle_1 \mid \alpha$ is a propper initial factor of $w \}$

$$\delta (\langle\alpha\rangle_i , x) = \begin{cases} \langle\alpha x\rangle_i & \text{if } \langle\alpha x\rangle_i \in S \\ \langle\alpha'\rangle_{1-i} & \text{if } \alpha x = w \\ \langle\alpha'\rangle_i & \text{otherwise} \end{cases}$$

where $i \in \{0,1\}$, $x \in [2]$, α' is the longest proper terminal factor of αx being a proper initial factor of w

$s = 0^q$, where 0^q, $q \geq 0$, is the longest initial factor of w not containing 1

$F_i = \{ \langle\alpha\rangle_i \}$, $i \in \{0,1\}$.

Then A_w is minimal among the sorting automata with the sorting sequence $\underline{a}_w$.

Let $T_w = (S, s, \sigma, \{0,1\}, \tau)$ be the 2-uniform tag system corresponding to the automaton A_w from Lemma 1 according to Proposition 1. Hence $seq_{T_w} = sort_{A_w} = \underline{a}_w$. Denote $\underline{b}_w = b(0)b(1)b(2)\ldots = intseq_{T_w} = state_{A_w}$.

Lemma 2. σ is an injective mapping.

To obtain our main result we will first investigate the structure of the word $\underline{b}_w$, the results for $\underline{a}_w$ will follow directly as can be seen from the following Lemma 4.

Let $x \in S$, $x = \langle\alpha\rangle_i$. Denote $\bar{x} = \langle\alpha\rangle_{1-i}$. Elements $x,y \in S$ will be called associated ($x \sim y$) iff $x = y$ or $x = \bar{y}$.

<u>Remark 1.</u> If for some $x,y \in S$ we have $x \sim y$ and $\tau(x) = \tau(y)$ then $x = y$.

<u>Lemma 3.</u> For each $i \geq 0$ $b(i) \sim b(i+2^{|w|-1})$.

Proof. Based on Lemma 1.

<u>Lemma 4.</u> If $\underline{a}_w = \alpha u^{2^{|w|}} \ldots$ for some α, u $\{0,1\}^*$ then $\underline{b}_w = \alpha'(u')^2 \ldots$ for some $\alpha', u' \in S^*$ such that $|\alpha'| = |\alpha|$, $|u'| = |u^k|$, $k = 2^{|w|-1}$.

<u>Lemma 5.</u>

(i) Let $i > 0$. Then

$$\tau(b(2i)) + \tau(b(2i+1)) \equiv$$
$$\equiv \tau(b(2i+2^{|w|})) + \tau(b(2i+1+2^{|w|})) \pmod 2$$

(ii) There is exactly one $0 \leq j \leq 2^{|w|-1} - 1$ such that for all $i \geq 0$ and all $0 \leq q \leq 2^{|w|-1} - 1$

$$\tau(b(2^{|w|}.i+2q)) + \tau(b(2^{|w|}.i+2q+1)) \not\equiv$$
$$\not\equiv \tau(b(2^{|w|}.i+2q+2^{|w|-1})) + \tau(b(2^{|w|}.i+2q+1+2^{|w|-1})) \pmod 2$$

iff $k = j$.

Proof. The assertions follow from Proposition 2.

A word $x \in S^*$ will be called <u>m-block</u> ($m \geq 0$) iff $x = \sigma^m(d)$ for some $d \in S$. A word $x \in S^*$ is <u>m-factorizable</u> iff it is a (possible empty) concatenation of m-blocks. The set of all m-blocks will be denoted $\mathcal{B}_m$, the set of all m-factorizable words will be denoted $\mathcal{F}_m$.

<u>Remark 2.</u> Each m-block is of length 2^m.

Each initial factor of $\underline{b}_w$ is of a length divisible by 2^m iff it is m-factorizable.

An m-block x will be called <u>even</u> /<u>odd</u>/ iff for some $i \geq 0$ $x = \sigma^m(b(2i))$ / $x = \sigma^m(b(2i+1))$ /.

<u>Remark 3.</u> For each $m \geq 1$ each m-block is a concatenation of some even (m-1)-block with some odd (m-1)-block.

<u>Lemma 6.</u> For $m \geq 0$ no m-block can be both even and odd.

Proof. Based on Lemma 2 and Lemma 5.

Lemma 7. If $\underline{b}_w = xB\ldots$ where $B \in \mathcal{B}_m$ then $x \in \mathcal{F}_m$.

Proof. Induction on m.

Lemma 8. If $\underline{b}_w = x_1u\ldots = x_2u\ldots$ where $x_1 \in \mathcal{F}_m - \mathcal{F}_{m+1}$, $x_2 \in \mathcal{F}_{m+1}$ then u is a proper initial factor both of some even and some odd m-blocks.

Proof. u is an initial factor of an infinite word starting with an odd m-block, and of some other starting with an even m-block. Since no m-block can be both even and odd, $|u| < 2^m$.

Lemma 9. If $\underline{b}_w = xuBu\ldots$, B being a word of length divisible by 2^m, and $x \in \mathcal{F}_m$, then $u \in \mathcal{F}_m$.

Lemma 10. $\underline{b}_w$ contains no factors of the form uBuBu where $u \in S^*$, $B \in \mathcal{B}_m$, $m \geqslant 0$.

Proof. Induction on u.

Corollary 1. $\underline{b}_w$ contains no cubes.

Lemma 11. $\underline{b}_w$ contains no factor of the form xyBzxyBzx where $x \in S$, $B \in \mathcal{B}_m$, $zxy \in \mathcal{B}_m$, $m = |w|-2$.

Proof. Based on Lemma 5.

Lemma 12. $\underline{b}_w$ contains no factor of the form xyBzxyBzx where $x \in S$, $B \in \mathcal{B}_m$, $zxy \in \mathcal{B}_m$, $m \geqslant 0$.

Proof. Induction on m.

Lemma 13. $\underline{b}_w$ contains no factor of the form xyBzxyBzx where $x \in S$, $B \in \mathcal{B}_m$, $z, y \in S^*$.

Proof. Induction on $|zxy|$.

Lemma 13 and Lemma 4 imply the following properties of $\underline{b}_w$ and $\underline{a}_w$.

Theorem 1. $\underline{b}_w$ contains no factor of the form xvxvx, $x \in S$, $v \in S^*$.

Theorem 2. $\underline{a}_w$ contains no factor of the form $(xv)^{2^{|w|}}x$, $x \in \{0,1\}$, $v \in \{0,1\}^*$.

Theorem 2 does not exclude the possibility that $\underline{a}_w$ contains a factor of the form u^q , $q = 2^{|w|}$. Our next goal is to find some necessary conditions for appearing of such a factor in $\underline{a}_w$. First we will describe how do the squares in $\underline{b}_w$ look like.

Lemma 14. Let $\underline{b}_w = \alpha uBu\ldots$, $B \in \mathcal{B}_m$, $u \neq \varepsilon$.
Then $\alpha, u \in \mathcal{F}_m$.

Proof. Let $\alpha \in \mathcal{F}_{m'} - \mathcal{F}_{m'+1}$. Let $m' < m$. Lemma 7 implies $u \in \mathcal{F}_{m'}$. Since $\alpha \notin \mathcal{F}_{m'+1}$ the first m'-block of u is odd. Since $B \in \mathcal{F}_{m'+1}$ the same block is even - a contradiction. Thus $m' \geq m$, $\alpha \in \mathcal{F}_m$. Lemma 7 implies $u \in \mathcal{F}_m$.

Lemma 15. Let $\underline{b}_w = \alpha uBu\ldots$, $B \in \mathcal{B}_m$. Then $|u| = 2^q - 2^m$ for some $q \geq m$.

Proof. Induction on $|u|$.

Lemma 16. Let $\underline{b}_w = \alpha uu\ldots$, $u \neq \varepsilon$, $\alpha \in \mathcal{F}_m$. Then $|u| = 2^q$ for some $q \geq m+|w|-1$.

Lemma 17. Let $\underline{b}_w = \alpha uu\ldots$, $u \neq \varepsilon$, $\alpha \in \mathcal{F}_m - \mathcal{F}_{m+1}$. Then $|u| = 2^{m+|w|-1}$ and $u \in \mathcal{F}_m - \mathcal{F}_{m+1}$.

Corollary 2. If $\underline{b}_w = \alpha uu\ldots$, $u \neq \varepsilon$, $\alpha \in \mathcal{F}_m - \mathcal{F}_{m+1}$ for some $m \geq 0$ then an analogical assertion is true for $m = 0$.

Let now $\bar{d}$ denote the inverse of the rightmost digit of w and let $v(w)$ denote the integer with binary notation w .

Lemma 18. Let $\underline{b}_w = \alpha uu\ldots$, $u \neq \varepsilon$, $\alpha \in \mathcal{F}_0 - \mathcal{F}_1$.
Then either for $y = \alpha$ or for $y = \alpha u$

$$|y| = v(w) + \bar{d} \pmod{2^{|w|}}$$

Proof. Based on Lemma 17 and Lemma 5.

Our knowledge of the powers in $\underline{b}_w$ and $\underline{a}_w$ can now be summarize in the following theorems.

Theorem 3. If $\underline{b}_w = \alpha uu\ldots$, $u \neq \varepsilon$, $\alpha \in \mathcal{F}_m - \mathcal{F}_{m+1}$ then

(i) $|u| = 2^{m+|w|+1}$ and $u \in \mathcal{F}_m - \mathcal{F}_{m+1}$

(ii) $\underline{b}_w = \alpha' u' u' \ldots$ for some $u' \neq \varepsilon$, $\alpha' \in \mathcal{F}_0 - \mathcal{F}_1$ and either for $y = \alpha'$ or for $y = \alpha' u'$

$|y| = v(w) + \bar{d} \pmod{2^{|w|}}$.

Theorem 4. If $\underline{a}_w = \alpha u^{2^{|w|}}\ldots$, $u \neq \varepsilon$, $|\alpha|$ divisible by 2^m and not divisible by 2^{m+1} then

(i) $|u| = 2^m$

(ii) either for $z = |\alpha|/2^m$ or for $z = |\alpha|/2^m + 2^{|w|-1}$

$z \equiv v(w) + \bar{d} \pmod{2^{|w|}}$.

Using Corollary 2 one can show that $\underline{b}_{1101}$ does not contain squares, and consequently, that $\underline{a}_{1101}$ does not contain a factor of the form u^{16}. For each w of the form 1^k , $k > 1$, $\underline{a}_w$ contains the factor 0^{2^k} . As can be shown, each $\underline{a}_w$ contains the factor $0^{2^{|w|}-1}$.

REFERENCES

1. Černý, A., On generalized words of Thue-Morse. L.I.T.P. report 83-44, Université Paris VI et VII, 1983
2. Christol, G., Kamae, T., Mendes-France, M., Rauzy, G., Suites algébriques, automates et substitutions, Bull. Soc. math. France 108 (1980), 401 - 419
3. Cobham, A., Uniform tag sequences, Math. Syst. Theory 6 (1972), 164 - 192
4. Fife, E.D., Binary sequences which contain no BBb, Transactions of the Am. Math. Soc. 261 (1980), 1, 115 - 136
5. Morse, H.M., Recurrent geodesics on a surface of negative curvature, Transactions of the Am. Math. Soc., 22 (1921), 84 - 100
6. Pansiot, J.J., The Morse sequence and iterated morphisms, Inf. Proc. Letters, 12 (1981), 2, 68 - 70
7. Thue, A., Über die gegenseitige Lage gleichen Teile gewisser Zeichenreihen, Videnskapsselskapets Skifter, I. Mat. - naturv. Klasse, Kristiania 1912, 1, 1 - 67
8. Thue, A., Über unendliche Zeichenreihen, Videnskabs-Selskabets Skifter, Math. Naturv. Klasse, Kristiania 1906, 7, 1 - 22

NONDETERMINISM IS ESSENTIAL FOR TWO-WAY COUNTER MACHINES

Marek Chrobak
Institute of Mathematics
Polish Academy of Sciences
Sniadeckich 8, 00-950 Warsaw, Poland.

Definitions:

2dpda - two-way deterministic pushdown automaton,
2nc - two-way nondeterministic counter machine,
2dc - two-way deterministic counter machine,
2sdfa(2) - two-way two-head deterministic finite automaton such that the second head is blind: it can see only the endmarkers,
2DPDA, 2NC, 2DC, 2SDFA(2) are the corresponding classes of languages.

Z.Galil posed in [4] several open questions about two-way deterministic pushdown automata and related them to some famous problems of the theory of computations: P=NP, LBA, etc.. Two of these questions concerned counter languages:

- is 2DPDA = 2DC ?
- is 2DC = 2NC?

Duris and Galil proved in [2] that 2DPDA $\neq$ 2DC using an ingenious generalization of the crossing sequence argument (see [6]). We will adapt their method to prove the following theorem.

Theorem 1. 2DC $\neq$ 2NC.

Hence the technique of Duris and Galil turned out to be quite fruitful; it was also used by them in [3] to construct a hierarchy of reversal-bounded counter machine languages . So far it is probably the strongest method for proving that a given language cannot be recognized by some two-way devices, altough it applies only to counter machine languages. Other results of this kind concern usually restricted types of automata (real-time, reversal-bounded, etc.) or use indirect methods (diagonalization). For examples, see [1,2, 5].

<u>Proof of theorem 1.</u>

Let $L = \{ w_0 \# w_1 \# \ldots \# w_n : w_i \in \{0,1\}^* \text{ for } 0 \leq i \leq n, \text{ and for some } 0 < j \leq n \ \ w_j = w_0 \}$.

Duris and Galil proved that $L \notin$ 2DC, [2] . However, L does not seem to be in 2NC. Then to prove our result we must make L easier for 2nc's but still too difficult for 2dc's.

We define a function code : $\{0,1\}^* \rightarrow \{0,1,2\}^*$, such that

$\mathrm{code}(a_1a_2\ldots a_k) = a_1 2^{k+1} a_2 2^{k+2} \ldots a_{k-1} 2^{2k-1} a_k$,

where $a_1,\ldots,a_k \in \{0,1\}$.

Let $M = \{ w_0 \# \mathrm{code}(w_1) \# \mathrm{code}(w_2) \# \ldots \# \mathrm{code}(w_n) : w_0 \# w_1 \# \ldots \# w_n \in L \}$.

<u>Lemma 1.</u> $M \in$ 2NC.

<u>Proof.</u> We describe briefly actions of a 2nc A recognizing M. Assume for simplicity that #'s on the input are numbered $\#_1, \ldots, \#_n$.

Step 0. A checks if the input word is well formed, that is, if between every two # 's there is some value code(w). It can be easily done using the counter.

Step 1. Let $w_0 = a_1a_2\ldots a_k$. A stores a_1 in the finite memory and k on the counter and moves right until it stops nondeterministically on $\#_j$. Let $w_j = b_1b_2\ldots b_l$. Now A checks if $l = k$ and $b_1 = a_1$.

Step i. Suppose that $b_1 = a_1$, $b_2 = a_2$, ... , $b_{i-1} = a_{i-1}$, the counter stores 0, and the head of A is on the symbol b_{i-1} (fig.1). Let x be the distance between $\#_1$ and b_{i-1}. A moves left increasing the counter until it reaches the left endmarker.

Then the counter stores x+k. A must guess now which of the bits $a_1, a_2, \dots, a_k$ is a_i. In order to do this A moves right increasing the counter and stops nondeterministically on some symbol a_p. The value of the counter is then x+k+p. At last, A stores a_p, moves to $\#_1$, and again moves right, decreasing now the counter until it becomes 0. Let d be the symbol scanned by the head. The distance between the position of the head and b_{i-1} is k+p. Hence the following equivalence holds:

$p = i$ iff $d \in \{0,1\}$ iff $d = b_i$.

We have two cases now:

$d = 2$: wrong guess, A rejects the input,

$d \neq 2$: right guess, if $d \neq a_i$ A rejects the input, otherwise the computation continues.

Repeating step i, for i = 2,...,k, A can compare w_0 and w_j. After the k-th repetition A accepts the input.

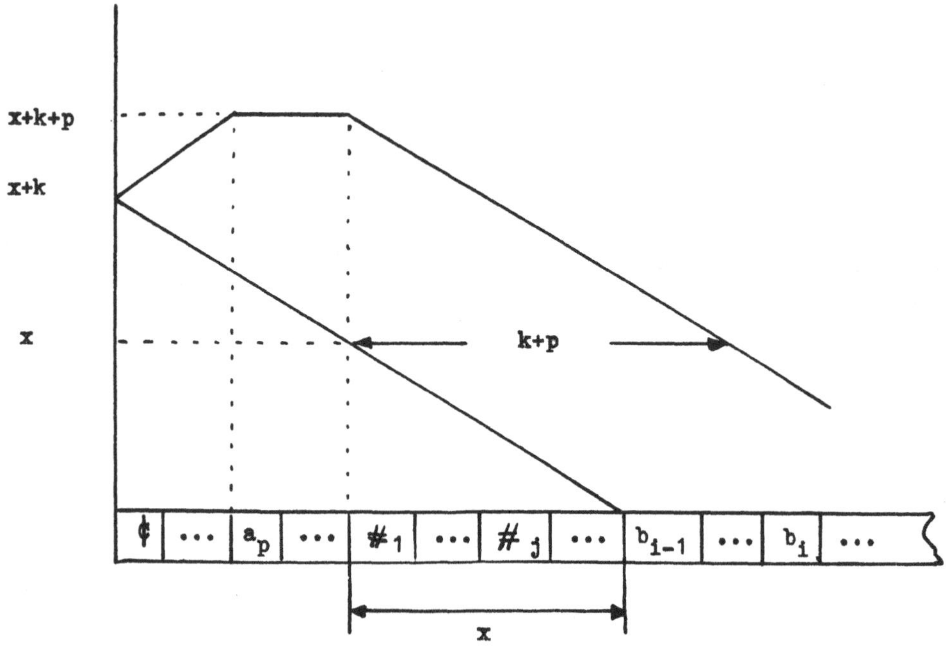

Fig.1. A guesses a_i and verifies the guess.

Lemma 2. $M \notin 2DC$.

Proof. The proof of Duris and Galil applies to M almost without any change. We only sketch the necessary modifications. A more detailed proof will appear elsewhere.

Since $2DC \subseteq 2SDFA(2)$, it is enough to show that $M \notin 2SDFA(2)$. Let B be any given 2sdfa(2). The idea is to construct languages $M_1 \in \{0,1\}^*$ and $M_2 \in \{\#,0,1,2\}^*$ such that:

(a) for every $u,v \in M_2$, $u \neq v$, there exists $y \in M_1$ such that exactly one of the words yu,yv belongs to M,

(b) there exist two words $\bar{u},\bar{v} \in L(B)$ such that for every $x \in M_1$, $x\bar{u} \in L(B)$ iff $x\bar{v} \in L(B)$.

From (a) and (b) follows $L(B) \neq M$.

Let L_1 and L_2 be the languages from the proof that $L \notin 2SDFA(2)$ in [2]. Then we take:

$M_1 = L_1$,

$M_2 = \{\#code(z_1)\#code(z_2)\#\ldots\#code(z_s) : \#z_1\#z_2\#\ldots\#z_s \in L_2\}$.

The construction in [2] remains valid when we take M_1 and M_2 instead of L_1 and L_2 since it does not depend on the actual structure of the strings in L_2. □

Theorem 1 follows easily from lemmas 1,2. □

Acknowledgment. I would like to thank W.Rytter for many discussions on automata and complexity.

References

1. T-h Chan, Reversal complexity of counter machines, Proc. 13th Annual STOC, Milwaukee, 1981, 146-157.
2. P.Duris, Z.Galil, Fooling a two-way automaton or one pushdown is better than one counter for two-way machines, Theoret. Comput. Sci. 21(1982), 39-53.
3. P.Duris, Z.Galil, On reversal-bounded counter machines and on pushdown automata with a bound on the size of the pushdown store, Automata, Languages and Programming, 9-th Colloq., Aarus, 1982,

Lect. Notes Comput. Sci. 140(1982), 166-175.

4. Z.Galil, Some open problems in the theory of computation as questions about two-way deterministic pushdown automata languages, Math. Systems Theory 10(1977), 211-228.
5. E.M.Gurari, O.H.Ibarra, Two-way counter machines and Diophantine equations, J. Assoc. Comput. Mach. 29(1982), 863-873.
6. J.E.Hopcroft, J.D.Ullman, Formal languages and their relation to automata, Addison-Wesley, Reading, MA, 1969.

WEAK AND STRONG FAIRNESS IN CCS*

Gerardo Costa** and Colin Stirling
Dept. of Computer Science
Edinburgh University

INTRODUCTION

It is desirable that models of parallel systems imply nothing about the relative speeds of the concurrent subcomponents but also imply that each subcomponent always eventually proceeds (unless it has terminated). Such models reflect the principle of fairness [Pa1,M2]. Complications arise if a subcomponent can not always proceed autonomously: other components may prevent it from proceeding or it may not be able to proceed without interacting. In such circumstances authors distinguish between weak and strong fairness [Pa2, P1, GPSS]. A weak fairness assumption implies that if a subcomponent can almost always proceed then eventually it must proceed while a strong fairness assumption implies that a subcomponent which can proceed infinitely often does proceed infinitely often.

Fairness assumptions have been examined from disparate viewpoints. Examples include [AO,DK,GPSS,HO,HS,M2,OL,Pa1,P1,QS]. Here, we are concerned with an operational approach, the problem of defining and generating the admissible execution sequences of a concurrent language when weak or strong fairness is assumed. One general method, described for instance in [LPS], is to invoke two semantic levels. One level prescribes the finite and infinite execution sequences without assuming fairness whilst the other filters out the inadmissible ones. The first level is given via a set of generative rules whereas the second is encoded as a definition of admissible (or inadmissible) execution sequence. The inadmissible sequences generated by the rules are discounted.

A more interesting solution, a single level approach, is simply to offer rules which generate just the admissible sequences. A second semantic level is then unnecessary. One standard approach is to appeal to random assignment. This has been done in [AO,Pa2] for a guarded command language and in [P1] for a concurrent while language. Technical problems occur when these approaches are extended to cover the language we consider in this paper, Milner's CCS [M1]. Moreover, the use of random assignment for capturing fairness assumptions is more a simulation than a description

* This work was supported by the UK SERC.
** On leave from Istituto di Matematica, Università di Genova, Italy.

because it involves predictive choice [M2]. This forcing in advance may also have unforeseen consequences.

In this paper we offer an alternative single level approach for Milner's CCS. We define the notions of weak-admissible and strong-admissible CCS execution sequence. To do this we make use of a labelled CCS syntax. We show that the weak-admissible execution sequences (the admissible sequences assuming weak fairness) can be generated by a finite set of rules which do not, unlike random assignment approaches, involve predictive choice. This is because weak-admissibility can be 'locally' characterized. However, this is not the case for strong-admissibility. Consequently, although we show that strong-admissible sequences can be generated by a finite set of rules predictive choice is involved. In both cases the set of rules we offer are an extension of the standard CCS rules. The results presented here extend [CS1].

In section 1 we describe Milner's CCS and in section 2 we define the weak and strong-admissible CCS execution sequences. Sections 3 and 4 contain the rules for weak and strong fair CCS. Proofs together with more details are given in the full version [CS2].

1. CCS

Milner's CCS is a calculus whose <u>closed expressions</u> stand for processes. Let $Act = \Delta \cup \bar{\Delta}$ be a set of <u>atomic actions</u> where $\bar{\Delta}$ is a set of co-actions disjoint from Δ and in bijection with it. The bijection is $\bar{\ }$: $\bar{a} \in \bar{\Delta}$ stands for the co-action of $a \in \Delta$, and $\bar{\bar{a}} = a$. The calculus allows synchronization of co-actions which is represented by τ, a silent or internal action. Let $Move = Act \cup \{\tau\}$ and let a,b,c range over Act, m,n over Move and let X,Y,Z be process variables. The syntax of the language is:

$$E ::= X \mid NIL \mid m\,E \mid E + E \mid fix\ X.E \mid E\backslash a \mid E|E\ .$$

NIL is the nullary process which does nothing; + represents nondeterministic choice; | concurrency; fix recursion; and $\backslash a$ represents restriction (prevention) of a and $\bar{a}$ actions. We assume that in fix X.E <u>X is guarded in</u> E : that is, every free occurrence of X in E is within a subexpression m F of E.

The behaviour of a process E is determined by the following rules where $E \xrightarrow{m} F$ means process E becomes F by performing the move m :

Move $\quad m\,E \xrightarrow{m} E$

+R $\quad \dfrac{E \xrightarrow{m} E'}{E + F \xrightarrow{m} E'}$

+L $\quad \dfrac{F \xrightarrow{m} F'}{E + F \xrightarrow{m} F'}$

fix $\quad \dfrac{E[fixX.E/X] \xrightarrow{m} E'}{fix\ X.E \xrightarrow{m} E'}$

Res $\quad \dfrac{E \xrightarrow{m} E'}{E\backslash a \xrightarrow{m} E'\backslash a} \quad m \notin \{a,\bar{a}\}$

$$|R \quad \frac{E \xrightarrow{m} E'}{E|F \xrightarrow{m} E'|F} \qquad |L \quad \frac{F \xrightarrow{m} F'}{E|F \xrightarrow{m} E|F'} \qquad |Syn \quad \frac{E \xrightarrow{a} E' \quad F \xrightarrow{\bar{a}} F'}{E|F \xrightarrow{\tau} E'|F'}$$

Three points are worth mentioning. Sometimes the + rules do not allow choice: $(a\,E + b\,F)\backslash b$ can only become E by performing a move. The | rules do not, in general, compel synchronization: $a\,E|\bar{a}\,F$ can perform a or can perform $\bar{a}$ as well as τ. On the other hand, $(a\,E|\bar{a}\,F)\backslash a$ can only perform τ because of the restriction. Finally, the number of conucrrent subprocesses may increase as moves are performed.

A <u>derivation</u> is any finite or infinite sequence of the form $E_0 \xrightarrow{m_0} E_1 \xrightarrow{m_1} \ldots$. An <u>execution sequence</u> is a maximal derivation: if it is finite its last term will be unable to make a move. We let h,k range over finite sequences of moves. If $E_0 \xrightarrow{m_0} E_1 \xrightarrow{m_1} \ldots \xrightarrow{m_i} \ldots \xrightarrow{m_n} E_{n+1}$ we use $E_0 \xrightarrow{h} E_{n+1}$ where $h = m_0 m_1 \ldots m_n$ as an abbreviation.

We have offered Milner's CCS without value passing and renaming: they can be dealt with within the framework we develop. For a full discussion of CCS see [M1].

2. FAIRNESS AND CCS

Fairness imposes the constraint that concurrent subprocesses always eventually proceed unless they are deadlocked or have terminated. Such a constraint may affect the behaviour of processes: for instance it precludes the CCS execution sequence $E|F \xrightarrow{a} E|F \xrightarrow{a} \ldots$ when $E = fixX.aX$ and $F = fixX.bX$ because F does not continually contribute a move. By ruling out execution sequences fairness affects liveness properties [OL,Pa1]. The process $((fixX(aX + bNIL))|\bar{b}NIL)\backslash b$ always eventually terminates (becomes $(NIL|NIL)\backslash b$) when fairness is assumed because the subcomponent $\bar{b}NIL$ must eventually proceed (since it is always able to). The possibility that subcomponents may not be able to proceed autonomously in CCS means that weak and strong fairness are distinguishable [Pa2, Pl]: the <u>weak</u> fairness constraint states that if a subcomponent can <u>almost always</u> proceed then eventually it must do so while the <u>strong</u> fairness constraint states that if a subcomponent can proceed <u>infinitely often</u> then it must proceed infinitely often. Thus, the sequence $E \xrightarrow{a} G \xrightarrow{c} E \xrightarrow{a} G \xrightarrow{c} E \xrightarrow{a} \ldots$ where $E = (F|\bar{b}NIL)\backslash b$, $F = fixX.a(cX + bNIL)$ and $G = ((cF + bNIL)|\bar{b}NIL)\backslash b$ is admissible under weak but not under the strong fairness constraint.

We formally define the two fairness constraints in terms of admissible execution sequences. To this end we appeal to a labelled CCS syntax: a labelling which is a slight modification of that associated with a tree representation of CCS expressions. For any expression distinct occurrences of all operators receive different labels. The slight modification we make means that unicity of labels is preserved both under substitution and derivation. Variable occurrences are also labelled (except for those X in the binder fix X) and substituted expressions inherit the label of the variable occurrences they replace: under the labelling fixX.aX is equivalent to an 'expression' $a_{u_1} a_{u_2} \ldots a_{u_n} \ldots$ where each label u_i is distinct. The calculus is unaffected. We assume that the moves performed are unlabelled: the Mov rule becomes $m_u E \overset{m}{\dashrightarrow} E$. Moreover, we assume that $(\text{fix}X)_u$ binds labelled occurrences of X and that $(\setminus b)_u$ restricts b and $\bar{b}$ actions. From now on, we assume, usually implicitly, that CCS expressions are labelled and when examples are given we omit those labels which are inessential to the surrounding discussion.

We need to define the notion of a live concurrent subprocess - a concurrent subcomponent that can perform or can contribute (via synchronization) to the performance of a move. First, we let P(E) be the set of labelled concurrent subprocesses of E irrespective of liveness. This is defined inductively letting labels represent processes:

$$P(X_u) = \emptyset \qquad P(NIL_u) = \{u\} \qquad P(m_u E) = \{u\}$$

$$P(E +_u F) = \begin{cases} \{u\} \text{ if } P(E) \cup P(F) \neq \emptyset \\ \emptyset \text{ otherwise} \end{cases} \qquad P(\text{fix } X.E) = P(E\setminus a) = P(E)$$

$$P(E|F) = P(E) \cup P(F)$$

The concurrent subprocesses in $E|F$ consist of those in E together with those in F. Note that we identify a subprocess by the label of its main combinator except in the case of $\setminus a$ (restriction) and fix X. Hence $(F|G) + bH$ is a single subprocess despite the presence of the parallel operator, $|$, in it. Moreover, the notion of subcomponent is dynamic: if $E = a_u a_v$ NIL and E performs a then the resulting subcomponents do not include u. (And by the method of labelling, once a label is 'lost' in this way it is never regained under derivation.)

We now define Act(E), the set of unlabelled actions in E (excluding τ) which can happen autonomously:

$$\text{Act}(X) = \text{Act}(NIL) = \text{Act}(\tau E) = \emptyset \qquad \text{Act}(a_u E) = \{a\}$$

$$\text{Act}(\text{fix } X.E) = \text{Act}(E) \qquad \text{Act}(E\setminus a) = \text{Act}(E) - \{a,\bar{a}\}$$

$$\text{Act}(E+F) = \text{Act}(E|F) = \text{Act}(E) \cup \text{Act}(F)$$

A simple consequence of this definition is: $a \in \text{Act}(E)$ iff $\exists F.\ E \overset{a}{\dashrightarrow} F$.

Next, we define LP(E,A) to stand for the set of live concurrent subprocesses in E when the environment prevents the set of actions $A \subseteq \text{Act}$. We let $\overline{\text{Act}}(E)$ be the set $\{\bar{a} : a \in \text{Act}(E)\}$: that is, the set of co-actions of Act(E).

$$LP(X_u,A) = LP(NIL_u,A) = \emptyset \qquad LP(\text{fix } X.E,A) = LP(E,A)$$

$$LP(m_u E,A) = \begin{cases} \{u\} \text{ if } m \notin A \\ \emptyset \text{ otherwise} \end{cases} \qquad LP(E +_u F,A) = \begin{cases} \{u\} \text{ if } LP(E,A) \cup LP(F,A) \neq \emptyset \\ \emptyset \text{ otherwise} \end{cases}$$

$$LP(E\setminus a, A) = LP(E, A \cup \{a,\bar{a}\})$$

$$LP(E|F, A) = LP(E, A - \overline{\text{Act}}(F)) \cup LP(F, A - \overline{\text{Act}}(E)) .$$

The set of live subprocesses in E is defined as $LP(E,\emptyset)$ which we abbreviate to $LP(E)$. For example, $LP(((aE +_u bF)|\bar{a}_v G)|b_w H)\backslash a\backslash b) = \{u,v\}$ for any E,F,G,H.

Let $\gamma = E_0 \xrightarrow{m_0} E_1 \xrightarrow{m_1} \ldots$ be a finite or infinite CCS execution sequence then we say that:

Definition 2.1 i. γ is weak-admissible iff $\neg(\exists u \exists i \; \forall k \geq i \,.\, u \in LP(E_k))$
ii. γ is strong-admissible iff $\neg(\exists u \; \forall i \; \exists k \geq i \,.\, u \in LP(E_k))$.

Here weak-admissible (strong-admissible) means admissible under the weak (strong) fairness constraint. A sequence is weak-admissible just in case no subcomponent becomes live and remains live throughout. A sequence is strong-admissible just in case no subcomponent is live infinitely often. Notice that a component can not be live infinitely often and proceed infinitely often because of the dynamic labelling: as soon as a component contributes a move it 'disappears' .

3. WEAK FAIR CCS

In this section we offer weak fair CCS rules: rules for CCS which generate just the weak-admissible execution sequences.

Our starting point is an alternative, more local, formulation of weak-admissible than definition 2.1. Let $\gamma = E_0 \xrightarrow{m_0} E_1 \xrightarrow{m_1} \ldots$ be a CCS execution sequence then

i. γ is w-admissible at i if $\exists j \geq i \;\; LP(E_i) \cap LP(E_{i+1}) \cap \ldots \cap LP(E_j) = \emptyset$
ii. γ is w-admissible if $\forall i : i \leq$ length of γ . γ is admissible at i.

Clearly, γ is w-admissible at i iff $\forall j \leq i$ γ is w-admissible at j.

Theorem 3.1 γ is w-admissible iff γ is weak-admissible.

The significance of this alternative definition of weak-admissibility is that it allows us to think of admissibility in terms of a localizable property and not just a property of complete execution sequences. Moreover, it suggests a way of generating the admissible execution sequences.

When B is a finite set of (process) labels we say that $E_0 \xrightarrow{m_0} E_1 \xrightarrow{m_1} \ldots \xrightarrow{m_n} E_{n+1}$ is a B-step just in case $B \cap LP(E_0) \cap \ldots \cap LP(E_{n+1}) = \emptyset$. We abbreviate $E_0 \xrightarrow{m_0} E_1 \xrightarrow{m_1} \ldots \xrightarrow{m_n} E_{n+1}$ is a B-step to $E_0 \xrightarrow[B]{h} E_{n+1}$ where $h = m_0 \ldots m_n$.

Lemma 3.2 i. if $LP(E) \neq \emptyset$ then $\forall$ finite set of labels $B \; \exists F,h . E \xrightarrow[B]{h} F$
ii. if $E \xrightarrow[B]{h} F$ and $F \xrightarrow{k} G$ then $E \xrightarrow[B]{h} F \xrightarrow{k} G$.

The definition of w-admissible at i means that a B-step from E when $B = LP(E)$ is of special interest. An $LP(E)$-step from E is 'locally' admissible: the live subprocesses of E have lost their liveness at some point in the step. For instance the following is an LP(E)-step from E where $E = (a_u baF \mid (\bar{a}_v \bar{b}\, G \mid \bar{a}_y H)) \setminus a$: $E \xrightarrow{\tau\tau} (a\, F \mid (G \mid \bar{a}_y H)) \setminus a$.

A w-sequence from E_0 is any maximal sequence of steps of the form $E_0 \underset{LP(E_0)}{\overset{h_0}{\dashrightarrow}} E_1 \underset{LP(E_1)}{\overset{h_1}{\dashrightarrow}} \ldots$. A w-sequence is simply a concatenation of 'locally' admissible steps. If δ is a w-sequence then its associated CCS execution sequence is the sequence which drops all reference to the $LP(E_i)$s.

<u>Theorem 3.3</u> A CCS execution sequence is weak-admissible iff it is the sequence associated with a w-sequence.

Rules are now offered for a transition relation $\xrightarrow[B]{h}$ by building on Milner's rules: $\xrightarrow[B]{h}$ satisfies the condition that $E \xrightarrow[B]{h} F$ iff $E \underset{B}{\overset{h}{\dashrightarrow}} F$. (Like $E \overset{h}{\dashrightarrow} F$, $E \xrightarrow{h} F$ will be an abbreviation of a sequence of moves $E \xrightarrow{m_0} E_1 \xrightarrow{m_1} \ldots \xrightarrow{m_n} F$ with $h = m_0 \ldots m_n$.) The w-sequences will then be generable from such rules. Note that the significance of lemma 3.2i in terms of rules involving $\xrightarrow[B]{h}$ is that it expresses a 'no stuck' property: for any live E and finite set of labels B the rules must allow a derivation of the form $E \xrightarrow[B]{h} F$. Moreover, 3.2ii says that any extension of a B-step is still a B-step.

We assume Milner's rules given in section 1 except that the expressions of the language are labelled. We merely add two new rules involving $\xrightarrow[B]{h}$ to give a weak fair CCS. We assume that B,C are finite sets of labels.

$$\text{Sh} \quad \frac{E \overset{m}{\dashrightarrow} F}{E \xrightarrow[B]{m} F} \quad B \cap LP(E) \cap LP(F) = \emptyset \qquad \text{Tr} \quad \frac{E \xrightarrow[B]{h} F \quad F \overset{m}{\dashrightarrow} G}{E \xrightarrow[B \cup C]{hm} G} \quad C \cap LP(G) = \emptyset$$

In the Tr(ans) rule $E \xrightarrow[B \cup C]{hm} G$ abbreviates $E \underset{B \cup C}{\xrightarrow{h} F \xrightarrow{m}} G$. The Sh(ift) rule allows one to derive any B-step involving a single move only. The Trans rule, on the other hand, allows one to generate B-steps involving a sequence of moves. This is all we need to get the required result.

<u>Theorem 3.4</u> $E \underset{B}{\overset{h}{\dashrightarrow}} F$ iff $E \xrightarrow[B]{h} F$

To generate all the admissible sequences from E it is not sufficient to be 'rid' of the processes in LP(E) as soon as possible. The following example admissible sequence makes this clearer where E = fix X.aX

$$E \mid b_u NIL \overset{a}{\dashrightarrow} E \mid b_u NIL \overset{a}{\dashrightarrow} \ldots \overset{a}{\dashrightarrow} E \mid b_u NIL \overset{b}{\dashrightarrow} E \mid NIL \overset{a}{\dashrightarrow} \ldots$$

There is no finite upper bound on how many a actions can be performed before u loses its liveness. The Shift and Trans rules allow for this freedom: one can 'freely' generate steps at any point. Previous choices do not constrain future possibilities. This is in contrast to those analyses or simulations of admissibility which appeal to random assignment.

We now come to the main result of this section. A WFCCS execution sequence is any maximal sequence of the form: $E_0 \overset{h_0}{\underset{B_0}{\longrightarrow}} E_1 \overset{h_1}{\underset{B_1}{\longrightarrow}} \ldots$ where $B_i = LP(E_i)$. The CCS execution sequence associated with it is the sequence which omits the B_is and replaces $\overset{m}{\dashrightarrow}$ for $\overset{m}{\longrightarrow}$ throughout. An immediate consequence of previous results is the following theorem.

Theorem 3.5 A CCS execution sequence is weak-admissible iff it is the sequence associated with a WFCCS execution sequence.

4. STRONG FAIR CCS

We now offer rules for strong fair CCS. As far as possible we develop an analysis similar to the previous section. However, immediately there is a problem: unlike weak-admissibility there is not an alternative localized equivalent definition of strong-admissibility. The following definition includes the possibility that if δ is s-admissible at i then this can only be known by inspecting the complete future segment after i. This appears to be intrinsic to the difference between weak and strong fairness. Let $\gamma = E_0 \overset{m_0}{\dashrightarrow} E_1 \overset{m_1}{\dashrightarrow} \ldots$ be a CCS execution sequence then

i. γ is s-admissible at i iff $\exists j \geq i \ \forall k \geq j \ P(E_i) \cap LP(E_k) = \emptyset$

ii. γ is s-admissible iff $\forall i \leq$ length of γ. γ is s-admissible at i

γ is s-admissible at i just in case no subcomponent at i, whether live or not, becomes infinitely often live. It is straightforward to show that γ is s-admissible at i iff $\forall j \leq i$. γ is s-admissible at j.

Lemma 4.1 γ is s-admissible iff γ is strong-admissible.

The s-admissibility condition does not immediately suggest an analogue of a B-step. In fact, one can give analogues but they always appear to involve predictive choice: that is, one can 'localise' strong-admissiblity at the expense of predictive choice. We here present one based on a finite queue of labels. (The idea of using queues for capturing strong admissibility is not new: see [Pa2] for instance.) First we need some notation. If B is a finite set of labels then let Seq(B) be the set of permutations of elements in B: if $T \in Seq(B)$ then each element of B appears exactly once in T. Conversely, if $T \in Seq(B)$ then $Set(T) = B$. We use . to denote concatenation of sequences. If u,v are labels then $u \subseteq_T v$ iff u = v or u precedes v in T. Finally, $T \upharpoonright B$ is the sequence which results from deleting any label not in B from T.

The analogue of a B-step we use is a T-step. The following definition does not enforce a strict queue discipline because it is too restrictive. Instead, the discipline used is given by the 'min' condition in the definition. For simplicity we define a T-step from E_0 when $T \in Seq(P(E_0))$ and not for arbitrary T. When $T \in Seq(P(E_0))$ we say that $E_0 \xrightarrow{m_0} E_1 \xrightarrow{m_1} \dots \xrightarrow{m_n} E_{n+1}$ is <u>a T-step</u> just in case the following 'min' condition holds:

$$\forall u \in Set(T).\ \forall i : 0 \le i \le n\ \ (\text{if}\ \ u \in LP(E_i)\ \ \text{then}$$
$$\exists v.\ v \subseteq_T u\ \ \exists j \le i.\ \ v \in LP(E_j)\ \ \text{and}\ \ v \notin P(E_{j+1}))$$

The min condition states that the live subcomponent of E_0 which is earliest in the queue must contribute a move in preference to any other process and, moreover, if at a later point say E_j in the derivation an even earlier process u in the queue becomes live then either E_j is E_{n+1} or u contributes to the move m_j. We abbreviate $E_0 \xrightarrow{m_0} E_1 \xrightarrow{m_1} \dots \xrightarrow{m_n} E_{n+1}$ is a T-step to $E_0 \xrightarrow[T]{h} E_{n+1}$ where $h = m_0 \dots m_n$. Let $E_0 = ((b_u caF \mid d_v eG) \mid \bar{a}_y H) \backslash a$ and let $T = \langle y,v,u \rangle$. Then $E_0 \xrightarrow{b} ((caF \mid d_v eG) \mid \bar{a}_y H) \backslash a$ is not a T-step. Whereas $E_0 \xrightarrow{d} E_1$ is where $E_1 = ((b_u caF \mid eG) \mid \bar{a}_y H) \backslash a$. Moreover so is the following where $E_2 = ((caF \mid e\,G) \mid \bar{a}_y H) \backslash a$:

$$E_0 \xrightarrow{d} E_1 \xrightarrow{b} E_2 \xrightarrow{c} ((aF \mid e\,G) \mid \bar{a}_y H) \backslash a.$$

<u>Lemma 4.2</u> i. if $LP(E) \neq \emptyset$ then $\forall T \in Seq(P(E))\ \exists F,h.\ \ E \xrightarrow[T]{h} F$

ii. if $E \xrightarrow{h} F$ is not a T-step and $F \xrightarrow{k} G$ then $E \xrightarrow{h} F \xrightarrow{k} G$ is not a T-step.

A <u>s-sequence</u> from E_0 is any maximal sequence of steps of the form $E_0 \xrightarrow[T_0]{h_0} E_1 \xrightarrow[T_1]{h_1} \dots$ where for $i \ge 0$ $T_{i+1} = (T_i \upharpoonright P(E_{i+1})) \cdot U$ and $U \in Seq(P(E_{i+1}) - P(E_i))$. It is easy to check that $T_i \in Seq(P(E_i))$ for every i.

This linking together of steps in the definition of s-sequence means that any process in E_i which does not contribute a move will move towards the front of the queue. Thus, if it repeatedly becomes live then eventually by the min condition it must contribute to a performance of a move. If δ is an s-sequence then its associated CCS execution sequence is the sequence which drops all reference to the T_is

<u>Theorem 4.3</u> A CCS execution sequence is strong-admissible iff it is a sequence associated with an s-sequence.

We now offer operational rules for strong fair CCS based upon the notion of a T-step. Our method is analogous to weak fair CCS: we add two extra rules to standard CCS involving a transition relation $\overset{h}{\underset{T}{\Longrightarrow}}$ which coincides with $\xrightarrow[T]{h}$. If U is a non-empty sequence then Hd(U) is the first member of U. We assume that

$T \in Seq(P(E))$:

$$\mathrm{Sh}\ \frac{E \overset{m}{-\!\!-\!\!>} F \qquad Hd(T \upharpoonright LP(E)) \notin P(F)}{E \underset{T}{\overset{m}{\Longrightarrow}} F} \qquad \mathrm{Tr}\ \frac{E \underset{T}{\overset{h}{\Longrightarrow}} F \quad F \overset{m}{-\!\!-\!\!>} G \qquad \forall u \in (LP(F) \cap P(E))\ \exists v. v \subseteq_T u \wedge v \notin P(G)}{E \underset{T}{\overset{hm}{\Longrightarrow}} G}$$

In the Tr(ans) rule $E \underset{T}{\overset{hm}{\Longrightarrow}} G$ abbreviates $E \overset{h}{\Longrightarrow} \underset{T}{F} \overset{m}{\Longrightarrow} G$. The Sh(ift) rule guarantees that the earliest live process in T contributes to the performance of m. The Trans rule on the other hand guarantees that if a process in P(E) becomes live at F and is earlier in T than any process which has contributed to the sequence of moves h then it must contribute to the move m. Thus, these two rules guarantee that the min condition always holds. (Note lemma 4.2i expresses, like 3.2i, a 'no stuck' property.)

Theorem 4.4 $E \underset{B}{\overset{h}{-\!\!-\!\!>}} F$ iff $E \underset{T}{\overset{h}{\Longrightarrow}} F$.

Lemma 4.2ii says that if a derivation is not a T-step then it cannot be extended to become one. Thus the particular choice of T may prescribe in advance the performance of a process. This means that unlike the rules for weak fairness these rules do not allow one to 'freely' generate steps: the choice of T constrains future possibilities.

We now come to the main result of this section which is analogous to theorem 3.5. A SFCCS execution sequence is any maximal sequence of the form: $E_0 \underset{T_0}{\overset{h_0}{\Longrightarrow}} E_1 \underset{T_1}{\overset{h_1}{\Longrightarrow}} \ldots$ where the T_is satisfy the conditions given in the definition of an s-sequence. The CCS execution sequence associated with it is the expected one.

Theorem 4.5 A CCS execution sequence is strong-admissible iff it is the sequence associated with a SFCCS execution sequence.

Acknowledgements

We would like to thank Matthew Hennessy, Robin Milner and Gordon Plotkin for many illuminating discussions. We would also like to thank Dorothy McKie for typing.

References

[AO] K. Apt and E. Olderog. 'Proof rules and transformations dealing with fairness'. Theoretical Computer Science pp.65-100 (1983).

[CS1] G.Costa and C.Stirling. 'A fair calculus of communicating systems'. LNCS Vol.158 pp.94-105 (and extended version, Technical Report CSR-137-83, Dept. of Computer Science, Edinburgh). (1983)

[CS2] G. Costa and C. Stirling. 'Weak and strong fairness in CCS', to appear as a technical report.

[DK] Ph. Darondeau and L. Kott. 'On the observational semantics of fair parallelism' LNCS Vol.154 pp.147-159 (1983).

[GPSS] D. Gabbay, A. Pnueli, S. Shelah and J. Stavi. 'On the temporal analysis of fairness'. Proc. 7th ACM POPL, Las Vegas (1980).

[H] M. Hennessy. 'Modelling finite delay operators', Technical Report CSR-153-83, Dept. of Computer Science, Edinburgh (1983).

[HS] M. Hennessy and C. Stirling. 'The power of the future perfect in program logics', in this volume.

[LPS] D. Lehmann, A. Pnueli and J. Stavi. 'Impartiality, justice and fairness: the ethics of concurrent termination'. LNCS Vol.115, pp.264-77, (1981).

[M1] R. Milner. 'A Calculus of Communicating Systems'. LNCS Vol.92, (1980).

[M2] R. Milner. 'A finite delay operator in synchronous CCS'. Technical Report CSR-116-82, Dept. of Computer Science, Edinburgh University (1982).

[OL] S. Owicki and L. Lamport. 'Proving liveness properties of concurrent programs'. ACM Transactions on Programming Languages and Systems pp.455-495 (1982).

[Pa1] D. Park. 'On the semantics of fair parallelism'. LNCS Vol.85, pp.504-26 (1980).

[Pa2] D. Park. 'A predicate transformer for weak fair iteration', in Proceedings 6th IBM Symposium on Mathematical Foundations of Computer Science, Hakone, Japan (1981).

[P1] G. Plotkin. 'A powerdomain for countable nondeterminism', LNCS Vol.140 pp.418-28 (1982).

[QS] J. Queille and J. Sifakis. 'Fairness and related properties in transition systems', Acta Informatica 19, pp.195-220 (1983).

ON THE COMPLEXITY OF INDUCTIVE INFERENCE†

(Preliminary Report)

Robert P. Daley
Department of Computer Science
University of Pittsburgh
Pittsburgh, PA 15260

Carl H. Smith
Department of Computer Science
University of Maryland
College Park, Maryland

§1 Introduction

Inductive inference, also known as the algorithmic synthesis of programs given examples of their intended input/output behavior, has been the subject of several recent survey papers [2,16]. The abstract study of inductive inference has focused on distinguishing various criteria for successful inference by a given class of machines. Herein a notion of the complexity of the inference process is presented with some examples of trade-offs between the complexity of an inference and the accuracy of the result of the inference. Then an axiomatization of the concept of the complexity of inductive inference is intoduced. Our axiomatization parallels the approach made by Blum [6] for the complexity of computations. Earlier studies of the complexity of inference were concerned with showing that certain inference problems were members of well known complexity classes like P and NP [1,11,17,18]. Freivald in [9] also attempted an axiomatization of a notion of complexity of inductive inference. The basic approach employed below is similar to his on a number of points.

The class of all total recursive functions is denoted by R. We use $f(x)\downarrow$ and $f(x)\uparrow$ to indicate that $f(x)$ is defined and $f(x)$ is undefined respectively. Also, we use $f|n$ to denote the finite initial segment of f whose domain consists of $\{x \mid x \leq n\}$. For any $h \in R$, h^0 is the identity function and $h^{n+1} = h(h^n)$. We use $\phi_1 =^n \phi_2$ to mean that the cardinality of $(\{x \mid \phi_1(x) \neq \phi_2(x)\}) \leq n$, and $\phi_1 =^* \phi_2$ to mean that $\{x \mid \phi_1(x) \neq \phi_2(x)\}$ is finite. The quantifiers $\overset{\infty}{\exists}$ and $\overset{\infty}{\forall}$ stand for "there exist infinitely many" and "for all but finitely many" respectively. The sequence $\{\phi_i\}$ denotes an arbitrary *acceptable programming system* (see [16]), also known as an *acceptable numbering* of all and only the partial recursive functions [20,21].

†) Supported by NSF Grants MCS 7803617, 8017332, 7903912, 8105214 and 8301536.

Inferences will be performed by *Inductive Inference Machines* (IIMs) as defined recursion theoretically in [5] and used in essence previously in [10,19]. An IIM M operates in a limiting recursive manner as follows: M is presented with successively larger segments of the graph of some function f and in response produces a sequence $p_1, p_2, \ldots$ of hypothesized programs for f. We say that M EX^n identifies f (and write $f \in EX^n(M)$) if there is some program p such that $\phi_p =^n f$ and either M produces a finite sequence of programs and p is the last one, or M produces an infinite sequence of programs all but finitely many of which are p. The class of sets of functions $EX^n = \{S \mid (\exists M)[S \subseteq EX^n(M)]\}$, consists of all the sets of functions which can be successfully inferred by some IIM with respect to EX^n type inference. There are several assumptions which can be made about IIMs without loss of generality with respect to inferrability. For example, M can be assumed to be total, i.e., defined on any finite segment of the graph of any function. Also, it can be assumed that the function f is presented to M (or querried by M) in increasing order with respect to its domain (i.e., f is presented as $f \mid 0, f \mid 1, f \mid 2, \ldots$). Although such assumptions do not effect inferrability, they can alter the complexity of some inferences. Consequently, rather than adopt the convention that each IIM is total we will assume that the domain of each IIM is *closed under subfunctions*, i.e., if $\phi_i(\tau)\downarrow$, then $\phi_i(\tau')\downarrow$ for all $\tau' \subseteq \tau$, where τ denotes a finite function (encoded as an integer) whose domain is not necessarily an initial segment of the integers. Each acceptable programming system $\{\phi_i\}$ can be effectively transformed into an enumeration $\{\psi_i\}$ of the class of all partial recursive functions whose domains are closed under subfunctions. Such an enumeration includes all the total recursive functions, and will be called an *prefixed numbering*.

Each IIM M will be chosen from a prefixed numbering and, hence, will be an effective device. Consequently, $\phi_M(f \mid n)$ will be used to denote the output value (if it exists) of M given the intital segment $f \mid n$ of the graph of f as input. We will use $\phi_M(f)$ to denote the limit (if it exists) $\lim_{n\to\infty} \phi_M(f \mid n)$. This limit exists if and only if for some n, either $\phi_M(f \mid m)\uparrow$ for all $m \geq n$, or $\phi_M(f \mid m) = \phi_M(f \mid n)$ for all $m \geq n$. In this way we see that the (partial) map $\phi_M: R \to N$ is a limiting recursive functional. Moreover, we view the input $f \mid n$ as being encoded as an integer, so that ϕ_M is simply a partial recursive function with certain limit properties, viz., $\phi_M(f)$ is the limit of ϕ_M on the sequence of integers $f \mid 0$, $f \mid 1$, $f \mid 2$, ... Most of the constructions below will involve functions of finite support (i.e., functions which are non-zero at only finitely many points). The set of functions of finite support will be denoted by S_* and defined formally by $S_* = \{f \in R \mid f =^* 0\}$.

A limiting recursive functional ψ is *total* on a set of functions X if and only if for all $f \in X$ and for all $n \in N$ $\psi(f \mid n)$ is defined and

$\psi(f) \equiv \lim_{n\to\infty} \psi(f \mid n)$ exists. In the case where X is the set of total recursive functions R we will simply say that ψ is a total limiting recursive functional. A limiting recursive functional ψ is called *nondecreasing* if $\psi(\sigma) \geq \psi(\sigma')$ for all σ and $\sigma' \subseteq \sigma$. The *modulus function* μ is defined by

$$\mu(\psi, f) = \begin{cases} \min\{n \mid (\forall m > n)[\psi(f \mid m) = \psi(f \mid n)]\}, \\ \qquad \text{if } \psi(f)\downarrow \text{ and } (\forall m)[\psi(f \mid m)\downarrow], \\ \max\{n \mid \psi(f \mid n)\downarrow\}, \\ \qquad \text{if } \psi(f)\downarrow \text{ and } (\exists m)[\psi(f \mid m)\uparrow], \\ \uparrow, \text{ if } \psi(f)\uparrow. \end{cases}$$

The modulus function specifies the point of convergence. We will use $\mu_i(f)$ to denote $\mu(\phi_i, f)$. It is clear that if ϕ_i is a limiting recursive functional, then so is μ_i. The number of mind changes of ψ on the input f is denoted by $\delta(\psi, f)$ and is defined by the cardinality of the set $(\{n \mid \psi(f \mid n) \downarrow \neq \psi(f \mid n-1)\})$. We use $\delta_i(f)$ to denote $\delta(\phi_i, f)$. Finally, the number of distinct hypotheses produced by an IIM M on input f is given by the cardinality of the set $(\{\phi_M(f \mid n) \mid \phi_M(f \mid n)\downarrow\})$.

§2 Complexity of Inference

For ordinary computations the complexity of a computation is synonymous with the complexity of the mechanism performing the computation. IIMs, however, are continually receiving inputs and reevaluating their most recent conjecture as to a program which computes the input function. Any IIM will then use an infinite amount of time to attempt any inference independently of whether or not the IIM eventually converges. The notion of inference complexity introduced below will measure only the amount of computation resources used up to the point of convergence.

Suppose $\{\tilde{\Phi}_i\}$ is any Blum computational complexity measure [5] for $\{\phi_i\}$ which is a prefixed numbering, so that

1) $\tilde{\Phi}_i(x)\downarrow \iff \phi_i(x)\downarrow$

2) $\tilde{\Phi}_i(x) = y$ is a recursive predicate in i, x, and y.

We extend this definition to limiting recursive functionals, and define the *inference complexity* of an IIM M on input f by

$$\Phi_M(f) = \sum_{n=0}^{\mu_M(f)} \tilde{\Phi}_M(f \mid n).$$

Suppose that all the functions in the underlying Blum measure only take on strictly positive values, then $\Phi_M(f) \geq \mu_M(f)$. Intuitively, the above measure is the sum of the computation resources used by M on input f until the point at which M converged to a particular program, and thus is the "area under the curve" of effort during the active period of inference by M. Accordingly, we will call such an inference complexity measure an *a.u.c. measure*. Observe that so long as M converges the complexity of M is defined whether or not the program to which M converged is a correct program for f.

Our first result shows the existence of a precise trade-off between complexity and accuracy.

Theorem 1: *There exists a $g \in R$ and IIMs $M_0, M_1, \ldots$ such that for all $h \in R$ and for all $n \in N$ there exists $S_{n,h} \in EX^n$ and*

1) for all $k < n$, $S_{n,h} \subseteq EX^{n-k}(M_k)$ and

$$(\overset{\infty}{\forall} f \in S_{n,h})[h(\Phi_{M_{k-1}}(f)) \leq \Phi_{M_k}(f) \leq g(h(\Phi_{M_{k-1}}(f)))],$$

2) for all M and for all $k \in N$ and for all $m < k$, if $S_{n,h} \subseteq EX^m(M)$ then

$$(\forall f \in S_{n,h})\ [\Phi_M(f) > h^{k-m}(\Phi_{M_k}(f))].$$

Our next theorem reinforces our contention that such strong trade-offs as depicted in Theorem 1 are indeed rare. Let $\overline{S}_{n,h} = \bigcup_{k=1}^{n} S_{k,h}$, where $S_{k,h}$ is as in Theorem 1.

Theorem 2: *There exist IIMs $\overline{M}_1, \overline{M}_2, \ldots$ such that for all $f \in R$ and for all $n \in N$ there exists a constant c such that*

$$\overline{S}_{n,h} \subseteq EX^k(\overline{M}_k) \text{ and } (\overset{\infty}{\exists} f \in \overline{S}_{n,h})[\Phi_{\overline{M}_k}(f) \leq c].$$

Theorem 2 demonstrates the existence of easy to infer sets of functions. We conclude this section with a result which emphasizes this point.

Theorem 3: *There exists a set of total recursive functions S and an IIM M and $h \in R$ such that $S \subseteq EX(M)$ and $(\forall f \in S)[\Phi_M(f) \leq h(f(0))]$ and*

$$(\forall f \in R)(\exists g \in S)[f =^1 g].$$

§3 Axiomatic Approach

In [6] Blum initiated an axiomatic approach to the complexity of computations of the partial recursive functions. His approach has proved to be a tremendous success in understanding the nature of computations and their complexity. In this section we formulate an analagous axiomatization of the complexity of inductive inference. In some sense the a.u.c. measures of the previous section could be considered an axiomatization since they were based on an arbitrary Blum computational complexity measure. However, to restrict attention only to such measures would exclude from consideration the modulus function and the number of mind changes as possible measures of inference complexity, which apriori seem to represent reasonable notions of measure. Let

$$\mu_M(f\,|\,n) = \max\{m \le n \mid \phi_M(f\,|\,m) \downarrow\ne \phi_M(f\,|\,m-1)\},$$

and

$$\Phi_M(f\,|\,n) = \sum_{m=0}^{\mu_M(f\,|\,n)} \tilde{\Phi}_M(f\,|\,m).$$

Hence, $\mu_M(f) = \lim_{n\to\infty} \mu_M(f\,|\,n)$ and $\Phi_M(f) = \lim_{n\to\infty} \Phi_M(f\,|\,n)$. Observe further that μ_M has a domain which is closed under subfunctions and that $\mu(\mu_M, f) = \mu_M(f)$ and $\mu(\Phi_M, f) = \mu_M(f)$.

We now present our axiomatization of an inference complexity measure. We say that a set of limiting recursive functionals $\{\Phi_i\}$ is an inference complexity measure for the prefixed numbering $\{\phi_i\}$ if and only if all of the following are satisfied:

1) a) $\Phi_i(f)\downarrow \iff \phi_i(f)\downarrow$,

b) $\Phi_i(\sigma)\downarrow \iff \phi_i(\sigma)\downarrow$,

2) $\Phi_i(x) = y$ is a limiting recursive predicate in i, x, and y,

3) a) $\Phi_i(f) \ge \delta(\phi_i, f)$,

b) $\mu(\Phi_i, f) = \mu(\phi_i, f)$.

It is not difficult to see that any a.u.c. measure satisfies these axioms as well as the modulus function itself. It is easy to see that the modulus function $\{\mu_i\}$ satisfies the axioms of an inference complexity measure. However, since $\mu_i(f\,|\,n)$ has an average value bounded above by 1, it follows from the recursive relatedness of any two computational complexity measures and the existence of arbitrarily difficult to compute recursive functions that μ_i cannot be an a.u.c. measure. Thus not every inference complexity measure can be an a.u.c. measure. Note that the number of mind changes will also satisfy Axiom 1 and so will constitute a measure of inference complexity, but the number of distinct hypotheses will not be a measure of inference complexity in our sense.

We now establish some basic results for inference complexity measures.

Lemma 4: *Φ_i is a limiting recursive functional.*

One of the very rich areas of work in computatinal complexity was in the study of complexity classes. We analagously define for any limiting (partial) recursive function ψ the inference complexity class named by ψ by

$$C_{\psi} = \{S \mid (\exists M)(\overset{\infty}{\forall} f \in S)[\Phi_M(f) \le \psi(f)]\}.$$

Only total limiting recursive functionals are used for names of complexity classes below. One of the most useful and pleasing properties of computational complexity measures is their recursive relatedness. Below is our analog of recursive relatedness for inference complexity measures.

Lemma 5: *If* $\{\Phi_i\}$ *and* $\{\hat{\Phi}_i\}$ *are two inference complexity measures for the prefixed numbering* $\{\phi_i\}$, *then there exists a total limiting recursive functional* Ψ *such that for all* $i \in N$ *and for all* $f \in R$ $\Phi_i(f) \le \Psi(f, \hat{\Phi}_i(f), i)$ *and* $\hat{\Phi}_i(f) \le \Psi(f, \Phi_i(f), i)$.

Observe that the recursive relatedness of Φ_i and $\hat{\Phi}_i$ depends on i in contrast to the case for compuational complexity measures. The following result shows that even though the modulus measure $\{\mu_i\}$ cannot be an a.u.c. measure, every a.u.c. measure can be levelled into the modulus measure, so that in some sense the modulus measure is a canonical measure.

Lemma 6: *For every a.u.c. measure there exists program transformation* $\alpha \in R$ *such that for all* $i \in N$ *and for all* $f \in R$ $\phi_{\alpha(i)} = \phi_i$ *and* $\delta(\phi_{\alpha(i)}, f) = \delta_i(f)$ *and* $\mu(\phi_{\alpha(i)}, f) = \Phi_i(f)$.

The following shows that even though there are arbitrarily difficult to infer subsets of the functions of finite support (Theorem 8 below) the set S_* can be inferred by an IIM M_* within complexity bounded by a partial limiting recursive functional. The IIM M_* simply constructs for each $f \in S_*$ a table of the finitely many non-zero points of the graph of f which it has seen thus far.

Lemma 7: *There exists a limiting recursive functional* Ψ *such that* $(\forall f \in S_*)[\Phi_{M_*}(f) \le \Psi(f)]$.

We have the following analogs with the computational complexity of recursive functions.

Theorem 8: *(Arbitrarily Difficult Sets) For every total limiting recursive functional* ψ *there exists a set* S_ψ *of total recursive functions such that* $S_\psi \in EX$ *and* $(\forall M)(\overset{\infty}{\forall} f \in EX(M))[\Phi_M(f) > \psi(f)]$.

Theorem 9: *(Compression Theorem) There exists a total limiting recursive functional* Ψ *such that for every increasing total limiting recursive functional* ψ *there exists a set* S_ψ *of total recursive functions such that* $S_\psi \in EX$ *and* $S_\psi \notin C_\psi$ *and* $S_\psi \in C_{\Psi \circ \psi}$.

Theorem 10: *(Speed-up Theorem) For every strictly increasing total limiting recursive functional* ψ *there exists a set S of total recursive functions such that* $S \in EX$ *and for all M if* $S \subseteq EX(M)$ *then there exists an* M' *such that* $S \subseteq EX(M')$ *and* $(\overset{\infty}{\forall} f \in S)[\psi(f, \Phi_{M'}(f)) \leq \Phi_M(f)]$.

If we contrast Lemma 7 with Theorems 8, 9 and 10, we see that the properties of the inference complexity classes with partial limiting recursive functionals as names merit further study, since they may be more natural than classes with total limiting recursive functionals as names. We see from the form of the Compression Theorem 9 above that there cannot exist an analog of the Gap Theorem. Moreover, we have

Lemma 11: *There does not exist a total limiting recursive functional* ψ *such that* $(\forall n \in N)(\overset{\infty}{\forall} f \in R)[\psi(f) > n]$.

Observe that if such a limiting recursive functional were to exist then the Gap Theorem would hold.

§4 Further Considerations

In this section we briefly examine the influence which the order of presentation has on the complexity of inference. Changing the order of presentation, of course, will have no effect on inferrability itself, but it can have an effect on the complexity of inference as we shall see below. In [14] Kinber also demonstrated that changing the order of presentation can speed-up the inference process. We will restrict our attention here to effective presentations of the input function. If $g \in R$ is a one-to-one function, then we denote by $f:g$ the *g-presentation* of f, and $f:g|n$ will denote the initial segment $\langle (g(0), f(g(0))), \ldots, (g(n), f(g(n)))\rangle$ of this presentation. Similarly, $\tau:g$ will denote the g-presentation of the finite function τ. Also, $\phi_M(f:g) = \lim_{n\to\infty} \phi_M(f:g|n)$, and whenever $\phi_M(f:g)\downarrow$ then $\mu_M(f:g) = \min\{n \mid (\forall m \geq n)[\phi_M(f:g|m) = \phi_M(f:g|n)]\}$. We denote by $f:g \in EX(M)$ that $\phi_M(f:g) = p$ and $\phi_p = f$. By our assumptions regarding the prefixed numbering it is clear that for all $g \in R$ if $\phi_M(\tau:g)\downarrow$ then $\phi_M(\tau':g)\downarrow$

for all $\tau':g \subseteq \tau:g$.

We begin with an example involving the functions of finite support S_*, the natural IIM M_* for S_*, and the modulus complexity measure, which illustrates the dramatic increases and reductions which are possible when the order of presentation is changed. Here we will suppose that M_* has been extended to arbitrary enumerations in the obvious way so that $f:g \in EX(M_*)$ for all $f \in S_*$ and all $g \in R$.

Theorem 12: *For every* $h \in R$ *and for every IIM* M *such that* $S_* \subseteq EX(M)$,

1) $(\overset{\infty}{\exists} g \in R)$ *such that*

a) $(\overset{\infty}{\exists} f \in S_*)[\mu_M(f:g) > h(\mu_{M_*}(f))]$,

b) $(\overset{\infty}{\exists} f \in S_*)[\mu_M(f) > h(\mu_{M_*}(f:g))]$,

2) $(\overset{\infty}{\exists} f \in S_*)$ *such that*

a) $(\overset{\infty}{\exists} g \in R)[\mu_M(f:g) > h(\mu_{M_*}(f))]$,

b) $(\overset{\infty}{\exists} g \in R)[\mu_M(f) > h(\mu_{M_*}(f:g))]$.

Theorem 12 shows that there can in general be no recursive bound for the increase (or decrease) in inference complexity when the order of presentation is changed. We now see that a limiting recursive upper bound does exist.

Theorem 13: *For every inference complexity measure* $\{\Phi_i\}$ *there exists a total limiting recursive functional* Ψ *such that for all* M *there exists an* M' *such that for all* $g \in R$ *and for all* $f \in \mathbf{dom}\ \phi_M$, $\phi_{M'}(f:g) = \phi_M(f)$ *and* $\Phi_{M'}(f:g) \le \Psi(f, g, \Phi_M(f))$.

There is another type of inductive inference called BC inference in Case and Smith [8], where the IIM is not required to converge to a fixed program for the input function, but only required to produce a (possibly infinite) sequence of programs whose behavior converges to that of the input function. Thus the convergence is only second order. Barzdin in [3] also studied the notion of BC inference which he called GN^∞ inference. More formally, we say that an IIM M BC^n identifies (and write $f \in BC^n(M)$) if and only if there exists and integer k such that $(\forall m \ge k)[\phi_{\phi_M(f|m)} =^n f]$. We also define

$$BC^n = \{S \mid (\exists M)[S \subseteq BC^n(M)]\}.$$

Clearly, in the case of BC inference the number of mind changes doesn't make any sense as a measure of complexity, but a (second order) modulus function μ^2 does exist and can be used as a basis for defining an a.u.c. type complexity measure. We can define

$$\mu^2(f,k) = \mathbf{min}\{n \mid (\forall m \geq n)[\phi_{\phi_M(f|m)} =^k f]\}.$$

Observe that in contrast with EX inference where $\phi_M(f)\downarrow$ and $f \notin EX^*(M)$ is possible, since the convergence is only second order, the criterion for success must be a part of the notion of convergence. Thus, the intuitive notion for the complexity of BC inference is well founded and we could proceed to develop an axiomatization analagous to that for EX inference above. However, we instead conclude with a strengthening of Theorem 1, which follows from the construction of $f_{n,h,j}$ in the proof since the diagonalization there was against any program produced by any IIM $M \leq j$.

Theorem 14: *There exists a* $g \in R$ *and IIMs* $M_0, M_1, \ldots$ *such that for all* $h \in R$ *and for all* $n \in N$ *there exists* $S_{n,h} \in EX^n$ *and*

1) for all $k < n$ $S_{n,h} \subseteq EX^{n-k}(M_k)$ *and*

$$(\overset{\infty}{\forall} f \in S_{n,h})[h(\Phi_{M_{k-1}}(f)) \leq \Phi_{M_k}(f) \leq g(h(\Phi_{M_{k-1}}(f)))]],$$

2) for all M *and for all* $k \in N$ *and for all* $m < k$ *if* $S_{n,h} \subseteq BC^m(M)$ *then*

$$(\forall f \in S_{n,h})\ [\Phi_M(f) > h^{k-m}(\Phi_{M_k}(f))]].$$

References

1) Angluin, D., *On the complexity of minimum inference of regular sets*, ***Information and Control*** 39 (1978), 337-350.
2) Angluin, D. and Smith, C., *Inductive inference: theory and methods*, ***Computing Surveys***, to appear.
3) Barzdin, J., *Two theorems on the limiting synthesis of functions*, ***Theory of Algorithms and Programs I***, Latvian State University, Riga, U.S.S.R. (1974), 82-88.
4) Barzdin, J. and Freivald, R., *On the prediction of general recursive functions*, ***Soviet Math. Doklady*** 13 (1972), 1224-1228.
5) Blum, L. and Blum, M., *Toward a mathematical theory of inductive inference*, ***Information and Control*** 28 (1975), 125-155.
6) Blum, M., *A machine-independent theory of the complexity of recursive functions*, ***Journal Assoc. Comput. Mach.*** 14 (1967), 322-326.

7) Case, J. and Ngo-Manguelle, S., *Refinements of inductive inference by Popperian machines*, Technical Report, SUNY Buffalo, Department of Computer Science (1979).
8) Case, J. and Smith, C., *Comparison of identification criteria for machine inductive inference*, ***Theoretical Computer Science*** 25 (1983), 193-220.
9) Freivald, R., *On the complexity and optimality of computation in the limit*, ***Theory of Algorithms and Programs II***, Lativan State University, Riga, U.S.S.R. (1975), 155-173.
10) Gold, E., *Language identification in the limit*, ***Information and Control*** 10 (1967), 447-474.
11) Gold, E., *Complexity of automaton identification from given data*, ***Information and Control*** 37 (1978), 302-320.
12) Hempel, C., ***Aspects of Scientific Explanation***, The Free Press (1965), New York.
13) Khodzhayev, J., *On the complexity of computation on Turing machines with oracles*, Ph.D. Dissertation, Tashkent (1970).
14) Kinber, E., *On speeding up the limiting identification of recursive functions by changing the sequence of questions*, ***Elektronische Informationsverarbeitung und Kybernetik*** 13 (1977), 369-383.
15) Klette, R. and Wiehagen, R., *Research in the theory of inductive inference by GDR mathematicians - a survey*, ***Information Sciences*** 22 (1980), 149-169.
16) Machtey, M. and Young, P., ***An Introduction to the General Theory of Algorithms***, North-Holland (1978), New York.
17) Pudlak, P., *Polynomially complete problems in the logic of automated discovery*, ***Lecture Notes in Computer Science*** 32 (1975), 358-361.
18) Pudlak, P. and Springsteel, F., *Complexity in mechanized hypothesis formation*, ***Theorectical Computer Science*** 8 (1979), 203-225.
19) Putnam, H., *Probability and confirmation*, ***Mathematics, Matter, and Method***, Cambridge University Press (1975). Originally appeared in 1963 as a Voice of America Lecture.
20) Rogers, H., *Gödel numberings of partial recursive functions*, ***Journal of Symbolic Logic*** 23 (1958), 331-341.
21) Rogers, H., ***Theory of Recursive Functions and Effective Computability***, McGraw Hill (1967), New York.
22) Selman, A., *Polynomial time enumeration reducibility*, ***SIAM Journal on Computing*** 7 (1978), 440-457.
23) Young, P., *Speed-ups by changing the order in which sets are enumerated*, ***Math. Systems Theory*** 5 (1971), 148-152.

MONOTONE EDGE SEQUENCES IN LINE ARRANGEMENTS AND APPLICATIONS
(extended abstract)

Herbert Edelsbrunner and Emmerich Welzl
Institutes for Information Processing, Technical University of Graz, Schießstattgasse 4a, A-8010 Graz, Austria

INTRODUCTION

A host of seemingly unrelated geometric tasks are solved algorithmically by reducing them to a common underlying idea dealing with arrangements of lines in the Euclidean plane. Let H denote a set of n lines in the plane. For the sake of simplicity we assume that H is in general position, that is no three lines of H are concurrent, not two are parallel, no line is vertical and no two intersection points of the lines lie on a common vertical line, see Figure 1-1.

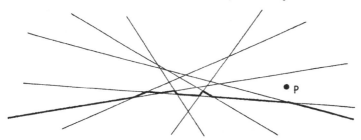

Figure 1-1. Seven lines with solid lower 3-curve.

H defines in a natural way a cell complex A(H), called the arrangement of H, whose faces edges, and vertices are also obviously defined. Let p be an arbitrary point in the plane. Then b(p) and a(p) denote the number of lines in H which lie strictly below and above p. In Figure 1-1, for example, b(p) = 4 and a(p) = 3. The lower k-curve of A(H) consists of the edges whose inner points p satisfy b(p) = k-1, see Figure

1-1 which shows the lower 3-curve. Similarly, the upper k-curve of A(H) consists of the edges whose inner points p satisfy a(p) = k-1.

Observation 1.1: The k-curve of an arrangement A(H) of n lines, for $1 \leq k \leq \lceil n/2 \rceil$, intersects every vertical line in exactly one point.

Another simple fact is important for the construction of k-curves:

Observation 1.2: Let e_1 and e_2 denote two edges of A(H) such that e_1 and e_2 share a vertex and they lie on the same k-curve. Then e_1 and e_2 are not parallel, that is, they come from different lines.

We will sketch two algorithms which construct k-curves of arrangements of lines. Both make use of the above observations and the complexity of both depends on the number of edges in the k-curve. The calculation of bounds for this number turns out to be a non-trivial task. We cite preliminary results on this problem. To this end let $b_k(H)$ denote the number of edges in the lower k-curve of A(H). Then we define $b_k(n)$ as the maximum of $b_k(H)$ for all sets H of n lines in the plane.

Proposition 1.3 [ELSS] and [EW1]: $b_k(n)$ is in $\Omega(n\log k)$ and in $O(nk^{1/2})$, for $1 \leq k \leq \lceil n/2 \rceil$.

We conjecture that the upper bound is far from being asymptotically tight. On the other hand, Paterson, Pippenger, and Pippenger [PPP] provide some indications that also the lower bound is not asymptotically tight.

2. CONSTRUCTING k-CURVES

This section sketches two algorithms for constructing k-curves of arrangements of lines. Let H denote a set of n lines in general position and let k be fixed with $1 \leq k \leq \lceil n/2 \rceil$. Without loss of generality, we consider computing the lower k-curve of A(H) only.

Algorithm TRIVIAL PLANE SWEEP:

Let V denote a vertical line conceptually sweeping from left to right. A dictionary D associated with V maintains the list of the lines sorted by y-coordinates of the intersections with V. At any moment, the k-th line in D is also the k-th bottommost line at the current position of V. The intersections of adjacent lines in D to the right of V are stored in a priority queue Q.

Let p denote the leftmost intersection point currently stored in Q. Then the sweep of V is continued by (1) swapping the positions in D of the two lines which define p, (2) inserting new intersection points into Q of lines which became adjacent, and (3) deleting old intersection points of lines which became non-adjacent. If one of the lines defining p was the k-th line in D before swapping, then p is a new vertex of the k-curve of A(H).

It is readily seen that each of the $\binom{n}{2}$ intersection points can thus be processed in O(logn) time yielding:

Theorem 2.1: Algorithm TRIVIAL PLANE SWEEP constructs a k-curve of A(H) in $O(n^2 \log n)$ time and $O(b_k(n))$ space.

The method can be improved using a result due to Overmars and van Leeuwen [OvL] on the maintenance of halfplanes. They give a method which maintains the intersection of n halfplanes in $O(\log^2 n)$ time per insertion or deletion. Let e be some edge of a k-curve in an arrangement, and let h be the line supporting e. The right endpoint of e is the rightmost point of the intersection of h with the intersection of halfplanes above the lines below e and below the lines above e. Identifying the right endpoint of e also gives the line h′ that supports the next edge to the right of e. To adjust the intersection of halfplanes accordingly, the halfplane bounded by h′ is deleted and a halfplane bounded by h is inserted. This finally leads to:

Theorem 2.2: A k-curve of A(H) can be constructed in $O(b_k(n)\log^2 n)$ time and $O(b_k(n))$ space.

3. APPLICATIONS

Surprisingly, k-curves of arrangements of lines and the two algorithms sketched above have various nice applications to a host of seemingly unrelated problems.

3.1 Minimum Area Triangle

Let S denote a set of n points in the plane. A minimum area triangle TR(S) is a triangle whose vertices are in S such that the area is a minimum. The following simple fact is the key idea of our solution:

Observation 3.1: Let p, q, and r denote the vertices of TR(S). Then r minimizes the distance to the line through p and q.

In a dual environment each point $p = (p_1, p_2)$ in S is transformed into the line T(p) whose points (x,y) satisfy $y = p_1x+p_2$. Observation 3.1 translates now to the fact that T(r) is either immediately below or above the intersection point defined by T(p) and T(q). These tripels, however, can be determined by an adapted version of Algorithm TRIVIAL PLANE SWEEP leading to:

Theorem 3.1: The minimum area triangle of a set S of n points in the plane can be determined in $O(n^2 \log n)$ time and O(n) space.

The time-bound can be improved to $O(n^2)$ (see [CGL] and [EOS]) if one is willing to construct the arrangement explicitely which costs $\Omega(n^2)$ space however.

3.2 Halfplanar Range Estimation

Let S denote a set of n point in the plane. A halfplanar range query consists of a halfplane h and asks for a rough idea of the number of points in S which lie in h. For the sake of concretness, we focus on the special case where "a rough idea" means the decision whether the number N(h) of points of S in h satisfies

$$N(h) < \lfloor n/3 \rfloor,$$
$$\lfloor n/3 \rfloor \leq N(h) \leq n-\lfloor n/3 \rfloor, \text{ or}$$
$$n-\lfloor n/3 \rfloor < N(h).$$

S is to be accommodated such that for a query halfplane h the decision can be found with little effort.

The problem is solved by transforming each point p in S into the line T(p) as described above. Similarly, a query halfplane h whose points (x,y) satisfy $y \leq h_1 x + h_2$ is transformed into the vertical ray T(h) whose upper endpoint is $(-h_1, h_2)$. This dualization process guarantees that p is in h if and only if T(p) intersects T(h), that is T(p) is below the point $(-h_1, h_2)$, see Figure 3-1.

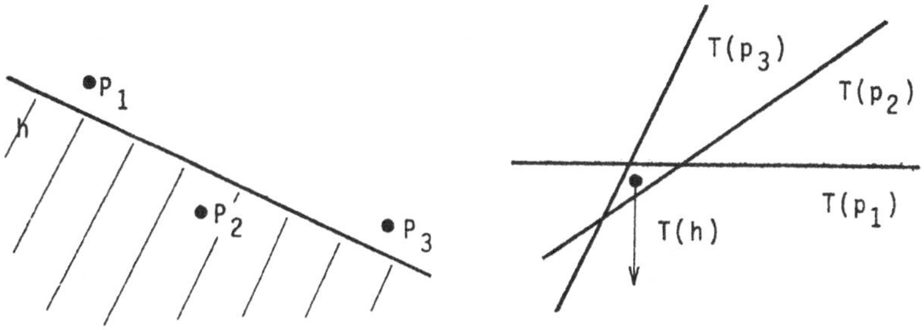

Figure 3-1. Transforming points and halfplane.

Thus, $N(h) < \lfloor n/3 \rfloor$ if and only if the endpoint of T(h) is below the lower $\lfloor n/3 \rfloor$-curve of T(S), $\lfloor n/3 \rfloor \leq N(h) \leq n-\lfloor n/3 \rfloor$ if and only if the endpoint is between the lower and the upper $\lfloor n/3 \rfloor$-curve, and $n-\lfloor n/3 \rfloor < N(h)$ if and only if the endpoint is above the upper $\lfloor n/3 \rfloor$-curve. This implies:

<u>Theorem 3.2:</u> There is a data structure for a set S of n points in the plane which requires $O(b_{\lfloor n/3 \rfloor}(n))$ space and $O(b_{\lfloor n/3 \rfloor}(n)\log^2 n)$ time for construcion such that a halfplanar range estimation query can be answered in O(logn) time.

3.3 Center-Point

Let S denote a set of n points in the plane. A centerpoint c(S) has the property that each line through c(S) has at least $\lfloor n/3 \rfloor$ points in both closed halfplanes. In general, c(S) is not in S and the existence of c(S) is a consequence of Helly's theorem. In the dual environment, c(S) is a line which is contained in the area between the lower and the upper $\lfloor n/3 \rfloor$-curve of T(S). This implies:

<u>Theorem 3.3:</u> The region of centerpoints of a set S of n points in the plane can be determined in $O(b_{\lfloor n/3 \rfloor}(n)\log^2 n)$ time and $O(b_{\lfloor n/3 \rfloor}(n))$ space.

The line corresponding to a centerpoint of S can be used for halfplanar range estimation which yields:

<u>Theorem 3.4:</u> Let S be a set of n points in the plane. There exists a data structure which requires O(1) space and $O(b_{\lfloor n/3 \rfloor}(n)\log^2 n)$ time for construction such that O(1) time suffices to decide for a query halfplane whether it contains at least $\lfloor n/3 \rfloor$ points or at most $n-\lfloor n/3 \rfloor$.

3.4 <u>Moving Points</u>

Let S denote a set of n moving points on a vertical line. Each point p moves with constant speed s(p) up or down. In the former case s(p) is positive, in the latter negative. A host of problems dealing with those point-sets is examined in Ottmann and Wood [OW]. One which relates to k-curves is:
Determine the sequence of the k-th topmost points, for some fixed k. When we use a horizontal axis for displaying time then S induces an arrangement T(S) of n lines and the required sequence is exactly the upper k-curve respectively the lower (n-k)-curve of T(S). This implies:

<u>Theorem 3.5:</u> The sequence of the k-th topmost points of a set S of n moving points on a vertical line can be computed in $O(b_k(n)\log^2 n)$ time and $O(b_k(n))$ space.

3.5 Constructing Polygon and Conjugation Trees

[W] and [EW] propose the polygon and the conjugation tree for the polygon range search problem which stores a set S of n points such that the number of those inside a convex query polygon with at most a constant number of edges can be determined efficiently. These trees require O(n) space and a polygon query can be answered in $O(n^{0.77})$ and $O(n^{0.69})$ time, respectively. A crucial step in constructing such a tree is the determination of a line that separates two finite sets of points into halves, respectively. This amounts to intersecting two k-curves of different arrangements, for suitable k. Theorem 2.2 then implies:

Theorem 3.6: There exists an algorithm which constructs the polygon or conjugation tree for a set S of n points in the plane in $O(b_{\lfloor n/2 \rfloor}(n)\log^3 n)$ time.

Trivial modifications of the algorithm leading to Theorem 2.2 even allow us to perform the construction in O(n) space. Details are given in [EW2].

References

[CGL] Chazelle,B., Guibas,L.J., and Lee,D.T. The Power of Geometric Duality. Proc. 24th Symp. Found. Comp. Sci. (1983), 217-225.

[ELSS] Erdös,P., Lovasz,L., Simmons,A., and Strauss,E.G. Dissection Graphs of Planar Point Sets. In: A Survey of Combinatorial Theory, J.N. Srivastava et. al. eds., North-Holland, Amsterdam, 1973.

[EOS] Edelsbrunner,H., O'Rourke,J., and Seidel,R. Constructing Arrangements of Lines and Hyperplanes with Applications. Proc. 24th Symp. Found. Comp. Sci. (1983), 83-91.

[EW1] Edelsbrunner,H. and Welzl,E. On the Number of Line-Separations of a Finite Set in the Plane. Report F97, IIG, TU Graz, Austria (1982).

[EW2] Edelsbrunner,H. and Welzl,E. Halfplanar Range Search in Linear Space and $O(n^{0.695})$ Query Time. Report F111, IIG, TU Graz, Austria (1983).

[OW] Ottmann,Th. and Wood,D. Dynamical Sets of Points. Manuscript (1982).

[OvL] Overmars,M.H. and van Leeuwen,J. Maintenance of Configurations in the Plane. J. of Comp. and Sys. Sci. 23 (1981), 166-204.

[PPP] Paterson,M., Pippenger,M., and Pippenger,N. Private Communication (1982).

[W] Willard,D.E. Polygon Retrieval. SIAM J. on Comp. 11 (1982), 149-165.

MANY-SORTED TEMPORAL LOGIC FOR MULTI-PROCESSES SYSTEMS *

P. ENJALBERT

LCR THOMSON-CSF
Domaine de Corbeville 91400 ORSAY FRANCE

M. MICHEL

CNET - 92131 - ISSY LES MOULINEAUX
38-40 Rue du Général Leclerc

INTRODUCTION

The present paper is concerned with temporal logic for parallel systems composed of separate communicating processes. Examples can be CSP programs [HOA]. Others are Kahn's or Boussinot's networks of sequential processes [BOU][KAH], the study of which is one of our first motivations (see Application II below).

Our approach rises from the following considerations :

In current work about temporal logic and parallelism, a system of concurrent processes is always considered as a whole; the time considered is the time of changes of configurations of the system. We do not find this stand point very satisfactory from a conceptual point of view . Nor do we find it convenient in practice, as an obstacle to modularity of specification and proof.

Oppositely, the main ideas of our approach can be summarized that way :

1) Each separate sequential process defines its own "local" time, the time of its state changes. Accordingly, each process defines a "local" temporal logic in a rather standard way.

2) The "global" time and logic are obtained by some kind of composition of the local times and logics.

3) Technically, a temporal logic is a "many-sorted" one; with each sequential process is associated a type, with typed propositionnal variables and modal operators.

Another aspect is the introduction of (modal) operators in order to express explicitely synchronisation properties.

These ideas are certainly related to Arnold-Nivat's approach [A-N] in the framework of language theory.

(*) Work partially supported by ADI (contrat n° 821682).

The paper is divided in two parts. In the first one we introduce the logical basis: the system TT of many-sorted temporal logic and its extension TTS which includes "explicit" synchronisation operators; first examples of applications to parallel programs are given. The second part is devoted to two more important applications: An axiomatic definition of "maximum Semantics" of [SAL]. An axiomatisation of the "channel" of [BOU1-2], and hence of communication in these networks; on this occasion we mention a third possible version for our logic, without global operators at all.

I - THE BASIC SYSTEM TT

1. Syntax

We consider k "sorts" or "types" denoted by the integers i = 1,..., k; plus the empty type denoted by the empty word.

For i = 1,..., k, we consider :

- $V^{(i)}$ the set of propositionnal variables of type i. $\forall i,j \; V^{(i)} \cap V^{(j)} = \{\text{true}\}$
- $N^{(i)}$, $G^{(i)}$ two modal operators of type i. ("Next" and "Always"). And, corresponding to the empty type: N , G two global modal operators.

We note [k] the set {1,...,k}. The set For of formulas is defined in the following way :

- First we define $For^{(i)}$, the set of formulas of sort i - $For^{(i)}$ is the smallest set s.t.:
  - $V^{(i)} \subseteq For^{(i)}$,
  - If $X,Y \in For^{(i)}$, $X \wedge Y$, $\sim X$, $N^{(i)} X$ and $G^{(i)} X \in For^{(i)}$

  We shall call these formulas "typed" or "locals"- $X^{(i)}$ means: "X of type i".
- Now, For is the smallest set s.t. :
  - $\forall i = 1,\dots, k \;\; For^{(i)} \subseteq For$,
  - If X, Y ∈ For, $X \wedge Y$, $\sim X$, N X and G X ∈ For. These are the "global" formulas.

Hence, we have k "local" temporal languages (of sort i ∈ [k]), each defined in a usual way; then, we get the whole language by considering local formulas as atomic, and combining them by means of the classical and global modal operators. For any type t, we define the operators $P^{(t)}$ and $F^{(t)}$ by: $P^{(t)} X \Leftrightarrow \sim N^{(t)} \sim X$ and $F^{(t)} X \Leftrightarrow \sim G^{(t)} \sim X$.

2. Semantics

a/ TT - Structures :

Definition 1 A TT-structure for k sorts is a k+1 - tuple: $\Sigma = < \sigma^1, \dots, \sigma^k, \sigma >$ where:

(TT1) For i = 1 ... k; σ^i is a temporal structure for the sub-logic of type i: $\sigma^i = < Z^i, Succ^i, \leq^i; v^i >$ with $v^i : V^{(i)} \to \mathcal{P}(Z^i)$ a truth function for propositionnal variables, and the rest of the structure isomorphic to N or some initial segment of it.

$\langle z^i(n) \rangle$ or $\sigma^i(n)$ will denote the nth element of the sequence σ^i.

Let us call configuration any k-tuple: $\vec{z} = < z_1, \dots, z_k >$ with $z_i \in Z^i$ for i ∈ [k]. We set $\vec{z}(i) = z_i$. Then:

(TT2) $\sigma = < Z, Succ, \leq >$ is also a sequence isomorphic to some initial segment of N or N itself, and satisfies :

TT2-1 : For all n : - $Z(n+1) \neq Z(n)$

- $\forall i \in [k] \;\; Z(n+1)(i) = Succ^i (Z(n)(i))$ or $Z(n)(i)$.

TT2-2 : $\forall z_i \in Z^i \; \exists n$: $Z(n)(i) = z_i$. $Z(o)$ is called the initial configuration.

Thus σ is also a temporal structure, the typed formulas acting as propositionnal variables.

Remark 1 : $\forall \vec{z}, \vec{z}', \in Z$

$z \leq z'$, iff $\forall i \in [k] \quad z_i \leq z'_i$.

b/ Validity of formulas

- For each $i \in [k]$, the relations : "$X^{(i)}$ is satisfied by $z_i \in Z^i$ in σ^i", denoted: $\sigma^i, z_i \models X^{(i)}$, is defined by the classical rules and :

$\sigma^i, z_i \models N^{(i)} X$ iff $\forall z'_i = \text{Succ}(z_i) \; \sigma^i, z'_i \models X$

$\sigma^i, z_i \models G^{(i)} X$ iff $\forall z'_i \geq z_i \; \sigma^i, z'_i \models X$.

we set: $\sigma^i \models X^{(i)}$ for: $\forall z_i \in Z^i \quad \sigma^i, z_i \models X^{(i)}$.

- similarly, for any configuration $\vec{z}$: $\sigma, \vec{z} \models X$ is defined by :

$\sigma, \vec{z} \models X^{(i)}$ iff $\sigma^i, \vec{z}(i) \models X^{(i)}$

and the usual rules for classical and modal connectives.

Again: $\sigma \models X$ stands for: $\forall \vec{z} \quad \sigma, \vec{z} \models X$

We remark that for typed $X^{(i)}$: $\sigma \models X^{(i)}$ iff $\sigma^i \models X^{(i)}$ (due to TT2-2) - This insures the coherence of our definitions. Finally, we set also $\Sigma \models X$ for $\sigma \models X$.

A theory T is any set of TT-formulas. We set :

- $\Sigma \models T$ iff $\forall X \in T \quad \Sigma \models X$

- $T \models X$ iff for all Σ, $\Sigma \models T \Rightarrow \Sigma \models X$

3. Systems of processes

Let $\Pi = < \Pi_1, \ldots, \Pi_k >$ a system of k processes. With each Π_i we associate a type, i. The general idea is that the specification of Π can be splitted in two parts :

- "Local" specification of each process Πi, by axioms of type i.
- Synchronisation conditions are expressed by "global" axioms.

A convenient convention is to consider that for each process Π_i the states can be of two kinds: "action-states" or "rest-states"; and we require only that the state following an action-state (if any) must be a rest-state. Accordingly, for each type we have a special propositional variable $A^{(i)}$, caracteristic of action-states, and the following local axiom of alternance :

$$(\text{Alt}^i) \quad A^{(i)} \rightarrow N^{(i)} \sim A^{(i)}$$

This is a good example of a "local" property, axiomatisation of which by means of global time and operators would be much tedious.

Example 1 : $\Pi = < \Pi_1, \Pi_2 >$. The sequential process Π_1 performs two actions a and b alternatively. Two actions are separated by a rest-state. With obvious notations the axioms of Π_1 are :

- (Alt^1), $A^1 \longleftrightarrow (a \vee b)$, $\sim (a \wedge b)$, $\sim A^1 \rightarrow N^1 A^1$

The process Π_2 is similar with actions c and d. The synchronisation conditions are: b and d are in mutual exclusion; a and c are always simultaneous. The corresponding axioms are :

- $\sim (b \wedge d)$, $a \Longleftrightarrow c$.

4. Axiomatics

Let M stand for any of $N^{(t)}$ or $G^{(t)}$. The axiom System TT-Ax is composed of :

Axioms: (CP) Enough classical toutologies

(M1) $(M\ X \wedge M\ Y) \rightarrow M(X \wedge Y)$

(M2) M True.

Linearity of the structure σ^t for every type t:

(L1) $P^t\ X \rightarrow N^t\ X$

(L2) $G^t\ X \Longleftrightarrow X \wedge N^t\ G^t\ X$

(L3) $G^t\ (X \rightarrow N^t\ X) \rightarrow (X \rightarrow G^t\ X)$

Axioms for the global sequence :

(G1) $P\ X^{(i)} \rightarrow X^{(i)} \vee P^{(i)}\ X^{(i)}$ for $i = 1,\dots, k$

(G2) $P\ (\bigwedge_{i=1,\dots,k} X^{(i)}) \rightarrow \bigvee_{i=1 \dots k} P^i\ X^{(i)}$

(G3) P^i true $\rightarrow$ P true.

(G4) $G^i\ X^{(i)} \Longleftrightarrow G\ X^{(i)}$

Inference rules :

M.P. $\dfrac{X \quad X \rightarrow Y}{Y}$ $\quad$ NEC^t $\dfrac{X}{G^t X}$, for any type t

All formulas above are supposed well formed, i.e. well typed.

Prop 1: TT-Ax is consistent for the semantic consequence operation: $T \vdash X \Rightarrow T \models X$

Proof: We just have to check the specification G1 - G4
G1 - G2 come from TT2-1 . G3 - G4 from TT2-2 and Remark 1 □

Prop 2: TT-Ax is complete and decidable relatively to finite theories.
This is a consequence of :

Prop 3: Let T be any finite theory, u a formula T-consistent (Not: $T \vdash \sim u$). There exists a TT-structure Σ with initial configuration $\vec{z}$ s.t. :

$$\Sigma \models T \quad \text{and} \quad \Sigma, \vec{z} \models u$$

The technique of the proof can be seen as an extension of that of [GAB], itself derived from [KOZ], to do with types and finite sequences . We also remark that the mentioned method can be extended to cope with finite theories (This remark also applies to PDL).

First we consider each sublogic of type i and define a "PDL-like" structure M_i, which is OK but for the Next relation which can be one-many. Then we have to draw linear sequences out of each M_i in a coherent way and to define simultaneously the global sequence of configurations.

II - TYPED LOGIC WITH EXPLICIT SYNCHRONISATION - THE EXTENDED SYSTEM TTS

In this section in addition to the previous operator we add for all $i,j \in [k]$ a new modality S^{ij}. The associated relation R^{ij} is a binary relation on $Z^i \times Z^j$ and $z_i\ R^{ij}\ z_j$ means that there is a Rendez-vous between processes i and j in state z_i and z_j^i respectively.

1. Syntax and Semantics

We keep the definitions of I.1 and add :

- A new operator S^{ij} "of type (i,j)" for all $i,j \in [k]$
- The clause : If $X \in For^j$, $S^{ij}\ X \in For^i$ in the definition of For^i. We set $U^{ij}\ X \longleftrightarrow \sim S^{ij} \sim X$.

Semantics :

A TTS structure is a :

$$\Sigma = < \sigma_1, \ldots, \sigma_k , \sigma , \{R^{ij}\}_{i,j \in [k]} >$$

where :

- $\sigma_1, \ldots, \sigma_k, \sigma$ satisfies TT1 and 2
- $\forall i,j \in [k]$, $R^{ij} \subseteq Z^i \times Z^j$
- the following properties hold :

(TT3) : (1) $\forall i,j,l, \in [k]$

R^{ij} is functional, $R^{ji} = (R^{ij})^{-1}$, and $R^{ij} \circ R^{jl} \subseteq R^{il}$

(2) $z_i\ R^{ii}\ z_i'$ iff $z_i = z_i'$ and $\exists j \neq i\ \ z_i R^{ij}\ zj$.

(3) If $Z(n) = < z_1, \ldots, z_k >$ and $z_i\ R^{ij}\ z_j'$, then $z_j' = z_j$

We set : $\sigma_i, z_i \models S^{ij}\ X^{(j)}$ iff $\exists z_j$ s.t. $z_i\ R^{ij}\ z_j$ and $z_j \models X^{(j)}$

Example 2: We can replace in ex. 1 the synchronisation axiom: $a \longleftrightarrow c$

by: $a \rightarrow S^{12}\ c$ (or equivalently, $c \rightarrow S^{21}\ a$)

Forgetting the global sequence σ, we can represent TTS models by "scale-like" diagrams. For instance in the case of ex. 2 :

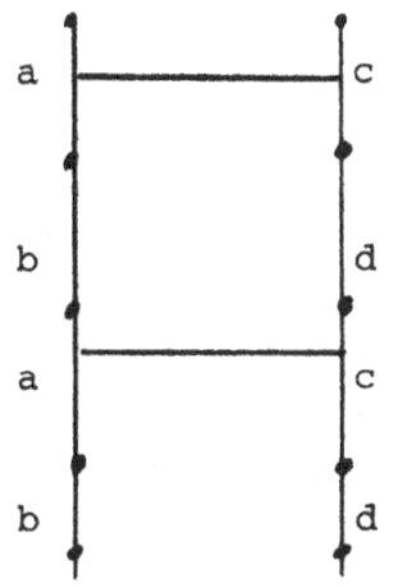

(The dots represent the rest-states and the edges action-states).

Example 3: Consider a resource X shared by k processes Π^i (for instance, X is a variable). Two kinds of operations can be performed relatively to X. The first ones are exclusive: if Π_i performs such an operation no other Π_j can access to X (f.i. Π_i modifies the content of variable X). The other kind includes operations that can be performed simultaneously by 2 processes. (f.i. reading X).

We can define a conflict set of instructions (see [SAL]) as being a set including two accesses to some resources X of which one is of the first "exclusive" kind.

Let us formalize X by a process Π^x (of "type" X). The other processes are Π^i for i $\in [k]$. Let $W^{(x)}$ be a propositional variable characteristic of the first kind of access, and $R^{(x)}$ of the second. Axioms for X can be :

- (Alt^x)
- $A^{(x)} \longleftrightarrow (R^{(x)} \vee W^{(x)})$
- $A^{(x)} \longleftrightarrow S^{x,x}$ True
- $(R^{(x)} \wedge S^{x,i}$ True$) \rightarrow \sim S^{x,j}$ True (for $i \neq j$).

Conflict sets of instructions are clearly excluded from all configurations.

From these first examples we see that the S^{ij} operators allow synchronisation specification "from a local point of view". This fact will be applied in ex. 5.

2. Axiomatics

The axiomatic system TTS-Ax is compound of :

1) TT-Ax with M1-M2 extended to $M = U^{ij}$
2) Synchronisation axioms :

(S1) $S^{ij}\, X \rightarrow U^{ij}\, X$

(S2) $X \rightarrow U^{ij}\, S^{ji}\, X$

(S3) $S^{ij}\, S^{jl}\, X \rightarrow X^{il}\, X$

(S4) $S^{ii}\,\text{True} \rightarrow \bigvee_{j \neq i} S^{ij}\,\text{True}$

(S5) $S^{ij}\, X \rightarrow X$

and rule NEC^{ij} : $\dfrac{X}{U^{ij}\, X}$

(In all formulas X is supposed to have the correct type)

Axioms (S1) - (S3) correspond to clauses TT3-1, (S4) to TT3-2 and (S5) to TT3-3.

Prop. 4: TTS is complete and decidable for finite theories. The scheme of the proof is the same as that of Prop. 2-3, with an appropriate definition of the closure operation on sets of formulas (cf. [KO2] and involves a few technical lemmas. It has been fully written in [ENJ2].

Part 2 : Applications

I - "MAXIMUM - SEMANTICS"

In [SAL], A. Salwicki introduces a distinction between different possible semantics for parallel programs. Among those, the so-called "Maximum Semantics" expresses the requirement that processes should not be lazy.

In max-semantics, at each step of the computation, the set of instructions in process must be a maximal non-conflict set (the notion of conflict set is as in Ex. 3). We can restate this condition in the following way :

(Max-Cond) . If some process Π_i is inactive, it must be the case that the next instruction in Π_i would have been in conflict with those presently performed by the other processes.

An alternative is "Arb-semantics" in which an arbitrary non conflict set of instructions is performed at each step. It is easy to see that the two semantics can lead to different computations and different results.

In Ex. 3, in order to express the notion of conflict set of instructions, we associated each resource with a process. Another solution is to suppose that we have a finite set of synchronisation axioms of the form :

$$\text{(Sync. Ax.)} \qquad \{p_m^{(i)} \rightarrow \bigvee_1 q_{m,1}^{(j)}\}_m$$

where $p_m^{(i)}$ and $q_{m,1}^{(j)}$ are propositionnal variables or negations of them.

Consider the following axiom :

$$\text{(MAX)} \quad (\sim A^{(i)} \wedge P \sim A^{(i)}) \rightarrow \bigvee_m (N^{(i)}\, p_m^{(i)} \wedge P \bigwedge_1 \sim q_{m,1}^{(j)})$$

The addition of (MAX) to (Sync-Ax) insures the requirement (Max-Cond) to be met. Note that the set of specific axioms keeps finite. Therefore, any property (expressible in the language) based on Max-Semantics will be procable from this new set of axioms.

II - NETS OF PROCESSES COMMUNICATING BY MEANS OF BUFFERS

In [BOU] a model for parallel computation, extension of G. KAHN's networks [KAH] is defined and studied which can briefly be described that way :

- A parallel program is a network of sequential processes, interconnected by means of buffers named "channels". Two processes can communicate or be synchronized only by sending or receiving messages.
- A channel is a FIFO-file where the processes can write or read values. A channel C must be fair :

(Fair-Cond): If a process X wants to write [resp. read] a value in C, and if C is infinitely often not full [resp. empty], the message from X will eventually be accepted [delivered] by C.

We give an axiomatisation for channels and a proof of liveness relative to the fairness hypothesis.

Formalization :

- Channels: A channel C is considered as a process managing a file and is assigned a type "C". $V^{(c)}$ is the following set: {Wait, Accept, Send, Empty, Full} whose meaning is respectively: C is inactive, accepts or sends a message; the file is empty or full.
- Processes: C is connected with Readers (R^i) and Writers (W^j). Each one of these processes is described as above. We just suppose that for any R^i [W^j], $V^{(i)}$[$V^{(j)}$] includes a variable Pre-Read^i(C) [Pre-Write^j(C)], the meaning of which is : R^i[W^j] asks for reading [writing] a value in C.

Axioms

- For the channel

(C1) OREXC (Wait, Accept, Send)
(C2) OREXC (Wait, $\{S^{c,i}\ \text{True}\}_i$, $\{S^{c,j}\ \text{True}\}_j$)
(C3) Wait $\rightarrow$ N^c (Accept v Send)
(C4) (Accept v Send) $\rightarrow$ p^c Wait

where OREXC (F_1, F_2,...) is the exclusive disjunction of F_1, F_2, ...

Managing of the file :

(C5) Accept $\rightarrow$ $P^c \sim$ Empty — Send $\rightarrow$ $P^c \sim$ Full
(C6) (Empty $\wedge$ $P^c \sim$ Empty) $\rightarrow$ Accept — (Full $\wedge$ $P^c \sim$ Full) $\rightarrow$ Send
(C7) ($\sim$ Full $\wedge$ P^c Full) $\rightarrow$ Accept — ($\sim$ Empty $\wedge$ P^c Empty) $\rightarrow$ Send
(C8) $\sim$ (Full $\wedge$ Empty)

- For the Readers and Writers

(R) Pre Read^i(C) $\rightarrow$ $N^i\ S^{i,c}$ Send
(W) Pre Write$^{(j)}$(C) $\rightarrow$ $N^j\ S^{j,c}$ Accept

- Fairness axioms

(F1) Pre Read^i(C) $\rightarrow$ (N^i false) $\rightarrow$ $F^c\ G^c$ Empty
(F2) Pre Write^j(C) $\rightarrow$ (N^j false) $\rightarrow$ $F^c\ G^c$ Full

They express (Fair-Cond).

Remark: This is a "weak" axiomatisation in this sense that :

1) The only properties of the file we consider is to be full or empty.

2) We express nothing about the transmission of values.

Still, we can prove the following "liveness" property :

Prop. 5: The following formula D is provable from the above axioms in TTS :

$$D \equiv N^c \text{ false} \rightarrow \quad (\bigwedge_i G^i \sim \text{Pre Read}^i(c) \wedge \bigwedge_j G^j \sim \text{Pre Write}^j(c))$$
$$\vee\ (\bigwedge_i G^i \sim \text{Pre Read}^i(c) \wedge \text{full})$$
$$\vee\ (\bigwedge_j G^j \sim \text{Pre Write}^j(c) \wedge \text{empty})$$

D means :

If C blocks, then :

- either there will be no demand for communication.
- or C is full [empty] and there will be no demand for reading [writing].

We have written a fully formal proof of D (cf. [ENJ1]).

The remarkable fact is that neither in the above axioms nor in the proof of D do we need global modalities. The proof makes use of the following Necessitation Rule :

$$\text{NEC}' : \quad \frac{X^{i_1} \quad \dots \quad X^{i_p} \rightarrow X^{i_{p+1}}}{G^{i_1} X^{i_1} \quad \dots \rightarrow G^{i_{p+1}} X^{i_{p+1}}}$$

In fact we can define a sublogic of TTS without global operators at all. Such a logical system represents an extreme standpoint in the discussion sketched in introduction.

Following this work, a complete axiomatisation and a tableau decision method has been proposed by E. Auduneau, L. Farinas, J. Henry [A-F-H].

CONCLUSION :

Several formalisms have been presented which introduce "local times" in temporal logics. To our sense, the results obtained for TTS indicate not only that this logic can be convenient and practicable, but also that there is no reason to do without the facilities of local times. Still, much work should be done, especially: compare the different formalism, apply them to synthesis of parallel programs, apply other decision methods like that of [MIC], develop a first order system.

REFERENCE :

[A-F-H] E. AUDUREAU, L. FARINAS, J. HENRY
"Logique multidimensionnable structurée" in "Logique Temporelle pour les réseaux de processus communiquant". Contrat ADI n° 82/682. Final Report.

[A-N] A. ARNOLD, M. NIVAT
"Comportement de processus" - Rapport LITP, University of Paris VII (1982).

[BOU] F. BOUSSINOT et al
"A language for formal description of real time systems". Safety of Computer Control System 1983, p. 119-126.

[ENJ1] P. ENJALBERT, L. FARINAS, M. MICHEL
"Logique Temporelle pour les Réseaux de Processus communiquant par file". T-R n° 2878 - LCR - Thomson-CSF, (1983).

[ENJ2] P. ENJALBERT
"Un théorème de complétude pour une logique multitemporelle typée". T-R n° 2917 - LCR - Thomson-CSF, (1983).

[GAB] D. GABBAY et al
"On the temporal analysis of fairness". 7th ACM Symposium on Principles of Programming Languages, 1980, p. 163-173.

[HOA] C.A.R. HOARE
"Communicating sequential processes". C.A.C.M. Vol. 21 n° 8, p. 666-677, (1978).

[KAH] G. KAHN, P.B. MAC QUEEN
"Coroutines and networks of parallel processes". Proc. IFIP 77.

[KOZ] D. KOZEN, R. PARIKH
"An elementary proof of the completeness of P D L"
TCS 14 (1981), p. 113-118.

[MIC] M. MICHEL
"Algèbre de Machines et Logique Temporelle". STACS 1984, LNCS n° 166.

[SAL] A. SALWICKI
"Semantics of parallel programs"
ICS-PAS Report - Polish Academy of Science.

ACKNOWLEDGEMENT: to L. FARINAS and E. AUDUREAU for many discussions and common work on this subject.

PROCESS LOGICS : TWO DECIDABILITY RESULTS

Z. Habasiński
Computer Centre, Technical
University of Poznań, Poznań,
Poland

Introduction.

There are many kinds and variants of logics designated for reasoning about programs. We introduce here the Dynamic Process Logic, DPL, which is a framework for logics designated for reasoning about events during regular programs computations. For this purpose they have path-formulae (interpreted over sequences of states) which tell us something about events along the path considered. The "proper" formulae are state formulae, interpreted over states. A criterion for decidability of the satisfiability problem is given, due to which any logic definable in DPL-framework and meeting a regularity condition is decidable in time $O(\exp cn^3)$. For instance some logics strictly stronger than PL from [H] are still decidable in that time. DPL^+ is an extension of DPL allowing boolean combinations of the elementary path-formulae. Any DPL^+-logic meeting the same regularity condition is decidable in time $O(\exp(n^3 \exp ck))$, where k is a number (usually appreciable) less than the length n of the given formula.
Temporal Process Logic, TPL, is a class of temporal branching time logics which is similar to DPL without the program connectives but with infinite sequences of states as the semantics of the atomic programs. A modification of the regularity condition gives the second criterion : any TPL-logic which fulfills the second condition is decidable in $O(\exp cn^2)$-time and any TPL^+-logic in time $O(\exp(n^2 \exp ck)$ This is an improvement of the theorem 8.5 from [EH], where $O(\exp^2(cn \cdot \log n))$ upper bound is given for CTL^+, the Computation Tree Logic with boolean combinations of the path formulae. Cf. [W] for similar results in the linear time temporal logic.

1. Tools from the automata theory

Besides the usual finite automata accepting finite strings we will alao use the Büchi automata on infinite strings. We review below their definition.

Definition 1.1

$\mathfrak{A}=(S, M, s_0, F)$ is a (deterministic) Büchi automaton on infinite strings from Σ^ω iff S is a finite set of states, $s_0 \in S$ is the initial state, $F \subseteq S$ and $M : S \times \Sigma \longrightarrow S$. Let $w=\sigma_0\sigma_1\ldots$ be an ω-sequence from Σ^ω. A run of $\mathfrak{A}$ on w is the function $r : \omega \longrightarrow S$ such that $r(0)=s_0$ and $r(i+1)=M(r(i), \sigma_i)$ for any natural i. $\mathfrak{A}$ accepts w iff $\{s \mid r(i)=s$ for infinitely many $i\} \cap F \neq \emptyset$, for the unique run r of $\mathfrak{A}$ on w.

Automata on trees are of crucial importance for our decision procedure :

Definition 1.2

The N-ary infinite tree T_N, $N \geq 1$, is the set $\{0,\ldots,N-1\}^*$. A path $(x_0, x_1, x_2,\ldots)$ in T_N is any (finite or not) sequence of nodes from T_N such that x_{i+1} is an immediate successor of x_i. A finite N-ary tree T is any finite subset of T_N such that :

- the empty string, root of T_N, belongs to T and
- if $x \in T$ then no successor of x in T_N belongs to T or all the immediate successors belong to T.

By $Fr(T)$, frontier of T, we denote $\{x \in T \mid x$ has no successors in $T\}$. N-ary infinite (finite resp.) Σ-tree is any function $f : T_N \longrightarrow \Sigma$ ($f : T \longrightarrow \Sigma$ for some finite tree $T \subseteq T_N$).

Now we fix N and omit it in subsequent definitions.

Definition 1.3

A special automaton on (infinite) Σ-trees, see [R], is a tuple $Aut=(S, M, s_0, F)$ where S, s_0 and F are defined as for Büchi automata and $M : S \times \Sigma \longrightarrow Powerset(S^N)$. r is a run of Aut on $f : T_N \longrightarrow \Sigma$ iff $r : T_N \longrightarrow S$ is such that :

- $r(\Lambda)=s_0$, Λ is the root of T_N, and
- $(r(x0),\ldots,r(x(N-1))) \in M(r(x), f(x))$ for any x in T_N.

Aut accepts $f : T_N \longrightarrow \Sigma$ iff there is a run r of Aut on the given Σ-tree such that : $\{s \mid r(x)=s$ for infinitely many $x \in \pi\}$ $\cap F \neq \emptyset$, for any infinite path π starting at the root $\Lambda \in T_N$.

One can prove the following theorem, cf. [R].

Theorem 1.4

The emptiness problem for special automata on N-ary tree is decidable in deterministic time proportional to $\mathrm{card}(\Sigma)\cdot N\cdot m^{N+3}$ (m is the number of states of the automaton tested).

□

2. DPL$^+$ and its decidability criterion

Assume we have given two countable disjoint sets of atomic programs $A_0, A_1, \ldots$ and atomic formulae $P_0, P_1, \ldots$ and a finite set of path connectives $E_0, \ldots, E_z$. Programs are defined exactly as in Propositional Dynamic Logic, see e.g. [FL],(with tests over state formulae, see below).

Formulae of DPL$^+$:

1. Any atomic formula is a state formula
2. If a is a program and $\underline{p}$ is a path formula then $\langle a \rangle \underline{p}$ is a stat formula (called a diament formula).
3. State formulae are closed under the usual boolean connectives
4. If E is a path connective of arity m and $p_1, \ldots, p_m$ are state formulae then $E(p_1, \ldots, p_m)$ is a path formula (called elementary on
5. Path formulae are closed under the usual boolean connectives

The set of DPL$^+$-formulae consists of all the state formulae defin by the above rules. DPL-formulae are those defined by 1.-4. For any natural k, DPL(k) denotes DPL$^+$ with the following restriction : any path formula may contain at most k occurrences of the elementary subformulae. As usual $[a]\underline{p}$ denotes $\neg\langle a \rangle \neg \underline{p}$.

Semantics of DPL$^+$:

Let S be a nonempty set. By $\underline{S}$ we denote the se of all nonempty finite sequences of elements from S. If $X,Y \subseteq \underline{S}$ the $X;Y$ is simply $\{(s_0,\ldots,s_m) \mid (s_0,\ldots,s_k) \in X$ and $(s_k,\ldots,s_m) \in Y$ fo some $0 \leqslant k \leqslant m\}$.

A <u>structure</u> is any triple $(S, \models, Tr)$ where S is a nonempty set of <u>states</u>, $\models$ is a <u>satisfiability relation</u> such that

$$\models \subseteq S \times \{\text{state formulae}\} \cup \underline{S} \times \{\text{path formulae}\}$$

$Tr : \{\text{programs}\} \longrightarrow \text{Powerset}(\underline{S})$, $Tr(a)$ is the set of <u>traces</u> of the program a. The above components should meet the following conditions:

C1. $s \models \langle a \rangle \underline{p}$ iff there is $\underline{s} \in Tr(a) : \underline{s} \models \underline{p}$ and the first element of $\underline{s}$ is s.

C2. $Tr(A) \cap \{(s) \mid s \in S\} = \emptyset$, for any atomic program A.

C3. $Tr(a;b) = Tr(a);Tr(b)$

C4. $Tr(a \cup b) = Tr(a) \cup Tr(b)$

C5. $Tr(a^*) = \{(s) \mid s \in S\} \cup \bigcup \{Tr(a^n) \mid n \geq 1\}$

C6. $Tr(p?) = \{(s) \mid s \models p\}$

C7 - C9 describe the standard behaviour of the satisfiability relation on the boolean connectives $\neg$, $\vee$ and $\wedge$.

Example 2.1

Let the set of path connectives consists of the six two-ary connectives : <u>until</u>, <u>while</u>, <u>before</u>, <u>pres</u>, <u>since</u> and <u>imp</u> . We write p <u>until</u> q instead of <u>until</u>(p,q) etc. Let $\underline{s} = (s_0, \ldots, s_k)$ for some $k \geq 0$. The semantics of the elementary formulae is given below.

$\underline{s} \models p$ <u>until</u> q iff $\exists i : s_i \models q$ and $\forall j \leq i \;\; s_j \models p$

$\underline{s} \models p$ <u>while</u> q iff $\forall i : (\forall j \leq i \;\; s_j \models q) \Rightarrow s_i \models p$

$\underline{s} \models p$ <u>before</u> q iff $\exists i : s_i \models p$ and $\exists j > i \;\; s_j \models q$

$\underline{s} \models p$ <u>pres</u> q iff $\forall i : s_i \models p \Rightarrow \forall j > i \;\; s_j \models q$

$\underline{s} \models p$ <u>since</u> q iff $\exists i : s_i \models q$ and $\forall j > i \;\; s_j \models p$

$\underline{s} \models p$ <u>imp</u> q iff $\forall i : s_i \models p \Rightarrow \exists j > i \;\; s_j \models q$

In this formalism some other constructs may be easily defined : <u>some</u> p is simply <u>true</u> <u>until</u> p, <u>all</u> p corresponds to p <u>while</u> <u>true</u> and <u>last</u> p to <u>false</u> <u>since</u> p. So Pratt's construct "during" written in [P] or [H] as $a \perp p$ is expressible in DPL$^+$ as $[a]$<u>some</u> p and in DPL as $\neg\langle a \rangle$<u>all</u>$\neg p$. Similary "preserves" from the quoted papers is simply $[a](p$ <u>pres</u> $p)$ in DPL$^+$ i.e. $\neg\langle a \rangle(p$ <u>before</u> $\neg p)$ in DPL. The formula $\langle a \rangle p$ from the Propositional Dynamic Logic is definable as $\langle a \rangle$<u>last</u> p. ψ-formula from [H], written there $\psi(a,p,q)$ is expressible in DPL$^+$ as $[a](p$ <u>imp</u> $q)$ or as $\neg\langle a \rangle(\neg q$ <u>since</u> $p)$ in DPL and is not definable in PL from [H]. So, DPL over the six path connectives is an actual extension of PL from Harel's paper. It follows from the theorem below that it is decidable in one-exponential time.

□

For any path or state formula of DPL^+ we define :
Sub(p) to be the set of all state subformulae of p and
Cl(p) as the set $Sub(p) \cup \{\neg q \mid q \in Sub(p)\}$.

$D \subseteq Cl(p)$ is a <u>description</u> <u>for</u> p iff the following hold :

- for any $q \in Sub(p)$: $q \in D$ iff $\neg q \notin D$
- if $q \vee r \in Sub(p)$ then $q \vee r \in D$ iff $q \in D$ or $r \in D$
- if $q \wedge r \in Sub(p)$ then $q \wedge r \in D$ iff $q \in D$ and $r \in D$

The set of all descriptions for p is denoted by Des(p).

Definition 2.2

A deterministic automaton $\mathfrak{A}$ accepting finite strings on the alphabet Des($\underline{p}$) <u>defines the path formula</u> $\underline{p}$ iff for any sequence $(s_o, \ldots, s_k)$ in any structure $(S, \models, Tr)$ we have :
$(s_o, \ldots, s_k) \models \underline{p}$ iff $\mathfrak{A}$ accepts $D_o \ldots D_k$ where each $D_i = \{q \in Cl(p) \mid s_i \models q\}$

Let E be a path connective of arity m.

E is <u>regular</u> iff there is a constant c : for any state formulae $q_1, \ldots, q_m$ there is an usual deterministic sequential automaton $\mathfrak{A}$ on the alphabet $Des(E(q_1, \ldots, q_m))$ such that :

- $\mathfrak{A}$ defines $E(q_1, \ldots, q_m)$
- the number of states in $\mathfrak{A}$ is less than c
- $\mathfrak{A}$ is constructable in time proportional to the cardinality of $Des(E(q_1, \ldots, q_m))$

□

Note that each path connective quoted in example 2.1 is regular. The following theorem gives a decidability criterion for DPL^+ logics. It says, roughly speaking, that any logic with regular path connective is decidable in less then two or one-exponential time depending on the presence of boolean combinations of the elementary path formulae. The parameter k occurring in the upper bound is usually much less then the length n of the given formula.

Theorem 2.3

Let us fix certain finite set of regular path connectives. For any state formula p_o of DPL^+ the satisfiability problem is decidable in time proportional to

$$\exp(n^3 \cdot \exp ck)$$

where :

n is the length of p_o, $k = \max\{k_1, \dots, k_N\}$

k_i is the number of occurrences of elementary subformulae in p_i, $(1 \leq i \leq N)$ and

$p_1, \dots, p_N$ are all the diament subformulae of p_o

Corollary 2.4

The same problem for DPL(k)- and DPL-logics having regular path connectives is decidable in time $O(\exp cn^3)$.

Proof of the theorem 2.3 (sketch).

$Des =_{df} Des(p_o)$. Note that if each path connective is regular then any path formula is definable by an automaton constructable in time exponential in the length of the formula in question. Let us make copies A', A'', A''' for any atomic program A in p_o in order to represent DPL^+-models by labelled trees. Define Pr as $\bigcup\{\{A, A', A'', A'''\} \mid A$ - atomic program in $p_o\}$. The set of <u>labells</u> (for p_o), Lab, is defined as $Pr^N \times Des \cup \{!\}$. For any labelled tree $f : T_N \longrightarrow Lab$ we define partial functions $f_o, \dots, f_{N-1} : T_N \longrightarrow Pr$, $f_N : T_N \longrightarrow Des$ by the condition : if $f(x) = !$ then $f(x) = (f_o(x), \dots, f_N(x))$ and a function $\langle\ \rangle_f : Pr \longrightarrow Powerset(T_N{}^2)$ by the requirement : $x\langle B\rangle_f y$ iff $y = xi$ for some $i < N$ and $f_i(x) = B$ and $f(y) \neq !$.

In the sequel <u>x</u> ranges over the set of finite paths from T_N. We also define the set of <u>traces</u> <u>in</u> f :

$Tr_f(A) = \{(x_o, \dots, x_k) \mid k \geq 1,\ x_o\langle A'\rangle_f x_1$ and $x_{k-1}\langle A''\rangle_f x_k$ and $x_i\langle A\rangle_f x_{i+1}$ for any $1 \leq i \leq k-2$ $\{(x,y) \mid x\langle A'''\rangle_f y$

$Tr_f(a;b) = Tr_f(a);Tr_f(b)$ $\qquad Tr_f(a \cup b) = Tr_f(a) \cup Tr_f(b)$

$Tr_f(a^*) = \bigcup\{Tr_f(a^n) \mid n \geq 1\} \cup \{(x) \mid f(x) \neq !\}$ $\qquad Tr_f(p?) = \{(x) \mid p \in f_N(x)\}$

By a <u>tree</u> <u>for</u> p_o we mean any $f : T_N \longrightarrow Lab$ such that for any

x in T_N we have : there is $m \geqslant 0$ such that $f(x0^m)=!$ and if $f(x)=!$ and y is a successor of x in T_N then $f(y)=!$.

If $\underline{p}$ is a path subformula of p_o or a negation of such then by $\mathfrak{A}(\underline{p})$ we denote the automaton defining $\underline{p}$ but working on the extended alphabet Des rather then on Des($\underline{p}$). For a given $f : T_N \longrightarrow$ Lab we will say that $\mathfrak{A}(\underline{p})$ accepts the path $(x_o, \ldots, x_k)$ iff $\mathfrak{A}(\underline{p})$ accepts the word $f_N(x_o)\ldots f_N(x_k) \in Des^+$.

Theorem 2.5

Assume each path connective is regular. p_o is satisfiable iff there is a tree for p_o $f : T_N \longrightarrow$ Lab such that for any x in T_N and $i < N$ we have :

T1. If $p_i \in f_N(x)$, $p_i = \langle a \rangle \underline{p}$ then there is a proper prefix $\underline{x}$ of the infinite path $(x, xi, xi0, xi0^2, \ldots)$ such that $\underline{x} \in Tr_f(a)$ and $\mathfrak{A}(p)$ accepts $\underline{x}$

T2. If $\neg \langle a \rangle \underline{p} \in f_N(x)$, $\underline{x} \in Tr_f(a)$ and the first element of $\underline{x}$ is x then $\mathfrak{A}(\neg p)$ accepts $\underline{x}$

T3. $p_o \in f_N(\Lambda)$

See [S] for the proof of a similar theorem. From the above theorem we know that for any given p_o there is a set of labelled trees X such that: p_o is satisfiable iff $X \neq \emptyset$. It is easy to construct a special automaton accepting exactly X. Moreover this may be done in time exponential in the length of p_o. Having already the special automaton one can use the theorem 1.4 in order to test it for emptiness. □

3. TPL$^+$ and its decision problem

The set of TPL$^+$-formulae is defined as for DPL$^+$ except that in this case we admit atomic programs only.

Semantics of TPL$^+$:

For any set $S \neq \emptyset$ let $\underline{S}$ denotes now the set of all (finite or not) sequences from S with at least two elements. A structure for TPL$^+$ is any triple $(S, \models, \langle\ \rangle)$ where S is a non-empty set of states, $\models$ is the satisfiability relation i.e.

$$\models \subseteq S\times\{\text{state formulae}\}\cup \underline{S}\times\{\text{path formulae}\}$$

$$\langle\ \rangle : \{\text{atomic programs}\} \longrightarrow \{\text{binary relations on } S\}$$

For any program A, $Tr(A)$ is called the set of A-<u>traces</u> and denotes $\{(s_0,\dots,s_k)\in\underline{S} \mid \forall i<k \ \ s_i\langle A\rangle s_{i+1}$ and there is no $t\in S$: $s_k\langle A\rangle t\}\cup$ $\cup\{(s_0,s_1,\dots)\in\underline{S} \mid \forall i : \ s_i\langle A\rangle s_{i+1}\}$. The components of any structure meet the following conditions :

S1. $s \models \langle A\rangle \underline{p}$ iff there is $\underline{s}\in Tr(A)$: $\underline{s} \models \underline{p}$ and the first element of $\underline{s}$ is s.

S2. $\langle A_0\rangle = \bigcup\{\langle A_i\rangle \mid i \geqslant 1\}$

S3 - S5 describe the usual behaviour of $\models$ on the boolean connectives

The sets $Sub(p)$, $Cl(p)$ and the notion of a description for any TPL^+-formula p are defined exactly as for DPL^+.

Definition 3.1

A deterministic Büchi automaton $\mathfrak{A}$ on $Des(p)\cup\{!\}$ <u>defines the path formula</u> $\underline{p}$ iff for any path $(s_0, s_1,\dots)$ in any structure $(S, \models, \langle\,\rangle)$ we have :

- if $\underline{s}=(s_0, s_1,\dots)$ is infinite then $\underline{s}\models\underline{p}$ iff $\mathfrak{A}$ accepts $D_0D_1\dots$
- if $\underline{s}=(s_0,\dots,s_k)$ then $\underline{s}\models\underline{p}$ iff $\mathfrak{A}$ accepts $D_0\dots D_k!!!\dots$

  where $D_i=\{q\in Cl(p) \mid s_i\models q\}$.

Let E be a path connective of arity m.
E is <u>definable</u> iff there is a constant c : for any $p_1,\dots,p_m$ there are Buchi automata $\mathfrak{A}^+$ and $\mathfrak{A}^-$ on the alphabet $Des(E(p_1,\dots$ $\dots,p_m))\cup\{!\}$ such that :

- $\mathfrak{A}^+$ defines $E(p_1,\dots,p_m)$, $\mathfrak{A}^-$ defines $\neg E(p_1,\dots,p_m)$
- the numbers of states in $\mathfrak{A}^+$ and $\mathfrak{A}^-$ are less then c
- both automata are constructable in time proportional to the cardinality of their input alphabet
- for any w in $(Des(E(p_1,\dots,p_m))\cup\{!\})^\omega$ we have :

  $\mathfrak{A}^-$ accepts w iff $\mathfrak{A}^+$ rejects w

Note that all the path connectives from example 2.1 as well as the connective <u>next</u> ($(s_0,s_1,\dots)\models$ <u>next</u> p iff $s_1\models p$) are definable. This justifies why the next theorem is an improvement of the theorem 8.5 from [EH]. The theorem 3.2 may be proved using methods similar to that presented in the previous paragraph.

Theorem 3.2

Let us fix certain finite set of definable path connectives For any state formula p_0 of TPL^+ the satisfiability problem is decidable in time proportional to

$$\exp(n^2 \cdot \exp ck)$$

where n and k are defined exactly as for the DPL^+-case. □

Acknowledgements

I would like to thank many people for their support in writing this paper. Among them : Marek Karpiński, Andrzej W. Mostowski, Ryszard Danecki, Robrt Knast and Peter Widmayer. The result presented here was obtained during my study at the Mathematical Institute of Polish Academy of Sciences.

References

[FL] Fischer M. J., Ladner R. E.
Propositional Dynamic Logic of regular programs
JCSS 18(2), 1979, pp. 194 - 211.

[H] Harel D.
Two results on Process Logic
Inf. Processing Letters 8(4), 1979, pp. 195 - 198.

[EH] Emerson E. A., Halpern J. Y.
Decision procedures and expressiveness in the Temporal Logic of Branching Time
STOC'82, pp. 169 - 180.

[P] Pratt V. R.
Process Logic : preliminary report
POPL'79, pp. 93 - 100.

[R] Rabin M. O.
Weakly definable relations and special automata
in : Math. Logic and Found. of Set Theory, (Y. Bar-Hillel ed.
North Holland Pub. Co., 1970, pp. 1 - 23.

[W] Wolper P. L.
Synthesis of Communicating Processes from Temporal Logic Specifications,
Report of the Dep. of Comp. Sci., Stanford Univ., 1982.

ON SEARCHING OF SPECIAL CLASSES OF MAZES AND FINITE EMBEDDED GRAPHS

A. Hemmerling
Sektion Mathematik der
rnst-Moritz-Arndt-Universität
DR-2200 Greifswald, Jahnstraße

K. Kriegel
Institut für Mathematik der
Akademie der Wissenschaften der DDR
DDR-1086 Berlin, Mohrenstraße

NTRODUCTION

This paper is the union of shortened versions of two papers sub-itted separately. In Part 1 the second author sketches his proof that, o every number k, there exists a 1-pebble automaton searching every abyrinth (maze) having no more than k barriers. This proof is announ-ed in /4/ and given in detail in /5/. Here labyrinths are considered s special finite, embedded graphs of degrees ≤ 4, but, moreover, the power f compass is essentially used. In Part 2 the first author generalizes he above result to the class of arbitrary finite, embedded, connected raphs with degrees bounded by a number d. More precisely, it is shown hat to every number k, there exists a 1-pointer automaton searching very finite, embedded d-graph which has no more than k vertices touch-ng three or more regions. An extended version of this proof will be ublished in /2/. In both cases, the number of internal states of the utomata can be bounded by exp(O(k)).

Our results show that the number of barriers in Hoffmann's con-truction of traps for 1-pebble automata /3/ cannot be bounded by a onstant. On the other hand, Müller showed that one can always construct raps with only three barriers for finite automata working in mazes /7/ nd with only two regions for finite automata working in embedded -graphs /6/.

A 1-pebble automaton searching every labyrinth with at most 3 arriers was already constructed by Szepietowski /8/.

PART 1

Let $D = \{n,e,s,w\}$ denote the set of compass directions. For any $r \in D$, $\bar{r}$ denotes the opposite direction.

A labyrinth L is a graph (Z,E) where Z is a set of points and $E \subseteq Z \times D \times Z$ a set of labelled arcs such that

(i) if $e = (P,r,Q) \in E$ then $\bar{e} = (Q,\bar{r},P) \in E$ ($\bar{e}$ is called the inverse arc of e),

(ii) if $e = (P,r,Q) \in E$ and $e' = (P,r,Q') \in E$ then $e = e'$.

The set $val_L P = \{r \in D: \exists (P,r,Q) \in E\}$ is called the valency of the point P. The degree of the point P, $deg_L P$, is defined to be the cardinality of $val_L P$.

We always claim that L is a connected graph.

A finite automaton α being able to move in labyrinths is a Mealy-type automaton. In each step, α determines the valency of the point it is visiting, and depending on its state α computes one of the free directions and moves one step in this direction.

A pebble automaton α may drop a pebble on a point it is visiting. After returning to that point α can sense the pebble's presence and if desired pick it up and move it to another point.

A standard example of a labyrinth is given by the integer grid $\mathbb{Z}^2$. A labyrinth L is called plane if it is isomorhic to a subgraph of $\mathbb{Z}^2$, and it is called fully plane if it is isomorphic to an induced subgraph of $\mathbb{Z}^2$. (Isomorphism means direction preserving isomorphism.) Note that a fully plane labyrinth L can be presented either by its set of points $Z \subseteq \mathbb{Z}^2$ or by its set of obstracles $M = \mathbb{Z}^2 \setminus Z$. The set of obstracles decomposes in components of connectedness called components of the labyrinth L.

For any labyrinth $L = (Z,E)$, there is defined a canonical rotation system π, that is a family $\{\pi_P: P \in Z\}$ of cyclic permutations of the set of arcs leaving a point P. The permutations π_P are defined by the clockwise orientation of the plane:

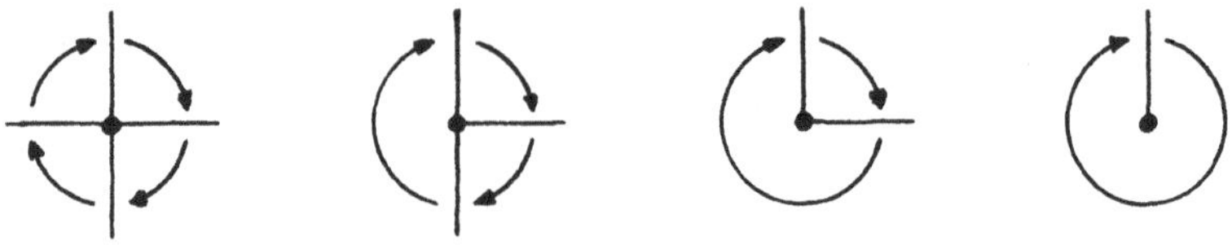

A face of a labyrinth is represented by a closed path

$e_1, e_2, \ldots, e_n = (P_1 \xrightarrow{r_1} P_2 \xrightarrow{r_2} \ldots \xrightarrow{r_n} P_{n+1} = P_1)$ such that, for any i $(1 \leq i < n)$, $e_{i+1} = \pi_{P_{i+1}}(\overline{e_i})$ and $e_1 = \pi_{P_1}(\overline{e_n})$.

The closed paths $e_1, e_2, \ldots, e_n$ and $e_j, e_{j+1}, \ldots, e_n, e_1, \ldots, e_{j-1}$ will be identified.

Let ι denote the cyclic permutation (e,n,s,w) of the set of directions D. We define a function $\chi : D^* \longrightarrow \mathbb{Z}$ in the following way:

$\chi(r) = 0$ for any $r \in D$,

$$\chi(rr') = \begin{cases} 1 & \text{if } r' = \iota(r), \\ 0 & \text{if } r' = r, \\ -1 & \text{if } r' = \iota^{-1}(r), \\ -2 & \text{if } r' = \overline{r}, \end{cases}$$

$$\chi(r_1 r_2 \ldots r_n) = \sum_{j=1}^{n-1} \chi(r_j r_{j+1}) .$$

The function χ is called <u>rotation index</u>. It is used to describe the curvature of paths in labyrinths. The rotation index of a face $f = (P_1 \xrightarrow{r_1} P_2 \xrightarrow{r_2} \ldots \xrightarrow{r_n} P_{n+1} = P_1)$ is defined to be the integer $\chi(r_1 r_2 \ldots r_n r_1)$.

Now we can formulate the main result of this part.

<u>Theorem:</u> For any natural number k, there is a 1-pebble automaton $\mathcal{O}\!\mathcal{L}(k)$ which is able to search any finite, fully plane labyrinth L with at most k components.

The proof of this theorem proceeds in two steps.

First step: To any finite, fully plane labyrinth L with k components we construct a finite, plane labyrinth L' with k faces such that the behaviour of automata in L' can be simulated in L. For a very simple construction see Part 2. Therefore, it is sufficient to prove the weak version of the theorem for plane labyrinths with k faces.

Second step: We prepare the proof of the weak version by some definitions and facts.

<u>Proposition 1:</u> Let L = (Z,E) be a finite, plane labyrinth with k faces. Then the following two conditions hold:

(i) The Euler-Poincaré characteristic
$$\mathcal{E}(L) = \mathrm{card}(Z) - \tfrac{1}{2}\mathrm{card}(E) + k$$
is equal to 2.

(ii) There is exactly one face - called the periphery of L - having rotation index -4, and all other faces have rotation index 4.

Definition: The greatest sublabyrinth R(L) which does not contain points of degree 1 (with respect to R(L)) is called the root of the labyrinth L. A point P in R(L) is called an R-point if its degree in R(L) is greater than 2.

Lemma 1: A finite, plane labyrinth with k faces has at most 2k-4 R-points.

Definition: Let P be a point of degree ≥ 3 in a labyrinth L.

- P is called $\mathcal{P}$-point ($\mathcal{P}$-periphery) if it belongs to the periphery of L.
- P is called T-point (T-tree) if there is no circle (= simply closed path) containing P.
- P is called A-point (A-action) if it is an R-point but neither $\mathcal{P}$- nor T-point.

Lemma 2: Any face of a finite, plane labyrinth contains a $\mathcal{P}$-point or an A-point. If two faces have a common point then they have a $\mathcal{P}$-point or an A-point in common. Any finite, plane labyrinth L can be represented as a digraph G_L the set of which is the union of all $\mathcal{P}$-points and all A-points. The arcs of G_L are segments of faces between two such vertices.

Proposition 2: Let L be a finite, plane labyrinth with k faces. Then the graph G_L is strongly connected, and for any A-point Q there is a path of length $\leq 2k-4$ in G_L from Q to a $\mathcal{P}$-point and, conversely, a path of length $\leq 2k-4$ from a $\mathcal{P}$-point to Q.

Taking into account Lemma 1 and Lemma 2, the proof of this proposition is straightforward. Note that paths in G_L can be represented by words of the monoid $\{0,1,2,3\}^*$ of the same length if we use the canonical cyclic ordering of arcs by the rotation system.

Example:

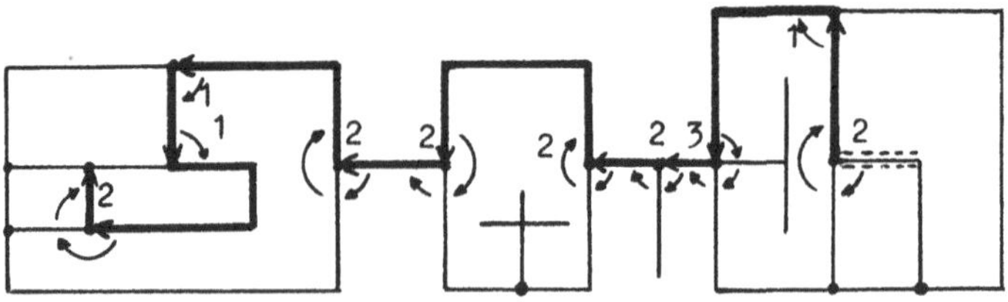

Now we can sketch the proof of the "weak version" of our theorem. One constructs a 1-pebble automaton which decides for a given point P in L the following three questions:

1) Is P a T-point ?
2) Is P a $\mathcal{P}$-point ?
3) Is P an A-point ?

Having such an automaton it is sufficient to describe an algorithm for searching the digraph G_L. This can be done as follows:

i) Initial part ("finding the periphery"): We search from a given A-point all paths of the length 2k-4 until we find the periphery,i.e. a $\mathcal{P}$-point.
ii) Periodical part ("searching the labyrinth"): For all $\mathcal{P}$-points, we search all paths of length 2k-4.

Note that this algorithm works correctly by Proposition 2. One needs an $O(k \cdot 4^{2k-4})$-state automaton to implement it.

PART 2

An embedded d-graph is a connected graph embedded in the plane and having no vertex of a degree greater than d (where $d \geq 3$). For brevity,then we identify the vertices and edges of the graph with their embeddings. Let the order of a vertex v,Ord(v),be the number of regions touched by v. Analogously,the order of an edge e,Ord(e),denotes the number of regions,the closures of which contain e. The order of a vertex is bounded by its degree,$Ord(v) \leq Deg(v)$. For an edge e, $Ord(e) \in \{1,2\}$,and e is a bridge of the graph iff Ord(e)=1.

Example:

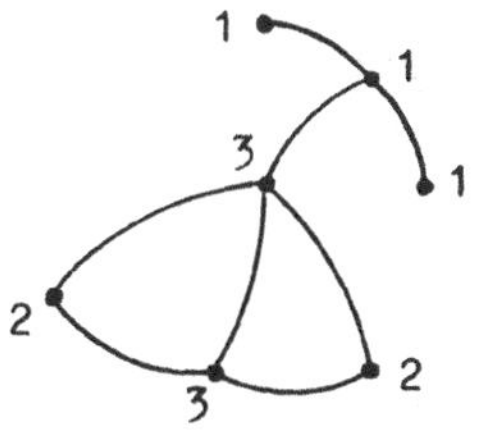

Orders of vertices

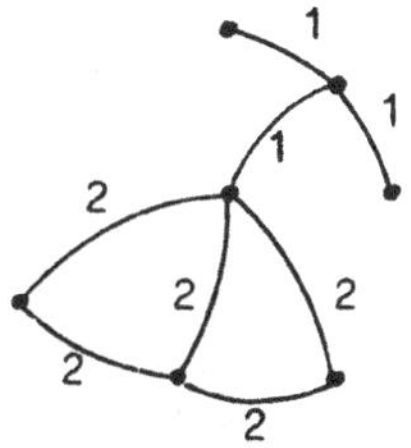

Orders of edges

We assume that the reader is familiar with the notation of a finite automaton which is able to walk in embedded d-graphs (see /1/). To improve their ability,finite automata are equipped with a so-called pointer which can be used to mark a directed edge of the graph. (Distinguish pointers from pebbles which serve to mark vertices of the graph.)

In our searching algorithm the pointer is needed only to compute the orders of vertices and edges.

Lemma: There are 1-pointer automata $\mathfrak{A}_V$ and $\mathfrak{A}_E$ which compute the order of their starting vertices and their starting edges, respectively. More precisely, starting with a position (v,u) on a finite, embedded d-graph, $\mathfrak{A}_V$ halts eventually in a state a_k at position (v,u) if k is the order of vertex v. On the same condition, $\mathfrak{A}_E$ halts eventually in a state a_k at position (v,u) if Ord({v,u})=k. Both automata use their pointer only to mark the starting position.

This can be easily proved.

Theorem: For every maximal degree $d \geq 3$ and every number k, one can construct a 1-pointer automaton which searches every finite, embedded d-graph having no more than k vertices with orders ≥ 3. In graphs having at least one vertex of an order ≥ 3 the automaton halts eventually. The number of internal states can be bounded by exp(O(k)).

In order to prove the theorem, we give a program for a 1-pointer automaton $\mathfrak{A}_{F,D,SEQ}$ working in embedded 3-graphs.

Program:

```
begin       FOL:= F; DIR:= D; IND:= 0;
      LOOP: (v,u):= current position;
            if Deg(v)=1 then  MOVE B  else
            if Deg(v)=2 then  MOVE R  else
            if Ord(v)=1 then  MOVE FOL  else
            if Ord(v)=2 then
               if Ord({v,u} )=2 and Ord( R(v,u) )=1 then
                  begin MOVE R; DIR:= R  end             else
               if Ord({v,u} )=2 and Ord( L(v,u) )=1 then
                  begin MOVE L; DIR:= L  end
               else MOVE DIR
            else MARK: if IND=Length(SEQ) then  STOP  else
                       begin  IND:= IND+1; MOVE SEQ(IND);
                              if SEQ(IND)=B then FOL:= ¬FOL else  ;
                              goto  LOOP
                       end
end
```

Here $F, D \in \{R,L\}$ are the initial values of variables FOL and DIR, respectively, and $SEQ \in \{R,L,B\}^*$. SEQ(i) denotes the i-th letter of the word SEQ if $1 \leq i \leq$ Length(SEQ). R,L and B mean the directions right, left and back, respectively. Let $\overline{R}$= L, $\overline{L}$= R and $\overline{B}$= B. For DIR $\in \{R,L,B\}$, MOVE DIR is the corresponding moving instruction for the automaton. DIR(v,u) denotes the edge of the position occupied after executing the instruction MOVE DIR from position (v,u). For example, B(v,u)= $\{u,v\}$. If Deg(v)=2 for the attached vertex v, MOVE R means to move foreward. STOP denotes the halting instruction.

From the lemma it follows that there exists a 1-pointer automaton $\alpha_{F,D,SEQ}$ with exp(O(Length(SEQ))) internal states and working according to our program in embedded 3-graphs.

On vertices of degree 1 or 2 or of degree 3 and order 1 the automaton walks by "keeping the right/left hand on the wall" if FOL= R/L. If a vertex has the degree 3 and order 2, there is exactly one incident bridge of the graph. When the automaton is arrived through this bridge, it continues the walk according to the direction stored in DIR. If it has reached such vertex through an edge of order 2, it has to walk through the bridge and to store in DIR the direction which secures the continuation of the walking in the case of returning to the vertex just attached. By the statement MARK the sequence SEQ controls the behaviour of the automaton at vertices of order 3.

<u>Agreement:</u> In the following considerations regarding the working of $\alpha_{F,D,SEQ}$ we shall neglect all steps which are used to compute the order of a vertex or of an edge according to the lemma.

Now we make up some essential facts concerning the behaviour of the automaton in finite, embedded 3-graphs G.

For a bridge $\{v,u\}$, let G_v denote that shore which contains the vertex v.

<u>Fact 1:</u> Let $\alpha_{F,D,SEQ}$ leave the vertex u through the bridge $\{v,u\}$. If G_v does not contain a vertex of order 3 then the automaton eventually returns through edge $\{v,u\}$, and by that time it has visited all edges in G_v and the variables FOL and DIR have the same values as before. If G_v contains a vertex of order 3 then the automaton eventually reaches such vertex.

For a sequence SEQ= $D_1D_2 \dots D_{n-1}D_n$, the <u>dual sequence</u> is defined by $\overline{SEQ}$= $\overline{D_n}\,\overline{D_{n-1}} \dots \overline{D_2}\,\overline{D_1}$. For example, $\overline{RRLBLR}$ = LRBRLL .

<u>Fact 2:</u> Let Length(SEQ) $\geq$ 1. If the graph G has a vertex of order 3, the automaton $\alpha_{F,D,SEQ\cdot B\cdot \overline{SEQ}\cdot B}$ halts eventually in that position in which it reached at first time a vertex of order 3.

Fact 3: Let G contain no more than k vertices of order 3. Then, for any two positions $(v_1,u_1),(v_2,u_2)$ in G, there exists a sequence $SEQ \in \{R,L,B\}^*$ such that $Length(SEQ) \leq k$ and that $\mathfrak{A}_{F,D,SEQ}$, starting from (v_1,u_1), eventually reaches the position (v_2,u_2).

Now let SEQ_i^k be the i-th sequence of the length k with respect to the lexicographical ordering in $\{R,L,B\}^*$ $(1 \leq i \leq 3^k)$, and

$$SEQ^k = SEQ_1^k \cdot B \cdot \overline{SEQ_1^k} \cdot B \cdot SEQ_2^k \cdot B \cdot \overline{SEQ_2^k} \cdot B \cdot \ldots \cdot B \cdot SEQ_{3^k}^k \cdot B \cdot \overline{SEQ_{3^k}^k} \cdot B \; .$$

Then the 1-pointer automaton $\mathfrak{A}_{F,D,SEQ^k}$ has all properties required in the theorem for d=3.

The theorem follows from its special case for d=3 by simulation based on replacements of the vertices with degrees >3 by suitable hypernodes of degree 3 (see /1/).

Let be given an embedded d-graph G. Each vertex of a degree ≤ 3 is replaced by itself. The vertices having higher degrees and orders $\neq 2$ are replaced by hypernodes in the following manner:

To prevent that vertices of order 3 arise by replacing from vertices of a degree ≥ 3 and the order 2, we need another replacement:

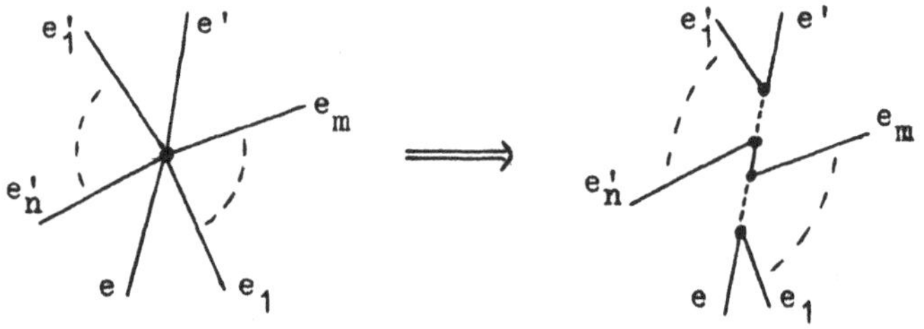

From the Jordan curve theorem it follows that every vertex of order 2 has exactly two incident edges e,e' of order 2 and, possibly, some further incident edges $e_1,\ldots,e_m,e_1',\ldots,e_n'$ of order 1 and arranged in the given manner.

r(G) denotes the embedded 3-graph obtained from G by the sketched replacements. Let $\mathfrak{A}_{dk}$ be a 1-pointer automaton searching all finite embedded 3-graphs having no more than dk vertices of order 3. Then one can construct a 1-pointer automaton $r(\mathfrak{A}_{dk})$ which, in every embedded d-graph G, simulates the working of $\mathfrak{A}_{dk}$ in the 3-graph r(G).

This automaton searches every finite,embedded d-graph with at most k vertices of orders ≥ 3,and the number of states can be bounded by $\exp(O(k))$.

Corollary: For every $d \geq 3$ and every number k,one can construct a 1-pointer automaton which searches every finite,embedded d-graph having no more than k regions. The number of internal states can be bounded by $\exp(O(k))$.

Indeed,from the Euler theorem it follows that $n \leq 2m$ if m is the number of regions and n is the number of vertices of degrees ≥ 3 of an embedded,finite d-graph.

Finally,we remark that the corollary implies the theorem of Part 1. Indeed,given a labyrinth (maze) L,let $G(L)$ denote the corresponding finite,embedded 4-graph. By removing all horizontal edges connecting two vertices which both have southern neighbours,we obtain an embedded 4-graph $\overline{G}(L)$ which has exactly k regions if k is the number of barriers (components) of the labyrinth L.

Example:

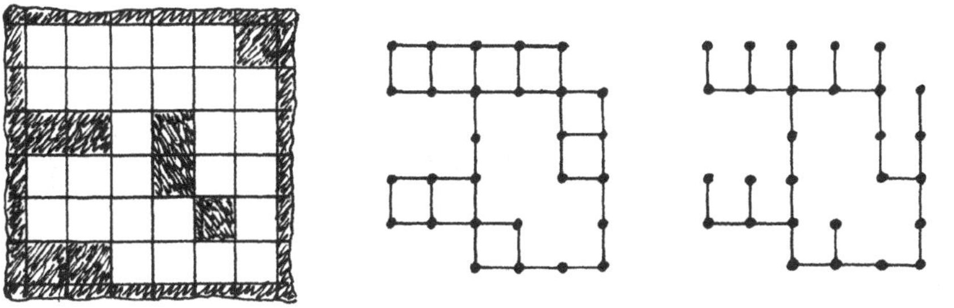

L (2 barriers) G(L) (11 regions) $\overline{G}(L)$ (2 regions)

To search a labyrinth L,a 1-pebble automaton can simulate the working of a 1-pointer automaton which searches the graph $\overline{G}(L)$.

REFERENCES

/1/ Blum,M.,Kozen,D., On the power of compass. 19th IEEE FOCS Conference,1978, 132-142

/2/ Hemmerling,A., 1-pointer automata searching finite embedded graphs. will be published in Zeitschr. f. math. Logik und Grundlagen d. Math.

/3/ Hoffmann,F., One pebble does not suffice to search planar labyrinths. Proc. FCT'81, Lect. Notes in Comp. Sc. 117 (1981), 433-444

/4/ Hoffmann,F.,Kriegel,K., Two more results on labyrinths. Short Communic. of the Internat. Congr. of Mathematicians, Warsaw 1983, XII, p. 52

/5/ Kriegel,K., Universelle 1-Kiesel-Automaten für k-komponentige Labyrinthe.
to appear as Report R-MATH-04/84, AdW der DDR,Berlin

/6/ Müller,H., Endliche Automaten und Labyrinthe.
EIK 7 (1971), 146-160

/7/ Müller,H., Automata catching labyrinths with at most three components.
EIK 15 (1979), 3-9

/8/ Szepietowski,A., Remarks on searching labyrinths by automata.
Proc. FCT'83, Lect. Notes in Comp. Sc. 158 (1983),457-464

THE POWER OF THE FUTURE PERFECT IN PROGRAM LOGICS*

M. Hennessy and C. Stirling
Dept. of Computer Science
University of Edinburgh
Edinburgh EH9 3JZ, Scotland

§1. Introduction

Many varieties of tense (temporal) logics have been suggested for describing properties of programs, [GPSS,CES,EH,BMP,Pn,MP,HKP,Ab,KVR]. This proliferation suggests that there is no simple criterion for judging the adequacy of such languages. They should be able to describe all properties which are commonly agreed to be of interest. However this class of properties is difficult to delineate and the most that one can hope for is to prove that language A is more expressive than language B in the sense that there is an interesting property expressible in A which is not expressible in B. There are of course other criteria for comparing these logics, such as the simplicity of their related proof systems. This paper will examine only their descriptive powers, i.e. their expressiveness.

One interesting question posed of such logics is whether they are adequate for expressing the various formulationsof fairness, [GPSS,La,EH]. Since this inevitably involves consideration of infinite sequences the models for these languages should state which infinite sequences are admissible. Most often these models are some form of transition system together with some criteria for admissible infinite sequences through the transition system. This is the approach taken, for example, in [QS,LPS,Mi4]. However much of the discussion is independent of the actual set of infinite sequences allowed. In §2 we introduce a generalisation of transition systems, called general transition systems, which are a natural generalisation of these models. They consist of a set of programs (or program points), or processes, and any collection of computations through these points satisfying simple criteria. We will use these are models throughout the paper.

In §3 we pose the problem of finding a tense logic which is adequate for expressing properties of general transition systems. We offer a natural candidate L_T which is a generalisation of the logic HML of [HM1,BR,HM2]. L_T is also a relativized branching time tense logic in the sense that the eventually operators are relativized to

* This work was partially supported by the SERC of the UK.

particular general paths into the future: instead of having an operator $\Diamond$ meaning in some future it will be that... we have an operator <u> meaning in some u-future it will be that... . We show that L_T is not sufficiently expressive. We do this by example; by exhibiting two general transition systems which intuitively have different computational behaviour but which cannot be distinguished by the language L_T. One may increase the power of L_T by adding new operators and thereby enable the language to express the intuitive difference between the two transition systems of this example.

There are two dangers in this approach. By adding more and more operators to cope with different examples the language becomes increasingly complicated. This criticism might be levelled at CTL* in [EH]. Moreover by simply adding a new operator one cannot be sure that the new language is sufficiently expressive to capture all the interesting phenomena of a general transition system. Criteria, independent of program logics, are required.

A simple criterion already exists for transition systems, namely bisimulation equivalence. Informally two programs in a transition system are bisimulation equivalent if no amount of experimentation will ever discover a difference between them. One may experiment on a program using the actions or transitions defined in the transition systems. This equivalence has been studied for various specific transition systems in [Si2,Pa1,Mi1,Mi2]. It is of interest to us because it serves as a simple characterisation of the expressive power of the language HML; in a finite branching transition system two programs enjoy the same properties expressible in HML if and only if they are bisimulation equivalent. In other words they can be distinguished by HML if and only if they are not bisimulation equivalent. We say that HML is expressive complete relative to bisimulation equivalence. This is a different definition of expressive completeness than that used in [EH,GPSS,Ka1].

Bisimulation equivalence is not of much interest for general transition systems since it is formulated in terms of experiments which only use finite transitions or computations. However by allowing experiments to consist of infinite computations one may define the notion of extended bisimulation equivalence on general transition systems. This, we claim, is a natural yardstick for the expressiveness of generalisations of HML. In §4 we give a very simple extension of L_T, J_T, which is expressive complete relative to extended bisimulation equivalence. The extra operator required is a relativised past tense operator ⓐ ; the property ⓐ A is true of a program if it has just executed the action a and immediately previous to its execution the property A is true. The expressive power of J_T stems from its ability to express the future perfect: it will have been the case that... .

In the full version we offer proofs and show that J_T is very stable - the addition of a wide variety of operators does not increase its expressiveness. We also show there that a modification of CTL* is as powerful as J_T.

§2. General Transition Systems

Transition systems have long been recognised as useful for modelling discrete systems [Ke,Si1,Pl]. Here we extend this concept to encompass the infinite behaviour of systems.

For any set X let X* be the set of finite sequences and X^ω the set of infinite sequences over X, and let $X^\dagger = X^* \cup X^\omega$. For $u \in X^\dagger$ let u(n) be the prefix of length n of u, if it exists and u[n] be the nth element of u, if it exists. Relative to a given set P of process names and a set A of action names, a computation is any non-empty finite or infinite sequence of the form:

$$p_0 \xrightarrow{a_0} p_1 \xrightarrow{a_1} p_2 \xrightarrow{a_2} \ldots\ldots \qquad (*)$$

This computation is called a u-computation from p_0 whenever $u \in A^\dagger$ and $u[n] = a_n$ for all n. It will sometimes be useful to allow computations of length zero from p which we write as $p \xrightarrow{\varepsilon} p$. We use c to range over C(A,P), the set of computations relative to P and A. If c is the computation (*) then term(c) = p_n if c is finite and of length n; $P(k,c) = p_k$ and $A(k,c) = a_k$ when defined; $A(c) = u \in A^\dagger$ such that $u[n] = a_n$ for all n; and c(n) = c if A(n,c) is not defined and $p_0 \xrightarrow{a_0} p_1 \xrightarrow{a_1} \ldots \xrightarrow{a_{n+1}} p_{n+1}$ otherwise. Finally composition of computations, $c_1 \cdot c_2$, is defined in the obvious way.

Definition 2.1 A general transition system, gts, is a triple <P,A,C> where

i. P is a set of process names

ii. A is a set of action names

iii. $C \subseteq C(A,P)$ is a set of computations satisfying

a. $c_1 \cdot c_2 \in C$ whenever defined and $c_1, c_2 \in C$

b. $c_1, c_2 \in C$ whenever $c_1 \cdot c_2 \in C$ and c_1 is finite. □

An immediate consequence of iii b is the condition:

c. $c(n) \in C$ for every $n>0$ whenever $c \in C$

The converse need not hold. When it does we say that a gts <P,A,C> is standard (or limit closed): $c \in C$ whenever $c(n) \in C$ for every $n>0$. Moreover, we say that a gts <P,A,C> is finite branching if for every $p \in P$ and $a \in A$ the set $S(a,p) = \{p': p \overset{a}{\to} p' \in C\}$ is finite.

General transition systems differ from transition systems in their account of infinite computations. Unlike transition systems they are built into the definition of general transition systems. In case of transition systems when fairness is not an issue, infinite computations are defined as the limits of sequences of finite computations: there is, then, a 1-1 correspondence between standard gtss and transition systems. Non-standard gtss have been used in [QS,LPS], and [CS,H2] contain finite branching instances. In the sequel we always assume that S(a,p), for any a and p, is countable. Labelled directed graphs can be used to represent gtss. If g is such a graph and I is a set of points through g then gts(g,I) is the gts <P,A,C> where P is the set of nodes in g, A the set of labels on the arcs of g and C is I together with the set of finite paths through g. We use this form of representation throughout the paper when giving examples.

§3. Relativized Branching Time Tense Logics

There is a variety of program logics for describing program properties. Tense (temporal) logics have become especially popular in recent years. They view a program essentially as a kind of transition system. Each formula of the logic expresses a property that the transition system may or may not possess. Many people have worried about the exact form these logics should have if they are to capture interesting programming properties such as fairness [GPSS,La,EH,Pn,OL]. Here, we address the question of which program logics are suitable for expressing properties of general transition systems and when these logics may be said to be sufficiently expressive.

In [HM1,HM2] a simple logic, HML, was introduced for expressing properties of finite branching transition systems. HML was further examined in [BR,St]. The first language we consider is a natural extension of HML to deal with general transition systems. When T = <P,A,C> is a gts $L_T^{\prime}$ is the language defined by:

$$\phi ::= Tr \mid \neg\phi \mid \vee\{\phi_i : i \in I\} \mid \langle u\rangle\phi \mid [u]\phi$$

where I is a non-empty set of indexes (which may be infinite) and where u ranges over non-empty members of $A^{\dagger}$. Intuitively, $\langle u\rangle\phi$ means ϕ at some point during some u-computation whereas $[u]\phi$ means ϕ at some point during every u-computation. (Thus <u> and [u] are not duals of each other.) We inductively define a satisfaction relation $\models \subseteq P \times L_T$: $p \models \phi$ is to be understood as process p satisfies (the property expressed by) the formula ϕ.

$p \models Tr$ for any $p \in P$

$p \models \neg\phi$ iff not($p \models \phi$)

$p \models \vee\{\phi_i : i \in I\}$ iff $p \models \phi_j$ for some $j \in I$

$p \models \langle u \rangle \phi$ iff for some u-computation c from p and for some n>0 $P(n,c) \models \phi$

$p \models [u]\phi$ iff for every u-computation c from p there is an n>0 $P(n,c) \models \phi$

L_T' extends HML in two ways: first, by allowing infinitary and not just finitary disjunction and secondly, by allowing u to range over $A^{\dagger}$ and not just A. (For different but finitary extensions of HML see [BR].) Let HML_T, therefore, be the sublanguage of L_T' where u is restricted to range over A only and where disjunction is finitary and let HML_T^{∞} be the result of admitting infinitary disjunction into HML_T.

The operators $\langle u \rangle$ and $[u]$ are also based on the two eventually operators of branching time: in some future it will be the case that and in every future it will be that [Pr,Bu]. They are, however, relativized eventually operators, relativized to particular paths into the future and when u is finite these paths are also bounded. Except when $u \in A$, $\langle u \rangle$ and $[u]$ are not duals of each other: $\neg\langle u \rangle\neg$ means 'in <u>every</u> u-future it will always be that' whereas $\neg[u]\neg$ means 'in <u>some</u> u-future it will always be that'. Thus, L_T' is a <u>relativized branching time tense logic</u> (and HML_T and HML_T^{∞} are relativized next logics).

To see the powers of L_T' consider the following example gts.

<u>Example 3.1</u>

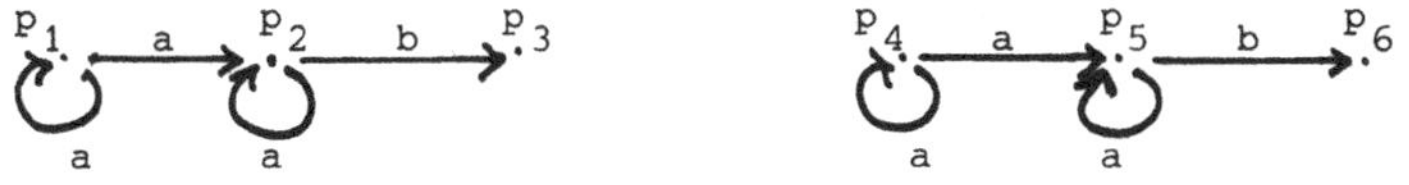

Let I_i be the set of infinite paths through this graph which almost always pass through p_i and let $I = I_2 \cup I_4 \cup I_5$. The graph together with I determine a gts. Note that $p_1 \xrightarrow{a} p_1 \xrightarrow{a} \ldots$ is not an infinite computation from p_1 whereas $p_4 \xrightarrow{a} p_4 \xrightarrow{a} \ldots$ is from p_4: hence $p_1 \models [a^{\omega}]\langle b \rangle Tr$ unlike p_4 because all a^{ω} computations from p_1 must eventually pass through p_2 whereas there is an a^{ω}-computation from p_4 which does not pass through p_5. It is easy to show that no formula of HML_T^{∞} distinguishes between p_1 and p_4. □

When defining a particular gts as in example 3.1 it is more

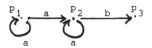

convenient to specify I in terms of subset J which generates it by virtue of the closure properties of paths given in definition 2.1 iii. In the above example I is completely specified by the three paths $p_2 \overset{a}{\rightarrow} p_2 \overset{a}{\rightarrow} \ldots$, $p_4 \overset{a}{\rightarrow} p_4 \overset{a}{\rightarrow} \ldots$, and $p_5 \overset{a}{\rightarrow} p_5 \overset{a}{\rightarrow} \ldots$. In future examples we specify the set of infinite paths in this way.

We now turn our attention to the expressibility of L'_T. Ideally, one wants L'_T to express all the interesting properties of T without, at the same time, being overly discriminatory. Formulas of L'_T may be used to distinguish processes in T. This ability to distinguish can be formalized by associating with L'_T an equivalence relation on $P \in T$:

$$p \sim_{L'_T} p' \text{ iff } \forall \phi \in L'_T .\ p \models \phi \text{ iff } p' \models \phi$$

Thus, $p \sim_{L'_T} p'$ holds just in case p and p' are indistinguishable by any formula of L'_T. In example 3.1 $p_2 \sim_{L'_T} p_5$ and $p_3 \sim_{L'_T} p_6$ but not $(p_1 \sim_{L'_T} p_4)$.

The following lemma shows that <u> is redundant in L'_T. Let L_T be the sublanguage of L'_T without the operator <u>.

<u>Lemma 3.2</u> $p \sim_{L'_T} p'$ iff $p \sim_{L_T} p'$. □

If T is standard then the powers of distinguishing between infinitary computations given by the operator [u], $u \in A^\omega$, are also redundant. Moreover, if T is finite branching there is no need for infinitary disjunction ([Mi3] shows the need for infinitary disjunction when T is not finite branching.)

<u>Lemma 3.3</u> a. If T is standard then $p \sim_{L_T} p'$ iff $p \sim_{HML^\infty_T} p'$

b. If T is standard and finite branching then $p \sim_{L_T} p'$ iff $p \sim_{HML_T} p'$. □

The logic L_T will not be expressively rich enough if there is a gts T and two processes which differ in their interesting properties but are indistinguishable by L_T. The following is such an example.

<u>Example 3.4</u>

I is generated by $\{p_2 \overset{a}{\rightarrow} p_2 \overset{a}{\rightarrow} \ldots;\ q_2 \overset{a}{\rightarrow} q_2 \overset{a}{\rightarrow} \ldots;$ $p_2 \overset{a}{\rightarrow} p_3 \overset{a}{\rightarrow} p_2 \overset{a}{\rightarrow} \ldots\}$. Let T be the gts described by this example. Then

<u>Lemma 3.5</u> $p_i \sim_{L_T} q_i$ for i=1,2 and 3. □

The language L_T, therefore, does not distinguish between p_2 and q_2.

However they do appear to have different interesting properties: all a^{ω}-computations from q_2 pass through q_2 almost always whereas all a^{ω}-computations from p_2 only pass through p_2 infinitely often and at p_2 and q_2 (unlike p_3 and q_3) there is the possibility of termination by moving to p_1 and q_1. This difference is bound up with distinctions between weak-fair and strong-fair computations [LPS,Pa2]. If we want a logical language to express such subtle distinctions then L_T needs to be extended. Analogous expressibility problems exist for branching time future tense logics with unrelativized eventually operators: see, for instance [La] and, in particular, [EH] which offers extensions which are more expressive.

Finding a suitable extension of L_T, which distinguishes p_2 from q_2 of example 3.4 will not guarantee sufficient expressibility. There is always the danger that a new example will be found showing it is still not expressive enough (or it is overly expressive). Indeed, in the full version we offer an extension of L_T which distinguishes the processes of example 3.6 but which is still not expressive enough. The main problem is, of course, that we want a logic which precisely captures the interesting properties of a gts. But we don't have a formal criterion for what an interesting property is: nor is it clear that we could ever have one given the open ended nature of what counts as interesting.

The way out of this impasse is to offer relative expressibility results. We need a yardstick, independent of the logic, for measuring expressibility. This independent criterion should be justifiable. For example, there is Kamp's result that linear time sentential tense logic with the two dyadic operators, since and until is completely expressive with respect to complete linear orders [Ka1]. (This is not the last word because of first-order tense logics - see [Ka2], for instance.) In the next section we offer expressive completeness results relative to a more programming oriented criterion, a criterion which arose independently of considerations of program logics.

§4. Bisimulations and Relative Expressibility

Let $T = \langle P,A,C\rangle$ be a gts. A <u>bisimulation</u> on T is a relation $R \subseteq P \times P$ satisfying

if $\langle p,p'\rangle \in R$ then for every $a \in A$

(1) i. if $p \xrightarrow{a} p_1$ then $p' \xrightarrow{a} p_1'$ for some p_1' such that $\langle p_1,p_1'\rangle \in R$

ii. if $p' \xrightarrow{a} p_1'$ then $p \xrightarrow{a} p_1$ for some p_1 such that $\langle p_1,p_1'\rangle \in R$

These relations give rise to a natural equivalence on P in T:

$p \sim_T p'$ if there exists a bisimulation R such that $\langle p,p'\rangle \in R$. It is straightforward to check that $\sim_T$ is indeed an equivalence and that it is also the maximal bisimulation (under inclusion). If $p \sim_T p'$ then their computational potentials are very similar. In fact the definition of $\sim_T$ is one attempt at formalising the idea that processes are equivalent if no amount of experimentation will ever discover a difference between them. (In a transition system one experiments on a process by asking it to perform particular actions.) The relations $\xrightarrow{a}$ formalise the effect that this has on processes. For more details on bisimulations see [Pa1,Mi3,Mi4,Si2].

The equivalence $\sim_T$ is a natural yardstick for measuring the expressibility of program logics. If Δ is a family of general transition systems then we say that a program logic M_T is <u>expressive complete</u> with respect to Δ and with respect to an equivalence E_T on processes in T whenever:

$$\forall T \in \Delta.\ \forall p,p' \in P \text{ in } T.\ p\ E_T\ p' \text{ iff } p \sim_{M_T} p'$$

The import of this stipulation is that the distinguishing powers of M_T coincide with E_T. The notion of expressive complete is different from that used in [EH,Ka1,GPSS]. We examine the difference in the full paper. The following theorem, a generalization of [HM1], is an expressive completeness result relative to standard gtss.

<u>Theorem 4.1</u> a. For any standard gts T, $p \sim_T p'$ iff $p \sim_{HML_T^{\infty}} p'$

b. For any finite branching standard gts T, $p \sim_T p'$ iff $p \sim_{HML_T} p'$. □

As previously noted, standard gtss are essentially transition systems. We now offer a natural extension of $\sim_T$ for gtss in general.

Let $T = \langle P,A,C\rangle$ be a gts. A relation $R \subseteq P \times P$ is an <u>extended bisimulation</u> on T if it satisfies:

if $\langle p,p'\rangle \in R$ then for every u-computation c

i. if c is from p then there is a u-computation c' from p' s.t. for all $n>0$, $\langle P(n,c), P(n,c')\rangle \in R$.

ii. if c is from p' then there is a u-computation c' from p s.t. for all $n>0$, $\langle P(n,c'), P(n,c)\rangle \in R$.

As with bisimulations, extended bisimulations give rise to a natural equivalence on processes in T:

$p \approx_T p'$ if there is an extended bisimulation R on T s.t. $\langle p,p'\rangle \in R$.

It is easy to show that $\approx_T$ is an equivalence relation and enjoys many of the properties of $\sim_T$: for instance, $\approx_T$ is the maximal extended bisimulation on T. Moreover, it is the natural counterpart of the relation 'partial extended bisimulation' defined in [H1]. It is also a very natural extension of $\sim_T$: $\approx_T$ becomes $\sim_T$ when computations are restricted to be of length one. Furthermore, the following lemma holds.

Lemma 4.2 If T is a standard gts then $p \sim_T p'$ iff $p \approx_T p'$. □

This lemma gives us further expressibility results: L_T and HML_T^∞ are expressive complete relative to standard gtss and $\approx_T$. However, the interesting issue is to find a logic which is complete relative to $\approx_T$ and arbitrary gtss. L_T is not such a logic. Example 3.4 is a counter-example; $p_2 \sim_{L_T} q_2$ whereas not($p_2 \approx_T q_2$) because of the a^ω-computation from p_2; $p_2 \overset{a}{\to} p_3 \overset{a}{\to} p_2 \overset{a}{\to} \ldots$.

In the previous section we remarked that similar expressiveness problems occur for standard branching time future tense logic. We cited [EH] where a more expressive logic is offered. Rather than extend L_T in this way these authors would suggest - an extension which not only complicates the semantics but also the syntax - we offer an alternative which is more in keeping with standard branching time tense logics. Instead, we add to L_T a relativized past tense operator ⓐ, $a \in A$, with intended meaning: it was the case at the last moment, just before a, that. (Thus, it is a past analogue of a relativized next operator.) The addition of this operator allows us to express a relativized future perfect tense; in every u-future it will have been the case just before s where s is a non-empty member of A*. (For instance, if $s=a_0,\ldots a_n$ then [u] ⓐ$_n$ ... ⓐ$_0$ expresses this tense.) Consider again example 3.4 and let ϕ be the L_T formula <a>[a]False:ϕ is true only of p_2 and q_2. The process q_2 satisfies : in every a^ω-future it will be that (ϕ and at the last moment, just before a,ϕ). Because of the a^ω-computation $p_2 \overset{a}{\to} p_3 \overset{a}{\to} p_2 \overset{a}{\to} \ldots$ p_2 fails this.

Where T = <P,A,C> is a gts then J_T is the language:

$\phi ::= Tr \mid \neg\phi \mid \bigvee\{\phi_i : i \in I\} \mid [u]\phi \mid$ ⓐ ϕ

where u is a non-empty member of $A^\dagger$ and $a \in A$. The presence of ⓐ in J_T complicates its semantics. The satisfaction relation $\models$ is inductively defined between finite computations d in C and J_T:

$d \models Tr$	forevery d
$d \models \neg\phi$	iff not($d \models \phi$)
$d \models \bigvee\{\phi_i : i \in I\}$	iff $d \models \phi_j$ for some $j \in I$
$d \models [u]\phi$	iff for every u-computation c from term(d) there exists $n \geq 0$ such that $d \cdot c(n) \models \phi$
$d \models$ ⓐ ϕ	iff d is $d' \cdot p \overset{a}{\to} p'$ and $d \models \phi$.

J_T induces an equivalence relation on finite computations in T:

$$d \sim_{J_T} d' \text{ iff } \forall\phi \in J_T.\ d \models \phi \text{ iff } d' \models \phi$$

We extend $\sim_{J_T}$ to processes in T in the obvious way:

$$p \sim_{J_T} p' \text{ iff } p \overset{\varepsilon}{\Rightarrow} p \sim_{J_T} p' \overset{\varepsilon}{\Rightarrow} p'.$$

It is easy to check that $q_2 \models [a^\omega](\phi \wedge ⓐ \phi)$ unlike p_2 in example 3.4 where ϕ is <a>[a]False.

We now come to the central theorem.

Theorem 4.3 If T is a gts then $p \sim_{J_T} p'$ iff $p \approx_T p'$. □

References

[Ab] K. Abrahamson, 'Modal logic of concurrent nondeterministic programs' LNCS Vol. 70, pp. 21-33 (1979).

[BMP] M. Ben-Ari, Z. Manna, A. Pnueli, 'The temporal logic of branching time', pp. 164-176, POPL (1981).

[BR] S. Brookes, W. Rounds, 'Behaviour equivalence relations induced by programming logics', LNCS Vol. 154, pp. 97-108 (1983).

[Bu] R.A. Bull, 'An approach to tense logic', pp. 282-300, Theoria (1970).

[CES] E.M. Clarke, E.A. Emerson, A.P. Sistla, 'Automatic verification of finite state concurrent systems using temporal logic specifications: a practical approach', pp. 107-126, POPL Proc. (1983).

[CS] G. Costa, C. Stirling, 'A fair calculus of communicating systems' LNCS Vol. 158, pp. 94-105 (1983).

[EH] E.A. Emerson, J.Y. Halpern, 'Sometimes and not never revisited: on branching versus linear time', pp. 127-140, POPL Proceedings (1983).

[GPSS] D. Gabbay, A. Pnueli, S. Shelah, J. Stavi, 'On the temporal analysis of fairness', pp. 163-173, POPL Proceedings (1980).

[H1] M. Hennessy, 'Axiomatizing finite delay operators', Technical Report CSR-124-82, Dept. of Computer Science, Edinburgh (1982). To appear in Acta Informatica.

[H2] M. Hennessy, 'Modelling finite delay operators', Technical Report CSR-153-83, Dept. of Computer Science, Edinburgh (1983).

[HKP] D. Harel, D. Kozen, R. Parikh, 'Process logic: expressiveness, decidability, completeness', JCSS, pp. 144-170 (1982).

[HM1] M. Hennessy, R. Milner, 'On observing nondeterminism and concurrency', LNCS Vol. 85, pp. 299-309 (1980).

[HM2] M. Hennessy, R. Milner, 'Algebraic laws for Nondeterminism and Concurrency, Technical Report CSR-133-83, Dept. of Computer Science, Edinburgh (1983).

[Ka1] H. Kamp, Tense logic and the theory of linear order Ph.D. thesis UCLA (1968).

[Ka2] H. Kamp, 'Formal properties of now', pp. 227-273, Theoria (1971).

[Ke] R. Keller, 'A fundamental theorem of asynchronous parallel computation' in Parallel Processing ed. T.Y. Feng, Springer (1975).

[KVR] R. Koymans, J. Vytopil, W. de Roever, 'Real time programming and asynchronous message passing', Technical Report RUU-CS-83-9, Rijksuniversiteit, Utrecht (1983).

[La] L. Lamport, '"Sometime" is sometimes "Not Never"', POPL Proc., pp. 174-185 (1980).

[LPS] D. Lehmann, A. Pnueli, J. Stavi, 'Impartiality, justic and fairness: the ethics of concurrent termination', LNCS Vol. 115, pp. 264-277 (1981).

[Mi1] R. Milner, A calculus of communicating systems, LNCS Vol. 92 (1980).

[Mi2] R. Milner, 'A modal characterisation of observable machine-behaviour' LNCS Vol. 112, pp. 25-34 (1981).

[Mi3] R. Milner 'Calculi for synchrony and asynchrony', TCS, pp. 267-310 (1983).

[Mi4] R. Milner, 'A finite delay operator in synchronous CCS', Technical Report CSR-116-82, Dept. of Computer Science, Edinburgh (1982).

[Mo] E. Moore, 'Gedanken-experiments on sequential machines', Automata Studies, edited by C.E. Shannon and J. McCarthy, Princeton University Press, pp. 129-153 (1956).

[MP] Z. Manna, A. Pnueli, 'How to cook a temporal proof system for your pet language', POPL Proceedings, pp. 141-154 (1983).

[OL] S. Owicki, L. Lamport, 'Proving liveness properties of concurrent programs', ACM Transactions on Programming Languages and Systems, pp. 455-495 (1982).

[Pa1] D. Park, 'Concurrency and automata on infinite sequences', LNCS Vol. 104 (1981).

[Pa2] D. Park, 'A predicate transformer for weak iteration', in Proceedings 6th IBM Symposium on Mathematical Foundations of Computer Science, Hakone, Japan (1981).

[PL] G. Plotkin, 'A structural approach to operational semantics', Lecture Notes, Aarhus University (1981).

[Pn] A. Pnueli, 'The temporal semantics of concurrent programs' LNCS Vol. 70, pp. 1-20 (1979).

[Pr] A. Prior Past, Present and Future, Oxford (1967).

[QS] J.P. Queille, J. Sifakis, 'Fairness and related properties in transition systems', Technical Report RR292, IMAG (1982).

[Si1] J. Sifakis, 'A unified approach for studying the properties of transition systems', TCS pp. 227-258 (1982).

[Si2] J. Sifakis, 'Property preserving homomorphisms of transition systems', Technical Report, IMAG (1982).

[St] C. Stirling, 'A proof theoretic characterization of observational equivalence'. **Proc. of** FCT-TCS Bangalore (1983).

HIERARCHY OF REVERSAL AND ZEROTESTING BOUNDED MULTICOUNTER MACHINES *

J. Hromkovič
Department of Theoretical Cybernetics
Comenius University
842 15 Bratislava
Czechoslovakia

ABSTRACT.

We consider one-way partially blind multicounter machines introduced by Greibach [2], and one-way multicounter machines. We study the reversal and zerotesting bounded versions of these machines, where for the input words of the length n the number of reversals or zerotests is bounded by a function f(n). It is established that time, nondeterminism, and counter as the resources, cannot compensate for a substantial decrease of the number of reversals allowed. Several hierarchy results are consequences of this fact.

Another group of results involves the hierarchy results according to zerotest number bound, and relates the reversal complexity and the zerotest complexity.

INTRODUCTION

A multicounter machine is a multipushdown machine whose pushdown stores operate as counters, i.e. have a single - letter alphabet. Unrestricted multicounter machines accept all recursively enumerable sets. So far various types of restricted multicounter machines have been considered to define proper subclasses [1, 3, 4, 5, 6]. In this paper we

* This work was supported in the part by the grant SPZV I-5-7/7.

shall consider one-way deterministic and nondeterministic multicounter machines with different types of restrictions. Let us first informally define these machines, the formal definitions can be found in Greibach [2] and Jantzen [7].

A multicounter machine consists of a finite state control, a reading head which reads the input from the input tape and a finite number of counters. We can regard a counter as an arithmetic register containing an integer which may be positive or zero. In one step, a multicounter machine may increment or decrement a counter by 1. The action or the choise of actions of the machine is determined by the input symbol currently scanned, the state of the machine and the sign of each counter: positive or zero. The machine starts with all counters empty and accepts if it reaches a final state with all counters empty. The class of multicounter machines without time limitation will be denoted by COUNTER, the deterministic version by DCOUNTER. The class of multicounter machines working in quasirealtime (for each machine there exists such a constant d that the length of each part of any computation, in which the reading head is stationary, is bounded by d) will be denoted by QR-COUNTER, the deterministic version by QR-DCOUNTER.

A partially blind multicounter machine is a multicounter machine which has no information about the contents of its counters, i.e. it does not known whether its counters are empty or nonempty. If by the computation should any counter go negative, no further transitions are allowed and the machine does not accept the input word. The machine accepts the input word iff it ends its computation in a final state with all "pblind" counters empty. The class of one-way (deterministic) partially blind multicounter machines will be denoted by PBLIND (DPBLIND), the class of quasirealtime (deterministic) partially blind multicounter machines will be denoted by QR-PBLIND (QR-DPBLIND).

Let M be a class of multicounter machines introduced. Then $\mathcal{L}(M)$ denotes the class of languages accepted by the machines of the class M. Let A be a multicounter machine from the class M and L(A) be the language accepted by A. Let f be a real function defined on natural numbers. Then $L_{Rf}(A)$ denotes the set of all words in L(A) for which there is some accepting computation containing at most f(n) reversals (i.e. changes from increasing to decreasing contents of a counter or vice versa), where n is the length of the input word. Let F be the class of all functions q from naturals to positive reals such that for all naturals n $f(n) \geq q(n)$. Then we define the classes of languages:

$$\mathcal{L}_{Rf}(M) = \bigcup_{B\in M} L_{Rf}(B) \quad \text{and} \quad \mathcal{L}(M\text{-}R(f)) = \bigcup_{q\in F} \mathcal{L}_{Rq}(M).$$

Furthermore we shall consider the following machines class $\mathcal{L}(M\text{-}RC) = \bigcup_{c\in N} \mathcal{L}_{Rc}(M)$, where N is the set off all naturals.

Now, we shall introduce the zerotesting restriction. Let f be a function from naturals to positive reals and let A be a multicounter machine from the class M. Then $L_{Zf}(A)$ is the set of all words in L(A) for which there is some accepting computation having its zero test numbe at most f(n) [i.e. the machine A empties the counters at most f(n) times in the computation], where n is the lenght of the input word. Let F be the class of all functions q from naturals to positive reals such that for all naturals n $f(n) \geq q(n)$. We define the following classes of languages:

$$\mathcal{L}_{Zf}(M) = \bigcup_{B\in M} L_{Zf}(B),\ \mathcal{L}(M\text{-}Z(f)) = \bigcup_{q\in F} \mathcal{L}_{Zq}(M)\ ,\ \text{and}\ \mathcal{L}(M\text{-}ZC) = \bigcup_{c\in N} \mathcal{L}_{Zc}(M)\ .$$

In what follows we will often consider computations in which a multicounter machine reads a group of identical symbols whose number is greater than the number of states. Clearly, there has to be a state q which will be entered twice (or more) in different configurations in thi part of computation. If these two occurrences of the state q are adjacen [no futher state q and no two equal states different from q occur in-between] we say that this part of the computation is a cycle with state characteristic q, reading head characteristic - the number (positive or zero) of symbols over which the reading head moves to the right in this cycle, and counter characteristic for each counter which is the difference between the counter contents at the beginning and at the end of the cycle. Thus, the counter characteristic can be positive, if the machine increases the contents of the counter in this cycle, can be negative if the counter contents is decreased in this part of the computation and obviously it can be zero.

We call attention to the fact that if a cycle occures in a computation of some one-way deterministic multicounter machine reading a group of identical symbols it follows that this cycle will be executed repeatedly until the reading head reaches some different symbol, i.e. until the reading head reads through the whole group of identical symbol

Let s be the number of states of an arbitrary multicounter machine A and k be the number of A counters, Then we can bound the number of all cycles with different characteristics by the following constant

$$s \cdot s \cdot (2s+1)^k \ .$$

Now, we introduce the following notation which we shall use in this paper. Let d be a real number. Then $\lceil d \rceil$ is the smallest natural number k such that $d \leq k$ and $\lfloor d \rfloor$ is the greatest natural number m such that $d \geq m$. Let f and g be arbitrary functions defined on naturals. Then the fact $\lim_{n\to\infty} f(n)/g(n) = 0$ we shall denote by $f(n) = o(g(n))$, and the fact that there exists a constant c such that $\lim_{n\to\infty} f(n)/g(n) = c$ we shall denote $f(n) = C(g(n))$.

This paper consists of two Sections. In Sect. 1 we shall study the hierarchy of the classes of languages accepted by multicounter and partially blind multicounter machines according to the bound on the number of reversals. First we prove some assertions for constant bounds which supplement the result obtained by Greibach [2]. Next we shall consider bounds given by increasing functions. We obtain a generalization of the results in Chan[7] for the hierarchy of reversal bounded one-way multicounter machines and we also prove this results for partially blind multicounter machines. In Sect. 2 we shall study the hierarchy of zero-testing multicounter machines and the relation between these machines and reversal bounded multicounter machines.

1. REVERSAL BOUNDED MULTICOUNTER MACHINES

First, we shall study the reversal bounded multicounter and partially blind multicounter machines, where the reversal bound is a constant. In the paper [2] it was shown

$\mathcal{L}$(QR-PBLIND-RC) $\subseteq$ $\mathcal{L}$(QR-COUNTER-RC) $\subsetneq$ $\mathcal{L}$(QR-PBLIND) $\subseteq$ $\mathcal{L}$(QR-COUNTER).

Now, we shall show the equality between $\mathcal{L}$(QR-PBLIND-RC) and $\mathcal{L}$(QR-COUNTER-RC). Whether equality holds between $\mathcal{L}$(QR-PBLIND) and $\mathcal{L}$(QR-COUNTER) remains open.

Theorem 1. Let c be a natural number. Then

$$\mathcal{L}_{Rc}(\text{QR-PBLIND}) = \mathcal{L}_{Rc}(\text{QR-COUNTER}) .$$

Idea of the proof. Each counter of a QR-COUNTER machine A can be simulated by $\lceil c/2 \rceil$ pblind counters of a QR-PBLIND machine B. During the computation B can nondeterministicly decide whether a simulated A's counter is empty. If B decides that s is empty B will work as A with the information that s is empty. Besides this B lets one of its $\lceil c/2 \rceil$ pblind counters simulating s stationary till the end of the computation in order to verify its nondeterministic decision. It is no problem to solve some local difficulties which origine in the proof of the fact $L_{Rc}(A) = L_{Rc}(B)$.

Corollary 1. $\mathcal{L}$(QR-PBLIND-RC) = $\mathcal{L}$(QR-COUNTER-RC)

Now, we shall formulate a theorem whose proof is not given here because it is too long but not very hard.

Theorem 2. Let c be a positive real number. Then

$$\mathcal{L}(\text{QR-PBLIND}) \subseteq \mathcal{L}_{Rcn}(\text{QR-PBLIND})$$

$$\mathcal{L}(\text{QR-COUNTER}) \subseteq \mathcal{L}_{Rcn}(\text{QR-COUNTER})$$

$$\mathcal{L}(\text{QR-DPBLIND}) \subseteq \mathcal{L}_{Rcn}(\text{QR-DPBLIND})$$

$$\mathcal{L}(\text{QR-DCOUNTER}) \subseteq \mathcal{L}_{Rcn}(\text{QR-DCOUNTER})$$

Now, we shall study the hierarchy of multicounter machines and partially blind multicounter machines according to reversal bound which is a function of the input word length. We shall show stronger hierarchi for multicounter machines, as in [8] and extend them for partially bli multicounter machines. We shall obtain these results as simple consequences of the following main theorem.

Theorem 3. Let $C(n) \geq g(n) \geq f(n)$ be arbitrary increasing functions defined on naturals. Let $f(n) = o(g(n))$. Then

$$\mathcal{L}(\text{QR-DPBLIND-R}(g)) - \mathcal{L}(\text{COUNTER-R}(f)) \neq \emptyset .$$

Proof. We shall prove this result dividing the proof in two parts according as $g(n) = C(n)$ or $g(n) = o(n)$.

I. Let $g(n) = C(n)$. Then we shall consider the language $L_1 = \{w \text{ in } \{a,b\}$ $| \#a(w) = \#b(w)$ and for all u, x in $\{a,b\}^*$ such that $w = ux$ it follows $\#a(u) \geq \#b(u)\}$, where $\#a(w)$ is the number of symbols a in the word w. Realizing the assertion of Theorem 2 we have L_1 is in $\mathcal{L}$(QR-DPBLIND-R(g)) We shall show by contradiction that L_1 does not belong to the class $\mathcal{L}$(COUNTER-R(f)) now.

Let h be a function from naturals to positive reals, such that $h(n) \leq f(n)$. Let there exist a COUNTER machine A with k counters and s states, for some natural k and s, such that $L_{Rh}(A) = L_1$. Then we consider the accepting computation y on the input word

$$x = \left(a^{s+1}b^{s+1}\right)^{d_1 d_2 (k+1)\cdot h(n)}$$

in L_1 for some constants d_1 and d_2 such that $d_1 > s$ and $d_2 > s^2(2s+1)^k$. Obviously, n can be choosen in such a way that $|x| = 2(s+1)d_1d_2(k+1)\cdot h(n) \leq n$.

In the following we shall construct a word x' not in L_1 such that A accepts x' if it accepts x. Clearly, this will prove our assertion.

It is easy to see that there exists such a subword

$$x_1 = \left(a^{s+1}b^{s+1}\right)^{d_1 d_2 (k+1)}$$

of x that the machine A computing on x_1 does not reverse its counters. The fact, that no reversal of counters is done in the part y_1 of the accepting computation y on the subword x_1, implies that each counter can be emptied at most once in the computation y_1 on x_1. So, we can assume that there exists such subword

$$x_2 = \left(a^{s+1}b^{s+1}\right)^{d_1 d_2}$$

of the word x_1 that no counter is reversed or emptied in the part y_2 of the accepting computation y on x_2.

Now, we shall consider all cycles with reading head characteristic 0 in the computation y_2 on x_2. Let the number of all such cycles according the different state characteristic be a constant c which can be clearly bounded by s. Let $y = u_1y_2u_2 = u_1v_1r_1v_2r_2v_3u_2$, where r_1 and r_2 are some cycles with the same state characteristic and reading head characteristic 0, and v_1, v_2, v_3, u_1, u_2 are some parts of computation. Considering the fact no counter is reversed or emptied in y_2 it can be easy seen that if $y = u_1v_1r_1v_2r_2v_3u_2$ is an accepting computation on $x = w_1x_2w_2$ then $y' = u_1v_1r_1r_2v_2v_3u_2$ is an accepting computation on x too. So we can assume that the accepting computation $y = u_1y_2u_2 = u_1z_1o_1z_2o_2\ldots z_co_cz_{c+1}u_2$, where o_j consists of all cycles with the reading head characteristic 0 and some state characteristic q_j (obviously the reading head is stationary in the part o_j of the computation) and z_j involves no cycle with reading head characteristic 0 for $j = 1,\ldots,c+1$.

Using this assumption we obtain that there exists such a subword

$$x_3 = \left(a^{s+1}b^{s+1}\right)^{d_2}$$

of the word x_2 that the part y_3 of the accepting computation y on x_3 involves no cycle with reading head characteristic 0 and no counter is reversed or emptied in the computation y_3 on x_3. There exists at least one cycle on each subword a^{s+1} of x_3 and since the number of cycles with different characteristics is bounded by $s^2(2s+1)^k < d_2$ there exist some cycles p_1 and p_2 with the same characteristics which are situated in two parts of the computation y_3 on two different groups of a's of the word x_3. Let the reading head characteristic of p_1 and p_2 be m. Clearly $m > 0$.

Choosing m symbols a from the first group of a's and pumping a^m to the second group of a's we obtain a word x' which does not belong to L_1. Now, we shall construct an accepting computation on the word x' what proves our assertion for $g(n) = C(n)$.

Let $y = uy_3v = us_1p_1s_2p_2s_3v$ be the accepting computation on the word $x = wx_3w'$, where $y_3 = s_1p_1s_2p_2s_3$ is the part of the computation on x_3 and u, v, s_1, s_2, s_3 are some parts of the computation. Then the accepting computation on x' will be $us_1s_2p_1p_2s_3v$, because the machine A is after the initial part of the computation $us_1p_1s_2p_2s_3$ on x in the same configuration (it means in the same state, with the same contents of all counters, and with the same postfix w' of the words x and x' on the input tape) as after the initial part of the computation $us_1s_2p_1p_2s_3$ on x'.

We call attention to the fact that no counter can be emptied in the computation $s_1s_2p_1p_2s_3$ what is the essential point in our consideration (No counter can reverse in the computation $s_1s_2p_1p_2s_3$ and no counter can be emptied in the computation $s_1p_1s_2p_2s_3$ imply this fact).

II. Let $g(n) = o(n)$. Then we shall consider the language $L(g) = \{w = u_1u_2\ldots u_p \mid \#a(w) = \#b(w),\ u_j \in a^+b^+$ for $j = 1,\ldots,p$, $p \leq \lceil g(n) \leq 2\rceil$, and for all v, y in $\{a,b\}^*$ such that $w = vy$ it follows $\#a(v) \geq \#b(v)\}$. It can be easy seen that this language belongs to $\mathfrak{L}$(QR-DPBLIND(g)).

Now, we shall show by contradiction that L(g) does not belong to $\mathfrak{L}$(COUNTER-R(f)). Let h be a function from naturals to positive reals such that $h(n) \leq f(n)$. Let there exist a COUNTER machine C with k counters, for a natural k, such that $L_{Rh}(C) = L(g)$. Let us consider an input word

$$x = (a^m b^m)^{\lceil g(n)/2\rceil}$$

where m is greater than the states number of A. Since $g(n) = o(n)$ there exists such large n that $2m\lceil g(n)/2\rceil \leq n$. So, we can assume x is in L(g). As a consequence of the assumption $h(n) \leq f(n) = o(g(n))$ we have, for a sufficiently large n, $\lceil g(n)/2\rceil \geq g(n)/2 \geq d(h(n)+1)$, where d is an arbitrarly large constant. Now using the same idea as in the case I it can be constructed a word x', which does not belong to L(g), but the machine C accepts it.

Using Theorem 3 we obtain several hierarchy results which we shall formulate in Theorem 4.

Theorem 4. Let $C(n) \geq g(n) \geq f(n)$ be increasing functions defined on naturals such that $\lim_{n\to\infty} f(n)/g(n) = 0$. Then

$$\mathfrak{L}(\text{QR-PBLIND-R}(f)) \subsetneq \mathfrak{L}(\text{QR-PBLIND-R}(g))$$
$$\mathfrak{L}(\text{PBLIND-R}(f)) \subsetneq \mathfrak{L}(\text{PBLIND-R}(g))$$
$$\mathfrak{L}(\text{COUNTER-R}(f)) \subsetneq \mathfrak{L}(\text{COUNTER-R}(g))$$
$$\mathfrak{L}(\text{QR-COUNTER-R}(f)) \subsetneq \mathfrak{L}(\text{QR-COUNTER-R}(g)) .$$

We note that the hierarchy results obtained in Theorem 4 can be obviously formulated for deterministic versions of the introduced machines and for different versions of time restricted machines too.

Finishing Sect. 1 we shall formulate some open problems referring to the hierarchy of reversal bounded multicounter machines.

Open problem 1. Is the class $\mathfrak{L}$(QR-PBLIND) a proper subclass of $\mathfrak{L}$(QR-COUNTER)? A solutation of this problem can be interesting for the versions of the machines with k counters, for $k \geq 2$, too.

Open problem 2. Is the class of languages $\mathfrak{L}_{Rc}$(QR-DPBLIND) equal to the class $\mathfrak{L}_{Rc}$(QR-DCOUNTER) for arbitrary natural c ?

Open problem 3. What is the relation between $\mathfrak{L}$(QR-PBLIND-R(f)) and $\mathfrak{L}$(QR-COUNTER-R(f)), where f is an arbitrary function ?

We note that one of the referee´s of MFCS´84 gave the negative answer to the first question of Open problem 1.

2. ZEROTESTING BOUNDED MULTICOUNTER MACHINES

We shall study the hierarchy of zerotesting bounded multicounter machines and their relation to reversal bounded multicounter machines in this Section. First we obtain a hierarchy result for deterministic zerotesting bounded machines.

Theorem 5. Let f and g be increasing functions from naturals to positive reals such that $g(n) \geq f(n)$ for all natural n. Let $f(n) = o(g(n))$, and

and $g(n)=o(\log_2 n)$. Then $\mathcal{L}$(QR-DCOUNTER-Z(f)) $\subsetneq$ $\mathcal{L}$(QR-DCOUNTER-Z(g)).

Proof. Let us consider the language $L_g = \{u_1u_2\ldots u_p \mid p \leq g(|u_1u_2\ldots u_p|)$ and $u_i = a^{n_i}b^{n_i}$ for all $i=1,\ldots,p$, where $n_i \geq 0\}$. Clearly, this language can be accepted by QR-DCOUNTER machine B with one counter such that $L_{Zg}(B) = L_g$.

Now, we shall show by contradiction that the language L_g does not belong to $\mathcal{L}$(QR-DCOUNTER-Z(f)). Let h be a function from naturals to positive reals such that $h(n) \leq f(n)$ and let there exist a QR-COUNTER machine A such that $L_{Zh}(A) = L_g$. Let A have s states and let the length of each part of any computation, in which the reading head is stationary be bounded by a constant c. We shall consider the word

$$x = a^{n_1}b^{n_1}a^{n_2}b^{n_2}\ldots a^{n_{g(n)}}b^{n_{g(n)}},$$

where $n_1 \geq s+1$, $n_i \geq 2sc\sum_{j=1}^{i-1} n_j$ for $i=1,\ldots,g(n)$. Since $g(n)=o(\log_2 n)$ we can assume that such a word exists and belongs to L_g.

Since $f(n)=o(g(n))$ there exists such subword

$$y = a^{n_j}b^{n_j}\ldots a^{n_k}b^{n_k}$$

of the word x that k-j is greater than the number of all cycles of A with different characteristics and the machine empties no counter during the computation on y. It is easy to see that, for $i=j,j+1,\ldots,k$, the counter characteristics of all cycles used in the computation on a^{n_i} must be equal or greater than 0.

Since k-j is greater than the number of all cycles of A there exist such two numbers d, z : $j \leq d < z \leq k$ that A works on a^{n_d} in a cycle s_1 and on a^{n_z} in a cycle s_2, where s_1 and s_2 have the same characteristic. Let m be the reading head characteristic of these cycles. Clearly, we can assume $m \geq 0$ because no finite computation of a deterministic machine can involve a cycle with the reading head characteristic 0. It can be easy seen that if A accepts the word $x = v_1yv_2$ then A must accept the word $x' = v_1a^{n_j}b^{n_j}\ldots a^{n_d+m}b^{n_d}\ldots a^{n_z-m}b^{n_z}\ldots a^{n_k}b^{n_k}v_2$ what proves our assertion because x' does not belong to L_g.

Now, we shall show some relation between zerotesting bounded machines and reversal bounded machines.

Theorem 6. Let $f(n)=o(n)$ be an increasing function from naturals to positive reals. Then $\mathcal{L}$(QR-DCOUNTER-Z(1)) - $\mathcal{L}$(COUNTER-R(f)) $\neq \emptyset$.

<u>Proof.</u> Let us consider the language $L' = \{w$ in $\{a,b\}^* \mid \#a(w) = \#b(w)$ and for all x,y in $\{a,b\}^+$, $w = xy$ implies $\#a(x) \geq \#b(x)\}$. L' can be accepted by a QR-DCOUNTER machines with one counter. We shall not prove that L' is not in $\mathcal{L}$(COUNTER-R(f)) since the proof is very similar to the proof of Theorem 3.

<u>Corollary 2.</u> Let $f(n) = o(n)$ be an increasing function defined on natural numbers. Then

$$\mathcal{L}(\text{QR-COUNTER-R}(f)\text{-Z}(1)) \subsetneq \mathcal{L}(\text{QR-COUNTER-Z}(1))$$
$$\mathcal{L}(\text{COUNTER-R}(f)\text{-Z}(1)) \subsetneq \mathcal{L}(\text{COUNTER-Z}(1)) .$$

Finishing this paper we note that the hierarchy results formulated in Corollary 2 can be introduced for deterministic versions and different time restricted versions of machines considered too.

ACKNOWLEDGEMENTS

I would like to thank Pavol Ďuriš for many helpful and interesting discussions and Branislav Rovan for his comments concerning this work.

REFERENCES

1. Ginsburg, S.: Algebraic and Automata - Theoretic Properties of Formal Languages. North-Holland Publ. Comp., 1975
2. Greibach, S.A.: Remarks on Blind and Partially Blind One-Way Multicounter Machines. TCS 7, 311-324 (1978)
3. Greibach, S.A.: The Hardest Context-free Language. SIAM J. of Comp. 2, 304-310 (1973)
4. Greibach, S.A.: Remarks on the Complexity of Nondeterministic Counter Languages. TCS 1, 269-288 (1976)
5. Hack, M.: Petri Net Languages. Computation Structures Group Memo 124, Project MAC, MIT, 1975
6. Ibarra, O.H.: Reversal - Bounded Multicounter Machines and Their Decision Problems. JACM 25, 116-133 (1978)
7. Jantzen, M.: On Zerotesting - Bounded Multicounter Machines. Proc. of 4th GI Conference, pp. 158-169, Berlin,Heidelberg,New York:Springer, 1979.
8. Chan, T.: Reversal Complexity of Counter Machines. Proc. of Annual Symposium on Theory of Computing 1981, 146-157

ON THE POWER OF ALTERNATION IN FINITE AUTOMATA*

Juraj Hromkovič
Department of Theoretical Cybernetics
Comenius University
842 15 Bratislava
Czechoslovakia

ABSTRACT

We shall deal with the following three questions concerning the power of alternation in finite automata theory:

1. What is the simplest kind of device for which alternation adds computational power ?

2. What are the simplest devices (according to the language family accepted by them) such that the alternating version of these devices is as powerful as Turing machines ?

3. Can the number of alternations in the computations of alternating devices be bounded by a function of input word length without the loss of the computational power ?

We give a partial answer to the Questions 1 and 2 , i.e. we find the simplest known devices having the required properties according to alternation (multihead simple finite automata and one-way multicounter machines with blind counters respectively) . Besides this considering one-way multicounter machines whose counter contents is bounded by the input word length we find a new characterisation of P (the class of languages accepted by deterministic Turing machines in polynomial time). For one-way alternating multihead finite automata we show that the number of alternations in computations cannot be bounded by $n^{1/3}$ for input words of length n.

* The research was supported by SPZV I-5-7/7 grant.

NOTATION

1dfa(k) [1nfa(k)] - one-way deterministic (nondeterministic) finite automaton with k (read only) heads,

1sdfa(k) [1snfa(k)] - one-way simple deterministic (nondeterministic) k-head finite automaton which is the same as above except that k-1 heads are blind, i.e. they can see only the endmarkers ¢ and $,

1dsefa(k) [1nsefa(k)] - one-way deterministic (nondeterministic) sensing k-head automaton, the same device as 1dfa(k) except that it can recognize the coincidence of its heads,

1sdsefa(k) [1snsefa(k)] - one-way simple deterministic (nondeterministic) sensing k-head finite automaton,

1cm(k) - one-way nondeterministic machine with with k counters,

1bm(k) - one-way nondeterministic machine with k blind counters, the 1bm(k) is a 1cm(k) which knows nothing about the contents of its blind counters during the computation, but it accepts only in a final state with all its blind counters empty (the formal definition can be found in Greibach [2]),

2dfa(k), 2nfa(k), 2sdfa(k), etc. are two-way extensions of 1dfa(k) , 1nfa(k), 1sdfa(k) respectively,

1afa(k), 2afa(k), 1asfa(k), 2asfa(k), 1acm(k), 1abm(k), etc. are alternating 1nfa(k), 2nfa(k), 1snfa(k), 2snfa(k), 1cm(k), 1bm(k) respectively,

1DFA(k) - the family of languages recognized by 1dfa(k), similarly 2DFA(k), 1NFA(k) , 2NFA(k), etc. are the corresponding families of languages.

R - the family of regular languages,

CFL - the family of context-free languages,

P - the class of languages accepted by deterministic Turing machines in polynomial time,

TM - the family of languages accepted by deterministic Turing machines,

1A(f(n))FA(k) - the subclass of 1AFA(k), where every language is accepted by a 1afa(k) which uses in its accepting computations at most f(n) alternations (n is the input word length),

1AB(k)-S(f(n)) - the subclass of 1AB(k), where every language is accepted by a 1ab(k) which has the contents of each of its blind counters bounded by f(n) for the input word length n.

Let w be a word in Σ^*, for some alphabet Σ, and let a be in Σ. Then $\#a(w)$ denotes the occurrence number of symbol a in w.

INTRODUCTION

The alternation was introduced by Chandra, Kozen, and Stockmeyer [1] in order to obtain a theoretical model of parallel computations. It has proved to be useful in many areas of complexity theory giving some new characterisations of basic complexity classes.

The alternating multihead finite automata were defined in King [4,5] as a generalization of nondeterministic multihead finite automata (in the same way as the alternating Turing machines were introduced as a generalization of nondeterministic Turing machines). Studying alternating finite automata King [4,5] obtained several new results about multihead pushdown automata and several new characterizations of the basic complexity classes. Some of the main results established are $P = \bigcup_{k \in N} 2AFA(k)$, $CFL \subseteq 1AFA(3)$. Some equivalent problems to the PN - P problem are given in [4,5] too.

This paper consists of three Sections. In Section 1 we shall deal with the problem to find the simplest kind of finite automata for which alternation adds computational power. King showed that $R = 2AFA(1)$ and so the simplest device considered in [4] for which alternation helps are 1nfa(2) automata. We show that alternation increases the computational power of 1snfa(2) automata too (we note $1SNFA(2) \subsetneq 1NFA(2)$). Especially, we prove that $1ASFA(k) = 1ASSEFA(k) = 1AFA(k) = 1ASEFA(k)$, and $2ASFA(k) = 2ASSEFA(k) = 2AFA(k) = 2ASEFA(k)$ for all natural k.

In Section 2 we shall find the simplest known devices (according to language family accepted) such that the alternating version of these devices is as powerful as Turing machines. We shall consider 1bm(k) machines. The simplicity of these machines can be seen in what follows. It is known [2] that each 1bm(k) can be simulated by a 1cm(k) with constant bound of counter reversals. The simple languages $L_1 = \{wcw \mid w \text{ in } \{a,b\}^*\}$ in 1DFA(2) and $L_2 = \{w \text{ in } \{a,b\}^* \mid \text{for all } x,y \text{ in } \{a,b\}^+ \text{ such that } w = xy, \text{ it follows } \#a(x) > \#b(x)\}$ in 1SDSEFA(2) do not belong to 1BM(k) for any natural k. But we shall prove $1ABM(2) = TM$ through the equality $1ABM(2) = 1ACM(2)$. Besides this we show $1ABM(k)\text{-}S(n) \subseteq 2AFA(k+1)$ and $2AFA(k) \subseteq 1ABM(k)\text{-}S(n)$, what follows $P = \bigcup_{k \in N} 1ABM(k)\text{-}S(n)$.

In Section 3 we shall prove that there exists a language L_R which belongs to 1AFA(3) and 2DFA(k), and does not belong to $\bigcup_{k \in N} 1A(n^{1/3})FA(k)$ for any natural k.

1. ALTERNATING SIMPLE MULTIHEAD FINITE AUTOMATA

In spite of the fact that multihead finite automata are more powerful than simple multihead finite automata we shall now show that 1ASFA(k) = 1AFA(k) and 2ASFA(k) = 2AFA(k).

King [4,5] showed that 1ASEFA(k) = 1AFA(k) and 2ASEFA(k) = 2AFA(k). First we shall show this result for simple multihead finite automata.

Lemma 1. For all natural k: 1ASFA(k) = 1ASSEFA(k)
2ASFA(k) = 2ASSEFA(k).

Outline of the proof. Let A be a 1asfa(k) [2asfa(k)]. Then a 1assefa(k) [2assefa(k)] A′ can be simulated by A in the following way. In each step during the A-computation, A can nondeterministically decide whether a group of heads coincide, and in parallel computation verifies its nondeterministic decision. Let

be subtrees of A′-computation tree if the considered group of heads coincide and if it does not coincide resp. (s, p, q, r, v are states of A′ and P, Q, R, V are computation subtrees from states p, q, r, v resp.). A simulates this step of A′ using the following computation subtree:

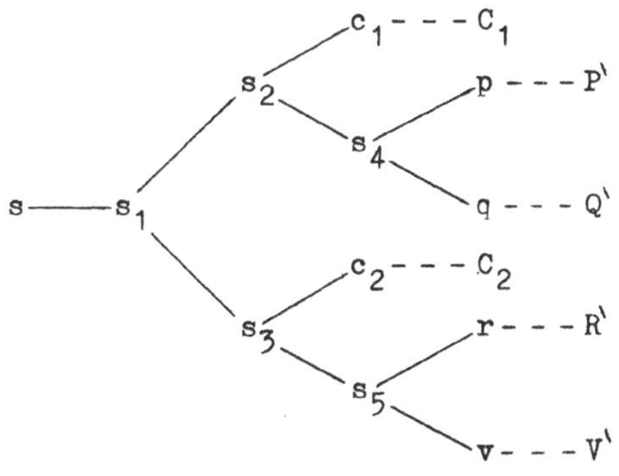

where s_1 is an existential state, s_2 and s_3 are universal states, and s_4, s_5 are existential (universal) states if s is existential (universal) state of A'. In the computation C_1 (C_2) A moves the whole considered group of heads to the right endmarker in order to verify which of its two nondeterministic decisions holds.

Clearly, C_1 (C_2) ends in an accepting (unaccepting) state iff all heads of the considered group reach the right endmarker in the same moment. Obviously, P', Q', R', V' are simulations of A of the computational subtrees P, Q, R, V of A'.

Theorem 1. For all natural k: 1AFA(k) = 1ASFA(k)
2AFA(k) = 2ASFA(k).

Outline of the proof. Let A be a 1afa(k) [2afa(k)] with k reading heads $H_1,\dots,H_k$. Since A can recognize coincidence of its heads we can assume that A computes in such way that in each configuration if H_i is on the j_i-th position of the input word (for $i = 1,\dots,k$) then $j_m \geq j_k$ for $m < k$. A 1asfa(k) [2asfa(k)] automaton A' with one reading head H'_k and k-1 blind heads $H'_1,\dots,H'_{k-1}$ can simulate A as follows. The head H'_i simulates the movement of H_i, i.e. the heads H_i in the computation of A and H'_i in the simulating computation of A' have the same position. A' can nondeterministically decide which symbol is occured on the H'_i-position of the input word for all $i = 1,\dots,k-1$. Since the reading head H'_k follows the blind heads $H'_1,\dots,H'_{k-1}$ A' using one-way parallel computation can verify its nondeterministic decision. Clearly, a similar tree as in the proof of Lemma 1 can be outlined to show the simulation of one step of the automaton A.

Combining this result with the known results mentioned in Introduction we obtain the following.

Corollary 1. $P = \bigcup_{k\in N} 2ASFA(k) = \bigcup_{k\in N} 2AFA(k)$.

Corollary 2. For all natural k: $1NFA(k) \subsetneq 1ASFA(k)$.

Corollary 3. $CFL \subseteq 1ASFA(3)$.

2. ONE-WAY BLIND MULTICOUNTER MACHINES

Now, we show that 1bm(2) machines adds using alternation so many computational power that they are as powerful as Turing machines.

Theorem 2. For all natural k: 1ABM(k) = 1ACM(k).

Outline of the proof. A 1abm(k) machine M' can simulate a 1acm(k) machine M nondeterministically deciding whether its blind counters are empty, and verifying in parallel its nondeterministic decision.

Corollary 4. $TM = 1ABM(2)$.

Proof. $TM = 1NCM(2) = 1ACM(2)$.

Theorem 3. For all natural k: $2AFA(k) \subseteq 1ABM(k)-S(n)$.

Outline of the proof. Let A be a 2afa(k) automaton with k two-way reading heads $H_1, \ldots, H_k$. Then A can be simulated by a 1abm(k) machine M with one-way reading head H and counters $C_1, \ldots, C_k$ whose contents is bounded by the input word length. The situation in which H_i is on the j_i-th position of the input word, for $i = 1,\ldots,k$, can be simulated by M's configuration in which H is on the left endmarker and the contents of C_i is equal to j_i. For each H_i, M can nondeterministicaly decide which symbol is read by H_i, and using its one-way reading head and the counter C_i (we can assume M recognizes whether or not C_i is empty) can in a parallel computation verify its nondeterministic decision.

Theorem 4. For all natural k: $1ABM(k)-S(n) \subseteq 2AFA(k+1)$.

Proof. Let M be a 1abm(k) machine with one-way reading head H and k blind counters $C_1, \ldots, C_k$ with their contents bounded by the input length. A 2afa(k+1) A can simulate M using one head to simulate H and k other heads $H_1,\ldots,H_k$ to simulate the counters $C_1,\ldots,C_k$ (the position of H_i is equal to the contents of C_i).

The following results we obtain as simple consequences of Theorems 3 and 4, and of the fact $P = \bigcup_{k \in N} 2AFA(k)$ in [4].

Corollary 5. $P = \bigcup_{k \in N} 1ABM(k)-S(n) = \bigcup_{k \in N} 1ACM(k)-S(n) = \bigcup_{k \in N} 2AFA(k)$.

Corollary 6. For all natural k: $1ABM(k)-S(n) \subsetneq 1ABM(k+2)-S(n)$.

Proof. In [4] it was shown $2AFA(k) \subsetneq 2AFA(k+1)$.

3. ONE-WAY MULTIHEAD AUTOMATA AND ALTERNATION NUMBER BOUND

In this Section we shall deal with the problem whether the number of alternations in the computations of alternating devices can be bounded by a constant or by a function of input word length without the loss of the computational power (similar problem was studied according to the number of nondeterministic decisions in nondeterministic computations). For one-way alternating multihead automata, we shall show the alternation number cannot be bounded by $n^{1/3}$.

We shall consider the following languages introduced by Yao and Rivest [6].

$L_f = \{w_1cw_2c\ldots cw_fdw_fc\ldots cw_2cw_1 \mid w_i \text{ in } \{a, b\}^* \text{ for } i = 1,2,\ldots,f\}$ for all natural f, where $L_f \subseteq \{a,b,c,d\}^*$. Let $L_R = \bigcup_{f\in N} L_f$. It is no problem to show that L_R belongs to 2DFA(2) and 1AFA(3).

<u>Theorem 5.</u> The language L_R does not belong to $1A(n^{1/3})FA(k)$ for any natural k.

<u>Outline of the proof.</u> We prove this result by contradiction. Let there exist an 1afa(k) automaton A accepting L_R whose accepting computation tree contains at most $n^{1/3}$ universal states for each input word w (of the length n) in L_R.

We shall consider the language

$L(m,hk^2m^r) = \{w_1cw_2c\ldots cw_fdw_fc\ldots cw_2cw_1 \mid |w_i| = m,\ f = hk^2\lfloor m^r\rfloor\} \subseteq L_R$,

where h is the maximal possible branching in A computations from a universal state, and $0 < r < 1/2$. We shall show that if A accepts all words in $L(m,hk^2m^r)$, for sufficiently large m and suitable r, then A has to accept a word x not in L_R.

The length of words in $L(m,hk^2m^r)$ is $n = 2hk^2\lfloor m^r\rfloor(m+1)-1$. For sufficiently large m and suitable, positive real number $r < 1/2$ the inequality $n^{1/3} < m^r$ hols. So, it is sufficient to show that m^r alternations do not suffice to recognize L_R.

We can assign to each A - accepting computation tree on an input word y in $L(m,hk^2m^r)$ a pattern of computation on y which is an arrangement (not tree) of special configurations. This can be done in a similar way as in [3]. It can be shown that there exist two different words

$y_1 = w_1cw_2c\ldots cw_{j-1}cvcw_{j+1}c\ldots cw_fdw_fc\ldots cw_{j+1}cvcw_{j-1}c\ldots cw_2cw_1$

$y_2 = w_1cw_2c\ldots cw_{j-1}cucw_{j+1}c\ldots cw_fdw_fc\ldots cw_{j+1}cucw_{j-1}c\ldots cw_2cw_1$

in $L(m,hk^2m^r)$ with the same pattern of computations, in which no pairs of reading heads compare the two words v in y_1 (u in y_2).

Then, it can be constructed an accepting computation tree of A (in a way which is a little more complicated as in [3,6]) on the word

$$x = w_1cw_2c\ldots cw_{j-1}cvcw_{j+1}c\ldots cw_fdw_fc\ldots cw_{j+1}cucw_{j-1}c\ldots cw_2cw_1 ,$$

which does not belong to L_R.

Corollary 7. For each natural $k \geq 3$:

$$1NFA(k) \subsetneq \bigcup_{c \in N} 1A(c)FA(k) \subseteq 1A(n^{1/3})FA(k) \subsetneq 1AFA(k).$$

Corollary 8. $\bigcup_{k \in N} 1A(n^{1/3})FA(k) \subsetneq \bigcup_{k \in N} 1AFA(k)$.

ACKNOWLEDGEMENTS

I am grateful to Pavol Ďuriš who pointed out the questions considered in this paper to me, and Branislav Rovan for their comments concerning this work.

REFERENCES

1. Chandra,A.K., Kozen,D.C., and Stockmeyer,L.J.: Alternation. Journal of ACM 28, 1981, 114-133.

2. Greibach,S.A.: Remarks on Blind and Partially Blind One-Way Multicounter Machines. TCS 7, 1978, 311-324.

3. Hromkovič, J.: One-Way Multihead Deterministic Finite Automata. Acta Informatica 19, 1983, 377-384.

4. King, K.N.: Alternating Multihead Finite Automata. Proc. 8th Inter. Coll. on Automata, Languages, and Programming, Lecture Notes in Comp. Science 115, Berlin,Heidelberg,New York: Springer 1981.

5. King,K.N.: Alternating Finite Automata. Doctoral disser., University of California, Berkeley.

6. Yao,A.C., and Rivest,R.L.: K+1 Heads are Better than K. Journal of ACM 25, 1978, 337-340.

THE EQUIVALENCE PROBLEM AND CORRECTNESS FORMULAS FOR A SIMPLE CLASS OF PROGRAMS

(Extended Abstract)

Oscar H. Ibarra[1] and Louis E. Rosier[2]

ABSTRACT

This paper is concerned with the semantics (or computational power) of very simple loop programs over different sets of primitive instructions. Recently, a complete and consistent Hoare axiomatics for the class of {x←0, x←y, x←x+1, x←x∸1, *do* x...*end*} programs which contain no nested loops, was given, where the allowable assertions were those formulas in the logic of Presburger arithmetic. The class of functions computable by such programs is exactly the class of Presburger functions. Thus, the resulting class of correctness formulas has a decidable validity problem. In this paper, we present simple loop programming languages which are, computationally, strictly more powerful, i.e. which can compute more than the class of Presburger functions. Furthermore, using a logical assertion language that is also more powerful than the logic of Presburger arithmetic, we present a class of correctness formulas over such programs that also has a decidable validity problem. In related work, we examine the expressive power of loop programs over different sets of primitive instructions. In particular, we show that an {x←0, x←y, x←x+1, *do* x ... *end*, *if* x=0 *then* y←z}-program which contains no nested loops can be transformed into an equivalent {x←0, x←y, x←x+1, *do* x ... *end*}-program (also without nested loops) in exponential time and space. This translation was earlier claimed, in the literature, to be doable in polynomial time, but then this was subsequently shown to imply that PSPACE=PTIME. Consequently, the question of translatability was left unanswered. Also, we show that the class of functions computable by {x←0, x←y, x←x+1, x←x∸1, *do* x ... *end*, *if* x=0 *then* x←c}-programs is exactly the class of Presburger functions. When the conditional instruction is changed to "*if* x=0 *then* x←y+1", then the class of computable functions is significantly enlarged, enough so, in fact, as to render many decision problems (e.g. equivalence) undecidable.

1. INTRODUCTION

In the previous ten or so years there has been a tremendous interest in the topic of program correctness—both from a theoretical and practical point of view[3,10,17,23]. Work in the theory of program schemata has dealt with the concept of identifying classes of programs for which various assertions about programs are mechanically verifiable. What actually needs to be verified is a correctness formula. A correctness formula is a logical formula of the form {p}S{q}, where S is a program and p and q are logical assertions about the variables in the program S. The interpretation is that, if p is true before the execution of S, then q will be true following the execution of S, assuming S terminates. Now once results are established for a particular class of programs (a program schemata) the procedures developed can be applied to any instance in the class. Much of the literature has concentrated on classes of simple programming languages but with using general assertions from a particular logic. Unfortunately, this work has produced mostly negative results. That is, for many classes of programs (and simple assertions) these questions are computationally unsolvable.

A positive result in this area is the result of Cherniavsky and Kamin[3]. They present a simple programming language and an assertion language for which they were able to provide a complete and consistent axiom system. An axiom system includes the programming language, the assertion language and a set of axioms or proof rules from which one can derive (or prove) certain logical assertions about the programs. An axiom system is said to be consistent and complete if the true correctness formulas coincide exactly with the provable ones. (We ignore here any discussion of a model for the assertion language as we expect it to be implicitly defined within each system.)

The programming language of Cherniavsky and Kamin [3] is the loop language L_1(x←0, x←y, x←x+1, x←x∸1). A program P in this language has the form:

---

[1]Department of Computer Science, University of Minnesota, Minneapolis, MN 55455. This research was supported in part by NSF Grant MCS 83-04756.

[2]Department of Computer Sciences, The University of Texas at Austin, Austin, Texas 78712. This research was supported in part by The University Research Institute, The University of Texas at Austin and the IBM Corporation.

P : input($x_1,...,x_k$)
A
output($y_1,...,y_t$)

where A is a block of instructions from the set {x←0, x←y, x←x+1, x←x∸1, *do* x...*end*}, k and t are constants and no nesting of loop structures is permitted in A. In general, a L_i(BB)-program will be defined similarly. In this case, however, the instructions in the block A must be taken from those in BB ∪ {*do* x...*end*} and the maximum level of nesting allowed for loop structures is i. Note then that L_i(x←0, x←y, x←x+1) are the loop languages in the subrecursive hierarchy of Meyer and Richie[21]. In particular, L_1(x←0, x←y, x←x+1) is the language shown to compute exactly the "simple" functions[24]. The assertion language used in the system of Cherniavsky and Kamin[3] is composed of those formulas in the logic of Presburger arithmetic. The computational power of L_1(x←0, x←y, x←x+1, x←x∸1) is quite limited. In fact programs over this language were shown to be capable of computing exactly those functions which are Presburger[2,3,7]. Actually the decidability of correctness formulas is reduced to the problem of deciding the validity of a formula in Presburger arithmetic. The equivalence problem for L_1(x←0, x←y, x←x+1, x←x∸1)-programs was also reduced to the same problem[2,3,7].

In any class of programming languages where the equivalence problem is not decidable, finding interesting correctness formulas whose validity is decidable is clearly not possible. In fact, for almost any class of programs where the equivalence problem is undecidable, the validity of correctness formulas of the form, {true}S{x=y}, cannot be mechanically verified. The equivalence problem for a class of programs is, given two programs in the class, to decide if these programs produce the same outputs when they are given identical inputs. Much is known about classes of simple programming languages and the corresponding classes of functions which they compute, as well as the difficulty of the respective equivalence problems. As mentioned earlier, Meyer and Richie[21] exhibited the hierarchy of simple programming languages L_i(x←0, x←y, x←x+1) whose union is a class of programs, which is capable of computing exactly the class of primitive recursion functions. The programming language classes, L_1(x←0, x←y, x←x+1) and L_2(x←0, x←y, x←x+1), constitute the lower two levels of this hierarchy. Programs in these classes compute the "simple" functions of Tsichritzis[24] and the elementary recursive functions, respectively. The programming language used by Cherniavsky and Kamin[2,3] is a slight generalization of the L_1(x←0, x←y, x←x+1) language. The class of L_1(x←0, x←y, x←x+1) (L_2(x←0, x←y, x←x+1)) programs has a decidable (undecidable) equivalence problem. Hence possible languages of interest, in terms of computational power, would include those that reside somewhere between L_1(x←0, x←y, x←x+1, x←x∸1) and L_2(x←0, x←y, x←x+1). Many examples of programming languages are known whose computational power is equivalent to that of L_1(x←0, x←y, x←x+1, x←x∸1)[2,3,7,12]. Until recently, however, few examples of programming languages, whose computational power lies properly within this range, appeared in the literature. Recent work by the authors has contributed in this area[16]. A possible approach for further research in this area then would be to examine various simple languages over different instruction sets. Programming languages of interest would have computational power greater than the language of Cherniavsky and Kamin, and yet still have interesting classes of correctness formulas that are mechanically verifiable. Additional work would be required to find suitable assertions. Another approach would be to use a simple assertion language and allow nontrivial, but limited, predicates. In [23], the predicate PERM(M,N), (indicating array M is a permutation of array N), was added to the assertion language of a simple system, and the resulting system still had interesting decidable correctness formulas.

In this paper we introduce simple loop programming languages S and T which are computationally more powerful than L_1(x←0, x←y, x←x+1, x←x∸1). Subsequently, we show that the classes of S and T programs have a decidable equivalence problem. We also show a decidable class of correctness formulas for such programs.

We begin by looking at the language L_1(x←0, x←y, x←x+1, x←x∸1) and see what constructs we can add and still have a decidable equivalence problem. L_1(x←0, x←y, x←x+1, x←x∸1) computes exactly the Presburger functions[2] and those are precisely the functions which are computable by straight-line programs over the instruction set:

(1) x←1
(2) x←x+y
(3) x←x∸y
(4) x←x/k, where / denotes integer division with truncation and k is a positive integer constant.
(5) x←x *mod* k, where for positive integers x and y, x *mod* y is defined to be x-y⌊x/y⌋ if y ≠ 0 and x otherwise[19]. Again k is a positive integer constant.
(6) *if* x=0 *then* I

where I denotes any instruction of type 1-5[12]. In fact, an L_1(x←0, x←y, x←x+1, x←x∸1) program can be converted into an equivalent straight-line program over these instructions in polynomial time[7]. Thus addition and proper subtraction are allowed but multiplication, division, and the modulo operation are only allowed by positive integer constants (i.e., $x \leftarrow k^*x$, $x \leftarrow x/k$, $x \leftarrow x$ *mod* k). Now what we want is to extend the language $L_1(x \leftarrow 0, x \leftarrow y, x \leftarrow x+1, x \leftarrow x \dot{-} 1)$ so that the resulting language becomes equivalent to a computationally more powerful class of straight-line programs, but which still has a decidable equivalence problem. Clearly, a more powerful class of straight-line programs would result if we allowed any one of the following constructs: $x \leftarrow x/y$, $x \leftarrow x^*y$, $x \leftarrow x$ *mod* y. Unfortunately, it is known (see [13]) that programs over $\{x \leftarrow 1, x \leftarrow x+y, x \leftarrow x/y\}$ or over $\{x \leftarrow 1, x \leftarrow x+y, x \leftarrow x \dot{-} y, x \leftarrow x^*y\}$ have an undecidable zero-equivalence problem. (The zero-equivalence problem for a class of programs is deciding for a member of that class whether the program outputs a zero for all possible inputs.) The only other case worth considering then is the addition of the construct $x \leftarrow x$ *mod* y, and for this we can show that equivalence is decidable.

Throughout, U will denote the class of programs over the following instruction set:

(1) $x \leftarrow 1$	(4) $x \leftarrow x/k$
(2) $x \leftarrow x+y$	(5) $x \leftarrow x$ *mod* y
(3) $x \leftarrow x \dot{-} y$	(6) *if* x=0 *then* I

where I denotes any instruction of type 1-5.

We now define the classes of loop programs S and T which are (polynomially) equivalent to U. For ease of explanation we describe the programming languages S and T over the following seven instruction types:

(1) $x \leftarrow 0$
(2) $x \leftarrow x+1$
(3) $x \leftarrow x \dot{-} y$
(4) $x \leftarrow y$
(5) *for* ι=u *to* v *by* c *do*
A
end
(6) $x \leftarrow y$ *mod* z
(7) *if* x|y *then* $z \leftarrow 0$, where '|' means 'divides'

where A is a block of primitive instructions (1-4, 6-7), ι is called the loop control variable and c is a constant. The interpretation is that ι is assigned the value of u on the 1st pass, u+c on the 2nd pass,..., and v-(v-u) *mod* c on the last pass, where changes to variables u, v and ι inside the loop do not affect the number of loop executions or the assignment made to variable ι preceding each pass of the loop. (If v<u, then the loop is not executed.)

The language S allows only the primitive instructions 1-4 and 6, and the language T allows only primitive constructs 1-4 and 7. Restrictions are also placed on loop structures which are allowable in S and T programs. The restrictions are syntactic restrictions placed on the blocks of instructions A that are allowed inside *for* loops. First, however, we need the following definitions.

Consider a block of instructions $A=I_1;...;I_l$, where each I_j $(1 \leq j \leq l)$ is an instruction of type 1-4 or 6-7. Then the functions p_A and b_A are defined to be for $1 \leq j \leq l$,

$$p_A(j,x) = \begin{cases} y & \text{if } I_j \text{ is "}x \leftarrow y\text{" or "}x \leftarrow x \textit{ mod } y\text{"} \\ x & \text{otherwise} \end{cases} \qquad \underline{\text{and}} \quad b_A(j,x) = p_A(1,...p_A(j\text{-}1,p(j,x))...),\ 1 \leq j \leq l.$$

Thus, $p_A(j,x)$ is the name of the variable before the execution of I_j, which is x after the execution of I_j (i.e. the value of x, after the execution of I_j, is derived from the value of $p_A(j,x)$ before the execution of I_j), and $b_A(j,x)$ is the name of the variable, before I_1 is executed, from which the value of x is derived, after the execution of $I_1;...;I_j$.

Note that the functions p_A and b_A are defined with respect to a block of instructions, A. Consider a loop structure of the form "*for* ι=u *to* v *by* c *do* A; *end;*". Then this is an allowable loop structure for S (or T)-programs if A contains only the instructions allowable in S (or T)-programs and restrictions 1 and 2 hold for A.

Restriction 1. If I_j is the instruction "*if* x|y *then* $z \leftarrow 0$" or "$x \leftarrow y$ *mod* z", then $b_A(j,y)=\iota$.

Restriction 2. If I_j is the instruction "*if* x|y *then* $z \leftarrow 0$" or "$z \leftarrow y$ mod x" then x may not be altered (appear on the left hand side of an assignment statement) within A.

Note that for any block of instructions of types 1-4 and 6-7 restrictions 1 and 2 can be syntactically checked.

A S (T)-program is a program over instruction types 1-6 (1-5,7) that allows no nesting of loop structures and where restrictions 1 and 2 hold for each block of instructions, A, which is enclosed within a loop structure.

Clearly S (T)-programs are more powerful than L_1(x←0, x←y, x←x+1, x←x∸1)-programs since an S (T)-program can compute x *mod* y. We shall show that the class of S (T)-programs has a decidable equivalence problem. In each case, we illustrate a polynomial time procedure that converts an S (T)-program into an equivalent U-program.

One might question whether the restrictions are necessary for S and T-programs. While we are not able to provide proofs in either case we provide evidence that indicates, probably so. For example, if restrictions 1 and 2 were not imposed on S-programs then such programs would be capable of computing the function *gcd*(x,y). First, our proof that the equivalence problem is decidable seems to fail if such functions as *gcd*(x,y) are allowed. Secondly, suppose that the introduction of the *gcd* function (i.e. the instruction z←*gcd*(x,y)) adds no computational power to the class of U-programs. Then the *gcd* function would not be harder to compute than multiplication. This would answer an open question in [1] in a very surprising way. In the case of T-programs we show that the removal of restriction 1 or 2 implies that the resulting class of programs has a PSPACE-complete 0-evaluation problem. (The 0-evaluation problem for a class of programs C is given a C-program P with one output variable, does P, when all input variables are initially zero, output a zero?) See [14]. Clearly the class of T-programs has a polynomial time 0-evaluation problem by virtue of the polynomial time translation of a T-program to a U-program.

Another point of comparison for S and T-programs is the class of DL-programs introduced in [6]. In [6] it was shown that the class of functions computable by DL-programs properly include the class of Presburger functions and that the class of DL-programs has a decidable equivalence problem. The essential difference between DL-programs and classes of programs that compute Presburger functions seems to be solely that a DL-program can perform an unbounded number of I/O operations. (In fact it was shown in [6], that any DL-program with a bounded number of inputs computed a Presburger function.) Thus it is clear that the class of functions computable by DL-programs is not comparable with the class of functions computable by S, T or U-programs, although both properly include the class of Presburger functions.

In Section 3, we show that the class of U-programs has a decidable equivalence problem. We then generalize this by looking at a class of unquantified correctness formulas. We show that this class of correctness formulas is decidable. Lastly we mention how this work can be used to extend the class of decidable correctness formulas in [17,23].

In the remaining section, we investigate claims made (without proof) in [18], concerning L_1(BB)-programs where BB ⊆ {x←0, x←y, x←x+1, x←x∸1, *if* x=0 *then* A *else* B}, and where A and B are (finite blocks) of the other primitive instructions in the set BB. We paraphrase the following definitions from [18]. Let L and L' be classes of programs, and $\mathcal{C}$ and $\mathcal{C}'$ the corresponding classes of functions they compute. L is effectively translatable into L' if for every program P in L there is a constructive way to obtain a program P' in L' such that P and P' compute the same function. If there is such a translation we write L -*-> L', where the "*" may be replaced by "⊂", "l", "p" or "e", according to whether the translation is the trivial inclusion map or produces a program P' in L' which is of length at most "linear", "polynomial", or "exponential", in the length of P. Also for our results as well for the claims made in [18], whenever the translation procedure given is "l", "p", or "e", it is also the case that it will take at most "linear", "polynomial", or "exponential" time, respectively (as a function of the size of the source program P).

Let "if" denote the instruction "*if* x=0 *then* A *else* B". Let BB1={x←0, x←y, x←x+1}. Let BB2 denote the set BB1 ∪ {x←x∸1}. The following theorem was claimed without proof in [18]. To make the notation less cumbersome the set brackets have been dropped in expressing the sets BB, of primitive instructions.

Theorem. Let γ_1 and γ_2 be subsets of {x←y, x←x∸1, if}. All possible translations from L_1(x←0, x←x+1, γ_1) to L_1(x←0, x←x+1, γ_2) can be read off the following diagram:

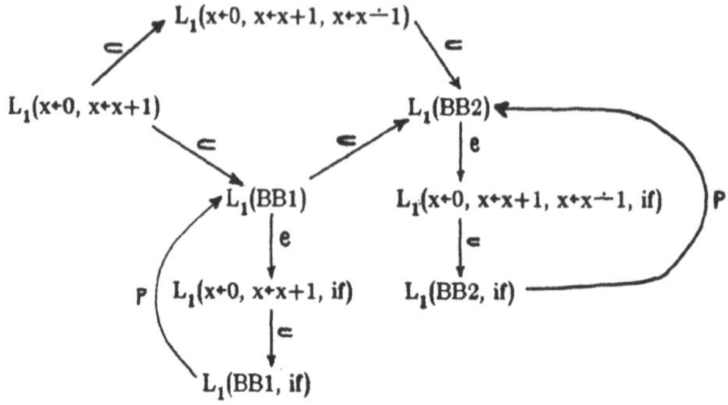

If an omitted arrow in the above diagram cannot be obtained by composition from the arrows already drawn, then it is the case of non-translatability, which also requires some proof. □

This theorem probably contains an error concerning the translations: L_1(BB1) --e--> L_1(x←0, x←x+1, if) --⊂--> L_1(BB1, if) --p--> L_1(BB1); and definitely contained an error concerning: L_1(BB2) --e--> L_1(x←0, x←x+1, x←x∸1, if) --⊂--> L_1(BB2, if) --p--> L_1(BB2). The remaining claims of the theorem are correct. From results in [14], it can be noted that L_1(BB1, if) --p--> L_1(BB1) implies PSPACE=PTIME. In [16], it was noted that L_1(x←0, x←x+1, x←x∸1, if)-programs were computationally more powerful than L_1(BB2)-programs. This is an important jump in terms of computational power for two reasons. First the class of functions computable by such programs is no longer Presburger. Secondly, the jump is so great that most decision problems for such programs are now undecidable, e.g. the equivalence problem is undecidable and hence there no longer exists a decision procedure to decide the validity of even very simple correctness formulas. For example, the validity of correctness formulas of the form {true}S{x=y} is no longer decidable. Left unanswered then is the question of translatability between L_1(BB1, if) and L_1(BB1).

In Section 4, we consider this problem as well as examine the computational gap between L_1(BB2)-programs and L_1(BB2, if)-programs. We concentrate on allowing the instructions "y←c" and "y←y∸1" to be conditionally executed. That is, we introduce the constructs "*if* x=0 *then* y←y+1", "*if* x=0 *then* y←y∸1" and "*if* x=0 *then* y←c", where c is any nonnegative integer constant. We are then able to show that: L_1(BB1, if) --e--> L_1(BB1), which is perhaps not surprising but nevertheless had not been confirmed. This should be contrasted with the corresponding situation for BB2, where the inclusion of the "if" construct provides an increase in computational power. We also show that: L_1(BB2, *if* x=0 *then* y←c) --e--> L_1(BB2). This should be contrasted with the result in [16], showing that L_1(BB2, *if* x=0 *then* y←y+1) is strictly more powerful than L_1(BB2). L_1(BB2, *if* x=0 *then* y←y+1)-programs were shown to be computationally equivalent to L_1(x←0, x←x+1, x←x∸1, if)-programs, in [16]. If both constructs, "*if* x=0 *then* y←c" and "*if* x=0 *then* y←y∸1" are concurrently considered it is easy to show that L_1(BB2, *if* x=0 *then* y←1, *if* x=0 *then* y←y∸1) is computationally equivalent to L_1(BB2, *if* x=0 *then* y←y+1). Unfortunately, we are unable to resolve the computational power of L_1(BB2, *if* x=0 *then* y←y∸1)-programs. The "*if* x=0 *then* y←y∸1" construct seems similar to the "*if* x=0 *then* y←y+1" construct, but as pointed out in [14], functions of one variable computed over BB2 ∪ {*if* x=0 *then* y←y∸1} are monotonic. Thus the proof techniques used in [16] do not seem to work with this language. This same problem was apparent in [14], where the authors were able to show that the 0-evaluation problem for this language is PSPACE-complete. Unfortunately, these techniques do not seem to work either. Lastly we note that the addition of the construct "*if* x=0 *then* y←z" to the set BB2 poses a difficult question. (It is easy to show that L_1(BB2, *if* x=0 *then* y←z)-programs are computationally equivalent to L_1(BB2, if)-programs.) If L_1(BB2, *if* x=0 *then* y←z)-programs are computationally more powerful than L_1(BB2, *if* x=0 *then* y←y+1)-programs, then it would imply that 0(n) space bounded Turing machines are more powerful than Turing machines operating simultaneously in 0(n) space and $0(2^{\lambda n})$ time, $\lambda<1$. This problem seems very difficult. The answer is not known even for the case when the time restriction is reduced to a polynomial. (See [16].) Other corrections to errors in [18] can be found in [15]. Although the proofs of the theorems are not included in this extended abstract, they can also be found in [15].

2. THE RELATIONSHIP BETWEEN S, T AND U-PROGRAMS

In this section, we show that there is a polynomial time procedure which translates an S or T-program into an equivalent U-program. Thus when in the next section we show that the class of U-programs has a decidable equivalence problem, the same can be said for the class of S and T-programs, respectively. The time required for the decision procedure is $O(2^{p(n)})$ for U-programs, and thus also for S and T-programs, where p is a polynomial. The last part of this section is concerned with showing that if we relaxed the class of T-programs by removing either restriction 1 or 2, then the resulting class of programs would have a PSPACE-complete 0-evaluation problem.

Theorem 1. Given an S-program P, we can construct a U-program P', in time polynomial in the length of P, such that P' is equivalent to P.

Theorem 2. Given a T-program P, we can construct a U-program P', in time polynomial in the length of P, such that P' is equivalent to P.

As indicated in the introduction, the restrictions are probably necessary for S (T)-programs. In [14] sufficient conditions were given, with respect to loop programs that do not allow nesting of loop structures that ensure that the 0-evaluation problem is PSPACE-hard. These results can be used to show that the 0-evaluation problem for T-programs without restrictions 1 or 2 is PSPACE-complete. It should be clear, however, that Theorems 1 and 2 imply that the 0-evaluation problem for S and T-programs can be done in polynomial time. Note that this says nothing about the decidability of the equivalence problem for classes of such programs.

It is interesting to note that if T-programs allowed the instruction "*if* x|y *then* I", where I is an instruction of the form $u \leftarrow u+1$, $u \leftarrow u \dot{-} 1$ or $u \leftarrow v$, then the equivalence problem becomes undecidable (even with both restrictions). This follows from the fact that such programs can compute integer division. For example, the program

$w \leftarrow x$
for $\iota = 1$ *to* x *by* 1 *do;*
  if $y|\iota$ *then* $w \leftarrow w \dot{-} 1$
end
$w \leftarrow x \dot{-} w$

computes x/y. The undecidability of the equivalence problem then follows from the undecidability of Hilbert's tenth problem[20]. The equivalence problem also becomes undecidable for either S or T-programs if the increment (c) is not constrained to be a constant.

3. THE EQUIVALENCE PROBLEM AND CORRECTNESS FORMULAS FOR U-PROGRAMS

In this section we look at the decidability problem for the class of U-programs as well as the decidability of a restricted class of correctness formulas. First we define a class of unquantified logical formulas, $\mathcal{F}$. The formulas are unquantified because we do not permit quantifiers, either $\forall$ or $\exists$ in any formula. Such a formula is valid, however, if it is true for all possible assignments. This then amounts to considering every variable to be universally quantified.

Definition. The class of logical formulas $\mathcal{F}$ is composed of unquantified logical formulas of the form: $F(x_1,...,x_n)$ where, $F(x_1,...,x_n)$ is any logical expression built up from integer constants and the variables $x_1,...,x_n$ such that: (1) the only arithmetic operators are + and -, (2) the relational operators are $<$, $=$, $\neq$, $\leq$ and | (divides), and (3) the logical operators are $\wedge$, $\vee$ and $\neg$.

The following two lemmas show the relationship of the formulas of $\mathcal{F}$ to the class of U-programs (and therefore S and T programs).

Lemma 1. Let $F(x_1,...,x_n)$ be a formula in $\mathcal{F}$. Then there exists a U-program P such that P is equivalent to the zero program on n-inputs (i.e. the U-program with n input variables that outputs 0 for all possible inputs) if and only if $\forall x_1...\forall x_n\ F(x_1,...,x_n)$ is true (i.e., F is valid). Furthermore, P can be found in polynomial time and the length of P is linear in the length of F.

Lemma 2. Let P_1 and P_2 be U-programs. Then there exists a formula F in $\mathcal{F}$ such that F is valid if and only if P_1 and P_2 are equivalent. Furthermore, F can be found in polynomial time and the length of F is linear in the lengths of P_1 and P_2.

Lemmas 1 and 2 have shown that the decidability of the equivalence problem for U-programs is exactly that of deciding the validity of $\mathcal{F}$ formulas. In [9] an algorithm to decide the truth of logical formulas of the form $\exists x_1 \ldots \exists x_n\ F(x_1,\ldots,x_n)$ (over the nonnegative integers), where F is in $\mathcal{F}$, was given. Since any formula of the form $\forall x_1 \ldots \forall x_n\ F(x_1,\ldots,x_n)$ is true if and only if $\exists x_1 \ldots \exists x_n \neg F(x_1,\ldots,x_n)$ is true, this also yields a decision procedure for the validity of formulas in $\mathcal{F}$. The procedure given in [9] runs in polynomial space. This seems to be the best we can do at this time. It should be noted, however, that the inequivalence problem is NP-hard and hence an exponential time algorithm is the best we can hope for. The NP-hardness follows from results in [4].

Next we go a step further and explore the question of decidability for simple correctness formulas of the form $\{p\}S\{q\}$, where p and q are formulas in $\mathcal{F}$ and S is a U-program. Using the strongest post condition calculus of [10], one can derive a formula $F(x_1,\ldots,x_n)$ such that $\forall x_1 \ldots \forall x_n\ F(x_1,\ldots,x_n)$ is true if and only if $\{p\}S\{q\}$ is true. Unfortunately the length of F in general is exponential in the length of $\{p\}S\{q\}$. This results since the strongest post condition (SPC) of "*if* x=0 *then* y←z" and the formula p(x,y,z), for example would be: $(x=0 \wedge SPC(y \leftarrow z, p(0,y,z))) \vee (x>0 \wedge p(x,y,z))$. Hence we provide an alternate proof.

Theorem 3. Let $\{p\}S\{q\}$ be a correctness formula where p and q are in $\mathcal{F}$ and S is a U-program. There exist a formula F in $\mathcal{F}$ such that F is valid if and only if $\{p\}S\{q\}$ is valid. Furthermore, F can be found in polynomial time and the length of F is linear in the length of $\{p\}S\{q\}$.

Let $\mathcal{P}$ be a class of programs and $\mathcal{L}$ a class of assertions. Then the validity of correctness formulas over $\mathcal{P}$ and $\mathcal{L}$ is equivalent to the equivalence problem for $\mathcal{P}$, if it is the case that for each formula F in $\mathcal{L}$, that there is a program P such that F is true (for a given input) if and only if P outputs a zero (for that input), and vice versa. This was actually the case in [3], since in the earlier paper [2] it was shown that the class of Presburger formulas was realized by he class of $L_1(x \leftarrow 0,\ x \leftarrow y,\ x \leftarrow x+1,\ x \leftarrow x \dot{-} 1)$-programs. The size of such a realizing program for a given Presburger formula, however, is at least doubly exponential in the length of the formula. If in [3] only unquantified assertions were allowed, the problem of deciding the truth of a correctness formula would be Co-NP complete. Hence an exponential time procedure would more than likely be required.

The previous theorem can be generalized somewhat in that the formula p (q) can also be of the form $p \wedge P$ or $p \vee P$ ($q \wedge Q$ or $q \vee Q$) where P (Q) is a Presburger formula. It follows from results in [2] that one can construct a U-program for a Presburger formula (with n-free variables) that is equivalent to the zero program on n-inputs if and only if the Presburger formula is valid. The constructed program need not contain any instructions of the form "x←x *mod* y". This is almost Lemma 1. Unfortunately, the length of the resulting program is at least double exponential in the length of the formula [2]. Hence the complexity of this problem cannot be the same. This follows from the complexity of deciding the validity of Presburger formulas[5].

Consider once again only formulas in $\mathcal{F}$. Then we can observe that for quantified formulas in $\mathcal{F}$, the validity problem is undecidable even if we limit the formulas to a single occurrence of the $\exists$ quantifier[22]. In fact, this is very easy to see using the strongest post condition calculus and adding the instruction "z←*lcm*(x,y)" to U-programs. We have $SPC(z \leftarrow lcm(x,y), Q(x,y,z)) = \exists w\ Q(x,y,w) \wedge z = lcm(x,y) = \exists w\ Q(x,y,w) \wedge x|z \wedge y|z \wedge \forall v\ ((x|v \wedge y|v) \supset z|v)$; hence the undecidability of such correctness formulas follows in a straightforward manner from [20], since multiplication can be simulated using the "z←*lcm*(x,y)". This follows because $lcm(x,x+1)=x^2+x$. Those readers familiar with the SPC calculus should note that the logical formulas derivable from such correctness formulas, $\{p\}S\{q\}$ where p and q are formulas in $\mathcal{F}$ and S is a U-program, are of the form: $\forall \bar{x}[\exists \bar{w} W(\bar{w}, \bar{x}) \supset q(\bar{x})]$ or equivalently, $\forall \bar{x} \forall \bar{w}[\neg W(\bar{w}, \bar{x}) \vee q(\bar{x})]$. It is then the universal quantifier appearing in the SPC(z←*lcm*(x,y), Q(x,y,z)) that is the problem. An interesting question then, is whether the same results for the instruction "z←*gcd*(x,y)" hold. The strongest post condition calculus produces a similar formula as it did for the "z←*lcm*(x,y)" instruction but we are unable to answer the decidability of the validity problem for such correctness formulas, at this time.

In [17,23], it was shown that unquantified Presburger array formulas have a decidable validity problem. In fact it was shown that special predicates could be added to the formulas (in a limited way) and the validity problem remains decidable. Such predicates considered were those concerning properties of the arrays such as orderedness or the property that one array is a permutation of another. The logical formulas in Presburger array theory are equivalent to correctness formulas of the form {p}S{q}, where p and q are logical formulas similar to those in $\mathcal{F}$ although they allow array elements as terms and do not allow the "|" predicate, and S is a {x←1, x←x+y, x←x∸y, x←x/k, x←A(i), A(i)←x, if x=0 then I}-program (I can be any of the other types of instructions). The proofs in [17,23] reduce the Presburger array formulas to equivalent unquantified Presburger formulas. Thus the validity problem for such formulas is decidable. The reduction is such that, even if an additional function f(x) were allowed in the Presburger array formula, the resulting formula is unaffected except that it too contains occurrences of f(x). Hence it is the case that even if we add the *mod* function to the theory in [17,23] it still yields a decidable validity problem.

4. L_1(BB)-Programs

In this section, we consider the computational power of L_1(BB)-programs over different sets of primitive instructions, BB. Most of our results consider problems considered in [14-16,18]. Our first result shows that L_1(BB1) and L_1(BB2)-programs can be converted into L_1(x←0, x←x+1, *if* x=0 *then* y←y+1)-programs and L_1(x←0, x←x+1, x←x∸1, *if* x=0 *then* y←y+1)-programs, respectively, in polynomial time. This is an improvement over the exponential time needed in [18].

Theorem 4. Given a L_1(BB1)-program P, one can construct in polynomial time, a L_1(x←0, x←x+1, *if* x=0 *then* y←y+1)-program P' such that P' is equivalent to P.

The proof for L_1(BB2)-programs is similar. One merely allows the additional instruction x←x∸y.

Our next result considers whether L_1(BB1, if)-programs can be converted into equivalent L_1(BB1)-programs. In [18] it was claimed without proof that this could be done in polynomial time. However in [14], it was shown that this was only possible if PSPACE=PTIME. Thus the question of convertibility seems to be in doubt. Here we provide an exponential algorithm. This result should be contrasted with the corresponding case for the set BB2, where the addition of the "if" construct provided an increase in the computational power of the language [16].

Theorem 5. Let P be an L_1(BB1, if)-program. Then an equivalent L_1(BB1)-program P' can be constructed in exponential time (and space).

Next we consider the computational power of L_1(BB2)-programs when allowing the instructions "y←c" and "y←y∸1" to be conditionally executed, where c can be any nonnegative constant. In our next result, we show that L_1(BB2, *if* x=0 *then* y←c)[3]-programs compute Presburger functions. For such a program we construct a nondeterministic reversal bounded multicounter machine (CM) to, in some sense, simulate the programs computation. The result then follows from the results concerning nondeterministic reversal bounded CM's in [11].

Theorem 6. Every L_1(BB2, *if* x=0 *then* y←c)-program computes a Presburger function.

This result should be contrasted with the result in [16], showing that L_1(BB2, *if* x=0 *then* y←y+1) is strictly more powerful than L_1(BB2). If both constructs, "*if* x=0 *then* y←c" and "*if* x=0 *then* y←y∸1" are concurrently considered it is easy to show that L_1(BB2, *if* x=0 *then* y←c, *if* x=0 *then* y←y∸1) is computationally equivalent to L_1(BB2, *if* x=0 *then* y←y+1), since the instruction "*if* x=0 *then* y←y+1" can be simulated by the following sequence of instructions from BB2 ∪ {*if* x=0 *then* y←c, *if* x=0 *then* y←y∸1}: w ← 0; *if* x = 0 then w ← 1; y ← y + 1; *if* w = 0 then y ← y ∸ 1, where w is a new variable. (The converse was shown in [16].) Unfortunately, we have been unable to resolve the computational power of L_1(BB2, *if* x=0 *then* y←y∸1). The "*if* x=0 *then* y←y∸1" construct seems similar to the "*if* x=0 *then* y←y+1" construct, but as pointed out in [14] functions of one variable computed over BB2 ∪ {*if* x=0 *then* y←y∸1} are monotonic. Thus the proof techniques used in [16] (as well as those presented in the last theorem) do not seem to work with this language. This same problem arose in [14], where the authors were able to show that the 0-evaluation problem for this language is PSPACE-complete.

---

[3]Any constant can be substituted for c. In fact each instance of such a statement can have a different constant.

REFERENCES

[1] Alt, H., Functions Equivalent to Integer Multiplication, *Lecture Notes in Computer Science, No. 85: Automata, Languages and Programming (ICALP 80)*, Springer Verlag, 1980.

[2] Cherniavsky, J., Simple Programs Realize Exactly Presburger Formulas, *SIAM J. Comput.* 5 (1976), pp. 666-677.

[3] Cherniavsky, J. and Kamin, S., A Complete and Consistent Hoare Axiomatics for a Simple Programming Language, *J. ACM*, 26 (1979), pp. 119-128.

[4] Constable, R., Hunt, H. and Sahni, S., On the Computational Complexity of Scheme Equivalence, Proc. 8th Ann. Princeton Conf. on Information Sciences Systems, Princeton, NJ, 1974.

[5] Fischer, M. and Rabin, M., Super-Exponential Complexity of Presburger Arithmetic, *Project Mac. Tech. Memo 43*, MIT, Cambridge, 1974.

[6] Gurari, E., Decidable Problems for Powerful Programs, to appear in *J. ACM.*

[7] Gurari, E. and Ibarra, O., The Complexity of the Equivalence Problem for Simple Programs, *J. ACM* 28, 3 (July 1981), pp. 535-560.

[8] Gurari, E. and Ibarra, O., The Complexity of the Equivalence Problem for Two Characterizations of Presburger Sets, *Theor. Computer Science*, 13 (1981) pp. 295-314.

[9] Gurari, E. and Ibarra, O., Two-Way Counter Machines and Diophantine Equations, *J. ACM* 29, 3 (July 1982), pp. 863-873.

[10] Hoare, C., An Axiomatic Basis of Computer Programming, *CACM*, Vol. 12, No. 10, pp. 576-580, 1969.

[11] Ibarra, O., Reversal-Bounded Multicounter Machines and their Decision Problems, *J. ACM*, Vol. 25, No. 1, January 1978, pp. 116-133.

[12] Ibarra, O. and Leininger, B., Characterizations of Presburger Functions, *SIAM J. Comput.*, Vol. 10, No. 1, pp. 22-39, February, 1981.

[13] Ibarra, O. and Leininger, B., On the Equivalence and Simplification Problems for Simple Programs, *J. ACM*, 30, 3 (July 1983), pp. 641-656.

[14] Ibarra, O., Leininger, B. and Rosier, L., A Note on the Complexity of Program Evaluation, accepted for publication in *Math. Systems Theory.*

[15] Ibarra, O. and Rosier, L., The Equivalence Problem and Correctness Formulas for a Simple Class of Programs, University of Texas at Austin, Department of Computer Sciences, Tech. Rep. No. 83-23 (1983).

[16] Ibarra, O. and Rosier, L., Simple Programming Languages and Restricted Classes of Turing Machines, *Theoretical Computer Science*, Vol. 26, No. 1 and 2, pp. 197-220, September 1983.

[17] Jefferson, D., Type Reduction and Program Verification (Ph.D. thesis), Department of Computer Science, Carnegie-Mellon University, 1980.

[18] Kfoury, A., Analysis of Simple Programs Over Different Sets of Primitives, *7th ACM SIGACT-SIGPLAN Conference Record*, 1980, pp. 56-61.

[19] Knuth, D., The Art of Computer Programming: Vol. 1, Fundamental Algorithms, Addison-Wesley, Reading, MA, 1973.

[20] Matijasevic, Y., Enumerable Sets are Diophantine, *Dodl. Akad. Nauk.*, SSSR 191 (1970), pp. 279-282.

[21] Meyer, A. and Richie, D., The Complexity of Loop Programs, in *Proc. 22nd Nat. Conf. of the ACM*, Thompson Book Co., Washington, DC, 1976, pp. 465-469.

[22] Robinson, J., Definability and Decision Problems in Arithmetic, *J. Symbolic Logic*, 14, pp. 98-114 (1949), MR 11, 151.

[23] Suzuki, N. and Jefferson, D., Verification Decidability of Presburger Array Programs, *J. ACM* 27, 1 (Jan. 1980), pp. 191-205.

[24] Tsichritzis, D., The Equivalence Problem of Simple Programs, *J. ACM* 17, 4 (Oct. 1970), pp. 729-738.

LOWER BOUNDS FOR POLYGON SIMPLICITY TESTING AND OTHER PROBLEMS

Jerzy W. Jaromczyk

Institute of Informatics

Warsaw University, PKiN VIII p.

00-901 Warsaw, Poland

Abstract

The new non-trivial lower bounds of time complexity for some problems of computational geometry such as:

- polygon simplicity testing (computational version)
- finding diameter of a set of points in R^2
- finding maximal distance between two sets of points in R^2

are derived. Some attractive geometrical constructions e.g., a curve with a constant width are used while proving the lower bounds.

1. Introduction

For many interesting algorithms, especially in computational geometry, the question of optimality of these algorithms is open. It is an effect of difficulties with proving lower bounds even for restricted models of computation.

Now the situation is changed a bit owning to the recent powerful results of Steele, Yao [10] (decision tree model) and Ben-Or [2] (computation tree model). These results, depending heavily on the Milnor's inequality [7], enable one to establish a lower bound for the following membership problem for a set W: " given a set $W \subseteq R^n$ - for $x \in R^n$ decide if $x \in W$ ". The lower bound for membership problem can be expressed on a ground of a number of disjoint connected components of the set W.

A method giving lower bounds was introduced by Jaromczyk [6]. The method, exploiting greately the techniques of Steele, Yao [10] and Ben-Or [2] considers both geometrical and topological properties of the set W; it works even for the problems with very few connected components (see [6]).

In the present paper we will prove $\Omega(n\log n)$ lower bounds in a model of decision trees for the following geometrical problems:

1. given a polygon $p=p_0\dots p_{k-1}$ - report all edges which are intersected (with certain edge of p);
2. given a plane set A of n points in R^2 - find a diameter of A;
3. given two sets A, B of n points in R^2 each - find a maximal distance between A and B.

Precise definitions of the above problems will be given in sections 3, The lower bounds for problem 2 and 3 settle an open question of Toussaint [11]. The lower bound for problem 1 reffers to the question of Edelsbrunner [4].

All the above presented problems are of great practical interest e.g., problems of finding diameter and maximal distance appear in pattern recognition and scene analysis.

This paper is composed in the following way. In section 2 the definition of the model is given and basic theorem of Jaromczyk [6] is quoted. Sections 3 and 4 are devoted to proving lower bounds for problems 1-3. Some concluding remarks pertaining mainly the method and further work are also provided.

2. Basic notions

Let $W\subseteq R^n$ be any set. The problem "given $x\in R^n$ determine if $x\in W$" is reffered to as a membership problem for the set W.

A d-th order polynomial decision tree T (shortly d-tree) for testing if $x\in W$ is a ternary tree. Each internal node of T contains a test of the form $p(x) : 0$ where p is polynomial of degree $\leq d$. The leaves of T are labeled with the answer "YES" or "NO". For a given input x the procedure starts from the root and proceeds down the tree branching at each internal node accordingly to the output of the polynomial test at that node. Eventually a leaf with the answer is reached. We say that T solves the membership problem for W if for any x the answer is given correctly with respect to "$x\in W$?".

Let $C_d(T)$ stand for the height of T. $COST_d(W)$ denotes minimal $C_d(T)$ over all algorithms T solving the membership problem for W.

Thus $COST_d(W)$ measures the worst case complexity of the membership problem for W; it is called lower bound of time complexity for the (membership) problem W.

Further results are based upon the theorem of Jaromczyk [6] so, for readers convenience, we quote this theorem (without proof) with needed notations and notions.

We call $\varphi : D \subseteq R^p \to R^m$ a d-th order polynomial (transform) if $\varphi(z_1,\dots,z_p) = (\varphi_1(z_1,\dots,z_p),\dots,\varphi_m(z_1,\dots,z_p))$ and φ_i, $i=1..m$, are polynomials in $z_1,\dots,z_p$ of degree $\leq d$.

φ is called S-slicer for set $W \subseteq R^m$ if the number of disjoint connected components of $W \cap \varphi(D)$ is at least S.

(we will use $\# W$ to denote number of disjoint connected components of W)

The following theorem holds

<u>Theorem 1.</u> Let $W \subseteq R^m$, $b,d,S \in N$, $\max(b,d) \geq 2$. Let φ be a b-polynomial S-slicer for W, $\varphi : D \subseteq R^p \to R^m$.

Then $COST_d(W) + COST_{b \cdot d}(D) \geq c_1 \log_2 S - p$

where $c_1 = (2\log_2(2bd - 1))^{-1}$.

(this theorem is included in [6], in asymptotic version).

As it turns out in applications the crucial notion for the method which theorem 1 yields is S-slicer.

3. Simple polygon problem

Given $p_o,\dots,p_{n-1} \in R^2$, $p_i=(x_i,y_i) \in R^2$, $i=0..n-1$. These points form an open polygon $p=p_o,\dots,p_{n-1}$ with edges $p_i p_{i+1}$, $i=0..n-2$. If we add an edge $p_{n-1}p_o$ we will obtain closed polygon.

<u>Simple polygon</u> (Jordan polygon) is a closed polygon homeomorphic to the unit circle. Only neighboring edges of the simple polygon can have points in common i.e., their endpoints. In the other words we can say that no two edges of the simple polygon <u>intersect</u> one another.

A convex simple polygon is called convex polygon. Without any confusion simple polygon will be meant as a plane figure or a border of this figure.

We will study the following computational problem SP_n:
"given points $p_o,\dots,p_{n-1} \in R^2$ - for polygon $p=p_o,\dots,p_{n-1}$ report all intersected edges"

Observe that the number of possible output answers to SP_n problem is at most 2^n thus an information theoretic lower bound is trivial.

The simple polygons take a significiant place incomputational geometry (see Shamos [9]). It is know that many computational problems in the field of computational geometry are easier to solve while a given list of points forms a simple polygon e.g.,convex hull problem, finding diameter of a set, finding distance between two sets of points and many others.

In order to derive lower bound for SP_n problem let us introduce an auxiliary membership problem of <u>half-simple polygon</u>.
Let $m=3k+1$ and $p=z_0,\dots,z_{k-1},a_0,w_0,\dots,a_{k-1},w_{k-1},a_k$ be a closed polygon. We say that this polygon is half-simple if no edge $z_i z_{i+1}$, $i=0..k-2$, is intersected.

<u>Half-simple polygon problem</u> is a membership problem for a set

$$W\text{-}SP_m = \{ p=(z_0,\dots,z_{k-1},a_0,w_0,\dots,a_{k-1},w_{k-1},a_k) \in R^{2m}\text{: p is half-simpl}$$

We will need the following geometrical construction.
Let $C= \{(x,x^2): x\in R\}$ be a parabola. Fix the points $t_i \in R^2$, $i=0..k-1$, $a_i \in R^2$, $i=0..k$, on such the way that two following properties hold:

R1. the points $t_0,\dots,t_{k-1}$ are the vertices of a convex polygon Q (in anti-clockwise order) such that each edge $t_0 t_1,\dots,t_{k-1}\, t_{k-2}$ intersect the parabola C in exactly two points

R2. the points $a_0,\dots,a_k$ lie inside the polygon Q.

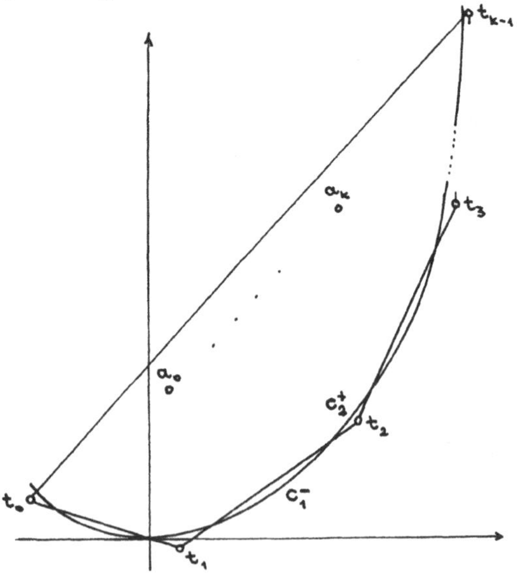

fig. 1

The existance of the points satisfying both properties R1 and R2 is intuitively obvious (see fig.1); simple calculations are skiped.

Observe that the convex polygon Q divides the parabola C onto disjoint open arcs $c_0^+,\dots,c_{k-1}^+$ lying inside Q (in interior intQ), and closed arcs $c_{-1}^-,\dots,c_{k-1}^-$ lying outside Q with both endpoints in a certain edge $t_i t_{i+1}$ (see fig.1.)
The polygon Q and the parabola C will be decisive for the construction of a slicer φ for $W\text{-}SP_m$.

<u>Lemma 2.</u> For any $d \geqslant 2$, $m=3k+1$

$$COST_d(W\text{-}SP_m) \geqslant c\log_2 k^k - k$$

where c 0 depends only on d.
<u>Proof</u>: Let us take $D=R^k$ and define slicer $\varphi : D \to R^m$

$$\varphi(x_0,\dots,x_{k-1}) = (t_0,\dots,t_{k-1},a_0,w_0,\dots,w_{k-1},a_k)$$

where $w_i=(x_i,x_i^2)$, $i=0..k-1$, and the points satisfy properties R1, R2. Observe that the points w_i while moving along the parabola C pass alternately through the arcs c_j^- and c_j^+ making polygon $t=t_0,\dots,t_{k-1}$, $a_0w_0,\dots,w_{k-1},a_k$ non-half-simple or half-simple respectively (see fig.2).
It is not difficult to see that all connected components of the set $\varphi(D) \cap W\text{-}SP_m$ are of the shape

$$W(j_0,\dots,j_{k-1}) = \{(t_0,\dots,t_{k-1},a_0,w_0,\dots,w_{k-1},a_k) : w_i \in c_{j_i}^+ , c_{j_i}^+ \in \{c_0^+,\dots,c_{k-1}^+\} , i=0..k-1 \}$$

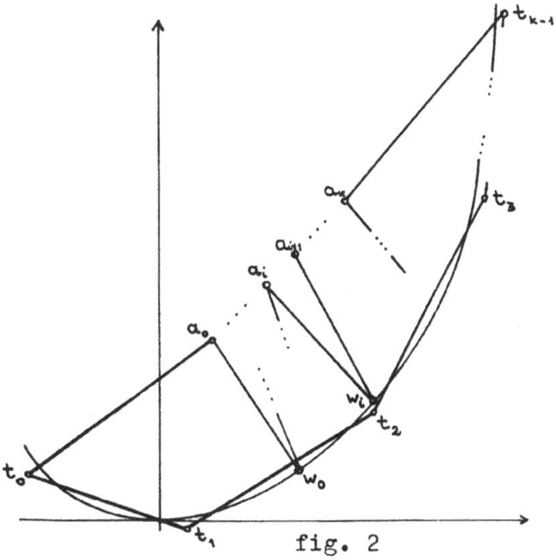

fig. 2

A number of sets $W(j_0,\dots,j_{k-1})$ is equal to k^k thus $\# \varphi(D) \cap W\text{-}SP_m \geq k^k$ and φ is k^k-slicer for $W\text{-}SP_m$.
Now the lemma follows from theorem 1 since $COST_d(D)=0$ for $D=R^k$. □

Theorem 3. For any fixed $d \geq 2$

$$COST_d(SP_n) = \Omega(n\log n)$$

Proof: For a fixed d $COST_d(SP_n)$ is a non-decreasing function of n since any algorithm for (n+1)-polygon yields an algorithm of the same cost for n-polygon by an appropriate setting of one vertex.
Now the theorem immediately follows by virtue of lemma 2 since $COST_d(W\text{-}SP_m) \leq COST_d(SP_m)$. □

Problem SP_n is a computational version of a decision problem DSP_n: "given polygon p - decide if p is simple polygon". (i.e. no witness in a form of intersected edge is needed). Obviously $COST_d(DSP_n) \leq$ $\leq COST_d(SP_n)$. The question regarding a lower bound for DSP_n was posed by Edelsbrunner [4]. The same lower bound is expected.

4. Diameter of a set and distance between two sets

Let $d(a,b)$ stand for the usual metric in R^2.
A (maximal) distance between two sets $A,B \subset R^2$ of points is $dist(A,B) = \max(\ d(a,b) : a \in A, b \in B\ )$.
A diameter of a set $A \subset R^2$ of points is $diam(A)=\max(d(a,b): a,b \in A)$.

Let $S_1=\{p_0,\dots,p_{n-1}\}$, $S_2=\{q_0,\dots,q_{n-1}\}$ be two sets in R^2.

A distance problem $DIST_n$ is: "given $S_1,S_2 \subset R^2$ - give a pair (i,j) of indices such that $d(p_i,q_j)=dist(S_1,S_2)$"

A diameter problem $DIAM_n$ is: "given $S_1 \subset R^2$ - give a pair (i,j) of indices such that $d(p_i,p_j)=diam(S_1)$".

Remark that we require only one pair of points yielding distance or diameter respectively.
In order to prove lower bounds for $DIST_n$, $DIAM_n$ two auxiliary membership problems are needed.
Let $h>0$ and $m=2k$ where k is odd natural number. Define two sets

$$h\text{-}DIAM_m = \{(p_0,\dots,p_{m-1}) \in R^{2m} : diam(\{p_0,\dots,p_{m-1}\}) \leq h\}$$

$$h\text{-}DIST_m = \{(p_0,\ldots,p_{m-1},q_0,\ldots,q_{m-1}) \in R^{4m} : dist(\{p_0,\ldots,p_{m-1}\}, \{q_0,\ldots,q_{m-1}\}) \leq h\}$$

The following easy fact holds.

<u>Fact 4</u>. $COST_d(h\text{-}DIAM_m) \leq COST_d(DIAM_m) + 1$
$COST_d(h\text{-}DIST_m) \leq COST_d(DIST_m) + 1$

We will need the following geometrical construction in order to prove the lower bounds for $DIAM_n$ and $DIST_n$ problems.

Let $h > 0$ and k be an odd natural number. Consider k-polygon i.e. regular polygon with k vertices and a length of diagonal lines equal to h. Denote this polygon by P. From each vertex of P we circumscribe an arc of a circle of radius h joining two neighbor vertices of polygon P (see fig. 3). The union of all these arcs forms a curve G with a constant width equal to h (see Jaglom,Boltianski [5], problem 79).
Let $z_0,\ldots,z_{k-1}$ stand for the vertices of P. Without any confusion P will be meant as a plane figur or a border of this figure. Similary G denotes a plane figure or the curve G.
To end the construction we draw a circle K of a radius r such that G and K possess two properties:

P1. the circle K intersects each arc (joining two neighbor vertices of the polygon P) of the curve G in two points. Therefore the curve G divides the circle K on two sets of arcs:
$L^- = \{l_0^-,\ldots,l_{k-1}^-\}$ of disjoint open arcs lying outside G
$L^+ = \{l_0^+,\ldots,l_{k-1}^+\}$ of disjoint closed arcs lying inside G with endpoints on the curve G.

P2. for any point $x \in l^-$, $l^- \in L^-$ there exists a vertex z_p of P such that $d(x,z_p) > h$.

The existance of circle K satisfying both properties P1 and P2 is intuitively obvious (see fig.3). We can for example put $r := h/2$. An easy proof that such r is good is omitted.

The polygon P, curve G and the circle K will be decisive for the construction of a slicer φ of the sets $h\text{-}DIAM_m$ and $h\text{-}DIST_m$.

<u>Lemma 5.</u> For any $d \geq 2$ and odd natural number k

$$COST_d(h\text{-}DIAM_{2k}) \geq ck\log_2 k - 3k$$

where $c > 0$ depends only on d.

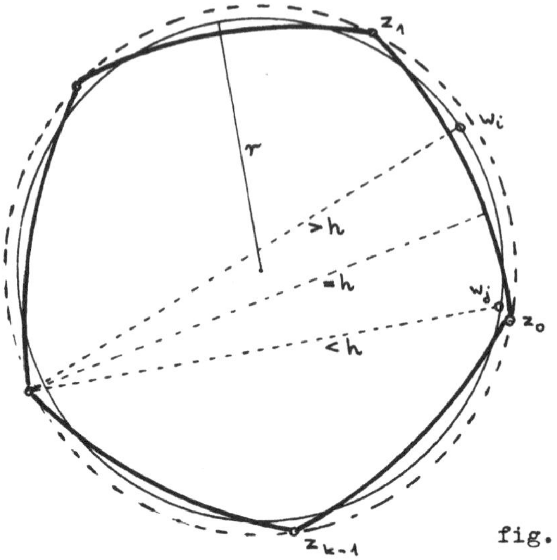

fig. 3

Proof: Let $w_i=(x_i,y_i) \in R^2$, $i=0..k-1$. Take

$D= \{(w_0,\dots,w_{k-1}) \in R^{2k}: x_i^2+y_i^2=r^2,\ i=0..k-1\}$ where r is radius of the circle K. A slicer φ, $\varphi: D \to R^{4k}$ is defined as

$\varphi(w_0,\dots,w_{k-1}) = (z_0,\dots z_{k-1},w_0,\dots,w_{k-1})$ where $z_0,\dots,z_{k-1}$ are the vertices of the polygon P. Observe that all connected components of $\varphi(D) \cap h\text{-}DIAM_{2k}$ are of the shape

$W(j_0,\dots,j_{k-1})= \{(z_0,\dots,z_{k-1},w_0,\dots,w_{k-1}) : w_i \in l^+_{j_i} \in L^+, i=0..k-1\}$.

Hence φ is k^k-slicer of the set $h\text{-}DIAM_{2k}$ and lemma follows(since $COST_d(D) \leq k$) by virtue of theorem 1. □

The above lemma and fact 4 yield

Theorem 6. For any fixed $d \geq 2$

$$COST_d(DIAM_m) = \Omega(m\log m)$$

Following the similar reasoning we come to effect that

Theorem 7. For any fixed $d \geq 2$

$$COST_d(DIST_m) = \Omega(m\log m)$$

The above theorems settle the open question (see Toussaint [11]) concerning the optimality of known, for $DIAM_m$ and $DIST_m$ problems, algorithms (Shamos [9] , Toussaint, McAlear [12]).

The above lower bounds are also true for computation tree model.

5. Concluding Remarks

We have shown new non-trivial lower bounds for some interesting problems in computational geometry. These examples give evidence that the method formulated in theorem1 (with the notion of slicer) delivers a strong tool for proving lower bounds of time complexity. While applying the method we are able to shed light on intrinsic complexity of the problem through an appropriate choosing of the transform φ (see theorem 1), thus constructing a topological space in which the complexity of the problem becomes evident.
On the other hand the method includes, as a special case, the known technique of reduction one computational problem to another one (it is done through an appropriate choosing of the domain D).
Therefore the further results in form of new lower bounds can be expected. Some of them have been already presented in [6], others e.g., for finding extremal polygon (see Boyce et al. [3]) are also known. Finally as it was mentioned in [6] the method is useful for polyhedral problems considered by Moravek [8], Yao, Rivest [13], Avis [1].

References

1. Avis, D., Lower bounds for geometric problems. Allerton Conference, October 1980.
2. Ben-Or, M., Lower bounds for algebraic computation trees, Proc. of 15-th STOC, Boston,1983, 80-86.
3. Boyce, J.E., Dobkin, D.P.,Drysdale, R.L., Guibas, L.J., Finding extremal polygons, Proc. of 14-th STOC, 1982, 282-289.
4. Edelsbrunner, H., Bulletin of EATCS, No.21, October 1983.
5. Jaglom, I.M., Boltianski, V.G.,Convex Figures,W-wa 1955 (polish)
6. Jaromczyk, J.W., A complexity bounds and euclidean space partitionig, Coll. on Alg.,Comb.& Logic, Györ-Hungary, September 83.
7. Milnor, J., On the betti numbers of real algebraic varieties, Proc. AMS 15, 1964, 275-280.
8. Morávek, J., A lokalization problem in geometry and complexity of discrete programming,Kybernetika 8(6), 1972, 498-516.
9. Shamos, M.I., Computational geometry, Ph.D.Thesis, Yale Uni. 1978.
10. Steele, J.M., Yao, A.C., Lower bounds for algebraic decision trees, J.Algorithms 3, 1982, 1-8.
11. Toussaint, G.T., Pattern recognition and geometrical complexity, Conference on Pattern Recognition, 1980, 1324-1347.
12. Toussaint, G.T., McAlear, J.A., A simple O(nlogn) algorithm for finding the maximum distance between finite planar sets, Tech.Rep.No.SOCS-82-6, March 1982, McGill Uni.
13. Yao, A.C., Rivest, R., On the polyhedral decision problem, SIAM J.on Comput. 9(2), 1980, 343-347.

A UNIFORM INDEPENDENCE OF INVARIANT SENTENCES

M. I. Kanovič
Kalinin University, USSR

Beginning with [1] a number of independence results in computer science has been stated [1,2,3 and many others]. The majority of these results is based on refined partial recursive representations of recursive functions. A reduction to simpler subrecursive representations is desirable.

On the other hand, independent formulas are disconnected, namely, in these formulas different representations of the same set are used. For example, there did exist m for which

$$P^{L_m} = NP^{L_m}$$

is independent but there was no answer whether

$$NP^{L_m} = co\text{-}NP^{L_m} \quad \text{or} \quad \text{"}L_m \text{ is empty"}$$

(with the same m) were independent of a given theory S.

In this paper general theorems of an uniform independence for subrecursive levels are stated. The main working notion of the theorem is the notion of a representation invariance. In a sense the paper could be considered as an uniform solution of the equation:

the undecidability of the halting problem is to the Rice's theorem as [1] is to x.

Let us denote by

$$R = \{ G_0, G_1, \ldots , G_i, \ldots \}$$

a class of representations ("machines" or "programs" or "operator terms").

In the paper we consider the following classes:

1) the class R_0 of all Turing machines

$$M_0, M_1, M_2, \ldots , M_i, \ldots$$

or, equivalently, the class of all partial recursive operator terms (built from the function symbols of the simplest functions and the operator symbols of the operations of superposition, primitive recursion and minimization),

2) the class of all primitive recursive operator terms,

3) the class of all elementary (according to Kalmar) operator terms,

4) the class of all E^n-terms (here E^n is the Grzegorchyk's class, n = 0, 1, 2, ...)

5) the class of all context-sensitive grammars,

6) the class of all codes of non-deterministic Turing machines with the space complexity bounded above by a function s(n), where $s(n) \geq n+1$,

7) the class of all codes of one-tape deterministic Turing machines with the time complexity bounded above by a function t(n) and the space complexity bounded above by a function s(n), where $t(n) \geq n \cdot \log_2 n$, $s(n) \geq n+1$, etc.

L(G) denotes the set defined by representation G,

f_G denotes the function defined by representation G.

For a standard notation,

$\varphi_i = f_{M_i}$, $L_i = L(M_i)$,

$T_i(n)$ is the time complexity of M_i,

$S_i(n)$ is the space complexity of M_i.

DEFINITION 1. An unary predicate Q(x) with the domain R is called invariant over R if, for any representations G and H of R for which $f_G = f_H$, the equivalence Q(G) = Q(H) is valid.

DEFINITION 2. A predicate Q(x) is called ω-invariant over R if, for any representations G and H of R for which the functions f_G and f_H coincide at all inputs except for a finite set of inputs, the equivalence Q(G) = Q(H) is valid.

E. g., the predicate

$$P^{L(x)} = NP^{L(x)}$$

is ω-invariant over an arbitrary R. The most interesting predicates in computer science, especially the predicates with oracles, are ω-invariant because of oracles are not sensitive to their finite alterations.

Remark. As to the quantifier complexity, an arbitrary ω-invariant over R predicate requires more than $\Sigma_2 \cap \Pi_2$-form (see [4]).

The main difficulty is to establish the non-triviality of the predicate in question, e.g., in this connection the existence of Turing machines G and H such that $P^{L(G)} = NP^{L(G)}$ is true and $P^{L(H)} = NP^{L(H)}$ is false is required [5] .

DEFINITION 3. A sequence of predicates

$$Q_0(x), Q_1(x), \ldots , Q_k(x), \ldots$$

is said to be effectively non-trivial over R if there exists an algorithm α such that, for any integer k, $\alpha(k)$ is a pair of representations (G,H) of R for which $Q_k(G)$ is true and $Q_k(H)$ is false.

If there exists an algorithm α (of definition 3) with a finite range of valuesthen we shall say that the sequence

$$Q_0(x),\ \ldots,\ Q_k(x),\ \ldots$$

is finitely non-trivial.

Let S be a fixed formal theory (like Peano arithmetic or Zermelo-Fraenkel set theory). We introduce the weakest variant of the notion: "a predicate $Q(x)$ is formally expressed in a formal theory S" .

DEFINITION 4. A predicate $Q(X)$ is said to be formally expressed w.r.t. R by means of a formula $F(x)$ if, for any representation G of R,

1) if $\vdash_S F(G)$ then $Q(G)$ is true,

2) if $\vdash_S \neg F(G)$ then $Q(G)$ is false.

Let us recall that a sentence F is independent of a given theory S if neither $\vdash_S F$ nor $\vdash_S \neg F$.

THEOREM 1. Let R be a class of the list of the beginning of the paper.Let

$$Q_0(x),\ \ldots,\ Q_k(x),\ \ldots$$

be a finitely non-trivial over R sequence of ω-invariant over R predicates and

$$F_0(x),\ \ldots,\ F_k(x),\ \ldots$$

be a computable sequence of formulas formally expressing (w.r.t. R) the predicates $Q_0(x),\ \ldots, Q_k(x), \ldots$ respectively.Then, for any representation G of R, there exists a representation H of R such that

1) $f_H = f_G$, $L(H) = L(G)$,

2) all the sentences

$$F_0(H),\ F_1(H),\ \ldots\ ,\ F_k(H),\ \ldots$$

are independent of the given theory S.

Proof. Diagonal arguments are used. A general scheme of the proo is given.

Let t be a partial recursive function such that, for any i,

1) if all the sentences

$$F_0(\tilde{G}),\ F_1(\tilde{G}),\ \ldots,\ F_k(\tilde{G}), \ldots$$

are independent of S then $t(i)$ is divergent, (here $\tilde{G} = G_{i.h(i)}$, h is described below),

2) if $t(i) = 2p+1$ then $\vdash_S F_p(\tilde{G})$,

3) if $t(i) = 2p$ then $\vdash_S \neg F_p(\tilde{G})$.

Let us denote by $\alpha_j(k)$ the jth element of the pair $\alpha(k)$. For a given G, we construct representations $G_i{'}$ such that

1) if $t(i)$ is divergent then $f_{G_i{'}} = f_G$,

2) if $t(i) = 2p+1$ then $f_{G_i{'}}$ is equal to $f_{\alpha_2(p)}$ except for a

finite set,

3) if $t(i) = 2p$ then $f_{G_{i'}}$ is equal to $f_{\alpha_1(p)}$ except for a finite set.

The key idea is to estimate the succinctness of $G_{i'}$ by $\log_2 h(i) + \text{Const.}$ where h is used for encoding integers by means of R.

For a suitable enumeration of R, we have $G_{i'}$ such that

$$i' = C_0 \cdot h(i)$$

for some C_0.

A desired representation H is $G_{C_0 \cdot h(C_0)}$.

If $t(C_0)$ were convergent then, for some p, we had the contradiction

$$Q_p(H) = \neg Q_p(H) .$$

In fact, $t(C_0)$ is divergent.It gives the desired effect.

Remark. If implications both

$$Q_0(x) \,\&\, Q_1(x) \,\&\, \dots \,\&\, Q_k(x) \longrightarrow Q_{k+1}(x)$$

and

$$Q_0(x) \,\&\, Q_1(x) \,\&\, \dots \,\&\, Q_k(x) \longrightarrow \neg Q_{k+1}(x)$$

are non-trivial over R then in addition we can state that the sentence $F_{k+1}(H)$ is independent of the theory

$$S + F_0(H) + F_1(H) + \dots + F_k(H).$$

Numerous corollaries are obvious.

COROLLARY 1. Let S be powerful enough to express predicates mentioned below. Then

1) for any elementary set L, there exists an elementary operator term H such that

$$L(H) = L$$

and

a) all the sentences

" $P^{L(H)} = NP^{L(H)}$ ",

" $NP^{L(H)} = co\text{-}NP^{L(H)}$ ",

" $P^{L(H)} = NP^{L(H)} \cap co\text{-}NP^{L(H)}$ ",

" L(H) is in P ",

" L(H) is in NP ",

" f_H is in E^2 ", etc.,

are independent of the theory S,

b) the sentence

" $NP^{L(H)} = co\text{-}NP^{L(H)}$ "

is independent of the theory $S +$ " $P^{L(H)} \neq NP^{L(H)}$ ",

2) for any context-sensitive language L, there exists a con-

text-sensitive grammar H such that

$$L(H) = L$$

and all the sentences

"L(H) is finite",

"L(H) is regular", etc.,

"L(H) is deterministic context-sensitive language"

are independent of the theory S (the latter sentence is independent provided that DCSL $\neq$ CSL),

3) for any function t of E^0 , there exists an E^0-term H such that

$$f_H = t$$

and the sentence

$$"TIME[f_H] = TIME[f_H^2]"$$

is independent of the theory S, etc.

The condition of finiteness in the non-triviality is essential, e.g., for any primitive recursive operator term H, one can find an integer k such that the sentence

$$"f_H \text{ is in } E^k "$$

is provable in Peano arithmetic.

At the high level R_0 the situation is simpler.

<u>THEOREM 2</u>. Let

$$Q_0(x), \ldots, Q_k(x), \ldots$$

be an effectively non-trivial over R_0 sequence of ω-invariant over R_0 predicates and

$$F_0(x), \ldots, F_k(x), \ldots$$

be a computable sequence of formulas formally expressing (w.r.t. R_0) the predicates $Q_0(x),\ldots,Q_k(x),\ldots$ respectively.Then, for any Turing machine M_i, there exists a Gödel number m such that

$$\varphi_m = \varphi_i, \quad L_m = L_i ,$$

$$T_m(n) \leq \max (n\log_2 n, T_i(n)),$$

$$S_m(n) \leq S_i(n)$$

and all the sentences

$$F_0(M_m), F_1(M_m), \ldots , F_k(M_m), \ldots$$

are independent of the given formal theory S.

Proof. Similarly to theorem 1.

<u>COROLLARY 2</u>. One can find a Turing machine M_m such that

$$\varphi_m = 0, \; L_m \text{ is empty},$$

$$T_m(n) \leq n\log_2 n,$$

$$S_m(n) \leq n+1$$

and

a) all the sentences

"L_m is finite",

"L_m is recursive",

"the time complexity of M_m is bounded by a total recursive function",

"$P^{L_m} = NP^{L_m}$", etc.,

are independent of the given theory S,

b) all the sentences

"$NP^{L_m} = \text{co-}NP^{L_m}$",

"every infinite set in NP^{L_m} contains an infinite subset in P^{L_m}",

"there exists an NP^{L_m}-simple set"

are independent of the theory

$$S + \text{" } P^{L_m} \neq NP^{L_m} \text{ "}.$$

For invariant predicates we have the following general theorem on an uniform independence.

THEOREM 3. Let

$$Q_0(x), \ldots , Q_k(x), \ldots$$

be an effectively non-trivial over R_0 sequence of invariant over R_0 predicates and

$$F_0(x), \ldots , F_k(x), \ldots$$

be a computable sequence of formulas formally expressing (w.r.t. R_0) the predicates $Q_0(x), \ldots, Q_k(x), \ldots$ respectively. Then there exists a Turing machine M_m such that

$$\varphi_m = 0, \ L_m \text{ is empty}$$

and all the sentences

$$F_0(M_m), F_1(M_m), \ldots , F_k(M_m), \ldots$$

are independent of the given theory S.

Proof. By means of the recursion theorem.

An uniform Gödel's incompleteness is given by theorem 3.

COROLLARY 3. Let S be powerful enough to express predicates mentioned below. Then there exists a Turing machine M_m such that $\varphi_m = 0$, L_m is empty and all the sentences

"L_m is empty",

"0 is in L_m", "1 is in L_m", ... , "k is in L_m", ... ,

"the theory S + L_m constructed by adding formulas of L_m as new axioms is consistent",

"the sentence "the theory S is consistent" is independent of the theory S + L_m", etc.

are independent of the theory S.

Close upper and lower bounds are stated for a size of "independent" representations H and m.

General theorem 1 is valid for an arbitrary class of subrecursive representations R having properties of a "weak universality" and being closed under branching and so on. We do not cite this theorem because of the awkwardness of its conditions. For simplicity, we confine ourselves to an enumeration of interesting and representative classes .

References

1 J.Hartmanis and J.E.Hopcroft, Independence results in computer science, SIGACT News 8 (4) , (1976) 13 - 24.

2 J.Hartmanis, Feasible computations and provable complexity properties, SIAM, 1978.

3 P.Hajek, Arithmetical hierarchy and complexity of computation, Theoret. Comp. Sc., 8 (1979) 227 - 237.

4 M.I.Kanovič, Complex properties of context-sensitive languages, Soviet Math. Dokl., 18 (1977), No. 2, 383 - 387.

5 T.Baker,J.Gill and R.Solovay, Relativizations of the P = NP question SIAM J. Comput. 4 (1975) 431 - 442.

6 H.Rogers Jr., Theory of recursive functions and effective computability, N.Y., 1967.

ON THE EQUIVALENCE OF COMPOSITIONS OF MORPHISMS AND INVERSE MORPHISMS ON REGULAR LANGUAGES

J. Karhumäki
Department of Mathematics,
University of Turku,
Turku, Finland

H.C.M. Kleijn
Institute of Applied Mathematics
and Computer Science
University of Leiden,
P.O. Box 9512
The Netherlands

ABSTRACT

We establish as our main results the following two theorems on compositions of morphisms and inverse morphisms. It is undecidable whether or not two transductions of the form $h_2h_1^{-1}$, where h_1 and h_2 are morphisms, are equivalent (word by word) on a given regular language, while the same problem for transductions of the form $h_1^{-1}h_2$ is decidable. Consequently, a sharp borderline between decidable and undecidable problems is found.

1. INTRODUCTION

Among the most natural problems in formal language theory are different kinds of equivalence problems. A typical example is the question of whether or not two transductions of a certain type are equivalent on their domain, cf. [B]. We consider this problem in a very simple set-up, namely assuming that the transductions are compositions of morphisms and inverse morphisms and that they are restricted to regular languages.

It was proved in [G] that the equivalence problem for 1-free nondeterministic sequential mappings is undecidable. Consequently, the equivalence problem for rational transductions is also undecidable, cf. [B]. On the other hand, this problem becomes decidable when the single-valued rational transductions are considered, cf. [BH].

The problem of whether or not two morphisms are equivalent (word by word) on a given language of certain type was raised in [CS], where the problem was also shown to be decidable for context-free languages. The topic of this paper, i.e., to study the equivalence of more complicated mappings on languages of certain type, was suggested in [KW].

As we saw the problem of whether or not two morphisms are equivalent on a regular language is decidable. On the other hand, recent characterization results of rational transductions, cf. [KL], [LL] or [T1], imply that for suitable compositions of morphisms and inverse morphisms the problem of whether or not such compositions are equivalent on a regular language becomes undecidable. That leads one to look for the borderline between the decidability and the undecidability.

The purpose of this note is to point out this borderline. We show, using the previously mentioned result of Griffiths and a recent result of Turakainen, cf. [T2], that it is undecidable whether or not two transductions of the form $h_2h_1^{-1}$ are equivalent on a regular language. Furthermore, using the Cross Section Theorem of Eilenberg, cf. [E] or [B], we prove that the same problem for transductions of the form $h_1^{-1}h_2$ is decidable. Consequently, we have found a "well-defined" borderline between decidability and undecidability for this particular problem setting.

To emphasize that the above undecidability result is not due to the fact that our family of languages is too complicated but rather because of the properties of morphisms, we also show that this problem remains undecidable if regular languages are replaced by languages of the form F^*, where F is finite. Hence, it is also undecidable whether or not two transductions of the form $h_3h_2^{-1}h_1$ are equivalent on Σ^*.

Finally, using a result from [BH], we conclude that for arbitrary compositions of morphisms and inverse morphisms, such that either all morphisms or all inverse morphisms in at least one of the compositions are injective, their equivalence on a regular language can be decided.

2. DEFINITIONS AND BASIC RESULTS

In this note we adopt the terminology of [B] and we use it also as general reference on basic results on formal languages. We now recal the notions and results needed later on.

Let Σ^* be the free monoid generated by a finite alphabet Σ. The identity of Σ^* is denoted by 1 and $\Sigma^+ = \Sigma^* - \{1\}$. A <u>transduction</u> $\tau: \Sigma^* \to \Delta^*$ is a mapping from Σ^* into the set of subsets of Δ^*. For two transductions τ and τ' their composition (if defined) is denoted by $\tau' o \tau$, or simply $\tau'\tau$. The domain of a transduction τ is denoted by $\mathrm{dom}(\tau)$. The transduction determined by the inverse of a transduction τ is denoted by τ^{-1}. Transductions $\tau, \tau': \Sigma^* \to \Delta^*$ are said to be <u>equivalent on a language</u> $L \subseteq \Sigma^*$, in symbols $\tau \overset{L}{\equiv} \tau'$ if $\tau(x) = \tau'(x)$ for all x in L. They are said to be <u>equivalent</u> if they are equivalent on Σ^*.

Let L be a family of languages and Θ a family of transductions (defined on suitable alphabets). We denote by

$$EP_{\forall}(\Theta, L)$$

the problem of deciding whether or not two given transductions from Θ are equivalent on a given language of L. We shall use the notations H and H^{-1} for the families of all morphisms and inverse morphisms, respectively. By HoH^{-1}, for instance, we mean the family of transductions of the form $h_2 h_1^{-1}$, where h_1 and h_2 are morphisms. The family of all regular languages is denoted by Reg.

In the next few lines we recall some results and terminology concerning rational transductions. If necessary the reader may consult [B]. A transduction $\tau: \Sigma^* \to \Delta^*$ is <u>rational</u> if and only if it is "realized" by a transducer, i.e., by a sixtuple $(\Sigma, \Delta, Q, q_0, F, E)$, where Σ is an input alphabet, Δ is an output alphabet, Q is a set of states, q_0 is the initial state, F is a set of final states, and $E \subseteq Q \times \Sigma^* \times \Delta^* \times Q$ is a set of transitions of T.

A transducer is called <u>1-free</u> if $E \subseteq Q \times \Sigma^* \times \Delta^+ \times Q$ and <u>simple</u> if $F = \{q_0\}$. Further we call a transducer <u>1-output</u> if $E \subseteq Q \times \Sigma^* \times \Delta \times Q$. By a <u>nondeterministic sequential</u> transducer we mean a transducer satisfying $F = Q$ and $E \subseteq Q \times \Sigma \times \Delta^* \times Q$. Of course, a rational transduction is called <u>1-free</u>, <u>simple</u>, <u>1-output</u> or <u>nondeterministic sequential</u> if it is realized by such a transducer. Finally, a transduction $\tau: \Sigma^* \to \Delta^*$ is called <u>single-valued</u> if, for each x in Σ^*, $\tau(x)$ contains at most one element, i.e., τ defines a partial function from Σ^* into Δ^*.

The following characterization result for rational transductions is given in [KL] and [T1] (cf. also [LL] and [CFS]).

<u>PROPOSITION 1.</u> Each rational transduction $\tau: \Sigma^* \to \Delta^*$ admits a factorization

$$\Sigma^* \xrightarrow{\ \cdot\$\ } (\Sigma \cup \{\$\})^* \xleftarrow{\ h_1\ } \Gamma_1^* \xrightarrow{\ h_2\ } \Gamma_2^* \xleftarrow{\ h_3\ } \Gamma_3^* \xrightarrow{\ h_4\ } \Delta^*$$

where each h_i is a morphism and $\cdot\$$ denotes the marking, i.e., the mapping which associates with each word x a new word $x\$$, where $\$$ is a new symbol not in Σ.

For simple transductions the above result was recently generalized by Turakainen, cf. [T2]. For our purposes his result can be stated as follows:

<u>PROPOSITION 2.</u> Each 1-output simple rational transduction $\tau\colon \Sigma^* \to \Delta^*$ admits a factorization

$$\Sigma^* \xleftarrow{\ h_1\ } \Gamma_1^* \xrightarrow{\ h_2\ } \Gamma_2^* \xleftarrow{\ h_3\ } \Delta^*$$

where, for any large enough m, the morphisms h_i can be chosen such that, for all $a \in \Delta$, h_3 is defined by $h_3(a) = a\$^m$, where $\$$ is a new symbol not in Δ.

We shall need not only Proposition 2 but also properties of the construction needed to prove the proposition. Hence, for the sake of completeness, we repeat this construction (implicitly presented already in [LL]). If τ is realized by 1-output simple transducer $T = (\Sigma,\Delta,\{s_0,\dots,s_n\},s_0,\{s_0\},E)$, then the morphisms are defined as follows:

$$h_1\colon E^* \to \Sigma^*, \qquad h_1(s_i,u,v,s_j) = u,$$
$$h_2\colon E^* \to (\Delta \cup \{\$\})^*, \qquad h_2(s_i,u,v,s_j) = \$^i v \$^{m-j},$$
$$h_3\colon \Delta^* \to (\Delta \cup \{\$\})^*, \qquad h_3(a) = a\$^m,$$

where m is any natural number $\geqslant n$, and $\$$ is a new symbol not in Δ.

The following remarks on the proof of Proposition 2 will be useful in our later considerations. Firstly, h_3 is injective and can be chosen to be the same for arbitrary two 1-output simple rational transductions. Secondly, the morphisms of the proposition satisfy

$$(1)\qquad h_1h_2^{-1}h_3h_3^{-1}h_2h_1^{-1}(\mathrm{dom}(\tau)) \subseteq \mathrm{dom}(\tau)\ .$$

Next we state two more known results used in our later considerations. The first one is due to Griffiths, cf. [G], and it is, in our terms, as follows:

PROPOSITION 3. It is undecidable whether or not two 1-free nondeterministic sequential transductions are equivalent.

Our last proposition, due to Blattner and Head, cf. [BH], is as follows:

PROPOSITION 4. It is decidable whether or not a rational transduction is single-valued. Moreover, the equivalence of two single-valued rational transductions is decidable.

Observe that Proposition 4 implies that the equivalence of two rational transductions, one of which is single-valued, is decidable, too.

3. RESULTS

By Propositions 1 and 3 we conclude that it is undecidable whether or not two given rational transductions of the form $h_4 h_3^{-1} h_2 h_1^{-1}$ are equivalent on $\Sigma^*\$$, where $\$$ is a new symbol not in Σ. This observation can be strengthened as follows:

THEOREM 1. $EP_\forall(H^{-1} \circ H \circ H^{-1}, Reg)$ is undecidable.

Proof. By Proposition 2, for 1-output simple rational transductions τ and τ' from Σ^* into Δ^* there exist morphisms h_1, h_2, h_3, g_1, g_2 and g_3 (in suitable alphabets) such that $\tau = h_3^{-1} h_2 h_1^{-1}$ and $\tau' = g_3^{-1} g_2 g_1^{-1}$. Hence, Theorem 1 follows from Proposition 3 and the following lemma.

LEMMA 1. For each 1-free nondeterministic sequential rational transduction $\tau_1 : \Sigma^* \to \Delta^*$ there exists a 1-output simple rational transduction $\tau_s : (\Sigma \cup \{\$\})^* \to (\Delta \cup \{\$\})^*$, with $\$ \notin \Sigma \cup \Delta$, such that $\tau_1(x)\$ = \tau_s(x\$)$ for all x in Σ^*.

Proof of Lemma 1. Let τ_1 be realized by a 1-free transducer $T_1 = (\Sigma, \Delta, Q, q_0, Q, E)$. Then we define a 1-output simple transducer $T_s' = (\Sigma \cup \{\$\}, \Delta \cup \{\$\}, Q', q_0, \{q_0\}, E')$, where $\$$ is a new symbol not in $\Sigma \cup \Delta$ and the transitions (and states) of T_s' are defined as follows. Let $e = (q,u,v,q')$, with $v = b_1 \dots b_n$, $n \geq 2$ and each b_i in Δ, be a transition in E. Then define a set $\psi(e)$ of transitions as follows: $\psi(e) = \{(q,u,b_1,q_1),(q_{n-1},1,b_n,q')\} \cup \{(q_i,1,b_{i+1},q_{i+1}) \mid i = 1,\dots,n-2\}$, where the states $q_1,\dots,q_{n-1}$

are new not in Q. We also require that the sets of new states obtained from different transitions as above are mutually disjoint. Further, for a transition $e = (q,u,v,q')$, with $v \in \Delta$, in E let $\psi(e) = \{e\}$. Then we set $E' = \bigcup_{e \in E} \psi(e) \cup \{(q,\$,\$,q_0) \mid q \in Q\}$.

Then, clearly, the statement of Lemma 1 and hence also Theorem 1 follows.

□

Observe that in Theorem 1 the language on which the equivalence of transductions is considered can be assumed to be of the form $\Sigma^*\$$, or, as is easy to see, of the form $\mathrm{dom}(h_3^{-1}h_2h_1^{-1})$.

In the next theorem we still strengthen the result of Theorem 1.

THEOREM 2. $EP_\forall(\mathcal{H}\circ\mathcal{H}^{-1}, \mathcal{R}eg)$ is undecidable.

Proof. Let $(h_3^{-1}h_2h_1^{-1}, g_3^{-1}g_2g_1^{-1}, L)$ be an instance of the problem of Theorem 1. Without affecting the undecidability we may assume, as is straightforward to see, that this instance satisfies the following conditions. Firstly, $L = \mathrm{dom}(h_3^{-1}h_2h_1^{-1}) = \mathrm{dom}(g_3^{-1}g_2g_1^{-1})$. Secondly, $h_3 = g_3$ and moreover h_3 is injective. Thirdly, the morphisms satisfy

$$(2)\qquad \begin{cases} h_1h_2^{-1}h_3h_3^{-1}h_2h_1^{-1}(L) \subseteq L \text{ and} \\ g_1g_2^{-1}h_3h_3^{-1}g_2g_1^{-1}(L) \subseteq L \end{cases}$$

The last two sentences follow from remarks after Proposition 2.

We shall show

$$(3)\qquad \begin{cases} h_3^{-1}h_2h_1^{-1} \stackrel{L}{\equiv} h_3^{-1}g_2g_1^{-1} \quad \text{if and only if} \\ h_2h_1^{-1}(L) = g_2g_1^{-1}(L) \quad \text{and} \quad h_1h_2^{-1}(L) \stackrel{L'}{\equiv} g_1g_2^{-1} \\ \text{where } L' = h_3h_3^{-1}h_2h_1^{-1}(L) \end{cases}$$

The equivalence (3) together with the known properties of regular languages imply Theorem 2.

To prove (3) we first observe, by (2), that $h_3^{-1}h_2h_1^{-1} \stackrel{L}{\equiv} h_3^{-1}g_2g_1^{-1}$ if and only if $L_1 = h_3^{-1}h_2h_1^{-1}(L) = h_3^{-1}g_2g_1^{-1}(L)$ and $h_1h_2^{-1}h_3 \stackrel{L_1}{\equiv} g_1g_2^{-1}h_3$. Hence, the injectiveness of h_3 yields (3). □

Let $\mathcal{C}_w$ denote the family of morphisms h satisfying $|h(a)| \leq 1$ for all letters a and $\mathcal{H}_1$ a family of nonerasing morphisms. Then, by a careful analysis of our proof of Theorem 2, one can see that the family $\mathcal{H}\circ\mathcal{H}^{-1}$ of transductions can actually be replaced by the family

$C_W \circ H_1^{-1}$. Indeed, in the proof of Lemma 1 the transitions of T_s are in $Q\times(\Sigma \cup \{1\})\times\Delta\times Q$ and for such transductions the morphisms h_1 and h_2 in the proof of Proposition 2 are in C_W and H_1, respectively.

To emphasize that the above undecidability results are mainly due to powerful properties of morphisms and not because the equivalence is restricted to complicated enough languages, we still strengthen our result slightly. In order to be able to do this let F denote the family of languages of the form F^* where F is finite.

<u>THEOREM 3.</u> $EP_\forall(H \circ H^{-1}, F)$ is undecidable.

<u>Proof.</u> According to the proof of Theorem 2, $EP_\forall(H \circ H^{-1}, Reg)$ remains undecidable even if only the regular star languages, i.e., regular languages of the form L_1^*, are considered. Indeed, $dom(h_3^{-1}h_2h_1^{-1})$ is always a star language. Further by a result in [LL] (cf. also [T2]) for each regular star language L_* there exist an <u>injective</u> morphism h and a finite language F such that $L_* = h^{-1}(F^*)$. Hence, the result follows from Theorem 2.
□

The remark following Theorem 2 applies to Theorem 3, too. As another remark we state the following interesting corollary of Theorem 3.

<u>THEOREM 4.</u> It is undecidable whether or not two transductions from $H \circ H^{-1} \circ H$ are equivalent.
□

Our next result shows that the order of the morphisms and the inverse morphisms in our previous results is crucial.

<u>THEOREM 5.</u> $EP_\forall(H^{-1} \circ H, Reg)$ is decidable.

<u>Proof.</u> Let $(h_2^{-1}h_1, g_2^{-1}g_1, L)$ be an instance of the problem and let $L' = h_2^{-1}h_1(L)$. If $L' \neq g_2^{-1}g_1(L)$, then the transductions $h_2^{-1}h_1$ and $g_2^{-1}g_1$ are not equivalent on L . Since L' is effectively regular this can be decided. So we may assume that $h_2^{-1}h_1(L) = g_2^{-1}g_1(L) = L'$. By the similar arguments we may also assume that $L \subseteq dom(h_2h_1^{-1}) = dom(g_2^{-1}g_1)$.

We define a partition of L induced by h_1 , in symbols $\sim_{L,h_1}$, as follows: $x \sim_{L,h_1} x'$ if and only if $h_1(x) = h_1(x')$. Similarly, we define partitions $\sim_{L',h_2}$, $\sim_{L,g_1}$ and $\sim_{L',g_2}$. Furthermore, let L_{cs} be a cross section of L with respect to h_1 , i.e., L_{cs} is a regular subset of L such that h_1 maps L_{cs} bijectively onto $h_1(L)$. Such a cross section can be effectively found, cf. [E] or [B]. Similarly, let L'_{cs} be a cross section of L' with respect to h_2.

With the above notation we establish the following diagram:

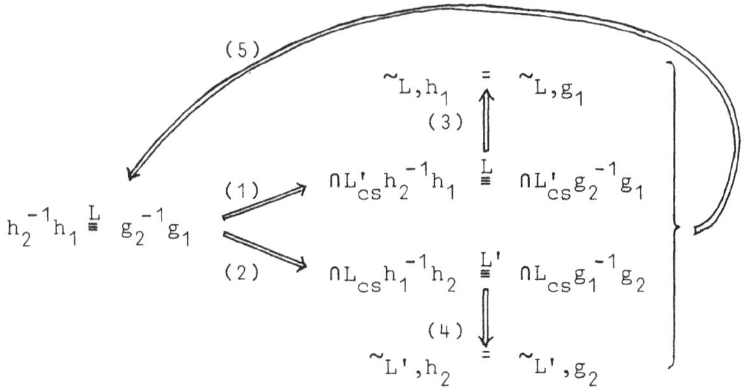

Diagram 1.

Before proving Diagram 1 we point out how Theorem 5 follows from it. By definition, rational transductions $\cap L'_{cs} h_2^{-1} h_1 \cap L$ and $\cap L_{cs} h_1^{-1} h_2 \cap L'$ are single-valued. Consequently, by Proposition 4, it is decidable whether or not the transductions $\cap L'_{cs} h_2^{-1} h_1$ and $\cap L'_{cs} g_2^{-1} g_1$ are equivalent on L, as well as whether or not the transductions $\cap L_{cs} h_1^{-1} h_2$ and $\cap L_{cs} g_1^{-1} g_2$ are equivalent on L'. Hence Diagram 1 implies Theorem 5.

Now, we turn to prove the implications in Diagram 1. Implication (1) is trivial, and also (2) is easy to see. To prove (3) let $\cap L'_{cs} h_2^{-1} h_1 \overset{L}{\equiv} \cap L'_{cs} g_2^{-1} g_1$ and $x \sim_{L,h_1} x'$. Then $h_1(x) = h_1(x')$ and so $\cap L'_{cs} g_2^{-1} g_1(x) = \cap L'_{cs} g_2^{-1} g_1(x')$. This together with our assumption $L \subseteq \mathrm{dom}(h_2 h_1^{-1}) = \mathrm{dom}(g_2 g_1^{-1})$ imply that $g_1(x) = g_1(x')$, i.e., $x \sim_{L,g_1} x'$. Therefore (3) follows by symmetry. Implication (4) can be proved in the same way as (3). Finally, implication (5) follows from the fact that conditions $h_2^{-1} h_1 \overset{L}{\equiv} g_2^{-1} g_1$ and $\cap L'_{cs} h_2^{-1} h_1 \overset{L}{\equiv} \cap L'_{cs} g_2^{-1} g_1$ are equivalent under the assumptions $\sim_{L,h_1} = \sim_{L,g_1}$ and $\sim_{L',h_2} = \sim_{L',g_2}$. Hence, we have proved Diagram 1 and so also Theorem 5. □

It is worth noting that if we consider a weaker equivalence, i.e., the existential equivalence of [KW], of transductions then the problem of Theorem 5 becomes undecidable. We have even the following stronger result: It is undecidable whether or not, for a given triple (h^{-1}, g^{-1}, L) where h and g are morphisms and L a regular language, the relation $h^{-1}(x) \cap g^{-1}(x) \neq \phi$ holds for all x in L. The details can be found in [KM].

We conclude this note with another decidability result.

THEOREM 6. It is decidable whether or not transductions τ and τ' are equivalent on L, where τ and τ' are compositions of morphisms and inverse morphisms such that either all the morphisms of τ or all the inverse morphisms of τ are injective and L is a regular language.

Proof. Firstly, we assume that all the inverse morphisms of τ are injective. Then τ is single-valued and therefore the equivalence of τ and τ' on L can be decided by Proposition 4. Secondly, we assume that all the morphisms of τ are injective. Then $\tau^{-1}\tau(L) \subseteq L$ and therefore τ and τ' are equivalent on L if and only if τ^{-1} and $\cap L\tau'^{-1}$ are equivalent on $\tau(L)$. Now, τ^{-1} is single-valued and therefore, since $\tau(L)$ is effectively regular, the result follows again from Proposition 4. □

As a consequence of Theorem 6 we note that the undecidability problem of Theorem 2 becomes decidable if one assumes that at least one of the four morphisms is injective.

ACKNOWLEDGEMENT. The authors are grateful to Y. Maon, G. Rozenberg and A. Salomaa for useful comments and to the Academy of Finland for financial support.

REFERENCES

[B] Berstel,J., Transductions and Context-Free Languages (B.G. Teubner, Stuttgard, 1979).

[BH] Blattner,M. and Head,T., Single-valued a-transducers, J. Comput. System Sci. 15 (1977) 310-327.

[CFS] Culik,K.II, Fich,F.E. and Salomaa,A., A homomorphic characterization of regular languages, Discrete Appl. Math. 4 (1982) 149-152.

[CS] Culik,K.II and Salomaa,A., On the decidability of homomorphism equivalence for languages, J. Comput. System Sci. 17 (1978) 163-175.

[E] Eilenberg,S., Automata, Languages and Machines, Vol. A (Academic Press, New York, 1974).

[G] Griffiths,T.V., The unsolvability of the equivalence problem for λ-free nondeterministic generalized machines, J. Assoc. Comput. Mach. 15 (1968) 409-413.

[KM] Karhumäki,J. and Maon,Y., A simple undecidable problem: Existential agreement of inverses of two morphisms on a regular language, manuscript (1983).

[KL] Karhumäki,J. and Linna,M., A note on morphic characterization of languages, Discrete Appl. Math. 5 (1983) 243-246.

[KW] Karhumäki,J. and Wood,D., Inverse morphic equivalence on languages, manuscript (1983).

[LL] Latteux,M. and Leguy,J., On the composition of morphisms and inverse morphisms, Lecture Notes in Computer Science 154 (1983) 420-432.

[T1] Turakainen,P., On homomorphic characterization of principal semi AFL's without using intersection with regular sets, Inform. Sci. 27 (1982) 141-149.

[T2] Turakainen,P., A machine-oriented approach to compositions of morphisms and inverse morphisms, EATCS Bull. 20 (1983) 162-166.

SOME CONNECTIONS BETWEEN PRESENTABILITY OF COMPLEXITY CLASSES AND THE POWER OF FORMAL SYSTEMS OF REASONING

Wojciech Kowalczyk
Institute of Mathematics
Warsaw University
PKiN IX p.
00-901 Warsaw

INTRODUCTION

In this paper we consider the following problem: "How do NP$\cap$coNP-complete sets, if any exist, look like?". Sipser [5] has shown that for some oracle X $NP^X \cap coNP^X$ has no complete sets, whereas for any oracle X most of the complexity classes /as NP, PSPACE etc./ relativized to X have complete sets. Then he suggested that if NP$\cap$ coNP-complete sets do exist then they must have quite different nature than complete sets in other complexity classes.

We show that NP$\cap$ coNP has complete sets if and only if NP$\cap$ coNP is presentable in a suitable way. Moreover, having any such presentation we can effectively construct an NP$\cap$ coNP-complete set.

In the case when NP$\cap$ coNP has no complete sets /thus is not suitably presentable/ we can prove that for any reasonable system of reasoning T /as Peano Arithmetic, Zermelo-Fraenkel Set Theory etc./ there exist nondeterministic Turing machines M and N, working in polynomial time, such that $L(M)=\overline{L(N)}$ but $T \nvdash$ "$L(M)=\overline{L(N)}$", where $\overline{L(N)}$ denotes the complement of the set accepted by N.

Thus the assumption that NP$\cap$ coNP has no suitable presentation implies that some sentences are independent of any fixed theory T. We show that this phenomenon takes place for other complexity classes which are not presentable. For instance, knowing that NP-P is not recursively presentable /unless P=NP/ we can prove that some true sentences of the form "$L(M)=L(N)$" or "$L(M)\notin P$", where M /N/ denotes a nondeterministic /deterministic/ Turing machine working in polynomial time, cannot be proved in T.

NOTATION

All machines are represented by sequences of letters from the fixed, finite alphabet Σ which contains 0,1,# etc. If M is a nondeterministic Turing machine then L(M) denotes the set accepted by M /$L(M) \subset \Sigma^*$/, if M is a deterministic Turing machine then f_M denotes the function computed by M /$f_M: \Sigma^* \longrightarrow \Sigma^*$/. If $L \subset \Sigma^*$ then $\bar{L} = \Sigma^* - L$. By the nondeterministic /deterministic/ Turing machine working in polynomial time /NDTMP /DTMP// we mean the usual one tape nondeterministic /deterministic/ Turing machine, which has installed the "clock" which stops computations after n^k steps /$k+n^k$ steps/, where k is a natural number. Let us observe that if M is a DTMP then M may be treated as an acceptor or as a transducer /i.e. a function from Σ^* to Σ^*/; however it will be always clear from the context which case should be taken. Under this convention

$$P = \{L(M) : M \text{ is a DTMP}\},$$

$$NP = \{L(M) : M \text{ is an NDTMP}\},$$

$$coNP = \{\bar{L} : L \in NP\}.$$

We say that $L \subset \Sigma^*$ is $NP \cap coNP$-complete /or complete in $NP \cap coNP$/ if $L \in NP \cap coNP$ and for each $L' \in NP \cap coNP$ there exists a DTMP M such that for each $x \in \Sigma^*$ $x \in L'$ iff $f_M(x) \in L$.

COMPLETNESS IN $NP \cap coNP$

We begin with the proof of Lemma 1 which connects the existence of $NP \cap coNP$-complete sets with the existence of a suitable presentation of $NP \cap coNP$.

Lemma 1.

The following statements are equivalent:

(i) There exists a complete set in $NP \cap coNP$.

(ii) There exists a recursive sequence (M_i, N_i) such that

(a) M_i, N_i are NDTMPs for i=1,2,... ,

(b) $L(M_i) = \overline{L(N_i)}$ for i=1,2,... ,

(c) $NP \cap coNP = \{L(M_i) : i=1,2,\dots\}$.

Proof.

(i)→(ii). Let L be an $NP \cap coNP$-complete set, M and N NDTMPs working in time n^k such that $L(M) = \overline{L(N)} = L$ and let $R_1, R_2, \dots$ be a recursive list of DTMPs which compute all polynomial time computable functions from Σ^* into Σ^*. Now we define the sequence (M_i, N_i), i=1,2,... as follows:

M_i accepts x iff M accepts $f_{R_i}(x)$

and

$$N_i \text{ accepts } x \quad \text{iff} \quad N \text{ accepts } f_{R_i}(x).$$

It is clear that the function $i \longmapsto (M_i, N_i)$ is recursive, and M_i and N_i work in time $(s(i)+n^{s(i)})^k$, where $s(i)+n^{s(i)}$ is the time needed by R_i to compute f_{R_i}. Slightly modifying the construction of M_i and N_i /to assure that words of length 1 are accepted in one step/ we get the desired presentation of $NP \cap coNP$.

(ii)$\rightarrow$(i). Let us suppose that there exists a Turing machine G such that $f_G(0^i) = (M_i, N_i)$, where M_i, N_i are NDTMPs, $L(M_i) = \overline{L(N_i)}$, for $i=1,2,\ldots$, and $NP \cap coNP = \{L(M_i) : i=1,2,\ldots\}$. Let $t(i)$ denotes the number of steps needed by G to compute successively $f_G(0), f_G(0^2), \ldots, f_G(0^i)$ and let $s(i)$ be such that M_i and N_i work in time less than $n^{s(i)}$, for $i=1,2,\ldots$. Clearly, t and s are recursive. Define $L \subset \Sigma^*$ in the following way:

$$L = \{(M_i, N_i, 0^{t(i)}, x, 0^{|x|^{s(i)}}) : x \in L(M_i)\} .$$

We will show that L is $NP \cap coNP$-complete. Let M be the NDTMP which works on input w as follows:

M deterministically checks whether w is in the form

$$(M_i, N_i, 0^{t(i)}, x, 0^{|x|^{s(i)}}) ;$$

if so then M simulates M_i on the input x, otherwise M rejects w.

M does this in no more than $c|w|^2$ steps, where c is a constant which depends only on Σ . Analogously, let N be the NDTMP which for given input w checks whether w is in "good" form; if so then N simulates N_i on the input x, otherwise N accepts w. Thus $L = L(M) = \overline{L(N)}$ and $L \in NP \cap coNP$. Moreover, for any $L' \in NP \cap coNP$ which is recognized by M_i the function

$$f(x) = (M_i, N_i, 0^{t(i)}, x, 0^{|x|^{s(i)}})$$

is the polynomial time reduction of L' to L, so L is $NP \cap coNP$-complete.□

Let T denotes any fixed formal system of reasoning such that:

/1/ the set of all theorems of T is recursively enumerable subset of Σ^*,

/2/ some basic facts about Turing machines are exprssible and provable in T,

/3/ only true sentences about Turing machines can be proved in T.

For example Basic Number Theory, Peano Arithmetic, Zermelo-Fraenel Set Theory satisfy these conditions.

Now we are ready to formulate

Theorem 2.

If there are no complete sets in $NP \cap coNP$ then there exists $L \in NP \cap coNP$ such that for any NDTMPs M and N satisfying $L = L(M) = \overline{L(N)}$

$$T \not\vdash \text{"} L(M) = \overline{L(N)} \text{"}.$$

Proof.

Suppose the opposite. Thus $(NP \cap coNP)_T = NP \cap coNP$, where $(NP \cap coNP)_T$ denotes the following set

$$\left\{ L \subset \Sigma^* : \begin{array}{l} \text{there exist NDTMPs } M \text{ and } N \text{ such that} \\ L(M) = L \text{ and } T \vdash \text{"} L(M) = \overline{L(N)} \text{"} \end{array} \right\}.$$

But the set $\{(M,N) : M \text{ and } N \text{ are NDTMPs and } T \vdash \text{"} L(M) = \overline{L(N)} \text{"}\}$ is recursively enumerable so $NP \cap coNP$ has recursive presentation by pairs (M_i, N_i) such that M_i and N_i are NDTMPs, $L(M_i) = \overline{L(N_i)}$, for $i = 1,2,\ldots$ and $NP \cap coNP = \{L(M_i) : i=1,2,\ldots\}$. This yields a contradiction with Lemma 1. □

Let us observe that $(NP \cap coNP)_T$ is a very interesting class of sets. It lies between P and $NP \cap coNP$ and, for T strong enough, it contains all known sets from $NP \cap coNP$. It may be treated as the "constructive part of $NP \cap coNP$". Moreover it has "almost" complete set. In fact, let

$$L_T = \left\{ (M, N, P(M,N), x, 0^{|x|^k}) : \begin{array}{l} M, N \text{ are NDTMPs working in time } n^k, \\ P(M,N) \text{ is a formal proof of "} L(M) = \overline{L(N)} \text{"} \\ \text{and } x \in L(M) \end{array} \right\}$$

/Without loss of generality we may assume that correctness of proofs can be verified in deterministic polynomial time./

Then we have

Theorem 3.

1/ $L_T \in NP \cap coNP$.

2/ Each $L \in (NP \cap coNP)_T$ is reducible /in polynomial time/ to L_T.

Proof.

Similar to the proof of Lemma 1. □

Thus L_T has the following property:

Corrolary 4.

If there exist complete sets in $NP \cap coNP$ then either L_T is one of them or for any such set and NDTMPs M, N recognizing it and its complement

$$T \not\vdash \text{"} L(M) = \overline{L(N)} \text{"}.$$

Proof.
If for an NP$\cap$ coNP-complete set L there exist NDTMPs M, N such that $L = L(M) = \overline{L(N)}$ and T$\vdash$"$L(M) = \overline{L(N)}$" then L is reducible to L_T, so L_T is NP$\cap$ coNP-complete.□

Moreover, if NP$\cap$ coNP has complete sets then we can choose a theory T making L_T complete in NP$\cap$coNP:

Corrolary 5.

If NP$\cap$ coNP has complete sets then there exists a theory T, satisfying /1/ - /3/, such that L_T is NP$\cap$ coNP-complete.

Proof.
Suppose that there exists an NP$\cap$ coNP-complete set. Thus, by Lemma 1, there exists a recursive sequence (M_i,N_i) of NDTMPs such that NP$\cap$ coNP = $\{L(M_i) \ : \ i=1,2,\dots\}$ and $L(M_i) = \overline{L(N_i)}$, for i=1,2,... . Taking

$$T = \text{Peano Arithmetic} + \{"L(M_i) = \overline{L(N_i)}" \ : \ i=1,2,\dots\}$$

we get the desired theory.□

INDEPENDENCE RESULTS

As we have seen, the assumption that the class NP$\cap$ coNP has no suitable presentation implies that some sentences are independent of T. We will show that this phenomenon occurs for other complexity classes. Landweber, Lipton and Robertson [3] have shown that if P $\neq$ NP then NP-P is not recursively presentable by Turing machines computing characteristic functions /i.e. there is no recursive list $M_1, M_2, \dots$ such that M_i computes a function $f_{M_i}: \Sigma^* \to \{0,1\}$, i=1,2,... and NP-P = $\{f_{M_i}^{-1}(\{1\}) : i=1,2,\dots\}$/. Using this fact we can prove

Theorem 6.

If P $\neq$ NP then

a/ there exist L$\in$P and NDTMP M such that L(M) = L but for any DTMP N T$\not\vdash$"L(M) = L(N)",

b/ there exists L$\in$NP-P such that for any NDTMP M recognizing L T$\not\vdash$ "L(M)$\notin$ P".

/T still denotes any fixed theory satisfying /1/-/3//.

Proof.
a/. Let us suppose that for each L$\in$P and NDTMP M recognizing L there exist the DTMP N and the proof in T of "L(M) = L(N)". We will construct a recursive presentation of NP-P. Let $M_1, M_2, \dots$ be a recursive list of all NDTMPs. For i=1,2,... let M_i' denote the following NDTMP:

M_i' on input x tries to find /in no more than $|x|$ deterministic steps/ a DTMP N and a proof in T of "$L(M_i) = L(N)$" ; if such a proof and N are found then M_i' accepts x iff $x \in SAT$, otherwise M_i' simulates M_i on input x. It is clear that the function $i \mapsto M_i'$ is recursive and

$$L(M_i') = \begin{cases} \text{SAT almost everywhere,} & \text{if } L(M_i) \in P \\ L(M_i) & \text{otherwise.} \end{cases}$$

Thus $NP-P = \{L(M_i') : i=1,2,\ldots\}$ and $M_1', M_2', \ldots$ is the recursive presentation of NP-P which gives a contradiction.

b/. If for each $L \in NP-P$ there exists an NDTMP M such that $L(M) = L$ and $T \vdash$ "$L(M) \notin P$" then NP-P is recursively presentable since the set $\{M : M \text{ is an NDTMP and } T \vdash \text{"}L(M) \notin P\text{"}\}$ is recursively enumerable.□

An analogous theorem also holds for other complexity classes which are not recursively presentable. For the list of such classes see Schöning [4].

Let us observe that sets, which existence is guaranteed by the above theorem, do not explicitly involve any diagonal process. Thus it is quite possible that some problems of practical importance /e.g. PRIMES/ have fast, deterministic algorithms, but the correctness of any such algorithm cannot be proved in T.

REFERENCES

[1] Hartmanis, J. and Hopcroft, J., Independence results in computer science, SIGACT News, vol. 8, no. 4 /1976/ pp. 13-24.

[2] Hartmanis, J., Relations between diagonalization, proof systems, and complexity gaps, Theoretical Computer Science 8 /1979/ 239-253.

[3] Landweber, L., Lipton, R., and Robertson, E., On the structure of sets in NP and other complexity classes, Theoretical Computer Science 15 /1981/. 181-209.

[4] Schöning, U., A uniform approach to obtain diagonal sets in complexity classes, Theoretical Computer Science 18 /1982/ 95-103.

[5] Sipser, M., On relativization and the existence of complete sets, 9th International Colloquium on Automata, Languages and Programing, Springer-Verlag, 1982.

FINDING A MAXIMUM FLOW IN /S,T/-PLANAR NETWORK IN LINEAR EXPECTED TIME

L. Kučera
Charles University
Prague, Czechoslovakia

INTRODUCTION

Since the pioneer work of Ford and Fulkerson [4] much effort has been devoted to the problem of finding a maximum flow in a network. Recently, the development of algorithms for the general problem has apparently gor into a blind alley. All efficient algorithms known are based on the Dinic´s decomposition [2], which has a lower time bound $\Omega(nm)$, where n,m resp. denote the number of vertices, edges, resp. of a network On the other hand, there is an algorithm of Karzanov [7] with the time complexity $O(n^3)$ /which is optimal up to a multiplicative constant among Dinic´s decomposition based algorithms for dense networks with $m \sim n^2$ / and $O(nm \log n)$ algorithm of Sleator and Tarjan [11]. It does not seem that the factor log n in the latter bound /corresponding to the use of a very sophisticated data structure/ could be eliminated and therefore the future progress requires a completely new approach to finding a maximum flow in a general network without the use of the Dinic´s decomposition.

However, the problem of finding a maximum flow is much simpler in some special cases. One of the most important among them is represented by the class of /s,t/-planar networks. /A network is /s,t/-planar if it is planar and if the source and terminal belong to the boundary of the same face/.

We know that the number of edges of any planar directed network is at most 6n, where n is the number of vertices of the network. Thus, the /very complicated/ algorithm of Sleator and Tarjan requires $O(n^2 \log n)$ time when applied to a /s,t/-planar network. On the other hand, the classical algorithm proposed by Berge [1] and Ford and Fulkerson [3], though very simple, has $O(n^2)$ time bound. Using a concept of priority queue,

Itai and Shiloach [6] have given an O(n log n) implementation of this algorithm. In the same paper an Ω(n log n) lower bound was proved for any implementation of the Berge-Ford-Fulkerson algorithm and it is widely believed that this bound applies to every algorithm for finding a maximum flow in a /s,t/-network.

All above time bounds concern the worst case time complexity. An obvious question arises concerning the time bounds for the average case time complexity. The aim of this paper is to show that the maximum flow in a /s,t/-planar network can be found in the linear average time. More precisely, we are going to prove the next

THEOREM: There is an algorithm A and a constant C such that for each positive real number b, each directed planar graph G with n vertices and for each two different vertices s,t of G belonging to the same face of the graph G the next stament holds:
If we choose randomly a real number c(e) for every edge e of the graph G in such a way that $0 < c(e) \leq b$ and the probability distributions of c(e)´s are uniform on $\langle 0,b\rangle$ and independent for different edges, then the algorithm A finds a maximum flow in the network G with the source s, the terminal t and the capacity c in at most C.n steps in the average case and in at most C.n.log n steps in the worst case.

It is known [5] that the Itai-Shiloach algorithm for finding a maximum flow in a /s,t/-planar network G is nothing else than the Dijkstra algorithm for one-to-many shortest path problem in the network dual to G. Thus, the theorem gives simultaneously a linear expected time algorithm for one-to-many shortest path problem for planar graphs.

A construction of the algorithm A of the theorem is quite simple. We shall use the Itai-Shiloach algorithm with an O(1) average case implementation of the priority queue used. A priority queue of this type has been described in [12]. The construction is based on hash techniques or, more correctly, on distribution techniques, because a monotone "hash" function has been used. Unfortunately, the proof of the O(1) average performance postulates a time-independent distribution of keys, while the key of an edge e in the Itai-Shiloach algorithm is equal to $c(e) + f_e$, where f_e is an immediate value of the flow in the moment of insertion of the edge e into the queue and therefore the key distribution "shifts to the right on the number axis". A priority queue described in [9] is constructed with the use in the Itai-Shiloach algorithm in mind. Using this queue, a maximum flow in a /s,t/-planar network can be found in the average time $O(n) + t_{sh}$, where t_{sh} is the time necessary to update a pointer to the location of the minimum item in the queue. It is shown [9] that t_{sh} is O(n) for certain classes of networks, but for some graph t_{sh} grows quadratically even in the average case.

We shall show that t_{sh} is O(n) even in the worst case if the queue used in [9] is dynamized in the sense of [10,12].

The paper is divided into four parts. In the part 2 we give a brief description of the Itai-Shiloach algorithm. Part 3 presents an implementation of priority queue and the last part of the paper gives an analysis of the time complexity of the resulting algorithm.

ITAI-SHILOACH ALGORITHM

A maximum flow f in a /s,t/-planar network G with the source s, the terminal t and capacity c /assumed to be embedded into the plane in such a way that s,t, resp. is the leftmost, rightmost, resp. point of the picture/ can be found using the next algorithm [6] in which
F is the size of the constructed flow,
$c_m(e)$ denotes so called modified capacity of the edge e
and Q is a priority queue with operations INITIALIZE, INSERT(e), DELETE(e) and MIN /MIN is a function returning the edge e minimizing $c_m(e)$ among all edges contained in the queue/.

```
procedure MAX-FLOW;
begin  F:= Ø; INITIALIZE;
for all edges e do f(e):=Ø;
find the uppermost path P from s to t in G;
loop
   for all edges e of the path P not contained in Q do
      begin c_m(e):= c(e) + F; INSERT(e) end;
   e_o:= MIN; F:=c_m(e_o);
   for all edges e in Q such that c_m(e)= F do
      begin DELETE(e); delete e from the graph G; f(e):=F-c_m(e) end;
   if there is no path from s to t in G
   then begin for all edges e in Q do f(e):=F-c_m(e);
                 stop the computation end
   else find the new uppermost path P from s to t in G;
   for all edges e in Q but not on the path P do
      begin  DELETE(e); f(e):=F-c_m(e)  end;
end of loop
end.
```

For the correctness of the algorithm and implementation details see [6]. It was also proved in [6] that the computation stops after

$O(n) + t_Q$ steps, where n is the number of vertices of the network and t_Q is the time necessary to perform all INSERT, DELETE and MIN operations. Thus, the time complexity of the algorithm depends only on the efficiency of the priority queue.

PRIORITY QUEUE

In the rest of the paper we assume the capacities of edges to be less than b , where b is a positive real constant. The number of vertices of a considered network will be denoted by n.

We are going to describe a possible implementation of the priority queue of the procedure MAX-FLOW. As can be easily seen, the next proposition is true during the computation:

If an edge e is contained in the queue Q or is to be inserted into it then $c_m(e)$ is defined and it is $F < c_m(e) \leq F + b$.

Based on this, the next implementation of the queue can be used: the queue consist of two integers p,q such that $0 \leq q < p$ and of doubly linked ordered lists /see e.g. [8],vol.1/ $Q_0, Q_1, \ldots, Q_{p-1}$. /We can use any priority queue implementation with constant time DELETE and MIN and linear time INSERT and INITIALIZE instead of doubly linked ordered lists/.

The operations of the queue are defined as follows / $\lceil x \rceil$ denotes the smallest integer greater or equal to x/ :

```
INIT: begin  p:=1; q:=Ø; r:=Ø; initialize Q_Ø end;

MIN:  begin while Q_q is empty do shift;
            MIN:= the first item in the list Q_q  end;

INSERT:  begin k:=(⌈c_m(e)*p/b⌉ -1+q )  mod p;
               insert e into Q_k; expand end;

DELETE:  begin k:=(⌈c_m(e)*p/b⌉ -1+q )  mod p;
               delete e from Q_k; compress end;
```

The procedure shift shifts a pointer q cyclically one position to the right:

```
procedure shift; begin q:= q+1; if q ≥ p then q:=Ø end;
```

Procedures expand and compress are used to keep the average number of elements of the lists Q_i between 1/2 and 2:

```
procedure expand;
begin r:=r+1; if r ≥ 2*p then
   begin p:= 2*p; q:= 2*q;
   for i:=Ø to p/2 - 1 do
     split the original Q_i into the new Q_2i and Q_2i+1 in such a way that
      Q_2i contains exactly those edges e of the original Q_i satisfying
      the inequality (⌈c_m(e)*p/b⌉-1+q) mod p ≤ 2*i
   end
end;
procedure compress;
begin r:=r-1; if Ø < r ≤ p/2 then
   begin p:=p/2; q:=⌊q/2⌋;
   for i:= Ø to p-1 do
      merge the original Q_2i and Q_2i+1 into the new Q_i
   end
end;
```

It is not very difficult to prove the correctness of the implementation, provided the queue is used in the algorithm described as the procedure MAX-FLOW.

TIME COMPLEXITY ANALYSIS

We shall call an execution of the expand or compress statement to be active if the conditions $r \geq 2p$ or $\emptyset < r \leq p/2$ of the if statement hol i.e. if the queue is restructured.

We are going to estimate the time t_Q spent by operations with the queue Q during computation of a maximum flow by the procedure MAX-FLOW using the above described implementation of a priority queue. The time t_Q can be partitioned as follows:

$$t_Q = t_{ini} + t_{ins} + t_{del} + t_{min} + t_{sh} + t_{mod}$$

where

t_{mod} is the time of the execution of expand and compress procedures together with the first executions of the procedure shift following the active execution of expand /note that an active compress follc wed by an active expand could move a pointer q one position back/;

t_{sh} is the time necessary to perform all shifts statements with the exception of those included into t_{mod};

$t_{ini}, t_{ins}, t_{del}, t_{min}$, resp. is the time necessary to perform the queue operations INITIALIZE,INSERT,DELETE,MIN,resp. with the exception of the time included either in t_{mod} or in t_{sh} .

It is easy to see that $t_{ini} = O(1)$, $t_{del} + t_{min} = O(n)$ in the worst case.

Lemma 1: It is $t_{ins} = O(n)$ in the average case.
Proof: Each edge is inserted into the queue at most once. Owing to the probability distribution of capacities of edges, the probability of inserting an edge into a particular Q_i is the same for all i. Since an insertion into Q_i has linear time complexity, the average time to execute a single INSERT is proportional to the average size of Q_i´s. The use of the expand statement guarantees that the latter quantity is bounded by 2.

Lemma 2: It is $t_{mod} = O(n)$ in the worst case.
Proof: After an active execution of expand or compress procedure the queue Q contains exactly p edges and therefore at least p/2 following execution of these procedures will be inactive. It follows that the time cost of active executions of the expand and compress procedures can be redistributes in such a way that each execution receive a constant amount of time.

The crucial point of the proof of the Theorem is a time bound for t_{sh}. While the number of auxiliary queues $Q_o, \dots, Q_{p-1}$ should be large in order to minimize the average value of t_{ins}, a small value of p is necessary to obtain a small value of t_{sh}, because one execution of the shift procedure corresponds to the increase of the flow by approximately b/p .

Lemma 3: It is $t_{sh} = O(n)$ in the worst case.
Proof: Let us denote steps of the computation of the procedure MAX-FLOW by numbers 0,1,2,... . We shall use the following notation:

$p(\tau)$ is the immediate value of p in the τ-th step;

$\nu(\tau)$ is the first step of computation when all edges contained in the queue Q in the moment τ have already been deleted;

$\varkappa(\tau)$ is the first step $\tau' > \tau$ such that $p(\tau') > p(\tau)$, provided it exists, otherwise $\varkappa(\tau) = \infty$;

$$\mu(\tau) = \begin{cases} \nu(\tau) & \text{if it is } \nu(\tau) < \varkappa(\tau), \\ \mu(\varkappa(\tau)) & \text{otherwise;} \end{cases}$$

$$\lambda(\tau) = \begin{cases} \tau & \text{if it is } \nu(\tau) < \varkappa(\tau), \\ \lambda(\varkappa(\tau)) & \text{otherwise.} \end{cases}$$

The last two definitions are correct, because $p(\tau) \leq n$ for each τ and if $\varkappa(\tau) \leq \nu(\tau)$ then $p(\varkappa(\tau)) = 2\,p(\tau)$ and therefore the recursion involved in the definition of μ and λ is to be repeated at most $\log_2 n$ times.

Let $\tau_o, \dots, \tau_k$ be the longest sequence of steps of computation such that

τ_o is the first step when the initial path from s to t is determined and its edges are insertes into the queue Q,

τ_i, $i \geq 1$, is the first step after $\mu(\tau_{i-1})$ when the new path from s to t is determined and all edges that belong to it /or do not, resp./ are inserted into /deleted from, resp./ the queue Q.

Note that this sequence is uniquely determined. Let ϱ_i be the number of executions of the shift procedure that are involved in t_{sh} and occurs between τ_i and $\mu(\tau_i)$. As t_{sh} increases only during these time intervals, we have

$$t_{sh} = O(\varrho_o + \dots + \varrho_k).$$

Let us prove that it is $\varrho_i \leq 2\,p(\lambda(\tau_i))$. Given a step τ, we know that for each τ', such that $\tau \leq \tau' < \min(\nu(\tau), \varkappa(\tau))$ the queue Q contains at least one edge contained in the queue Q in the step τ and it is $p(\tau') \leq p(\tau)$. It follows that only $p(\tau)$ executions of the shift statement involved in t_{sh} occurs between τ and $\min(\nu(\tau), \varkappa(\tau))$. Especially, the equality

$$\mu(\tau_i) = \nu(\tau_i) \quad \text{implies} \quad \varrho_i \leq p(\tau_i) \leq p(\lambda(\tau_i)).$$

If it is $\varkappa(\tau_i) < \nu(\tau_i)$ then there is a sequence

$$\tau_i = \omega_o, \dots, \omega_m \quad \text{such that}$$

$$\omega_j = \varkappa(\omega_{j-1}) \leq \nu(\omega_{j-1}) \quad \text{for } j = 1, \dots, m$$

$$\mu(\tau_i) = \mu(\omega_m) = \nu(\omega_m) < \varkappa(\omega_m) .$$

As we have shown, it is $\lambda(\tau_i) = \omega_m$ and

$$\varrho_i \leq p(\omega_o) + p(\omega_1) + \dots + p(\omega_m) = (2^{-m} + \dots + 1)\, p(\omega_m) < 2p(\lambda(\tau_i)) .$$

Let E_i be the set of edges contained in the queue Q in the $\lambda(\tau_i)$. The sets E_i are pairwise disjoint and it follows from the description of the comprese statement that the cardinality $|E_i|$ of E_i is at least $p(\lambda(\tau_i))/2$. If the number of edges of the network is n' then

$$\varrho_o + \dots + \varrho_k \leq 2\left(p(\lambda(\tau_o)) + \dots + p(\lambda(\tau_k))\right) \leq 4\left(|E_o| + \dots + |E_k|\right) \leq 4n',$$

which implies $t_{sh} = O(n)$.

We have shown that the only part of the computation of the maximum flow which has a superlinear worst case time complexity is an inserting of edges into the auxiliary queues $Q_o, \dots, Q_{p-1}$. Therefore the first part of the Theorem follows from the lemma 1. In order to obtain an O(n log worst case bound it is sufficient to modify our algorithm in the following way:

If the computation does not stop after n.log n steps, replace the queue Q described in this paper by arbitrary O(log n) implementation of the priority queue /see [8], vol.3/ and finish the computation.

References:

1. Berge,C., Ghouila-Houri,A. Programming, Games and Transportation Networks, Methuen, Agincourt, Ontario .
2. Dinic,E.A. Algorithm for solution of a problem of maximal flow in a network with power estimation. Soviet Math.Dokl. 11 /1970/,1277-1280.
3. Ford,L.R.,Fulkerson,D.R. Maximal flow through a network. Canad.J.Math. 8 /1956/, 399-404.
4. Ford,L.R.,Fulkerson,D.R. Flows in Networks. Princeton University Press, New Jersey, 1962.
5. Hassin,R. Maximum flow in /s,t/-planar networks. Info.Proc.Letters, 13,3 /1981/, 107.
6. Itai,A.,Shiloach,Y. Maximum flow in planar networks. SIAM J.COmput. 8 /1979/,135-150.
7. Karzanov,A.V. Determining the maximal flow in a network by the method of preflows. Soviet Math.Dokl. 15 /1974/, 434-437.
8. Knuth,D.E. The Art of Computer Programming. Vol.1.Basic Algorithms. Vol.3.Sorting and Searching. Addison-Wesley, Reading, 1968 and 1973.
9. Kučera,L. Maximum flow in planar networks. In Mathematical Foundation of Computer Science.Gruska,J.and Chytil,M.eds.,Lecture Notes in Computer Science 118, Springer Verlag, Berlin, 1981, 418-422.
10. Overmars,M.H. The Design of Dynamic Data Structures, Lecture Notes in Computer Science 158, Springer Verlag, Berlin, 1983.
11. Sleator,D,D. An O(nm log n) algorithm for maximum network flow. Ph.D.Thesis, Stanford Univ., 1980.
12. Wiederman,J. Preserving total order in constant expected time. In Mathematical Foundation of Computer Science, Gruska,J. and Chytil,M. eds., Lecture Noters in Computer Science 118, Springer Verlag, Berlin, 1981, 554-562.

NONDETERMINISTIC LOGSPACE REDUCTIONS

Klaus - Jörn Lange
Fachbereich Informatik, Universität Hamburg
Schlüterstraße 70, 2000 Hamburg 13, West Germany

ABSTRACT

Nondeterministic logspace-bounded reductions are introduced. A set in DSPACE(log n), NP-complete with respect to these reductions, is exhibited, which makes the transitive closure of the NLOG(·)-operation quite unlikely, unless NSPACE(log n) = NP is assumed. In contrast to this the NLOG-closure of NLOG(NLOG(·))-classes is shown.

1. INTRODUCTION

The notion of reduction strongly influenced the field of computational complexity. Probably the most important types of reductions are the many-one reductions with a polynomial time or logarithmic space bound. The later one seems to be more useful since it permits us to extend the results to classes below *P* , e.g. NSPACE(log n), LOG(*DCF*), and LOG(*CF*). In fact, the one-way log-tape reduction of [HIM] is even stronger (in the sense of [LLS]) than LOG reductions.
Some results concerning LOG reductions and formal languages are contained in sections 2 and 3.

It is a natural idea to generalize the notion of reduction by using nondeterminism as it was done with languages. In this way several kinds of nondeterministic polynomial time reductions were introduced in [LLS]. In this paper we consider nondeterministic logspace many-one reductions.

Since a logarithmic space bound on the worktape does not ensure polynomial length of the output, we impose in addition a polynomial time bound on the turing transducer. The resulting deifinition of the NLOG reduction is contained in section 4.

In section 5 the nondeterminism of NLOG reductions is related to nonerasing homomorphisms by decomposing a NLOG reduction into a LOG reduction followed by a nonerasing homomorphism followed by an one-way LOG **reduction.** (We use a slightly different definition of one-way LOG reduction than that used in [HIM] ; but the results of section 5 are not affected).

In section 6 we exhibit some sets in DSPACE(log n) complete with respect to NLOG reductions for various (nondterministic) complexity classes. The results make it unlikely - unless *P* = *NP* is assumed, that NLOG reductions fit in between LOG reductions and POL reductions as in the language case, where DSPACE(log n) $\subseteq$ NSPACE(log n) $\subseteq$ *P* holds.

In section 7 relations of NLOG reductions to some families of formal languages are investigated. In particular it is remarkable that the families *LIN* and *EDTOL* , which are not distinguished by LOG reductions, i.e. LOG(LIN) = LOG(EDTOL) = = NSPACE(log n), behave differently with respect to NLOG reductions: while still NLOG(LIN) = NSPACE (log n) we have NLOG(EDTOL) = NP!

2. PRELIMINARIES

Throughout this paper the standard notation from [Ho Ul] or [Su 3] is used. In addition we use some further notions and definitions.
The paper deals with the following complexity classes:

L: = DSPACE(log n), NL: = NSPACE(log n), C: = $APDA_{PT}$(log n), NC: = $NAPDA_{PT}$(log n)
P: = DTIME(POLY), NP: = NTIME(POLY), DLBA: = DSPACE(n), CS: = NSPACE(n), and
PSPACE: = DSPACE(POLY), where Poly indicates the union over all polynomials.

The known relations of these classes are:

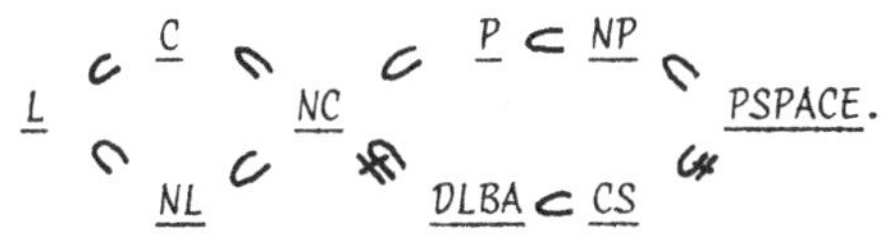

With REG, LIN, DCF, CF, and CS we denote the class of regular, linear, deterministic context-free context-free, and context-sensitive languages. For the definition of EOL, EDTOL, and ETOL we refer to [Ro Sa] .

We call a nonerasing homomorphism $c:\Sigma^* \longrightarrow \Pi^*$ a coding, iff $c(\Sigma) \subset \Pi$. A homomorphism $h:\Sigma^* \longrightarrow \Pi^*$ is called polynomially erasing on a language $L \subset \Pi^*$, iff there exists a polynomial p such that $/v/ \leq p(/h(v)/)$ holds for all v in Σ^*, where /v/ denotes the length of v. For a class of languages $\mathfrak{A}$ set $C(\mathfrak{A}) := \{c(L) \mid L \in \mathfrak{A}, c$ is a coding$\}$ and $H_{pe}(\mathfrak{A}) := \{h(L) \mid L \in \mathfrak{A}$, h is an on L polynomially erasing homomorphism$\}$.
Let TRIO ($\mathfrak{A}$) denote the smallest family of languages containing $\mathfrak{A}$ and closed under nonerasing homomorphism, inverse homomorphism and inersection with regular sets.
Let $TRIO_{pe}(\mathfrak{A})$ be the smallest family containing $\mathfrak{A}$ and closed under polynomially erasing homomorphism, inverse homomorphism and intersection with regular sets.

Remark: $TRIO_{pe}(D_2)$ = CF and $TRIO_{pe}(S_2)$ = LIN, where D_2 is the semidyck-set and S_2 the symmetric language of degree two. (see [Be]).

3. DETERMINISTIC LOG-SPACE REDUCTIONS

One of the main tools in complexity theory is the notion of reduction. In this paper we consider logspace-bounded reductions both for two-way and for one-way input.
We say that an off-line Turing machine A (as defined in [Ho Ul] , but) with an

one-way input tape is logspace-bounded iff for every input v and every prefix u of v = = uw A has visited at most $\lceil \log(|u|) \rceil$ cells of its working tape before reading the fi symbol of w, where $\lceil x \rceil$ denotes the smallest integer greater or equal x.

Definition 3.1: Let $h:\Sigma^* \longrightarrow \Pi^*$ be a recursive function, $L \subset \Sigma^*$; and $M \subset \Pi^*$.

a) h is a reduction from L to M iff $v \in L \Leftrightarrow h(v) \in M$ holds for all v in Σ^*, or equivalently $h^{-1}(M) = L$.

b) A reduction h from L to M is a LOG-reduction iff h is computable by a logspace-bounded deterministic Turing machine. In this case we write $L \leq_{LOG} M$.

c) A reduction h from L to M is an one-way LOG-reduction iff h is computable by a one-way logspace-bounded Turing machine. In this case we write $L \leq_{1\text{-}LOG} M$.

d) The LOG-closure of a laguage family $\mathfrak{A}$ is the class of all languages LOG-reducible to elements of $\mathfrak{A}$: $LOG(\mathfrak{A}) := \{L \mid \exists M \in \mathfrak{A} : L \leq_{LOG} M\}$.

If $\mathfrak{A}$ is unary, i.e. $\mathfrak{A} = \{M\}$ we drop the parentheses using LOG(M).

e) The one-way LOG-closure of $\mathfrak{A}$ is defined by

$1\text{-}LOG(\mathfrak{A}) := \{L \mid \exists M \in \mathfrak{A}: L \leq_{1\text{-}LOG} M\}$.

f) A family $\mathfrak{A}$ is LOG-closed if $LOG(\mathfrak{A}) = \mathfrak{A}$.

g) A set $L \in \mathfrak{A}$ is $\mathfrak{A}$-complete (with respect to LOG-reductions) iff $LOG(L) \supset \mathfrak{A}$.

Remark: In order to avoid confusions it should be remarked that the 'directions' of t C(·)- and the LOG(·)-operations are opposite to each other. If, for instance, $L \in LOG(C(M))$ then there exists a language N, a coding c and a logspace computable function h such that $c(M) = N$ and $h^{-1}(N) = L$.

The importance of the notion of LOG-reducibility is reflected by the fact that except *DLBA* and *CS* al complexity classes mentioned above are LOG-closed and possess complete sets (see [Ha] and [Su 3]). A decisive point in this connection is the transitivity of LOG-reductions (see [St Me]), i.e. $LOG(LOG(\mathfrak{A})) = LOG(\mathfrak{A})$ and $1\text{-}LOG(1\text{-}LOG(\mathfrak{A})) = 1\text{-}LOG(\mathfrak{A})$ for arbitrary $\mathfrak{A}$.

We close this section with some well-known facts about LOG-reductions.

Proposition 3.2: *L* is the smallest LOG-closed class and each nonempty element of *L* is *L*-complete.

LOG-closure relates complexity classes to formal languages:

Proposition 3.3: ([Jo Sk] , [Su 2], [Su 3] , [vL]):

L = LOG(*REG*), *NL* = LOG(*LIN*) = LOG(*EDTOL*), *C* = LOG(*DCF*), *NC* = LOG(*CF*) = LOG(*EOL*), and *NP* = LOG(*ETOL*).

Question: Is there a formal language class $\mathfrak{A}$ defined by some rewriting system such that $LOG(\mathfrak{A}) = $ *P* or $LOG(\mathfrak{A}) = $ *PSPACE*?

4. NONDETERMINISTIC LOGSPACE REDUCTIONS

Nondeterminism is a basic concept both in formal language theory and in complexity theory. It is now a natural idea to use this concept for reducibility as it was used for recognition of languages.

Definition 4.1: A set-valued mapping $\tau:\Sigma^{*}\longrightarrow 2^{\Pi^{*}}$ (where 2^{N} denotes the powerset of a set N) is called a nondeterminsitic reduction from $L \subset \Sigma^{*}$ to $M \subset \Pi^{*}$, iff $\forall\ v \in \Sigma^{*}$: $v \in L \Leftrightarrow \tau(v) \cap M \neq \emptyset$ or equivalently $\tau^{-1}(M) = L$.

Remark: In Definition 4.1 we used the standard definition of an inverse, set-valued mapping, which is the 'existential' version. If we define it in the 'universal' way, i.e. $\tau^{-1}(M):=\{v \in \Sigma^{*} \mid \tau(v) \subset M\}$, (compare this with the notion of weak and strong control set in [Gi Ro]), then in the following we would come to complements of nondeterministic complexity classes.

If we want to consider complexities of nondeterministic reductions, we first have to determine our machine model, which is characterized by the following postulations:
A transforming Turing machine has

- one two-way read-only input tape (sometimes we consider one-way input, which will be stated explicitly),
- one two-way read/write working tape,
- one one-way write-only output tape, and
- a finite, nondeterministic control with accepting and rejecting states.(These we use only to ease the handling. The entering of a final rejecting state could be simulated by printing a garbage symbol on the output tape).

We say that the machine is logspace-bounded, iff for every terminating computation (not necessarily accepting) on an input v the number of used cells on the working tape is bounded by $\lceil \log(/v/) \rceil$.
Remark: There is no difference in postulating the logspace-bound for accepting computations only, because the log-function is fully space-constructible. The definition of logspace-bounded one-way machines corresponds to that of the deterministic case.

Now, how to restrict the complexity of nondeterminstic reductions? An obvious idea would be: A nondeterministic reduction τ from $L \subset \Sigma^{*}$ to $M \subset \Pi^{*}$ is a NLOG-reduction, iff there exists a logspace-bounded nondeterministic Turing machine A, such that for every v in Σ^{*} $w \in \tau(v)$ iff there is an accepting computation of A on v with output w. But in that way each recursive enumerable set would be reducible to $\underline{L}$ (or any other complexity class) by 'unbounded padding': If B is a deterministic Turing machine accepting a language $L \subset \Sigma^{*}$, set $\tau(v) := v\#^{*}$, where # is a padding symbol not in Σ, and set $M:= \{v\#^{n} \mid$ B accepts v using $\lceil \log(n) \rceil$ tapesquares$\}$. Then τ is a nondeterministic reduction from L to M computable without workingtape and with one-way input, i.e. by an a-transducer.
The solution we choose is to restrict the transforming Turing machine A to have polynomial running time. (It would be equivalent to demand only polynomial output

length, because the output tape is one-way, the number of configurations of a log-space-bounded workingtape is bounded by a polynomial, and polynomials are fully time-constructible.)

Definition 4.2: a) A nondeterministic reduction τ from $L \subset \Sigma^*$ to $M \subset \Pi^*$ is called a (one-way)NLOG-reduction, iff there exist a polynomial p and a nondeterministic logspace-bounded (one-way) transforming Turing machine A such that

i) Each computation of A on input $v \in \Sigma^*$ terminates within $p(/v/)$ steps and

ii) $w \in \tau(v)$ iff there exists an accepting computation of A on v with output w.

This case we denote by $L \leq_{NLOG} M$ $(L \leq_{1\text{-}NLOG} M)$.

b) $NLOG(\mathfrak{A}) := \{L \mid \exists\, M \in \mathfrak{A}\colon L \leq_{NLOG} M\}$, and

c) $1\text{-}NLOG(\mathfrak{A}) := \{L \mid \exists\, M \in \mathfrak{A}\colon L \leq_{1\text{-}NLOG} M\}$.

Obviously we have

Proposition 4.3: $LOG(\mathfrak{A}) \subset NLOG(\mathfrak{A})$ and $1\text{-}LOG(\mathfrak{A}) \subset 1\text{-}NLOG(\mathfrak{A})$ for arbitrary $\mathfrak{A}$.

Let SAT be the set of satisfiable Boolean expressions and QBF the set of true, quantified Boolean formulae.

Theorem 4.4: a) $NLOG(\underline{REG}) = \underline{NL}$,

b) $NLOG(SAT) = NLOG(\underline{ETOL}) = \underline{NP} = NLOG(\underline{NP})$, and

c) $NLOG(QBF) = \underline{PSPACE} = NLOG(\underline{PSPACE})$.

Proof: a) Each nondeterministic logspace-bounded algorithm A for recognition of some L in $\underline{NL}$ can be regarded as a NLOG-reduction from L to the set {ACCEPT} (or {+}, {True}, {Yes}, {1} ,...), because we can enforce polynomial running time by augmenting A with a clock. Hence $\underline{NL} = NLOG(\{ACCEPT\}) \subset NLOG(\underline{REG})$. On the other hand each NLOG-reduction to a regular set R can be converted into a nondeterministic logspace-bounded algorithm by dropping the output tape and instead encoding some finite automaton accepting R into the finite control of the transforming machine.

b) By [vL] and Proposition 4.3 we have $\underline{NP} \subset NLOG(SAT) \subset NLOG(\underline{ETOL}) \subset NLOG(\underline{NP})$. But it is easy to see, that $\underline{NP}$ is closed under NLOG-reductions, i.w. $NLOG(\underline{NP}) = \underline{NP}$.

c) similar to part b).

If we would know

(*) $\forall\, \mathfrak{A}$: $LOG(\mathfrak{A}) = NLOG(\mathfrak{A})$

then we would have by Pr. 3.3 and Th. 4.4 $\underline{L} = \underline{NL}$ and hence the LBA-problem would be solved. Below we will see, that (*) even implies $\underline{L} = \underline{NL} = \underline{C} = \underline{NC} = \underline{P} = \underline{NP}$.

5. DECOMPOSITION OF NLOG-REDUCTIONS

In this section we look for the closure of complexity classes under NLOG-reductions and hence for the transitivity of the $NLOG(\cdot)$-operation. First we decompose NLOG-reductions.

Theorem 5.1: $NLOG(\mathfrak{A}) = LOG(1\text{-}NLOG(\mathfrak{A}))$ for arbitrary class $\mathfrak{A}$.

Proof: "$\subset$" Let $M \in \mathfrak{A}$ and $L \in NLOG(\mathfrak{A})$. Further, let τ be a NLOG-reduction from $L \subset \Sigma^*$

to $M \subset \Pi^*$. Then there exist a polynomial p and a nondeterministic logspace-bounded TM A, which fulfill the conditions of Def. 4.2. Now we construct a nondeterministic one-way logspace-bounded machine B as follows: B only accepts words (more precisely allows accepting computations on inputs) in $(\Sigma^*\$)^*$, where \$ is a new symbol not in Σ. B reads and counts at first symbols of Σ until it reaches the first \$ symbol. It stores the binary representation of the number n of read Σ symbols on its working tape. Then B initiates a binary counter i with 1, which can take values between 1 and n. Now B does the following actions (if possible; whenever any condition is not satisfied, B stops in a rejecting state):

1 B reads i symbols in Σ from the input tape.

2 B simulates (nondeterministically) one step of A applied on the symbol last read on the input tape. The result on the workingtape and the actual state are stored (or simulated) on the working tape and the finite control of B, directly.

The movement of the (two-way) input head of A is simulated by either increasing or decreasing the counter i.

3 If A has not reached a final state, B proceeds to the next \$ symbol and repeates step 1. Otherwise B finishes the input and stops according to A. Obviously B is a nondeterministic one-way logspace-bounded transforming machine.

Let $\sigma: (\Sigma \cup \{\$\})^* \longrightarrow 2^{\Pi^*}$ be the mapping realized by B and set $N := \sigma^{-1}(M) =$ $= \{u \in (\Sigma^*\$)^* \mid \sigma(u) \cap M \neq \emptyset\}$. Clearly we have $N \in$ 1-NLOG($\mathfrak{A}$). Now define $f: \Sigma^* \longrightarrow (\Sigma^*\$)^*$ by $f(v) := (v\$)^t$, where $t := p(/v/)$. Then f is logspace-computable and it is easy to see that $f^{-1}(N) = L$ holds because of $\tau^{-1}(M) = L$. Hence $L \in$ LOG(1-NLOG($\mathfrak{A}$)).

"$\supset$" Let $N = \sigma^{-1}(M)$ for some $M \in \mathfrak{A}$, where σ is a 1-*NLOG*-reduction realized by a machine B, and $L = f^{-1}(N)$, where f is a LOG-reduction realized by a machine A. Then simulate B and A simultaneously; whenever B demands an input symbol, continue the simulation of A until it prints the next output symbol on its (one-way!) output tape. This symbol use as the next input for B. The whole construction is clearly logspace-bounded.

Theorem 5.1 proves the existence of *NL*-complete sets in 1-*NL*, as shown by Sudborough in [Su 1]. In Corollary 5.4 we will see, that this holds even for the class C(1-*L*), which is a subclass of 1-*NL*.

Because of the transitivity of LOG-reductions we get the LOG-closure of NLOG-classes:

Corollary 5.2: LOG(NLOG($\mathfrak{A}$)) = NLOG($\mathfrak{A}$) for arbitrary $\mathfrak{A}$

Proof: LOG(NLOG($\mathfrak{A}$)) = LOG(LOG(1-NLOG($\mathfrak{A}$))) = LOG(1-NLOG($\mathfrak{A}$)) = NLOG($\mathfrak{A}$).

In the following we characterize the amount of nondeterminism used by (one-way) NLOG-reductions in terms of nonerasing homomorphisms:

Theorem 5.3: 1-NLOG($\mathfrak{A}$) = 1-LOG(C(1-LOG($\mathfrak{A}$))) for arbitrary $\mathfrak{A}$.

Proof: "$\subset$" Assume $L \in$ 1-NLOG($\mathfrak{A}$), hence $L = \tau^{-1}(M)$ for some $M \in \mathfrak{A}$ and a τ realized by a nondeterministic one-way logspace-bounded machine A. Without loss of generality A has at each step at most two choices for its next step. Let k be the number of states of A and m the number of working tape symbols. If v is an element of Σ^*,

$v = a_1a_2...a_n$, $a_i \in \Sigma$, then A, working on v, has used at most $\lceil \log(i-1) \rceil$ cells on its workingtape before reading the symbol a_i. Hence A can perform at most $m^{\lceil \log i \rceil} \cdot k \cdot \lceil \log i \rceil$ steps without loop between reading the symbols a_i and a_{i+1}. Now set $d := \lceil \log m \rceil + 1$ and $q(i) := k \cdot 2^d \cdot i^d$. Then clearly $m^{\lceil \log i \rceil} k \cdot \lceil \log i \rceil \leq q(i)$. Therefore each possible computation of A on v is determinable by strings over $\{0,1\}$ of length q(i) inserted after each a_i. We construct a deterministic machine B, which accepts inputs from $(\Sigma\{0,1\}^*)^*$. While possible B does the following actions: B reads a symbol from Σ and simulates A for at most q(i) steps, where i is a counter in B for the number of read Σ symbols. The simulation is deterministic controlled by 0,1-symbols. If A is to read the next input symbol, i is increased by one and B skips elements from $\{0,1\}$ until an element of Σ is reached. If A stops accepting, B finishes the input. If anything else happens (overflow of the q(i)-counter, rejecting by A, not enough control symbols,...) B finishes the input and prints a 'garbage' symbol ¶ not in Π on the output tape.

Obviously, B is a deterministic one-way logspace-bounded machine realizing a mapping $g: (\Sigma \cup \{0,1\})^* \longrightarrow (\Pi \cup \{¶\})^*$. Then $N := g^{-1}(M) \in$ 1-LOG(M). Define the coding $c: (\Sigma \cup \{0,1\})^* \longrightarrow (\Sigma \cup \{0\})^*$ by $c(0) := c(1) := 0$ and $c(a) := a$ for all a in Σ. Further on define $f: \Sigma^* \longrightarrow (\Sigma \cup \{0\})^*$ by $f(a_1a_2...a_n) := a_1 0^{q(1)} a_2 0^{q(2)} ... a_n 0^{q(n)}$, which is a 1-LOG-reduction from L to f(L). Now we put all parts together by observing that $g(u) \in M$ iff $u \in N$ iff $c(u) \in f(L)$, because of $\tau^{-1}(M) = L$. That is $L = f^{-1}(c(g^{-1}(M)))$ and hence $L \in$ 1-LOG(C(1-LOG($\mathfrak{A}$))).

"$\supset$" Essentially the same idea as in the proof of Th.5.1. "$\supset$".

Corollary 5.4: NLOG($\mathfrak{A}$) = LOG(C(1-LOG($\mathfrak{A}$))) for arbitrary $\mathfrak{A}$.

Now we can treat the question of NLOG-closure or transitivity of NLOG-reductions. The one-way case behaves similar to the deterministic case.

Proposition 5.5: 1-NLOG(1-NLOG($\mathfrak{A}$)) = 1-NLOG($\mathfrak{A}$) for arbitrary $\mathfrak{A}$.

Proof: The idea is parallel simulation as in Th.5.1.

This contrasts the two-way case, where we will show that in general NLOG($\mathfrak{A}$) $\neq$ NLOG(NLOG($\mathfrak{A}$)), unless $\underline{NL} = \underline{NC} = \underline{P} = \underline{NP}$. But we can prove:

Theorem 5.6: NLOG(NLOG($\mathfrak{A}$)) = NLOG(LOG($\mathfrak{A}$)) for arbitrary $\mathfrak{A}$.

Proof: Only "$\subset$" has to be shown. Assume $L \subseteq \Sigma^*$, $M \subseteq \Pi^*$, $N \subseteq \Gamma^*$, $N \in \mathfrak{A}$ and let σ and be NLOG-reductions such that $L = \sigma^{-1}(M)$ and $M = \tau^{-1}(N)$. Further let τ be realized by a q-time-bounded NTM A. We have to put the nondeterminism of A into σ. This is done in a way similar to the proof of Th. 5.3. So we give only a sketch. We simulate A deterministically on input $v \in \Pi^*$ with the help of a 'controlstring' α in $\{0,1\}^*$ of length q(/v/). This controlstring can be given in one block because we now work in two-way mode and logspace is enough to have 'book-marks'. We modify σ to σ' defined by $\sigma'(u) := \{v\$\alpha \mid v \in \sigma(u), \alpha \in \{0,1\}^*, /\alpha/ = q(/v/)\}$ for all u in Σ^*.

Similar to Th. 5.3. we construct a deterministic two-way machine B, which on input $v \in (\Pi \cup \{0,1\})^*$ first checks $v = v'\$\alpha$ for some $v' \in \Pi^*$ and $\alpha \in \{0,1\}^*$ and then $/\alpha/ = q(/v'/)$ or else rejects by printing a garbage symbol not in Γ. Then B simulates at most $q(/v/)$ steps of A on v' according to α, where rejecting or break-up of the simulation is again done by printing a garbage symbol. It should be clear that for the resulting mapping f we have $L = \sigma'^{-1}(f^{-1}(N)), f^{-1}(N) \in \text{LOG}(N)$ and $L \in \text{NLOG}(f^{-1}(N)) \subset \text{NLOG}(\text{LOG}(N) \subset \text{NLOG}(\text{LOG}(\mathfrak{A}))$.

Corollary 5.7: $\text{NLOG}(\text{NLOG}(\text{NLOG}(\mathfrak{A}))) = \text{NLOG}(\text{NLOG}(\mathfrak{A}))$ for arbitrary $\mathfrak{A}$.

Proof: $\text{NLOG}(\text{NLOG}(\text{NLOG}(\mathfrak{A}))) = \text{NLOG}(\text{LOG}(\text{NLOG}(\mathfrak{A}))) = \text{NLOG}(\text{NLOG}(\mathfrak{A}))$ by Cor. 5.2.

Hence $\text{NLOG}(\text{NLOG}(\mathfrak{A}))$ is closed under *NLOG*-reductions. It is interesting to compare the following consequence of Th. 5.6. with Cor. 5.4:

Corollary 5.8: $\text{NLOG}(\text{NLOG}(\mathfrak{A})) = \text{LOG}(C(\text{LOG}(\mathfrak{A})))$ for arbitrary $\mathfrak{A}$.

Proof: By Th. 5.6., Cor. 5.4. and the transitivity of LOG-reductions we get $\text{NLOG}(\text{NLOG}(\mathfrak{A})) = \text{NLOG}(\text{LOG}(\mathfrak{A})) = \text{LOG}(C(1\text{-LOG}(\text{LOG}(\mathfrak{A})))) = \text{LOG}(C(\text{LOG}(\mathfrak{A})))$.

6. NLOG-REDUCTIONS AND COMPLETENESS

In this section some sets complete for well-known complexity classes with respect to NLOG-reductions will be exhibited. To do so, we first derive some results on TRIO operations.

Lemma 6.1: $\text{TRIO}_{pe}(\mathfrak{A}) \subset \text{NLOG}(\mathfrak{A})$ for arbitrary $\mathfrak{A}$.

Proof: By standard arguments each $L \in \text{TRIO}_{pe}(\mathfrak{A})$ is representable as $L = f(g^{-1}(M) \cap R)$ for some $M \in \mathfrak{A}$, $R \in \underline{REG}$, g a homomorphism, and f a homomorphism, the erasing of which is bounded on $g^{-1}(M) \cap R$ by a polynomial p. Let Γ be the alphabet of R. We have to compute nondeterministically $v \longmapsto \tau(v) := \{g(u) \mid \exists\ u \in \Gamma^* : f(u) = v, u \in R\}$ within log-space, since $\tau^{-1}(M) = L$. If f would be nonerasing, τ would be computable even by an a-transducer. Now set $\Gamma' := f^{-1}(\lambda) \cap \Gamma$, where λ denotes the empty word. The set $f^{-1}(v)$ can be build up by first generating all words u in $(\Gamma \setminus \Gamma')^*$ with $f(u) = v$ and then shuffling these words with elements of Γ'^*. Since f is polynomial erasing on $g^{-1}(M) \cap R$ it is sufficient to shuffle at most $p(/v/)$ symbols from Γ' into these u's. By augmenting the a-transducer with a p(n)-counter, we get the result.

With the help of L.6.1. it is now possible to show the $\underline{NP}$-completeness w.r.t. NLOG-reductions of the set $\text{COPY} := \{(\$v)^n \mid n \in \mathbb{N}, v \in \{0,1\}^*\}$.

Theorem 6.2: a) $\text{NLOG}(\text{COPY}) = \underline{NP}$ and

b) $\text{LOG}(\text{COPY}) = \underline{L}$.

Proof: By [Fl St] we know $\text{LOG}(\text{TRIO}(\text{COPY})) = \underline{NP}$. Now Lemma 6.1 and Theorem 4.4 imply $\text{NLOG}(\text{COPY}) = \underline{NP}$. $\text{COPY} \in \underline{L}$ and Proposition 3.2 imply $\text{LOG}(\text{COPY}) = \underline{L}$.

Remark 1: If $\mathrm{NLOG}(\mathfrak{A}) = \mathrm{LOG}(\mathfrak{A})$ for all $\mathfrak{A}$, then we would have $\underline{L} = \underline{NL} = \underline{C} = \underline{NC} = \underline{P} = \underline{NP}$ and hence $\underline{DLBA} = \underline{CS}$.

Remark 2: If $\mathrm{NLOG}(\mathrm{NLOG}(\mathfrak{A})) = \mathrm{NLOG}(\mathfrak{A})$ for all $\mathfrak{A}$, then we would have $\underline{NL} = \mathrm{NLOG}(\underline{REG}) = \mathrm{NLOG}(\mathrm{NLOG}(\underline{REG})) = \mathrm{NLOG}(\underline{NL}) = \underline{NP}$ because of COPY $\in \underline{NL}$. which would imply $\underline{NL} = \underline{NC} = \underline{P} = \underline{NP}$.

Question: Is there a family $\mathfrak{A}$ such that $\mathrm{NLOG}(\mathrm{NLOG}(\mathfrak{A})) = \underline{PSPACE}$ holds, but probably not $\mathrm{NLOG}(\mathfrak{A}) = \underline{PSPACE}$? (With probably we mean 'as probably as $\underline{NL} = \underline{NP}$').

Remark 3: If POL(·) denotes the closure under polynomial time reductions ([Ho Ul]), we see $\mathrm{NLOG}(\underline{REG}) = \underline{NC} \subset \underline{P} = \mathrm{POL}(\underline{REG})$ and

$\mathrm{NLOG}(\mathrm{COPY}) = \underline{NP} \supset \underline{P} = \mathrm{POL}(\mathrm{COPY})$, where both inclusions are conjectured to be strict. This corresponds in a way with the existence of certain oracle sets in connection with the relativized '$\underline{NL} \subset \underline{P}$'-question . ([La Ly]).

Another consequence of Lemma 6.1 is the existence of simple complete sets for $\underline{NL}$ and $\underline{NC}$, what we state without proof.

Theorem 6.3: a) $\mathrm{NLOG}(D_2) = \underline{NC}$ and

b) $\mathrm{NLOG}(S_2) = \underline{NL}$. (Compare with the remark in the second section)

Contrast this with the fact $\mathrm{LOG}(D_2) = \mathrm{LOG}(S_2) = \underline{L}$.

Remark: Let L_{ECF} denote the set of all context-free grammars, which generate a non-empty language. Then $\mathrm{LOG}(L_{ECF}) = \underline{P}$ is well-known. Using the machine model of [Ga Jo] it is easy to show $\mathrm{NLOG}(L_{ECF}) = \underline{NP}$.

7. FORMAL LANGUAGES AND NLOG-REDUCTIONS

In this section we give a short survey on nondeterministic analogies to proposition 3.3.

We already showed $\underline{NL} = \mathrm{NLOG}(\underline{REG})$ and because of COPY $\in \underline{EDT0L}$ also $\underline{NP} = \mathrm{NLOG}(\underline{EDT0L}) = \mathrm{NLOG}(\underline{ET0L})$. Note that $\underline{EDT0L} \subset \underline{NL}$!

In contrast to this the classes $\underline{LIN}$ and $\underline{CF}$ behave less dramatically.

Proposition 7.1: $\mathrm{NLOG}(\underline{CF}) = \underline{NC}$.

Proof: Only "$\subset$" has to be shown. If $M \in \underline{CF}$, $L = \tau^{-1}(M)$, and τ is realized by a NTM A which is time-bounded by a polynomial p, and if M is accepted in realtime by a PDA B, then simulate B on the output of A, which is possible because B has one-way input. The whole construction works in polynomial time and hence $L \in \mathrm{NAPDA}_{PT}(\log n) = \underline{NC}$.

Since $\underline{CF} = \mathrm{TRIO}_{pe}(\underline{DCF})$ we get

Corollary 7.2: $\mathrm{NLOG}(\underline{DCF}) = \underline{NC} = \mathrm{NLOG}(\underline{CF}) = \mathrm{LOG}(\underline{CF})$.

Sudborough showed in [Su 1] $\mathrm{LOG}(\underline{E0L}) = \mathrm{LOG}(\underline{CF})$. It is not hard to show with his result $\mathrm{NLOG}(\underline{E0L}) = \mathrm{LOG}(\underline{CF})$.

Finally we get

Theorem 7.3: NLOG(*LIN*) = *NL*.

Sketch of proof: We have to show NLOG(*LIN*) ⊂ *NL*, only. For each L ∈ NLOG(*LIN*) there is some k ∈ N, such that L is accepted by a PDA A with k two-way input heads, whose push-down head is only allowed to make one turn. We construct an nondeterministic finite automaton B with 2·k two-way input heads accepting L: B guesses the 'turn-configuration' of A, except for the content of the push-down store, i.e. the position of all k heads and the state of the finite control. Similar to the proof method for *LIN* ⊂ *NL* [Fl St] , B simulates the reading of the pushdown store forward to an endconfiguration with k heads, where the content of the push-down store is generated by synchronous backward simulation of the writing of the push-down store, again with k heads.

REFERENCES

[Be] J. Berstel, Transductions And Context-Free Languages, Teubner, Stuttgart, 1979

[Fl St] P. Flajolet, J.M. Steyart, Complexity Of Classes Of Languages And Operators, IRIA Rapport De Recherche N° 92, 1974

[Ga Jo] M. Garey, D. Johnson, Computers and Intractability: A Guide to the Theory of NP-Completeness, W.H. Freeman and Company, San Fransisco, 1979

[Gi Ro] S. Ginsburg, G. Rozenberg, TOL schemes and control sets, Inform. and Control 27 (1974), 109-125

[Ha] J. Hartmanis, Feasible Computations and Provable Complexity Properties, SIAM Monographie, Philadelphia, 1978

[H I M] J. Hartmanis, N. Immerman, S. Mahaney, One-way Log-Tape Reductions, Proc. of the 19th Symposium on Foundations of Computer Science (1978), 65-71

[Ho Ul] J.E. Hopcroft, J.D. Ullman, Introduction to Automata Theory, Languages, and Computation, Addison-Wesley Publ., Reading, 1979

[Jo Sk] N.D. Jones, S. Skyum, Recognition of deterministic ETOL Languages in logarithmic space, Inform. and Control 35 (1977), 177-181

[La Ly] R.E. Ladner, N. Lynch, Relativization of questions about Log space computability, Math. Sys. Theory 10 (1976), 19-32

[L L S] R. Ladner, N. Lynch, A. Selman, A Comparison of Polynomial Time Reducibilities, Theor. Comput. Sci. 1 (1975), 103-123

[Mo Su] B. Monien, I.H. Sudborough, The interface between language theory and complexity theory, in 'Formal Language Theory' (R.V. Book ed.), Academic Press, New York, 1980, 287-324

[Ro Sa] G. Rozenberg, A. Salomaa, The Mathematical Theory Of L Systems, Academic Press, New York, 1980

[St Me] L.J. Stockmeyer, A.R. Meyer, Word problems requiring exponential space, Proc. Fifth. Annual ACM Symposium on the Theory of Computing, 1-9, 1973

[Su 1] I. Sudborough, On tape bounded complexity classes and multihead finite automata, J. Comput. Syst. Sci 10 (1975), 62-76

[Su 2] I.H. Sudborough, The complexity of the membership problem for some extensions of context-free languages, Intern. J. Computer Math. 6 (1977), 191-215

[Su 3] I.H. Sudborouth, On the tape complexity of deterministic context-free languages, J. Assoc. Comput. Mach. 25 (1978) 405-414

[vL] J. van Leeuwen, The membership question for ETOL languages is polynomially complete, Inform. Process. Lett. 3 (1975) 138-143

Factoring multivariate polynomials over algebraic number fields

Arjen K. Lenstra
Centrum voor wiskunde en informatica
Kruislaan 413
1098 SJ Amsterdam
The Netherlands

Abstract

We present an algorithm to factor multivariate polynomials over algebraic number fields that is polynomial-time in the degrees of the polynomial to be factored. The algorithm is an immediate generalization of the polynomial-time algorithm to factor univariate polynomials with rational coefficients.

1. Introduction

We show that the algorithm from [7] to factor univariate polynomials with rational coefficients can be generalized to multivariate polynomials with coefficients in an algebraic number field. As a result we get an algorithm that is polynomial-time in the degrees and the coefficient-size of the polynomial to be factored.

An outline of the algorithm is as follows. First the polynomial $f \in \mathbf{Q}(\alpha)[X_1, X_2, ..., X_t]$ is evaluated in a suitably chosen integer point $(X_2 = s_2, X_3 = s_3, ..., X_t = s_t)$. Next, for some prime number p, a p-adic irreducible factor $\tilde{h}$ of the resulting polynomial $\tilde{f} \in \mathbf{Q}(\alpha)[X_1]$ is determined up to a certain precision. We then show that the irreducible factor h_0 of f for which $\tilde{h}$ is a p-adic factor of $\tilde{h}_0$, belongs to a certain integral lattice, and that h_0 is relatively short in this lattice. This enables us to compute this factor h_0 by means of the so-called *basis reduction algorithm* (cf. [7: Section 1]).

As [7] is easily available, we do not consider it to be necessary to recall the basis reduction algorithm here; we will assume the reader to be familiar with this algorithm and its properties.

Although the algorithm presented in this paper is polynomial-time, we do not think it is a useful method for practical purposes. Like the other generalizations of the algorithm from [7], which can be found in [8; 9; 10; 11], the algorithm will be slow, because the basis reduction algorithm has to be applied to huge dimensional lattices with large entries. In practice, a combination of the methods from [6], [14], and [15] can be recommended (cf. [6]).

2. Preliminaries

In this section we introduce some notation, and we derive an upper bound for the coefficients of factors of multivariate polynomials over algebraic number fields.

Let the algebraic number field $\mathbf{Q}(\alpha)$ be given as the field of rational numbers $\mathbf{Q}$ extended by a root α of a prescribed *minimal polynomial* $F \in \mathbf{Z}[T]$ with leading coefficient equal to one; i.e. $\mathbf{Q}(\alpha) \simeq \mathbf{Q}[T]/(F)$. Similarly, we define $\mathbf{Z}[\alpha] = \mathbf{Z}[T]/(F)$ as a ring of polynomials in α over $\mathbf{Z}$ of degree $< l$, where l denotes the degree δF of F.

Let $f \in \mathbf{Q}(\alpha)[X_1, X_2, ..., X_t]$ be the polynomial to be factored, with the number of variables $t \geq 2$. By $\delta_i f = n_i$ we denote the degree of f in X_i, for $1 \leq i \leq t$. We often use n instead of n_1. We put $N_i = \prod_{k=i}^{t}(n_k + 1)$, and $N = N_1$. Let $lc_0(f) = f$. For $1 \leq i \leq t$ we define $lc_i(f) \in \mathbf{Q}(\alpha)[X_{i+1}, X_{i+2}, ..., X_t]$ as the leading coefficient with respect to X_i of $lc_{i-1}(f)$, and we put $lc(f) = lc_t(f)$. Finally, we define the *content* $\text{cont}(f) \in \mathbf{Q}(\alpha)[X_2, X_3, ..., X_t]$ of f as the greatest common divisor of the coefficients of f with respect to X_1. Without loss of generality we may assume that $2 \leqslant n_i \leqslant n_{i+1}$ for $1 \leqslant i < t$, that f is *monic* (i.e. $lc(f) = 1$), and that $\delta_i \operatorname{cont}(f) = 0$ for $2 \leqslant i \leqslant t$.

Let $d \in \mathbf{Z}_{>0}$ be such that $f \in \frac{1}{d}\mathbf{Z}[\alpha][X_1, X_2, ..., X_t]$, and let discr$(F)$ denote the discriminant of F. It is well-known (cf. [15]) that if we take $D = d\,|\operatorname{discr}(F)|$, then all monic factors of f are in $\frac{1}{D}\mathbf{Z}[\alpha][X_1, X_2, ..., X_t]$ (in fact it is sufficient to take $D = d \cdot s$, where s is the largest integer such that s^2 divides discr(F), but this integer s might be too difficult to compute).

We now introduce some notation, similar to [8: Section 1]. Suppose that we are given a prime number p such that

(2.1) $\quad p$ does not divide D.

For $G = \sum_i a_i T^i \in \mathbf{Z}[T]$ we denote by G_l or $G \bmod p^l$ the polynomial $\sum_i (a_i \bmod p^l)T^i \in (\mathbf{Z}/p^l\mathbf{Z})[T]$, for any positive integer l. Suppose furthermore that we are given some positive integer k, and that p is chosen in such a way that a polynomial $H \in \mathbf{Z}[T]$ exists such that

(2.2) $\quad H$ has leading coefficient equal to one,

(2.3) $\quad H_k$ divides F_k in $(\mathbf{Z}/p^k\mathbf{Z})[T]$,

(2.4) $\quad H_1$ is irreducible in $(\mathbf{Z}/p\mathbf{Z})[T]$,

(2.5) $\quad (H_1)^2$ does not divide F_1 in $(\mathbf{Z}/p\mathbf{Z})[T]$.

Clearly H_1 divides F_1 in $(\mathbf{Z}/p\mathbf{Z})[T]$, and $0 < \delta H \leqslant I$. In the sequel we will assume that conditions (2.1), (2.2), (2.3), (2.4), and (2.5) are satisfied.

By $\mathbf{F}_q$ we denote the finite field containing $q = p^{\delta H}$ elements. From (2.4) we have $\mathbf{F}_q \simeq (\mathbf{Z}/p\mathbf{Z})[T]/(H_1) \simeq \{\sum_{i=0}^{\delta H - 1} a_i \alpha_1^i : a_i \in \mathbf{Z}/p\mathbf{Z}\}$, where $\alpha_1 = T \bmod (H_1)$ is a zero of H_1. Furthermore we put $W_k(\mathbf{F}_q) = (\mathbf{Z}/p^k\mathbf{Z})[T]/(H_k) = \{\sum_{i=1}^{\delta H - 1} a_i \alpha_k^i : a_i \in \mathbf{Z}/p^k\mathbf{Z}\}$, where $\alpha_k = T \bmod (H_k)$ is a zero of H_k. Notice that $W_k(\mathbf{F}_q)$ is a ring containing q^k elements, and that $W_1(\mathbf{F}_q) \simeq \mathbf{F}_q$. For $a \in \mathbf{Z}[\alpha]$ we denote by $a \bmod (p^l, H_l) \in W_l(\mathbf{F}_q)$ the result of the canonical mapping from $\mathbf{Z}[\alpha] = \mathbf{Z}[T]/(F)$ to $W_l(\mathbf{F}_q) = (\mathbf{Z}/p^l\mathbf{Z})[T]/(H_l)$ applied to a, for $l = 1, k$. For $\tilde{g} = \sum_i \frac{a_i}{D} X_1^i \in \frac{1}{D}\mathbf{Z}[\alpha][X_1]$ we denote by $\tilde{g} \bmod (p^l, H_l)$ the polynomial $\sum_i (((D^{-1} \bmod p^l) a_i) \bmod (p^l, H_l)) X_1^i \in W_l(\mathbf{F}_q)[X_1]$ (notice that $D^{-1} \bmod p^l$ exists due to (2.1)).

We derive an upper bound for the height of a monic factor g of f. As usual, for $g = \sum_{i_1} \sum_{i_2} \cdots \sum_{i_t} \sum_j a_{i_1 i_2 \ldots i_t j} \alpha^j X_1^{i_1} X_2^{i_2} \ldots X_t^{i_t} \in \mathbf{Q}(\alpha)[X_1, X_2, ..., X_t]$, the *height* $g_{\max}$ is defined as $\max |a_{i_1 i_2 \ldots i_t j}|$, and the *length* $|g|$ as $(\sum a_{i_1 i_2 \ldots i_t j}^2)^{1/2}$. Similarly, for a polynomial h with complex coefficients, we define its height $h_{\max}$ as the maximum of the absolute values of its complex coefficients.

For any choice of $\alpha \in \{\alpha_1, \alpha_2, ..., \alpha_I\}$, where $\alpha_1, \alpha_2, ..., \alpha_I$ are the conjugates of α, we can regard g as a polynomial g_α with complex coefficients. We define $\|g\|$ as $\max_{1 \leqslant i \leqslant I} (g_{\alpha_i})_{\max}$. From [3] we have

$$\|g\| \leqslant e^{\sum_{i=1}^t n_i} \|f\|.$$

In [8: Section 4] we have shown that this leads to

$$g_{\max} \leqslant e^{\sum_{i=1}^t n_i} \|f\| I (I-1)^{(I-1)/2} |F|^{I-1} |\operatorname{discr}(F)|^{-1/2}. \tag{2.6}$$

From [13] we know that the length $|F|$ of F is an upper bound for the absolute value of the conjugates of α, so that

$$\|f\| \leqslant f_{\max} \sum_{i=0}^{I-1} |F|^i,$$

which yields, combined with (2.6),

$$(2.7)\qquad g_{\max} \leqslant e^{\sum_{i=1}^{t} n_i} f_{\max} I (I-1)^{(I-1)/2} |F|^{I-1} |\mathrm{discr}(F)|^{-½} \sum_{i=0}^{I-1} |F|^{i}.$$

The upper bound for the height of monic factors of f, as given by the right hand side of (2.7), will be denoted by B_f. Because $|\mathrm{discr}(F)| \geqslant 1$, we find

$$(2.8)\qquad \log B_f = O\left(\sum_{i=1}^{t} n_i + \log f_{\max} + I \log(I\,|F|)\right).$$

3. Factoring multivariate polynomials over algebraic number fields

We describe an algorithm to compute the irreducible factorization of f in $\mathbf{Q}(\alpha)[X_1, X_2, ..., X_t]$.

Let $s_2, s_3, ..., s_t \in \mathbf{Z}_{>0}$ be a $(t-1)$-tuple of integers. For $g \in \mathbf{Q}(\alpha)[X_1, X_2, ..., X_t]$ we denote by $\tilde{g}_j$ the polynomial $g \bmod ((X_2 - s_2), (X_3 - s_3), ..., (X_j - s_j)) \in \mathbf{Q}(\alpha)[X_1, X_{j+1}, X_{j+2}, ..., X_t]$; i.e. $\tilde{g}_j$ is g with s_i substituted for X_i, for $2 \leqslant i \leqslant j$. Notice that $\tilde{g}_1 = g$ and that $\tilde{g}_j = \tilde{g}_{j-1} \bmod (X_j - s_j)$. We put $\tilde{g} = \tilde{g}_t$.

Suppose that a polynomial $\tilde{h} \in \mathbf{Z}[\alpha][X_1]$ is given such that

(3.1) $\tilde{h}$ is monic,

(3.2) $\tilde{h} \bmod (p^k, H_k)$ divides $\tilde{f} \bmod (p^k, H_k)$ in $W_k(\mathbf{F}_q)[X_1]$,

(3.3) $\tilde{h} \bmod (p, H_1)$ is irreducible in $\mathbf{F}_q[X_1]$,

(3.4) $(\tilde{h} \bmod (p, H_1))^2$ does not divide $\tilde{f} \bmod (p, H_1)$ in $\mathbf{F}_q[X_1]$.

We put $l = \delta_1 \tilde{h}$, so $0 < l \leqslant n$. By $h_0 \in \frac{1}{D}\mathbf{Z}[\alpha][X_1, X_2, ..., X_t]$ we denote the unique, monic, irreducible factor of f such that $\tilde{h} \bmod (p^k, H_k)$ divides $\tilde{h}_0 \bmod (p^k, H_k)$ in $W_k(\mathbf{F}_q)[X_1]$ (cf. (3.2), (3.3), (3.4)).

(3.5) Let $m = m_1, m_2, m_3, ..., m_t$ be a t-tuple of integers satisfying $l \leqslant m < n$ and $0 \leqslant m_i \leqslant \delta_i lc_{i-1}(f)$ for $2 \leqslant i \leqslant t$, and let $M = 1 + I \sum_{i=1}^{t} m_i N_{i+1}$ (where of course $N_{t+1} = 1$). We define $L \subset (\frac{\mathbf{Z}}{D})^M$ as the lattice of rank M, consisting of the polynomials $g \in \frac{1}{D}\mathbf{Z}[\alpha][X_1, X_2, ..., X_t]$ for which

(i) $\delta_1 g \leqslant m$ and $\delta_i g \leqslant n_i$ for $2 \leqslant i \leqslant t$;

(ii) If $\delta_j lc_{j-1}(g) = m_j$ for $1 \leqslant j \leqslant i$, then $\delta_{i+1} lc_i(g) \leqslant m_{i+1}$ for $1 \leqslant i < t$;

(iii) If $\delta_i lc_{i-1}(g) = m_i$ for $1 \leqslant i \leqslant t$, then $lc(g) \in \mathbf{Z}$;

(iv) $\tilde{h} \bmod (p^k, H_k)$ divides $\tilde{g} \bmod (p^k, H_k)$ in $W_k(\mathbf{F}_q)[X_1]$.

Here M-dimensional vectors and polynomials satisfying conditions (i), (ii), and (iii), are identified in the usual way (cf. [8: (2.6); 11: (2.2)]). For notational convenience we only give a basis for L in the case that $m_i = n_i$ for $2 \leqslant i \leqslant t$; the general case can easily be derived from this:

$$\{\tfrac{1}{D} p^k \alpha^j X_1^i : 0 \leqslant j < \delta H,\ 0 \leqslant i < l\}$$

$$\cup\ \{\tfrac{1}{D} \alpha^{j-\delta H} H(\alpha) X_1^i : \delta H \leqslant j < I,\ 0 \leqslant i < l\}$$

$$\cup\ \{\tfrac{1}{D} \alpha^j \tilde{h} X_1^{i-l} : 0 \leqslant j < I,\ l \leqslant i \leqslant m\}$$

$$\cup\ \{\tfrac{1}{D} \alpha^j X_1^{i_1} \prod_{r=2}^{t} (X_r - s_r)^{i_r} : 0 \leqslant j < I,\ 0 \leqslant i_1 \leqslant m,\ 0 \leqslant i_r \leqslant n_r$$

$$\text{for } 2 \leqslant r \leqslant t,\ (i_2, i_3, ..., i_t) \neq (0, 0, ..., 0),$$

$$\text{and } (i_1, i_2, i_3, \ldots, i_t) \neq (m, n_2, n_3, \ldots, n_t)\}$$

$$\cup \ \{X_1^m \prod_{r=2}^{t} (X_r - s_r)^{n_r}\}$$

(cf. [8: (2.6); 11: (2.19)], (2.2), and (3.1)).

(3.6) **Proposition.** *Let b be a non-zero element of L and let*

$$\tilde{B}_j = f_{\max}^m b_{\max}^n (n+m)! \left[DN_2(1+F_{\max})^{I-1} \prod_{i=2}^{j} s_i^{n_i} \right]^{n+m}, \tag{3.7}$$

for $1 \leqslant j \leqslant t$, where $f_{\max}^m$ denotes $(f_{\max})^m$.

Suppose that

$$s_j \geqslant ((n+m)n_j + 1)^{1/2} \tilde{B}_{j-1} \tag{3.8}$$

for $2 \leqslant j \leqslant t$, and

$$p^{k\delta H} \geqslant |F|^{I-1} (I^{1/2} \tilde{B}_t)^I. \tag{3.9}$$

Then $gcd(f, b) \neq 1$ in $\mathbf{Q}(\alpha)[X_1, X_2, \ldots, X_t]$.

Proof. Denote by $R = R(Df, Db) \in \mathbf{Z}[\alpha][X_2, X_3, \ldots, X_t]$ the resultant of Df and Db (with respect to the variable X_1). An outline of the proof is as follows. First we prove that an upper bound for $(\tilde{R}_j)_{\max}$ is given by $\tilde{B}_j$. Combining this with (3.8), we then see that $X_j = s_j$ cannot be a zero of $\tilde{R}_{j-1}$ if $\tilde{R}_{j-1} \neq 0$, for $2 \leqslant j \leqslant t$. This implies that the assumption that $R \neq 0$ (i.e. $gcd(f, b) = 1$) leads to $\tilde{R} \neq 0$. We then apply a result from [6], and we find with (3.9) that $\tilde{R} \bmod (p^k, H_k) \neq 0$. But this is a contradiction, because $\tilde{h} \bmod (p^k, H_k)$ divides both $\tilde{f} \bmod (p^k, H_k)$ and $\tilde{b} \bmod (p^k, H_k)$ in $W_k(\mathbf{F}_q)[X_1]$. We conclude that $R = 0$, so that $gcd(f, b) \neq 1$ in $\mathbf{Q}(\alpha)[X_1, X_2, \ldots, X_t]$.

If a and b are two polynomials in any number of variables over $\mathbf{Q}(\alpha)$, having l_a and l_b terms respectively, then

$$(a \cdot b)_{\max} \leqslant a_{\max} b_{\max} \min(l_a, l_b)(1+F_{\max})^{I-1}. \tag{3.10}$$

From (3.10) we easily derive an upper bound for $(\tilde{R}_j)_{\max}$, because $\tilde{R}_j \in \mathbf{Z}[\alpha][X_{j+1}, X_{j+2}, \ldots, X_t]$ is the resultant of $D\tilde{f}_j$ and $D\tilde{b}_j$:

$$(\tilde{R}_j)_{\max} \leqslant (D\tilde{f}_j)_{\max}^m (D\tilde{b}_j)_{\max}^n (n+m)! N_{j+1}^{n+m-1} (1+F_{\max})^{(I-1)(n+m-1)}. \tag{3.11}$$

It follows from $\tilde{f}_j = \tilde{f}_{j-1} \bmod (X_j - s_j)$, that $(\tilde{f}_j)_{\max} \leqslant (\tilde{f}_{j-1})_{\max}(n_j+1)s_j^{n_j}$, so that

$$(\tilde{f}_j)_{\max} \leqslant f_{\max} \prod_{i=2}^{j} (n_i+1) s_i^{n_i}. \tag{3.12}$$

Combining (3.11), (3.12), and a similar bound for $(\tilde{b}_j)_{\max}$, we obtain

$$(\tilde{R}_j)_{\max} < f_{\max}^m b_{\max}^n (n+m)! (DN_2 \prod_{i=2}^{j} s_i^{n_i})^{n+m} (1+F_{\max})^{(I-1)(n+m-1)}, \tag{3.13}$$

for $1 \leqslant j < t$. (Remark that (3.13) with "$<$" replaced by "$\leqslant$" holds for $j = t$.)

Now assume, for some j with $2 \leqslant j \leqslant t$, that $\tilde{R}_{j-1}$ is unequal to zero. We prove that $\tilde{R}_j \neq 0$. Because $\tilde{R}_j = \tilde{R}_{j-1} \bmod (X_j - s_j)$, the condition $\tilde{R}_j = 0$ would imply that all polynomials in $\mathbf{Z}[X_j]$ that result from $\tilde{R}_{j-1}$ by grouping together all terms with identical exponents in α and X_{j+1} up to X_t, have $(X_j - s_j)$ as a factor. These polynomials have degree (in X_j) at most $(n+m)n_j$, so that we get, with the result from [12], that

$$|s_j| \leqslant ((n+m)n_j + 1)^{1/2} (\tilde{R}_{j-1})_{\max}.$$

Combined with (3.13) and (3.7) this is a contradiction with (3.8). We conclude that $\tilde{R}_j \neq 0$ if $\tilde{R}_{j-1} \neq 0$ for any j with $2 \leqslant j \leqslant t$, so that the assumption $gcd(f, b) = 1$ (i.e. $R \neq 0$) leads to $\tilde{R} \neq 0$.

Assume that $H_k(T)$ divides $\tilde{R}(T) \in \mathbf{Z}[T]$ in $(\mathbf{Z}/p^k\mathbf{Z})[T]$, i.e. $\tilde{R} \bmod (p^k, H_k) = 0$. The polynomial $H_k(T)$ is also a divisor of $F(T)$ in $(\mathbf{Z}/p^k\mathbf{Z})[T]$, so that $gcd(F(T), \tilde{R}(T)) = 1$ and [6: Theorem 2] lead to

$$p^{k\delta H} \leq |F|^{I-1}(I^{1/2}\tilde{R}_{max})^I.$$

With the remark after (3.13) and (3.7) this is a contradiction with (3.9), so that $\tilde{R} \bmod (p^k, H_k) \neq 0$. This concludes the proof of (3.6). □

(3.14) **Proposition.** *Let $b_1, b_2, ..., b_M$ be a reduced basis for L (cf. [7: Section 1]), where L and M are as in (3.5), and let*

(3.15) $$B_j = (n+m)!(M2^{M-1})^{n/2}\left\{B_f DN_2(1+F_{max})^{I-1}\prod_{i=2}^{j} s_i^{n_i}\right\}^{n+m},$$

for $2 \leq j \leq t$, where B_f is as in Section 2. Suppose that

(3.16) $$s_j \geq ((n+m)n_j+1)^{1/2}B_{j-1}$$

for $2 \leq j \leq t$, that

(3.17) $$p^{k\delta H} \geq |F|^{I-1}(I^{1/2}B_t)^I,$$

and that f does not contain multiple factors. Then

(3.18) $$(b_1)_{max} \leq (M2^{M-1})^{1/2}B_f$$

and h_0 divides b_1, if and only if $h_0 \in L$.

Proof. If h_0 divides b_1, then $h_0 \in L$, because $b_1 \in L$; this proves the "if"-part.

To prove the "only if"-part, suppose that $h_0 \in L$. Because h_0 is a monic factor of f, we have from (2.7) that $(h_0)_{max} \leq B_f$. With [7: (1.11)] and $h_0 \in L$ this gives $|b_1| \leq (M2^{M-1})^{1/2}B_f$, so that (3.18) holds, because $(b_1)_{max} \leq |b_1|$. Because of (3.18), (3.16), (3.17), (3.15), and the definition of B_f, we can apply (3.6), which yields $gcd(f, b_1) \neq 1$.

Now suppose that h_0 does not divide b_1. This implies that h_0 also does not divide $r = gcd(f, b_1)$, where r can be assumed to be monic. But then $\tilde{h} \bmod (p^k, H_k)$ divides $(\tilde{f}/\tilde{r}) \bmod (p^k, H_k)$, so that Proposition (3.6) can be applied with f replaced by f/r. Conditions (3.8) and (3.9) are satisfied because $(f/r)_{max} \leq B_f$ (cf. (2.7)) and because of (3.16), (3.17), and (3.15). It follows that $gcd(f/r, b_1) \neq 1$, which contradicts $r = gcd(f, b_1)$ because f does not contain multiple factors. □

(3.19) We describe how to compute the irreducible factor h_0 of f. Suppose that f does not contain multiple factors, and that the polynomial $\tilde{h}$, the $(t-1)$-tuple $s_2, s_3, ..., s_t$, and the prime power p^k are chosen such that (3.1), (3.2), (3.3), (3.4), (3.16), and (3.17) are satisfied with, for (3.16) and (3.17), m replaced by $n-1$. Remember that we also have to take care that conditions (2.1), (2.2), (2.3), (2.4), and (2.5) on p and H are satisfied.

We apply the basis reduction algorithm (cf. [7: Section 1]) to a sequence of M_j-dimensional lattices as in (3.5), where the $M_j = 1 + I\sum_{i=1}^{t} m_i N_{i+1}$ run through the range of admissible values for $m_1, m_2, ..., m_t$ (cf. (3.5)), in such a way that $M_j < M_{j+1}$. (So, for $m = l, l+1, ..., n-1$, and $m_i = 0, 1, ..., \delta_i lc_{i-1}(f)$ for $i = t, t-1, ..., 2$ in succession.) According to (3.14), the first vector b_1 that we find that satisfies (3.18) equals $\pm h_0$ (remember that b_1 belongs to a basis for the lattice), so that we can stop if such a vector is found. If for none of the lattices a vector satisfying (3.18) is found, then h_0 is not contained in any of these lattices according to (3.14), so that $h_0 = f$.

(3.20) **Proposition.** *Assume that the conditions in (3.19) are satisfied. The polynomial h_0 can be computed in $O((\delta_1 h_0 IN_2)^4 k \log p)$ arithmetic operations on integers having binary length $O(INk \log p)$.*

Proof. Observing that $\log(INp^{2k}) = O(k \log p)$ (cf (3.17), (3.15), and (2.8)), the proof immediately follows from (3.19), (3.5), and [7: (1.26), (1.37)]. □

(3.21) We now show how $s_2, s_3, ..., s_t$ and p can be chosen in such a way that the conditions in (3.19) can be satisfied. The algorithm to factor f then easily follows by repeated application of (3.19).

We assume that f does not contain multiple factors, so that the resultant $R = R(df, df')$ of df and its derivative df' with respect to X_1 is unequal to zero. First we choose $s_2, s_3, ..., s_t \in \mathbf{Z}_{>0}$ minimal such that (3.16) is satisfied with m replaced by $n-1$. It follows from (3.16), (3.15), (2.8) and $\log D = O(\log d + I \log(I|F|))$ (because $D = d|\text{discr}(F|)$, that

$$\begin{aligned}\log s_j &= O(\log((n+m)n_j) + \log B_{j-1}) \\ &= O(InN + n(\log B_f + \log D + I \log(1+F_{\max}) + \sum_{i=1}^{j-1} n_i \log s_i)) \\ &= O(n(IN + \log(df_{\max}) + I \log(I|F|) + \sum_{i=1}^{j-1} n_i \log s_i))\end{aligned}$$

for $2 \leqslant j \leqslant t$, so that

$$\log s_j = O(n(IN + \log(df_{\max}) + I \log(I|F|)) \prod_{i=2}^{j-1} (1 + nn_i))$$

and

(3.22)
$$\sum_{i=2}^{t} n_i \log s_i = O(n^{t-2} N(IN + \log(df_{\max}) + I \log(I|F|))).$$

From the proof of (3.6) it follows that, for this choice of $s_2, s_3, ..., s_t$ the resultant $R \in \mathbf{Z}[\alpha]$ of $d\tilde{f}$ and $d\tilde{f}'$ is unequal to zero.

Next we choose p minimal such that p does not divide D or discr(F), and such that $\tilde{R} \not\equiv 0 \bmod p$. Clearly

$$\prod_{q \text{ prime}, q<p} q \leqslant d \, \text{discr}(F) \tilde{R}_{\max}$$

which yields, together with

$$\prod_{q \text{ prime}, q<p} q > e^{Ap}$$

for all $p > 2$ and some constant $A > 0$ (cf. [4: Section 22.2]), that

(3.23)
$$p = O(\log d + I \log(I|F|) + \log \tilde{R}_{\max}).$$

Similar to (3.13) we obtain

$$\tilde{R}_{\max} \leqslant f_{\max}^{2n-1} n^n (2n-1)! \left[dN_2 \prod_{i=2}^{t} s_i^{n_i} \right]^{2n-1} (1+F_{\max})^{(I-1)(2n-2)},$$

so that we get, using (3.22)

$$\log \tilde{R}_{\max} = O(n^{t-1} N(IN + \log(df_{\max}) + I \log(I|F|))).$$

Combining this with (3.23) we conclude that

(3.24)
$$p = O(n^{t-1} N(IN + \log(df_{\max}) + I \log(I|F|))).$$

Notice that (2.1) is now satisfied. In order to compute a polynomial $H \in \mathbf{Z}[T]$ satisfying (2.2), (2.4), (2.5), and (2.3) with k replaced by 1, we factor $F \bmod p$ by means of Berlekamp's algorithm [5: Section 4.6.2] and we choose H as an irreducible factor of $F \bmod p$ for which $\tilde{R} \bmod (p, H_1) \neq 0$; such a polynomial H exists because $\tilde{R} \bmod p \neq 0$. Conditions (2.4) and (2.3) with k replaced by 1 are clear from the construction of H, and because we may assume that H

has leading coefficient equal to one, (2.2) also holds. The condition that $\mathrm{discr}(F) \bmod p \neq 0$, finally, guarantees that $F \bmod p$ does not contain multiple factors, so that (2.5) is satisfied.

We choose k minimal such that (3.17) holds, so that

$$k \log p = O(I(InN + n\log(df_{\max}) + In\log(I|F|) + n\sum_{i=2}^{t} n_i \log s_i) + \log p)$$

(cf. (3.15) and (2.8)), which gives, with (3.22) and (3.24)

(3.25) $$k \log p = O(In^{t-1}N(IN + \log(df_{\max}) + I\log(I\,|F|))).$$

Now we apply Hensel's lemma [5: Exercise 4.6.22] to modify H in such a way that (2.3) holds for this value of k (this is possible because (2.3) already holds for $k = 1$), and finally we apply Berlekamp's algorithm as described in [1: Section 5] and Hensel's lemma as in [14] to compute the irreducible factorization of $\tilde{f} \bmod (p^k, H_k)$ in $W_k(\mathbb{F}_q)[X_1]$. Condition (3.4) is satisfied for each irreducible factor $\tilde{h} \bmod (p^k, H_k)$ of $\tilde{f} \bmod (p^k, H_k)$ because $\tilde{R} \bmod (p, H_1) \neq 0$, and (3.1), (3.2), and (3.3) are clear from the construction of h.

We have shown how to choose $s_2, s_3, \ldots, s_t$ and p, and how to satisfy the conditions in (3.19). We are now ready for our theorem.

(3.26) **Theorem.** *Let f be a monic polynomial in $\frac{1}{d}\mathbb{Z}[\alpha][X_1, X_2, \ldots, X_t]$ with $t \geq 2$, of degree n_i in X_i, and $2 \leq n = n_1 \leq n_2 \leq \ldots \leq n_t$. The irreducible factorization of f can be found in $O(n^{t-1}(IN)^5(IN + \log(df_{\max}) + I\log(I\,|F|)))$ arithmetic operations on integers having binary length $O(n^{t-1}(IN)^2(IN + \log(df_{\max}) + I\log(I\,|F|)))$, where $N = \prod_{i=1}^{t}(n_i + 1)$.*

Proof. If f does not contain multiple factors, then f can be factored by repeated application of (3.19). In that case (3.26) follows from (3.21), (3.20), (3.25), and the well-known estimates for the application of Berlekamp's algorithm and Hensel's lemma (cf. [5; 1] and [16]).

If f contains multiple factors, then we first have to compute the monic *gcd* g of f and its derivative with respect to X_1, and the factoring algorithm is then applied to f/g. The cost of factoring f/g satisfies the same estimates as above, because $(f/g)_{\max} \leq B_f$ (cf. (2.7)), and this dominates the costs of the computation of g, which can be done by means of the subresultant algorithm (cf. [2]). □

References

1. E.R. Berlekamp, Factoring polynomials over large finite fields, *Math. Comp.* **24** (1970), 713-735.

2. W.S. Brown, The subresultant PRS algorithm, *ACM Transactions on mathematical software* **4** (1978), 237-249.

3. A.O. Gel'fond, Transcendental and algebraic numbers, Dover Publ., New York 1960.

4. G.H. Hardy, E.M.Wright, An introduction to the theory of numbers, Oxford University Press 1979.

5. D.E. Knuth, The art of computer programming, vol. 2, Seminumerical algorithms, Addison-Wesley, Reading, second edition 1981.

6. A.K. Lenstra, Lattices and factorization of polynomials over algebraic number fields, Proceedings Eurocam 82, LNCS 144, 32-39.

7. A.K. Lenstra, H.W. Lenstra, Jr., L. Lovász, Factoring polynomials with rational coefficients, *Math. Ann.* **261** (1982), 515-534.

8. A.K. Lenstra, Factoring polynomials over algebraic number fields, Report IW 213/82, Mathematisch Centrum, Amsterdam 1982 (also Proceedings Eurocal 83, LNCS 162, 245-254).

9. A.K. Lenstra, Factoring multivariate polynomials over finite fields, Report IW 221/83, Mathematisch Centrum, Amsterdam 1983 (also Proceedings 15th STOC, 189-192).

10. A.K. Lenstra, Factoring multivariate integral polynomials, Report IW 229/83, Mathematisch Centrum, Amsterdam 1983 (also Proceedings 10th ICALP, LNCS 154, 458-465).

11. A.K. Lenstra, Factoring multivariate integral polynomials, II, Report IW 230/83, Mathematisch Centrum, Amsterdam 1983.

12. M. Mignotte, An inequality about factors of polynomials, *Math. Comp.* **28** (1974), 1153-1157.

13. J. Stoer, Einführung in die numerische Mathematik I, Springer, Berlin 1972.

14. P.S. Wang, Factoring multivariate polynomials over algebraic number fields, *Math. Comp.* **30** (1976), 324-336.

15. P.J. Weinberger, L.P. Rothschild, Factoring polynomials over algebraic number fields, *ACM Transactions on mathematical software* **2** (1976), 335-350.

16. D.Y.Y. Yun, The Hensel lemma in algebraic manipulation, MIT, Cambridge 1974; reprint: Garland Publ. Co., New York 1980.

GÖDEL NUMBERINGS, PRINCIPAL MORPHISMS, COMBINATORY ALGEBRAS

A Category-theoretic characterization of functional completeness.

G. Longo, E. Moggi
Dip. Informatica
Università di Pisa
Corso Italia 40, I-56100 Pisa

Abstract. Functional languages are based on the notion of application: programs may be applied to data or programs. By application one may define algebraic functions and a programming language is functionally complete when any algebraic function $f(x_1,\ldots,x_n)$ is representable (i.e. there is a constant a such that $f(x_1,\ldots,x_n) = ax_1\cdot\ldots\cdot x_n$). Combinatory Logic (C.L.) is the simplest type-free language which is functionally complete.

In a sound category-theoretic framework the constant a above may be considered an "abstract gödel-number" for f, as gödel-numberings are generalized to "principal morphisms". By this, models of C.L. are categorically characterized and their relation is given to λ-calculus models within Cartesian Closed Categories.

Finally, the partial recursive functionals in any finite higher type are shown to yield models of C.L..

1. INTRODUCTION

There is a natural category-theoretic generalization of the notion of Gödel-numbering:

1.1 Definition. Let X,Y be objects in a category C. A morphism $f \in C(X,Y)$ is principal if

$$\forall g \in C(X,Y)\ \exists h \in C(X,X)\quad g = f \circ h$$

As a matter of fact, consider the category EN of numbered sets whose objects are pairs (X,e_X), with $e_X : \omega \to X$ (onto) and morphisms defined by $f \in EN(X,Y)$ if $\exists f' \in R$ (the recursive functions) $f \circ e_X = e_Y \circ f'$. Then, for $e_\omega = id$ and any Gödel numbering of PR (the partial recursive functions), one has: $\phi \in EN(\omega,PR)$ is principal iff ϕ is an acceptable Gödel-numbering.

Principal morphisms were introduced in Longo (1982) and Longo & Moggi (1983) (implicitly) and explicitly in Longo & Moggi (1984) for the purposes of higher types computability. The significance of this notion is confirmed in this paper, by the fact that in Cartesian Closed Categories (CCC) principal morphisms, plus two simple conditions (see §.3), characterize combinatory algebras, i.e. models of Combinatory Logic (CL).

Category-theoretic characterizations of models of λ-calculus (λ-models) and of the purely equational theory (λ-algebras) were given by Berry (1979) and Scott, Lambek, Koymas (see Barendregt (1983) for a survey). The weakest theory, CL was still missing a semantic characterization (besides the definition: see Hindley & Longo (1980), Meyer (1982)). Notice also that, by gödel-numbering PR, ω may be turned into a partial combinatory algebra, namely Kleene's $(\omega,\cdot)$ where $n\cdot m = \{n\}(m)$ $(=\phi(n)(m)$ in the notation above). Then the comparative study of partial applicative structures and of the conditions for combinatory algebras, as total applicative structures, via the present approach, should give some insight on the nature of gödel-numberings and on their relations to combinatory algebras.

§.2 gives a semantic characterization of CL in the general setting of Cartesian Categories.

§.3 deals with that characterization within Cartesian Closed Categories, as the natural setting for models of a functional language such as λ-calculus and its sub-theories. Combinatory algebras and λ-models turn out to be tidily related in these categories.

In a concluding remark some hints are given for applications in the recursion theoretic hierarchy of functionals, which actually motivated the present paper. As a matter of fact, it turns out that the set of the partial recursive functions or the effective functionals at any higher finite type yield combinatory algebras (actually, λ-models).

§.2 Combinatory Algebras and Cartesian Categories

2.1 Definition. Let $A = (X, \cdot)$ be an applicative structure.

(i) $f: X^n \to X$ is algebraic if it can be defined from $x_1, x_2, \ldots, x_n$ and constants from X using application.

(ii) The set of representable functions of n-arguments $F^n(A)$ is defined by $F^n(A) = \{f: X^n \to X \mid \exists a \in X\ \forall \vec{x} \in X^n\ f(\vec{x}) = a \cdot x_1 \ldots x_n\}$

(iii) A is functionally (or combinatorially) complete if any algebraic function is representable.

(See Barendregt (1984) for a formal definition of algebraic function. If there is no ambiguity, write F^n for $F^n(A)$ and F for F^1.)

2.2 Theorem. (Curry-Shoenfinkel) $(X, \cdot)$ is functionally complete iff $K = \lambda xy.x$ and $S = \lambda xyz.xz(yz)$ are representable.

Call Combinatory Algebra any non trivial functionally complete applicative structure. As well known, Combinatory Logic, the theory of combinatory algebras, is powerful enough to be a functional programming language for computing all partial recursive functions.

2.3 Definition. Let $A = (X, \cdot)$ be a combinatory algebra. The category of representable morphisms $\underline{A}$ is the sub-category of Set (the category of sets) given by

(i) $\{X^n / n \in \omega\}$ are the objects of $\underline{A}$;

(ii) $f \in \underline{A}(X^n, X^m)$ if $f: X^n \to X^m$ and $\forall i < m\ \ p_i^m \circ f \in F^n$, where p_i^m is the i-th projection.

To prove that $\underline{A}$ is a category note that:

1) $\forall i < n\ \ p_i^n = \lambda x_0 \ldots x_{n-1}.x_i \in F^n$ and, hence $id_{X^n} \in \underline{A}(X^n, X^n)$.

2) If $f \in F^n$ and $g_i \in F^m$ for all $i < n$, then $f \circ \langle g_0, \ldots, g_{n-1}\rangle = \lambda x.f(g_0\vec{x}, \ldots, g_{n-1}\vec{x})$ where $\langle -,- \rangle$ is the usual cartesian tupling of morphisms.

Thus, for $f \in \underline{A}(X^n, X^r)$ and $g \in \underline{A}(X^m, X^n)$, one has $p_i^r \circ (f \circ g) = (p_i^r \circ f) \circ g = (p_i^r \circ f) \circ \langle p_0^n \circ g, \ldots, p_{n-1}^n \circ g\rangle \in F^m$ and, hence, $f \circ g \in \underline{A}(X^m, X^r)$.

A category is Cartesian if it has all finite products, i.e. there is a terminal object T s.t. $\forall X\ \exists! f\ X \xrightarrow{f} T$ and, for all objects X,Y there is an object $X \times Y$, the product, and, for all Z, an isomorphism $\langle -,- \rangle : C(Z,X) \times C(Z,Y) \to C(Z, X \times Y)$

s.t.

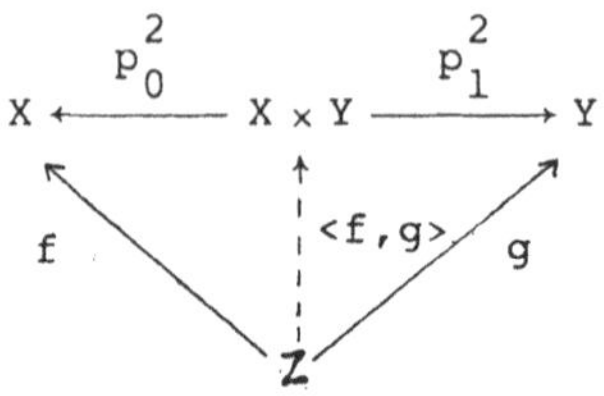

It has enough points if, informally, its morphisms behave like set-theoretic functions, in extenso (see Barendregt (1983)).

2.4 <u>Lemma</u>. Let $\mathcal{A} = (X,\cdot)$ be a Combinatory Algebra. Then $\underline{\mathcal{A}}$ is cartesian; indeed $X^n \times X^m = X^{n+m}$ and X^0 is the terminal object. Moreover $\underline{\mathcal{A}}$ ha enough points.

<u>Proof</u>. Standard. Δ

2.5 <u>Definition</u>. Let C be a category. Define then

1) for $f,g \in C(X,Y)$, f is <u>reducible</u> to g ($f \leqslant_X g$) if $\exists h \in C(X,X)\ f = g \circ h$.

2) Assume also that C is cartesian. Then $u \in C(X \times Y, Z)$ is <u>Kleene-universal</u> (K-universal) if $\forall f \in C(X \times Y, Z)$ $\exists s \in C(X,X)\ f = u \circ (s \times id)$, i.e.

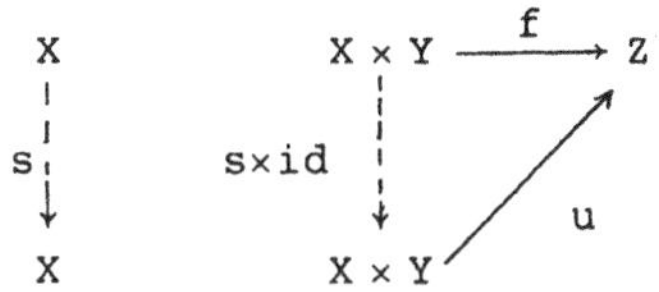

2.6 <u>Remark</u>. Both notions in 2.5 have an obvious recursion theoretic meaning. Principal morphisms in 1.1 are the largest element in $C(X,Y)$ w.r.t. the preorder $\leqslant_X$ in 2.5(1). See then the discussion in §.1 to look at this factorization in terms of recursive reducibility. Similarly, K-universality generalizes the s-m-n (interation) theorem. Just assume that in the category of numbered sets (EN) a <u>partial</u> morphism $u \in EN_p(\omega \times \omega, \omega)$ is universal iff the corresponding total morphism $\bar{u} \in EN(\omega \times \omega, \omega^\perp)$ is K-universal (where $\omega^\perp$ is the completion of ω by a bottom element, i.e. $\forall z\ EN_p(z,\omega) \cong EN(z,\omega^\perp)$, Longo & Moggi (1984)

2.7 Definition. Let C be cartesian and $u \in C(X \times X,X)$. Then $u^{(n)} \in C(X \times X^n,X)$ is inductively defined by:
$u^{(1)} = u$, $u^{(n+1)} = u^{(n)} \circ (u \times id^n)$, i.e.

$$u^{(n+1)} : X \times X \times X^n \xrightarrow{u\times id^n} X \times X^n \xrightarrow{u^{(n)}} X, \text{ for } X^{n+1} = X \times X^n.$$

Notation: a retraction $X \triangleleft Y$ between objects X,Y in a category C, is a pair $i \in C(X,Y)$ and $j \in C(Y,X)$ s.t. $j \circ i = id_X$.

2.8 Lemma. Let C be cartesian. Assume that, for some U in C, $U \times U \triangleleft U$ and there is a K-universal $u \in C(U \times U,U)$. Then $\forall n\ u^{(n)} \in C(U \times U^n,U)$ is K-universal.

Proof. By assumption, this is true for $n = 1$.
Let $U \times U \triangleleft U$ via (i,j) and $f \in C(U \times U^{n+1},U)$. Then, by the inductive hypothesis, for some $s^{(n)} \in C(U,U)$ the following diagram commutes:

$$\begin{array}{ccccc} U \times U^n & \xrightarrow{j\times id^n} & U \times U \times U^n & \xrightarrow{f} & U \\ \Big\downarrow{\scriptstyle s^{(n)}\times id^n} & & & \nearrow{\scriptstyle u^{(n)}} & \\ U \times U^n & & & & \end{array} \qquad (1)$$

By assumption, for some $s \in C(U,U)$ one also has:

$$\begin{array}{ccccc} U \times U & \xrightarrow{i} & U & \xrightarrow{s^{(n)}} & U \\ \Big\downarrow{\scriptstyle s\times id} & & & \nearrow{\scriptstyle u} & \\ U \times U & & & & \end{array} \qquad (2)$$

Then compute
$$\begin{aligned} f &= f \circ (j \times id^n) \circ (i \times id^n) & \\ &= u^{(n)} \circ (s^{(n)} \times id^n) \circ (i \times id^n) & \text{by (1)} \\ &= u^{(n)} \circ (u \times id^n) \circ (s \times id^{n+1}) & \text{by (2)} \\ &= u^{(n+1)} \circ (s \times id^{n+1}) & \end{aligned}$$
Δ

We are now in the position to prove the main theorem of this section. Let's first express in suitable category-theoretic terms the notion of applicative structure.

2.9 Definition. Let C be cartesian and U an object in C, with $T \triangleleft U$ and $u \in C(U \times U,U)$. The applicative structure associated to u,

$A(u)$, is given by $A(u) = (C(T,U),\cdot)$, where $a \cdot b = u \circ \langle a,b\rangle$.

2.10 Theorem. Let C be a cartesian category. Assume that for some object U, one has $T \triangleleft U$, $U \times U \triangleleft U$ and there exists a K-universal $u \in C(U \times U,U)$. Then $A(u)$ is a combinatory algebra.

Proof. Let $T \triangleleft U$ via (i_T, j_T). Then, by 2.8, $\forall n\ \forall f \in C(U^n,U)$ $\exists s \in C(U,U)$ s.t. the following diagram commutes, with $[f] = s \circ i_T$:

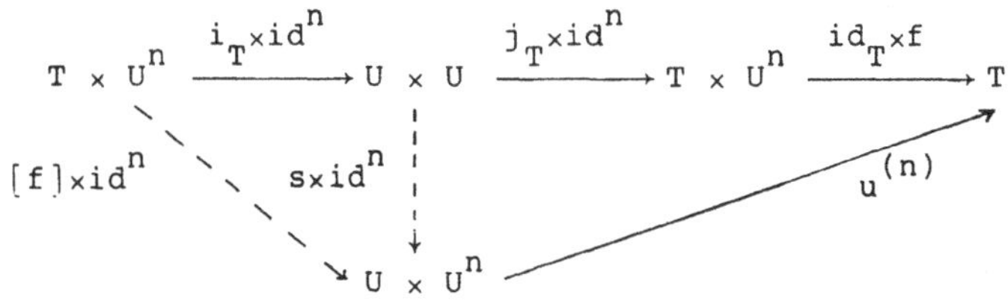

Notice now that $u^{(n)}$ is simply the application, from left to righ of its n+1 arguments and that $[f]$ "represents" f, similarly as in definition 2.1(i). Thus, by 2.2, we only need to write down $f \in C(U^2,U$ and $g \in C(U^3,U)$ such that $[f]$ and $[g]$ represent K and S, respec tively.

Just take $f = p_0^2$ and $g = u \circ \langle\langle u \circ \langle p_0^3, p_2^3\rangle, u \circ \langle p_1^3, p_2^3\rangle\rangle$. Δ

It is very simple to prove the converse of 2.10: it just formally says the well known fact that the (η expansion of) the identity is the universal function in Combinatory Logic (λ-calculus). The point is that by applying the construction in 2.9, one gets back to the given combinatory algebra.

2.11 Theorem. Let $A = (X,\cdot)$ be a combinatory algebra. Then $T \triangleleft X$, $X \times X \triangleleft X$ and $u = \lambda xy.x\cdot y$ is K-universal in the category $\underline{A}$ (see def. 2.3). Moreover $A(u) = A$.

Proof. $T \triangleleft X$ trivially holds, for $X \neq \emptyset$. Let now $c, c_1, c_2 \in X$ represent $\lambda xyz.zxy$, $\lambda xy.x$, $\lambda xy.y$, respectively, in the sense of 2.1(i). Then, for $\langle x,y\rangle = cxy$ and $p_i^2(x) = xc_i$, $\langle , \rangle \in \underline{A}(X^2,X)$, $p_i^2 \in \underline{A}(X,X)$ and $X \times X \triangleleft X$ via $(\langle , \rangle, \langle p_1, p_2\rangle)$. Note that $X \times X$ exists by 2.4.

Finally, assume that $f \in \underline{A}(X^2,X)$ and that $a \in X$ represents f. Then $f = u \circ ((\lambda x.ax) \times id)$, hence u is universal.

It is easy to check from the definition that $A(u) = A$. Δ

§.3 From Cartesian to Cartesian Closed Categories

A Cartesian Closed Category (CCC) C is a Cartesian Category such that for all objects Y,Z there is an object Z^Y, which "corresponds" to C(Y,Z), and, for all X, an isomorphism $\Lambda: C(X \times Y, Z) \cong C(X, Z^Y)$. The correspondence of Z^Y to C(Y,Z) is given by the following diagram:

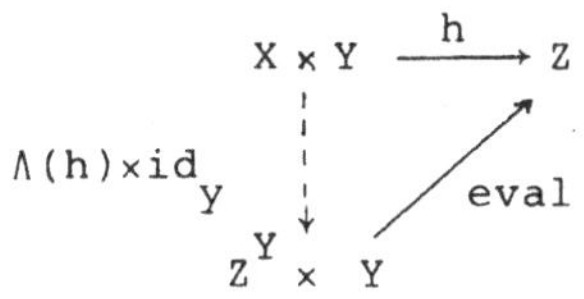

3.1 Proposition. Let C be a CCC and Λ the isomorphism $C(X \times Y, Z) \cong C(X, Z^Y)$. Then $u \in C(X \times Y, Z)$ is K-universal iff $\Lambda(u) \in C(X, Z^Y)$ is principal.

Proof. The isomorphism Λ implies, by definition, the equivalence of the following diagrams:

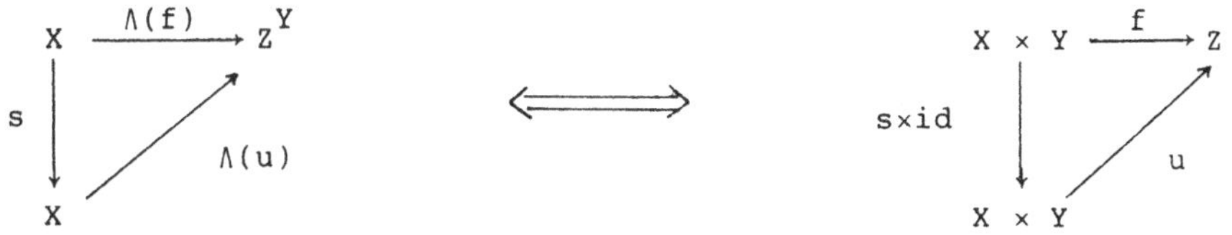

And we are done. Δ

The following fact characterizes combinatory algebras in CCC's.

3.2 Corollary. Let C be a CCC; U an object in C and $g \in C(U, U^U)$. Then $T \triangleleft U$, $U \times U \triangleleft U$ and g is principal iff $A(\Lambda^{-1}(g))$ is a combinatory algebra.

Proof. By using 2.10 and 2.11. Δ

As summarized in Barendregt (1983), models of the purely equational theory of λ-calculus, λ-algebras, are characterized as reflexive objects in a CCC, i.e. as U such that $U^U \triangleleft U$.

λ-algebras, originally called pseudo-λ-models in Hindley & Longo (1980), are exactly the models of Combinatory Logic with strong reduction (see Hindley & Longo (1980) or Hindley & Seldin (198?)).

Moreover, one gets exactly the "first order" models of λ-calculus, λ-modles (Meyer (1982), Scott (1980)), if U has enough points (see

2.4; Barendregt (1983)).

3.3 <u>Lemma</u>. Let C be a category. Then

(i) if $Y \lhd X$ via (i,j), then j is principal;

(ii) if $Y \lhd X$ and $f \in C(X,Y)$ is principal, then there exists $g \in C(Y,X)$ s.t. $Y \lhd X$ via (f,g).

<u>Proof</u>. (i) Just notice that the following diagram commutes, for all $h \in C(X,Y)$, where $Y \lhd X$ via (i,j):

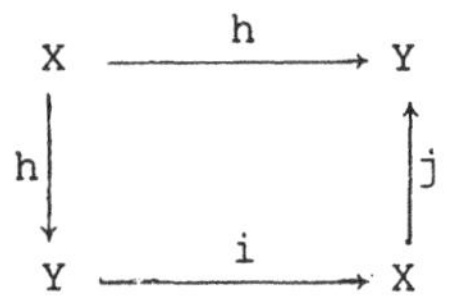

Thus $h = j \circ (i \circ h)$.

(ii) Since f is principal, $\exists s \in C(X,X)\ j = f \circ s$. Then, for $g = s \circ i$, one has $f \circ g = j \circ i = id_X$. As a diagram:

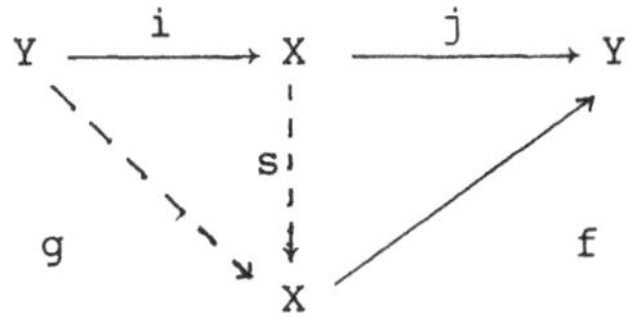

Δ

3.4 <u>Theorem</u>. Let C be a CCC. Assume that, for some U in C, $U^U \lhd U$ via (i,j). Then one has

(1) $T \lhd U$, $U \times U \lhd U$ and j is principal.

(2) Let $u \in C(U \times U, U)$ be universal, then also $A(u)$ can be turned into a λ-algebra.

<u>Proof</u>. (1) Clearly $T \lhd U$, for $T \lhd U^U \lhd U$. Write now $[M]$ for the interpretation of the λ-term M (see Barendregt (1983)). If, say, M has n free variables this determines a morphism in $C(U^n,U)$. Then consider $\langle,\rangle \in C(U^2,U)$ and $p_1, p_2 \in C(U,U)$, given by $\langle x,y\rangle = [\lambda z.zxy]$ and $p_i(x) = [x(\lambda y_1 y_2 . y_i)]$, i.e. consider the usual pairing and projections in λ-calculus.

Clearly one has $U \times U \lhd U$ via $(\langle,\rangle, \langle p_1,p_2\rangle)$.

Finally, j is principal by 3.3(i).

(2) Immediate by 3.1 and 3.3(ii). Δ

3.5 Warning: In general, the converse of 3.4(2) does not hold. Namely, one may have a CCC C and U such that, for some K-universal $u \in C(U \times U, U)$, $A(u)$ can actually be turned into a λ-algebra (or even a λ-model, if there are enough points); nevertheless $U^U \triangleleft U$ in C may fail, for the inverse of the principal morphism $\Lambda(u) \in C(U, U^U)$ need not be a morphism in C. Note, though, that by definition of $A(u)$ one does have that the representable functions F over U coincide with U^U.

Remark (summary of §.4: applications)

A constructive domain is the collection of all computable elements in an effectively given domain (see Scott (1982)). The category CD constructive domains is a sub-CCC of EN.

4.1 Theorem. Let X,Y be in CD with $\omega^{\perp} \triangleleft X$ and $X \times X \triangleleft X$. Then there exists a principal $f \in CD(X,Y)$.

Consider then X in CD such that $\omega^{\perp} \triangleleft X$ and $X \times X \triangleleft X$. By 4.1 and 3.2, X may be turned into a combinatory algebra $(X,\cdot)$ by taking $f \in C(X, X^X)$ principal and by setting, for $x,y \in X$, $x \cdot y = f(x)(y)$. In particular, consider the constructive domain PR of the partial recursive functions and the CCC $\{PR^{\sigma}\}_{\sigma \in T}$ generated by PR in CD, where T are the cartesian type symbols (i.e. the higher type partial computable functionals, see Longo & Moggi (1983/4)). Since, for $(0) \neq \sigma \in T$, $\omega^{\perp} \triangleleft PR^{\sigma}$ and $PR^{\sigma} \times PR^{\sigma} \triangleleft PR^{\sigma}$, one obtains a combinatory algebra $(PR^{\sigma},\cdot)$ in any finite higher type. By 3.3(ii) and results in Plotkin (1978), $(PR^{\sigma},\cdot)$ may be actually turned into a λ-model. Note, however, that

1) $\exists f \in EN(\omega^{\perp}, PR)$ principal (e.g. a gödel-numbering), but $\omega^{\perp} \times \omega^{\perp} \triangleleft \omega$ fails

2) $\omega \times \omega \triangleleft \omega$, but $\exists f \in EN(\omega, \omega^{\omega})$ principal clearly fails;

this is why one cannot turn Kleene's $(\omega,\cdot)$, i.e. type (0), into a combinatory algebra.

REFERENCES

Barendregt H. (1983) "Lambda-calculus and its models" in Proceedings of Logic Colloquium 82 (Lolli, Longo, Marcja eds), North-Holland.

Barendregt H. (1984) The lambda-calculus: its syntax and semantics, revised and expanded edition, North-Holland.

Berry G. (1979) "Some Syntactic and Categorial Constructions of λ-calculus models" INRIA, Valbonne.

Hindley R., Longo G. (1980) "Lambda-calculus models and extensionality" Zeit. Math. Logik 26 (289-310).

Hindley R., Seldin J. (198?) Introductory book on lambda-calculus and CL (in preparation).

Longo G. (1982) "Hereditary Partial Effective Functionals in any finite type" (Preliminary note), Forschungsinst. Math. ETH, Zürich.

Longo G., Moggi E. (1983) "The Hereditary Partial Effective Functionals and Recursion Theory in Higher Types" J. Symb. Logic (to appear).

Longo G., Moggi E. (1984) "Cartesian Closed Categories for effective type structures" Part I Symp. Semantics Data Types, LNCS, Springer-Verlag (to appear).

Meyer A. (1982) "What is a model of lambda-calculus?" Info. Contr. 52, 1 (87-122).

Plotkin G. (1978) "T_ω as a universal domains" JCSS 17,2 (209-236).

Scott D.S. (1980) "Relating theories of lambda-calculus" in To H.B. Curry: essays... (Hindley, Seldin eds.), Academic Press (403-450).

Scott D.S. (1982) "Some ordered sets in Computer Science" Ordered sets (Rival ed.), Reidel.

Research partially by Min. P.I. (Fondi 40%: Comitato per la Matematica).

REPRESENTATIONS OF INTEGERS AND LANGUAGE THEORY

Aldo de Luca

Dipartimento di Matematica
e Applicazioni
Università di Napoli
Italy .

Antonio Restivo

Istituto di Matematica
Università di Palermo
Italy.

0. *Introduction.*

The relationships between numbers and automata date from the origins of the theory of computability,i.e.from Turing machines and recursive function theory.Later the study of different families of computing devices were often related to their ability to "recognize" sets of numbers.In particular,in the theory of finite automata, many papers were devoted to the study of so called recognizable sets of integers (*see, for instance,* [4]).In this framework the problem of the representation of non negative integers plays an important role.

Given an integer $k \geq 2$, a set I of non negative integers is k-recognizable if there exists a recognizable language over the alphabet $\{0,1,\ldots,k-1\}$ whose words represent the elements of I in base k .It must be remembered that a k-recognizable set is not,in general, k'-recognizable for $k' \neq k$:in other words ,the recognizability of a set of integers strongly depends on its representation. A fundamental result of Cobham [1] characterizes the sets of non negative integers k-recognizable for all $k \geq 2$.

In this paper we consider the general problem of obtaining a "simpler" recognizability of a given set of non negative integers by changing the representation. By "simpler" we mean here that,given a k-recognizable set,the language obtained by changing the representation is contained in a proper subclass of recognizable languages as,for instance,*aperiodic* (or "non-counting regular sets")or *locally testable languages* [3,6,8].

In this framework we consider the notion of a *number system* (*see* [2]) which is a generalization of the usual representation in base k ,and we introduce the notion of an S-recognizable set of integers for a given number system S .As a first result we prove (cf.Theorem 1) that,if S is a number system, any S-recognizable set of non negative integers is k-recognizable for a suitable $k \geq 2$.

Given a set I of non negative integers recognizable in base k, we ask the question whether one can obtain an aperiodic (recognizable) language by representing I over another base or another number system. A positive answer is given in some particular cases. If the set I is recognizable, i.e. it is k-recognizable for all $k \geq 2$, we prove, by using Cobham's theorem, that there exist an integer $h \geq 1$ and a set of words representing I in base h, which is a *generalized definite language*, and hence aperiodic (cf. Theorem 2). If the set of words representing I in base k is a *bounded* recognizable language [5,7], we prove that there exists an integer h such that the set of all expansions of the numbers of I in base h is an *aperiodic* (recognizable) language (cf. Theorem 3). In the case in which the set of words representing I in base k is a *finitely generated submonoid*, satisfying a certain supplementary condition, we construct a number system S and a *strictly locally testable* language (and hence aperiodic) representing I in the system S (cf. Theorem 5).

These results are remarkable since the bounded recognizable languages and finitely generated submonoids have a *star-height (unrestricted)* ≤ 1 whereas the aperiodic recognizable languages, as a consequence of a theorem of Schützenberger [9], have a star-height equal to 0. In other words the aperiodic recognizable languages over a finite alphabet A can be generated from the finite subsets of A^* by using only a finite number of Boolean operations and products (without star-operations). Thus as regards to the star operation the set of words representing I in the new number system, looks to be more simple than the original one.

1. *Preliminaries.*

Let N denote the set of non negative integers. For any $k \geq 2$, we consider the alphabet $A_k = \{0,1,\ldots,k-1\}$. If $w = w_1w_2\ldots w_n$, $w_i \in A_k$, $1 \leq i \leq n$, the *standard interpretation* of w (in the k-ary system) is the integer $t = [w]_k$ defined as:

$$[w]_k = \sum_{i=1}^{n} w_i \, k^{n-i} .$$

If $k = 1$, i.e. $A_1 = \{a\}$ then the standard interpretation of a word $w \in A_1^*$ is simply given by its length $|w|$. For any $t \in N$ the set of all the words over A_k whose standard interpretation is t contains a unique element $\langle t\rangle_k$ of minimal length called the *expansion of* t *in base* k.

If $L \subseteq A_k^*$ is a language over the alphabet A_k, let $[L]_k$ denote the subset of N defined as:

$$[L]_k = \{ [w]_k \in N \mid w \in L \} .$$

If I is a subset of N we say that $L \subseteq A_k^*$ *represents* I *in base* k if $[L]_k = I$. We say that $I \subseteq N$ is k-*recognizable* if there exists a recognizable language $L \subseteq A_k^*$ which represents I in base k.

In [2] Culik and Salomaa generalize the notion of representation of an integer in base k by introducing the notion of *number system*. We consider here the following slightly modified version. A number system S is a $(k+1)$-tuple:

$$S = (n, m_1, m_2, \ldots, m_k)$$

of non negative integers such that $n \geq 2$, $k \geq 1$, $0 \leq m_1 < m_2 < \ldots < m_k$. The integer n is called the *base* of the system S and $A_S = \{m_1, \ldots, m_k\}$ is the *alphabet*, or set of *digits*, of S. A word $w = w_1 w_2 \ldots w_r$, $w_i \in A_S$, $1 \leq i \leq r$, *represents* the number $t = [w]_S$ defined as:

$$[w]_S = \sum_{i=1}^{r} w_i \, n^{r-i} .$$

If S is of the form $S = (n, 0, 1, \ldots, n-1)$ one obtains the usual interpretation in the n-ary system. If $S = (n, 1, \ldots, n)$ then S is the so-called n-adic system. It is known (*see* [2]) that an arbitrary number system S is, in general, *ambiguous* (i.e. an integer may have many different representations in S) and *non complete* (i.e. the set of integers representable in S is a proper subset of N). If $L \subseteq A_S^*$ is a language over the alphabet A_S, let $[L]_S$ denote the subset of N defined as:

$$[L]_S = \{ [w]_S \in N \mid w \in L \} .$$

We say that $L \subseteq A_S^*$ *represents* a subset I of N in the number system S if $[L]_S = I$. A subset I of N is called S-*recognizable* if there exists a recognizable language $L \subseteq A_S^*$ such that $[L]_S = I$. The following theorem shows that this definition does not extend the class of sets which are k-recognizable for some k.

Theorem 1. Let $I \subseteq N$ be an S-recognizable set with $S = (n, m_1, \ldots, m_k)$. Then I is also n-recognizable.

The proof is essentially obtained by using an argument of Culik and Salomaa [2] who show that the application $\alpha : \{m_1, \ldots, m_k\}^* \to \{0, \ldots, n-1\}^*$, which maps the words of $\{m_1, \ldots, m_k\}^*$ into their expansions in base n, is a rational transduction. The theorem is then a consequence of the fact that the image of a recognizable language by a rational transduction is again recognizable [4]. □

2. Main results.

The problem we now consider is the following: given a k-recognizable set I of non negative integers,do there exist a base h (*resp.* a number system S) and an aperiodic(recognizable) language L over the alphabet A_h (*resp.* A_S) such that $[L]_h = I$ (*resp.* $[L]_S = I$) ? Let us illustrate the question by the following example:

Example. Let $I = \{2^{2n} \mid n \geq 0\}$. I is 2-recognizable: $I = [L]_2$ with $L = 1(00)^*$. L is not aperiodic. However, in base 4, there exists an aperiodic language $L' = 10^*$ such that $[L']_4 = I$.

Next theorems give an affirmative answer to the question in some particular cases.

Theorem 2. If a subset I of N is k-recognizable for all $k \geq 2$ then there exists a base $h \geq 1$ and a generalized definite language $L \subseteq A_h^*$ such that $I = [L]_h$.

Proof: If a set I of non negative integers is k-recognizable for all $k \geq 2$ then from a result of Cobham [1] (cf.[4]) I is of the kind $I = B \cup (C + p^*)$,where B and C are finite subsets of N , $p \in N$ and $p^* = 0 \cup p \cup 2p \cup \ldots$. If $p = 0$ the the result is trivial since I is a finite set. If $p = 1$ then $p^* = N$ so that I is the union of a finite set and the set of all the integers greater than or equal to a fixed non negative integer. In this case one can take h=1,i.e. $A_1 = \{a\}$ and find an $L \subseteq a^*$ such that $[L]_1 = I$ by assuming $L = F \cup a^* a^q$, where F is a finite subset of a^* and q a suitable element of N . If $p \geq 1$ let us set $A = \{0,1,\ldots,p-1\}$, $n_o = \max\{ n \mid n \in B \cup C \}$, $|n_o| = |\langle n_o \rangle_p|$, $I' = \{ n \in I \mid n < p^{|n_o|+1} \}$ and $R = \{ r \in A \mid [r]_p = \mathrm{res}(c)$ for some $c \in C \}$,where res(c) denotes the residual of c modulus p .

We can then consider the language $L = F \cup (A \setminus 0) A^* E$ where $E = A^{|n_o|} R$ and F is a finite subset of A^* such that $[F]_p = I'$. L is a generalized definite language over A such that $[L]_p = I$. This is a consequence of the fact that for $n \in N$ and $c \in C$ one has that : $n \equiv c \pmod p$ and $n \geq p^{|n_o|+1}$ if and only if $\langle n \rangle_p \in (A \setminus 0) A^* E$. □

We remark that if one refers to k-adic (instead of k-ary) systems then one can prove,by making use of Theorem 1 and Cobham's theorem,in a way similar to that of Theorem 2, the stronger result: *A subset* I *of* N *is recognizable in a* k-*adic system for all* $k \geq 2$ *if and only if there exist an* h-*adic system* $A = \{1,\ldots,h\}$ *and a reversed definite language* $L = E \cup A^*F$ (E *and* F *being finite subsets of* A^*) *such that* $[L]_h = I$.

Our main result concerns bounded and aperiodic recognizable languages. A language $L \subseteq A^*$ is *bounded* if there exists a finite number of words $w_1, w_2, \ldots w_p$ of A^* such that $L \subseteq w_1^* w_2^* \ldots w_p^*$. A language $L \subseteq A^*$ is *aperiodic* if there exists a positive integer k such that for all $n \geq k$, $f \in A^*$, $u, v \in A^*$ one has:

$$u f^n v \in L \quad \text{if and only if} \quad u f^{n+1} v \in L .$$

Let us explicitly remark that a bounded recognizable language,in general,is not aperiodic.For instance on the alphabet $A = \{a,b\}$,the language $L = (a^2)^* b$ is bounded but not aperiodic.Vice versa an aperiodic recognizable language needs not to be bounded.For instance the language $\{ab,ba\}^*$ is aperiodic but it is not bounded (cf.[7]).

In the following a bounded language L over the alphabet A will be called *standard* if there exist a positive integer h and words w_i $(i=1,\ldots,h)$, u_i $(i= 1,\ldots,h+1)$ such that

$$L = u_1 w_1^* u_2 w_2^* \ldots u_h w_h^* u_{h+1} .$$

We say that each w in L is a *loop* of L of length $|w|$. The standard language L will be called *elementary* if all the loops of L have length 1 ,i.e. they are letters.

Theorem 3 . Let I be a subset of N , $k \geq 2$ a base and L a bounded recognizable language over A_k such that $[L]_k = I$. There exist then a base $h \geq 2$ and an aperiodic recognizable language L' over A_h such that $[L']_h = I$.

Proof (Outline) . The proof is based on some lemmas (cf.Lemmas 1,2,3).First one shows,by making use of a result of Ginsburg and Spanier [5],that any bounded recognizable language over a given alphabet can be expressed as a finite union of standard languages.Then one proves that there exist some bases on which a standard language "becomes"(via the set of the represented integers) an aperiodic recognizable language.Finally one derives that it is possible to choose the same base for all the standard languages in the preceding union, on which they become aperiodic and recognizable.

The proofs of the following lemmas are omitted for the sake of brevity:

Lemma 1. A language L over the alphabet A is a recognizable bounded language if and only if L is a finite union of standard languages.

Lemma 2. An elementary standard language L over the alphabet A is aperiodic.

Lemma 3. Let $L \subseteq A_k^*$ be a finite union of standard languages L_m, $m \in M$, and r a positive integer. If the loops in all L_m, $m \in M$, have a same length $d > 0$, which does not depend on m, then there exists a language L' over the alphabet A_h, $h = k^{rd}$, which is a finite union of elementary standard languages and such that $[L']_h = [L]_k$.

We are now in the position to prove Theorem 2. Let us first observe that if L is a standard language over A_k, i.e. $L = u_1 w_1^* u_2 w_2^* \dots u_p w_p^* u_{p+1}$, with $|w_i| \geq 1$ $(i = 1,\dots,p)$ then L can be expressed as a finite union of standard languages whose loops have all a same length which can be any fixed multiple of the lengths of the loops in L. In fact let $d = \text{m.c.m.}(|w_1|,\dots,|w_p|)$, r a positive integer and $p_i = rd/|w_i|$ $(i = 1,\dots,p)$. One can express w_i^* $(i=1,\dots,p)$ as:

$$w_i^* = \bigcup_{0 \leq j < p_i} (w_i^{p_i})^* w_i^j .$$

If one replaces all loops in L by the preceding expressions then one obtains that L can be written as a finite union of standard languages whose loops have all a same length equal to

$$|w_i^{p_i}| = p_i |w_i| = r d \qquad (i = 1,\dots,p) .$$

Now let L be any bounded recognizable language over the alphabet A_k. From Lemma 1, L can be expressed as a finite union

$$L = \bigcup_{m \in M} L_m$$

of standard languages L_m, $m \in M$. For each $m \in M$ we denote by d_m the least common multiple of the lengths of the loops in L_m. As we have previously seen each language L_m can be expressed as a finite union of standard languages whose loops have a same length $r_m d_m$, where r_m is an arbitrary positive integer. Let us set $d = \text{m.c.m.}(d_m \mid m \in M)$ and take for any m, $r_m = d/d_m$. By making use of Lemma 3 there exists then, for any m, a unique base $h = k^d$ and a language $L'_m \subseteq A_h^*$ which is union of elementary standard languages and such that $[L'_m]_h = [L_m]_k$, $m \in M$. Setting

$$L' = \bigcup_{m \in M} L'_m$$

one has $[L']_h = [L]_k$. Moreover by Lemma 2 it follows that L' is aperiodic. □

We explicitly note that if L' is an aperiodic recognizable language such that $[L']_h = I$ then the set of all the expansions of I in base h given by $L' \setminus 0A_h^*$,is still an aperiodic (recognizable) language.

Next theorems (cf.Theorems 4 and 5) give an answer to the initially posed problem in the case of finitely generated submonoids. A finitely generated (f.g.) submonoid of A^* is a language L of the form $L = X^*$ with $X = \{x_1, \ldots, x_n\}$ a finite subset of A^* . If the elements of X have the same length , i.e. $|x_1| = |x_2| = \ldots = |x_n|$,we say that L is *uniform* . If X is such that for a given letter $a \in A$, $a^n \in X \Rightarrow n=1$, we say that L is a-*pure*.

Theorem 4. Let I be a subset of N , $k \geq 2$ a base and L a uniform f.g; submonoid of A_k^* such that $I = [L]_k$. Then there exist a base $h \geq 2$ and a f.g. strictly locally testable submonoid L' of A_h^* such that $[L']_h = I$.

Proof : Let $L = X^* \subseteq A_k^*$,where $X = \{x_1, \ldots, x_n\}$ and $|x_1| = \ldots = |x_n| = d$. We take $h = k^d$ and define $Y = \{y_1, \ldots, y_n\} \subseteq A_h$ where $y_i = [x_i]_k$, $1 \leq i \leq n$. The language $L' = Y^*$ is then clearly a strictly locally testable language of A_h^* . Moreover for any word $w = x_{i_1} \ldots x_{i_r} \in L$ one has:

$$[w]_k = \sum_{s=1}^{r} [x_{i_s}]_k \quad k^{d(r-s)} = [y_{i_1} \ldots y_{i_r}]_h$$

so that $[L']_h = [L]_k = I$. □

Theorem 5 . Let I be a subset of N , $k \geq 2$ a base and L a 0-pure f.g. submonoid of A_k^* such that $[L]_k = I$. There exist then a number system S and a f.g. strictly locally testable submonoid L' of A_S^* such that $[L']_S = I$.

Proof: Let $L = X \subseteq A_k^*$ with $X = \{x_1, \ldots, x_n\}$ .We consider then the number system S whose base is k and whose alphabet A_S is given by $A_S = \{0, b_1, \ldots, b_n\}$ where $b_i = [x_i]_k$, $1 \leq i \leq n$. Let us set

$$Y = \{0^{|x_1|-1} b_1, \ldots, 0^{|x_n|-1} b_n\} \subseteq A_S^* \quad \text{and} \quad L' = Y^* .$$

For any word $w = x_{i_1} \dots x_{i_r} \in L$ one has:

$$[w]_k = [x_{i_r}]_k + [x_{i_{r-1}}]_k \, k^{|x_{i_r}|} + [x_{i_{r-2}}]_k \, k^{|x_{i_{r-1}} x_{i_r}|} + \dots + \dots =$$

$$= [0^{|x_{i_1}|-1} b_{i_1} 0^{|x_{i_2}|-1} b_{i_2} \dots 0^{|x_{i_r}|-1} b_{i_r}]_S ,$$

so that $[L]_k = [L']_S$. If one supposes that L is 0-pure then so will be Y^* . Moreover one can easily verifies that X is a "locally parsable" code. Hence L' is a strictly locally testable language (cf. [3]) . □

Acknowledgments

The work for this paper was partly supported by the Italian Ministry of the education (M.P.I.)

REFERENCES

[1] . Cobham,A., On the base-dependence of sets of numbers recognizable by finite automata, *Mathematical System Theory*, 3 (1969) 186-198.

[2] . Culik,II,K. and A.Salomaa, Ambiguity and decision problems concerning number systems,*Lecture Notes in Computer Science,vol.154,Springer Verlag,1983,pp.137-146.*

[3] . De Luca,A. and A.Restivo, A characterization of strictly locally testable languages and its application to subsemigroups of a free semigroup,*Information and Control*, 44(1980) 300-319.

[4] . Eilenberg,S., "Automata,Languages and Machines",*Academic Press,vol.A,1974, vol.B,1976.*

[5] . Ginsburg,S.and E.H.Spanier,Bounded regular sets,*Proc.American Math.Soc.* 17 (1966)1043-1049.

[6] . McNaughton,R.and S.Papert, "Counter-Free Automata",*MIT Press,New York,1971.*

[7] . Restivo,A., Mots sans répétitions et langages rationnels bornés,R.A.I.R.O.,I.T. 11(1977)197-202.

[8]. Restivo,A., Codes and aperiodic languages,*Lecture Notes in Computer Science, vol.2,Springer Verlag,1973,pp.175-181.*

[9]. Schützenberger,M.P.,On finite monoids having only trivial subgroups,*Information and Control,8(1965)190-194.*

NEW LOWER BOUND FOR POLYHEDRAL MEMBERSHIP PROBLEM WITH AN APPLICATION TO LINEAR PROGRAMMING

Jaroslav Morávek and Pavel Pudlák
Mathematical Institute
Czechoslovak Academy of Sciences
115 67 Praha 1 / Czechoslovakia

I. INTRODUCTION

Many computational problems can be formulated as a polyhedral membership problem (abbrev.: PMP) with respect to a given (convex) polyhedron $P \subseteq R^n$ (cf. eg. [1], [2]):

INSTANCE : $x \in R^n$;
QUERY : Decide whether $x \in P$. //

Our model of computation is the linear comparison tree (cf. eg. [1-6]) a formal definition of which is given in Section 2. Intuitively, we are interested in the minimal number of comparisons of the form $\lambda(x) \gtreqless 0$, where λ is a nonconstant linear function, needed to determine, in the worst case, whether $x \in P$. We call this number linear comparison complexity and denote it by lcc(P).

We shall prove that lcc(P) can be bounded from below using the number φ_d of all d-dimensional faces of P as follows:

Theorem 1 Let $P \subseteq R^n$ be an n-dimensional convex polyhedron, and let $d \in \{0,1,\dots,n-1\}$. Then

$$2^{\,\mathrm{lcc}(P)} \cdot \binom{\mathrm{lcc}(P)}{n-d} \geq \varphi_d \cdot 2^{n-d} \quad . \blacksquare$$

(Here and in the sequel the combination number $\binom{a}{b}$ is defined to be 0 if $a < b$).

This lower bound improves the previously known lower bound

$$2^{\,\mathrm{lcc}(P)} \cdot \binom{\mathrm{lcc}(P)}{n-d} \geq \varphi_d$$

proved for d = 0 in [2] (1972) and in the general case in [7] and [1].

Theorem 1 will be proved in Section 3. In the proof we generalize an idea from [8] in two respects: 1. Instead of the n-cube we consider arbitrary full-dimensional polyhedra. 2. Instead of the vertices we consider arbitrary faces of a given dimension.

In Section 4 we apply the lower bound from Theorem 1 to obtain a lower bound on the number of certain partially linear comparisons required for solving the following decision problem, closely related to the linear programming (n, k are given integers, $3 \leq n \leq k$):

INSTANCE : A real $n \times k$-matrix A and a real row vector x with n components;

QUERY : Decide whether there exists a nonnegative row vector y such that $A \cdot y^T = x^T$ (symbol T denotes the transposition). //

2. GENERALITIES

Terms 'open', 'closed', 'closure', 'interior', 'relative interior', 'dimension', etc. are meant w.r.t. the usual topology of R^n, dimension of the empty set is -1. Symbols fr(.), int(.) and relint(.) denote respectively the boundary, the interior and the relatitive interior, the closure of $M \subseteq R^n$ is denoted by $\overline{M}$. The least affine subset of R^n containing a given nonempty set $M \subseteq R^n$ will be denoted by aff(M) (the affine hull of M).

By a <u>polyhedron</u> we shall mean the intersection of a finite (possibly empty) family of (closed or open) halfspaces in R^n . Thus R^n and $\emptyset$ are polyhedra, and every polyhedron is convex.

In this paper we consider (relatively open!) faces of a polyhedron: A nonempty subset $F \subseteq R^n$ will be called a <u>face</u> of a polyhedron Q if F = int(Q) or $F = \mathrm{relint}(\overline{Q} \cap H)$ for some supporting hyperplane H of Q. (Let us observe that our relatively open faces are in a one-to-one correspondence to the usual (closed) faces [9]). The following fact can be easily verified:

<u>Lemma 1</u> Let a nonempty polyhedron Q be expressed as a finite intersection of halfspaces, $Q = X_1 \cap X_2 \cap \ldots \cap X_r$. Then every face F of Q can be expressed in the form $F = X_1' \cap X_2' \cap \ldots \cap X_r'$, where for each $j \in \{1,2,\ldots,r\}$: $X_j' = \mathrm{int}(X_j)$ or $X_j' = \mathrm{fr}(X_j)$.∎

Lemma 1 will be used as an equivalent definition of the face. In particular, it follows from it that the faces of Q partition $\overline{Q}$.

<u>Definition:</u> A <u>linear comparison tree</u> (in R^n) is an indexed family $\mathcal{T} = (\lambda_{i,t_1,\ldots,t_i} \mid t_1,\ldots,t_i \in \{-1, 0, 1\};\ i \in \{0,1,\ldots,h\})$, where $\lambda_{i,t_1,\ldots,t_i} : R^n \to R$ are linear (in general nonhomogeneous) functions; h is called the <u>height</u> of $\mathcal{T}$, tuples $(i,t_1,\ldots,t_i)$ are called <u>vertices</u> of $\mathcal{T}$, for i = h <u>output vertices</u>.

Given a linear comparison tree $\mathcal{T}$ and a vertex $(i,t_1,\ldots,t_i)$ of it we define the <u>polyhedron generated at the vertex</u> by

$$G(i,t_1,\ldots,t_i) \overset{\text{df}}{=} \bigcap_{j=0}^{i-1} \{x\in R^n \mid \operatorname{sign}(\lambda_{j,t_1,\ldots,t_j}(x)) = t_{j+1}\}.$$

In case of $i = h$ this polyhedron will be called an <u>output set.</u>

<u>Definition:</u> We say: $\mathcal{T}$ <u>solves</u> PMP <u>with respect to</u> a polyhedron P if every output set of $\mathcal{T}$ is a subset of P or $R^n \setminus P$. (Put otherwise, after ommitting the empty sets, the partition given by the output sets is a refinement of the partition $\{P, R^n \setminus P\}$).

<u>Definition:</u> For a polyhedron $P \subseteq R^n$ let us set

$$\operatorname{lcc}(P) \overset{\text{df}}{=} \min\{h \mid \text{there exists a linear comparison tree } \mathcal{T} \text{ having height } h \text{ and solving PMP w.r.t. } P\}.$$

<u>Observation:</u> In the sequel we may assume w.l.o.g. that all linear functions occurring in the definition of $\mathcal{T}$ are nonconstant.

3. PROOF OF THEOREM 1

For proving Theorem 1 we shall use some additional definitions and lemmas. (The idea of the proof is to assign suitable weights for d-dimensional faces and then to choose a path on which the sum of weights decreases as slowly as possible ; adversary or oracle approach. The weight assigned to a d-dimensional face is 0 or a power of 2 and, roughly speaking, expresses how close is the face to the polyhedron generated at the vertex.)

<u>Lemma 2</u> Let Y, $Q \subseteq R^n$ be nonempty polyhedra. If $\dim(Y\cap\overline{Q}) = \dim(Y)$ then there exists exactly one face G of Q such that

$$\dim(G\cap Y) = \dim(Y) .$$

<u>Proof:</u> 1. Existence: Let $\{F_1,F_2,\ldots,F_t\}$ be the set of all faces of Q. Since $\{F_1,\ldots,F_t\}$ is a partition of $\overline{Q}$ we have:

$\{Y\cap F_1,\ldots,Y\cap F_t\}$ is a partition of $Y\cap\overline{Q}$.

Thus, $\dim(Y\cap F_\tau) = \dim(Y\cap\overline{Q}) = \dim(Y)$ for some τ.

2. Uniqueness: Let G_1, G_2 be faces of Q such that $\dim(G_j\cap Y) = \dim(Y)$ for $j\in\{1,2\}$. We have to prove that $G_1 = G_2$.

We may assume w.l.o.g. that $\dim(G_1) = \dim(G_2)$, and let us consider the following two cases: (i) $\dim(G_1) = n$; (ii) $\dim(G_1) < n$.

In the (i) case, we want to prove that $\dim(G_2) = n$. Indeed, assuming on the contrary that $\dim(G_2) < n$ we obtain $\dim(Y) = \dim(G_2\cap Y) \leq \dim(\operatorname{aff}(G_2)\cap\operatorname{aff}(Y)) \leq \dim(\operatorname{aff}(Y)) = \dim(Y)$, hence $Y \subseteq \operatorname{aff}(Y) \subseteq$

$\subseteq$ aff(G_2). This yields $Y \subseteq H$ for some supporting hyperplane H of Q. On the other hand, G_1 = int(Q) $\neq \emptyset$. Now, combining the last two facts with dim($G_1 \cap Y$) = dim(Y) ≥ 0 we have: int(Q) $\cap H \neq \emptyset$, a contradiction. Thus, dim(G_2) = n, which yields

$$G_1 = G_2 = \text{int}(Q).$$

In the (ii) case, we have similarly as in the (i) case: $Y \subseteq \text{aff}(Y) \subseteq \text{aff}(G_j) \subseteq H_j$ for some supporting hyperplane H_j of Q with the property $G_j = \text{relint}(\overline{Q} \cap H_j)$ ($j \in \{1,2\}$).

This yields successively $\emptyset \neq Y \cap \overline{Q} \subseteq H_j \cap \overline{Q}$ ($j \in \{1,2\}$), and $\emptyset \neq$ $\neq \text{relint}(Y \cap \overline{Q}) \subseteq \text{relint}(H_1 \cap \overline{Q}) \cap \text{relint}(H_2 \cap \overline{Q}) = G_1 \cap G_2$, hence

$$G_1 = G_2. \blacksquare$$

<u>Definition:</u> Let Y, Q be polyhedra, $Y \neq \emptyset$. The <u>incidence degree of</u> Y <u>with respect</u> to Q, id(Y,Q), will be defined as follows:

$\text{id}(Y,Q) = 0$ if $\dim(Y \cap \overline{Q}) < \dim(Y)$,

$\text{id}(Y,Q) = 2^{\dim(G)-\dim(Y)}$ otherwise, where G is the face of Q determined uniquely by the condition $\dim(G \cap Y) = \dim(Y)$ (Lemma 2).

<u>Lemma 3</u> Let Y,Q be polyhedra, $Y \neq \emptyset$, H_1,H_2 two opposite open halfspaces, bounded by a given hyperplane H in R^n. Let $Q_i = Q \cap H_i$ ($i \in \{1,2\}$). Then $\text{id}(Y,Q) \leq \text{id}(Y,Q_1) + \text{id}(Y,Q_2)$. Moreover, if $\text{id}(Y,Q_1) > \text{id}(Y,Q_2)$ then $\overline{Q} \cap H_1 \cap Y \neq \emptyset$.

<u>Proof:</u> Let $F_1,F_2,\ldots,F_r$ be the faces of Q. The faces of Q_i are the nonempty members of the sequence $F_1 \cap H_i$, $F_2 \cap H_i, \ldots, F_r \cap H_i$, $F_1 \cap H$, $F_2 \cap H, \ldots, F_r \cap H$. Further we have:

$$\dim(F_t \cap H_i) = \dim(F_t) \qquad \text{if } F_t \cap H_i \neq \emptyset \ ; \tag{1}$$

$$\dim(F_t \cap H) \geq \dim(F_t) - 1 \quad \text{if } F_t \cap H \neq \emptyset \ . \tag{2}$$

For the proof of $\text{id}(Y,Q) \leq \text{id}(Y,Q_1) + \text{id}(Y,Q_2)$ we consider the following two cases: (a) $\dim(Y \cap \overline{Q}) < \dim(Y)$; (b) $\dim(Y \cap \overline{Q}) = \dim(Y)$.

In the (a) case we have id(Y,Q) = 0 and the inequality holds. Moreover, in this case we also have $\dim(Y \cap \overline{Q}_i) < \dim(Y)$ ($i \in \{1,2\}$) and thus $\text{id}(Y,Q_1) = \text{id}(Y,Q_2) = 0$.

In the (b) case let us consider the (unique) face F_t of Q such that $\dim(Y \cap F_t) = \dim(Y)$ (Lemma 2), and let us deal with the following two 'subcases': (b1) $Y \cap F_t \cap H_i \neq \emptyset$ for some i ;

(b2) $Y \cap F_t \cap H_i = \emptyset$ for $i \in \{1,2\}$.

In the (b1) case we have $Y \cap \overline{Q} \cap H_i \supseteq Y \cap F_t \cap H_i \neq \emptyset$. Further, observe that $\dim(Y \cap F_t \cap H_i) = \dim(Y \cap F_t) = \dim(Y)$, and hence $F' \overset{df}{=} F_t \cap H_i$ is a face of Q_i such that $\dim(Y \cap F') = \dim(Y)$. Now, it follows from (1) that $\dim(F') = \dim(F_t)$, and we have

$$\text{id}(Y,Q_i) = 2^{\dim(F')-\dim(Y)} = 2^{\dim(F_t)-\dim(Y)} = \text{id}(Y,Q) \ ,$$

which proves the inequality $id(Y,Q) \leq id(Y,Q_1) + id(Y,Q_2)$.

In the (b2) case we have $Y \cap F_t \subseteq H$, and hence

$$\dim(Y \cap F_t \cap H) = \dim(Y \cap F_t) = \dim(Y). \tag{3}$$

Further, $F_t \cap H \neq \emptyset$ is a common face of Q_1 and Q_2, and using (2) and (3) we have: $id(Y,Q_1) = id\ (Y,Q_2) =$

$$= 2^{\dim(F_t \cap H)-\dim(Y)} \geq 2^{\dim(F_t)-\dim(Y)-1} = \frac{1}{2}\, id(Y,Q)\ ,$$

which completes the proof of the first part of the Lemma.

For proving the second part observe that $id(Y,Q_1) > id(Y,Q_2)$ only if the (b1) case takes place with $i = 1$. So we have $Y \cap \overline{Q} \cap H_1 \supseteq Y \cap F_t \cap H_1 \neq \emptyset$. This completes the proof. ■

From Lemma 3 it follows immediately:

<u>Lemma 4</u> Let P, $Q \subseteq R^n$ be polyhedra, $int(P) \cap int(Q) \neq \emptyset$, H a hyperplane in R^n, H_1, H_2, Q_1, Q_2 introduced as in Lemma 3, F a face of P. If $id(F,Q_1) > id(F,Q_2)$ then $int(P) \cap int(Q) \neq \emptyset$. ■

<u>Definition:</u> A vertex $(i,t_1,t_2,\ldots,t_i)$ of a linear comparison tree is said to be <u>dichotomic</u> if $t_1 \neq 0,\ t_2 \neq 0, \ldots, t_i \neq 0$.

<u>Lemma 5</u> Let $\mathcal{F}$ be a set of faces of a polyhedron P, and let $(i,t_1,\ldots,t_i)$ be a dichotomic vertex of a linear comparison tree $\mathcal{T}$ where $i <$ height of $\mathcal{T}$. Suppose $G(i,t_1,\ldots,t_i) \cap int(P) \neq \emptyset$.

Then there exists $t_{i+1} \in \{-1,1\}$ such that

$$G(i+1,t_1,\ldots,t_{i+1}) \cap int(P) \neq \emptyset\ ,$$

and

$$\sum_{F \in \mathcal{F}} id(F,G(i,t_1,\ldots,t_i)) \leq 2 \cdot \sum_{F \in \mathcal{F}} id(F,G(i+1,t_1,\ldots,t_i,t_{i+1})).$$

<u>Proof:</u> Put $Q = G(i,t_1,\ldots,t_i)$, $Q_j = G(i+1,t_1,\ldots,t_i,(-1)^{j-1})$ $(j \in \{1,2\})$. Then by Lemma 3, for every F

$$id(F,Q) \leq id(F,Q_1) + id(F,Q_2),$$

and hence $\sum_{F \in \mathcal{F}} id(F,Q) \leq \sum_{F \in \mathcal{F}} id(F,Q_1) + \sum_{F \in \mathcal{F}} id(F,Q_2)$.

Now, if

$$S_1 \overset{df}{=} \sum_{F \in \mathcal{F}} id(F,Q_1) = S_2 \overset{df}{=} \sum_{F \in \mathcal{F}} id(F,Q_2)\ ,$$

choose an index $j \in \{1,2\}$ such that $P \cap Q_j \neq \emptyset$.

On the other hand, if $S_1 \neq S_2$, say $S_1 > S_2$, then we have for some $F \in \mathcal{F}$: $id(F,Q_1) > id(F,Q_2)$, and Lemma 4 yields $int(P) \cap Q_1 \neq \emptyset$. The proof is completed. ■

<u>Lemma 6</u> Let P and Q be polyhedra, $Q \subseteq P \neq \emptyset$. Then for every face F of P it holds $id(F,Q) \leq 1$. Moreover, $id(F,Q) = 1$ iff there exists a face G of Q such that $G \subseteq F$ and $\dim(G) = \dim(F)$.

Proof: It is sufficient to show that if F meets a face G of Q then $G \subseteq F$.

We may assume w.l.o.g. that $P = X_1 \cap X_2 \cap \dots \cap X_r$ and $Q = X_1 \cap X_2 \cap \dots \cap X_r \cap Y_1 \cap Y_2 \cap \dots \cap Y_s$ for some halfspaces $X_1, X_2, \dots, X_r, Y_1, Y_2, \dots, Y_s \subseteq R^n$. Our assertion now follows immediately from Lemma 1. ■

Proof of Theorem 1: Let $\mathcal{T}$ be a linear comparison tree for solving PMP with respect to P, and let h = height of $\mathcal{T}$. Let $\mathcal{F}_d$ be the set of all d-dimensional faces of P. Starting with the root (0) of $\mathcal{T}$ (where $G(0) = R^n$) and applying successively Lemma 5 we have

$$\varphi_d \cdot 2^{n-d} = \sum_{F \in \mathcal{F}_d} id(F,R^n) \leq 2^h \cdot \sum_{F \in \mathcal{F}_d} id(F,G(v)) \, , \tag{4}$$

where v is some dichotomic output vertex of $\mathcal{T}$ with

$$G(v) \cap int(P) \neq \emptyset \, .$$

Then $G(v) \subseteq P$ ($\mathcal{T}$ is a linear comparison tree solving our PMP) and Lemma 6 gives

$$\sum_{F \in \mathcal{F}_d} id(F,G(v)) \leq \gamma_d \, , \tag{5}$$

where γ_d is the number of all d-dimensional faces of G(v).

Finally,

$$\gamma_d \leq \binom{h}{n-d} , \tag{6}$$

which follows from the fact that every d-dimensional face of G(v) is determined by some (n - d)-tuple of boundary hyperplanes of G(v).

Combining (4) - (6) we complete the proof. ■

4. A LOWER BOUND FOR LINEAR PROGRAMMING

Let $3 \leq n \leq k$ be given integers, and let $M(n \times k)$ denote the set of all real matrices of the size $n \times k$. By $\Pi(n \times k)$ we denote the decision problem introduced in the end of Section 1, and by $\Pi(n \times k;\ A)$ the decision computational problem introduced as above except that $A \in M(n \times k)$ is fixed.

Problem $\Pi(n \times k;\ A)$ can be evidently stated as PMP with respect to the polyhedron $P = \{x = y \cdot A^T |\ y \geq 0\}$ (a closed convex polyhedral cone). Thus the linear comparison complexity of problem $\Pi(n \times k;\ A)$ is defined; we denote it by $lcc(n \times k;\ A)$.

On the other hand, it is well-known that problem $\Pi(n \times k)$ cannot be solved by linear comparison trees with comparisons of the form

$\lambda(x, A) \gtreqless 0$, where $\lambda(x, A)$ are linear both in x and A, cf. e.g. [10] or [11]. For solving $\Pi(n \times k)$ we use in our paper a more general decision tree model:

Definition: A partially linear comparison tree in $R^n \times M(n \times k)$ is an indexed family of functions of $R^n \times M(n \times k)$ into R,

$$\mathcal{B} = (\mu_{i,t_1,\ldots,t_i} \mid t_1,\ldots,t_i \in \{-1,0,1\};\ i = 0,1,\ldots,h),$$

such that $\mu_{i,t_1,\ldots,t_i}(x,A) = c_1(A)\cdot x_1+\ldots+c_n(A)\cdot x_n + c(A)$

($x \in R^n$, $A \in M(n \times k)$), where $c_1(A),\ldots,c_n(A)$, $c(A)$ are arbitrary (in practice: computable) functions of A.

All concepts concerning partially linear comparison trees are introduced just as for linear comparison trees. In particular, the set generated at the vertex $(i,t_1,\ldots,t_i)$ by $\mathcal{B}$ is

$$\bigcap_{j=0}^{i-1} \{(x,A) \in R^n \times M(n \times k) \mid \operatorname{sign}(\mu_{j,t_1,\ldots,t_j}(x,A)) = t_{j+1}\},$$

and $\mathcal{B}$ solves $\Pi(n \times k)$ if every output set of $\mathcal{B}$ is a subset of $\{x = y\cdot A^T \mid y \geqq 0\}$ or of its complement.

It is easy to see that for each ordered pair of integers n, k with $3 \leqq n \leqq k$ there exists a partially linear comparison tree solving $\Pi(n \times k)$. Indeed, we may assume w.l.o.g. that $x \geqq 0$ (in the contrary case, we multiply the equations from $A\cdot y^T = x^T$ with negative coordinates of x by -1) and reduce $\Pi(n \times k)$ to the following linear programming problem: Minimize $z_1 + z_2 +\ldots+ z_n$ subject to $A\cdot y^T + z^T = x^T$, $y \geqq 0$, $z = (z_1,\ldots,z_n) \geqq 0$.

Now, each simplex iteration of the ordinary simplex method applied to the above linear programming problem requires in general many 'nonlinear' arithmetic operations (multiplications and divisions) but only 'linear' operations (additions, subtractions and multiplications by rational expressions in the components of A, as indeterminates) are applied to the components of x. Thus, from a formal point of view, the simplex algorithm can be described as a partially linear comparison tree for solving $\Pi(n \times k)$, and we see that the class of all partially linear comparison trees for solving $\Pi(n \times k)$ contains nontrivial computing methods.

As the measure of complexity of $\Pi(n \times k)$ with respect to the class of all partially linear comparison trees we shall use

$$pl(n \times k) \overset{df}{=} \text{minimum height of a partially linear comparison tree } \mathcal{B} \text{ solving } \Pi(n \times k).$$

If we 'fix' $A \in M(n \times k)$ in a partially linear comparison tree solving $\Pi(n \times k)$ we obtain from it a linear comparison tree solving

$\Pi(n \times k;\ A)$ and having the same height. Thus we have:

Lemma 7 For every $A \in M(n \times k)$ it holds: $pl(n \times k) \geq lcc(n \times k;\ A)$. ■

Theorem 2 For each triple of integers d,n,k with $3 \leq n \leq k$ and $1 \leq d \leq \frac{1}{2}(n-1)$ it holds

$$2^{\,pl(n \times k)} \cdot \binom{pl(n \times k)}{n-d} \geq \binom{k}{d} \cdot 2^{n-d} \quad . ■$$

Theorem 2 improves the lower bound from [6]. Using this theorem we can obtain various corollaries concerning the asymptotic behaviour of $pl(n \times k)$, as e.g. :

Corollary Let (n_k) be a sequence of integers such that $3 \leq n_k \leq k$ $(k = 3,4,\ldots)$ and $n_k \to \infty$ as $k \to \infty$. If $\log n_k = o(\log k)$ then

$$pl(n_k \times k) \gtrsim \frac{1}{2}\, n_k \cdot \log k \ \ (k \to \infty) \quad . ■$$

For proving Theorem 2 we shall use the following:

Lemma 8 Let d,n,k be integers as in Theorem 2. The maximum number of d-dimensional faces of a convex polyhedral cone in R^n spanned by exactly k extreme rays is $\binom{k}{d}$.

Proof: Since the number of all d-dimensional faces of the cone in question is, at most, $\binom{k}{d}$ (each d-dimensional face is determined by exactly d extreme rays) it is sufficient to construct a convex polyhedral cone with k extreme rays and $\binom{k}{d}$ faces of dimension d.

Let H be an arbitrary hyperplane in R^n that does not pass through the origin; H is an (n - 1)-dimensional affine set. Let $Q \subseteq H$ be the cyclic polytope (cf. [9], pp. 61-63) with k vertices, and let C be the pointed convex polyhedral cone generated by Q. It is easy to see that C has $\binom{k}{d}$ faces of dimension d. ■

Proof of Theorem 2: Let C be the cone from Lemma 8. We construct a matrix $A \in M(n \times k)$ such that $C = \{y \cdot A^T \mid y \geq 0\}$. (For the rows of A we can put e.g. arbitrary generating elements of the extreme rays of C). For the completion of the proof it is sufficient to apply Lemma 7 and then Theorem 1 for P = C. ■

REFERENCES

1. A.C. Yao, R.L. Rivest, On the polyhedral decision problem. Siam J. Comput., Vol. 9, No. 2, May 1980.
2. J. Morávek, A localization problem in geometry and complexity of discrete programming. Kybernetika(Prague)8: 498-516(1972).

3. J. Morávek, On the complexity of discrete programming problems. Aplikace matematiky 14: 442-474 (1969).

4. J. Morávek, A note upon minimal path problem. Journal of Math. Analysis and Appl. 30: 702-717 (1970).

5. J. Morávek, A geometrical method in combinatorial complexity. Aplikace matematiky 26: 82-96 (1981).

6. J. Morávek, Decision trees and lower bound for complexity of linear programming, in Graphs and Other Combinatorial Topics. M. Fiedler, Ed., Proc. of the Third Czechoslovak Symposium on Graph Theory, held in Prague, 1982.

7. E. Kalinová, The localization problem in geometry and Rabin-Spira linear proofs. (Czech), M.Sci. thesis, Universitas Carolina, Prague, 1978.

8. E. Györi, An n-dimensional search problemwith restricted questions. Combinatorica 1(4) (1981) 377-380.

9. B. Grünbaum, Convex Polytopes. John Wiley, 1967.

10. D. Avis, Comments on a lower bound for convex hull determination: "On the Ω (n log n) lower bound for convex hull and maximal vector determination" by van Emde Boas. Inform. Process. Lett. 11 (1980), No. 3, 126.

11. J.W. Jaromczyk, Linear decision trees are too weak for convex hull problem. Infor. Process. Lett. 12 (1981), No. 3, 138-141.

DECIDABILITY OF THE EQUIVALENCE PROBLEM FOR SYNCHRONOUS DETERMINISTIC PUSHDOWN AUTOMATA

A.Sh. Nepomnjashchaja
Computing Center, Siberian Division,
USSR Academy of Sciences
Novosibirsk 630090, USSR

INTRODUCTION

The equivalence problem for deterministic pushdown automata (dpda) is very important both for the theory and practice. This problem is closely related with the questions regarding monadic recursion schemes /1/. Though the problem is open for general dpda's there are different subclasses of dpda's for which the equivalence problem is decidable. So, the equivalence problem is decidable for dpda's accepting by empty stack in real time (class R_o) /2/,/3/, for deterministic finite-turn pushdown automata /4/, for two dpda's one of which is a finite-turn or a one-counter machine /5/.

The purpose of this paper is to define a new subclass of dpda's called synchronous for which the equivalence problem is decidable.

DEFINITIONS

A synchronous deterministic pushdown automaton has an input tape over Σ and an m-track pushdown stack over $\Gamma=(\Gamma_1\cup\{v\})\times(\Gamma_2\cup\{v\})\times\ldots\times(\Gamma_m\cup\{v\})$ where for any $i\neq j$ $(i,j=1,2,\ldots,m)$ $\Gamma_i\cap\Gamma_j=\{Z_{01}\}$ v is a special symbol. Each symbol Z_i in Γ is m-tuple $[\bar{Z}_{i1}, \bar{Z}_{i2},\ldots, \bar{Z}_{im}]$ where $\bar{Z}_{ij}=v$ or $\bar{Z}_{ij}=Z_{ij}$ and $Z_{ij}\in\Gamma_j$. $Z_o=[Z_{01}, Z_{01},\ldots, Z_{01}]$ is the initial pushdown symbol. Symbols Z_{ij} are called essential symbols. The synchronous automaton M has a set of states $Q=\{q_\alpha\}$ where $\alpha=<\alpha_1, \alpha_2,\ldots,\alpha_m>$ and for each $j(1\leq j\leq m)$ $\alpha_j=0$ or $\alpha_j=1$, and there exists k for which $\alpha_k=1$. The initial state of M is $q_o=<1,1,\ldots,1>$. Let M be in a state q_α. The tracks of the pushdown stack are called active tracks if they are associated with $\alpha_i=1$ in $\alpha=<\alpha_1,\alpha_2,\ldots,\alpha_m>$.

The synchronous automaton M works in the state q_α synchronously on the active tracks of the pushdown stack. Suppose that $q_\alpha = \bar{q}_i$ iff a track i is active in the state q_α and other tracks are non-active.

We denote the projection of a symbol $Z_i \in \Gamma$ on a track j as $Pr_j(Z_i) = \bar{Z}_{ij}$ and the projection $\Phi \in \Gamma^+$ on a track j as a string φ_j ($Pr_j(\Phi) = \varphi_j$). Suppose that $Pr_j^-(\Phi) = \bar{\varphi}_j$ where $\bar{\varphi}_j$ is obtained from φ_j by deletion of all the symbols v.

Let $q_\alpha \in Q$, $Z_j \in \Gamma$. We define a subset $q_\alpha \times Z_j$ consisting of such symbols $\bar{Z}_{jl}$ of m-tuple Z_j for which $\alpha_l = 1$.

Let $\Phi \in \Gamma^+$, $\Phi = (\varphi_1, \varphi_2, \dots, \varphi_m)$ where for each $1 \le k \le m$ $\varphi_k \in (\Gamma_k \cup \{v\})^+$. Suppose that the string Φ is correctly constructed in the state q_α if the following conditions are satisfied: (1) there exists such an active track j that φ_j has no symbols v; (2) if on the active track i in the state q_α $(1 \le i \le m)$ φ_i has the symbols v, then $\varphi_i = \varphi_{i1}\varphi_{i2}$ where φ_{i1} is a string consisting only of the symbols v and $\varphi_{i2} \in \Gamma_i^+$.

Informally, we use the non-essential symbol v on the different active tracks to equalize the length of the strings.

Each rule of the automaton M has the following form: $(q_\alpha, c, Z_j) \rightarrow (q_\beta, \Phi, d)$ where $q_\alpha, q_\beta \in Q$, $c \in \Sigma$, $Z_j \in \Gamma$, $\Phi \in (\Gamma^+ \cup \{e\})$ (e is a special empty stack symbol), $d \in \{0,1\}$ and d denotes a move on the input tape (if d = 1 then M makes a move, otherwise M does not make a move). The automaton M has four types of rules (for the rules of the types (1)-(3), the subset $q_\alpha \times Z_j$ has only essential symbols):

(1) the rules of the type $WRITE(q_\alpha, c, Z_j) = (q_\alpha, \Phi, 1)$ where Φ is correctly constructed in the state q_α;

(2) the rules of type $ERASE(q_\alpha, c, Z_j) = (q_\alpha, e, 1)$;

(3) the rules of the type $BLOCK_i(q_\alpha, c, Z_j) = (q_\beta, Z_j, 0)$ where $\alpha_i = 1$ in the m-tuple $\alpha = \langle \alpha_1, \alpha_2, \dots, \alpha_m \rangle$ and there exists such $j \ne i$ that $\alpha_j = 1$, $\beta_i = 0$ in m-tuple $\beta = \langle \beta_1, \beta_2, \dots, \beta_m \rangle$ and for any $k \ne i$ $\beta_k = \alpha_k$ $(1 \le k \le m)$;

(4) the rules of the type $ERASE_v(\bar{q}_i, ¢, Z_j) = (\bar{q}_i, e, 0)$ where $Pr_i(Z_j) = v$.

So, using the rules WRITE and ERASE the automaton M works synchronously on all the active tracks in the state q_α such that at any time the automaton M makes a move and the state of M is not changed. It should be noted that the rules WRITE and ERASE can be used in the case when $q_\alpha = \bar{q}_i$. Then for each symbol $Z_j \in \Gamma$ from the left part of these rules $Pr_i(Z_j) \in \Gamma_i$. The rule $BLOCK_i$ turns the active track i of the pushdown stack into a non-active track and the automaton M goes from the state q_α to the state q_β. The rule $ERASE_v$ is used in the case when in the state $\bar{q}_i$ for the top symbol Z_j on the pushdown stack $Pr_i(Z_j) = v$.

The program of M is the set Δ of rules which satisfies the following three restrictions.

(1) The set Δ has no rules $WRITE(q_\alpha, c, Z_j)=(q_\alpha, \Phi, 1)$ and $WRITE(q_\beta, c, Z_p)=(q_\beta, \Omega, 1)$ with the same $c \in \Sigma$ if there exists such i that $\alpha_i=\beta_i=1$, $Pr_i(Z_j)=Pr_i(Z_p)$ and $Pr_i^{-}(\Phi) \neq Pr_i^{-}(\Omega)$.

(2) The set Δ has no rules $WRITE(q_\alpha, c, Z_j)=(q_\alpha, \Phi, 1)$ and $ERASE(q_\beta, c, Z_p)=(q_\beta, e, 1)$ with the same $c \in \Sigma$ if there exists such i that $\alpha_i=\beta_i=1$ and $Pr_i(Z_j)=Pr_i(Z_p)$.

(3) The set Δ has no rules $BLOCK_i(q_\alpha, c, Z_j)=(q_\beta, Z_j, 0)$ and $WRITE(q_\sigma, c, Z_p)=(q_\sigma, \Phi, 1)$ (or $ERASE(q_\sigma, c, Z_p)=(q_\sigma, e, 1)$) with the same symbol $c \in \Sigma$ if $\sigma_i=1$ and $Pr_i(Z_j)=Pr_i(Z_p)$.

We shall explain the sense of these restrictions.

Let two rules of M have the same input symbol $c \in \Sigma$ and the same top symbol Z_{ji} be on the same active track i of the pushdown stack. Then these rules are either the rules of the type WRITE in which the symbol Z_{ji} will be changed by the same string over Γ_i (restriction 1) or the rules of the type ERASE (restriction 2). Note that neither of these two rules can be a rule of the type $BLOCK_i$ (restriction 3).

A configuration of an automaton M is a 4-tuple $K=(q_\alpha, w\dashv, j, \Phi)$ where $q_\alpha \in Q$, w is an input string with the endmarker $\dashv$, j is a non-negative integer $(1 \leq j \leq |w|+1)$, Φ is a pushdown stack. The mode of a configuration K is a 3-tuple (q_α, c, Z_j) where $q_\alpha \in Q$, $c \in \Sigma$, $Z_j \in \Gamma$ that describes the state, the input symbol and the top stack symbol of K, respectively. There is a distinguished initial configuration $K_o=(q_o, w\dashv, 1, Z_o)$. On any track of the tape Z_{o1} is assumed to appear only on the bottom of the stack. The empty stack is denoted by the special symbol Λ. The accepting configuration has a form $(q_\alpha, w\dashv, |w|+1, \Lambda)$ where $w \in \Sigma^+$. A string w is accepted by the automaton M if M goes from the initial configuration $K_o=(q_o, w\dashv, 1, Z_o)$ to the accepting configuration. The language L(M) accepted by M is the set of the strings accepted by M. The automata M_1 and M_2 are equivalent, $M_1 \equiv M_2$, iff $L(M_1)=L(M_2)$.

The following proposition immediately follows from restrictions (1)-(3).

PROPOSITION 1. The automaton M is deterministic.

EXAMPLE 1. For the automaton M_o let $\Sigma=\{a,b,c,d\}$, $Q=\{q_o, \bar{q}_1, \bar{q}_2\}$, $\Gamma=\{Z_o=[Z_{o1}, Z_{o1}], Z_1=[Z_{11}, Z_{12}], Z_2=[v, Z_{22}], Z_3=[Z_{31}, Z_{32}], Z_4=[Z_{41}, v]\}$ and Δ be the set of rules:

$WRITE(q_o,a,Z_o)=(q_o,Z_1,1)$, $WRITE(q_o,a,Z_1)=(q_o,Z_2Z_1,1)$,
$WRITE(q_o,b,Z_1)=(q_o,Z_3,1)$, $WRITE(q_o,b,Z_3)=(q_o,Z_4Z_3,1)$,
$BLOCK_2(q_o,c,Z_3)=(\bar{q}_1,Z_3,0)$, $ERASE(\bar{q}_1,c,Z_3)=(\bar{q}_1,e,1)$,
$ERASE(\bar{q}_1,c,Z_4)=(\bar{q}_1,e,1)$, $ERASE_v(\bar{q}_1,\dashv,Z_2)=(\bar{q}_1,e,0)$,
$BLOCK_1(q_o,d,Z_3)=(\bar{q}_2,Z_3,0)$, $ERASE(\bar{q}_2,d,Z_3)=(\bar{q}_2,e,1)$,
$ERASE_v(\bar{q}_2,d,Z_4)=(\bar{q}_2,e,0)$, $ERASE(\bar{q}_2,d,Z_2)=(\bar{q}_2,e,1)$.

It can be verified that $L(M_o)=\{a^nb^kc^k/k,n\geqslant 1\}\cup\{a^nb^kd^n/k,n\geqslant 1\}$.

We shall informally describe the work of the automaton M_o. Let the symbol a be read on the input tape, then M_o will write the symbol v and the symbol Z_{22} corresponding to the symbol a on the first and the second tracks of the pushdown stack, respectively. Let the symbol b be read on the input tape, then M_o will write the symbol Z_{41} corresponding to the symbol b and the symbol v on the first and the second track of the pushdown stack, respectively. For the first time let the symbols c or d be read on the input tape, then the automaton M_o will go to the states $\bar{q}_1$ or $\bar{q}_2$ and the first or the second track will be active, respectively. If the symbol c is read in the state $\bar{q}_1$ the automaton M_o will erase the symbol Z_{41} from the first track and check equality of the number of all the symbols b to the number of all the symbols c in the input string. After that the automaton M_o in the state $\bar{q}_1$ will erase all the symbols Z_2 from the pushdown stack using the rule $ERASE_v$. If the symbol d is read in the state $\bar{q}_2$ the automaton M_o will erase all the symbols Z_4 from the pushdown stack using the rule $ERASE_v$. Let the symbol d be read in the state $\bar{q}_2$ on the input tape, then the automaton M_o will erase the symbol Z_{22} from the second track of the pushdown stack. So, M_o will check equality of the number of all symbols d to the number of all symbols a in the input string.

It should be noted that non-essential rules of the program are omitted.

PROPOSITION 2. The context-free language $\{a^nb^kc^k/k,n\geqslant 1\}\cup\{a^nb^kd^n/ /k,n\geqslant 1\}$ is accepted by the synchronous automaton (see example 1) and cannot be accepted by any machine of class R (real time machines /6/) or of class C (one-counter machines /6/).

PROPOSITION 3. The context-free language $\{a^nd/n\geqslant 1\}\cup \cup\{a^{n_1}b^{n_1}*a^{n_2}b^{n_2}*\ldots*a^{n_i}b^{n_i}c/i\geqslant 1,\ n_i\geqslant 1\}$ is accepted by the synchronous automaton and cannot be accepted by any automaton of the class R_o (real time machines which accept the input strings by the empty stack /6/)or of the class of finite-turn machines /4/.

Notice that the synchronous automata accept the context-free languages which are the union of the S-languages.

THE ALGORITHM OF CHECKING THE EQUIVALENCE

Let M_1 and M_2 be two synchronous automata. Without loss of generality we can assume that M_1 and M_2 have different sets of states and stack symbols. Let an automaton M_k (k = 1,2) be fixed. We define a function $f(q_\alpha, Z_j) = T_{\alpha,j}$ on such pairs q_α, Z_j that the subset $q_\alpha \times Z_j$ has only essential symbols and there exist such integers i, j $(i \neq j)$ that $\alpha_i = \alpha_j = 1$ in m-tuple $\alpha = <\alpha_1, \alpha_2, \dots, \alpha_m>$. $T_{\alpha,j}$ is the set of the shortest strings ε such that the automaton M_k goes from the configuration $(q_\alpha, \varepsilon \dashv, 1, Z_j)$ to the configuration $(q_\alpha, \varepsilon \dashv, |\varepsilon|+1, \Lambda)$. We define a set $R_k = \bigcup_{(\alpha,j)} T_{\alpha,j}$. Let $R = R_1 \cup R_2$.

PROPOSITION 4. R is a finite set.

Suppose that $r = \max\{ |\varepsilon| \,/\, \varepsilon \in R\}$.

For the automaton M_k we define a set $B = \{\varphi\}$ of strings over Γ from the right side of all the rules WRITE. We define a function $\omega(\bar{q}_i, \varphi) = C_{i,\varphi}$ where $\varphi \in B$, $Pr_i^-(\varphi)\, \Gamma_i^+$, $C_{i,\varphi}$ is the set of the shortest strings ξ_i such that the automaton M_k goes from the configuration $(\bar{q}_i, \xi_i \dashv, 1, Pr_i^-(\varphi))$ to the configuration $(\bar{q}_i, \xi_i \dashv, |\xi_i|+1, \Lambda)$. Suppose that $H_k = \bigcup_{(i,\varphi)} C_{i,\varphi}$ where $1 \leq i \leq m$, $\varphi \in B$. Let $H = H_1 \cup H_2$. Suppose that $t \doteq \max\{|\xi_i| \,/\, \xi_i \in H\}$. We define a constant $c = 2m(t^2 + 5t)(rt^2+1)$.

Let M_1 and M_2 be two synchronous automata the equivalence of which is to be tested. We suppose that M_1 has the sets $Q_1 = \{q_\alpha\}$, $\Gamma_1 = \{Z_r\}$ and M_2 has the sets $Q_2 = \{p_\beta\}$, $\Gamma_2 = \{y_s\}$. We define a deterministic pushdown automaton M which accepts an empty language iff M_1 and M_2 are equivalent. An algorithm of constructing M resembles that from /4/. The stack of M has two storeys. The work of M_1 is modelled on the left storey and the work of M_2 is modelled on the right storey. The stack of M consists of segments separated by ceilings. The length of the segment is limited by the constant c which has been defined above. Each ceiling has a form: $(q_\alpha, Z_r, p_\beta, y_s)$ which means that at the time of the creation of the ceiling the automaton M_1 in the state q_α has the symbol Z_r on the top of the stack and the automaton M_2 in the state p_β has the symbol y_s on the top of the stack. The initial configuration of M contains the initial configurations of M_1 and M_2 on the left and the right storeys, respectively. The automaton M uses the following rules without moving on the input tape.

(1) Let both the storeys in the top segment contain more than one symbol. Then a new ceiling is placed below the top symbol of each storey which divides the segment into two smaller ones.

(2) Let one of the storeys in the top segment be empty. Then the ceiling immediately below the top segment is removed. The left storey of the top segment is fused with the left storey of the segment below and result is a new left storey. The right storey of the top segment is fused with the right storey of the segment below and the result is a new right storey.

(3) Let the left storey in the top segment contain exactly one symbol and the right storey in the state $\bar{p}_j$ contain the string $\sigma' Y_s$ with the length being greater than $\frac{c}{2}$. Then the string $\sigma' Y_s$ in the top segment is replaced by the string σY_s where σ is received from σ' by removing all the symbols Y_t for which $Pr_j(Y_t) = v$.

(4) Let the left storey in the top segment in the state $\bar{q}_i$ contain the string $\omega' Z_r$ with the length being greater than $\frac{c}{2}$ and the right storey contain exactly one symbol. Then the string $\omega' Z_r$ in the top segment is replaced by the string ωZ_r where ω is received from ω' by removing all the symbols Z_t for which $Pr_i(Z_t) = v$.

It should be noted that if neither of the rules (1)-(4) can be applied, one symbol will be in the top segment on one of the storeys. If both automata M_1 and M_2 make a move or both automata do not make a move on the input tape, the automaton M makes a move or does not make a move, respectively. Let one of the automata M_1 and M_2 for example, M_1, make a move and the other not make a move. Then the automaton M does not make a move and executes the work of M_2 on the right storey. Let one of the automata M_1 and M_2 reject the input string until this string will be read completely. Then the automaton M will execute on the other storey the work of that automaton M_1 or M_2 which will continue its work on this string.

We consider that the context-free languages $L(M_1)$ and $L(M_2)$ are nonempty and the automata M_1 and M_2 have only living configurations (in the sense of /6/). A string w over Σ is accepted by M in the following cases: (a) when the size of the top segment of M is greater than c; (b) when the input string is accepted on one storey and is rejected on the other storey.

This algorithm differs from that given in /4/ on the rules (3), (4).

THE MAIN RESULT

The following theorem proves the correctness of the algorithm.

THEOREM. The equivalence problem for synchronous deterministic pushdown automata is decidable.

It should be noted that the following fact follows directly from the algorithm: if the automata M_1 and M_2 are inequivalent then M will accept the nonempty set. So, the following fact remains to be proved: if M_1 and M_2 are equivalent, the size of the top segment is restricted by the constant c.

The proof of the theorem directly follows from the following lemmas.

LEMMA 1. Let in the top segment of M a symbol Z_k be on the left storey and a string γ in the state $\bar{p}_j$ be on the right storey. If $M_1 \equiv M_2$ then $|\gamma| \leq \frac{c}{2}$.

LEMMA 2. Let in the top segment of M a symbol Z_k be on the left storey and a string γ in the state p_o be on the right storey. If $M_1 \equiv M_2$ then $|\gamma| \leq \frac{c}{2}$.

LEMMA 3. Let in the top segment of M a string Ψ in the state $\bar{q}_i$ be on the left storey and a string γ in the state p_o be on the right storey ($|\Psi| \geq 2$, $|\gamma| > |\Psi|$). If $M_1 \equiv M_2$ then $|\gamma| \leq c$.

LEMMA 4. Let in the top segment of M a string Ψ in the state $\bar{q}_i$ be on the left storey and a string γ in the state $\bar{p}_j$ be on the right storey ($|\Psi| \geq 2$, $|\gamma| > |\Psi|$). If $M_1 \equiv M_2$ then $|\gamma| \leq c$.

LEMMA 5. Let in the top segment of M a string Ψ in the state q_o be on the left storey and a string γ in the state p_o be on the right storey ($|\Psi| \geq 2$, $|\gamma| > |\Psi|$). If $M_1 \equiv M_2$ then $|\gamma| \leq c$.

We shall explain, for example, the idea of the proof of the lemma 1. Let $L(\bar{q}_i, \delta)$ be a set of all strings w_k such that the automaton M_1 goes from the configuration $(\bar{q}_i, w_k \dashv, 1, \delta)$ to an accepting configuration. Let $\tau_i(\delta)$ be the length of the shortest string in the set $L(\bar{q}_i, \delta)$. For the automaton M_2 we use the same notation $L(p_o, \gamma)$, $L(\bar{p}_j, \gamma)$, $\tau_j(\gamma)$. To prove the lemma 1 let us assume the contrary, that is, $M_1 \equiv M_2$ but $|\gamma| > \frac{c}{2}$. Let $(\bar{q}_i, Z_r, p_o, Y_s)$ be the ceiling immediately below the top segment. Since $M_1 \equiv M_2$, for any ω_1, ω_2 we have $L(\bar{q}_i, \omega_1 Z_r) = L(p_o, \omega_2 Y_s) \rightarrow L(\bar{q}_i, \omega_1 Z_k) = L(\bar{p}_j, \omega_2 \gamma)$.

Since $|\tau_i(\omega_1 Z_k) - \tau_i(\omega_1 Z_r)| < t$, we have $|\tau_j(\omega_2 \gamma) - \tau_j(\omega_2 Y_s)| < t$. Hence it follows that $|Pr_j^-(\gamma)| \leq 2t$. On the other hand, we obtain $|\gamma| = |Pr_j^-(\gamma)|$ by the rule (3) of the algorithm. Since $\frac{c}{2} > 2t$, we have $|\gamma| > 2t$. So, we receive that $|\gamma| > 2t$ and $|\gamma| \leq 2t$ which is impossible.

REMARK. It should be noted that we can define such a class of context-free grammars $H = \bigcup_{(m)} H_m$ (class H_m from /7/) that the class of languages which are accepted by synchronous automata coincides with the class of languages which are generated by grammars from H.

REFERENCES

1 Friedman E.P. Equivalence Problems for Deterministic Context-Free Languages and Monadic Recursion Schemes. - J.Comput.Syst.Sci., 1977, v.14, No 3.

2 Oyamaguchi M., Honda N., Inagaki Y. The Equivalence Problem for Real-Time DPDA. - Inform.Contr., 1980, v.45, No 1.

3 Romanovskij V.U. The Equivalence Problem for Strict Deterministic Pushdown Automata Working in Real-Time. - Kibernetica, 1980, No 5 (in Russian).

4 Valiant L.G. The Equivalence Problem for Deterministic Finite-Turn Pushdown Automata. - Inform.Contr., 1974, v.25, No 2.

5 Oyamaguchi M., Honda N., Inagaki Y. The Equivalence Problem for Two DPDA's, One of Which is a Finite-Turn or One-Counter Machine. - J.Comput.Syst.Sci., 1981, v.23, No 3.

6 Valiant L.G. Decision Procedures for Families of Deterministic Pushdown Automata. - Department of Computer Science, University of Warwick, Coventry, England. (A dissertation). July 1973.

7 Nepomnjashchaja A.Sh. The Decidability of Equivalence Problem for the Union of S-Grammars. - In: Translation and Optimization of Programs. - Computing Center of the Siberian Division of the USSR Academy of Sciences, Novosibirsk, 1983.

MODELS AND OPERATORS FOR NONDETERMINISTIC PROCESSES

Rocco De Nicola
Istituto di Elaborazione della Informazione
Consiglio Nazionale delle Ricerche - Pisa (ITALY)
and
Department of Computer Science
University of Edinburgh - Edinburgh (U.K.)

1. Background and Motivations

Programming languages in which concurrently active processes interact by some form of synchronized communication (CSP's handshake /12/, Ada's rendezvous /15/) are more and more widespreadly used. At the same time the need to reason effectively about programs written in such languages is growing. Hence the necessity to define mathematical models which allow to describe and to prove properties of languages with primitives for communication. To serve these scopes the models have to satisfy somewhat conflicting demands, it should be possible to use them to describe a wide range of systems while at the same time they need to be simple enough to offer a tractable theory. Various models for parallel and/or communicating systems have been put forward in the last decade; lately, mainly because of simplicity requirements, a class of models which offer a characterization of parallel computations in terms of their nondeterministic interleavings, is the object of growing interest /10,13,14,-17,18/.

Unfortunately even after the common choice to reduce parallelism to nondeterministic interleaving there are disagreements on what classes of mathematical entities are suitable for modelling systems (processes). The heterogeneity arises mainly from the different emphasis on the particular behavioural aspects one is willing to capture. So we have, just to mention few examples, that:

a. The traces model /13/ identifies a process with the set of its possible sequences of communications (the traces). It turns out to be suitable for reasoning about potential communications but insensitive to deadlock.

b. The refusals set model /14/ in addition to traces associates to every process the set of communications a process is able to refuse after every trace. This model has problem in handling divergent or underspecified processes /5/.

c. A synchronization tree /17/ is a tree whose nodes represent the states of a process and whose arcs are labelled to denote potential communications with the environment. Synchronization trees are a basic and very concrete model: tree semantics arise naturally when concurrency is simulated by nondeterministic interleaving; unfortunatetly they turn out to discriminate too much and need to be factored by certain equivalences /1,11,17/.

In /6/ we put forward another class of models for nondeterministic processes called Representation Trees and show how they can be motivated in terms of success or failure of particular experiments. In fact we have three different models depending on the chosen way of handling divergence (ignored, considered catastrophic and considered underspecification). Roughly speaking every Representation Tree associated with a particular process contains the following information:

1. The possible sequences of communications the process may perform.
2. Associated with every sequence there is a finite set of sets of communications which represents the possible futures of the process after the sequence. In general the future will depend on internal nondeterministic choices and is characterized by the sets of communications the process must accept at that stage of its progress.

However in /6/ we did not define any operation which apply directly to Representation Trees. In the present paper we define various operators on one of the three models and offer further motivations and intuitions connected with it. It will be characterized as a continuous Σ-algebra /4,7/ and two of its reducts /2,3/ will be used as semantic domains /19,20/ for significative subsets of the language for a Theory of Communicating Sequential Processes (TCSP) /14/ and of the Calculus of Communicating Systems (CCS) /17/. These are two languages which have greatly influenced theoretical research. They start from a common philosophy which presuppose that the notion of process (active entity) and of event (communication) are to play a prominent role; but their semantics is given in very different styles: operational (based on transition systems) for CCS and denotational (based on refusals sets) for TCSP. This and the very different sets of primitive operators have hindered various attempts at a direct comparison. In /1/ synchronization trees are used as semantic domain for both languages: the comparison is hampered by the different equivalences used to quotient the trees. In /16/ the normal forms of /5/ and /6/ are used to define translations from one language to the other and to compare their expressiveness: no additional insight into the interrelations between the various operators is gained. In the sequel we will use SRT as semantic domain for both CCS and TCSP without quotienting. This will allow a direct comparison of the two languages. The soundness of the results will be checked by relating the SRT-semantics to the operational semantics for CCS of /6/ and to the denotational semantics for TCSP of /14/.

The rest of the paper is organized as follows. In section 2 we introduce the SRT domain as a continuous Σ-algebra whose carriers are cpo's and whose operations are (ω)-continuous /7/. In section 3 we discuss some general techniques of algebraic semantics /8/ and use them to give the semantics of TCSP and CCS by using SRT as semantic domain. In section 4 we discuss the results and use them to sketch a taxonomy of operators for nondeterminism. We omit all proofs from this presentation, they will appear in the complete version of the paper.

2. Strict Representation Trees

2.1. Very often we may wish to think of machines, programs, in general of processes as black boxes. This either because we are not allowed to open machines up or to read the code of programs or because the components are so unfamiliar to us that nothing can be gained by looking at them. In these cases we may try to find out the "behaviour" of a process by proposing suitable experiments to it and by analysing its reactions to them. When the processes involved may be nondeterministic it is important to know not only whether given a particular experiment a process responds favourably or unfavourably to it but also whether the process responds consistently all the times the experiment is proposed.

This approach to semantics has been taken in /6/ and /9/ and can be characterized operationally in terms of transitions and communications in a very natural way. In the present paper we will define a particular kind of rooted trees which carry all the information which could be obtained by proposing external experiments from a particular class. The trees which we will call Strict Representation Trees and denote by SRT, will have in general labels on both arcs and nodes. The arcs will be labelled by actions from a set L while the nodes will be labelled by sets of subsets of L. If we think of an element of SRT as describing how a given process would react to external experiments we can see the labels on the arcs as standing for the set of experiments the process may accept and the labels associated to the nodes as standing for the sets from which the process will nondeterministically choose the set of experiments it must accept. For a precise definition of may and must see /6/.

Since we want to elicit the nondeterminism exhibited by a process we will have that all the information about the possible nondeterministic choices will be held by the labels of the nodes and that all the arcs outgoing from a particular node will have a distinct label. To use SRT as a semantic model we need to have a theory of partially defined trees. Because of this we introduce trees with nodes which can be "improved". In the sequel these nodes will be called open nodes and will be denoted by o. Fully specified nodes will be called closed nodes by constrast and will be denoted by •. For simplicity we will require that every open node is a leaf and that the set of outgoing arcs from every node is finite; these requirements amount to considering divergence as catastrophic. There are languages which introduce actions invisible to the external environment, these actions give a process the possibility to "withdraw" some visible actions and to force the environment to consider only some particular subprocess. From an experimental point of view internal actions turn out to be important expecially when they are among the initial actions of a process. To distinguish between processes with and without initial invisible actions, we need to allow a slight disomogeneity in the model. While we require all closed nodes to be labelled by a set of sets which contains the set of labels of the outgoing arcs we allow closed roots to be labelled by the empty set.

In the next page few examples show a possible use of SRT as semantic domain for nondeterministic processes. Trees from SRT are in fact used to give a meaning to various synchronization trees of /17/; the examples evidence some of the above mentioned choices. Figures 1a and 1b show how we differentiate between deterministic and nondeterministic trees: a representation tree used to describe a deterministic process is such that if there is a label associated to a node then it is equal to the set of the outgoing arcs. Figure 2 shows how we identify two processes whose internal structures are different but such that their reactions to external experiments are the same. Figure 3 shows the way we treat silent moves and finally figure 4 shows how we treat divergence, underspecification is treated similarly.

2.2. We are now ready to give a formal characterization of the SRT model. In the sequel we will denote the set of nodes, closed nodes and open nodes of a tree t respectively by Nodes(t), CNodes(t) and ONodes(t). Note that the set of open nodes can be obtained by Nodes(t) and CNodes(t) as follows: ONodes(t) = Nodes(t) - CNodes(t). Moreover we will use Succ(t,s) to denote the set of successors of the node of the tree t determined by the string s and $\mathcal{A}$(t,s) to denote the set of labels associated to the node s of the tree t, and FPOW(L) to denote the set of finite subsets of L. Note that the condition that for every node there exists at most one outgoing arc labelled by a particular action implies that, if L is a non empty set of labels, any strict representation tree t is uniquely determined by:

1. A non empty, prefix closed set of strings from L* which determines the set of nodes of t.
2. A subset of Nodes(t) which determines the set of closed nodes.
3. A total mapping $\mathcal{A}$: CNodes $\longrightarrow$ FPOW(L) U $\emptyset$

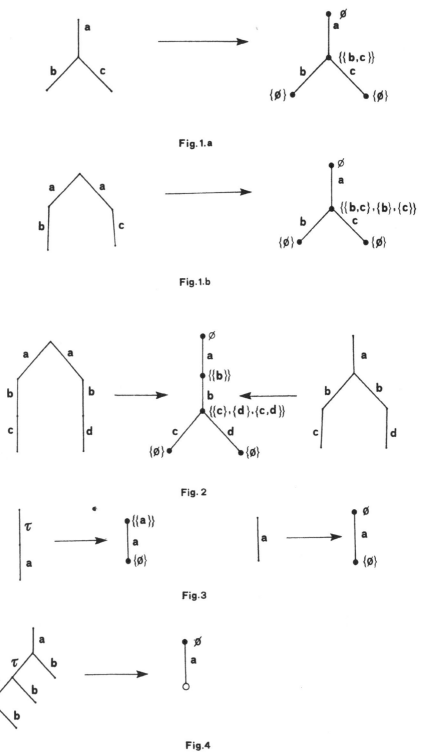

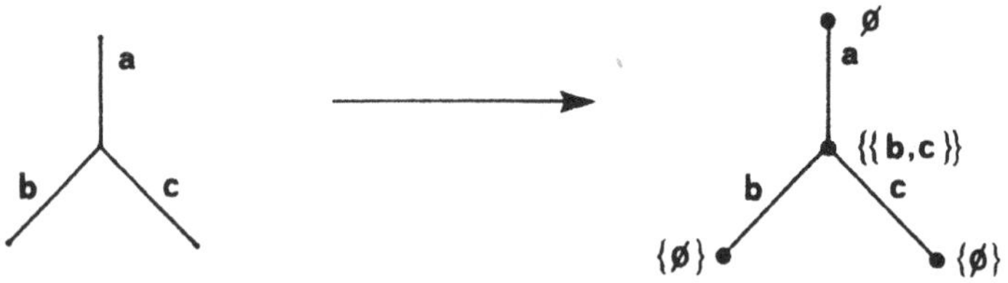

Fig. 1.a

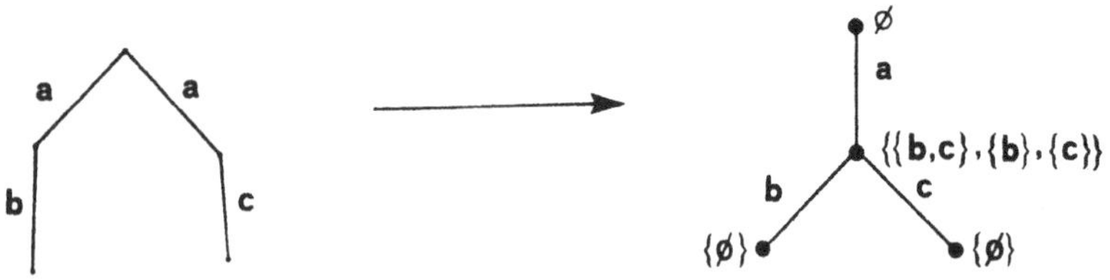

Fig. 1.b

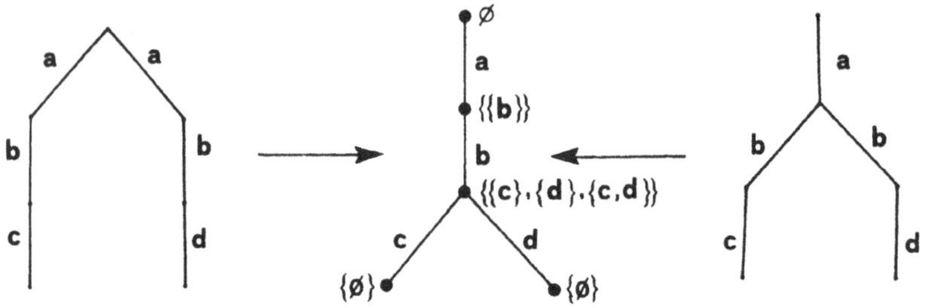

Fig. 2

Fig. 3

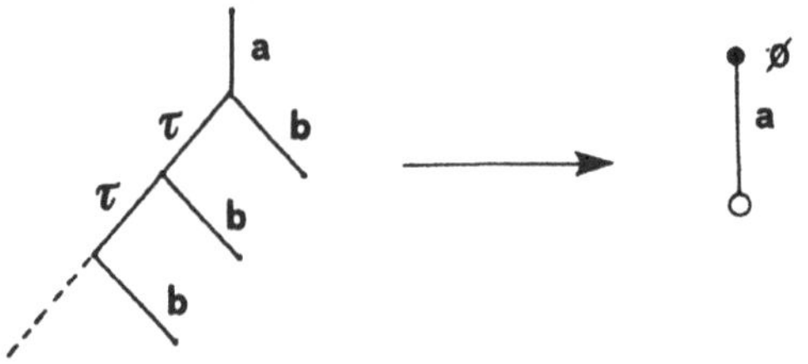

Fig. 4

Definition 2.1: A set of sets S is right closed with respect to A if and only if $X \in S$ and $X \subseteq Y \subseteq A$ implies $Y \in S$.

Definition 2.2: Let L be a set of labels, then SRT_L is the set of rooted, finitely branching trees such that $t \in SRT_L$ implies

1. Every branch is labelled by an element a of L and for every node there is at most one outgoing branch labelled by a.
2. Every node is either open (o) or closed (●).
3. If a node is open then it is a leaf.
4. Every closed node, t(s), is labelled either by $X \in FPOW(L)$ where X is right closed with respect to Succ(t,s) or by the empty set.
5. If $\mathcal{A}(t,s)$ is empty then t(s) is the root.

To use SRT_L as a semantic domain and apply standard methods of denotational semantics we need to prove that these trees enjoy particular properties. In particular we will prove that the structure of infinite trees can be entirely deduced from the structure of a particular set of finite trees. We need first of all to define an ordering on Strict Representation Trees. The main idea behind this ordering is that if t_1 and t_2 are members of SRT_L then t_1 is less defined than t_2 if t_1 is more nondeterministic than t_2.

Definition 2.3.: For $t_1, t_2 \in SRT_L$, $t_1 < t_2$ iff for every $s \in Nodes(t_2)$ we have that $t_1(s)$ closed implies

1. $t_2(s)$ is closed
2. $\mathcal{A}(t_2,s) \subseteq \mathcal{A}(t_1,s)$
3. $\mathcal{A}(t_2,s) = \emptyset$ implies either $\mathcal{A}(t_1,s) = \emptyset$ and $Succ(t_1,s) = Succ(t_2,s)$ or $Succ(t_2,s) \in \mathcal{A}(t_1,s)$

Note that $\mathcal{A}(t_2,s) \subseteq \mathcal{A}(t_1,s)$ and condition 4. of definition 2.2 imply that $Nodes(t_2) \subseteq Nodes(t_1)$.

Theorem 2.4.: $(SRT_L, <)$ is an w-algebraic c.p.o. whose finite elements are the trees with a finite number of nodes.

2.3. It is convenient to define a class of w-continuous operators over SRT_L. The set of operators introduced may seem a bit ad hoc and one may easily imagine more natural ones; they have been chosen mainly to have a direct correspondence with the operators of the two languages we want to give a meaning, since we aim at a comparison of the latters. We will first give an informal description of the various operators and then describe them formally. We will introduce two unary operators ($a\rightarrow$ and $\tau\rightarrow$) and three binary operators ($\oplus$, $\Box$, $+$).

The operator "$a\rightarrow$" prefixes a process with the elementary action a, while "$\tau\rightarrow$" gives a tree the possibility to change its internal state silently without the external environment having any control over it. The three binary operators are such that if P and Q are processes then a. $P \oplus Q$, b. $P \Box Q$ and c. $P + Q$ are processes which behaves like P or like Q, the difference between them consists in the amount of control over the choice left to the external environment. In case a. the choice is wholly nondeterministic, the environment does not have any control over it. In case b. the choice is completely determined by the first communication the environment offers to the process. In case c. the choice is again determined by the first communication the external environment offers unless one of the summands has the possibility to change its initial state silently, in this case P + Q will nondeterministically decide whether to leave the choice to the environment or to perform the silent move. The three binary operators and the operator "$\tau\rightarrow$" allow us to describe nondeterministic behaviours of processes; their analysis and comparison are among the principal aims of the paper. We will not consider any parallel operator, this mainly because the languages we are interested in reduce parallel composition to nondeterministic interleavings.

We are now ready to give a formal definition of the various operators. In order to define every operation on Representation Trees it is sufficient to describe three functions on Nodes, CNodes and $\mathcal{A}$. In the sequel we will let RC(S,X) denote the right closure of the set of sets S with respect to the set X.

ACTION

The first operation is the prefixing operation $a\rightarrow$; given any tree t, $a \rightarrow t$ produces a new tree whose root has a single branch labelled by a and after a is the same as t. Formally we have that for $a \in L$, $t \in SRT_L$, $a \rightarrow t$ is described by:

1. $Nodes\,(a \rightarrow t) = \{\varepsilon\} \cup \{as \mid s \in Nodes(t)\}$
2. $CNodes\,(a \rightarrow t) = \{\varepsilon\} \cup \{as \mid s \in CNodes(t)\}$
3. $\mathcal{A}(a \rightarrow t, \varepsilon) = \emptyset$
   $\mathcal{A}(a \rightarrow t, a) = Succ(t,\varepsilon)$ if $\mathcal{A}(t,\varepsilon) = \emptyset$
   $= \mathcal{A}(t,\varepsilon)$ otherwise
   $\mathcal{A}(a \rightarrow t, as) = \mathcal{A}(t,s)$ for all $s \in L^+$ and $s \in CNodes(t)$

INTERNAL ACTIONS

We have another unary operator $\tau\rightarrow$ which, given any tree labels its root, with the set of initial moves in case the root is labelled by the empty set. For all $t \in SRT_L$, $\tau \rightarrow t$ is described by:

1. $Nodes\,(\tau \rightarrow t) = Nodes(t)$
2. $CNodes\,(\tau \rightarrow t) = CNodes(t)$
3. $\mathcal{A}(\tau \rightarrow t, \varepsilon) = \mathcal{A}(t,\varepsilon)$ if $\mathcal{A}(t,\varepsilon) \neq \emptyset$
   $= Succ(t,\varepsilon)$ if $\mathcal{A}(t,\varepsilon) = \emptyset$ and $t(\varepsilon) \in CNodes(t)$
   $\mathcal{A}(\tau \rightarrow t, s) = \mathcal{A}(t,s)$ for all $s \in L^+$ and $s \in CNodes(t)$

CHOICE

Given two trees t_1 and t_2, the choice operator "+" glues them togheter at their roots. Particular care is needed to guarantee that the resulting tree does satisfy conditions 1 to 5 of definition 2.2, i.e. it is a Strict Representation Tree. We have to distinguish various cases depending on whether the label of the root of t_1 and t_2 is 0 or not and on whether the root is open or closed.

Case 1 $\quad \varepsilon \in ONodes(t_1)$ or $\varepsilon \in ONodes(t_2)$

Case 2 $\quad \varepsilon \in CNodes(t_1)$ and $\varepsilon \in CNodes(t_2)$
   a. $\mathcal{A}(t_1,\varepsilon) = \mathcal{A}(t_2,\varepsilon) \neq \emptyset$
   b. $\mathcal{A}(t_1,\varepsilon) \neq \emptyset$ and $\mathcal{A}(t_2,\varepsilon) \neq \emptyset$
   c. $\mathcal{A}(t_1,\varepsilon) = \emptyset$ and $\mathcal{A}(t_2,\varepsilon) \neq \emptyset$
   d. $\mathcal{A}(t_1,\varepsilon) \neq \emptyset$ and $\mathcal{A}(t_2,\varepsilon) = \emptyset$

In case 1 $\quad t_1 + t_2 = o$ (the trivial diverging tree)

In case 2 $\quad t = t_1 + t_2$ is defined by

1. $Nodes(t) = Nodes(t_1) \cup Nodes(t_2) - \{su \mid s \in ONodes(t_i),\ u \in L^+,\ i=1,2\}$
2. $CNodes(t) = \{s \mid s \in Nodes(t)$ and $s \in Nodes(t_i)$ implies $s \in CNodes(t_i),\ i=1,2\}$
3. $\mathcal{A}(t,) = \emptyset$ in case a
   $= RC(\mathcal{A}(t_1,\varepsilon) \cup \mathcal{A}(t_2,\varepsilon),\ Succ(t,\varepsilon))$ in case b
   $= RC(\mathcal{A}(t_2,\varepsilon) \cup \{Succ(t_1,\varepsilon) \cup X \mid X \in \mathcal{A}(t_2,\varepsilon)\},\ Succ(t,\varepsilon))$ in case c
   $= RC(\mathcal{A}(t_1,\varepsilon) \cup \{Succ(t_2,\varepsilon) \cup X \mid X \in \mathcal{A}(t_1,\varepsilon)\},\ Succ(t,\varepsilon))$ in case d
   $\mathcal{A}(t,s) = RC(\mathcal{A}(t_1,s) \cup \mathcal{A}(t_2,s),\ Succ(t,s))$ for all $s \in CNodes(t) \cap L^+$.

EXTERNAL CHOICE

Let $t = t_1 \Box t_2$

1. $Nodes(t) = Nodes(t_1 + t_2)$
2. $CNodes(t) = CNodes(t_1 + t_2)$
3. $\mathcal{A}(t,s) = \mathcal{A}(t_1 + t_2, s)$ for all $s \in CNodes(t) \cap L^+$

$\mathcal{A}(t,\varepsilon) = \emptyset$ in case a
$= \{X_1 \cup X_2 \mid X_1 \in \mathcal{A}(t_1,\varepsilon),\ X_2 \in \mathcal{A}(t_2,\varepsilon)\}$ in case b
$= \{Succ(t_1,\varepsilon) \cup X \mid X \in \mathcal{A}(t_2,\varepsilon)\}$ in case c
$= \{X \cup Succ(t_2,\varepsilon) \mid X \in \mathcal{A}(t_1,\varepsilon)\}$ in case d

INTERNAL CHOICE

$t = t_1 \oplus t_2$ is defined by

1. $Nodes(t) = Nodes(t_1 + t_2)$
2. $CNodes(t) = CNodes(t_1 + t_2)$
3. $\mathcal{A}(t,s) = \mathcal{A}(t_1 + t_2, s)$ for all $s \in Nodes(t),\ s \neq \varepsilon$

$\mathcal{A}(t,\varepsilon) = RC(Succ(t_1,\varepsilon) \cup Succ(t_2,\varepsilon),\ Succ(t,\varepsilon))$ in case a
$= RC(\mathcal{A}(t_1,\varepsilon) \cup \mathcal{A}(t_2,\varepsilon),\ Succ(t,\varepsilon))$ in case b
$= RC(\mathcal{A}(t_2,\varepsilon) \cup Succ(t_1,\varepsilon),\ Succ(t,\varepsilon))$ in case c
$= RC(\mathcal{A}(t_1,\varepsilon) \cup Succ(t_2,\varepsilon),\ Succ(t,\varepsilon))$ in case d

Together with the operations defined above we will use:

a. The nullary operator $\mho$ to denote the trivial tree $\bullet$
b. The nullary operator Ω to denote the trivial tree o

In the sequel we will let Σ to denote the set of operators over SRT_L

$\{\mho, \Omega, a\rightarrow \text{ (for all } a \in L),\ \tau\rightarrow,\ \oplus,\ \Box,\ +\}$.

Theorem 2.5. If op is an operation from Σ then it is continuous over SRT_L.

3. Using Strict Representation Trees as a Semantics

Guessarian /8/ and the ADJ group /7/ have shown how to interpret a language which only uses the operations of a given Σ-cpo. The idea does come out of the usual techniques of denotational semantics /19,20/. In general the semantic equations define the meaning of the various syntactic constructs as functions of their components. These functions can be seen as a homomorphism between algebras /2,3/. We have in fact that any BNF-like definition of a programming languages yields a free Σ-algebra and the ideal completion of the latter yields a free complete Σ-algebra: given any other Σ-algebra, initiality guarantees the existence of the semantic homomorphism.

Let X be a set of variables, ranged over by x. The set of recursive terms over Σ, REC_Σ, ranged over by t, can be defined by the following BNF-like schema:

$$t ::= x \mid op(t_1,\ldots,t_k),\ op \in \Sigma^k \mid rec\ x.\ t\ . \qquad (*)$$

Given any Σ-cpo D and a set of mapping from X to D, denoted by ENV_D and ranged over by e, we can define the semantic function $\mathcal{M}_D$ as follows:

Definition 3.1 If $\mathcal{M}_D : REC_\Sigma \rightarrow (ENV_D \rightarrow D)$ then

1. $\mathcal{M}_D(x)e = e(x)$
2. $\mathcal{M}_D(op\ (t_1,\ldots, t_k)e = op_D(\mathcal{M}_D(t_1)e, \ldots, \mathcal{M}_D(t_k)e)$ for all $op \in \Sigma$
3. $\mathcal{M}_D(rec\ x.\ t)e = \underline{fix}\ \lambda d. \mathcal{M}_D(t)e(d/x)$ where fix represents the least fixed point

operator and e(d/y) denote the environment which coincides with e except at y where it is d.

Pure CCS, i.e. CCS without value passing /17/, and the language for a theory of CSP (TCSP), /14/, may be defined by choosing a particular set of operators Σ_{ccs} and Σ_{csp}. In the rest of the section we will apply the sketched techniques to significative subsets of these two languages.

Let A be a set of elementary actions ranged over by a,b,c,... we will have that the syntax of TCSP and CCS are defined by (*) where Σ is replaced by Σ_{csp} and Σ_{ccs} respectively and

$$\Sigma^{0}_{ccs} = \{NIL, \Omega\} \quad \Sigma^{1}_{ccs} = \{\tau. \,,\ a.\} \quad \Sigma^{2}_{ccs} = \{+\} \quad \Sigma^{n}_{ccs} = \emptyset \quad \text{for } n \geq 3$$

$$\Sigma^{0}_{csp} = \{Stop,\ CHAOS\} \quad \Sigma^{1}_{csp} = \{a\rightarrow\} \quad \Sigma^{2}_{csp} = \{\sqcap, \Box\} \quad \Sigma^{n}_{csp} = \emptyset \text{ for } n \geq 3$$

Let (SRT, Σ) be the continuous Σ-cpo defined in section 2, the index L is omitted for simplicity. We have that the operations and SRT itself are too powerful to express the semantics of the two languages, two reduct of (SRT, Σ) are sufficient.

<u>Definition 3.2</u> Let SRT' be a subset of SRT obtained by replacing conditions 4 and 5 of definition 2.2 by:
4. Every closed node, t(s), is labelled by $X \in$ FPOW(L) which is right closed with respect to Succ(t,s).

We can now define two reducts of (SRT, Σ) which are obtained from SRT by just forgetting some of its sorts and some of its operators.

<u>Theorem 3.3</u> If $\Sigma' = \{\Phi, \Omega, a\rightarrow, \oplus, \Box\}$ then (SRT', Σ') is a reduct of (SRT, Σ) and is a continuous Σ'-cpo.

<u>Theorem 3.4</u> If $\Sigma'' = \{\Phi, \Omega, a\rightarrow, \tau\rightarrow, +\}$ then (SRT", Σ'') is a reduct of (SRT, Σ) and is a continuous Σ''-cpo.

If we take Σ_{csp} and Σ_{ccs} and rename $\sqcap$ by $\oplus$, $\tau.$, and a. by $\tau\rightarrow$ and $a\rightarrow$, Stop and NIL by Φ and finally CHAOS by Ω we can immediately apply definition 3.1 to obtain new semantics for TCSP and CCS which use, respectively, (SRT', Σ') an (SRT", Σ'') as semantic domains. These two reducts have interesting equational characterizations which allow us to relate the new semantics to existing ones. The equational characterization are summarized in Table 1 and Table 2 where X, Y are used to denote processes, a to denote actions and μ to denote both actions and silent moves.

$\Omega \sqsubseteq X$

$X \Box X = X$
$X \Box Y = Y \Box X$
$X \Box (Y \Box Z) = (X \Box Y) \Box Z$
$X \Box \Phi = X$
$X \Box \Omega = \Omega$

$X \oplus X = X$
$X \oplus Y = Y \oplus X$
$X \oplus (Y \oplus Z) = (X + Y) \oplus Z$

$(a\rightarrow X) \oplus (a\rightarrow Y) = a\rightarrow (X \oplus Y)$
$(a\rightarrow X) \Box (a\rightarrow Y) = a\rightarrow (X \oplus Y)$
$X \oplus (Y \Box Z) = (X \oplus Y) \Box (X \oplus Z)$
$X \Box (Y \oplus Z) = (X \Box Y) \oplus (X \Box Z)$

Table 1

$\Omega \sqsubseteq X$

$X + X = X$
$X + Y = Y + X$
$X + (Y + Z) = (X + Y) + Z$
$X + \Phi = X$
$X + \Omega = \Omega$

$(\mu\rightarrow X) + (\mu\rightarrow Y) = \mu\rightarrow((\tau\rightarrow X) + (\tau\rightarrow Y))$
$(\mu\rightarrow X) + (\tau\rightarrow((\mu\rightarrow Y) + Z)) =$
$= \tau \rightarrow ((\mu\rightarrow X) + (\mu\rightarrow Y) + Z$
$X + (\tau\rightarrow Y) \sqsubseteq \tau\rightarrow(X + Y)$
$(\tau\rightarrow X) + (\tau\rightarrow Y) \sqsubseteq X$

Table 2

Proposition 3.5 If A' is the set of axioms in table 1 then (SRT', Σ') is isomorphic to the initial Σ'-cpo which satisfies A'.

Propositions 3.6 If A" is the set of axioms in table 2 then (SRT", Σ'') is isomorphic to the initial Σ''-cpo which satisfies A".

Now we have that the Σ-cpo of /5/, (Bounded Refusal Set) which is a slight modification of the model based on refusal sets of /14/, is also isomorphic to the initial Σ'-cpo which satisfies A'. Moreover the set of axioms A" is fully abstract with respect to the precongruence $\sqsubseteq^c_{\sim 2}$ defined operationally in /6/. These considerations point out the relationship between our observational denotational semantics based on SRT and others well known semantics for CCS and TCSP.

4. Operators for Nondeterminism

From the definitions of section 2 it is possible to derive informations about the interrelations among the various SRT operators which given one or two trees as argument transform them into a tree which is more nondeterministic than the starting ones. The derived interrelations would allow us to get informations about the corresponding TCSP and CCS operators. The operators for nondeterminism are of vital interest whenever we choose the interleaving approach to model concurrency.

Theorem 4.1: If L is a set of labels, P, Q $\in$ SRT_L, $<$ is the preorder of definition 2.3, and P = Q if and only if P $<$ Q and Q $<$ P, then

$$(\tau \rightarrow P) + (\tau \rightarrow Q) = P \oplus Q \quad < \quad P + Q \quad < \quad P \,\Box\, Q \qquad (**)$$

We have that $<$ is based on the "amount" of nondeterminism exhibited by processes ; the observational approach taken implies that this is determined by the "amount" of control over the progression of a process left to the external environment. For these reasons theorem 4.1 suggests criteria to consider whenever choosing nondeterministic operators for describing systems. We have that expressiveness is only one of the requirements a set of operators of a model has to meet; it is important for them to be orthogonal to each other and that their interactions can be characterized algebraically. TCSP and CCS contain only a subset of the operators considered in theorem 4.1 just in order to meet some of the above criteria. Indeed we can see that $\tau\rightarrow$ and $\oplus$ and $\Box$ and + are not orthogonal at all. The first equality of (**) suggests that there is a close relationship between internal moves and internal choices. In fact once we choose to take the observational approach, implied by SRT, we have that any internal move can be rendered as an internal choice between performing the move and waiting for the decisions of the external environment. Indeed we can always replace $\tau\rightarrow$ with $\oplus$ in fact togheter with the equality of (**) we have:

$$P + (\tau \rightarrow Q) = (\tau \rightarrow (P + Q)) + (\tau \rightarrow Q) = (P + Q) \oplus Q \qquad (R1)$$

and

$$P \,\Box\, (\tau \rightarrow Q) = \tau \rightarrow (P \,\Box\, Q) = (P \,\Box\, Q) \oplus (P \,\Box\, Q) \qquad (R2)$$

Note that P $\oplus$ P = P does not hold in general in SRT, it does hold only if we restrict ourselves to SRT', the subset of SRT of definition 3.2.

There is another noteworthy consequence of the definitions of the various operators of section 2: in case the roots of both arguments are labelled by the empty set we have that

$$P \,\Box\, Q = P + Q \qquad (R3)$$

Now by using (R2) as many times as necessary, it is possible to have denotations of representation trees such that all the operands of $\Box$ and + denote trees whose roots are labelled by the empty set.

These previous considerations seem to suggest that the operators $\tau\rightarrow$ and + are more primitive than ⊕ and □; on the other side the equation of table 1 show that ⊕ and □ can have an interesting algebraic characterization. This duality will be the subject of further investigations. Anyway it is the author's conviction that only using the various operators to model "real life" systems we can get a definite answer, if any.

Acknowledgements

The present work is a direct result of the stimulating environment of the Department of Computer Science of the University of Edinburgh. In particular I would like to thank I. Castellani, M. Hennessy and C. Stirling for many helpful discussions. I would also like to thank Silvia Giannini and Stefania Soldani for typing the difficult manuscript.

References

LNCS n denotes Lecture Notes in Computer Science Volume n, Springer-Verlag, Berlin.

/1/ Brookes S.D. "On the relationship of CCS and CSP" Proc. ICALP '83, LNCS 154, pp. 83-96; (1983).

/2/ Burstall, R.B. and Goguen, J.A. "Algebras, Theories and Freeness: an Introduction for Computer Scientists" in Theoretical Foundations of Programming Methodology (M. Broy and G. Schmidt eds), Reidel, Dodrecht, (1982).

/3/ Cohn, P.M. "Universal Algebra" Harper and Row, New York (1965).

/4/ Courcelle, E. and Nivat, M. "Algebraic Families of Interpretations", Proc. 17th FOCS Annual Symposium, Houston, (1976).

/5/ De Nicola, R. "Two complete Axiom Systems for a Theory of Communicating Sequential Processes". Internal Report CSR-154-83, University of Edinburgh, (1983). A short version in Proc. FCT'83, LNCS 158, pp. 115-126.

/6/ De Nicola, R. and Hennessy, M. "Testing equivalences for processes", Proc. ICALP '83, LNCS 154, (1983). Ext. version to appear in Theoretical Computer Science.

/7/ Goguen, J.A., Thatcher, J.W., Wagner, E.G. and Wright, J.B. "Initial algebra semantics and continuous algebras" Journal of ACM 24, N°1, 68-95, (1977).

/8/ Guessarian, I. "Algebraic Semantics", LNCS 99, (1981).

/9/ Hennessy, M. "Synchronous and Asynchronous Experiments on Processes", Technical Report CSR-125-82, University of Edinburgh, (1982).

/10/ Hennessy, M. "A Model for Nondeterministic Machines", Technical Report CSR-135-83, pp. 299-309, (1983).

/11/ Hennessy, M., Milner, R. "On Observing Nondeterminism and Concurrency", LNCS 85, pp. 299-309, (1980).

/12/ Hoare, C.A.R. "Communicating Sequential Processes", CACM Vol. 21, N°8, (1978).

/13/ Hoare, C.A.R. "A Model for Communicating Sequential Processes", Technical Monograph Prg-22, Computing Laboratory, University of Oxford, (1982).

/14/ Hoare, C.A.R., Brookes, S.D. and Roscoe, A.D. "A Theory of Communicating Sequential Processes", Technical Monograph Prg-16, Computing Laboratory, University of Oxford (1981). To appear in Journal of ACM.

/15/ Ichbiach, J.D. et al. "Reference Manual for the ADA programming language", MIL - STD - 18 15A, (1983).

/16/ Margaria, I. and Zacchi, M. "Modelli per processi comunicanti: confronto tra la teoria dei CSP e CCS", Cnet project Technical Report N° 96, (1983).

/17/ Milner, R. "A Calculus of Communicating Systems", LNCS 92, (1980).

/18/ Rounds, W.C. and Brookes, S.D. "Possible futures, acceptances, refusals and communicating processes", Proc. 22nd FOCS Nashville, (1981).

/19/ Scott, I.S. "Data types as lattices", SIAM J. on Computing, vol. 5, N°3, (1976).

/20/ Stoy, J. "Denotational semantics: the Scott-Strachey approach to programming language theory", MIT Press, (1977).

ALGORITHMS FOR STRING EDITING WHICH PERMIT ARBITRARILY COMPLEX EDIT CONSTRAINTS+

B. John Oommen
School of Computer Science
Carleton University
Ottawa, Ontario, Canada, K1S 5B6

ABSTRACT

Let X and Y be any two strings of finite length. We consider the problem of transforming X to Y using the edit operations of deletion, insertion and substitution. The optimal transformation is the one which has the minimum edit distance associated with it. The problem of computing this distance and the optimal transformation using no edit constraints has been studied in the literature. In this paper we consider the problem of transform X to Y using any arbitrary edit constraints involving the number and type of edit operations to be performed. An algorithm has been presented to compute the minimum distance associated with editing X to Y subject to the specified constraint. The algorithm requires $O(|X|.|Y|.\min(|X|,|Y|))$ time and space. The technique to compute the optimal-transformation has also been presented.

I. INTRODUCTION

In the study of the automatic correction of garbled text a question that has interested researchers is that of quantifying the dissimilarity between two strings. A review of such distance measures and their application to string correction is given by Hall and Dowling [1] and Peterson [2].

+ Supported by the Natural Sciences and Engineering Research Council of Canada.

The most promising of all the measures introduced seems to be the one that relates the strings using various edit operations. The edit operations usually considered are substitution, deletion, insertion and the reversal of adjacent letters. The studies of various researchers (see the literature surveys of references 1-4) indicate that most of the common errors that occur in garbled text consist of a single substitution, deletion, insertion or reversal error and hence the quantification of the dissimilarity between two strings using the above edit operations seems reasonable. Further, since a single reversal error can be modelled as a sequence of an insertion and a deletion error, the vast majority of research that has been conducted deals with merely the edit operations of substitution, deletion and insertion.

The problem of editing one string, say X, to transform it to Y has been much studied in the literature [1-6]. All these papers deal with the problem leaving the edit process unconstrained. The only algorithm known to us that considers constrained editing is the one due to Sankoff [7]. The latter algorithm is a longest common subsequence algorithm and involves a simple constraint that has its application in the comparison of Amino acid sequences.

In this paper we consider the problem of editing X to Y subject to far more general edit constraints. These edit constraints can be arbitrarily complex so long as they specified in terms of the number and type of the edit operations to be included in the optimal edit transformation. We shall present an $O(|X|.|Y|.\min(|X|,|Y|))$ time and space algorithm to perform the editing.

I.1 Notation

Let A be any finite alphabet and A* be the set of strings over A. θ is the null symbol, $\theta \notin A$. Let $\tilde{A} = A \cup \{\theta\}$. $\tilde{A}$ is referred to as the Appended Alphabet. A string $X \in A^*$ of the form $X = x_1 \ldots x_N$, where each $x_i \in A$, is said to be of length $|X| = N$. Its prefix of length i will be written as X_i, $i \leq N$. Upper case symbols represent strings and lower case symbols, elements of the alphabet under consideration.

Let Z' be any element in $\tilde{A}^*$, the set of strings over $\tilde{A}$. The Compression Operator, C, is a mapping from $\tilde{A}^*$ to A*. C(Z') is Z' with all occurrances of the symbol θ removed from Z'. Note that C preserves the order of the non-θ symbols in Z'. For example, if $Z' = f\theta o\theta r$, C(Z') = for.

d(.,.) is a function whose arguments are a pair of symbols belonging to $\tilde{A}$, the appended alphabet, and whose range is the set of positive real numbers. $d(\theta,\theta)$ is undefined and is not needed. The elementary distance d(a,b) can be interpreted as the distance associated with transforming 'a' to 'b', for a, b $\in \tilde{A}$.

For every pair (X,Y), X,Y $\in$ A*, the finite set $G_{X,Y}$ is defined by means of the compression operator C, as a subset of $\tilde{A}^* \times \tilde{A}^*$

$$G_{X,Y} = \{(X',Y') | (X',Y') \in \tilde{A}^* \times \tilde{A}^*, \text{ and obeys (i) - (iii)}\} \quad (1)$$

(i) $C(X') = X,\ C(Y') = Y$

(ii) $|X'| = |Y'|$

(iii) In no (X',Y') is $x_i' = y_i' = \theta,\ 1 \le i \le |X'|$.

By definition, if $(X',Y') \in G_{X,Y}$, $\text{Max}[|X|,|Y|] \le |X'| = |Y'| \le |X| + |Y|$.

The meaning of the pair $(X',Y') \in G_{X,Y}$ is that is corresponds to one way of editing X into Y, using the edit operations of substitution, deletion and insertion. The edit operations themselves are specified for all $i=1, \ldots, |X'|$ by (x_i',y_i'), which represents the transformation of x_i' to y_i'. Note that $G_{X,Y}$ is an exhaustive enumeration of the set of all the ways by which X can be edited to Y. For example, if X=f and Y=go, then, $G_{X,Y} = \{(f\theta, go), (\theta f, go), (f\theta\theta, \theta go), (\theta f\theta, g\theta o), (\theta\theta f, go\theta)\}$. In this case, the pair $(\theta f, go)$ represents the edit operations of inserting the 'g' and replacing the 'f' by an 'o'.

II. EDIT CONSTRAINTS

II.1 Permissible and Feasible Edit Operations

Consider the problem of editing X to Y, where $|X| = N$ and $|Y| = M$. Suppose we edit a prefix of X into a prefix of Y, using exactly i insertions, e deletions (or erasures) and s substitutions. Since the number of edit operations are specified, this corresponds to editing $X_{e+s} = x_1 \ldots x_{e+s}$, the prefix of X of length e+s, into $Y_{i+s} = y_1 \ldots y_{i+s}$, the prefix of Y of length i+s.

Observe that if $r = e + s$ and $q = i + s$, the variables i, e, s, r and q will have to obey the following obvious constraints.

(i) $\text{Max}[0,M-N] \le i \le q \le M$ (ii) $0 \le e \le r \le N$ and

(iii) $0 \le s \le \text{Min}[M,N]$ (2)

Values of triples (i,e,s) which satisfy these constraints are termed as the feasible values of the variables. Let,

$H_i = \{j | \text{Max}[0,M-N] \le j \le M\}$, $H_e = \{j | 0 \le j \le N\}$, and

$H_s = \{j | 0 \le j \le \text{Min}[M,N]\}$ (3)

H_i, H_e and H_s are called the set of permissible values of i, e and s. Observe that a triple (i,e,s) is feasible if apart from $i \in H_i$, $e \in H_e$, $s \in H_s$, the inequalities $i + s \le M$ and $e + s \le N$ are satisfied.

II.2 Specification of Edit Constraints

An edit constraint is specified in terms of the number and type of edit operation that are required in the process of transforming X to Y. It is expressed by formulat the number and type of edit operations in terms of three sets Q_e, Q_i and Q_s which are subsets of the sets H_i, H_e and H_s defined in (3). For example, to edit X to Y performing exactly k_i insertions, k_e deletions and k_s substitutions is equivalent to specifying,

$$Q_i = \{k_i | k_i \in H_i\}, \qquad Q_e = \{k_e | k_e \in H_e\}, \text{ and } \qquad Q_s = \{k_s | k_s \in H_s\}.$$

THEOREM I.

Every edit constraint specified for the process of editing X to Y can be written merely as a subset of H_i. The proof is included in [8].

Remarks:

1. The set referred to above which describes the constraint and which is the subset of H_i shall in future be written as T.
2. The edit constraint can just as easily be written as a subset of H_s or H_e. We have however chosen to describe T as a subset of H_i. This choice is absolutely subjective.
3. In [8] we have shown that T is a function involving intersections and unions of Q_i, $\bar{Q}_e$ and $\bar{Q}_s$, where, $\bar{Q}_e = \{M-N+j | j \in Q_e\}$ and $\bar{Q}_s = \{M-j | j \in Q_s\}$ (4)

EXAMPLE I.

Let X = for and Y = ga. Suppose we want to transform X to Y by performing at least 1 insertion, at most 1 substitution and exactly 2 deletions. Then,

$Q_i = \{1,2\}$, $Q_e = \{2\}$ and $Q_s = \{0,1\}$. Hence,

$\bar{Q}_e = \{1\}$ and $\bar{Q}_s = \{1,2\}$. Thus $T = Q_i \cdot \bar{Q}_e \quad \bar{Q}_s = \{1\}$.

This means that the optimal transformations must contain exactly one insertion. Some candidates edit transformations are given by the following subset of $G_{X,Y}$.

$\{(\theta for, g\theta\theta a), (f\theta or, ga\theta\theta), (fo\theta r, g\theta a\theta), (for\theta, g\theta\theta a)\}$

Note that every pair in the above subset corresponds to at least one insertion, at most 1 substitution and exactly 2 deletions.

We shall refer to the edit distance subject to the constraint T as $D_T(X,Y)$. By definition, $D_T(X,Y) = \infty$ if $T = \emptyset$. This is merely a simple way of expressing that it is impossible to edit X to Y subject to the constraint T. We shall now consider the computations of $D_T(X,Y)$.

III. W: THE ARRAY OF CONSTRAINED EDIT DISTANCES

Let W(i,e,s) be the constrained edit distance associated with editing X_{e+s} to Y_{i+s} subject to the constraint that exactly i insertions, e deletions and s substitutions are performed in the process of editing. Let r=e+s and q=i+s. Let $G_{i,e,s}(X,Y)$ be the subset of the pairs in G_{X_r,Y_q} in which every pair corresponds to i insertions, e deletions and s substitutions. Since we shall consistently be referring to the strings X and Y, we refer to this set as $G_{i,e,s}$. Thus, using the notation of (1), W(i,e,s) has the expression,

$$W(i,e,s) = \operatorname*{Min}_{(X'_r,Y'_q)\ \varepsilon\ G_{i,e,s}} \left[\sum_{j=1}^{|X'_r|} d(x'_{rj}, y'_{qj}) \right] \quad \text{if } i,e \text{ or } s > 0 \tag{5}$$

THEOREM II.

Let W(i,e,s) be the quantity defined as in (5) for any two strings X and Y. Then,

$$W(i,e,s) = \text{MIN}\Big[\{W(i-1,e,s) + d(\theta,y_{i+s})\},\{W(i,e-1,s) + d(x_{e+s},\theta)\}, \{W(i,e,s-1) + d(x_{e+s}, y_{i+s})\}\Big]$$

for all feasible triples (i,e,s). The theorem is proved in [8].

The computation of the distance $D_T(X,Y)$ from the array W(i,e,s) only involves combining the appropriate elements of the array using T, the set of the number of insertions permitted. This is stated in the following theorem which is proved in [8].

THEOREM III.

The quantity $D_T(X,Y)$ is related to the elements of the array W(i,e,s) as below:

$$D_T(X,Y) = \operatorname*{Min}_{i\varepsilon T} \left[(W(i,N-M+i,M-i) \right]$$

Remark:

In an unconstrained editing problem i can take on any value ranging from Max[0,N-M] to M. Thus the unconstrained edit distance (which is commonly known as the Generalized Levenshtein Distance) D(X,Y), has the expression:

$$D(X,Y) = \operatorname*{Min}_{i\ \varepsilon\ [\text{Max}[0,M-N],M]} \Big[[W(i,N-M+i,M-i)] \Big]$$

Using the results of the above two theorems, we now propose a computational scheme for $D_T(X,Y)$.

IV. THE COMPUTATION OF W(.,.,.) AND $D_T(X,Y)$

To compute $D_T(X,Y)$ we shall first compute the index W(.,.,.) which is recursively computable, for all permissible values if i, e and s. Subsequently, using the array W(i,e,s) as the input, we get $D_T(X,Y)$ by comparing the contributions of the pertinent elements in W(.,.,.) as specified by Theorem III.

The computation of the array W(.,.,.) has to be done in a systematic manner, so that any quantity W(i,e,s) is computed before its value is required in any further computation. This is easily done by considering a three-dimensional coordinate system whose axes are i, e and s respectively. Initially the weight associated with the origin W(0,0,0) is assigned the value zero and the weights associated with the vertices on the axes are evaluated. Thus, W(i,0,0), W(0,e,0) and W(0,0,s) are computed for all permissible values of i, e and s. Subsequently, the i-e, e-s and i-s planes are traversed, and the weights associated with the vertices on these planes are computed using the previously computed values. Finally, the weights corresponding to feasible positive values of the variables are computed. Algorithms I and II compute the array W(.,.,.) and the quantity $D_T(X,Y)$ respectively.

ALGORITHM I.

Input: The strings $X = x_1x_2...x_N$, $Y = y_1y_2...y_M$, and the set of elementary edit distances defined by d(.,.). Let R=Min[M,N].

Output: The array W(i,e,s) for all feasible values of i, e and s.

Method: W(0,0,0)=0

```
        for i=1 to M do    W(i,0,0) = W(i-1,0,0) + d(θ,y_i)
        for e=1 to N do    W(0,e,0) = W(0,e-1,0) + d(x_e,θ)
        for s=1 to R do    W(0,0,s) = W(0,0,s-1) + d(x_s,y_s)
        for i=1 to M do
           for e=1 to N do
              W(i,e,0) = Min[W(i-1,e,0)+d(θ,y_i),W(i,e-1,0)+d(x_e,θ)]
           end
        end
        for i=1 to M do
           for s=1 to M-i do
              W(i,0,s) = Min[W(i-1,0,s)+d(θ,y_{i+s}),W(i,0,s-1)+d(x_s,y_{i+s})]
           end
        end
        for e=1 to N do
           for s=1 to N-e do
```

```
            W(0,e,s) = Min[W(0,e-1,s)+d(x_{s+e},θ),W(0,e,s-1)+d(x_{s+e},y_s)]
         end
      end
      for i=1 to M do
         for e=1 to N do
            for s=1 to Min[(M-i), (N-e)] do
         W(i,e,s) = Min[{W(i-1,e,s)+d(θ,y_{i+s})},{W(i,e-1,s)+d(x_{e+s},θ)},
                         {W(i,e,s-1)+d(x_{e+s},y_{i+s})}]
            end
         end
      end
      END ALGORITHM I
```

Algorithm II computes the quantity $D_T(X,Y)$ using the array W(.,.,.).

ALGORITHM II.

Input: The array W(.,.,.) computed using Algorithm I, and the constraint set, T.

Output: The constrained distance $D_T(X,Y)$.

Method:

```
$D_T(X,Y) = \infty$
for all i ε T do
   D_T(X,Y) = Min[D_T(X,Y), W(i,N-M+i,M-i)]
end
END ALGORITHM II.
```

Remark: From the last set of for-loops in Algorithm I it is easy to see that the time required to compute the array W(.,.,.) is of O(MNR), where R=Min[M,N]. Thus it has a worst case complexity which is cubic in time. Algorithm II clearly requires linear time. Thus the overall time required to compute $D_T(X,Y)$ is O(MNR).

IV.1 Computing the Best Edit Sequence

Once the quantity $D_T(X,Y)$ has been computed the optimal edit sequence can be obtained by backtracking through the array W(.,.,.) and printing out the actual edit sequence traversed, in the reverse order. The technique is well known in dynamic programming problems and has been used extensively for edit sequences [3,4,5,6] and longest common subsequences. Without further comment we now present Algorithm III which has as its input the distance $D_T(X,Y)$ and the optimal element of W(.,.,.), i.e., the element W(I,E,S) which is equal to $D_T(X,Y)$.

ALGORITHM III.

Input: The indices, I, E and S for which $D_T(X,Y) = W(I,E,S)$.

Output: The optimal sequence of edit operations subject to the specified constrained. The sequence is given in the reverse order and in the following notation:

(a) The pair (x_i, y_j) means the substitution of x_i by y_j.

(b) The pair (x_i, θ) means the deletion of x_i.

(c) The pair (θ, y_j) means the insertion of y_j.

Method:
```
i = I;      e = E;      s = S  where W(I,E,S) = D_T(X,Y)
while (i ≠ 0 and e ≠ 0 and s ≠ 0) do
begin
  if (W(i,e,s) = W(i-1,e,s) + d(θ,y_{i+s})) then
begin
     print (θ,y_{i+s})
     i = i-1
end
else if (W(i,e,s) = W(i,e-1,s) + d(x_{e+s},θ)) then
     begin
          print (x_{e+s},θ)
          e = e-1
     end
     else
     begin
          print (x_{e+s},y_{i+s})
     end
end
END ALGORITHM III.
```

Remark: Obviously Algorithm III is performed in $O(Max(M,N))$ time.

CONCLUSIONS

In this paper we have considered the problem of editing a string X to a string Y subject to a specified edit constraint. The edit constraint is fairly arbitrary and can be specified in terms of the number and type of edit operations desired in the optimal transformation. The way by which the constraint, T, can be specified has been proposed. Also the technique to compute $D_T(X,Y)$, the edit distance subject to the constraint T, has been presented. A final algorithm has been given which has as its input the quantity $D_T(X,Y)$ and the outputs the optimal edit transformation subject to the specified constraint.

Given the strings X and Y, $D_T(X,Y)$ and the array of constrained edit distances, W(.,.,.), can be computed in $O(|X|.|Y|.Min(|X|,|Y|))$ time. If $D_T([,Y)$ is given the optimal edit transformation can be obtained by backtracking through W(.,.,.) in $O(max(|X|,|Y|))$ time. The application of constrained string editing towards the recognition of noisy strings has been discussed in [8].

REFERENCES

[1] Hall, P.A.V., and Dowling, G.R., "Approximate String Matching", Computing Surveys, Vol.12, 1980, pp.381-402.

[2] Peterson, J.I., "Computer Programs for Detecting and Correcting Spelling Errors", C-ACM, Vol.23, 1980, pp.676-687.

[3] Kashyap, R.I., and Oommen, B.J., "A Common Basis for Similarity and Dissimilarity Measures Involving Two Strings", The International Journal of Computer Mathematics, Vol.13, March 1983, pp.17-40.

[4] Kashyap, R.I., and Oommen, B.J., "Similarity Measures for Sets of Strings", The International Journal of Computer Mathematics, Vol.13, May 1983, pp.95-104.

[5] Wagner, R.A., and Fischer, M.J., "The String to String Correction Problem", J-ACM, Vol.21, 1974, pp.168-173.

[6] Masek, W.J., and Paterson, M.S., "A Faster Algorithm Computing String Edit Distances", J. of Comp. and Syst. Sci., Vol.20, 1980, pp.18-31.

[7] Sankoff, D., "Matching Sequences Under Deletion/Insertion Constraints", Proc. of the Nat. Acad. Science., USA, Vol.69, January 1972, pp.4-6.

[8] Oommen, B.J., "Constrained String Editing", submitted for publication. Also available as School of Computer Science Technical Report, SCS-TR-48, Carleton University, Ottawa, Ontario, Canada.

THE STRUCTURE OF POLYNOMIAL COMPLEXITY CORES†

(Extended Abstract)

Pekka Orponen
Department of Computer Science
University of Helsinki
SF-00250 Helsinki 25, Finland

Uwe Schöning
Institut für Informatik
Universität Stuttgart
D-7000 Stuttgart 1, West Germany

1. INTRODUCTION

A polynomial complexity core for a language A is a set of strings C such that the running time of any deterministic Turing machine recognizing A exceeds any polynomial almost everywhere on C. That is, C is a set of strings uniformly hard for any algorithm for A. This notion was introduced by Lynch [8], who proved that every language not in P has an infinite complexity core. It is our purpose here to investigate further the structure and interrelationships of the cores for a language. We will be mostly concerned with proper cores, i.e., cores that are subsets of the languages, but many of our results could also be proved without this condition.

In Section 3 we study how dense the (proper) complexity cores for intractable languages can be. We prove that if each algorithm for a language has a nonsparse set of "hard" inputs, then in fact the language has a nonsparse proper complexity core. (A set A is nonsparse if for any polynomial p there is an n such that A contains more than p(n) strings of length n.) As applications of this result we show that any language that is NP-hard or "paddable" has a nonsparse proper complexity core. From a general point of view, the result is an interesting example of making a structural property of the hard input sets absolute by finding a single uniformly hard set with this property.

Except in trivial cases, a language never has a unique proper complexity core. In Section 4 we turn our attention to the interrelationships between the proper cores for a language and establish some connections between the complexity properties of a language and its global

†Part of this research was carried out while the authors were visiting the Department of Mathematics, University of California, Santa Barbara. This research was supported in part by the Emil Aaltonen Foundation, the Academy of Finland, the Deutsche Forschungsgemeinschaft, and the National Science Foundation under Grant No. MCS83-12472.

core structure. We characterize the languages with maximal proper complexity cores, and give a simple condition for two languages to have essentially the same set of proper cores.

2. PRELIMINARIES

All our languages will be over the alphabet $\Sigma = \{0,1\}$. For a string $w \in \Sigma^*$, $|w|$ denotes its length. The cardinality of a set A is denoted by $|A|$. For the finite initial segments of a language A we use the notation $A^{(n)} = \{x \in A \mid |x| \leq n\}$. A language A is (polynomially) sparse if for some polynomial p, $|A^{(n)}| \leq p(n)$ for all n. A language is co-sparse if its complement is sparse. We fix the standard sequence of polynomials: $p_j(n) = n^j + j$, $j = 0, 1, \ldots$.

The set of strings accepted by a Turing machine M is denoted by L(M). For a deterministic machine M, $time_M(x)$ is the number of steps M takes on input x ($time_M(x) = \infty$ if M doesn't halt on x). The familiar complexity classes P and NP are defined as usual. We define also EXPTIME = $\{L(M) \mid M$ is deterministic and $time_M(x) \leq 2^{c|x|}$ for some $c > 0$ and all $x \in \Sigma^*\}$.

A set $A \subseteq \Sigma^*$ is (polynomially one-way) paddable if there exists a polynomial time computable one-one function pad: $\Sigma^* \times \Sigma^* \longrightarrow \Sigma^*$ such that $pad(x,y) \in A \Longleftrightarrow x \in A$ for all $x, y \in \Sigma^*$. A paddable set is invertibly paddable if there exists also a polynomial time computable function decode: $\Sigma^* \longrightarrow \Sigma^*$ such that $decode(pad(x,y)) = y$, for all $x, y \in \Sigma^*$ [3,11].

A set A is polynomial time many-one reducible to a set B, denoted $A \leq_m^P B$, if there is a polynomial time computable function $f : \Sigma^* \longrightarrow \Sigma^*$ such that $x \in A \Longleftrightarrow f(x) \in B$, for all $x \in \Sigma^*$. For a class of sets $\mathcal{C}$, set A is $\mathcal{C}$-hard if $B \leq_m^P A$ for each $B \in \mathcal{C}$. If moreover $A \in \mathcal{C}$, A is $\mathcal{C}$-complete.

A (polynomial) complexity core for a language A is a set C such that for every deterministic Turing machine M with L(M) = A, and polynomial p, $time_M(x) > p(|x|)$, for all but finitely many $x \in C$. A complexity core C for A is proper if $C \subseteq A$. Lynch [8] has proved that any $A \notin P$ has an infinite complexity core; her proof can be easily modified to yield an infinite proper core. Even, Selman and Yacobi [5] have studied the general conditions needed for proofs of this type.

3. NONSPARSE COMPLEXITY CORES

In this section we show that intractable sets have not only infinite but even non-sparse complexity cores. We need a technical definition adapted from [10].

3.1 Definition. For a deterministic Turing machine M and a function f on natural numbers define the set of *f-hard inputs for M in L(M)* as $H(M,f) = \{x \in L(M) \mid time_M(x) > f(|x|)\}$. Let the class 1-APT (one-sided almost polynomial time) be defined as $\{L(M) \mid H(M,p)$ is sparse for some polynomial $p\}$.

3.2 Theorem. A recursive set A has a nonsparse proper complexity core if and only if $A \notin$ 1-APT.

Proof. The direction from left to right is obvious. For the converse direction, let $A \notin$ 1-APT, and $M_1, M_2, \ldots$ be an enumeration of all deterministic recognizers for A. For each $k \geq 1$, let $[M_1 \mid \ldots \mid M_k]$ be the Turing machine that simulates the machines $M_1, \ldots, M_k$ in parallel until the first of them comes to halt, and then accepts or rejects accordingly.

Now consider the following construction of a set C in stages where in stage k we extend the finite set C_{k-1} to C_k, and finally take C as $\cup\{C_k \mid k \geq 0\}$.

Stage 0: $C_0 := \phi;$

$n_0 := 0;$

Stage $k \geq 1$: Search for the smallest $n > n_{k-1}$ such that the the following holds:

(*) $|H([M_1 \mid \ldots \mid M_k], p_k)^{(n)}| > p_k(n)$.

$C_k := C_{k-1} \cup H([M_1 \mid \ldots \mid M_k], p_k)^{(n)};$

$n_k := n.$

It can be verified that if this construction can proceed at each stage then C becomes a non-sparse proper complexity core. Hence suppose there is some stage k such that the construction cannot proceed because for no n the condition (*) can be satisfied. But this would mean that $H([M_1 \mid \ldots \mid M_k], p_k)$ is sparse (via polynomial p_k) and hence would imply that $A \in$ 1-APT--contradicting the assumption. □

Observe that the proof technique is similar to forcing techniques in set theory or recursion theory.

Now we proceed to two specific applications of Theorem 3.2. First we consider paddable sets.

3.3 Lemma. The class 1-APT - P contains no paddable sets.

Proof. Suppose A is a paddable set in 1-APT. Let M witness that $A \in$ 1-APT. By applying M to sufficiently many (but still polynomially many) versions of the given input x, we obtain a polynomial time algorithm for A. This argument shows that each paddable set in 1-APT is already in P. □

3.4 Corollary. Each paddable set not in P has a non-sparse proper complexity core.

By extending Proposition 4.2 in [1], we can also get an upper bound on the complexity core densities of invertibly paddable sets.

3.5 Proposition. No invertibly paddable set can have a co-sparse complexity core.

Now we turn to complexity cores of NP-hard sets.

3.6 Lemma. If a set is in 1-APT then it is $\leq_m^P$-reducible to a sparse set.

Proof. Let A be in 1-APT. Hence for some M, L(M) = A, and polynomial p, H(M,p) is sparse. Let a be some fixed element of A. Then the following function f witnesses that $A \leq_m^P H(M,p) \cup \{a\}$.

$$f(x) = \begin{cases} a & \text{if } M \text{ accepts } x \text{ in } p(|x|) \text{ steps} \\ x & \text{otherwise.} \end{cases}$$ □

Using Mahaney's well known theorem [9] that no NP-hard set can be sparse unless P = NP, we immediately get the following conclusion:

3.7 Corollary. If P ≠ NP then each NP-hard set has a non-sparse proper complexity core.

Concerning the relationship between Corollaries 3.4 and 3.7, observe that it is not known whether each NP-complete set is paddable [3,11].

Now we show that for EXPTIME-hard sets these nonsparse core results can be strengthened to cores of exponential density.

3.10 Theorem. Any EXPTIME-hard set A has a proper complexity core C such that for some $\varepsilon > 0$ and almost every n, $|C^{(n)}| > 2^{n^\varepsilon}$.

Proof. In [3, Th. 13] a co-sparse set B in EXPTIME is constructed such that every $\leq_m^P$-reduction from B to some other set is one-one almost everywhere. Since it can be shown [7] that B is a core for itself, and since cores are preserved under $\leq_m^P$-reductions (which are in this case one-one a.e.), every set to which B is $\leq_m^P$-reducible has a core C such that for some $\varepsilon > 0$ and almost every n, $|C^{(n)}| > 2^{n^\varepsilon}$. This holds especially for every EXPTIME-hard set. □

4. GLOBAL STRUCTURE OF CORES

In this section we study the interrelationships between the proper complexity cores for a language. The obvious fact about this global structure is that subsets and finite unions of cores are again cores.

This observation leads us to use the language of lattice theory (see, e.g., [6]).

A lattice is a partially ordered set $\mathcal{L} = \langle \mathcal{L}, \leq \rangle$ satisfying the condition that any pair of elements in $\mathcal{L}$ has a least upper bound and a greatest lower bound with respect to $\leq$ in $\mathcal{L}$.

We will be considering the lattice $\Omega_F = \langle \Omega_F, \subseteq_F \rangle$ of languages modulo finite sets. The domain set of this lattice is $\Omega_F = \{\underline{A} \mid A \subseteq \Sigma^*\}$, where $\underline{A} = \{B \subseteq \Sigma^* \mid (A \setminus B) \cup (B \setminus A)$ is finite$\}$. The order relation is defined by $\underline{A} \subseteq_F \underline{B} \iff \exists A' \in \underline{A},\ B' \in \underline{B} : A' \subseteq B'$. It is easy to verify that Ω_F is indeed a lattice, where the l.u.b. and g.l.b. of elements $\underline{A}, \underline{B} \in \Omega_F$ are $\underline{A \cup B}$ and $\underline{A \cap B}$, respectively.

A subset $\mathcal{I}$ of a lattice $\mathcal{L}$ is an ideal in $\mathcal{L}$ if it satisfies the conditions (i) if $x, y \in \mathcal{I}$, then l.u.b. $(x,y) \in \mathcal{I}$; and (ii) if $x \in \mathcal{I}$ and $z \leq x$, then $z \in \mathcal{I}$. An ideal is principal if it has the form $\mathcal{I} = \{x \in \mathcal{L} \mid x \leq a\}$, for some $a \in \mathcal{L}$.

It is easy to see that for any $A \subseteq \Sigma^*$, the class $\Gamma_A = \{\underline{C} \mid C$ is a proper complexity core for $A\}$ forms an ideal in the lattice Ω_F. Our next result characterizes the languages A for which Γ_A is principal, i.e., the languages with maximal proper complexity cores.

4.1 Definition. A set is p-immune if it has no infinite subset in P. A set is almost p-immune if it is the disjoint union of a p-immune set and a set in P.

4.2 Theorem. For recursive A, Γ_A is principal if and only if A is almost p-immune.

We prove first a most useful little lemma.

4.3 Lemma. For recursive A, $C \subseteq A$ is a proper complexity core for A if and only if $C \cap E$ is finite for each $E \subseteq A$, $E \in P$.

Proof (of lemma). For each P-set E in A, A has an algorithm that runs in polynomial time on E. Hence if C is a core, $C \cap E$ is finite. Conversely, if C is not a core, some algorithm M for A runs in time bounded by a polynomial p infinitely often on C. But then $C \cap E$ is infinite for the P-set $E = \{x \mid M$ accepts x in time $p(|x|)\} \subseteq A$. □

Proof (of theorem). It is relatively straightforward to show that if A is almost p-immune, the p-immune part of A is a maximal (modulo finite sets) proper core for A, and so defines a largest element in Γ_A.

Assume then that Γ_A is principal, i.e., for some proper core C for A, any proper core C_1 for A is contained in C except for finitely many strings. Any proper core for a recursive set is p-immune, so it suffices to show that $A - C \in P$.

Assume not; we will construct an infinite complexity core C_1 for A in A - C, contradicting the maximality of C. Enumerate the P-subsets of A: $E_1, E_2, \ldots$. For $k \geq 1$ define $E_{(k)} = (\cup_{i\leq k} E_i) - C$. For each k, $E_{(k)} \in P$. But $E_{(k)} \subseteq A - C$ and $A - C \notin P$, so we can define an infinite sequence of strings in A - C by

c_1 = any string in A - C;

c_{k+1} = any string different from $c_1, \ldots, c_k$ in $(A-C) - E_{(k)}$.

Now $C_1 = \{c_1, c_2, \ldots\} \subseteq A - C$ has only finite interesection with each P-set $E_i \subseteq A$, hence it is a core for A. □

The lattice structure of Γ_A seems to get quite intricate as soon as we are out of the simplest case of A being almost p-immune. The study of these structures would be of some interest, because naturally occurring sets do not seem to have immunity properties. It has been proved by L. Berman [2] that no set complete for a deterministic time class can be p-immune. David Russo (personal communication) has observed that in fact such sets cannot be even almost p-immune. The immunity properties of NP-complete sets pose an intriguing open problem.

We now give another theorem relating complexity properties of sets to the structure of their core ideals.

4.4 Theorem. Let A and B be recursive. Then $\Gamma_A = \Gamma_B$ if and only if $A - B,\ B - A \in P$.

Proof. With the help of Lemma 4.3, verification of the "if" claim is not difficult. For the "only if" direction, let $\Gamma_A = \Gamma_B$. We show that $A - B \in P$, from which the result follows by symmetry.

We prove first that for some $E \in P$, $A - B \subseteq E \subseteq A$. Let $E_1, E_2, \ldots$ be the P-sets in A. If A - B is not covered by any E_i, we can define an infinite core $\{c_1, c_2, \ldots\}$ for A in A - B by

c_1 = any string in A - B;

c_{k+1} = any string different from $c_1, \ldots, c_k$ in $(A-B) - \cup_{i\leq k} E_i$.

But this is contrary to the assumption that $\Gamma_A \subseteq \Gamma_B$.

Let then E be as above, and define $E_1 = E \cap B$. We show that $E_1 \subseteq F$ for some $F \subseteq B$, $F \in P$, from which it follows that $A - B = E - F \in P$. Assume such a cover for E_1 does not exist in B. Then we can enumerate the P-sets in B and proceed just as above to obtain an infinite complexity core C for B in E_1. But because $C \subseteq E_1 \subseteq E \in P$, C is not a core for A, contradicting the assumption that $\Gamma_B \subseteq \Gamma_A$. □

Observe that having A - B and B - A in P is a very strong condition guaranteeing that A and B are of the same $\leq_m^P$-degree. It would be very interesting to obtain characterizations of reducibilities in terms of core lattice structures. Conversely, it would be interesting to know what happens if the notion of ideal similarity in the theorem is weakened from identity to, say, lattice isomorphism.

REFERENCES

1. J.L. Balcázar and U. Schöning, Bi-immune sets for complexity classes Math. Syst. Theory (1984), to appear.
2. L. Berman, On the structure of complete sets: almost everywhere complexity and infinitely often speedup. 17th FOCS (1976), 76-80.
3. L. Berman and J. Hartmanis, On isomorphism and density of NP and other complete sets. SIAM J. Computing 6 (1977), 305-322.
4. P. Berman, Relationship between density and deterministic complexity of NP-complete languages. 5th ICALP, Lecture Notes in Computer Science 62, 63-71, Springer-Verlag, 1978.
5. S. Even, A.L. Selman, and Y. Yacobi, Hard-core theorems for complexit classes. To appear.
6. G. Grätzer, Lattice Theory. Freeman & Co., 1971.
7. K. Ko and D. Moore, Completeness, approximation and density. SIAM J. Computing 10 (1981), 787-796.
8. N. Lynch, On reducibility to complex or sparse sets. Journal of ACM 22 (1975), 341-345.
9. S.R. Mahaney, Sparse complete sets for NP: solution of a conjecture by Berman and Hartmanis. J. Comput. Syst. Sci. 25 (1982), 130-143.
10. A.R. Meyer and M.S. Paterson, With what frequency are apparently intractable problems difficult? M.I.T. Technical Report TM-126, Feb. 1979.
11. P. Young, Some structural properties of polynomial reducibilities and sets in NP. 15th ACM STOC (1983), 392-401.

Additional note: After completing this paper, we have obtained the following results:

Theorem (with David Russo). If $P \neq NP$, then the NP-complete set SAT is not almost p-immune.

Theorem. If A and B are recursive sets that are not almost p-immune, then the ideals Γ_A and Γ_B are lattice-isomorphic.

Proofs of these results will appear elsewhere.

Solving Visibility Problems by Using Skeleton Structures †

Thomas Ottmann
Institut für Angewandte Informatik
und Formale Beschreibungsverfahren,
Universität Karlsruhe
Postfach 6380, D-7500 Karlsruhe, West Germany

Peter Widmayer
IBM Thomas J. Watson Research Center,
P.O. Box 218, Yorktown Heights, New York 10598, USA

ABSTRACT

This paper presents solutions to several visibility problems which can be considered as simplified versions of the hidden line elimination problem. We present a technique which allows to translate many line sweep algorithms involving sets of iso-oriented objects into algorithms for sets of non iso-oriented objects such that the asymptotic worst case time and space requirements do not change.

1. INTRODUCTION

Problems involving sets of iso-oriented rectangles or rectilinear polygons have become very popular in the blossoming area of computational geometry. Two basic reasons may be identified for this fact: First and above all these problems arise from applications in quite different areas of computer science like VLSI design, graphics and image processing. Second, for problems involving sets of iso-oriented objects there exist often efficient or even time and space optimal solutions in contrast to the available solutions of the corresponding problems involving sets of non iso-oriented objects.

In the iso-oriented case, without loss of generality, all objects in a given set have their sides parallel to orthogonal (vertical and horizontal) coordinate axes. (We will call objects of this kind *rectilinear* or *rectangular* as well.) Therefore, the projection of these objects to a vertical or horizontal scan line which sweeps over the set of objects can be represented efficiently by a semidynamic *skeleton structure*. The most popular and widely used structures of this kind are the segment tree introduced in [BW] and the tile tree of McCreight [McC] and its variant, the interval tree, independently discovered by Edelsbrunner [E]. These structures appear in many algorithms, cf. e.g. [Ov]. Modifications of the segment tree, the contracted segment tree and the visibility tree, have been used in algorithms to compute the contour of rectilinear polygons, cf. [Gü] and [W].

At first glance, segment and interval trees seem to have no analogue for sets of non iso-oriented objects. For, there is no obvious way to obtain a rastering induced by the set which can be used to build a skeleton structure in this case. Our major aim in the present paper is to show that semidynamic skeleton structures can be built and utilized for sets of non iso-oriented objects as well. This will be exemplified by a classical algorithmic problem arising in computer graphics, the hidden-line-elimination problem. However, for expository purposes we will restrict the problem considerably by ruling out a lot of "degenerate" cases which may appear in practice.

2. A RECTANGULAR VISIBILITY PROBLEM

† This work was supported by the grant Ot64/4-2 from the Deutsche Forschungsgemeinschaft

Suppose we are looking at a set of planar sheets of paper of different size. They have been laid on each other on a table but — by chance — in such a way that the projections of the sides (the edges) of the paper sheets are iso-oriented; i.e. they are all parallel to orthogonal x-y-coordinates. The visibility problem we are dealing with in this section is the problem of computing a "realistic" two-dimensional picture of such a three-dimensional scene. More precisely, we specify the input and the desired output of the rectangular visibility problem as follows:

Input: A set of n planar, iso-oriented rectangles in two-dimensional space considered to be two-dimensional images of planar three-dimensional objects which do not penetrate each other. Each rectangle r is given by the y-values of the lower and upper boundaries and by the x-values of the left and right boundaries. Furthermore, a three-dimensional plane equation of the form $ax + by + cz + d = 0$ is associated with each rectangle r. Thus, r may be identified with an 8-tuple $r = (x_l, x_r, y_b, y_t, a, b, c, d)$.

Output: A list of all visible edges in x-y-space where each edge is the two-dimensional projection of a maximal visible portion of a boundary edge of an input object.

For each point (x, y) belonging to a given object $r = (x_l, x_r, y_b, y_t, a, b, c, d)$, i.e. where $x_l \leq x \leq x_r$ and $y_b \leq y \leq y_t$, we can compute its distance to the projection plane $z = 0$ by evaluating the plane equation at the given point in constant time. (We think of an observer looking at this plane from far away, with all objects lying on the other side.)

$$dist(x, y) = \frac{1}{c}(-ax - by - d)$$

The function *dist* determines the relative visibility order "$\prec_{x,y}$" for the given set of objects at point (x, y). Observe that in general there is no total ordering of the objects which is compatible with the orderings "$\prec_{x,y}$" at all points (x, y) in the plane.

We make some assumptions of nondegeneracy: The x-values of left and right boundaries of rectangles are all pairwise distinct as are the y-values of lower and upper boundaries. This in particular implies that all points (the ones originally given as corner points of rectangles as well as the edge intersection points) are pairwise distinct.

We will now briefly sketch how to solve the above rectangular visibility problem using the plane sweep paradigm as explained e.g. in [NP]. The visibility of an edge can change only where the edge begins, ends, or intersects another edge. In order to further simplify the problem we treat the vertical and horizontal edges separately. Hence, we will only explain how to compute the visible parts of the *vertical* edges by sweeping a *horizontal* scan line from top to bottom. (The horizontal edges can be treated similarly by sweeping a vertical line, say, from left to right.) The scan line halts at exactly those points where the visibility of a vertical edge may change. An edge or a part of an edge is output as soon as it becomes invisible. Each active rectangle currently cut by the scan line is represented by its projection on the line, thus, by a closed interval $[e, e']$ with a pair of vertical edges e and e' as its boundaries. These intervals are stored in a structure L in such a way that updates and visibility tests can be carried out using L efficiently. Whenever we insert into the structure L a new interval $[e, e']$ or delete from L the interval $[e, e']$ representing a rectangle we have to determine all edges e'' such that e'' is the left or right boundary of some interval overlapping $[e, e']$. From [BO] we can see that these edges can be found in time proportional to $O(\log n + k)$ where k is the number of answers.

In the next section we will show how to build the structure L, to maintain it during the sweep, and how to use it for visibility tests.

3. SKELETON STRUCTURES FOR SETS OF ISO-ORIENTED OBJECTS

The left and right boundaries of all rectangles in the given set of n rectangles impose a *discrete raster* on the horizontal x-axis. We deal with a (normalized) raster over the subset $1,\ldots,2n$ of the integers. As far as visibility is concerned, this raster is equivalent to the raster given in the input, and it can be computed in $O(n \log n)$ time by sorting the boundaries and mapping them into their positions in sorted order. The horizontal projection of each of the n rectangles is a closed interval consisting of a contiguous sequence of elementary fragments. First we describe how these intervals are stored in a segment tree (cf. [BW]):

The *(empty) segment tree (skeleton)* is a binary tree of minimal height such that the i–th leaf represents both the i–th raster point and the closed open interval $[i, i+1)$, i.e. the i-th fragment. Each internal node represents the union of all fragments which belong to the leaves in the subtree defined by it. Each closed interval $[i, j]$, i.e. each sequence of consecutive fragments including the two raster points i and j of the boundary of the interval, are represented by the smallest set of nodes such that the union of the intervals represented by these nodes is exactly the desired interval; the right boundary of a closed interval is represented by a leaf. We say that these nodes *cover* the given interval. It is well known that $O(\log n)$ nodes always suffice to represent an arbitrary interval. Each interval and the corresponding rectangle is associated with all covering nodes, i.e. each node has its associated *node list* of intervals. While the skeleton of the segment tree remains fixed, these node lists change dynamically according to what intervals are currently active.

When implemented appropriately, the above structure has the following characteristics (cf. [BW], e.g.):

Space required to store n intervals $S(n) = O(n \log n)$

Insertion time $O(\log n)$.

Deletion time $O(\log n)$.

Query time $O(\log n + k)$, where k is the size of the answer, i.e. the number of elements reported when answering an inverse range query.

We could directly use this structure as the structure L in the above given algorithm to solve the rectangular visibility problem:What remains to be shown is that we can perform visibility tests. They can be carried out as follows. In order to check whether or not a vertical edge e is visible at some scan point we determine all currently active intervals (rectangles' projections) which contain the edge e. (This is an inverse range query with a raster point as query point.) These intervals represent exactly those rectangles which may hide e. Thus, it suffices to evaluate the plane–equations of these rectangles at the scan point and determine the rectangle with minimal distance to the observer. If and only if this rectangle has e as its left or right boundary the edge e is visible (at e's intersection with the scan line).

Performing visibility tests as just described can, of course, be quite time consuming. For, the node lists may have length up to $O(n)$ which implies that a single visibility test may take time $O(n)$ as well. We can improve this if we maintain the lists of names of intervals representing rectangles in sorted order according to the relative distance to the observer. Thus, the rectangle with minimum distance to the observer at a scan point can be determined in $O(\log n)$ steps.

Unfortunately, this speed-up of the query time is not obtained for nothing: The *insertion* of a new interval representing a rectangle at the appropriate position into the respective node list according to its relative distance to the observer requires more time. However, it should be clear that $O(\log n)$ steps *per node list* suffice if the node lists are implemented appropriately (as balanced trees, e.g.). Thus, $O(\log^2 n)$ steps per insertion are sufficient. Similarly, the *deletion* of the name of an interval representing a rectangle from all node lists in which it occurs can be carried out in time $O(\log^2 n)$. In order to solve the rectangular visibility problem at most n insertions and n deletions of intervals have to be performed. If there are k intersections of vertical and horizontal edges, a total number of $2n + k$ visibility tests have to be performed. We summarize the discussion by the following theorem.

Theorem 1: *The rectangular visibility problem for a set of n objects with k intersections can be solved in time $O(n \log^2 n + k \log n)$ and space $O(n \log n)$ using an appropriate version of segment trees.*

We will now describe how a version of the tile tree (or: interval tree of [E]) can be used to solve the rectangular visibility problem:

Again, the left and right boundaries of the set of given rectangles form a set of discrete raster points on the x–axis. First, a binary tree of minimal height storing the sorted set of these points is built. Each node has an associated list of left and right endpoints of intervals on the x–axis. Initially all lists of endpoints are empty. This is the skeleton of the interval tree storing the empty set of intervals.

Let a rectangle r have left and right boundaries e and e' on a horizontal scan line with the interval $[e, e']$ representing r. Interval $[e, e']$ is inserted into the above described skeleton structure as follows: Let $key(p)$ denote the raster point stored at node p. Starting at the root we check for each node p whether $key(p)$ is in $[e, e']$,i.e. whether or not $key(p)$ is an end point or an internal point of the interval $[e, e']$. If so, both end points e, e' representing the interval $[e, e']$ and the rectangle r are inserted into the sorted list of endpoints associated with p according to their respective x-value and the insertion is complete. If $key(p)$ is not in $[e, e']$ both endpoints of the interval lie either totally to the left or totally to the right of the raster point $key(p)$. Depending on which case occurs recursively apply the insertion procedure to the left or right son of p.

Deletion of an interval from this structure consists of determining the node in whose node list the endpoints of the interval are stored and in deleting these entries.

It should be clear how all intervals containing a raster point e can be reported: We search for e until the node p with $key(p) = e$ has been found; all intervals in the list of node p are reported and for each node q on the way from the root of the tree to p either a prefix or a suffix of the list associated to q is reported: Assume that p is in the right subtree of q. Then only intervals in the list associated to q should be reported which contain the raster point e which is, by assumption, to the right of the raster point $key(q)$ stored at q; therefore, scan the list associated with q from its right end and report all intervals containing e. These intervals form a suffix of the list associated with q. Similarly, if p is in the left subtree of q all intervals in a prefix of the list associated with q are reported.

Implementing the lists of intervals (associated with the nodes of the skeleton) as balanced trees obviously leads to a structure with the following characteristics: (cf.[E])

Space required to store n intervals $S(n) = O(n)$.

Insertion time per interval $O(\log n)$

Deletion time per interval $O(\log n)$

Query time, i.e. the time required to answer an inverse range query with a raster point as query point: $O(\log n + k)$ where k is the size of the answer.

In order to solve the rectangular visibility problem we have again to perform visibility tests and not just to answer inverse range queries whenever a vertical edge begins at or intersects a horizontal edge. Clearly, we can first determine all intervals containing a vertical edge e or, more precisely, their projections to the scan line. Then, for all these intervals we evaluate the plane equation of the associated three-dimensional objects at the scan position (i.e. at the point defined by the y-value of the scan line and the x-value of e) and determine for which interval the minimum is obtained. Thus, the time required to perform a *single* visibility test is of order $O(n)$ because there may be up to $O(n)$ intervals containing a given query point. Therefore, we augment the structure as follows in order to speed up a visibility test from time $O(n)$ to $O(\log^2 n)$:

The lists of left and right endpoints associated with a node p are stored at the leaves of two balanced leaf-search trees, such that the leaves are doubly linked. The internal nodes of these balanced trees contain routing information to guide the search for an element stored at some leaf and, moreover, store the interval with minimal distance at scan position in its subtree.

Observe that the interval having the minimum distance at scan position in some *prefix* of the list of (endpoints of) intervals can easily be determined when searching for the position among the leaves which is the (right) end of the prefix to be considered. Inspect the mindistance values stored at left brothers of the nodes on the search path and keep track of the minimum among these values. Similarly, the interval having the minimum distance to the observer at scan position appearing in a *suffix* of the endpoint list associated with a node p can be determined in time $O(\log l)$ where l is the length of the list. In order to perform a visibility test at most $O(\log n)$ prefixes or suffixes of lists, each of length at most $O(n)$ have to be inspected as just described. Hence, a single visibility test can be carried out in time $O(\log^2 n)$.

It should be clear that the just described structure can be maintained when inserting, respectively, deleting an interval in time $O(\log n)$. Thus we obtain the following theorem.

Theorem 2: *Given a set of n planar three-dimensional objects with rectangular, iso-oriented two-dimensional images having k intersections of horizontal and vertical boundaries in two-dimensional space, the rectangular visibility problem can be solved in time $O((n+k)\log^2 n)$ and space $O(n)$ using* *the (augmented) interval tree.*

In order to solve the rectangular visibility problem, $O(n)$ insertions and deletions of intervals and $O(n+k)$ visibility tests have to be performed. We have seen two different, semidynamic skeleton structures which support these operations and have different time– and space characteristics. We close this section by posing the open problem whether or not there exists a structure which allows to perform an insertion, deletion and visibility test in time $O(\log n)$ and which requires space $O(n)$ or at most $O(n \log n)$.

4. THE VISIBILITY PROBLEM FOR NON ISO-ORIENTED OBJECTS

In this section we are concerned with the following ***general visibility problem***: Given a set of planar three–dimensional objects which do not penetrate but which may partially or totally hide each other, compute the visible edges in two–dimensional space under orthographic projection. More precisely, we are dealing with a scene of planar polygonal objects in three–dimensional Euclidean $x-y-z$–space. The objects are not necessarily convex and may have holes. Following the standard terminology in computer graphics, cf.e.g. [FV], we call such objects ***surfaces*** or, simply, ***faces***. We assume that each face is bounded by edges which are oriented such that the interior lies to the right of a boundary edge. No two faces are allowed to intersect, that is, no point of the three–dimensional Euclidean space belongs to more than one face. The description of the three–dimensional scene is given as a collection of the descriptions of the faces. Each face is, in turn, described by a plane equation and a collection of descriptions of the edges constituting the boundary of the face. An edge is given by its starting and terminating point, implicitly giving the orientation, and the face it belongs to.

Because we want to compute a realistic image of the scene under orthographic projection, it is sufficient to assume that the two–dimensional images of the edges in $x-y$–space are given. For, the third z–coordinate can be computed for each endpoint by using the plane equation of the face to which the point belongs.

The output of an algorithm for solving the general visibility problem should yield a collection of all maximal visible parts of edges in $x-y$––space. A point of an edge is visible if and only if it has minimum distance to the observer among all points having the same two–dimensional coordinates (x, y) and belonging to some face in the given set of faces, i.e. its z-coordinate is minimal. A given set of faces with a total number of n edges as its boundaries may have up to $O(n^2)$ intersection points in two–dimensional space.

We make the following assumptions of nondegeneracy: There are no points (the ones originally given as endpoints of edges in two–dimensional space as well as the intersection points of edges in two–dimensional space), which belong to more than two edges. Furthermore, we assume that there are no two points (end points of edges and intersection points) having the same y–value. This latter assumption in particular implies that there are no horizontal edges.

As for the rectangular visibility problem the line sweep paradigm is an obvious method to solve the given problem: Sweep a horizontal line from top to bottom through the $x-y$–plane halting at each point where an edge starts or terminates or two edges intersect. These are the ***sweeping points***. The visibility status of an edge can change only at a sweeping point. By our assumptions about nondegeneracy each sweeping point falls into exactly one of the following two categories: It is either a point where two edges meet (both start, both end, or one starts and one ends) or an intersection point of two edges. Furthermore, all sweeping points have different y–values.

We assume that the k intersection points are either precomputed or computed "on the fly" during the sweep as described in [BO] in time $O((n+k)\log n)$ and space $O(n)$ (using the improvement of Brown [B]).

As in the rectangular case treated in Sections 2 and 3 each face will be represented by its projection on the scan line. Differing from the rectangular case, the projection of a face may now consist of several, noncontiguous intervals. Furthermore, the intervals may grow and shrink when sweeping the scan line downwards. The first difference stems from the fact that we do not assume convexity anymore. The second difference leads to our central question: How, if at all, can we build a skeleton structure which allows to insert and delete such intervals of nonfixed size and which supports visibility tests for non iso–oriented edges?

Recall how we proceed in the rectangular case: The knowledge about the entire set of objects to be processed is exploited to build once and for all in advance a structure that provides room for each object that will ever be present in the structure. When dealing with rectilinearly oriented rectangles in the plane, each ordering at scan line position of the currently active rectangles is a subordering of the total ordering of all rectangles' left and right boundaries. The latter, however, can be computed by a simple sort in advance. Once a left or right boundary has been inserted into the skeleton structure it does not change its position anymore until it disappears from the structure.

For non orthogonal objects, not only the computation of the initial places for the objects in some skeleton structure is no longer obvious. It may also happen that the relative position in the one–dimensional ordering of the interval boundaries at the scan line varies.

In order to solve the general visibility problem in the same way as the rectangular problem we will proceed as follows: We will first show how to define and compute an *initial placement order* for a set of non iso–oriented objects on which we can base a skeleton structure for a line sweep algorithm. Then the non iso–oriented analogues of segment and interval trees will be introduced. Finally, we will give a solution of the visibility problem using these structures.

5. INITIAL PLACEMENT ORDER OF LINE SEGMENTS

Line segments are the typical elementary kind of non orthogonal objects in the plane appearing in the general visibility problem of the previous section. As a first step we want to find an *initial placement order which is suitable for a line sweep* from top to bottom for the given set of line segments.

We call a total linear order $\cdot\!<$ on a given set of line segments an *initial placement order* iff the following algorithm *sweep* can be carried out and is correct:

Algorithm *sweep:*

S :=sequence of line segments in initial placement order $\cdot\!<$, all segments marked as not present; { S is the current placement order initialized as the initial placement order $\cdot\!<$ }
Q :=sequence of sweeping points (upper- and lower endpoints and intersection points of line segments) ordered according to decreasing y–value:
while Q not empty **do**
begin
p := next point from Q;
if p is an upper endpoint of segment s **then** mark s as present in S
else if p is a lower endpoint of segments s **then** mark s as not present in S
(*) **else** {p is an intersection point of segments s and t; both s and t are marked as present in S; there are no segments marked as present in between s and t}
exchange s and t in S, i.e. replace S by the ordering which results from S by interchanging s and t.
(**) {exactly the segments marked as present in the current placement order S appear in left to right order at scan line hight, i.e. at a horizontal line just below p}
end {while}
{all elements of S are marked as not present}
end of algorithm *sweep.*

We in particular require that the assertions marked (*) and (**) hold true. The most crucial one is (**) which says that the ordering among the currently active line segments induced by the scan line is exactly the subordering of the elements of the current placement order S which are marked as present.

Observe that it is not a trivial task to find the initial placement order $\cdot\!<$ for a given set of line segments. This ordering can be defined in a different way by considering a set of line segments as a set of *zigzags*. The notion of a zigzag has some similarity with the notion of *chain* introduced in [LePr] and [Pr].

The *zigzag of segment s* is the polygonal line starting at the upper end point of s, following s until the first intersection point of s with some other segment, say t, is met, from this intersection point following t in direction of decreasing y until the next intersection point of t with some other segments is met, and so on until the end point of a segment is met. Observe that the continuation of a zigzag at an intersection point is defined uniquely by the above description. Note that there is a one-to-one correspondence between zigzags and segments: the zigzag starting with segment s is associated with segment s is called the zigzag of s and will be denoted by z_s. A zigzag is a connected polygonal line. No two different points on a zigzag can have the same y-value. Zigzags can only touch each other (at intersection points of segments), but cannot cross.

We say that a zigzag z lies to the *left* of a zigzag z', iff z and z' have points p and p' with a common y-value, such that the x-value of p is less than the x-value of p'. Let $left^+$ denote the transitive closure of $left$.

Observe that there can still be pairs of zigzags unrelated within $left^+$. For all pairs of zigzags z and z' having no points with the same y-value, we say z is *above* z' iff all y-values of points of z are greater than all y-values of points of z'. Note that relation *above* is a partial order. Let us define the total order $left^+|above$ between zigzags z and z' as follows: Whenever z and z' are related in $left^+$ then they are related in $left^+|above$ in the same way; otherwise they are related in $left^+|above$ in the same way as in *above* (orderings of this type are sometimes called "$left^+$ *consulting above*", see [GY], [OW1]).

For a detailed proof that $left^+|above$ is really a linear order for the set of zigzags associated to an arbitrary given set of line-segments, see [OW2].

Once we have defined a linear order for the set of zigzags associated to a given set of line segments this order carries over to the set of line segments as follows:

For each two line segments s and t define

$$s \cdot\!< t \quad \text{iff} \quad z_s \ \ left^+|above \ \ z_t$$

It can be shown that $\cdot\!<$ is an initial placement order of the given set of line segments, i.e. the above algorithm *sweep* can be carried out and is correct with this order as initial placement order. Furthermore, it can be shown that the $left^+|above$ ordering for a set of zigzags, hence, the initial placement ordering $\cdot\!<$ for a given set of line segments can be computed efficiently by:

Theorem 3: *Given a set of n line segments with k intersection points, an initial placement order which is suitable for a line sweep from top to bottom can be computed in time* $O((n+k)\log n)$ *and space* $O(n)$.

Any improvement of the known algorithm to compute the k intersections of the n line segments to, say, a $O(n \log n + k)$ time algorithm would improve the algorithm for computing an initial placement order as well.

6. SKELETON STRUCTURES FOR POLYGONS IN THE PLANE

In this section we will describe how to apply the skeleton approach to sets of non iso-oriented polygons in the plane and how to adapt segment and interval trees.

The skeleton structures could be based on the initial placement order $\cdot\!<$ of *all* the line segments occuring in the given polygons. We can, however discard all those line segments which do not occur as topmost elements of zigzags. Here we deal with a slightly modified version of zigzags to utilize a fact which is valid for polygons, namely that often at an endpoint of a segment another segment starts. Hence, we consider zigzags to continue at such transition points. This has the immediate consequence that for a set of p *convex* polygons only $2p$ line segments and zigzags have to be considered. A set of nonconvex polygons, however, consisting of a total number of n line segments may decompose into $O(n)$ zigzags.

In general, assume that there is a total number of n line segments m of which appear as initial elements of zigzags. First, we compute the initial placement order $\cdot\!<$ for this set of m segments using the algorithm described in the previous section. Then we build the analogue of a segment or interval tree based on this ordering. Roughly speaking, these structures will be based on the *ranks* of the segments appearing as boundaries of intervals rather than on the boundaries themselves. For that purpose we use an array *rank* indexed by the n segments and initialized as follows:

$$rank(s) = \begin{cases} i, & \text{if } s \text{ is the } i\text{-th element in the order } \cdot< \\ undefined, & \text{otherwise} \end{cases}$$

Simultaneously, we maintain an array *current* indexed by the ranks which returns the element having a given rank:

$$current(i) = s \qquad \text{iff} \qquad rank(s) = i$$

Finally, with each polygon P we associate an ordered set of ranks of endpoints of intervals appearing on the scan line. This set is denoted by $endpoints(P)$. It allows us to find (the rank of) a left endpoint for a given right endpoint and (the rank of) a right endpoint for a given left endpoint of a currently active interval representing polygon P on the scan line. Furthermore, it allows us to determine the interval of ranks representing P which contains a given point (rank).

Observe that $endpoints(P)$ may contain up to $O(n)$ elements for nonconvex polygons but $endpoints(P)$ has at most two elements for convex polygons.

Both the segment tree and the interval tree structures are augmented in this way. Their skeletons are based on the ranks of elements as follows:

Segment tree skeleton: The nonorthogonal segment tree skeleton is a minimal–height binary tree with m leaves labelled $1, \ldots, m$ representing the m ranks. *Insertion* and *deletion* of a closed interval $[a, b]$ of segments a and b representing a polygon P on the scan line can be performed in exactly the same way as in the orthogonal case: Insert, respectively delete the corresponding interval $[rank(a), rank(b)]$.

Interval tree skeleton: The nonorthogonal interval tree skeleton is a minimal–height binary tree with m nodes labelled $1, \ldots, m$ representing the m ranks.Each node has an associated list of ranks of left and right endpoints of currently active intervals. These lists are ordered according to their ranks. Initially all lists are empty.*Insertion* and *deletion* of a closed interval is performed as usual by referring to ranks again.

A line sweep algorithm for a set of non iso–oriented polygons in the plane using a skeleton structure has the following rough structure:

Algorithm *h-sweep for non iso-oriented objects*

$Q :=$ sequence of sweeping points in decreasing y–order;
$S :=$ skeleton structure based on the initial placement order $\cdot<$ of segments and initialized such that it stores the empty set of intervals;
initialize *rank* and *current* arrays according to the initial placement order
for each polygon P **do** $endpoints(P) := \emptyset$;
while Q not empty **do**
  begin $y :=$ next point from Q;
    (∗) update $S, rank, current, endpoints$ according to the type of y;
    (∗∗) perform the problem-specific tasks and queries
end {*while*}
end of algorithm.

We will first disregard any problem–specific details, i.e. skip the the part marked (∗∗). However, we will explain the required update operations, i.e. the part marked (∗) for two sample cases:

Case 1: Let us assume that y is a concave cornerpoint of polygon P.

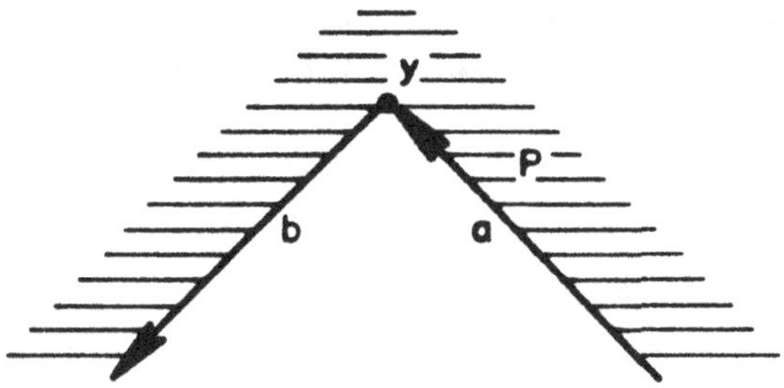

Search *endpoints*(P) for the successor of *rank*(a) and for the predecessor of the *rank*(b). Let us assume that segments r and s have these ranks. This means that the vicinity of sweeping point y looks as follows:

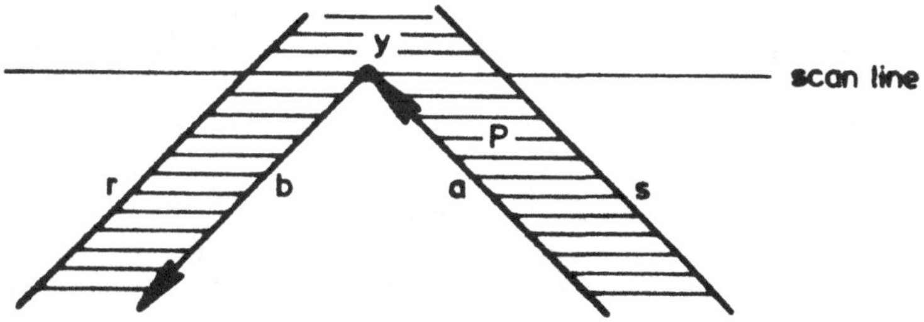

Just above sweeping points y the polygon P has the interval $[r, s]$ as a part of its projection on the scan line. Just below the sweeping point , the part of P is represented by the two intervals $[r, b]$ and $[a, s]$. Hence, it suffices to *delete* interval $[r, s]$ from the skeleton sructure S and insert $[r, b]$ and $[a, s]$ into S. Furthermore, (the ranks of) a and b have to be inserted into the set *endpoints*(P).

Case 2: Let us assume that y is an intersection point of the edges a of polygon P and b of polygon Q, and, that these two polygons overlap as depicted by the figure.

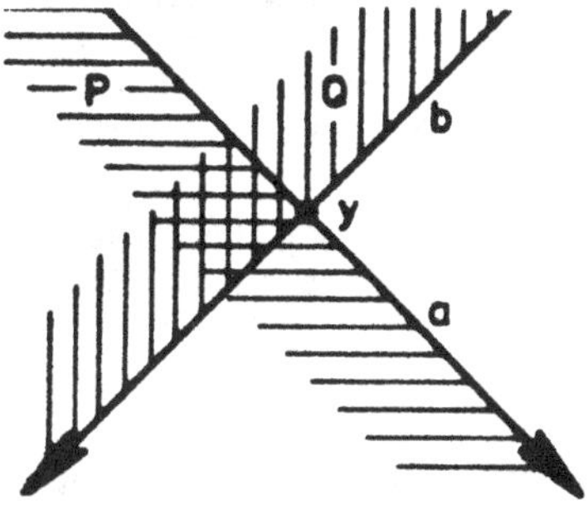

Again, we search the set *endpoints*(P) for the predecessor, say r, of a and the set *endpoints*(Q) for the predecessor, say s, of b. Then, above y, P and Q were represented by the intervals $[r, a]$ and $[s, b]$, respectively. Below y, P and Q are represented by the same intervals, but their ranks have been changed in between. This is reflected by the following updates:

delete $[r, a]$ from S;
delete $[s, b]$ from S;
exchange (*rank*(a), *rank*(b)) and update *current* accordingly;

insert $[r, a]$ into S;
insert $[s, b]$ into S;
It should be obvious how to treat the remaining cases analogously.

It should be clear from the above discussion that the following invariant holds true before each next iteration of the while loop starts:

(INV) The skeleton structure S stores exactly the currently active intervals (projections of polygons with the scan line) and for each P, *endpoints*(P) is the set of ranks of endpoints of these intervals.

Hence, as in the rectangular case, this structure can be used to answer inverse range queries: Given a currently active segment s. In order to determine all intervals containing s (or more precisely, the intersection point of s and the scan line) search the skeleton structure S for the raster point "*rank*(s)".

We did not explain so far how to implement the sets *endpoints*(P) and the algorithms for manipulating these sets; i.e. *insert*, *delete*, *successor* and *predecessor* operations. Furthermore, we have exactely the same alternatives for implementing the skeleton stucure S as in the iso-oriented case. However, it should be clear that it is possible to design the involved structures and tune the respective algorithms in such a way that $O(\log n)$ steps suffice to carry out all update operations which are necessary in one iteration of the while loop not counting the steps required to perform the problem specific tasks and queries.

The problem-specific tasks which have to be performed in order to solve the general visibility problem are logically the same as for the orthogonal case: One has to carry out visibility tests for all edges which meet at a sweeping point. An output has to be generated as soon as an edge (or a portion of it) becomes invisible. Using a buffer technique one can acchieve that always maximal visible portions of polygon edges are reported. Thus the additional overhead for solving the general visibility problem for a set of p polygons having n edges and k intersections in two-dimensional space compared with the orthogonal case is

a) the time and space required to compute the initial placement order $\cdot <$
b) the time and space required to maintain *rank*, *current*, *endpoints*(P) for all polygons P.

It is clear that this has no effect on the asymptotic worst case space requirement and overall runtime of the whole algorithm.

Let us summarize the above discussion on the application of non orthogonal skeleton structures to the hidden line elimination problem in the following.

Theorem 4: *The visibility problem for a set of p arbitrary planar polygons in three-dimensional space consisting of a total number of n line segments with k intersections in the projection plane can be solved in time $O(n \log^2 n + k \log n)$ and space $O(n \log n)$ using an appropriate version of segment trees, or in time $O((n+k)\log^2 n)$ and space $O(n)$ using a version of interval trees. In case the polygons are convex, the factors $\log n$ in the above time bounds can be replaced by $\log p$, and the segment tree space bound reduces to $O(n + p \log p)$.*

By using a new fully dynamic structure of [SL] it has recently been shown in [Nu] how to solve the visibility problem in time $O((n+k)\log n)$ and space $O((n+k)\log n)$ also.

8. CONCLUSION

In this paper we have discussed visibility problems as simplified versions of classical algorithmic problems in computer graphics. We have demonstrated how an algorithm that solves the iso-oriented case of a problem can be transferred to an algorithm for solving the non iso-oriented case as well. Our main contribution consists in adapting skeleton structures to sets of non iso-oriented objects appropriately. Skeleton structures have proved themselves as versatile and efficient data structures to manipulate sets of iso-oriented rectilinear objects; they appear in many algorithms. Using the technique described in this paper their advantages can be carried over to sets of non iso-oriented objects as well.

We presume to assert a *meta-theorem* which states the following:

A line sweep algorithm for solving a problem which involves iso-oriented rectilinear objects and which is based on variants of segment trees or interval trees, can be translated into an algorithm for solving the corresponding problem involving non iso-oriented objects such that both algorithms have essentially the same time and space requirements.

Because of its informal nature it is not possible to give a formal proof for the above meta-theorem. Instead we just furnished a "proof" by a typical and nontrivial example. Further examples can be provided easily. We just mention the report intersecting pairs problem as another, even simpler example: The iso-oriented version of the problem, which is the report intersecting pairs of rectangles problem, and the non iso-oriented version, which is the report intersecting pairs of polygons problem, can obviously be solved along the lines described in this paper.

There are two extreme possibilities for storing geometric data when performing a line sweep algorithm. On the one hand a fully dynamic structure like a balanced tree can be designed for storing exactly the currently active objects. This is an attempt to structure the data appropriately. On the other hand the whole universe from which the data are taken can be structured without taking into account which objects of the universe will occur in the algorithm. Nievergelt [N] provides us with many arguments why it might be better to follow the second approach. Skeleton structures fall in between these two extremes. It is still too early to express a considered opinion about whether or not they are a really good choice. One has to implement the algorithms and run experiments with real data in order to obtain the right impression. However, our first steps in this direction are quite promising.

Acknowledgement The authors wish to acknowledge the comments and criticism from R.H. Güting, O. Nurmi, F.P. Preparata, H.W. Six, and E. Soisalon-Soininen. We furthermore owe special thanks to M. Menzel and M. Schrapp for their TEXnical assistance in typesetting the paper.

REFERENCES

[B] Brown, K.Q., Comments on 'Algorithms for Reporting and Counting Geometric Intersections', *IEEE Transactions on Computers C-30*, (1981), 147–148

[BO] Bentley, J.L. and Ottmann, Th., Algorithms for Reporting and Counting Geometric Intersections, *IEEE Transactions on Computers C-28*, (1979), 643–647

[BW] Bentley, J.L. and Wood, D., An Optimal Worst-Case Algorithm for Reporting Intersections of Rectangles, *IEEE Transactions on Computers C-29*, (1980), 571–577

[E] Edelsbrunner, H., Dynamic Data Structures for Orthogonal Intersection Queries, *Technical University of Graz, Graz, Austria, Institute für Informationsverarbeitung, Report F59*, (1980)

[FV] Foley, J.D. and Van Dam, A., *Addison-Wesley Publishing Co., Reading, Mass.*, (1982)

[Gü] Güting, R.H., An Optimal Contour Algorithm for Iso-Oriented Rectangles, *Journal of Algorithms*, (1984), to appear

[GY] Guibas, L.J. and Yao, F.F., On Translating a Set of Rectangles, *Proceedings of the 12th Annual ACM Symposium on Theory of Computing*, (1980), 154–160

[LePr] Lee, D.T. and Preparata, F.P., Location of a Point in a Planar Subdivision and its Applications, *SIAM J. Comput. Vol.6, No.3*, (1977), 594–606

[LiPr] Lipski, W. and Preparata, F.P., Finding the Contour of a Union of Iso-Oriented Rectangles, *Journal of Algorithms, 1*, (1980), 235–246

[McC] McCreight, E.M., Efficient Algorithms for Enumerating Intersecting Intervals and Rectangles, *Report CSL-80-9, Xerox PARC*, (1980)

[N] Nievergelt, J., Trees as Data and File Structures, *Proceedings CAAP'81, LNCS 112*, (1981), 35–45

[NP] Nievergelt, J. and Preparata, F.P., Plane–Sweep Algorithms for Intersecting Geometric Figures, *Communications of the ACM 25*, (1982), 739–747

[Nu] Nurmi, O., A Fast Line-Sweep Algorithm for Hidden Line Elimination, *Universität Karlsruhe, Institut für Angewandte Informatik und Formale Beschreibungsverfahren, Report 134*, (1984)

[Ov] Overmans, M.H., The Design of Dynamic Data Structures, *Lecture Notes in Computer Science, 156, Berlin*, (1983)

[OW1] Ottmann, Th. and Widmayer, P., On Translating a Set of Line Segments, *Computer Vision, Graphics, and Image Processing 24*, (1983) 382–389

[OW2] Ottmann, Th. and Widmayer, P., On the Placement of Line Segments into a Skeleton Structure, *Universität Karlsruhe, Institut für Angewandte Informatik und Formale Beschreibungsverfahren, Report 114*, (1982)

[Pr] Preparata, F.P., A Note on Locating a Set of Points in a Planar Subdivision, *SIAM J., Comput., Vol.8, No.4*, (1979), 542–545

[SL] Swart, G. and Ladner, R., Efficient Algorithms for Reporting Intersections, *University of Washington, Seattle, Computer Science Department, Technical Report No. 83-07-03*, (1983)

[W] Wood, D., The Contour Problem for Rectilinear Polygons, *Report CS-83-20, University of Waterloo*, (1983)

ANOTHER LOOK AT PARAMETERIZATION USING ALGEBRAS WITH SUBSORTS

Axel Poigné
IMPERIAL COLLEGE *
London SW7 2AZ

0. INTRODUCTION

By now the theory of abstract data types is well developped. Especially modularization techniques have been a topic of interest (for example in /EKTWW 80/, /EKMP 82/, /EK 82/, /SW 82/, /OR 82/). The work ranges from a more syntactical approach of Ehrig & al /EKTWW 80/, /EKMP 82/ to the purely semantical argumentation of Lipeck /Li 82/. The work of Lipeck provides a deep insight into the mechanisms of modularization, especially the limitations are characterized. Nevertheless there seem to be at least two aspects which are not thoroughly discussed:

- It is a standard assumption that a 'construction step' is persistent. This seems to be too restrictive to deal for instance with error specifications. To our opinion one should assume conservativeness, i.e. preservation of distinctness of data.
- For sake of pragmatics reasonable syntactical criteria should be developped which guarantee correct application of modularization.

With the latter topic we have been concerned in /PV 83/. There we discussed the implementation of abstract data types by a (abstract functional) programming language, and we have been able to establish correctness of composition of implementations under certain even syntactical criteria. Trying to transfer the techniques to the more abstract framework of data types, we observed that the results are mainly due to a restricted interaction of the systems of equations envolved. We looked for a suitable modeling of this phenomenon in the language of algebraic specification, and at last learned about the algebraic specifications with partially ordered sorts of M.Gogolla /Go 83/. His approach, being originally designed to handle error specifications, seemed to be quite useful for our purposes as it allows to restrict the scope of equations to subclasses of terms. At this point we have started to reconsider the modularization techniques, the outcome being the notion of parameterization we present. The mathematics is a trifle more envolved than in the standard case, but we hope at least to model the situation we have in mind. Due to lack of space we restrict our attention to parameterization but give some hints how implementation is modeled. Moreover all proofs have been omitted which can be found in /Po 84/ where a full account of our approach is given. We have to assume familiarity with the notion of parameterization given in /EKTWW 80/ and the theory of abstract data types in general.

---

*The work was started while staying at the University of Dortmund.

]. MOTIVATION

A lot of motivation stems from handling of errors. For a stack over some parameter dat there is the well known question of the meaning of the top of the empty stack. A no so bad answer is to state that this should be interpreted as an error. Disregarding al the problems caused /ADJ 78/ we only like to point out that the parameter part of the stack specification must contain an error element, if persistency of parameterization is to be achieved, and that an error-element must be available in any suitable actual parameter.

We regard this situation as unsatisfactory as any specification must anticipate environments in which it may be used. This contradicts the idea of modularization where each specification should depend on its own. In our example the error element on the data component is caused by the stack mechanism and is not specific for the actual parameter specification. Hence the error should be added by the parameterization a posteriori and should not be assumed to exist in the actual parameter a priori. As a consequence operators of the actual parameter should not be applied to the error element because otherwise the effect of the application must be determined by the actual parameter.

Another, may be less natural, example occurs in case of higher type specification /Po 84a/. Higher type specifications include (non-extensional) functions as language constructs ('function spaces' are allowed as sorts). Then for instance a parameterization may add fixpoints for all endofunctions of a given specification, or, intuitively, the parameter is enriched by recursive definitions. To avoid technical overhead we will explain the problems occuring in this setting on the level of standard specifications. Assume that we add a fixpoint for just one function

```
spec  FIXPOINT is
sorts  $s
ops    fix: → s
       $f: s → s
eqns   f(fix) = fix
```

('$' indicates the parameters, notation is due to H.Ganzinger). The specification is conservative but not persistent as the element 'fix' is added. Updating by (in the sense of /EKTWW 80/) a specification BOOL for booleans with a certain set of equations, replacing 'f' by the operator 'not', we can conclude that

true = fix or not(fix) = fix or fix = fix = fix and fix = fix and not fix = false.

Thus the result specification identifies data (regarding BOOL as parameter), and conservativeness is not preserved by updating in general. We believe that parameter passing here does not reflect our intuition. The equations of BOOL are designed to hold for the actual parameter algebra but not for data added by updating. The observation is the same as in the preceeding example: Neither the actual parameter nor the parameterized specification can anticipate in which environment it will be used. Thus it should be possible to seperate between the equations of the actual parameter and those of the parameterized specification. This can be accomplished using

2. ALGEBRAS WITH SUBSORTS

Theories and algebras with subsorts are introduced in /Gog 78/ but the first neat formalization seems to be given in /Go 84/. The approach is based on two concepts, namely to use partial ordering on sorts to express subset relations on carriers of an algebra, and to overload operators in the sense that an operator name may be used with different sortings if the number of arguments is the same.

Example:

```
spec NUMBER is
sorts  nat < int
ops    0: → nat
       suc, _∸1:nat → nat, suc,pred:int → int
       _+_: nat nat → nat, _+_: int int → int
var    x,y:int, z:nat
eqns   pred(suc(x)) = x,  suc(pred(x)) = x
       pred(x) + y = pred(x + y),   suc(x) + y = suc(x + y)
       0 + y = y
       0 ∸ 1 = 0,  suc(z) ∸ 1 = z
```

While the notation is quite suggestive one has to be careful that the effect of operations only depend on the name and that the scope of equations is restricted to certain subclasses (to nat in case of suc(z) ∸ 1 = z).

The development of the theory of algebras with subsorts is similar to that for standard algebras. We slightly modify the approach of /Go 84/ to which we strongly refer for motivation.

2.1 *Definition*: (i) A *signature* (*with subsorts*) SIG consists of a partially ordered set $S = (S,<)$ of sorts, and a $S^* \times S$-sorted set of operators Σ such that

$\sigma: w \to s,\ \sigma: w' \to s' \varepsilon \Sigma \ \Rightarrow\ |w| = |w'|$.

We say that s is a *subsort* of s' if $s < s'$. The set of *operator names* is $\tilde{\Sigma} := \bigcup\{\Sigma_{w,s} \mid w \varepsilon S^*,\ s \varepsilon S\}$.

A SIG-*algebra* A consists of a S-sorted set A of elements, and a mapping $\sigma_A^{w,s}: A_w \to A_s$ for each $\sigma: w \to s \varepsilon \Sigma$ such that

(a) $s < s' \ \Rightarrow\ A_s \subseteq A_{s'}$

(b) $\sigma: w \to s,\ \sigma: w' \to s' \varepsilon \Sigma,\ \bar{a}:w,\ \bar{a}:w' \varepsilon A \ \Rightarrow\ \sigma_A^{w,s}(\bar{a}) = \sigma_A^{w',s'}(\bar{a})$.

A SIG-*homomorphism* $f:A \to B$ is a S-sorted mapping such that

(a) $h_s(\sigma_A^{w,s}(\bar{a})) = \sigma_B^{w,s}(h_w(\bar{a}))$ for $\sigma: w \to s \varepsilon \Sigma$ and $\bar{a}:w \varepsilon A$

(b) $a:s,\ a:s' \in A \Rightarrow h_s(a) = h_{s'}(a)$

The category of SIG-algebras and SIG-homomorphisms is denoted by SIG^b. Forgetting about the operators we shall refer to S-*ordered sets*, the respective category being denoted by S-set.

(For convience we use a notation similar to programming languages $x:s \in X$ stands for $x \in X_s$ and $\bar{x}:w \in X$ for $\bar{x} \in X_w$, relations are canonically extended to tuples)

Given a S-ordered set X construction of the *term algebra* is inductively defined by (a) $X_s \subseteq T_{SIG}(X)_s$ and (b) $\sigma:w \to s \in \Sigma,\ s < s',\ \bar{t}:w \in T_{SIG}(X) \Rightarrow \sigma(\bar{t}):s' \in T_{SIG}(X)$ with the standard definition of operations. As notation we use $T_{SIG}(X)$.

2.2 *Proposition*: $T_{SIG}(X)$ is a free SIG-algebra over a S-ordered set X (S being the sort component of SIG).

2.3 *Definition*: (i) A *presentation (with subsorts)* PRES consists of a signature SIG and a set of equations of the form '$\lambda\bar{x}:w.\ l =_s r$' with $l,r:s \in T_{SIG}(\{\bar{x}:w\})$ for some $s \in S$. $\bar{x}:w$ is used to denote the sequence '$x_0:s_0,\dots,x_{n-1}:s_{n-1}$', and $\{\bar{x}:w\}$ denotes the S-ordered set generated by this sequence.

(ii) A SIG-algebra *satisfies* an equation $\lambda\bar{x}:w.l =_s r$ if $I^{\#}_s(l) = I^{\#}_s(r)$ for all S-set-morphisms $I:\{\bar{x}:w\} \to A$ where $I^{\#}: T_{SIG}(\{\bar{x}:w\}) \to A$ is the unique homomorphic extension.

(iii) A PRES-*algebra* is a SIG-algbra which satisfies all the equations of PRES. With SIG-homomorphisms this defines a category $PRES^b$.

Remark: As we deal with heterogeneous algebras we have to be careful about the sorting of equations (compare /GoMe 81/). For practical purposes we proceed as in the example specification above and assume that an equation is abstracted by the variables occuring in the terms on both sides of an equation. It should be observed that the conditions ensure that application of equations is restricted by sorting of the variables.

As we are interested in modularization techniques we extend /Go 84/ and deal with several presentations and presentation morphisms.

2.4 *Definition*: A *morphism* h: PRES → PRES' *of presentations* consists of a monotone sort mapping $h: S \to S'$, i.e. $h(s) < h(s')$ if $s < s' \in S$, and a $S^* \times S$-sorted mapping $h: \Sigma \to \Sigma'$ such that (a) $h_{w,s}: \Sigma_{w,s} \to \Sigma'_{h(w),h(s)}$ and (b) $\sigma:w \to s,\ \sigma:w' \to s' \in \Sigma \Rightarrow h_{w,s}(\sigma) = h_{w',s'}(\sigma)$, and such that $\lambda\bar{x}:h(w).h(l) =_{h(s)} h(r) \in E'$ if $\lambda\bar{x}:w.l =_s r \in E$.

E,E' denote the respective sets of equations, h(l),h(r) are the terms canonically renamed by h.

Presentation morphisms induce forgetful functors which have a left adjoint.

In our case we are only interested in free constructions for presentation embeddings PRES ⊆ PRES'. In this case a free PRES'-algebra $T_{PRES'}(A)$ over a PRES-algebra A can be constructed as follows:

Let slightly ambiguously A denote as well the S'-ordered set $A_{s'} := \{A_s | s \in S, s < s'\}$.

On $\tilde{T}_{SIG'}(A) := \bigcup\{T_{SIG'}(A) | s \in S'\}$ we define a relation $t \overset{\cdot}{\sim} t'$ by

(i) $t = \sigma(\bar{a})$ and $t' = \sigma_A^{w,s}(\bar{a})$ with $\sigma: w \to s \in \Sigma$, $\bar{a}: w \in A$, or

(ii) $t = I_s^{\#}(l)$, $t' = I_s^{\#}(r)$ with $\lambda\bar{x}:w.l =_s r \in E'$ and $I: \{\bar{x}:w\} \to T_{SIG'}(A)$, or

(iii) $t = \sigma(\bar{t})$, $t' = \sigma(\bar{t}')$ with $\sigma: w \to s$, $\sigma: w' \to s' \in \Sigma'$, $\bar{t}:w$, $\bar{t}':w' \in T_{SIG'}(A)$ and $\bar{t} \overset{\cdot}{\sim} \bar{t}'$

or (iv) $t = t'$ or $t' \overset{\cdot}{\sim} t$.

Let $t \sim t'$ be the transitive closure of $t \overset{\cdot}{\sim} t'$.

2.5 *Proposition*: $T_{PRES'}(A) := T_{SIG'}(A)/_{\sim}$ with the canonical embedding $\eta: A \to (T_{PRES'}(A))_e$ $a \to [a]$ is a free PRES'-algebra over a PRES-algebra A where

- $[t] := \{t' \in T_{PRES'}(A) | t \sim t'\}$ and $\sigma^{w,s}_{T_{PRES'}(A)}([\bar{t}]) = [\sigma^{w,s}_{T_{SIG'}(A)}(\bar{t})]$, and
- A'_e is the restriction of a PRES'-algebra A to its PRES-part.

To conclude this section we informally state that *theories* (*with subsorts*) are defined in analogy to /GoMe 81/ but additionally terms may be compared which are not of the same sort. This is why we have to introduce the additional rules of

sort abstraction $\lambda\bar{x}:w.t =_s t' \vdash \lambda\bar{x}:w.\ t = t'$ and

sort concretion $\lambda\bar{x}:w.t = t' \vdash \lambda\bar{x}:w.\ t =_s t'$ if $t:s, t':s \in T_{SIG}(\{\bar{x}:w\})$ for $s \in S$.

2.6 *Proposition*: Theories with subsorts are sound and complete.

The results indicate that algebras with subsorts have the standard properties. While substitution is strictly sorted the difference is that congruences compare terms without caring for the sorting.

3. SPECIFICATION OF PARAMETERIZED DATA TYPES

With regard to abstract data types the notion of presentations is too general. We cannot discuss the technical problems here but will try to motivate the following

3.1 *Definition*: A presentation $((S,<),\Sigma,E)$ is said to be a *specification* (*with subsorts* if for all operator names σ there exists an operator $\sigma: w \to s \in \Sigma$ which is *maximal*, i.e. $\sigma: w' \to s' \in \Sigma$ implies $w' < w$, $s' < s$. We call w the *arity* and s the *coarity* of σ and use $a(\sigma)$, $c(\sigma)$ as notation. Thus $\sigma: a(\sigma) \to c(\sigma)$ denotes the maximal operator with name σ. We use SPEC as keyword for specifications.

A *specification morphism* h:SPEC $\to$ SPEC' is a morphism of presentations such that $h(a(\sigma)) = a(h(\sigma))$ and $h(c(\sigma)) = c(h(\sigma))$.

The definition reflects the phenomenon that for any operator name σ there implicitly exists a *maximally defined* operation, the *meaning* of σ,

$$\sigma_A: \bigcup_{\sigma:w \to s \in \Sigma} A_w \to \bigcup_{\sigma:w \to s \in \Sigma} A_s, \quad \bar{a} \to \sigma_A^{w,s}(\bar{a}) \text{ if } \bar{a}:w \in A$$

on the semantical level. For an arbitrary presentation σ_A may not have a syntactical counterpart, i.e. an operator $\sigma{:}w \to s \in \Sigma$ such that $\sigma_A^{w,s} = \sigma_A$. This is guaranteed for a specification. Intuitively then all operators with the same name are a restriction of the maximal one.

In chapter 1 we argued that parameterized specifications should
- restrict operators of the actual parameter to data of the actual parameter
- allow to add data to the actual parameter
- preserve distinctness of parameter data.

This can be achieved by the following conditions

3.2 *Definition*: (i) A *parameterized specification* SPEC = (S,Σ,E) (*with subsorts*) is a specification with a *formal parameter* PSPEC = $(PS,P\Sigma,PE)$ such that the following syntactical and semantical requirements hold:
- PSPEC is a specification and PSPEC $\subseteq$ SPEC as presentations
- *parameter completeness*, i.e.
  - $s \in S,\ s' \in PS,\ s < s'$ in $S \Rightarrow s \in PS$
  - $\sigma{:}w \to s \in \Sigma,\ s \in \tilde{PS} \Rightarrow \sigma{:}w \to s \in P\Sigma$
  - $\forall \sigma{:}w' \to s' \in \Sigma{:}\ \sigma \in P\Sigma \Rightarrow \exists\, \sigma{:}w \to s \in P\Sigma{:}\ w < w',\ s < s'$ and $\forall \hat{w} \in PS^*{:}\ \hat{w} < w' \Rightarrow \hat{w} < w$
- *parameter protection*, i.e. PSPEC $\subseteq$ SPEC is conservative

where a presentation embedding PRES $\subseteq$ PRES' is *conservative* if all the units of the corresponding adjunctions (2.5) are injective. If they are isomorphisms (identities) the embedding is called (*strongly*) *persistent*.

(ii) Given a parameterized specification SPEC with a formal parameter PSPEC and a parameterized specification ASPEC a *parameter passing* (*morphism*) is a specification morphism h:PSPEC $\to$ ASPEC. ASPEC is called the *actual parameter*.

Parameter completeness induce a seperation between parameter and body. The last condition appears to be rather technical but it guarantees that application of parameter operators (in the more general sense that the name occurs in the parameter) to parameter data is determined by the parameter algebra.

A - at a first sight counter intuitive - consequence is that parameterized specifications are persistent. For analysis let us consider the well known stack as example

```
spec PRIMITIVE STACK is
sorts  $pdata < data, stack
ops    error: → stack, push: stack data → stack,
       pop: stack → stack, top: stack → data, empty: → stack
var    d:pdata, s:stack
eqns   pop(push(s,d)) = s,  top(push(s,d)) = d
       pop(empty) = empty,  top(empty) = error
       push(s,error) = s
```

Conceptually data is a parameter sort as it contains parameter data. In this sense any sort $s \in S$ with a subsort $s' \in PS$ can be understood as a parameter sort (without occuring in the formal parameter). With this view parameterized specifications are only conservative as data may be added on 'conceptual' parameter sorts.

4. PARAMETER PASSING

In order to model parameter passing we proceed in analogy to /EKTWW 80/ and use pushouts in the category of presentations. According to the nature of presentations with subsorts we have to take care that all operators with the same name are renamed uniformly. To avoid unnecessary complications we only discuss pushouts of specific structure.

Throughout this chapter we consider a fixed parameterized specification SPEC with formal parameter PSPEC and an actual parameter specification ASPEC such that

$PSPEC = (PS,P\Sigma,PE)$ $\quad$ $SPEC = (S,\Sigma,E)$ $\quad$ $ASPEC = (AS,A\Sigma,AE)$

where $S = (PS+BS,<)$, $\Sigma = P\Sigma+B\Sigma$ and $E = PE+BE$. 'B' stands for body. Without restriction of generality we assume that $BS \cap AS = \emptyset$ and $B\Sigma \cap A\Sigma = \emptyset$. $h:PSPEC \to ASPEC$ is a specification morphism.

Let $RSPEC = (RS,R\Sigma,RE)$ be defined by $RS = (AS+BS,<)$, $R\Sigma = A\Sigma+h'(B\Sigma)$, $RE = AE+h'(BE)$ where h': $SPEC \to RSPEC$ is given by

$$h'(s) = \text{if } s \in PS \text{ then } h(s) \text{ else } s$$

$$h'_{w,s}(\sigma) = \text{if } \sigma \in \tilde{P\Sigma} \text{ then } h(\sigma) \text{ else } \sigma$$

and the ordering on RS is given as the transitive closure of

$$s_o < s_1 \quad \text{iff} \quad \begin{array}{ll} \text{(i)} & s_o < s_1 \text{ in } AS, \text{ or} \\ \text{(ii)} & s_o = h'(s'_o) \text{ and } s_1 = h'(s'_1) \text{ and } s'_o < s'_1 \text{ in } S. \end{array}$$

4.1 *Proposition*:

$$\begin{array}{ccc} PSPEC & \subseteq & SPEC \\ h\downarrow & & \downarrow h' \\ ASPEC & \subseteq & RSPEC \end{array}$$

is a pushout of presentations.

4.2 *Definition*: The *syntax* of parameter passing is given by the diagram

$$\begin{array}{ccccc} & & PSPEC & \overset{s}{\subseteq} & SPEC \\ & & h\downarrow & & \downarrow h' \\ PASPEC & \overset{s_A}{\subseteq} & ASPEC & \overset{s'}{\subseteq} & RSPEC \end{array}$$

the square being a pushout of presentations. The *result specification* is RSPEC with formal parameter PASPEC.

The *semantics* of parameter passing is given by the free functor

$$F_{s'} \circ F_{s_A} : PASPEC^b \to RSPEC^b.$$

Parameter passing is called *correct* if

- RSPEC is a parameterized specification
- *passing compatibility* holds, i.e. $F_s \circ V_h \circ F_{s_A} \simeq V_{h'} \circ F_{s'} \circ F_{s_A}$.

In general parameter passing is not correct due to identification of operators.

Example:

spec ACT is
sorts s
ops a,b: → s
f:s → s

spec PARA is
sorts \$s < s'
ops \$a,\$b: → s, c: → s'
\$f,\$g:s → s, f,g:s' → s'
eqns a = f(c), b = g(c)

Identification of f,g:s → s extends to f,g:s' → s'. Thus $a \neq b$ in the initial ACT-algebra but a = b in the initial algebra of the result specification.

We have to restrict parameter passing as well semantically as syntactically.

4.3 *Definition*: Parameter passing is *safe* if

(a) $h'(\sigma) = h'(\sigma') \Rightarrow h'(a(\sigma)) = h'(a(\sigma'))$ and $h'(c(\sigma)) = h'(c(\sigma'))$

(b) $SPEC \vdash \lambda\bar{x}:w.t = t'$ for all $w \in PS^*$ and all terms $t:s, t':s' \in T_\Sigma(\{\bar{x}:w\})$ such that $h'_s(t) = h'_{s'}(t')$

where - $T_\Sigma(X) \subseteq T_{SIG}(X)$ such that $t \in T_\Sigma(X)$ iff for any subterm $\sigma(\bar{t})$ of t with $\sigma \in P\Sigma$ there occurs a $\sigma' \in B\Sigma \setminus P\Sigma$ in $\bar{t}$,

- h' is the canonical renaming of terms.

A parameterized specification is *safe* if it is safe for all parameter passings.

The safeness condition states that only operators can be identified which have the same arity and coarity modulo h'. The second, semantical condition guarantees that parameter operators can only be identified if they behave in the same way on parameter data. The condition corresponds to the parameter completeness condition.

With all this preparation we are able to prove our

4.4 *Main Theorem*: Parameter passing is correct if parameter passing is safe.

The proof is rather elaborated. The conditions are sufficient, we do not know if they are necessary.

5. DISCUSSION AND FURTHER RESULTS

At a first glance the result may be disappointing because of the restrictive conditions. But with some minor encodings one can prove that the correctness result of /EKTWW 80/ is a special case of our result. The conditions are syntactical and easy to check except for the semantical condition for safeness which seems to be rather awkward. A closer look shows that this condition has to be ensured as well for the proof of persistency in the case of standard parameterized data types, and that one can design rather simple algorithms which check sufficient and reasonable conditions for safeness. This is done in /Po 84/ where as well hints for a pragmatics of parameterized specifications with subsorts are given.

In /Po 84/ we moreover discuss implementations of abstract data types. Analysis of the approach of /EKMP 82/ yields that the failure to establish correctness results in the general case is due to an unintended interaction of equations similar to the case of parameterizations. We demonstrate that one can exploit the theory of algebras with subsorts. Basically the idea is to guard the implementing data type and the data type to be implemented against the implementation specification in the same way as we did for formal parameters, the implementation specification then being a conservative extension of both data types. Composition of implementations can be defined by a pushout, and we can use our main theorem to establish correctness of composition (which is different to that of /EKMP 82/ but coincides in certain cases. Correctness is RI-correctness of /EKMP 82/). As the arguments depend on the same basic result one almost immediately gets compatibility of parameterizations and implementations.

REFERENCES

/ADJ 78/ ADJ: An Initial Algebra Approach to the Specification, Correctness and Implementation of Abstract Data Types, In: R.Yeh,Ed., Current Trends in Programming Methodology IV, Prentice-Hall 1978

/EK 82/ Ehrig,H.,Kreowski,H.-J.: Parameter Passing Commutes with Implementation of Parameterized Data Types. Proc. 9th ICALP, LNCS 140, 1982

/EKMP 82/ Ehrig,H.,Kreowski,H.-J.,Mahr,B.,Padawitz,P.: Algebraic Implementation of Abstract Data Types, TCS 20, 1982

/EKTWW 80/ Ehrig,H.,Kreowski,H.-J.,Thatcher,J.W.,Wagner,E.G.,Wright,J.B.: Parameter Passing in Algebraic Specification Languages,Techn. Rep. TU Berlin 1980, also Proc. Workshop on Progr. Spec., LNCS 134, 1982

/Gog 78/ Goguen,J.A.: Order sorted Algebras: Exception and Error Sorts, Coercion and Overloaded Operators, Comp. Dept. Rep. 14, UCLA 1978

/Go 84/ Gogolla,M.: Algebraic Specifications with Partially Ordered Sorts and Declarations, CAAP '84, LNCS 1 , 1984

/GoMe 81/ Goguen,J.A.,Meseguer,J.: Completeness of Many-Sorted Equational Logic, ACM SIGPLAN Notices 16,7, 1981

/Li 82/ Lipeck,U.: Ein algebraischer Kalkül für einen strukturierten Entwurf von Datenabstraktionen, Promotion Dortmund, 1982

/Or 82/ Orejas,F.: Characterizing Composability of Abstract Implementations, Rep. RR 82/08, Univ. Barcelona 1982

/Po 84a/ Poigné,A.: Higher Order Specifications, STACS'84, LNCS 1 , 1984

/Po 84/ Poigné,A.: Modularization Techniques for Algebraic Specifications Using Subsorts, To appear as Techn. Rep.

/PV 83/ Poigné,A.,Voss,J.: Programs over Abstract Data Types - On Implementation of Abstract Data Types. Techn. Rep. 171, Abt. Informatik, Univers. Dortmund 1983

/SW 82/ Sannella,D.,Wirsing,M.: Implementation of Parameterized Specifications. Proc. 9th ICALP, LNCS 140, 1982

A LOWER BOUND ON COMPLEXITY OF BRANCHING PROGRAMS

(Extended abstract)

P. Pudlák
Mathematical Institute
Czechoslovak Academy of Sciences
115 67 Praha 1, Czechoslovakia

1. Introduction

A branching program is a labelled directed acyclic graph with the following properties.

(i) There is exactly one source.

(ii) Every vertex has outdegree 2, one of the leaving edges is labelled by 0 and the other by 1.

(iii) The sinks are labelled by YES or NO and the other vertices are labelled by some numbers 1,...,n, where n is the length of inputs of the branching program.

The branching program computes a Boolean function $f : \{0,1\}^n \rightarrow \{0,1\}$ as follows. Given $\underline{a} = (a_1,\dots,a_n) \in \{0,1\}^n$,

(i) it starts at the source,

(ii) whenever it reaches a vertex labelled by i, $1 \leq i \leq n$, it continues along the edge labelled by a_i,

(iii) $f(\underline{a}) = 1$ (resp. = 0) iff the branching program accepts (resp. rejects) $\underline{a}$ iff it reaches a sink labelled by YES (resp. NO).

The complexity of a branching program is the number of vertices of the program. As usual this defines also a complexity measure for Boolean functions (namely, the complexity of f is the complexity of an optimal branching program that computes f). It is well-known that this measure is between cirquit size complexity and formula size complexity, in base $\{\neg, \wedge, \vee\}$. Actually, a branching program is, essentially, a special kind of a contact scheme, which has been investigated for quite a long time. Therefore also the bound $\Omega(n^2 \log^{-2} n)$ for an explicitly defined Boolean function, see [8], applies to branching programs.

The concept of the branching program that we are studying here originated as a description of the computations of automatons and Turing machines. In this description a vertex of the branching program is a pair $\langle$state, time$\rangle$ and the branching corresponds to divergence of computations caused by the symbol currently read by the automaton. It follows that the logarithm of branching program complexity (also called capacity) is a lower bound to the space complexity of Turing machines which use at least log n space, see e.g. [10]. Thus the research into branching programs is strongly motivated by LOG = NLOG? problem.

The width of a branching program is another important complexity parameter. Since we investigate subquadratic complexity it is important to have the definition of the width as general as possible. Therefore our definition is applicable to arbitrary branching programs (in contrast to the definition of [2],[3]). We call a sequence $L_0, L_1, \dots, L_t$ of mutually disjoint subsets of vertices levels if for every edge e there exist $0 \le i < j \le t$ such that e goes from L_i to L_j. The width of a level L_i is its cardinality plus the number of edges going from $\bigcup_{j<i} L_j$ to $\bigcup_{i<j} L_j$; the width with respect to $L_0, \dots, L_t$ is the maximum of the widths of the levels; the width of the branching program is the minimum of the widths over all possible sequences of levels.

For $\underline{a} \in \{0,1\}^n$, let $|\underline{a}|$ denote the number of ones among $a_1, \dots, a_n$. A Boolean function f is called symmetric if it is invariant under permutations of variables, i.e. if $f(\underline{a})$ depends only on $|\underline{a}|$. The most interesting applications of our lower bound theorem will be in symmetric functions. We shall show that most symmetric functions have nonlinear complexity. It should be pointed out that Nečiporuk's technique [8] does not give nonlinear lower bounds to symmetric functions.

For some particular functions, namely $E^n_{\lceil n/2 \rceil}$, nonlinear bounds have been found in [2] , [3] under the assumption of bounded width. The functions E^n_k, T^n_k, $C^n_{i,k}$, $C^n_k : \{0,1\}^n \to \{0,1\}$ are defined as follows. For $\underline{a} \in \{0,1\}^n$

$E^n_k(\underline{a}) = 1$ iff $|\underline{a}| = k$;

$T^n_k(\underline{a}) = 1$ iff $|\underline{a}| \ge k$;

$C^n_{i,k}(\underline{a}) = 1$ iff $|\underline{a}| \equiv i \bmod k$;

$C^n_k = C^n_{o,k}$.

We shall show that

(1) $T^n_{\lceil n/2\rceil}$, $E^n_{\lceil n/2\rceil}$ have nonlinear complexity <u>without</u> any restriction of width;

(2) for fixed k, T^n_m, E^n_{m-1} have width k linear complexity iff $m < k$;

(3) a similar result for $C^n_{i,k}$ and particular values of k.

(Notice that e.g. $C^n_6 = C^n_3 \wedge C^n_2$, hence has width 3 linear complexity). The actual nonlinear bounds for width ≥ 3 will be asymptotically better than the previously known ones [3].

2. Theorems

Theorem 1

$\forall k > 1\ \exists \varepsilon$ such that every Boolean combination of functions T^n_{m-1}, $C^n_{i,m}$, $1 < m \leq k$ can be computed by a branching program of width k and complexity $\leq \varepsilon \cdot n$.

The proof of this theorem is not difficult. We include Theorem 1 here only as a motivation for our lower bound theorem which shows that computations of the functions T^n_k, $C^n_{i,k}$ are inherently contained in any low complexity branching program.

We use the following notation. For a Boolean function $f(x_1,\dots,x_n)$ and a subset of variables $Y \subseteq \{x_1,\dots,x_n\}$, $f|Y$ denotes the function obtained from f by substituting zeros for every variable not contained in Y. We denote by

$$\log^{(2)} n = \log\log n, \quad \log^{(3)} n = \log\log\log n,$$

for every n for which it makes sense. All logarithms in this paper are to the base 2.

Theorem 2

There exists $\varepsilon > 0$ such that for every Boolean function $f(x_1,\dots,x_n)$, $n > 4$, which can be computed by a branching program of width k and complexity

$$N \leq \varepsilon \cdot n \frac{\log^{(2)} n - \log r}{\log^{(3)} n}$$

there exists a subset of variables $\{y_1,\dots,y_r\} \subseteq \{x_1,\dots,x_n\}$ of cardinality r such that $f|\{y_1,\dots,y_r\}$ is equivalent to a Boolean combination of functions

$$T^r_{m-1}(y_1,\dots,y_r),\ C^r_{i,m}\ (y_1,\dots,y_r),$$

where

$$1 < m \leq k \quad \text{and} \quad m \leq \max(\tfrac{2N}{n},\ 2).$$

For $n > 4$, let

$$S(n) = \log^{(2)} n \ / \ \log^{(3)} n.$$

Clearly $S(n) \to \infty$ for $n \to \infty$. For $k \geq 1$ let $k?$ denote the smallest positive integer divisible by all i, $1 \leq i \leq k$. Clearly $k?$ divides $k!$. For $\underline{a},\underline{b} \in \{0,1\}^n$ we shall write $\underline{a} \leq \underline{b}$ to denote that $a_i = 1 \Rightarrow b_i = 1$ for $i = 1,\dots,n$.

Corollary 3

$\exists\ \varepsilon > 0\ \forall k > 1\ \forall$ sufficiently large n the following holds. If a branching program P has width $\leq k$ and complexity $\leq \varepsilon.n.S(n)$, where n is the length of inputs, then for every input $\underline{a}, |\underline{a}| \leq n/2$, there exists $\underline{b}$, $\underline{c}$, $\underline{a} \leq \underline{b} \leq \underline{c}, |\underline{b}| = |\underline{a}|+k-1, |\underline{c}| = |\underline{b}| + k?$ such that P accepts $\underline{c}$ iff it accepts $\underline{b}$.

Corollary 4

$\exists\ \varepsilon > 0\ \forall k > 1\ \forall$ sufficiently large n the following holds.

(i) T^n_k, E^n_{k-1} cannot be computed by a branching program of width k and complexity $\varepsilon.n.S(n)$;

(ii) if $k = p^m$ for $m \geq 1$ and a prime number p, then $C^n_{i,k}$ cannot be computed by a branching program of width k-1 and complexity $\leq \varepsilon.n.S(n)$.

Corollary 5

$\exists\ \varepsilon > 0\ \forall k\ \exists K\ \forall$ sufficiently large n there are at most K symmetric Boolean functions of n variables which can be computed by branching programs of width $\leq k$ and complexity $\leq \varepsilon.n.S(n)$.

Remark

Those symmetric Boolean functions which can be computed by branching programs of width k and complexity $\leq \varepsilon.n.S(n)$ are, in fact, Boolean combinations of T^n_{m-1}, T^n_{n-m+2}, $C^n_{i,m}$, $1 < m \leq k$; thus, by Theorem 1, they can be computed even with complexity $c_k.n$, where c_k is a constant depending only on k. This shows that there is a gap in the complexity hierarchy of symmetric functions.

In the remaining two corollaries we consider branching programs with unrestricted width.

Corollary 6

$\exists \varepsilon > 0$ such that for every positive integers k, n, $2 \leq k \leq \varepsilon . S(n)$, the following holds. If a branching program has complexity $\leq 1/4\, k.n$, where n is the length of inputs, then for every input $\underline{a}$, $|\underline{a}| \leq n/2$, there exists $\underline{b}, \underline{c}$, $\underline{a} \leq \underline{b} \leq \underline{c}$, $|\underline{b}| = |\underline{a}| + k-1$, $|\underline{c}| = |\underline{b}| + k?$ such that it accepts $\underline{c}$ iff it accepts $\underline{b}$.

Corollary 7

$\exists \varepsilon > 0 \quad \forall$ n, k if

$$4\varepsilon . S(n) < k < n - 4\varepsilon . S(n),$$

then the branching program complexity of T^n_k is at least $\varepsilon . n . S(n)$.

3. Problems and related results

The largest lower bound to branching program complexity of a symmetric function that we know of is our $\Omega\,(n.\log^{(2)} n/\log^{(3)} n)$; the smallest upper bound is $c.n^2$. This leaves quite a bad gap.

For width 2, Borodin, Dolev, Fich and Paul [2] obtained a bound $\Omega\,(n^2/\log n)$ for $E^n{}_{\lceil n/2 \rceil}$. However, no polynomial width 2 upper bound is known for such a function.

The following strengthening of a theorem of Hodes and Specker was proved in [7].

Theorem

For every base Ω there exists $\varepsilon > 0$ such that for every Boolean function $f(x_1, \dots, x_n)$, if f can be realized by a formula of complexity $\leq \varepsilon . n . (\log^{(2)} n - \log r)$ in base Ω, then there exists a subset $\{y_1, \dots, y_r\} \subseteq \{x_1, \dots, x_n\}$ of cardinality r such that $f \restriction \{y_1, \dots, y_r\}$ is equivalent to a Boolean combination of $T^r_1(y_1, \dots, y_r)$ and $C^r_2(y_1, \dots, y_r)$.

It was also shown in [9] that the bound $\varepsilon . n . \log^{(2)} n$ has the best possible growth rate. Nothing is known about the optimality of the bound of Theorem 2.

Our lower bound theorem is applicable to many Boolean functions, however it gives very small bound. Therefore it would be very desirable to modify the theorem so that it gave larger bounds (if it were not possible to improve the bound in the theorem itself). Such a modification is known for Hodes-Specker theorem; it can be improved to ε .n.log n if we alow substitutions of ones too [4].

Other theorems of Hodes-Specker-type can be derived from proofs and results of Furst, Saxe, Sipser [5] and Ajtai [1]. Namely, the following is a corollary of a result of Ajtai:

<u>Theorem</u>

For every $\varepsilon < 1$, every polynomial p , every positive integer k and every sufficiently large n, if $f:\{0,1\}^n \to \{0,1\}$ can be computed by a cirquit of depth k and size p(n), then there exist $\underline{a},\underline{b} \in \{0,1\}^n$ such that $\underline{a} \leq \underline{b}, |\underline{b}| - |\underline{a}| \geq \varepsilon . n$ and f is constant on $\{ \underline{x} \in \{0,1\}^n \mid \underline{a} \leq \underline{x} \leq \underline{b} \}$.

4. Proofs

Here we only very briefly sketch some ideas of the proofs.

<u>Theorem 1</u> can be easily proved as follows. Let us say that a branching program has width k and property V if levels of width k can be defined so that all the sinks are in the last level. Then the theorem follows from the following claims.

1. Let $P_1,\ldots,P_t$ have width k and property V, let P_{t+1} have width k. Then the disjunction and the conjunction of the functions computed by $P_1,\ldots,P_{t+1}$ can be computed with width k and complexity equal at most to the sum of complexities of P_i, i = 1,...,t+1.

2. Let P have width k and Q width k-1 and both have property V, then similarly their conjunction can be computed with width k <u>and</u> property V.

3. T^n_{m-1}, $C^n_{i,m}$ can be computed with width m, property V and complexity $\leq$ 2.m.n.

4. Every Boolean combination of T^n_{m-1}, $C^n_{i,m}$, $m \leq k$, is equivalent to some combination of the following form:

$$\bigvee_a (T^n_{p_a} \wedge \neg T^n_{q_a}) \vee (T^n_{k-1} \wedge \bigvee_b (C^n_{i_b,k} \wedge \bigwedge_c C^n_{i_{b,c},m_c})),$$

where $p_a < k-1$; q_a, $m_c \leq k-1$.

The proof of Theorem 2 is much more involved. We shall state some lemmas without proofs. First we have to modify the concept of a branching program to include some additional structure which is irrelevant for computation but important for the proof. Namely we allow more sources in the graph, one of them being distinguished as the input source, we allow also double edges of the form $u \overset{0}{\underset{1}{\rightrightarrows}} v$ and add another labelling of nonterminal vertices so that every nonterminal vertex v is determined by the pair of its labels (i,a), (i is the label used in the computation). We shall not distinguish between v and (i,a). Let P be such a branching program for inputs of length n. Let $X \subseteq \{1,\dots,n\}$. Then PX denotes the "induced" branching program i.e.

1. $(i,a) \xrightarrow{\varepsilon} v$ is an edge in PX, $\varepsilon \in \{0,1\}$, if $i \in X$ and $(i,a) \xrightarrow{\varepsilon} v_1 \xrightarrow{0} v_2 \xrightarrow{0} \dots v_k \xrightarrow{0} v$ is the maximal path in P where the first labels of $v_1,\dots,v_k$ do not belong to X;
2. the vertices of PX are all the vertices of the form (i,a), $i \in X$ and the sinks incident with an edge described in 1.

Lemma 1. (i) $Y \subseteq X \Rightarrow PY = (PX)Y$;
(ii) if P computes f then PX computes $f|X$.

For $a,b,c,d \in \{1,\dots,n\}$, $a < b$, $c < d$ we define $P\{a,b\}$ be isomorphic to $P\{c,d\}$ if, after relabelling c by a and d by b, $P\{c,d\}$ turns out to be isomorphic to $P\{a,b\}$ as a labelled graph. We say that P is homogeneous if for each two such pairs the induced branching programs are isomorphic.

The most difficult part of the proof is a description of homogeneous branching programs. Because of the essential structural difference between branching programs and formulas we cannot make use of ideas of [9]. Now, let P be homogeneous and $n \geq 3$. Paths in P of the form

$$(1,a_1)\xrightarrow{0}(1,a_2)\xrightarrow{0}\dots\xrightarrow{0}(1,a_t)\xrightarrow{0}(2,a_1)\xrightarrow{0}(2,a_2)\xrightarrow{0}\dots\quad\dots\xrightarrow{0}(n,a_t)$$

resp.

$$(n,a_1)\xrightarrow{0}(n,a_2)\xrightarrow{0}\dots\xrightarrow{0}(n,a_t)\xrightarrow{0}(n-1,a_1)\xrightarrow{0}(n-1,a_2)\xrightarrow{0}\dots\xrightarrow{0}(1,a_t),$$

will be called + resp. - oriented states. We prove that any vertex of P which is relevant for computation either lies on a state or is a sink or lies on a path

$$(i,a)\xrightarrow{1}(i,b_1)\xrightarrow{1}(i,b_2)\xrightarrow{1}\ldots\xrightarrow{1}(i,b_k)\xrightarrow{1}v$$

connecting some element (i,a) of a state to an element of another state or sink v. We omit the technical description of how the states and sinks are connected and refer the reader to Figure 1 where some typical situations are shown. We shall call connections of states of the type a), b), e) regular. We use an <u>automaton</u> to describe the computation of P. The states of the automaton are the states defined above and some sinks and sets of sinks of P. The behaviour of the automaton can be easily determined from the list of possible connections of the states and sinks of P. Thus if the computation of P reaches a vertex (i,a) of some state, then the automaton is in this state and scans the i-th symbol of the input. If e.g. the state is oriented +, $i<n$ and the scanned symbol is 0, then it remains in the same state and moves to the right on the input. It turns out that no power is gained by allowing unequally oriented states. To link the width restriction with the properties of the computed function we use:

<u>Lemma 2</u>. If $s_1,\ldots,s_k$ are different states connected in a regular way $2k-1\leq n$, then the width of P is at least k.

Then we obtain the main lemma:

<u>Lemma 3</u>. If a homogeneous branching program has width $\leq k$ <u>or</u> has $\leq k$ states then it computes a Boolean combination of functions T^n_{m-1}, $C^n_{i,m}$, $m\leq k$.

The rest of the proof is short and follows the pattern of the proof of the lower bound theorem of [9]. Let $f:\{0,1\}^n\to\{0,1\}$ be the function computed by an arbitrary branching program P which has complexity $\leq 1/2S.n$, for some constant S, and width $\leq k$. We can assume that each of the first labels i, $1\leq i\leq n$, occurs at most S times, (if not, we substitute 0's for $<1/2$ of the variables whose index occurs too often). Hence we can choose S or less than S labels a so that the mapping $v\mapsto(i,a)$ is one-to-one. Then an upper bound to the number of nonisomorphic induced branching programs $P\{i,j\}$ is

(1) $\ell=\varepsilon_o^{S.\log S}$

for some constant ε_o. By a well-known bound to Ramsey theorem, there exists $Y\subseteq\{1,\ldots,n\}$ such that all $P\{i,j\}$'s with $i\neq j$, $i,j\in Y$ are isomorphic and $|Y|\geq r$, for every r which satisfies the inequality

(2) $\ell^{r.\ell} \leq n$.

By Lemma 1, PY is homogeneous and computes $f|Y$. Moreover PY has width $\leq k$ and has at most S states. Thus by Lemma 3 PY computes a Boolean combination of functions T^r_{m-1}, $C^r_{i,m}$ with $m \leq \min(k,S)$. Using the estimates (1), (2) we obtain a sufficient condition for S:

$$S \leq \varepsilon \frac{\log^{(2)} n - \log r}{\log^{(3)} n}$$

for some constant $\varepsilon > 0$.

The proofs of Corollaries 3,4,6,7 are easy. Corollary 5 can be proved from Theorem 2 in the same way as the analogical corollary for formula size is derived from Hodes-Specker theorem, [7].

References

1. M. Ajtai, Σ^1_1-formulae on finite structures, Annals of Pure and Applied Logic 24 (1983), pp. 1-48.
2. A. Borodin, D. Dolev, F.E. Fich and W. Paul, Bounds for width two branching programs, 15th Annual ACM Symposium on Theory of Computing, 1983, pp. 87-93.
3. A.K. Chandra, M.L. Furst and R.J. Lipton, Multiparty protocols, ibidem pp. 94-99.
4. M.J. Fischer, A.R. Meyer and M.S. Paterson, Ω(n log n) lower bounds on length of Boolean formulas, SIAM J. Comput. Vol. 11, No. 3, 1982, pp. 416-427.
5. M. Furst, J.B. Saxe, M. Sipser, Parity, cirquits and the polynomial-time hierarchy, 27nd Symposium on the Foundations of Computer Science, 1981, pp. 260-270.
6. L. Hodes and E. Specker, Lengths of formulas and elimination of quantifiers I, in Contributions to Mathematical Logic, H.A. Schmidt, K. Schütz, H.-J. Thiele, eds., North-Holland, 1968, pp. 175-188.
7. V.M. Khrapčenko, Complexity of realization of symmetric Boolean functions on finite basis, Problemy Kibernetiki 31, 1976, pp. 231-234, (Russian).

8. E.J. Nečiporuk, A Boolean function, Dokl. Acad. Nauk SSSR, Vol. 169, No. 4, 1966, pp. 765-766, (Russian), English translation in Sov. Math.-Dokl. Vol. 7, No. 4, pp. 999-1000.

9. P. Pudlák, Bounds for Hodes-Specker theorem, in Rekursive Kombinatorik, Münster, 1983, Springer-Verlag, to appear.

10. P. Pudlák and S. Žák, Space complexity of computations, manuscript.

11. J.E. Savage, The Complexity of Computing, John Wiley and Sons, 1976.

Figure 1

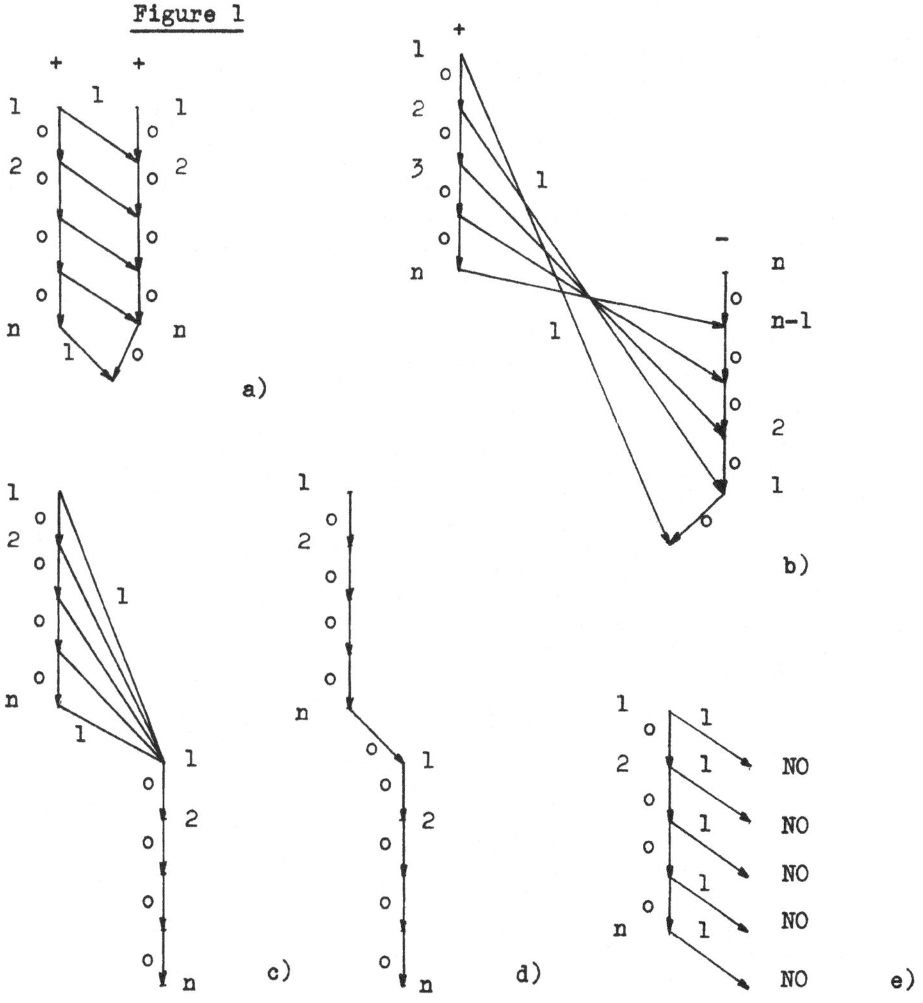

FROM DYNAMIC ALGEBRAS TO TEST ALGEBRAS

Jan Reiterman and Věra Trnková

Faculty of Nuclear Science and Technical Engineering, Technical University of Prague
Prague, Czechoslovakia

Faculty of Mathematics and Physics, Charles University, Prague
Prague Czechoslovakia

ABSTRACT. We generalize some results, known for dynamic algebras, to test algebras. Main results: Every free algebra in the equational class generated by separable test algebras is isomorphic to a Kripke test structure. Consequently, equational classes generated by separable test algebras and by Kripke test structures coincide. In contrary to dynamic algebras, free separable test algebras over finitely many generators do not exist. Epimorphisms in the equational class generated by separable dynamic or test algebras are shown not to be necessarily surjective.

I. INTRODUCTION

Dynamic algebras were introduced by Kozen [3], see also Pratt [11], to provide an algebraic range for semantics of the propositional dynamic logic PDL.

A dynamic algebra (as defined by Pratt [11]) is a two sorted algebra $(\mathcal{B}, \mathcal{R})$ with operations

$$\vee, -, 0 \quad \text{on } \mathcal{B} \text{ (of arities } 2, 1, 0 \text{ resp.)},$$
$$\cup, ;, * \quad \text{on } \mathcal{R} \text{ (of arities } 2, 2, 1 \text{ resp.)}$$
$$\Diamond : \mathcal{R} \times \mathcal{B} \longrightarrow \mathcal{B}$$

subject to identities

(1) $\vee, -, 0$ make $\mathcal{B}$ a Boolean algebra

(2) $a0 = 0$, $a(p \vee q) = ap \vee aq$

(3) $(a \cup b)p = ap \vee bp$

(4) $(ab)p = a(bp)$

(5) $p \vee aa^*p \leq a^*p \leq p \vee a^*(ap - p)$

where ; and $\Diamond$ are ommited for brevity, $a,b \in \mathcal{R}$ and $p,q \in \mathcal{B}$.

The basic examples of dynamic algebras are Kripke structures - the original models of PDL. A Kripke structure is a dynamic algebra $(\mathcal{B},\mathcal{R})$ where $\mathcal{B}$ is a Boolean algebra of subsets of a set W with the usual set theoretical operations; $\mathcal{R}$ is an algebra consisting of binary relations on W where $\cup$, ; are the union and composition of binary relations, respectively, and $*$ assignes to each relation $a \in \mathcal{R}$ its reflexive-and-transitive closure. Finally,

$$ap = \{ s \in W ; \exists t \in p , (s,t) \in a \} .$$

The set W appearing in the definition may be thought as the set of states of a computer; a proposition $p \in \mathcal{B}$ is identified with the set $p \subset W$ of all states in which the proposition p holds; elements of $\mathcal{R}$ are non-deterministic "actions" or "programs": each $(s,t) \in a$ represents a passage from the state s to the state t when the action a is applied. Operations $\cup$, ; , $*$ represent the program constructs "choice", "sequencing" and "iteration".

Program constructs if then else and while do cannot be obtained as combinations of the above; that is why tests are added to PDL; this leads to test algebras.

A test algebra (Pratt [11]) is a dynamic algebra $(\mathcal{B},\mathcal{R})$ with one operation $? : \mathcal{B} \longrightarrow \mathcal{R}$ more where

(6) $(?p)q = p \wedge q$.

A Kripke test structure is a test algebra $(\mathcal{B},\mathcal{R})$ which is a Kripke structure including relations

$$?p = \{ (s,s) ; s \in p \} \quad (p \in \mathcal{B}) .$$

Then if p then a else b is $((?p)a) \cup ((?-p)b)$ and while p do a is $((?p)a)^*(?-p)$.

Most of results on dynamic algebras can be extended to test algebras without essential changes of arguments; we list some of these results in the last section. This is not the case of Pratt´s theorem [11] on free separable algebras and of Nemeti´s result [9] showing that epis do not coincide with surjective homomorphisms. Both are formulated as open problems in [9]. We shall discuss them in the next two sections. Since the number of pages is limited, the full versions of proofs of Theorem 1 and Theorem 2 below will appear elsewhere.

II. THE EQUATIONAL HULL OF KRIPKE TEST STRUCTURES

A dynamic (or test) algebra $(\mathcal{B}, \mathcal{R})$ is called <u>separable</u> if for $a_1, a_2 \in \mathcal{R}$,

$$a_1 p = a_2 p \quad \text{for all} \quad p \in \mathcal{B} \quad \text{implies} \quad a_1 = a_2 .$$

The separability implies some evident identities for actions, e.g., the associativity for the composition of actions $(ab)c = a(bc)$, the commutativity for the union of actions $a \cup b = b \cup a$ for dynamic algebras and for test algebras, moreover,

$$(?p)\cup(?q) = ?(p \vee q), \quad (?p)(?q) = ?(p \wedge q), \quad (?p)^* = ?1 .$$

The class SDA (or STA) of separable dynamic algebras (or separable test algebras, respectively) is not equational. Still SDA admits <u>all</u> free algebras. This is an immediate consequence of the result that free algebras in HSP(SDA) are separable (due to Pratt [11] for non-void sets of Boolean generators and Andréka, Nemeti [1] in general case). (Here HSP(V) is the equational class generated by a class V of algebras ; it consists of <u>H</u>omomorphic images of <u>S</u>ubalgebras of <u>P</u>roducts of V-algebras.) This is no more true for test algebras:

<u>Theorem 1.</u> Let m,n be cardinal numbers. The following are equivalent.

(i) m+n is infinite or m = 0 ;

(ii) the free algebra $F_{m,n}$ in HSP(STA) over n Boolean and m action generators is separable;

(iii) the free separable algebra $F^s_{m,n}$ over n Boolean and m action generators does exists.

The essential part of the proof consists in exhibiting two actions c_1, c_2 in $F_{m,n}$ (m,n finite, $m \neq 0$) that are distinct but inseparable. Let $\alpha_1, \ldots, \alpha_m$ and $\pi_1, \ldots, \pi_n$ be action and Boolean generators, respectively. Put

$$\pi = \bigwedge_{i=1}^{n} \pi_i , \quad \alpha = \bigcup_{j=1}^{m} \alpha_j , \quad \rho = \pi - \alpha 1 .$$

Then the actions in question are

$$c_1 = \big((?\alpha^2\rho)\alpha \quad (?\rho)\big) \cup \alpha^2 , \quad c_2 = \alpha^2 .$$

The full version of the proof will appear elesewhere.

Though free algebras in HSP(STA) need not be separable, the following theorem is valid:

<u>Theorem 2.</u> Every free algebra in HSP(STA) can be embedded in a direct product of finite separable Kripke test structures.

The proof follows the scheme of Pratt's proof for HSP(SDA) involving the Fischer-Ladner closure [11] but it requires an essential refinement of arguments. The full version will appear elsewhere.

<u>Corollary 1.</u> The equational classes generated by separable finite Kripke test structures and by separable test algebras coincide.

This solves an open problem from [9].

<u>Corollary 2.</u> The Boolean part of the equational theory generated by finite Kripke test structures coincides with the Boolean part of the equational theory of test algebras.

This means that Boolean identities (i.e. identities in the Boolean part) which are valid in all finite Kripke test structures are the same as the Boolean identities valid in all test algebras. This yields an alternative proof of completeness of Segerberg's axioms including the test axiom [6] ; see Pratt [11] for dynamic logic without tests.

Another consequence is the following

<u>Proposition</u>. Every free algebra in HSP(STA) is <u>representable</u>, i.e. isomorphic to a Kripke test structure. (See [1] for HSP(SDA)).

III. EPIMORPHISMS

The problem whether epimorphisms coincide with surjective homomorphisms in various classes of dynamic or test algebras was considered in Németi [9] in connection with Beth definability property. This was shown in [9] not to be the case in SP(SDA) and left as an open problem for several classes of test algebras for the argument did not work in the presence of "?". The following shows that epimorphisms are not necessarily surjective in HSP(SDA) and in HSP(STA). A dynamic or test algebra $(\mathcal{B}, \mathcal{R})$ is called <u>associative</u> if

$$(ab)c = a(bc) \quad (a, b, c \in \mathcal{R}).$$

<u>Theorem 3.</u> Let K be any class of associative dynamic or test algebras containing separable Kripke (test) structures. Then K admits an epimorphism such that neither its action part nor its Boolean part are surjective.

Proof. Let $\exp Z$ be the power set algebra of the set Z of integers; let $\mathcal{B}_2$ be its subalgebra generated by all one-element sets $\{n\}$, $n \in Z$, and let $\mathcal{B}_1$ be the subalgebra of $\mathcal{B}_2$ generated by all $\{n\}$ with n positive. Let $(\mathcal{B}_1, \mathcal{R}_1)$ be the Kripke (test) structure such that $\mathcal{R}_1$ is generated by actions a_0, a_1 where

$$a_0 = \{(n,n) ; n \in Z\}, \quad a_1 = \{(n,n+1) ; n \in Z\}$$

and let $(\mathcal{B}_2, \mathcal{R}_2)$ be defined analogously with action generators a_0, a_1 and

$$a_{-1} = \{(n,n-1) ; n \in Z\} .$$

Clearly, $\mathcal{B}_2 - \mathcal{B}_1 \neq \emptyset$ and $\mathcal{R}_2 - \mathcal{R}_1 \neq \emptyset$. We prove that the embedding of $(\mathcal{B}_1, \mathcal{R}_1)$ into $(\mathcal{B}_2, \mathcal{R}_2)$ is an epimorphism in K. Let $h = (h_{\mathcal{B}}, h_{\mathcal{R}})$, $h' = (h'_{\mathcal{B}}, h'_{\mathcal{R}})$ be two homorphisms from $(\mathcal{B}_2, \mathcal{R}_2)$ to an associative dynamic (test) algebra; let these homorphisms coincide on $(\mathcal{B}_1, \mathcal{R}_1)$. To prove $h_{\mathcal{R}} = h'_{\mathcal{R}}$, it suffices to show $h_{\mathcal{R}}(a_{-1}) = h'_{\mathcal{R}}(a_{-1})$. We have $a_{-1}(a_1a_{-1}) = (a_{-1}a_1)a_{-1}$ and $a_1a_{-1} = a_{-1}a_1 = a_0$. Thus

$$\begin{aligned} h_{\mathcal{R}}(a_{-1}) &= h_{\mathcal{R}}(a_{-1}(a_1a_{-1})) = h_{\mathcal{R}}(a_{-1})h_{\mathcal{R}}(a_1a_{-1}) = \\ &= h_{\mathcal{R}}(a_{-1})h'_{\mathcal{R}}(a_1a_{-1}) = h_{\mathcal{R}}(a_{-1})h'_{\mathcal{R}}(a_1)h'_{\mathcal{R}}(a_{-1}) = \\ &= h_{\mathcal{R}}(a_{-1})h_{\mathcal{R}}(a_1)\, h'_{\mathcal{R}}(a_{-1}) = h_{\mathcal{R}}(a_{-1}\, a_1)\, h'_{\mathcal{R}}(a_{-1}) = \\ &= h'_{\mathcal{R}}(a_{-1}a_1)\, h'_{\mathcal{R}}(a_{-1}) = h'_{\mathcal{R}}(a_{-1}a_1a_{-1}) = h'_{\mathcal{R}}(a_1) , \end{aligned}$$

and $h_{\mathcal{B}} = h'_{\mathcal{B}}$ because $h_{\mathcal{B}}(\{n\}) = h'_{\mathcal{B}}(\{n\})$ for any n positive, hence

$$\begin{aligned} h_{\mathcal{B}}(\{0\}) &= h_{\mathcal{B}}(a_{-1}\{1\}) = h_{\mathcal{R}}(a_{-1})\, h_{\mathcal{B}}(\{1\}) = h'_{\mathcal{R}}(a_{-1})\, h'_{\mathcal{B}}(\{1\}) = \\ &= h'_{\mathcal{B}}(\{0\}) \end{aligned}$$

and we prove $h_{\mathcal{B}}(\{-n\}) = h'_{\mathcal{B}}(\{-n\})$ by induction.

<u>Remark</u>. The above theorem generalizes [9] for dynamic algebras and solves all problems for test algebras mentioned there except for the class of <u>all</u> dynamic algebras or test algebras where our proof does not work.

<u>Observation</u>. If $(f_{\mathcal{B}}, f_{\mathcal{R}})$ is an epimorphism in the class of <u>all</u> dynamic algebras then the action part $f_{\mathcal{R}}$ is surjective. If $(f_{\mathcal{B}}, f_{\mathcal{R}}) : (\mathcal{B}, \mathcal{R}) \to (\mathcal{B}', \mathcal{R}')$ is an epimorphism in the class of <u>all</u> test algebras then $f_{\mathcal{R}}(\mathcal{R}) \cup \{?p ; p \in \mathcal{B}'\}$ generates all of $\mathcal{R}'$.

<u>Open problem</u>. Does there exists an epimorphism in the class of <u>all</u>

dynamic or test algebras whose Boolean part is nor surjective?

IV. REMARKS

We give a list of some results on test algebras which are obtained from corresponding results on dynamic algebras using the same (or slightly modified) proofs. The references bellow are related to the latter.

Every separable test algebra can be represented by a <u>non-standard Kripke test structure</u> defined as the Kripke test structure except that the condition concerning * is replaced by a weaker condition " a^* is the least reflexive and transitive relation in $\mathcal{R}$ containing a",[3] .

There is a duality between separable test algebras and test modified topological dynamic spaces, Kozen [4] .

Every representable test algebra $(\mathcal{B}, \mathcal{R})$ is *-continuous, i.e.

$$a^*p = \bigvee_{i=0}^{\infty} a^i p \quad (a \in \mathcal{R},\ p \in \mathcal{B}) \text{ (Pratt [11])}.$$

Even, representable test algebras satisfy stronger continuity conditions - strong *-continuity (Kozen [3]) and relative σ-continuiuty (Reiterman, Trnková [13]). These conditions are not sufficient for a test algebra to be representable. E.g., let $(\mathcal{B}, \mathcal{R})$ be a full complete test algebra, i.e. $\mathcal{B}$ is a complete Boolean algebra and $\mathcal{R}$ is the set of all completely additive functions on $\mathcal{B}$ with operations defined in a standard way. If $(\mathcal{B}, \mathcal{R})$ is not the full Kripke structure and if the cardinality of $\mathcal{B}$ is not too large (specifically: non-measurable) then $(\mathcal{B}, \mathcal{R})$ is not representable (Reiterman, Trnková [14]). Another example [13] : if $\mathcal{B}$ = exp W with W of uncountable but non-measurable cardinality and $\mathcal{R}$ = all σ-additive functions on exp W then $(\mathcal{B}, \mathcal{R})$ is a separable *-continuous test algebra that even does not admit any homomorphism to any Kripke test structure. On the other hand, finite test algebras [11] and even test algebras with finite action part [13] are representable.

Test algebras can be equipped with one more unary operation $a \longrightarrow a^-$, called <u>reversion</u>, which is axiomatized by

$$(7)\ (ab)^- = b^-a^- \qquad (9)\ \ a^{--} = a$$
$$(8)\ (a \cup b)^- = a^- \cup b^- \qquad (10)\ \ p \leq -a(-a^-p)$$

The presence of the reversion ensures some nice properties: all actions in a test algebra with reversion are completely additive [14] . Every separable *-continuous test algebra with reversion can be represented as a subalgebra of a full complete test algebra [14]. Every countable separable $*$-continuous test algebra $(\mathcal{B},\mathcal{R})$ with reversion where $\mathcal{B}$ is atomic is representable (Kozen [8]).

We did not consider

- the validity of the results above for test algebras with reversion except for the result on epimorphisms which obviously remains valid if the reversion is present ,
- the problem of the existence of a countable non-representable separable test algebra [8, 13],
- the problem of the existence of a separable test algebra without reversion which cannot be embedded into a full complete test algebra [14],
- the non-equational logic aspects of test algebras [5, 9].

REFERENCES

[1] Andréka, H. and Németi, I.: Every free algebra in the variety generated by the separable dynamic algebras is separable and representable, to appear in Theoretical Comp. Sci.

[2] Fischer, M.J. and Lander, R.E.: Propositional modal logic of programs, Proc. 9th Ann. ACM Symp. on Theory of Computing, 286--294, Boulder, Col., May 1977.

[3] Kozen, D.: A representation theorem for models of $*$-free PDL, IBM Researche, July 1979.

[4] Kozen, D.: Dynamic algebras, in [10].

[5] Kozen, D.: On induction vs. $*$-continuiuty, IBM Research, Sept. 1980.

[6] Kozen, D.: On the duality of dynamic algebras and Kripke models, IBM Research, May 1979.

[7] Kozen, D.: On the representation of dynamic algebras, IBM Research Oct. 1979.

[8] Kozen, D.: On the representation of dynamic algebras II, IBM Research, May 1980.

[9] Németi, I.: Dynamic algebras of programs, Proc. FCT´81, Lecture notes Comp. Sci. 117, Springer 1981, 281-291.

[10] Parikh, R.: Propositional dynamic logic of programs: a survey, Laboratory for Comp. Sci., Mass.Inst. of Technology, Jan.1981.

[11] Pratt, V.R.: Dynamic algebras: examples, constructions, applications, Laboratory for Comp. Sci., Mass. Inst. of Technology, July 1979.

[12] Pratt, V.R.: Dynamic algebras and the nature of induction, Laboratory for Comp. Sci., Mass. Inst. of Technology, March 1980.

[13] Reiterman, J. and Trnková, V.: Dynamic algebras which are not Kripke structures, Proc. MFCS 1980.

[14] Reiterman, J. and Trnková, V.: On representation of dynamic algebras with reversion, Proc. MFCS 1981.

COMBINATORIAL GAMES WITH EXPONENTIAL SPACE COMPLETE DECISION PROBLEMS

J. M. Robson
Australian National University
Canberra, ACT, Australia

1. INTRODUCTION

In recent years, several results have been proved showing that, for a variety of games, it is an exponential time complete problem to decide, given an arbitrary configuration, whether a given player can force a win. The games include a number of artificial 'boolean formula' games [2,10] as well as generalisations to N x N boards of chess [3], checkers [9] and go [8].

The lower bound parts of these proofs have been obtained by reductions from the problem of deciding the outcome of a computation of a linear space bounded Alternating Turing Machine. [2] showed close relationships between Alternating and Deterministic Turing Machine complexities and started the reductions by giving a reduction to a boolean formula game called G_1 described later in this paper. The upper bound parts are trivially based on a 'brute force' method for solving such game problems in exponential time by classifying all configurations as won, lost or drawn and recording this information in a table.

Since the number of distinct configurations which can arise in these games is exponential in the input length (board size or formula length), complexity greater than exponential time will only arise if the outcome from a given configuration depends on the way in which the configuration arose. This happens in chess but only in unimportant ways (rules concerning castling, en passant capture and drawing rules). In the Chinese rules of go, there is a much more significant rule relating the outcome to past play; this rule forbidding a player to play so as to return to any configuration which has occurred previously is a generalisation of the Japanese ko rule. This rule means that the exponential time completeness result referred to above applies only to the Japanese version of go since the 'no-repetitions' rule invalidates both upper and lower bound parts of that proof. The realisation of this fact triggered the investigations reported in this paper which showed that this 'no-repetitions' rule in fact makes some games exponential space complete. It is not yet known whether this fact applies to the Chinese version of go.

Another way in which games with only an exponential number of physical configurations may still be harder than exponential time has been studied in [6,7]. It is shown there that games where the players have imperfect information of the physical configuration may be exponential space complete or harder depending on the number of players and the amount of hidden information. A survey of recent results in complexity of games appears in [4].

2. STATEMENT OF RESULTS

The following description of G_1 is paraphrased from [10]. 'G_1: a position is a triple (τ, F(X, Y, {t}), α) where $\tau \in$ {I, II} indicates which player is about to move, F is a boolean formula in 4CNF whose variables have been partitioned into disjoint sets X, Y and {t} and α is an assignment of values to the variables of F. Player I moves by setting t to true, and setting the variables in X to any values (thus producing a new position with τ changed from I to II, F unchanged and α changed according to the new values of X and t); player II moves by setting t to false and setting the variables in Y to any values. A player loses if the formula F is false after his move'.

The major result of this paper is:

Theorem 1

If G_1 is modified to G_1' by introducing the no-repetition rule: 'A player loses if, after his move, $\exists$ some earlier move M such that all variables have the same values as after M.', then deciding, for an arbitrary G_1' configuration, whether player I can force a win is complete in exponential space. The proof of this theorem also shows with a little extra work:

Theorem 2

If G_1 is modified to G_1^* by introducing the conditional no-repetition rule: 'Two special variables $x \in X$ and $y \in Y$ are specified. A player loses if, after his move, $\exists$ some earlier move M such that all variables have the same values as after M <u>and</u> at most one of x and y has been changed since M.', then deciding for an arbitrary G_1^* configuration, whether player I can force a win is complete in alternating exponential space.

Theorem 2 is equivalent to asserting that the decision problem for G_1^* is complete for double exponential time [2]. The lower bound parts of these proofs are described in the next section. The upper bound parts are relatively trivial. For G_1' a simple minimax algorithm runs in exponential space; for G_1^* a double exponential time decision procedure is obtained by applying the brute force method mentioned in the introduction to a game whose configurations are sequences of G_1^* configurations starting with a change of a special variable and not including either another change of that variable or a repeated G_1^* configuration.

3. THE LOWER BOUNDS

3.1 Overview

The lower bound parts of theorems 1 and 2 will be proved by two similar reductions, one from exponential time bounded Alternating Turing Machine (ATM) computations to G_1' and the other from exponential space bounded ATM computations to G_1^*. The result of [2], relating time f on an ATM to space f on a deterministic TM, completes the proof of theorem 1.

To simplify the presentation of the reductions, they will be split into three stages; (i) from a general ATM to a very restricted ATM, (ii) from the restricted ATM to games G_0' and G_0^*

(the variants of a new game G_0 even more basic than G_1) and (iii) from G_0 to G_1. It is hoped that this makes the critical central step from an ATM to a game as easy to understand as possible, though this approach does not give the simplest overall reduction.

The restrictions on the ATM are as follows:

(1) At each step the ATM will switch from an existential to a universal state or vice versa.

(2) The ATM has a single tape and a binary alphabet.

(3) The ATM starts at the left hand of its input, moves steadily right writing special markers at the left hand end of the tape and at some distance to the right and then never leaves the section of tape between the two markers. The states which may be used in this initial scan of the tape can never again be entered after it.

(4) The ATM's remaining states are partitioned into 'left moving' and 'right moving'. Once in a left (right) moving state the machine continues moving steadily left (right) until it encounters the marker at the left (right) end of the section of tape marked out for the computation which forces the ATM to enter one of two special states in which it must switch into a right (left) moving state.

The effect of restrictions (3) and (4) is that a computation of the ATM consists of alternate left and right scans, each of which covers the same section of tape, and that each tape cell in this section is either always scanned in an existential state or always scanned in a universal state. These restrictions will considerably simplify the expression F; it is fairly straightforward to show that an arbitrary ATM may be replaced by one subject to these restrictions and accepting the same inputs with at most a polynomial increase in its time and space complexity.

The new game G_0 differs from G_1 in that whether a move is legitimate depends on the values of the variables before the move as well as after. More rigorously, the formula F which determines whether a player has lost after his move may include, as well as X, Y and t, X^* and Y^* indicating the values of the corresponding variables before the move. The reduction from G_0 to G_1 will be described in section 3.4.

3.2 The Game Graph

3.2.1 The variables X and Y

The idea behind the reduction to $G_0^{(*)}$ is exactly the same as that of the reduction to G_1 in [10], namely that in each move of the game the players are expected to set their variables in a way which represents one step of the ATM; player I makes the moves corresponding to existential states and player II makes the moves corresponding to universal states; the formula F is designed so that (i) a player making a move not representing a possible ATM step loses very quickly, (ii) if the ATM computation reaches an accepting state, I wins and (iii) if the ATM computation reaches a rejecting state, II wins. Thus player I can force a win if and only if the ATM accepts its input.

In our case the variables are not sufficient in number to completely encode the current

ATM instantaneous description. Thus the formula F cannot ensure that a player loses instantaneously if he sets his variables to values inconsistent with the symbols written on the tape by the previous computation. Instead, if this happens, the opposing player can set in train a sequence of moves which quickly results in the transgressing player losing by the no-repetitions rule. The way in which this happens will be described first by showing a subgraph of the game graph corresponding to a single ATM step. Section 3.3 will describe the formula F which forces players to play according to this game graph.

The variables in each of the sets X and Y are as follows (subscripts X and Y will be used when necessary to distinguish them):

q_i $0 \leq i <$ log (number of ATM states)

p_i $0 \leq i <$ log (upper bound on space used by ATM)

S, A1, A2, A3, C, R

N (in the G_0^* case)

n_i $0 \leq i <$ log (upper bound on number of times ATM traverses its tape) (in the G_0' case)

The intended interpretation of these is: q_i are the bits of an ATM state number called Q; p_i are the bits of P, a position or address on the ATM tape; S is an ATM symbol; A(1,2,3) C, R stand for Accept(1,2,3), Claim, Refute; n_i (at G_0') are the bits of N, the number of times the ATM has switched from a left moving to a right moving state or vice versa; N (at G_0^*) : N_X and N_Y are the special variables which are mentioned in the conditional no-repetition rule of G_0^*, N_X the parity of the number of times the machine has switched from a left moving to a right moving state and N_Y the parity of the number of switches from right to left moving. A quadruple (N,Q,P,S) is interpreted as describing a single step of the ATM in which the symbol S is written and N,Q and P have the appropriate values <u>after</u> the step.

3.2.2 Moves at G_0'

At G_0' corresponding to an instant when the ATM has just made a universal (existential) step, is a G_0' configuration where both (N_X, Q_X, P_X, S_X) and (N_Y, Q_Y, P_Y, S_Y) describe this step and exactly one of the remaining variables in X and Y is true, namely $A3_{X(Y)}$. The game proceeds to the configuration corresponding to the instant one step later in the computation by a sequence of five moves illustrated in the left part of figure 1: first I clears $A3_X$, sets C_X and sets (N_X, Q_X, P_X, S_X) to what he CLAIMs are the correct new values (N′, Q′, P′, S′); then II, provided he ACCEPTs the CLAIM, sets $A1_Y$ and sets (N_Y, Q_Y, P_Y, S_Y) to the same new values; then I clears C_X and sets $A2_X$; II clears $A1_Y$ and sets $A3_Y$ then I clears $A2_X$ completing the sequence. The formula F is designed so that a player loses instantaneously, if he deviates from this pattern except in one of the two ways shown in figure 1.

The lower possible deviation (by I) is fatal, not instantaneously but very quickly: I can set R_X when he clears $A2_X$; II then clears $A3_Y$ and sets $A1_Y$; the formula F would then allow I only one move, namely to clear R_X and set $A2_X$, but the no-repetitions rule ensures that I cannot do this and so loses.

The upper deviation (by II) is more interesting. This deviation is fatal for II if I's CLAIM represented by (N′, Q′, P′, S′) is consistent with the previous computation of the ATM but is fatal for I otherwise. This fact is the crux of the correctness of the reduction.

(i) II REFUTEs I's CLAIM by setting (N_Y, Q_Y, P_Y, S_Y) to the description (N″, Q″, P″, S″) of the last step which wrote a symbol (S″) in the position on the tape read on this current step. F will allow II to do this if and only if the step (N″, Q″, P″, S″) would have made (N′, Q′, P′, S′) impossible, that is if N″ and P″ do describe the last step which wrote in the tape cell read by the supposed move (N′, Q′, P′, S′) and the supposed ATM transition from reading S″ in state Q to writing S′ and changing to state Q′ is not possible.

(ii) I's only possible move is now to set all the X variables to the same values as the corresponding Y variables.

(iii) II's only possible move is now to clear R_Y and set $A1_Y$. Thus the game has arrived at a configuration exactly like that already considered on the lower deviation but with the step described (N″, Q″, P″, S″) chosen by II instead of (N′, Q′, P′, S′) claimed by I.

If the step (N″, Q″, P″, S″) really did occur earlier in the computation, then I now loses by the no-repetition rule. Otherwise I clears R_X and sets $A2_X$; II is forced to clear $A1_Y$ and set $A3_Y$; I returns to the deviation by clearing $A2_X$ and setting R_X; II now loses by the no-repetition rule.

Thus the illustrated graph of possible moves and deviations ensures exactly as required that II can win by taking the deviation if I makes a false CLAIM but otherwise a player taking a deviation loses.

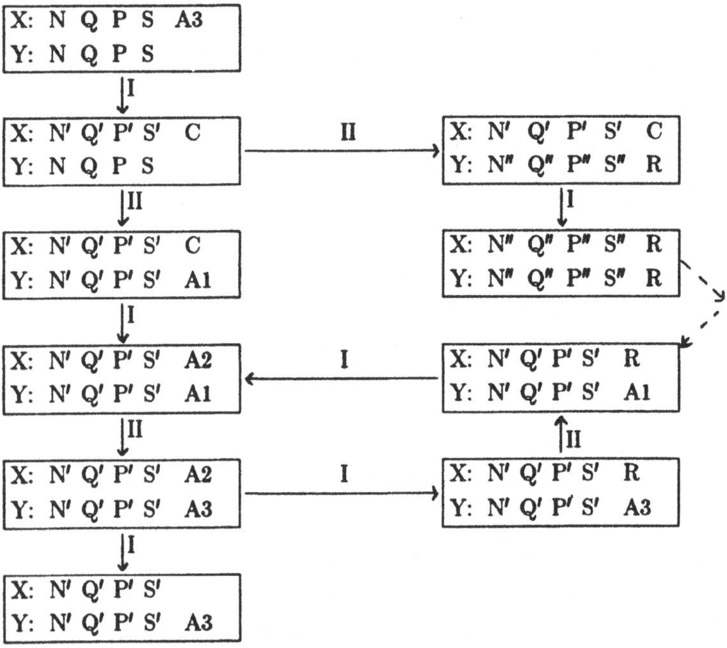

Figure 1 An existential step at G'_0

The graph of all possible moves at G_0' is obtained by taking copies of figure 1 for every possible CLAIM by I and every attempted REFUTATION by II, adding also copies of the corresponding figure for universal steps and finally identifying all nodes with equal values for all variables (including the turn variable t which is implicit in the figure).

3.2.3 Differences at G_0^*

Figure 2 illustrates the G_0^* moves corresponding to an ATM step together with possible deviations exactly as in figure 1. The discussion is identical except for the treatment of the special variables N_X and N_Y. N_X and N_Y are no longer required to be equal in the first and last configurations; instead I changes N_X or II changes N_Y at the appropriate end of the tape used. On the upper deviation II sets N_Y to the correct value for the step with which he tries to REFUTE I's CLAIM; then I is required to set N_X to the appropriate value (note that N_X is implied by N_Y and the knowledge of whether Q is a left or right moving state). Between two adjacent occasions when a tape cell is scanned, exactly one of N_X and N_Y changes whereas between two non-adjacent occasions, both N_X and N_Y change at least once. Thus with the conditional no-repetition rule of G_0^*, N_X and N_Y ensure that a refutation succeeds only if it relies on the last symbol written in the relevant tape cell, avoiding the need, as in G_0', to have enough n_i variables to count scans over the tape.

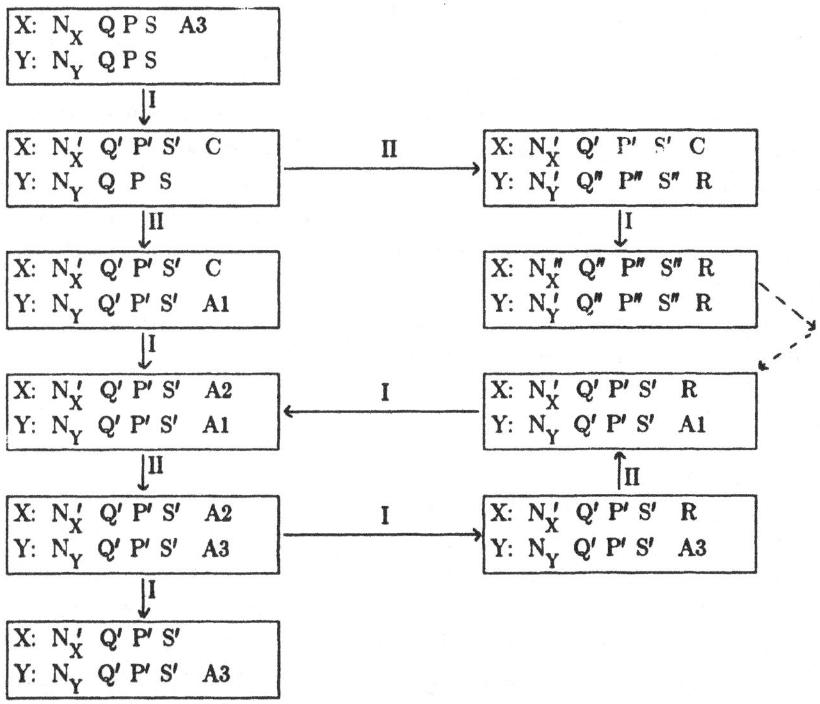

Figure 2 An existential step at G_0^*

3.3 The Boolean Formula

3.3.1 Semantics

The boolean formula F over the variables X, Y, t (in fact t is not needed in G_0), and X* and Y* is required mainly to ensure that any move not corresponding to the informal description above loses at once. (F must also ensure that the symbols read on the first scan of the tape do correspond to the input, that I loses if the computation rejects and that II loses if it accepts. These are trivial given the restriction (3) on the ATM and are not discussed further.) We distinguish between variables indicating the 'type' of a configuration, namely A(1,2,3) C and R and those imparting information, about the progress of a computation (N,Q,P,S).

Figure 1 (or 2) contains eleven arcs so that with the figure for universal steps, there are 22 types of move. With each move type i ($1 \leq i \leq 22$) we associate a type condition, namely an expression T_i indicating that the type variables before and after the move are correct for that type, and a semantic condition S_i. Then F can be written as $\forall i(S_i \vee \neg T_i) \wedge \exists i(T_i)$.

For nine of the eleven move types in figure 1, the semantic condition is simply, as indicated in the diagram, that $N_X=N_Y \wedge Q_X=Q_Y \wedge P_X=P_Y \wedge S_X=S_Y$.

For the tenth move type, where I sets C, the semantic condition simply states that the new values N_X and P_X are correctly related to the old ones N_Y and P_Y. Since, by the restrictions on the ATM, Q_Y determines whether the ATM is moving left or right and whether it is about to reverse direction, this can be written, for some expressions $f_i(Q_Y)$ $1 \leq i \leq 4$, as $(f_1(Q_Y) \vee N_X=N_Y) \wedge (f_2(Q_Y) \vee N_X=N_Y+1) \wedge (f_3(Q_Y) \vee P_X=P_Y+1) \wedge (f_4(Q_Y) \vee P_X=P_Y-1)$. An extra conjunct $f_5(Q_X,Q_Y)$ ensures correct state changes at ATM direction changes.

For the last move type, where II sets R attempting to refute I's claim by setting new values N'' etc., there are two semantic conditions: firstly N_Y and P_Y must refer to the previous occasion when this cell was scanned and secondly in state Q_Y^* scanning symbol S_Y the machine does not move into state Q_X and write symbol S_X. These can be written respectively as $(N_Y = N_X-1) \wedge (f_6(Q_Y^*) \vee P_Y = P_Y^*+2) \wedge (f_7(Q_Y^*) \vee P_Y = P_Y^*-2)$ and as $f_8(S_Y, Q_Y^*, S_X, Q_X)$.

For the eleven other move types, in the corresponding figure for universal steps, the conditions are obtained by interchanging subscripts X and Y.

3.3.2 The CNF form

To convert the expression F described above into CNF, we first convert various subexpressions to CNF. For some expressions such as the type conditions and the f_i whose exact form was not discussed, this clearly gives expressions whose length depends only on the ATM and not on the computation length or space. For the remaining arithmetic subexpressions, $N_X = N_Y$, for instance, is equivalent to $\forall i\ N_{X,i} = N_{Y,i}$ and $P_X = P_Y + 1$ is equivalent to $(P_{X,0} \neq P_{Y,0}) \wedge \forall(i>0)\{(P_{X,i}=P_{Y,i}) \equiv (P_{X,i-1} \vee \neg P_{Y,i-1})\}$. Next writing these simpler subexpressions involving only 0(1) variables in CNF and applying the distributive law, we obtain a CNF expression F whose length is bounded by a linear function of the number of variables (the linear function depending on the ATM).

Now F is a conjunction of disjunctions each of the form $E_1(X) \vee E_2(Y)$ or equivalently $(X\# = E_1(X)) \wedge (Y\# = E_2(Y)) \wedge (X\# \vee Y\#)$. Using the complement of the method of [1], a subexpression such as $(X\# = E_1(X))$ can be replaced by a 3CNF form $E_1'(X, X\#, X')$ which has a satisfying X′ assignment iff the original subexpression was true. Adding the X# and X′ variables into the set X and similarly for Y, gives a form of the expression F which is in 3CNF and has a length linear in the number of variables introduced in section 3.2.1. This completes the polynomial reduction from the original ATM problems to G_0' and G_0^*.

3.4 From G_0 to G_1

The reduction from G_0 to G_1 is most easily explained as a modification to the game graph described in section 3.2. The translation from the modified graph to the CNF form of F is then very similar to that described in section 3.3. The modification gives each player extra variables which can be used to record the state of the G_0 variables before the move as well as after and also three new 'timing' variables t^0, t^1 and t^2; at any time exactly one of these will be true and will indicate the number of moves made by the player (mod 3).

Figure 3 illustrates, for a move by player I, how a single arc in the G_0 game graph becomes a path of length 3 in the G_1 graph. Recall that the G_0 expression could use X* and Y* to refer to the values of the variables before the move. Initially I has two sets of copies of the initial G_0 variables X* and one set of copies of Y*; II has two sets of copies of Y* and one of X*. I moves by changing one of his copies of X* to the new value X; II responds by changing his single copy of X* to X and I finally changes his other copy. Note that if the G_0 move loses by causing a repetition, the repetition at G_1 occurs on I's second move so that the responsibility for the repetition remains correctly ascribed to I. In G_1^* the special variables are those corresponding to the G_0^* special variables in the second copy of the G_0^* variable sets; that is they are the ones changed in the third move of the sequence described.

Now the type conditions for a position in G_1^* can be written in a form like $t_X^2 \wedge t_Y^1 \wedge \neg t_X^0 \wedge \neg t_X^1 \wedge \neg t_Y^0 \wedge \neg t_Y^2 \wedge t$ for the last position in figure 3. (Remember t is the turn variable always set true by I).

The semantic conditions for the various positions consist of the obvious conditions that various copies of variable sets are in fact equal and the clauses inherited from G_0^* F(X[copy 1], X[copy 2], Y[copy 1], Y[copy 2]) in the two positions where one player has taken the first move of the three move sequence.

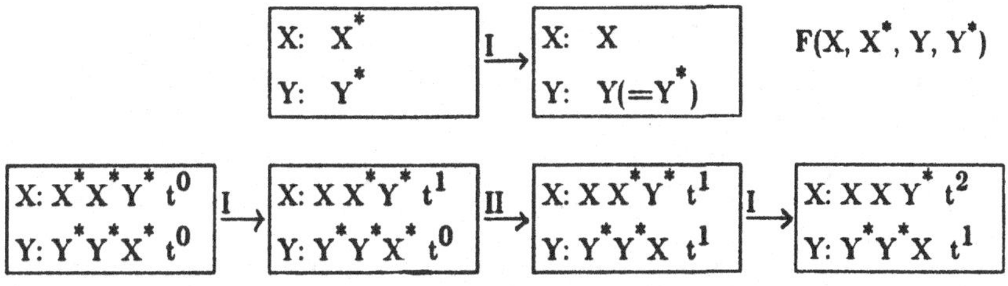

Figure 3 The reduction from G_0 to G_1

4. CONCLUSIONS

The reductions described in section 3, and the theorems of [2] on the relationships between deterministic and alternating time and space complexities complete the proof of the theorems 1 and 2.

Theorem 1 would apply also to the modified version X′ of any other game X, if there were a polynomial time reduction from G_1 to X satisfying a few simple conditions which force players at X′ to mimic play at G_1'.

In fact such reductions would exist using the already published reductions to G_2, G_3 [10], chess [3] and checkers [9] except for one minor snag: the original reduction from G_1 to G_2 produces multiple G_2 configurations for one at G_1. However this problem is easily removed (for each variable in the original set X, the reduction produces four variables $x_{i,1}$, $x_{i,2}$, $y_{i,1}$, and $y_{i,2}$; when player I changes $x_{i,1}$ or $x_{i,2}$, player II must change one of $y_{i,1}$ or $y_{i,2}$ but it is immaterial which he chooses; removing $y_{i,2}$ and requiring player II to change $y_{i,1}$ leaves a valid reduction; similar comments apply to the variables $x_{i,2}$ corresponding to original Y variables), establishing exponential space completeness for G_2', G_3', chess′ and checkers′.

The list of games whose ′ version is known to be exponential space complete does not contain go whose version go′ as played in China started this investigation. The reductions to go from G_3 make it easy for one player to force the other to break the no-repetition rule.

Finally it is of interest to note that instances of X′ for any game X are simply instances of the 'generalised geography game' [5], which is known to be complete in polynomial space. Thus the exponential space completeness results could be interpreted as implying that the geography game is still very hard (requiring space $\Omega(e^{\log^c x})$ for $c > 0$) even when restricted to apparently simple instances which can be described in logarithmic space and even when this short description is provided as part of the input.

REFERENCES

[1] Bauer M., Brand D., Fischer M., Meyer A. and Paterson M., A note on disjunctive form tautologies, SIGACT news, April 1973, pp 17-20.

[2] Chandra A.K., Kozen D.A. and Stockmeyer L.J., Alternation, J. ACM, vol. 28, 1981, pp 114-133.

[3] Fraenkel A.S. and Lichtenstein D., Computing a perfect strategy for n x n chess requires time exponential in n, J. Combinatorial Theory series A, vol. 31, 1981, pp 199-213.

[4] Johnson D.S., The NP-Completeness Column: an ongoing guide, Journal of Algorithms, vol. 4, 1983, pp 397-411.

[5] Lichtenstein D. and Sipser M., Go is polynomial-space hard, J. ACM, vol. 27, 1980, pp 393-401.

[6] Reif J.H. and Peterson G.L., Multiple person alternation, in Proceedings of 20th annual symposium on foundations of computer science, IEEE Computer Society, 1979, pp 348-363.

[7] Reif J.H., Universal games of incomplete information, in Proceedings of 11th annual ACM symposium of theory of computing, ACM, 1979, pp 288-308.

[8] Robson J.M., The complexity of go, in Information Processing 83, R.E.A. Mason ed., IFIP, 1983, pp 413-418.

[9] Robson J.M., N by N checkers is exptime complete, SIAM J. Comput., to appear.

[10] Stockmeyer L.J. and Chandra A.K., Provably difficult combinatorial games, SIAM J. Comput, vol. 8, 1979, pp 151-174.

FAST RECOGNITIONS OF PUSHDOWN AUTOMATON AND CONTEXT-FREE LANGUAGES

Wojciech Rytter
Institute of Informatics, Warsaw University
00-901 Warszawa, PKiN VIII p, Poland

Abstract

We prove: 1) every language accepted by two-way nondeterministic pushdown automaton can be recognized on RAM in $O(n^3/\log n)$ time; 2) every language accepted by two-way loop-free pushdown automaton can be recognized in $O(n^3/\log^2 n)$ time; 3) every context-free language can be recognized on-line in $O(n^3/\log^2 n)$ time. We improve the results of [1,7,4].

1. Introduction

2npda languages are languages accepted by two-way nondeterministic pushdown automata (2npda's, for short). The best known upper bound on time complexity for recognition of 2npda languages was $O(n^3)$. Recently this bound was improved for some subclasses of 2npda languages [7,8]. In this paper we introduce a compressed represenation of matrices and we prove that 2npda languages can be recognized in less than cubic time. A similar approach is possible in the case of the on-line recognition of context-free languages (cfl's, for short). Our model of the computation is a random access machine (RAM) with the uniform cost criterion for access operations. We assume that arithmetic operations have logarithmic cost (even $\log^2 n$ cost of aritnmetic operations does not change the order of time complexity of our recognition algorithms), however the cost of access to a table is constant.

2. The compressed representation of matrices

Let A be a matrix of the type array [0..m-1,0..m-1] of boolean, and let 1 denote the value true and 0 denote the value false. Define the operations: DIFCOL(i,j)= $\{k : A(k,i)=1$ and $A(k,j)=0\}$ (difference of columns i,j); DIFROW(i,j)= $\{k : A(i,k)=1$ and $A(j,k)=0\}$ (difference of rows); INSERT(k,j) - the effect of this operation, limited to the matrix A, is the same as A(k,j):=1, however this will affect other components of the designed later data structure.

Let $p = \lceil \log(m)/4 \rceil$, assume for simplicity that m is divisible by p and denote s=m/p. We maintain, together with A, two matrices R,C of the types R:array[0..m-1,0..s-1] of integer, C: array[0..s-1,0..m-1] of integer. R is the matrix of compressed rows, and C is the matrix of compressed columns of A. We divide rows and columns of A into sections of length p. The (i,j)th element of A "charges" to the t-th section of j-th column,where t =i div p, and to the l-th section of i-th row, where l = j div p. Define rowsection(i,r) to be r-th section of i-th row, and colsection(r,j) to be r-th section of j-th column.

rowsection(i,r)= (A(i,pr),A(i,pr+1),...,A(i,pr+p-1)),
colsection(r,j)= (A(pr,j),A(pr+1,j),...,A(pr+p-1,j)).

For a binary sequence $\bar{a} = (a_0,a_1,\ldots,a_{p-1})$ denote by code($\bar{a}$) the binary number represented by $\bar{a}$. The matrices R, C are related to A in the following way: for $0 \leq i,j < m, 0 \leq r < s$

R(i,r)=code(rowsection(i,r)), C(r,j)=code(colsection(r,j))

If $q=\mathrm{code}(a_0,a_1,\ldots,a_{p-1})$ then denote $\mathrm{set}(q)=\{k : a_k=1,\ 0 \leq k < p\}$. Hence the codes of sections can be treated as sets and we can define some set operations. Let $P = \{0,1,\ldots,2^p-1\}$. For q1, q2 $\in$ P define:

$\mathrm{dif}(q1,q2)= \mathrm{set}(q1)-\mathrm{set}(q2)$; $\mathrm{add}(q1,k)= \mathrm{set}^{-1}(\mathrm{set}(q1) \cup \{k\})$,
where $0 \leq k < p$.

Let $|P|$ denote the cardinality of P. We precompute and store in a table the values of dif(q1,q2), add(q1,k) for each q1,q2, k. This can be done easily with the cost $O(m^2)$ because $|P|^2=O(m^{1/2})$. Now we assume (disregarding the cost of preprocessing) that the costsof operations dif and add are constant. If the matrices R, C are computed then the operations DIFCOL, DIFROW, INSERT are implemented as follows:

```
function DIFCOL(i,j);
begin
  DIFCOL:=∅;
  for r:=0 to s-1 do
    for each q ∈ dif(C(r,i),C(r,j)) do
      DIFCOL:= DIFCOL ∪ {q+rp }
end;
```

The operation DIFROW is impemented analogously.

```
procedure INSERT(i,j);
begin
  A(i,j):=1; C(i div p, j):=add(C(i div p, j), i mod p);
  R(i,j div p):=add(R(i,j div p),j mod p)
end;
```

The following lemma follows directly from our impementation of A.

Lemma 1

Assume that the values of dif and add are precomputed. Then

1) The cost of DIFCOL(i,j) is $O(m/\log(m)+|DIFCOL(i,j)|\cdot\log m)$;

2) The cost of DIFROW(i,j) is $O(m/\log(m)+|DIFROW(i,j)|\cdot\log m)$;

3) The cost of INSERT(i,j) is $O(\log m)$.

3. Fast recognitions of 2npda languages

Let us fix a 2npda A and let w be a given input word of the length n. We want to decide if A accepts w. The size of our problem is n. The surface configuration of A (configuration, for short) is a vector (s,top,pos), where s is a state of A, top is an element of the stack alphabet and pos is a position of the input head. Let K be the set of all configurations (for a given w). Denote $m=|K|$. Observe that $m=O(n)$. Let us identify configurations with consecutive natural numbers and assume for simplicity that $K=\{0,\ldots,m-1\}$. Let 0 be the number of the initial configuration and m-1 the number of the accepting configuration. We assume that initially and in the moment of acceptance the stack is one-element and each move of A is a pop or a push move.

We say that a pair of configurations (k,l) is below (i,j) iff A can go from k to i in one push move, and from j to l in a pop move, moreover top symbols in k and l are the same. A pair (i,j) is said to be realizable iff A starting in the configuration i with one-element stack can go after some number (may be zero) of moves to the configuration j ending also with one-element stack. Let R denote the set of all realizable pairs and let below(i,j) denote the set of pairs which are below (i,j). Observe that $|below(i,j)|=O(1)$.

The next lemma is analogous to Lemma 3.2(a) in [9].

Lemma 2

a) A accepts w iff $(0,m-1)\in R$;

b) R is the minimal set satisfying the conditions:

1) $(i,i)\in R$ for each $i\in K$; 2) if $(i,j)\in R$ then $below(i,j)\subseteq R$;

3) if $(i,k)\in R$ and $(k,j)\in R$ then $(i,j)\in R$.

Now the simulation of A can be reduced to the computation of R.

Our first algorithm is similar to the algorithm MAIN in [9] and its correctness can be proved by a similar argument. We start with the set R consisting initially only of the pairs (i,i) and successively augment R according to Lemma 2(b). Let Q be a queue. The operation insert(x,S) inserts x into the set S. The operation delete(Q) deletes an element from Q, the deleted element is the value of this operation.

Algorithm 1

```
begin
1: Q:=R:= {(i,i) : i∈ K};
while Q ≠ ∅ do
  begin (i,j):=delete(Q)    {(i,j) is to be processed}
  2: for each (k,l)∈ below(i,j) do
       if (k,l)∉ R then  insert((k,l),R) and insert((k,l),Q);
  3: for each (k,i)∈R such that (k,j)∉ R do
       insert((k,j),R) and insert((k,j),Q);
  4: for each (j,k)∈ R such that (i,k)∉ R do
       insert((i,k),R) and insert((i,k),Q)
end;
 if (0,m-1)∈ R then ACCEPT
end.
```

In the above algorithm instructions 1,2 correspond to conditions 1,2, respectively, in Lemma 2(b), and instructions 3,4 correspond to the condition 3. We represent R by the boolean matrix A, such that (i,j) is an element of R iff A(i,j)=1. Assume that initially each entry of A contains 0. The algorithm below is a refinement of Algorithm 1.

Algorithm 2

```
begin
1: for each i∈ K do INSERT(i,i) and insert((i,i),Q);
while Q ≠ ∅ do
  begin (i,j):=delete(Q);
  2: for each (k,l) ∈below(i,j) do
       if A(k,l)=0 then INSERT(k,l) and insert((k,l),Q);
  3: for each k ∈DIFCOL(i,j) do
       INSERT(k,j) and insert((k,j),Q);
  4: for each k ∈DIFROW(j,i) do
       INSERT(i,k) and insert((i,k),Q)
  end;
 if A(0,m-1)=1 then ACCEPT
end.
```

We use here the implementation of INSERT, DIFROW, DIFCOL from the previous section.

<u>Theorem</u> 1

Every 2npda language can be recognized in $O(n^3/\log n)$ time.

Proof.

Algorithm 2 accepts iff the input w is accepted by 2npda A. The correctness of the algorithm follows from Lemma 2. Each pair (i,j) is inserted into Q at most once because before such insertion it is inserted into R . Hence the algorithm stops.Each of the instructions (2-4) is executed $O(m^2)$ times. The cost of operations DIFROW, DIFCOL is dominating. The cost of one of these operations is $O(m/\log(m) + v \log(m))$, where v is the number of computed (and inserted) elements. The terms $m/\log(m)$ give in total $O(m^3/\log m)$ cost, and the terms $v \log(m)$ give in total $O(m^2 \log m)$ cost, because each pair of configurations can be inserted in Q at most once. The number of inserted pairs in the instructions 3,4 is equal to the number of elements computed in DIFROW and DIFCOL. The total cost of the algorithm is $O(m^3/\log m)$, which is $O(n^3/\log n)$. This completes the proof.

We say that a 2npda A is loop-free iff there is no possible an infinite computation of A on any (finite) input word w.

<u>Theorem</u> 2

Every language accepted by a loop-free 2npda can be recognized in $O(n^3/\log^2 n)$ time.

Sketch of the proof.

We refer the reader to the algorithm in [7] simulating loop-free 2npda's in $O(n^3/\log n)$ time. The algorithm computes the function TERM(x), where TERM(x) is the set of terminators of a configuration x. We can represent sets TERM(x) by row-vectors of the size m consisting of zeros and ones (characteristic vectors), and redefine TERM(x) to be the compressed row (compressed version of the set of terminators). The unions of all possible sections of size p can be precomputed. After such preprocessing the (compressed) union of two (compressed) rows costs $O(m/\log m)$. The conversion from the compressed representation to the standard one is needed to compute the set S, however it is needed for each configuration at most once. We leave details to the reader. The algorithm is improved by the factor log n.

4.<u>A fast on-line recognition of cfl's</u>

The fastest known algorithm for the on-line recognition of cfl's had time complexity $O(n^3/\log n)$ on RAM, see [4]. We simplify this algorithm and improve its cost by the factor log n.

Remark

There is a simple alternative proof of the existence of a $O(n^3/\log n)$ time on-line algorithm. There is a Turing machine recognizing on-line cfl's in $O(n^3)$ time, see [6,page 440] . It was proved in[3] that if there is a Turing machine recognizing a given language in $O(n^s)$ time on-line, then an algorithm can be constructed which recognizes the same language on-line on RAM in $O(n^s/\log n)$ time, for $s > 1$.

Let $G=(V_N,V_T,P,S)$ be a context-free grammar in Chomsky normal form, where: V_N,V_T are the sets of, respectively, nonterminal and terminal symbols, P is the set of productions and S is the starting symbol. Whenever we write $A \to v$ this means that $A \to v$ is a production from P Let X denote the set of all subsets of V_N. We write $A \xrightarrow{*} v$ iff the string v can be derived from A in the grammar G. Let us fix G and let $w=a_1 \ldots a_n$ be a given input word of the length n.

We say that the triple (A,i,j) is realizable iff $A \xrightarrow{*} a_{i+1} \ldots a_j$, for $0 \leq i < j \leq n$. R is the set of all realizable triples.

It can be proved that R is the minimal set satisfying:

a) $(A,i,i+1) \in R$ for every $0 \leq i < n$, $A \to a_{i+1}$;

b) if $(B,i,k),(C,k,j) \in R$ and $A \to BC$ then $(A,i,j) \in R$.

We represent R by the (parsing) matrix T of the type array[0..n,0..n] of X, such that $T(i,j) = \{A : (A,i,j) \in R\}$.We use an algebraic approach. Let $N_i,M_i \in X$, for $0 \leq i \leq n$. Define:

$N_1 * N_2 = \{A : A \to BC,\ B \in N_1, C \in N_2\}$;

$(N_0,\ldots,N_n) * (M_0,\ldots,M_n) = (N_0 * M_0,\ldots,N_n * M_n)$;

$(N_0,\ldots,N_n) \cup (M_0,\ldots,M_n) = (N_0 \cup M_0,\ldots,N_n \cup M_n)$;

$(N_0,\ldots,N_n) * M_1 = (N_0 * M_1,\ldots,N_n * M_1)$.

Let COL(j) denote the j-th column of T.

Algorithm 3

```
begin {j-th output symbol is 1 iff a1..aj is generated by G}
for j:=1 to n do {compute COL(j)}
   begin
   aj:=next input symbol;
   T(j-1,j):={A : A→aj};
 1: for k:=j-1 downto 0 do
      COL(j):=COL(j) ∪ COL(k) * T(k,j);
   if S ∈ T(0,j) then write(1) else write(0);
   end
end.
```

Initially the value of each entry of T is the empty set.

We start with the above algorithm. Its cost is $O(n^3)$ and it is dominated by the total cost of instruction 1. This cost was improved in [4] by dividing columns into sections. Let $p=\lceil \log_c(n)/2 \rceil$, where $c=|X|$. Assume for simplicity that n+1 is divisible by p. There are $O(n^{1/2})$ vectors of the length p whose entries are elements of X. We write column-vectors and column-sections horizontally. For a column-vector $D=(D_0,\ldots,D_{p-1})$ whose entries are elements of X, and $0 \leq i \leq n$ define:

$$\text{PRODUCT}(i,D) = \bigcup_{D_k \neq \emptyset} \text{COL}(i+k) * D_k .$$

(if all D_k's are empty then the result is $\emptyset$).

Whenever we refer to COL(j) or colsection(r,j) we refer to the corresponding part of T.

<u>Algorithm</u> 4

```
begin
for j:=1 to n do
  begin
  a_j:= next input symbol;
  T(j-1,j):= {A : A -> a_j} ;
 {compute COL(j) section by section starting with the section con-
      taining entry T(j-1,j)}
  for r:=⌈(j-1)/p⌉ downto 0 do  {compute r-th section of COL(j)}
   begin
  1: for i:=rp+p-1 downto rp do
     if T(i,j) ≠ ∅ then
     colsection(r,j):=colsection(r,j) ∪ colsection(r,i)* T(i,j);
  2: H:= PRODUCT(i,colsection(r,j));  {i=rp}
  3: COL(j):= COL(j) ∪ H ;
   end;
  if S ∈ T(0,j) then write(1) else write(0);
  end
end.
```

Observe that immediately after executing instruction 1 we have i=rp. Algorithm 4 is a refinement of Algorithm 3. Instruction 1 in Algorithm 3 is replaced by the instruction computing COL(J) section by sections .We refer the reader to [4] where a similar technique was used, however we eliminated here the operation predict.After executing instruction 1 in Algorithm 4 colsection(r,j) is completely computed, now instruction 3 is equivalent to $\text{COL}(j):=\text{COL}(j) \cup \text{COL}(rp) * T(rp,j) \cup \text{COL}(rp+1) * T(rp+1,j) \cup \ldots \cup \text{COL}(rp+p-1) * T(rp\ p-1,j)$.

Instructions (1-3) are executed $O(n^2/\log n)$ times. The cost of instruction 1 is $O(\log^2 n)$, this gives in total $O(n^2 \log n)$ cost. The total cost of instruction 2 can be reduced by applying the tabulation method We store the computed values of PRODUCT(i,D) in the table TAB(i,D). Each entry of this table initially contains the special value "undefined". Replace instruction 2 by:

2′: {i=rp} D:=colsection(r,j);
    <u>if</u> TAB(i,D)="undefined" <u>then</u> TAB(i,D):=PRODUCT(i,D);
    H:=TAB(i,D);

In this manner the total cost of executed instructions 2 (after replacement) is $O(n^{2.5}\log n)$, because there are only $O(n^{1.5})$ possible pairs (i,D) and for each such pair we compute PRODUCT at most once.

Now the cost of instruction 3 is dominating. The total cost of executed instructions 3 is $O(n^3/\log n)$. The key to the improvement of the whole algorithm is a fast implementation of this instruction.

<u>Theorem</u> 3

Every context-free language can be recognized on-line in $O(n^3/\log^2 n)$ time.

Proof.

We encode each colsection(r,j) by a number in the range $[0..2^p-1]$ and obtain the matrix C of compressed columns of T analogously as in the case of boolean matrices. However the entries of T are subsets of V_N, while the entries of A are zeros and ones.

The operation INSERT is modified. The effect of INSERT(A,k,j) is the insertion of the nonterminal A into T(k,j) and the appropriate change of the compressed matrix C. If H is a column-vector of the size n+1 whose entries are elements of X then compr(H) denotes the compressed version of H (a column vector of the size (n+1)/p whose entries are small integers). We introduce the operation DIF with two arguments a compressed column vector and an index of a column of T such that:

$$\mathrm{DIF}(\mathrm{compr}(H),j) = \{(A,k) : A \in H_k \text{ and } A \notin T(k,j)\}, \text{for } H=(H_0,\ldots,H_n).$$

It can be proved (by an implementation similar to that from Lemma 1) that after a preprocessing the cost of INSERT(A,k,j) is $O(\log n)$ and the cost of $\mathrm{DIF}(\hat{H},j)$ is $O(n/\log n + |\mathrm{DIF}(\hat{H},j)|\cdot\log n)$, where $\hat{H}$ is a compressed column-vector. The cost of preprocessing is small.

Assume now that the values of PRODUCT(i,D) are compressed column-vectors. This can be done by changing slightly instruction 2′. The extra cost is small (we leave the details to the reader). Instead of the column H we have its compressed version $\hat{H}$ computed in (modified) instruction 2′. Now we replace instruction 3 by:

3': <u>for</u> <u>each</u> $(A,k) \in DIF(\hat{H},j)$ <u>do</u> INSERT(A,k,j).

The cost of instruction 3' is $O(n/\log(n) + v \log n)$, where v is the number of inserted triples (A,k,j). Every triple is inserted at most once, hence the numbers v give in total $O(n^2)$. We execute $O(n^2/\log n)$ instructions 3' .Hence the total cost of the algorithm (after modifications) is $O(n^3/\log^2 n)$.

Now we overcome the fact that n is not known in advance (the algorithm is on-line). Initially the algorithm assumes that n=2. In general, it assumes that $n=2^k$ and if the (2^k+1)th symbol is read the computation starts over with $n=2^{k+1}$ suppressing the initial outputs (see [3]). This does not change the order of time complexity. This completes the proof.

<u>References</u>

[1] A.Aho,J.Hopcroft,and J.Ullman. "Time and tape complexity of push-down automaton languages. Information and Control 13(1968) 186-206

[2] R.Bird. "Tabulation techniques for recursive programs".ACM.Computing Surveys 12(4) (1980)

[3] Z.Galil. "Two fast simulations which imply some fast string matching and palindrom recognition algorithms.Inf.Proc.Letters 4(1976) 85-87

[4] S.Graham,M.Harrison,and W.Ruzzo."On-line context-free recognition in less than cubic time". 8th ACM Symp. on the Theory of Computing (1976)

[5] S.Graham,and M.Harrison."Parsing of general context-free languages" Advances in Computers 14 (1976) 77-185

[6] M.Harrison."Introduction to formal language theory". Chapter 12. Addison-Wesley (1978)

[7] W.Rytter."Time complexity of loop-free two-way pushdown automata". Inf.Proc.Letters 16 (1983) 127-129

[8] W.Rytter."A note on two-way nondeterministic pushdown automata". Inf.Proc.Letters 15 (1982) 5-9

[9] W.Rytter."Time complexity of unambiguous path systems". Inf.Proc. Letters 15 (1982) 102-104

MULTIPROCESSOR SYSTEMS AND THEIR CONCURRENCY

Peter H. Starke
Sektion Mathematik
der Humboldt-Universität zu Berlin
DDR-1086 Berlin, PSF 1297

1. Introduction

Distributed systems are systems which consist of cooperating parts residing at different locations. These parts are connected by channels transmitting the messages sent and received by the processes realized by the different parts of the system. We can assume that at every location exactly one strictly sequential process is performed which interchanges messages with other processes in the system and which receives and processes inputs from its local environment.

At this level of abstraction there is only a very small difference between a distributed system and a multiprocessor system which is understood as a system consisting of several concurrently working sequential processors acting on a partially shared memory. The only difference between a channel on the one hand and a buffer on the other hand is that a channel leads somewhere, i.e. a channel is a memory item which can be read by only one processor. In this view, distributed systems are special multiprocessor systems.

PRINOTH (1),(2) has defined abstract distributed systems as collections of finite deterministic (incompletely specified) automata modelling the processors and of communication places (channels). Thereby the channels act as pre- and post-conditions for the state transitions of the automata. Here we propose a notion of an abstract multiprocessor system which is more general and more convenient to deal with. This notion includes REISIG's buffer synchronized systems of sequential machines (3) as a special case too (cf. also (7)).

The first point of our generalization is that we allow an arbitrary finite number of data units to be received from or to be stored into every channel or buffer during a single state transition instead of only one unit.

Our second point consists in allowing the state transitions of the processors to depend on a disjunction of preconditions referring to the channels rather than to depend on a single condition. This results in a certain amount of nondeterminism because now a single state transition can cause different changes in the memory content if two or more preconditions (with different postconditions) in this disjunction are fulfilled. We can show that the distribution problem is not solvable without nondeterminism.

If a real system is described as an abstract multiprocessor system, it is not hard to construct a Petri net such that the behaviour of the net sufficiently close describes the behaviour of the original system. Considering semilanguages (sets of semiwords) as the behaviour of multiprocessor systems and Petri nets we are, for the first time, in a position to compare such systems with respect to their concurrency and to measure the concurrency of a proposed behaviour.

2. Basic definitions

As mentioned above, at our level of abstraction the notions of a distributed system on the one hand and of a multiprocessor system on the other hand differ only in the so-called channel property.

Definition 1

The (n+4)-tuple $\Upsilon = (\underline{A}^1,\ldots,\underline{A}^n,C,K,k_0,\varkappa)$ is said to be an n-processor system iff the following conditions hold:

(a) For $i=1,\ldots,n$ $\underline{A}^i = (X_i,Z_i,f_i,z_{1i},M_i)$ is a deterministic finite automaton with input set X_i, state set Z_i, transition function $f_i\colon Z_i \times X_i \longrightarrow Z_i$, initial state z_{1i}, and the set M_i of designated final states;

(b) $\hat{C}$ is a finite set;

(c) The sets $X_1,\ldots,X_n,Z_1,\ldots,Z_n,C$ are mutually disjoint;

(d) $\varkappa$ is a mapping from the set $\bigcup_{i=1}^{n}(Z_i \times X_i)$ of all local situations into the set of all finite nonempty subsets of $\mathbb{N}^C \times \mathbb{N}^C$;

(e) $k_0 \in \mathbb{N}^C$ and K is a finite subset of $\mathbb{N}^C$.

Interpretation. The automata $\underline{A}^i$ are the local processors of the system Υ and the elements $c \in C$ are the channels or buffers of Υ; we prefer to call them channels. Every mapping $k\colon C \longrightarrow \mathbb{N}$ is called a channel content; $k(c)$ is the number of messages or data units the channel c actually contains. Hence, k_0 is the initial channel content. K is the set of designated final channel contents. The automata $\underline{A}^i$ are not synchronized by a common clock but work within independent time scales. Their behaviour differs from the behaviour of ordinary automata only in that they send and receive messages during their transition from state to state caused by certain inputs: If $\underline{A}^i$ is in its state $z \in Z_i$ and the input $x \in X_i$ is applied, then, in order to transit to its state $f_i(z,x)$, $\underline{A}^i$ has to select a pair (k^-,k^+) from $\varkappa(z,x)$ in such a way that $k^- \leq k$, i.e. $\forall c(c \in C \longrightarrow k^-(c) \leq k(c))$ holds, where k is the actual channel content. If this is not possible, the state transition is blocked (must wait). After selecting an appropriate pair $(k^-,k^+) \in \varkappa(z,x)$ the transition from z to $f_i(z,x)$ takes place and changes the channel content from k to $k' := k - k^- + k^+$ (componentwise), i.e. during this state transition for every c C $k^-(c)$ data units are received from the channel c and $k^+(c)$ messages are sent into c. We observe that the new channel content is not uniquely determined in general although the state transition itself is deterministic.

Definition 2

Let $\mathcal{T}$ be as in Def. 1.

1. $\mathcal{T}$ is said to be a distributed system iff $\mathcal{T}$ has the channel property, i.e. iff for all i,j with $1 \leq i < j \leq n$, all pairs $(k_i^-,k_i^+) \in \varkappa(Z_i \times X_i)$, $(k_j^-,k_j^+) \in \varkappa(Z_j \times X_j)$ and for all $c \in C$ it holds
$$k_i^-(c) \cdot k_j^-(c) = 0.$$
2. A global state of $\mathcal{T}$ is an (n+1)-tuple $(z_1,\ldots,z_n,k)$ where $z_i \in Z_i$ are the states of the local automata and where k is the channel content. The initial state of $\mathcal{T}$ is $(z_{11},\ldots,z_{1n},k_o)$ and the designated final states form the set $M_1 \times \ldots \times M_n \times K$.
3. The system $\mathcal{T}$ is said to be deterministic iff the mapping $\varkappa$ has only singleta as values.
4. We call $\mathcal{T}$ weakly deterministic iff for every global state $(z_1',\ldots,z_n',k')$ reachable from the initial state and for all $i=1,\ldots,n$, $x_i \in X_i$ the set $\varkappa(z_i',x_i)$ contains at most one pair (k^-,k^+) such that $k^- \leq k'$.
5. $\mathcal{T}$ is said to be weakly persistent iff for every reachable global state it holds: If two processors are enabled then they can perform their state transitions concurrently. $\mathcal{T}$ is called persistent iff it is weakly persistent and weakly deterministic.

The channel property demands that every channel supplies at most one processor (hence, it leads to a certain processor) although it may be fed by several automata. It seems reasonable to demand that distributed systems should have the channel property because in general different processors reside at different locations so that they cannot be supplied by the same channel. On the other hand, within an arbitrary multiprocessor system, buffers will be shared among certain processors in general. From Def. 2 it follows

Corollary 1:

Every distributed system is weakly persistent.

Translated into our terminology, PRINOTH (1) considered distributed systems, whose initial channel content k_o is identically zero, the set of designated final channel contents is the singleton containing k_o, and, which are deterministic and Boolean, i.e.

for all (k^-,k^+) (z,x), c C it holds $k^-(c),k^+(c)$ 0,1 . Therefore, PRINOTH's distributed systems are persistent, since the channel property ensures that there are no channel conflicts and determinism results in the absence of inner conflicts of the processors.

As a first example we consider the four-processor system = $(\underline{A}^1,\ldots$ $\ldots,\underline{A}^4,C,K,k_o,$) with $C =$ c_1,c_2,c_3 , $k_o = (0,0,0)$, $K =$ k_o which is represented by Figure 1. There e.g., the dotted arc from the arrow z_{11} --- z_{21} to c_1 means that during this state transition one message is send to c_1, i.e. $(z_{11},x_1) = ((0,0,0),(1,0,0))$.

Obviously, is deterministic and Boolean but not weakly persistent and the channel property is not fulfilled.

3. Multiprocessor systems and Petri nets

To describe the concurrent behaviour of systems which are modelled by Petri nets, in our paper (4) we developed the calculus of semiwords

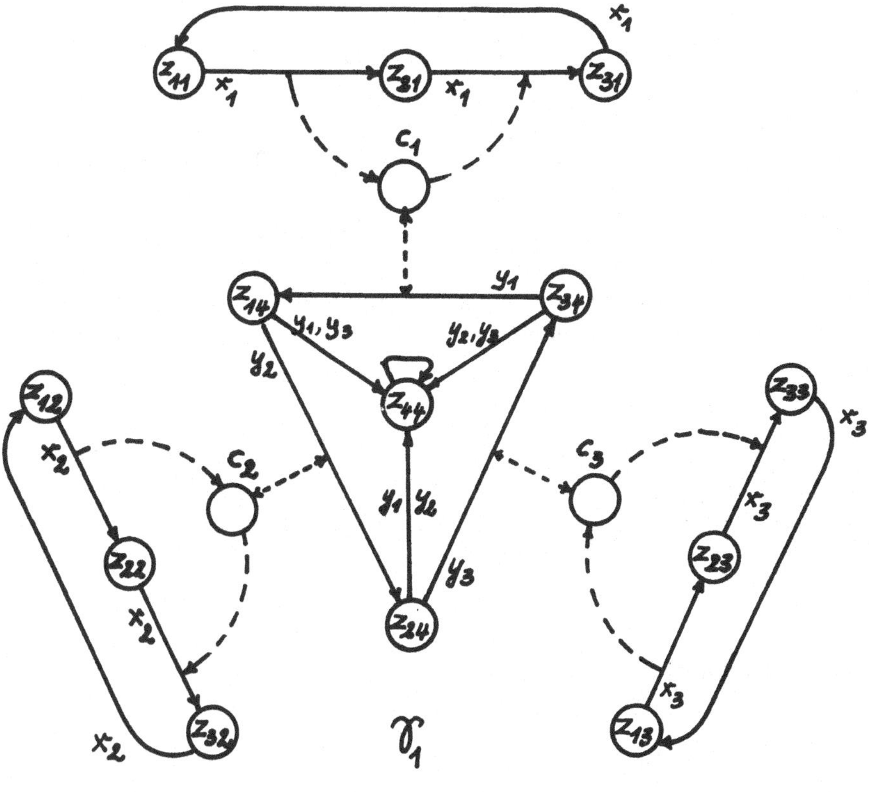

Fig. 1

and semilanguages. Now we adjoin a labelled Petri net with every multiprocessor system in order to apply this calculus. In the sequel we use the terminology and results of (4).

Let $\mathcal{T} = (\underline{A}^1,\dots,\underline{A}^n,C,K,k_0,\varkappa)$ be a multiprocessor system. The Petri net $N_{\mathcal{T}} = (P,T,F,V,m_0)$ corresponding to $\mathcal{T}$ is obtained by transforming the automata into Petri nets in the usual way (cf. (5)) while the buffers become (communication) places of the net. Hence,

$$P := C \cup \bigcup_{i=1}^{n} Z_i$$

and we put

$$T := \{(z,x,k^-,k^+) / (k^-,k^+) \in \varkappa(z,x)\} .$$

Every transition $t = (z,x,k^-,k^+)$ is labelled with the letter

$$h(t) := x,$$

i.e. $h: T \longrightarrow X := \bigcup_{i=1}^{n} X_i$ is an e-free labelling of $N_{\mathcal{T}}$. The initial marking m_0 is

$$m_0(p) := \begin{cases} 1, \text{ if } & p = z_{1i} \text{ for some } i \\ k_0(p), \text{ if } & p \quad C, \\ 0 \;, \text{ else,} & \end{cases}$$

and the set $M_{\mathcal{T}}$ of terminal markings of $N_{\mathcal{T}}$ consists of all reachable markings which on C coincide with an element of K and which mark only final states. The arcs F and their multiplicities V are determined by the transition functions f_i and by $\varkappa$; we omit the details but refer to Figure 2 where the net obtained from $\mathcal{T}$ is represented.

The nonterminal behaviour $SL_{N_{\mathcal{T}}}(m_0)$ and the terminal behaviour $SL_{N_{\mathcal{T}}}(m_0,M_{\mathcal{T}})$ are semilanguages over T, the set of all transitions of $N_{\mathcal{T}}$ which correspond to the state transitions of the system $\mathcal{T}$. Since the external behaviour of $\mathcal{T}$ consists only in concurrent reads of inputs at different locations by different processors we have to consider the images of these semilanguages under the projection h.

Let $q = (\!(A,R,\beta)\!)$ be a semiword over T, i.e. (A,R) is a finite irreflexive partial ordering, $\beta: A \longrightarrow T$ is a mapping such that for $a,b \in A$

$$\beta(a) = \beta(b) \longrightarrow aRb \vee bRa \vee a=b$$

holds and q is the class of all finite labelled partial orderings which are isomorphic with (A,R,β). Then $h(q)$ is defined as

$$h(\,(\!(A,R,\beta)\!)\,) := (\!(A,R,h\cdot\beta)\!).$$

The result of the application of h is a semiword again:

<u>Theorem 2</u>

1. If $q \in SL_{N_{\mathcal{T}}}(m_0)$ then $h(q)$ is a semiword over X.
2. If $\mathcal{T}$ is weakly deterministic then for every $w \in SW(X)$ there is at most one $q \in SL_{N_{\mathcal{T}}}(m_0)$ such that $h(q) = w$.

The first part of Theorem 2 is a special case of

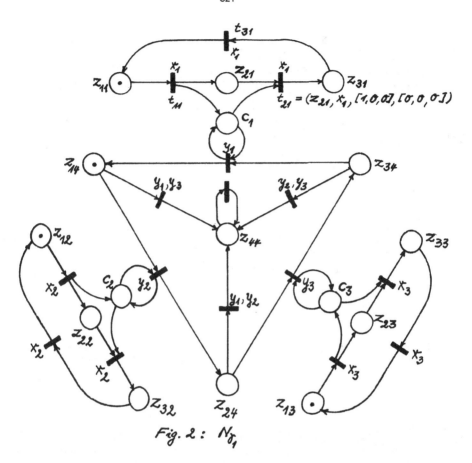

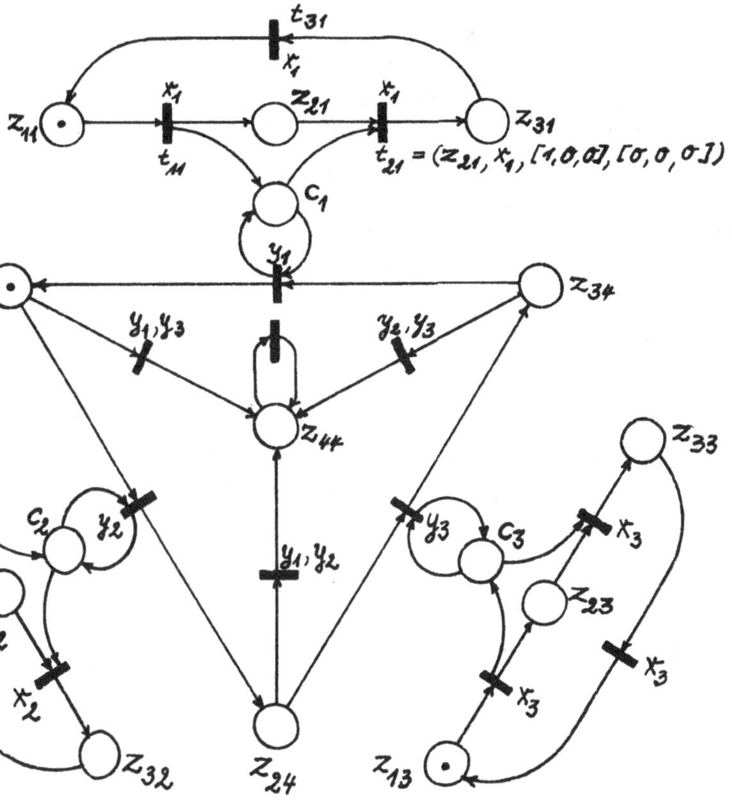

Fig. 2: N_{γ_1}

Theorem 3

If $N = (P,T,F,V,m_o)$ is a Petri net and $h\colon T \longrightarrow X$ is an e-free labelling such that the disjoint labelling condition (cf. (6))

$$\exists m(m \in R_N(m_o) \ \& \ t_1^- + t_2^- \leq m) \longrightarrow h(t_1) \neq h(t_2)$$

holds for all $t_1, t_2 \in T$ then $h(q)$ is a semiword for all $q \in SL_N(m_o)$.

Every multiprocessor system can be considered as an e-free disjointly labelled Petri net. The converse is also true:

Theorem 4.

Every e-free disjointly labelled Petri net is equivalent to a multiprocessor system.

In the proof a multiprocessor system is constructed which, for every label x of the Petri net, contains exactly one one-state automaton with the input set $\{x\}$.

Let γ be a multiprocessor system. Then

$$NSL(\mathcal{T}) := h(SL_{N_{\mathcal{T}}}(m_o))$$

is the nonterminal semilanguage of $\mathcal{T}$ and

$$TSL(\mathcal{T}) := h(SL_{N_{\mathcal{T}}}(m_o, M_{\mathcal{T}}))$$

is the terminal semilanguage of $\mathcal{T}$. By the preceding results the families NSL (resp. TSL) of all nonterminal (resp. terminal) semilanguages of multiprocessor systems coincide with the families of all nonterminal (resp. terminal) semilanguages of e-free disjointly labelled Petri nets. In our paper (4) we gave characterizations of the corresponding families of semilanguages of free (unlabelled) Petri nets which result in characterizations of NSL and TSL.

4. The distribution problem

In the paper (1) PRINOTH considered deterministic Boolean distributed systems and their behaviour - regular languages (not semilanguages). The distribution problem asks for an algorithm which for every finite deterministic automaton $\underline{A}$ and for every partition $\mathcal{Y} = \{X_1, \ldots, X_n\}$ of its input set X constructs a distributed system $\mathcal{T}$ containing as many processors as $\mathcal{Y}$ has classes and such that

$$L(\underline{A}) = L(\mathcal{T}) := TSL(\mathcal{T}) \cap W(X)$$

i.e. the terminal languages of $\underline{A}$ and $\mathcal{T}$ coincide.

Remembering that every finite deterministic automaton is equivalent to a disjointly labelled Petri net and using the method applied in the proof of Theorem 4 we are able to solve the distribution problem, i.e. to construct a Boolean weakly deterministic multiprocessor system with the same language.

Having semilanguages in mind, i.e. comparing not only the languages but also the semilanguages of the considered systems, a distributed system in general is not equivalent with an automaton because in this system some external events can be accepted and processed concurrently while the automaton can process them only sequentially.

At this point we should explain why we have introduced a certain amount of nondeterminism into our multiprocessor systems:

Theorem 5

There exist finite automata which are not equivalent to any deterministic multiprocessor system, i.e. which do not admit any deterministic nontrivial distribution.

We consider e.g. the automaton $\underline{A}$ given by Figure 3. The only nontrivial partition of $X = \{0,1\}$ is $\mathcal{Y} = \{X_1, X_2\}$ with $X_1 = \{0\}$, $X_2 = \{1\}$. Our method provides the two-processor system represented in Figure 4. One can show that every deterministic system processing the word 010 reaches the same state as if it processes 001 which contradicts its equivalence with $\underline{A}$.

5. The concurrency measure

We are now going to introduce a numerical measure of the concurrency of systems and of their behaviour.

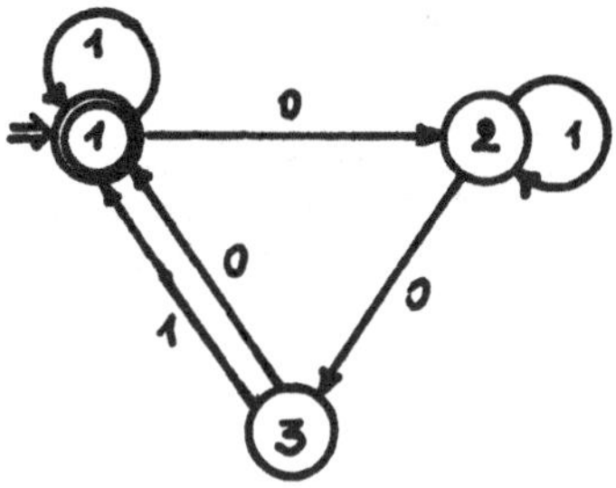

Fig. 3

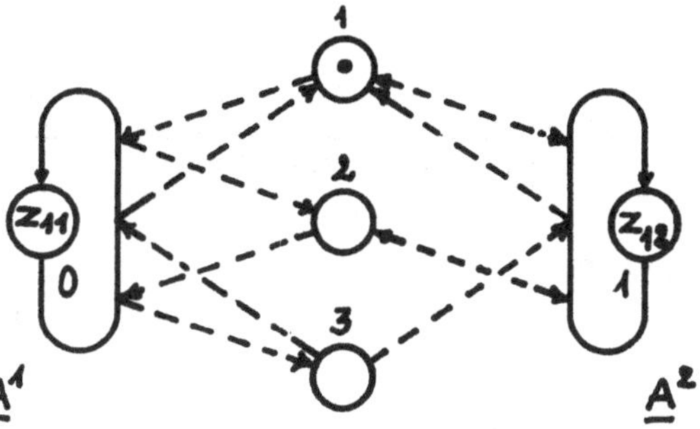

Fig. 4 $C = \{1,2,3\}$

Definition 3

1. If $S \in NSL$ is a nonterminal system semilanguage then
   $Co(S) := \min\{n \;/\;$ There is an n-processor system γ with $S = NSL(\gamma)\}$
   is the concurrency of S. The terminal concurrency Cot(S) of terminal system languages S is defined analogously.
2. If γ is a multiprocessor system then
   $Co(\gamma) := Co(NSL(\gamma))$ (resp. $Cot(\gamma) := Cot(TSL(\gamma))$)
   is the concurrency (resp. terminal concurrency) of γ.

It will become clear later that for $S \in NSL \cap TSL$ the equation $Co(S) = Cot(S)$ holds. For every regular language L we have $Cot(L) = 1$, hence sequential processes have, as desired, the concurrency 1. Next, one can show that the "width" of the processes realized by a system gives a lower bound to its concurrency:

Theorem 6

Let $S \subseteq SW(X)$ be a nonterminal (resp. terminal) system semilanguage. Then

$$card(X) \geq Co(S) \text{ (resp. } Cot(S)) \geq \max_{q \in S} \max\{card(B)/B \text{ antichain in } q\}$$

The inequalities in Theorem 6 are proper in general.

6. The reconfiguration problem

Given a multiprocessor system γ and a partition $\mathcal{Y} = \{Y_1,\ldots,Y_l\}$ of its input set X the reconfiguration problem demands a method to construct (if existing) an l-processor system $\hat{\gamma}$ equivalent to γ such that the processor $\underline{A}^j$ of $\hat{\gamma}$ has Y_j as its input set. From the preceding section we learned that it is not possible to distribute any system according to any partition of the input set. Therefore we first derive a necessary condition:

Definition 4

Let $S \subseteq SW(X)$ be a semilanguage and $x,y \in X$. Then x and y are said to be on a line with respect to S (abbr. $x \overline{\;S\;} y$) iff within all semiwords q from S, whenever x and y occur as labels in q, the corresponding actions are related, i.e.

$x \underset{S}{—} y :\longleftrightarrow \forall A \forall R \forall B(\ ((A,R,B)) \in S\ \&\ a,b \in A\ \&\ B(a)=x\ \&\ B(b)=y \longrightarrow$
$\longrightarrow aRb \vee bRa \vee a=b\ ).$

The relation "on a line" is reflexive and symmetric but in general not transitive, i.e. it is a compatibility relation. With respect to $S = \{x(y \times z)\} \subseteq SW(\{x,y,z\})$ we have $y \underset{S}{—} x$, $x \underset{S}{—} z$ but not $y \underset{S}{—} z$.

Corollary 7

If $S \subseteq S'$ then the relation $\underset{S'}{—}$ is contained in the relation $\underset{S}{—}$.

Theorem 8

Let γ be a multiprocessor system and $S = NSL(\gamma)$. Then every input set X_i of a processor $\underline{A}^i$ of γ is a compatibility class with respect to $\underset{S}{—}$

Hence, every input set of a processor of a multiprocessor system γ is a compatibility class with respect to $\underset{TSL(\gamma)}{—}$. This follows from $TSL(\gamma) \subseteq NSL(\gamma)$ and Corollary 7.

Let li(S) denote the minimal number of blocks of a partition of the alphabet of S such that every block is a compatibility class with respect to $\underset{S}{—}$.

Corollary 9

For $S \in NSL$ it holds $Co(S) \geq li(S)$ and for $S \in TSL$ we have $Cot(S) \geq li(S)$.

We have seen so far that a reconfiguration of a system γ according to a partition $\mathcal{Y}$ of its input alphabet is impossible if $\mathcal{Y}$ does not respect the on-a-line relation.

Theorem 10

Let γ be a multiprocessor system with the input set X, and let $\mathcal{Y}$ be a partition of X into compatibility classes with respect to $NSL(\gamma)$ ($TSL(\gamma)$ resp.). Then there exists a card($\mathcal{Y}$)-processor system $\hat{\gamma}$ such that $\mathcal{Y}$ is the set of the input alphabets of its components and such that $NSL(\gamma) = NSL(\hat{\gamma})$ ($TSL(\gamma) = TSL(\hat{\gamma})$ resp.).

Our method to prove Theorem 10 is to distribute the Petri net corresponding to γ according to $\mathcal{Y}$:

Theorem 11

Let $N = (P,T,F,V,m_0)$ be a Petri net, $M \subseteq \mathbb{N}^P$ a finite set of terminal markings, h: $T \longrightarrow X$ an e-free disjoint labelling of N, $S := h(SL_N(m_0))$, $S' := h(SL_N(m_0,M))$ and $\mathcal{Y}$ a partition of X into compatibility classes with respect to $\underset{S}{—}$ (to $\underset{S'}{—}$ resp.). Then there exists a card($\mathcal{Y}$)-processor system $\hat{\gamma}$ such that $\mathcal{Y}$ is the set of the input alphabets of its processors and such that $S = NSL(\hat{\gamma})$ ($S' = TSL(\hat{\gamma})$ resp.).

Corollary 12

For $S \in NSL$ it holds $li(S) = Co(S)$ and for $S \in TSL$ we have $li(S) = Cot(S)$.

Hence, the measures Co and Cot coincide on the intersection of their domain, if we apply them to semilanguages. Obviously, this is not the case if we measure systems, because e.g. the terminal system semilanguage may be empty, so its concurrency is 1 while the nonterminal system semilanguage of the same system may have an arbitrary amount of concurrency. From TSL($\mathcal{S}$) $\subseteq$ NSL($\mathcal{S}$), Cor. 7 and 12 we obtain

Corollary 13

For every multiprocessor system $\mathcal{S}$ it holds Cot($\mathcal{S}$) $\leq$ Co($\mathcal{S}$).

Hence, we have solved the reconfiguration problem: For a given multiprocessor system $\mathcal{S}$ a reconfiguration according to a given partition η of the input alphabet exists iff the partition η consists only of compatibility classes with respect to the on-a-line relation of the terminal (resp. nonterminal) behaviour of $\mathcal{S}$.

The full version of this paper is going to appear in the Journal of Information Processing and Cybernetics (EIK), Vol. 20 (1984).

7. References

(1) R. Prinoth: An algorithm to construct distributed systems from state-machines. Protocol Specification, Testing and Verification, C. Sunshine (ed.), North Holland Publ. Co. 1982, 261 - 282.

(2) R. Prinoth: Verteilungsalgorithmen für Zustandsmaschinen. Internal report GMD-IFV, 1981.

(3) W. Reisig: Deterministic buffer synchronization of sequential processes. Acta Informatica 18 (1982) 117 - 134.

(4) P.H. Starke: Processes in Petri nets. Elektron. Informationsverarbeitung und Kybernetik 17 (1981) 389 - 416.

(5) P.H. Starke: Petri-Netze. Grundlagen, Anwendungen, Theorie. DVW Berlin 1980.

(6) G. Rozenberg, R. Verraedt: Subset languages of Petri nets. Informatik-Fachberichte 66 (1983) 250 - 263.

(7) G. Berthelot, G. Memmi, W. Reisig: A control structure for sequential processes synchronized by buffers. Paper presented at the 4th European Workshop on Application ans Theory of Petri nets, Toulouse Sept. 1983.

FREE CONSTRUCTIONS IN ALGEBRAIC INSTITUTIONS

Andrzej Tarlecki*
Department of Computer Science, University of Edinburgh

Abstract

To provide a formal framework for discussing specifications of algebraic abstract data types we introduce the notion of an algebraic institution. Our main results concern the problem of the existence of free constructions in algebraic institutions. We review a characterization of logical specification systems that guarantee the existence of initial models for any consistent set of axioms given by Mahr and Makowsky in [MM 83a, MM 83b]. Then the more general problem of the existence of free functors (left adjoints to forgetful functors) for any theory morphism is analysed. We give a construction of a free model of a theory over a model of a subtheory (with respect to an arbitrary theory morphism) which requires only the existence of initial models.

1 Introduction

An abstract data type may be viewed as a pair: a signature and a set of axioms over this signature, which describes a class of many-sorted algebras that satisfy the specified axioms. Which formulae are actually accepted as axioms and what it means for an algebra to satisfy an axiom is determined by the logical system we use for our specifications. The informal notion of logical system has been formalized by Goguen and Burstall [GB 83], who introduced for this purpose the notion of institution (this generalizes the ideas of "abstract model theory" [Bar 74]). An institution consists of a collection of "abstract signatures" together with for any "signature" Σ a set of Σ-sentences, a collection of Σ-models and a satisfaction relation between Σ-models and Σ-sentences. The only "semantic" requirement is that when we change signatures, then the induced translations of sentences and models preserve the satisfaction relation.

In the present paper we consider an important special case of the abstract notion of institution (section 3). In an algebraic institution abstract signatures are many-sorted signatures in the usual algebraic sense and abstract models are just many-sorted algebras. We do not restrict the form of sentences, however. Besides the satisfaction condition, an algebraic institution is subject to the "abstractness condition" which reflects the fact that we are not interested in any particular representation of data objects. To guarantee a minimal specification power of algebraic institutions we also assume that in any algebraic institution all equations are expressible.

Specifications given in some standard institutions (e.g. in first-order logic) often are loose, i.e. admit many non-isomorphic models. This indicates that some mechanism for imposing additional constraints on the models admitted by a specification is necessary. The most widely accepted choice at this point is to require initiality (cf. [GM 83] for an extensive treatment of this notion). In this approach, from among all possible models of a set of axioms we choose as an acceptable realization of the abstract data type only the unique (up to isomorphism) initial model.

Unfortunately, not every class of algebras contains an initial algebra. Thus, if one wants

---

*On leave from Institute of Computer Science, Polish Academy of Sciences, Warsaw.

to avoid proving the existence of initial models for each specification separately he has to use an institution that guarantees the existence of initial algebras among the models of any consistent set of axioms. It is well-known that, for example, equational logic has this property but first-order logic has not.

In section 4 we briefly review a characterization of algebraic institutions which have this property due to Mahr and Makowsky [MM 83a, MM 83b]. Namely, in an algebraic institution the class of models of any specification contains an initial algebra with no "junk" (non-reachable elements) if and only if this algebraic institution is reducible to the institution of infinitary conditional equations and inequations.

It is well-known, however, that the initial algebra is not always the intended realization of an abstract data type. Quite often we want some parts of a data type to be interpreted loosely – and some others to be interpreted in a standard "initial" way given an interpretation of these "loose" parts. In other words we require that some part of a model must be a "free extension" of some other parts. This may be formally expressed using "initially restricting algebraic theories" [Rei 80] or, more generally, data constraints as introduced in [BG 80], cf. also [EWT 83, GB 83].

We investigate (section 5) the existence of the above-mentioned free extensions. Our basic result states that in a given algebraic institution these free extensions always exist if and only if the class of models of any set of axioms contains an initial algebra. Some further consequences of this fact are deduced from the Mahr and Makowsky result.

To define the meaning of data constraints Goguen and Burstall [GB 83] required the existence of the free extension for any interpretation of a subpart of an abstract data type. We point out (section 6) that this requirement may be slightly relaxed.

All the proofs, which due to space limitation are omitted here, may be found in [Tar 83].

2 Preliminaries

We assume that the reader is familiar with the notions of (algebraic) signature, (total) many-sorted algebra and homomorphism (see e.g. [ADJ 76]).

For any signature Σ, $Alg(\Sigma)$ denotes the category of Σ-algebras and Σ-homomorphisms. In the following we identify a class K of Σ-algebras with the full subcategory of $Alg(\Sigma)$ having objects K.

Let Σ be a signature with sorts S.

Let X be an S-sorted set. The algebra $T_\Sigma(X)$ of Σ-terms with variables X is defined in the standard way (see e.g. [ADJ 76]). We write T_Σ for $T_\Sigma(\emptyset)$, the algebra of ground terms.

For any Σ-algebra A by a valuation of variables X into A we mean an S-sorted function $v: X \to |A|$. For any valuation $v: X \to |A|$ there is exactly one Σ-homomorphism $v^{\#}: T_\Sigma(X) \to A$ that extends v. If X is empty then there is the unique (empty) function $\emptyset: X \to |A|$, and hence there is the unique homomorphism $\emptyset^{\#}: T_\Sigma \to A$. For a ground term t of a sort s we write t_A rather than $(\emptyset^{\#})_s(t)$.

We say that a Σ-algebra A is generated by an S-sorted set X, where $X \subseteq |A|$, if the homomorphism $\iota^{\#}: T_\Sigma(X) \to A$ induced by the inclusion $\iota: X \to |A|$ is surjective. An algebra is called reachable if it is generated by the empty set.

By an (explicitly quantified – cf. [GM 81] where the necessity of explicit quantification was pointed out) Σ-equation we mean any triple $\langle X, t1, t2 \rangle$, written as $\forall X. t1=t2$ (or $t1=t2$ if $X=\emptyset$), where X is an S-sorted set (of variables) and t1, t2 are Σ-terms of the same sort with variables X.

Let A be a Σ-algebra and $e=\langle X, t1, t2 \rangle$ be a Σ-equation. We say that e holds in A (or A satisfies e), written $A \models e$, if for any valuation $v: X \to |A|$, $v^{\#}(t1)=v^{\#}(t2)$. Note that any Σ-equation holds in the empty Σ-algebra (if it exists) and in any algebra containing only one element of every sort ("the" one-element algebra, for short).

Let $\Sigma 1=\langle S1,\Omega 1\rangle$ and $\Sigma 2=\langle S2,\Omega 2\rangle$ be signatures.

A signature morphism $\sigma:\Sigma 1\rightarrow\Sigma 2$ is a pair $\langle\sigma_{sorts},\sigma_{opns}\rangle$ where $\sigma_{sorts}:S1\rightarrow S2$ and σ_{opns} is an $S1^*\times S1$-sorted function $\sigma_{opns}:(\Omega 1_{u,s})_{u\in S1^*,s\in S1}\rightarrow(\Omega 2_{\sigma(u),\sigma(s)})_{u\in S1^*,s\in S1}$, where for $u=s1\ldots sn\in S1^*$ $\sigma(u)=\sigma(s1)\ldots\sigma(sn)$. We usually denote both components of a signature morphism σ by the same symbol as the signature morphism itself – just σ.

A signature morphism translates names of sorts and function symbols preserving their arities and sorts. This translation of "syntax" in a natural way determines a translation of semantic objects – algebras – but in the opposite direction.

Let $\sigma:\Sigma 1\rightarrow\Sigma 2$ be a signature morphism. For any $\Sigma 2$-algebra A, σ determines a $\Sigma 1$-algebra $U_\sigma(A)$ defined in the following way: for $s\in S1$ $|U_\sigma(A)|_s=|A|_{\sigma(s)}$ and for $f:s1\times\ldots\times sn\rightarrow s$ $f_{U_\sigma(A)}=\sigma(f)_A$. This translation amounts also to a translation of homomorphisms. It is easy to check that U_σ is a functor from the category of $\Sigma 2$-algebras to the category of $\Sigma 1$-algebras. We call it the forgetful functor for σ.

A signature morphism $\sigma:\Sigma 1\rightarrow\Sigma 2$ determines in a natural way a translation of $\Sigma 1$-terms to $\Sigma 2$-terms. We denote this translation just by σ. Then, this naturally extends to a translation of $\Sigma 1$-equations to $\Sigma 2$-equations (again denoted by σ) given by $\sigma(\forall X.\,t1=t2)=\forall\bar{X}.\,\sigma(t1)=\sigma(t2)$, where for any S1-sorted set X, $\bar{X}$ is an S2-sorted set defined by $\bar{X}_{s2}=\bigcup\{X_{s1}|\sigma(s1)=s2\}$. (N. B. this overloading of the symbol of signature morphism will never lead to confusion.)

3 Algebraic institutions

To formalize the notion of a logical system in a very general, abstract setting, Goguen and Burstall [GB 83] introduced the notion of institution. We concretize their ideas to the standard algebraic case. Our algebraic institutions are supposed to provide syntax and semantics to specify classes of algebras.

An algebraic institution INS consists of

- for each signature Σ a set (of Σ-sentences) $Sen_{INS}(\Sigma)$ and for each signature morphism $\sigma:\Sigma 1\rightarrow\Sigma 2$ a function $Sen_{INS}(\sigma):Sen_{INS}(\Sigma 1)\rightarrow Sen_{INS}(\Sigma 2)$ (a translation of $\Sigma 1$-sentences to $\Sigma 2$-sentences) such that the mapping $\sigma\longmapsto Sen_{INS}(\sigma)$ preserves the identities and the composition of signature morphism, i.e. this part of an institution is just a functor from the category of signatures to the category of sets.
- for each signature Σ a relation (called the satisfaction relation)
$$\models_{INS,\Sigma}\subseteq Alg(\Sigma)\times Sen_{INS}(\Sigma)$$

subject to the following conditions:

- (satisfaction condition)
for any signature morphism $\sigma:\Sigma 1\rightarrow\Sigma 2$, $A\in Alg(\Sigma 2)$ and $\psi\in Sen_{INS}(\Sigma 1)$
$$U_\sigma(A)\models_{INS,\Sigma 1}\psi \text{ iff } A\models_{INS,\Sigma 2}Sen_{INS}(\sigma)(\psi)$$
- (abstraction condition)
for any signature Σ, $\psi\in Sen_{INS}(\Sigma)$ and isomorphic $A,B\in Alg(\Sigma)$
$$A\models_{INS,\Sigma}\psi \text{ iff } B\models_{INS,\Sigma}\psi$$
- (expressibility of equations)
for any signature Σ and any Σ-equation e there is a set of Σ-sentences $\Psi\subseteq Sen_{INS}(\Sigma)$ such that for any $A\in Alg(\Sigma)$
$$A\models e \text{ iff for all } \psi\in\Psi\ A\models_{INS,\Sigma}\psi$$

Notational conventions:

- We omit subscripts (INS, Σ) whenever possible.
- For any signature morphism σ we write just σ instead of $Sen_{INS}(\sigma)$.

- For $\Psi \subseteq Sen(\Sigma)$ and $K \subseteq Alg(\Sigma)$ we write $K \models \Psi$ with the obvious meaning.
- For any signature Σ and $\Psi \subseteq Sen(\Sigma)$ by $Mod(\Psi)$ we mean the collection of all Σ-algebras that satisfy Ψ.

For the sake of compactness of the statement of some results we accept the following additional restrictions:

- If $\sigma: \Sigma1 \to \Sigma2$ is a signature inclusion then $\sigma: Sen(\Sigma1) \to Sen(\Sigma2)$ is an inclusion too.
- For any signature Σ, all Σ-equations are just Σ-sentences (elements of $Sen(\Sigma)$) rather than only being expressible via sets of Σ-sentences.

Note again that our algebraic institutions are institutions in the sense of [GB 83]. We concretize their abstract notion assuming that signatures are just standard algebraic signatures and models are just many-sorted algebras. Additional requirements (the abstraction condition and expressibility of equations) seem very natural from the algebraic point of view. However, the abstraction condition excludes some more general forms of specifications, which quite naturally arise in a study of specification languages or constraints (see [ST 84]). Nevertheless, in this paper we assume that hereafter any class of algebras which we consider is closed under isomorphism.

Our notion of an algebraic institution differs slightly from the notion of an algebraic specification language due to Mahr and Makowsky [MM 83a]. We use arbitrary signature morphisms rather than only renamings (bijective signature morphisms) and (implicit in [MM 83a]) signature inclusions. This seems to be essential for a study of specification languages, data constraints and parameterized specifications (cf. [BG 80, EWT 83, GB 83, ST 84]). We have not found enough justification for the requirement that "false" must be a sentence in an algebraic institution; after all, one should avoid inconsistent specifications rather than guarantee their existence.

Here are some examples of algebraic institutions which we use in the rest of the paper:

- EQ – the institution of equations. For any signature Σ, $Sen_{EQ}(\Sigma)$ is just the set of all Σ-equations with the satisfaction relation as defined in section 2.
- ICEQ – the institution of infinitary conditional equations. For any signature Σ, $Sen_{ICEQ}(\Sigma)$ consists of all formulae of the form $\forall X. \{t1_j = t2_j\}_{j \in J} \Rightarrow t1 = t2$ where X is an S-sorted set, J is an arbitrary set of indices and $t1_j$, $t2_j$ for $j \in J$ and $t1$, $t2$ are Σ-terms of pairwise the same sorts with variables X. The satisfaction relation is standard again: a Σ-algebra A satisfies the above formula if for any valuation $v: X \to |A|$ $v^{\#}(t1) = v^{\#}(t2)$ whenever $v^{\#}(t1_j) = v^{\#}(t2_j)$ for all $j \in J$.
- NEQ – the institution of universally quantified equations and inequations (i.e. negations of equations). For any signature Σ, $Sen_{NEQ}(\Sigma)$ consists of all Σ-equations (with the usual satisfaction relation) and all Σ-inequations, i.e. formulae of the form $\forall X. t1 \neq t2$ where X is an S-sorted set of variables and $t1$, $t2$ are Σ-terms of the same sort with variables X. A Σ-algebra A satisfies the above inequation if for any variable valuation $v: X \to |A|$, $v^{\#}(t1) \neq v^{\#}(t2)$.
- ICNEQ – the institution of infinitary conditional equations and inequations. For any signature Σ, $Sen_{ICNEQ}(\Sigma)$ consists of all infinitary conditional equations (as ICEQ) and all infinitary conditional inequations, i.e. formulae of the form $\forall X. \{t1_j = t2_j\}_{j \in J} \Rightarrow t1 \neq t2$ where X, J, $t1_j$, $t2_j$, $t1$, $t2$ are as in the definition of ICEQ. A Σ-algebra A satisfies the above formula if for any valuation $v: X \to |A|$, $v^{\#}(t1) \neq v^{\#}(t2)$ whenever $v^{\#}(t1_j) = v^{\#}(t2_j)$ for all $j \in J$.

Let INS be an arbitrary institution. For any signature Σ, we say that a class of Σ-algebras $K \subseteq Alg(\Sigma)$ is definable in INS if there is a set of Σ-sentences $\Psi \subseteq Sen(\Sigma)$ such that K consists of exactly those Σ-algebras that satisfy Ψ, i.e. $K = Mod(\Psi)$.

In the rest of the paper we try to characterize algebraic institutions that have some desirable properties. To do this we need some way to compare algebraic institutions; the most natural device seems to be the notion of reducibility (cf. [MM 83a]).

For any two algebraic institutions INS1 and INS2, we say that INS1 is reducible to INS2 if any class of algebras definable in INS1 is also definable in INS2.

Let INS be an algebraic institution.

By a theory in INS we mean a pair $\langle\Sigma,\Psi\rangle$, where Σ is a signature and Ψ is a set of Σ-sentences which contains all its logical consequences. A bit more formally: for any signature Σ and $K\subseteq Alg(\Sigma)$ let $Sen(K)$ denote the set of all Σ-sentences that hold in K, i.e. $Sen(K)=\{\psi\in Sen(\Sigma) \mid K\models\psi\}$. A theory is a pair $\langle\Sigma,\Psi\rangle$ where $\Psi=Sen(Mod(\Psi))\subseteq Sen(\Sigma)$. If $T=\langle\Sigma,\Psi\rangle$ is a theory, we use the notation Mod(T) for the class of all T-models, i.e. all Σ-algebras that satisfy Ψ.

For any two theories $T1=\langle\Sigma1,\Psi1\rangle$ and $T2=\langle\Sigma2,\Psi2\rangle$, by a theory morphism from T1 to T2, $\sigma: T1\rightarrow T2$, we mean a signature morphism $\sigma:\Sigma1\rightarrow\Sigma2$ such that $\sigma(\psi)\in\Psi2$ for any $\psi\in\Psi1$. Note that if $\sigma: T1\rightarrow T2$ is a theory morphism then the forgetful functor U_σ translates T2-models to T1-models, $U_\sigma: Mod(T2)\rightarrow Mod(T1)$.

4 Initiality

It is often the case that from among all admissible models of a theory we would like to pick out only the initial one(s).

Recall that a Σ-algebra A is initial in a class of Σ-algebras K if $A\in K$ and for any $B\in K$ there is exactly one Σ-homomorphism from A to B. By the initial model of a theory we mean "the" (up to isomorphism) initial algebra in the class of all models of this theory.

Unfortunately, in general the class of models of a theory need not contain an initial algebra. However, there are algebraic institutions in which any theory has an initial model. In this section we try to characterize these institutions.

We say that an algebraic institution admits initial semantics if any non-empty class of algebras definable in this institution contains an initial algebra.

Fact 1 The institutions EQ, NEQ, ICEQ and ICNEQ admit initial semantics.

Fact 2 For any two algebraic institutions INS1 and INS2, if INS1 is reducible to INS2 and INS2 admits initial semantics then INS1 admits initial semantics as well.

In [MM 83a] Mahr and Makowsky announced that an algebraic institution admits initial semantics iff it is reducible to ICEQ, the institution of infinitary conditional equations. Unfortunately, this is not true (cf. Fact 1). Following the proof given in [MM 83a], to get a correct result we have to exclude inequations and to restrict our considerations to initial algebras that are reachable.

We say that an algebraic institution strongly admits initial semantics if any non-empty class of algebras definable in it contains a reachable initial algebra.

Fact 3 The algebraic institutions EQ, ICEQ, NEQ and ICNEQ strongly admit initial semantics.

Fact 4 For any two algebraic institutions INS1 and INS2, if INS1 is reducible to INS2 and INS2 strongly admits initial semantics then INS1 strongly admits initial semantics as well.

Now, we have (a detailed proof is given in [Tar 83]):

Theorem 1 Let INS be an algebraic institution. INS strongly admits initial semantics and any class of algebras definable in INS contains the one-element algebra iff INS is reducible to the institution of infinitary conditional equations, ICEQ.

In fact, in [MM 83b] Mahr and Makowsky give an even stronger result:

Consider a class of Σ-algebras K. An algebra A is called EQ-free in K if $A\in K$, A is reachable and for any Σ-equation e if e holds in A then it holds in any algebra in K.

Now (cf. [MM 83b]) any class definable in an algebraic institution INS contains an EQ-free algebra if and only if INS is reducible to ICNEQ. (This holds only under the assumption implicit in [MM 83b] that only signatures Σ such that there are ground Σ-terms of any sort are considered - cf. [GM 81].)

Note, however, that it is not quite trivial to deduce a characterization of algebraic institutions which admit initial semantics from this fact. First, there is a slight technical problem: when empty carriers are admitted an initial algebra need not be EQ-free, since it may satisfy more equations than other algebras in the considered class (cf. [GM 81]). Then, an initial algebra need not be reachable, either.

Nevertheless, it follows from the above fact due to Mahr and Makowsky that an algebraic institution strongly admits initial semantics if and only if it is reducible to ICNEQ.

5 Liberality

As we have already mentioned, the choice of the initial model of a theory may be too restrictive. What we often need are models in which only some parts of a theory are interpreted in some standard way relative to some other parts which may be interpreted loosely. A formal definition of these more general constraints requires the existence of a standard "free" extension of any model of a subtheory to a model of the whole theory (cf. [BG 80, EWT 83, GB 83], also the next section). In this section we try to characterize algebraic institutions in which such extensions always exist.

Let INS be an algebraic institution, $T1=\langle\Sigma 1,\Psi 1\rangle$ and $T2=\langle\Sigma 2,\Psi 2\rangle$ be theories with sorts S1 and S2, respectively, and $\sigma: T1\rightarrow T2$ be a theory morphism.

For any $A\in Mod(T1)$, by a σ-free model over A we mean an algebra $F_\sigma(A)\in Mod(T2)$ together with a $\Sigma 1$-homomorphism $\eta_A: A\rightarrow U_\sigma(F_\sigma(A))$ such that the following "universality condition" holds: for any $B\in Mod(T2)$ and $\Sigma 1$-homomorphism $h: A\rightarrow U_\sigma(B)$ there is a unique $\Sigma 2$-homomorphism $h^{\#}: F_\sigma(A)\rightarrow B$ such that $\eta_A; U_\sigma(h^{\#})=h$.

Note that if a σ-free model over A exists for any $A\in Mod(T1)$ then the mappings $A\longmapsto F_\sigma(A)$ and $A\longmapsto\eta_A$ determine a functor $F_\sigma: Mod(T1)\rightarrow Mod(T2)$ which is a left adjoint to the forgetful functor $U_\sigma: Mod(T2)\rightarrow Mod(T1)$; η is the unit of the adjunction and its counit ϵ is determined by $\epsilon_B=(id_{U_\sigma(B)})^{\#}$ for $B\in Mod(T2)$ (cf. [MacL 71]).

In the following we try to construct free models using initial ones.

Consider $A\in Mod(T1)$.

By the diagram signature for A we mean the signature $\Sigma 1(A)$ which is the extension of $\Sigma 1$ by a constant $\underline{a}$ of sort s for each $s\in S1$ and $a\in|A|_s$.

Let $\Sigma 2(\sigma(A))$ be the extension of the signature $\Sigma 2$ by a constant $\underline{a}^s$ of sort $\sigma(s)$ for each $s\in S1$ and $a\in|A|_s$. Let $\bar\sigma$ be the obvious extension of σ to a signature morphism from $\Sigma 1(A)$ to $\Sigma 2(\sigma(A))$ given by $\bar\sigma(\underline{a})=\underline{a}^s$ for $s\in S1$ and $a\in|A|_s$.

Let E(A) be the obvious extension of A to a $\Sigma 1(A)$-algebra given by $\underline{a}_{E(A)}=a$ for any $s\in S1$ and $a\in|A|_s$.

By the diagram of A we mean the set $\Delta(A)$ of all ground $\Sigma 1(A)$-equations that hold in E(A). Let $\bar\sigma(\Delta(A))=\{\bar\sigma(e)\mid e\in\Delta(A)\}$.

Let $\iota:\Sigma 2\rightarrow\Sigma 2(\sigma(A))$ be the signature inclusion. For any $B\in Mod(T2)$ and $\Sigma 1$-homomorphism $h: A\rightarrow U_\sigma(B)$, by the h-extension of B we mean the $\Sigma 2(\sigma(A))$-algebra $E_h(B)$ defined by $U_\iota(E_h(B))=B$ (i.e. $E_h(B)$ is an extension of B) and $(\underline{a}^s)_{E_h(B)}=h_s(a)$ for $s\in S1$ and $a\in|A|_s$.

Finally, let $T3=\langle\Sigma 2(\sigma(A)),\Psi 3\rangle$ be the least $\Sigma 2(\sigma(A))$-theory such that $\Psi 2\cup\bar\sigma(\Delta(A))\subseteq\Psi 3$.

Lemma 1 For any $B\in Mod(T2)$ and $\Sigma 1$-homomorphism $h: A\rightarrow U_\sigma(B)$, $E_h(B)\in Mod(T3)$.

Lemma 2 Let $B\in Mod(T3)$. Define $h:|A|\rightarrow|U_\sigma(U_\iota(B))|$ by $h_s(a)=(\underline{a}^s)_B$ for any $s\in S1$ and $a\in|A|_s$. Then

- $U_\iota(B) \in Mod(T2)$
- h is a $\Sigma1$-homomorphism $h: A \to U_\sigma(U_\iota(B))$
- $B = E_h(U_\iota(B))$

Lemma 3 If $B1, B2 \in Mod(T2)$, $h1: A \to U_\sigma(B1)$ and $h2: A \to U_\sigma(B2)$ are $\Sigma1$-homomorphisms then there is a 1-1 correspondence between $\Sigma2(\sigma(A))$-homomorphisms from $E_{h1}(B1)$ to $E_{h2}(B2)$ and $\Sigma2$-homomorphisms $h: B1 \to B2$ such that $h1; U_\sigma(h) = h2$.

Using the above three lemmas one may easily prove our main theorem:

Theorem 2 A σ-free model over A exists iff theory T3 has an initial model. Moreover, if B is initial in Mod(T3) then a σ-free model over A, $\langle F_\sigma(A), \eta_A \rangle$, is defined by $F_\sigma(A) = U_\iota(B)$ and $(\eta_A)_s(a) = (\underline{a}^s)_B$ for any $s \in S1$ and $a \in |A|_s$.

We say that an algebraic institution is liberal if for any two theories T1 and T2 and theory morphism $\sigma: T1 \to T2$ there exists a σ-free model over any model of T1, or equivalently, if the forgetful functor $U_\sigma: Mod(T2) \to Mod(T1)$ has a left adjoint (cf. [GB 83]).

The above construction implies:

Corollary 1 An algebraic institution is liberal iff every class of algebras definable in it contains an initial algebra.

An algebraic institution is said to guarantee satisfiability if any theory in this institution is satisfiable (has a model).

Fact 5 Any liberal algebraic institution guarantees satisfiability.

Fact 6 If an algebraic institution guarantees satisfiability then any definable class of algebras contains a one-element algebra.

The above two facts allow us to use the Mahr and Makowsky result to characterize liberal algebraic institutions - again only under a generalized reachability condition.

An algebraic institution is called strongly liberal if for any two theories $T1 = \langle \Sigma1, \Psi1 \rangle$ and $T2 = \langle \Sigma2, \Psi2 \rangle$ with sorts S1 and S2, respectively, and a theory morphism $\sigma: T1 \to T2$, for any $A \in Mod(T1)$ there is a σ-free model over A, $\langle F_\sigma(A), \eta_A \rangle$, such that $F_\sigma(A)$ is generated by its $\sigma(T1)$-part, i.e. $F_\sigma(A)$ is generated by the S2-sorted set $X \subseteq |F_\sigma(A)|$, where for $s2 \in S2$ $X_{s2} = |F_\sigma(A)|_{s2}$ if $s2 = \sigma(s1)$ for some $s1 \in S1$ and $X_{s2} = \emptyset$ otherwise.

Corollary 2 An algebraic institution is strongly liberal iff it is reducible to ICEQ, the institution of infinitary conditional equations.

6 Data constraints

To pick out models with an interpretation of some parts of a theory that is standard relative to some other parts we introduce sentences called data constraints. The following definition is adapted directly from [GB 83].

Let us fix a liberal algebraic institution.

If $\sigma: T1 \to T2$ is a theory morphism and $B \in Mod(T2)$ then we say that B is σ-free if it is ("naturally" isomorphic to) a σ-free model over its T1-part, i.e., more formally, if the counit morphism $\epsilon_B = (id_{U_\sigma(B)})^{\#}: F_\sigma(U_\sigma(B)) \to B$ is an isomorphism.

By Σ-data constraint, for any signature Σ, we mean a pair $\langle \sigma: T1 \to T2, \theta: \Sigma2 \to \Sigma \rangle$ where T1 and T2 are theories, σ is a theory morphism, $\Sigma2$ is the signature of T2 and θ is a signature morphism.

We say that Σ-algebra A satisfies the above constraint if $U_\theta(A) \in Mod(T2)$ and $U_\theta(A)$ is σ-free.

Goguen and Burstall [GB 83] have proved that if we accept Σ-data constraints as additional Σ-sentences in a liberal institution then the resulting system is again an institution. It may also be proved that if the underlying institution is algebraic then the result institution is algebraic as well. It is worth noting that it need not be liberal. In fact, in our framework it is never liberal since it does not admit initial semantics.

Note that in order for the above satisfaction relation for data constraints to be well-defined we do not need σ-free models over all models of T1; we need them only for the models that are, roughly, T1-parts of models of T2.

Let σ: T1→T2 be a theory morphism. We say that A∈Mod(T1) is σ-consistent with T2 if A=U_σ(B) for some B∈Mod(T2), i.e. if A∈U_σ(Mod(T2)), where U_σ(Mod(T2)) is the range of the forgetful functor U_σ: Mod(T2)→Mod(T1).

We call an algebraic institution quasi-liberal if for any theory morphism σ: T1→T2 there is a σ-free model over any model of T1 that is σ-consistent with T2, i.e. if for any theory morphism σ: T1→T2 the forgetful functor U_σ: Mod(T2)→U_σ(Mod(T2)) has a left adjoint F_σ: U_σ(Mod(T2))→Mod(T1).

Corollary 3 An algebraic institution is quasi-liberal iff it admits initial semantics.

Note that if σ: T1→T2 is a theory morphism then the existence of σ-free models over models of T1 that are σ-consistent with T2 is sufficient to define σ-freeness of models of T2, and hence to use data constraints having σ as the first component. By Corollary 3, this allows us to use data constraints in any algebraic institution that admits initial semantics, even if it is not liberal. However, the following theorem indicates that this relaxation of requirements on the underlying institution is sometimes illusory.

Theorem 3 Let Σ be a signature. Any Σ-data constraint in NEQ (the institution of equations and inequations) is expressible by a Σ-data constraint in EQ (the institution of equations) and a set of Σ-sentences in NEQ, i.e. for any Σ-data constraint c in NEQ there is a Σ-data constraint c′ in EQ and a set Ψ of Σ-sentences in NEQ such that for any Σ-algebra A A⊨c iff A⊨$\Psi \cup$ {c′}.

7 Summary of results

We introduce the notion of an algebraic institution to formalize the concept of a logical system for specifying classes of algebras. In this framework we briefly review a characterization of algebraic specification languages that admit initial semantics given by Mahr and Makowsky [MM 83a, MM 83b]. The proof given in [MM 83a] leads to the result that an algebraic institution is reducible to the institution of infinitary conditional equations if and only if any theory in this algebraic institution has a reachable initial model as well as a model containing only one element of every sort (Theorem 1). A more general result given in [MM 83b] shows that any satisfiable theory in an algebraic institution has a reachable initial model if and only if this institution is reducible to the institution of infinitary conditional equations and inequations.

Considering the problem of the existence of free functors (left adjoints to forgetful functors) we give a basic construction of a free model of a theory over a model of a subtheory (w.r.t. an arbitrary theory morphism) which requires only the existence of initial models (Theorem 2). This leads to the result that an algebraic institution is liberal (i.e. guarantees the existence of a free functor for any theory morphism) if and only if any theory has an initial model (Corollary 1).

Another consequence of our basic construction is that the institution of infinitary conditional equations is the most general algebraic institution which is strongly liberal, i.e. which guarantees the existence of a model of a theory that is both free over and generated by a model of a subtheory w.r.t. any theory morphism (Corollary 2).

Finally, we point out that perhaps the requirement of liberality is too restrictive; we show that the meaning of data constraints remains well-defined in algebraic institutions which are only quasi-liberal, i.e. which guarantee the existence of a free model of a theory over a model of a subtheory provided that this model is consistent with the whole theory. Quasiliberality is equivalent to the existence of an initial model of any satisfiable theory (Corollary 3). For some standard algebraic institutions this relaxation turns out to be illusory (Theorem 3).

Acknowledgement

J. Cartmell, J. Goguen, G. Plotkin, O. Schoett, E. Robinson and D. Rydeheard were patient enough to listen to the story of the footnote to Mal'cev's theorem and helped me to understand its consequences. Special thanks to Rod Burstall for many instructive discussions and to Don Sannella for turning my attention to the papers by Mahr and Makowsky and for numerous useful comments and linguistic corrections. I am also grateful to B. Mahr for detailed comments on [Tar 83] and on the result due to him and J. A. Makowsky.

This research is supported by a grant from the (U. K.) Science and Engineering Research Council.

8 References

[ADJ 76] Goguen, J. A., Thatcher, J. W. and Wagner, E. G. An initial algebra approach to the specification, correctness, and implementation of abstract data types. Current Trends in Programming Methodology, Vol. 4: Data Structuring (R. T. Yeh, ed.), Prentice-Hall, pp. 80–149 (1978).

[Bar 74] Barwise, K. J. Axioms for abstract model theory. Annals of Math. Logic 7, pp. 221-265.

[BG 80] Burstall, R. M. and Goguen, J. A. The semantics of Clear, a specification language. Proc. of Advanced Course on Abstract Software Specifications, Copenhagen. Springer LNCS 86, pp. 292–332.

[EWT 83] Ehrig, H., Wagner, E. G. and Thatcher, J. W. Algebraic specifications with generating constraints. Proc. 10th ICALP, Barcelona. Springer LNCS 154, pp. 188–202.

[GB 83] Goguen, J. A. and Burstall, R. M. Introducing institutions. Proc. Logics of Programming Workshop, CMU.

[GM 81] Goguen, J. A., Meseguer, J. Completeness of many-sorted equational logic, SIGPLAN Notices 16(7), pp. 24–32, July 1981, extended version to appear in Houston Journal of Mathematics.

[GM 83] Goguen, J. A. and Meseguer, J. An initiality primer. to appear in Application of Algebra to Language Definition and Compilation (M. Nivat, J. Reynolds, editors), North Holland.

[MacL 71] MacLane, S. Categories for the Working Mathematician. Springer.

[MM 83a] Mahr, B. and Makowsky, J. A. Characterizing specification languages which admit initial semantics. Proc. 8th CAAP, L'Aquila, Italy. Springer LNCS 159, pp. 300–316.

[MM 83b] Mahr, B. and Makowsky, J. A. An axiomatic approach to semantics of specification languages. Proc. 6th GI Conf. on Theoretical Computer Science, Dortmund. Springer LNCS 145.

[Rei 80] Reichel, H. Initially restricting algebraic theories, in: Mathematical Foundations of Computer Science (Proc. 9th Symp. Rydzyna 1980, Poland, P. Dembinski, ed.), Lecture Notes in Computer Science 88, pp. 504–514, Springer-Verlag 1980.

[ST 83] Sannella, D. T. and Tarlecki, A. Building specifications in an arbitrary institution, Proc. International Symposium on Semantics of Data Types, Sophia-Antipolis, June 1984, to appear.

[Tar 83] Tarlecki, A. Free constructions in algebraic institutions. Report CSR-149-83, Dept. of Computer Science, Univ. of Edinburgh.

REMARKS ON COMPARING EXPRESSIVE POWER OF LOGICS OF PROGRAMS

Jerzy Tiuryn and Paweł Urzyczyn
Institute of Mathematics , University of Warsaw
PKiN, 00-901 Warszawa, Poland

1. INTRODUCTION

This paper is motivated by the following general problem: "Given two classes of program schemes, K_1 and K_2, compare the expressive power of programming logics based on these classes." In particular, we are interested in answers to the question: "Does nondeterminism add to the power of logics based on a given class K of (deterministic) program schemes ?"

A program scheme is understood as an arbitrary computing device, the behaviour of which may be divided into elementary steps, i.e. atomic first-order tests or assignments of depth at most 1. This is an obvious idea to assume that any class of program schemes should be closed under some natural operations on programs. Thus, all classes of program schemes considered below are assumed to form "acceptable programming languages" (apl's), where the definition of an apl follows the ideas of Lipton [L] and Clarke, German and Halpern [CGH].

For a given class K of program schemes, the logic of programs in K, L(K), is built up, in a standard way, from first-order connectives and program schemes from K, with help of the dynamic construct $\langle\ \rangle$, as in Dynamic Logic (see [H]). We write $L(K_1) \leqslant L(K_2)$ iff for every formula of $L(K_1)$ there is an equivalent formula in $L(K_2)$. $L(K_1) \equiv L(K_2)$ stands for $L(K_1) \leqslant L(K_2) \wedge L(K_2) \leqslant L(K_1)$.

We consider two important properties that may or may not be possessed by a given apl. The first one is "rich control", i.e. the ability to simulate a Turing machine. This is usually represented by allowing arithmetical operations (on counters) to occur in programs. Each class K of program schemes may be naturally extended to a new class KC, members of which, in addition to all features available in K, may also use counters. It turns out that adding counters is, in a sense, at least as powerful as adding nondeterminism. More formally, we show that L(KC) $\equiv$ L(NKC), where NK denotes the least nondeterministic apl containing the deterministic apl K (Theorem 2). In particular, each program scheme in NKC, computing a partial function, is equivalent to a program scheme in KC (Theorem 1). As an immediate corollary of these results we get: L(NFDC) $\equiv$ L(FDC) (see [T]) and L(NFDSC) $\equiv$ L(FDSC) (see [MT]), where FD is the class of all deterministic flow-diagrams, and FDS is the class of all deterministic flow-diagrams with a stack.

Another interesting property of apl´s is the ability to carry out a back-track search, i.e. to penetrate any data structure (more precisely: the part of data structure which is accessible from a given input, by means of algebraic operations). We reduce this condition to a weaker one, namely, it suffices to assume the existence of a program scheme determining whether the part of interpretation accessible from the input is finite or not (Lemma 1). Adding counters to such an apl provides a class of universal power, i.e. equivalent to FDSC (Theorem 3) thus we use the term "semi-universal" to denote apl´s of this kind. Another justification of the name "semi-universal" is Corollary 1, which states that all deterministic semi-universal apl´s are universal (and therefore equivalent) over all accessibly infinite interpretations, i.e. all interpretations where the input generates infinitely many elements. Therefore the problem of comparing expressive power of logics based on these apl´s leads to investigating their behaviours over accessibly finite interpretations. We have the following sufficient condition for elimination of nondeterminism from logics based on semi-universal apl´s:

"For any $P \in NK$, there is $Q \in K$ such that, for each accessibly finite interpretation I, P converges in I iff Q converges in I".

This follows from Theorem 4. Now we turn to the general problem of comparing logics of programs based on semi-universal apl´s. An apl is <u>divergence closed</u> iff, for each $P \in K$, there is $Q \in K$ such that, for each accessibly finite I, P converges in I iff Q does <u>not</u> converge in I. (Examples are FDS or flow-diagrams with arrays, cf. [TU].) Our last result (Theorem 5) says that comparing logics of semi-universal divergence closed apl´s is reducible to questions of comparative schematology. Namely, for divergence closed K_2, and arbitrary K_1, $L(K_1) \leqslant L(K_2)$ is equivalent to: " $\forall P \in K_1 \exists Q \in K_2$ $\forall$ accessibly finite I (P converges in I iff Q converges in I)".

The paper is organized as follows. Section 2 introduces some necessary definitions, while Section 3 is devoted to the formulation of results. Because of space limitations, the proofs are only briefly sketched.

2. PRELIMINARIES

Throughout the paper, σ denotes a fixed signature, i.e. a finite sequence of function and relation symbols (including =). A σ-interpretation is a pair $(A,\underline{a})$, where $\underline{a}$ is a finite sequence of elements of a σ-structure A. If t is a term in k variables and $\underline{a} \in A^k$, then $t_A(\underline{a})$ denotes the value of t for the valuation determined by $\underline{a}$. For any $A,\underline{a}$, the symbol $A(\underline{a})$ denotes the substructure generated by $\underline{a}$ in A. We say that the interpretation $(A,\underline{a})$ is accessibly (in-)finite iff $A(\underline{a})$ is (in-)finite.

An acceptable programming language (apl) is understood as a class of program schemes together with an operational semantics. We do not define the notion of a program scheme precisely, to make the definitions comprising various formalisms. What we assume is the following: a (k-ary) deterministic program scheme (over σ) is an arbitrary expression P which determines (effectively) a vector of input/output variables, $\underline{x} = (x_1,\dots,x_k)$, denoted Var(P), and an informal algorithm (like in [K2]). For a given interpretation $(A,\underline{a})$ with $\underline{a} \in A^k$, such an informal algorithm may perform the following operations:

- store the input values of $\underline{a}$ in distinguished memory registers, corresponding to the variables $x_1,\dots,x_k$;
- compute values of form $f_A(b_1,\dots,b_n)$, where f is an n-ary function symbol in σ , and $b_1,\dots,b_n$ are values already stored in the memory, and store the new values in arbitrary locations;
- determine the flow of the computation by testing conditions of form $r_A(b_1,\dots,b_n)$, where $b_1,\dots,b_n$ are as above, and r is an n-ary relation symbol in σ ;
- interchange the contents of memory registers.

We denote by $Val(P,A,\underline{a})$ the set of all values occurring (i.e. stored in memory registers) during the computation of P determined by $\underline{a}$ in A. For $b,c \in Val(P,A,\underline{a})$ we put $b \leqslant c$ iff b is stored in the memory earlier than c. Clearly, any program scheme P defines, in any structure A, a partial function $P_A: A^k \longrightarrow A^k$, such that $P_A(\underline{a})$, if defined, is the output of the computation of P in $(A,\underline{a})$. Two k-ary program schemes P,Q are equivalent ($P \equiv Q$) iff, for any σ-structure A, $P_A = Q_A$.

The reader may easily check that the well-known notions of flowcharts, recursive schemes etc (see [CG]) may be presented so that they satisfy the conditions above. In addition, it can be proved that any program scheme in the above sense is equivalent to a flow-chart (flow-diagram) with one stack and counters, as defined in [CG]. (Note that our program schemes can always test for equality.)

It should be clear how to generalize the above notions to the nondeterministic case. Note that, for a nondeterministic program scheme P the symbol P_A refers to an input-output relation rather than a partial function. Note also that the set Val(P,A,$\underline{a}$) cannot be ordered in the above way, since it refers to possibly many different computations.

A deterministic acceptable programming language (apl) over σ may be now defined as an arbitrary class K of program schemes, which satisfies certain structural conditions. We do not describe these conditions precisely, but we give an informal explanation of what should be assumed about an apl. First we assume that formal operations on a given scheme P, like replacements of variables in Var(P) or adding new input-output variables (but without changing the algorithm of P) are always possible within K. Further, we require that K (more precisely: the class of all algorithms defined by schemes in K) is closed under compositions "while...do" and "if...then...else"-constructs, with tests being open first-order formulas. Of course, we assume that assignments are in K.

An important observation is that each computation may be viewed as a sequence of steps. That is, the algorithm defined by a program scheme is not a "black box" with input-output information only - it is a prescription how to perform successive elementary actions. In particular, we assume that K is closed under the following constructs:

"run P until α" and "after each step of P do Q".

The meaning of the first construct is that the computation of P stops in the first state where α is true. As to the second one - Q is assumed not to change the contents of registers used by P. In the above, α is an open formula with variables which refer to memory registers.

A nondeterministic apl (napl) is defined as above, but it must be also closed under the construct "P or Q", which refers to a nondeterministic choice. If K is an apl then NK denotes the least napl containing K (as an algebra of algorithms).

If K_1, K_2 are (n)apl's then we write $K_1 \leqslant K_2$ iff, for any $P_1 \in K_1$, there is $P_2 \in K_2$ with $P_1 \equiv P_2$. Further, $K_1 \equiv K_2$ means "$K_1 \leqslant K_2 \wedge K_2 \leqslant K_1$".

Let K be a (n)apl over σ. The logic of programs over K, L(K), is the least set of expressions such that

- any open first-order formula over σ belongs to L(K);
- if $\alpha, \beta \in L(K)$ and x is a variable then $\alpha \vee \beta$, $\neg\alpha$, $\exists x \alpha \in L(K)$;
- if $P \in K$, and $\alpha \in L(K)$ then $\langle P \rangle \alpha \in L(K)$.

The semantics of L(K) is similar to that of predicate calculus except that $A, \underline{a} \models \langle P \rangle \alpha$ holds iff there is $\underline{b}$ such that $(\underline{a}, \underline{b}) \in P_A$ ($b = P_A(\underline{a})$ in the deterministic case) and $A, \underline{b} \models \alpha$. We abbreviate $\langle P \rangle$true by Halt(P).

Let $L_1 L_2$ be two logics. We say that L_1 is reducible to L_2 ($L_1 \leqslant L_2$) iff, for every $\alpha \in L_1$, there is $\beta \in L_2$, equivalent to α in all interpretations. L_1 and L_2 are of equal expressive power ($L_1 \equiv L_2$) iff $L_1 \leqslant L_2$ and $L_2 \leqslant L_1$. Of course, $K_1 \leqslant K_2$ implies $L(K_1) \leqslant L(K_2)$.

Let K be a (n)apl over σ and let $k \in N$, $\alpha \in L(K)$. If α has exactly k free variables, then the <u>k-spectrum</u> of α is the class <u>spec</u>$_k(\alpha) = \{ (A,\underline{a})$: $(A,\underline{a})$ is accessibly finite and $A = A(\underline{a})$ and $A,\underline{a} \models \alpha \}$, otherwise <u>spec</u>$_k(\alpha) = \emptyset$. For $L \subseteq L(K)$, we set <u>spec</u>$_k(L) = \{$ <u>spec</u>$_k(\alpha)$: $\alpha \in L\}$. If $P \in K$ then <u>spec</u>$_k(P)$ = <u>spec</u>$_k$(Halt(P)). Further, <u>spec</u>$_k(K) = \{$ <u>spec</u>$_k(P)$: $P \in K\}$. We write <u>spec</u>$(L) \subseteq$ <u>spec</u>(L') iff, for all k, <u>spec</u>$_k(L) \subseteq$ <u>spec</u>$_k(L')$.

3. LOGICS BASED ON SEMI-UNIVERSAL APL'S

We say that an apl K is <u>semi-universal</u> iff, for all $k \geqslant 1$, there exists a program scheme $FIN_k \in K$, with k variables, such that, for all interpretations $(A,\underline{a})$, with $\underline{a} \in A^k$, FIN_k converges in $(A,\underline{a})$ iff $(A,\underline{a})$ is accessibly finite. A napl is <u>semi-universal</u> iff it is of the form NK, where K is a semi-universal apl.

<u>Lemma 1</u>

Let K be a semi-universal apl. Then, for all $k \geqslant 1$, there exists a k-ary program scheme $SEARCH_k \in K$, such that, for all $A,\underline{a}$:

- $SEARCH_k$ converges in $(A,\underline{a})$ iff $(A,\underline{a})$ is accessibly finite;
- $Val(SEARCH_k, A, \underline{a}) = A(\underline{a})$.

For the proof of Lemma 1 we use the following:

<u>Lemma 2</u> (Lipton [L])

Let K be an apl and let $P \in K$ have k variables. There is a program scheme $SUCP \in K$, in k+1 variables, such that, for all $(A,\underline{a},b)$,

- SUCP converges in $(A,\underline{a},b)$ iff $b \in Val(P,A,\underline{a})$;
- the output of SUCP in $(A,\underline{a},b)$ is $(\underline{a},c)$, where c is:
  - equal to b, if b is the least element in $Val(P,A,\underline{a})$,
  - the element next to b in $Val(P,A,\underline{a})$, otherwise.

<u>Sketch of the proof of Lemma 1</u>

The key observation is the following. The algorithm given by FIN_k in every converging computation must reach all elements generated by the input. Similarly, in infinite computations, it must reach infinitely many values.

In accessibly infinite interpretations, $SEARCH_k$ uses $SUCFIN_k$ as a procedure to simulate arithmetical operations, where the role of integers is played by the elements of $Val(FIN_k, A, \underline{a})$. This enables it to ru a copy of FIN_k to reach all values of terms of depth at most n. For increasing n, all values are eventually computed. If the interpretation is accessibly finite, the algorithm is able to recognize it, since, fo some n, the Lipton´s successor fails to compute n+1. □

A program scheme with counters is an arbitrary program scheme P such that the informal algorithm defined by P may use special memory registers (counters) to store non-negative integers, and execute on them usual arithmetic operations and tests for zero and equality. For technical purposes we may think on schemes with counters as defining algorithms for inputs consisting of two parts: "algebraic" (corresponding to variables in Var(P)) and "arithmetic" (corresponding to distinguished counters). If K is a (n)apl then KC denotes the class "K with counters", i.e. the least (n)apl containing K and closed under constructs discussed in Section 2, with tests being Boolean combinations of equations on counters. Basic assignments on counters are also assumed to belong to KC.

The following theorems explain the use of counters for eliminating nondeterminism from program schemes and logics.

Theorem 1

Let K be an apl. For any k-ary $P \in NK$, there is a k-ary deterministic scheme $DETP \in KC$, such that, for all $(A, \underline{a})$:

- P converges in $(A, \underline{a})$ (i.e. there is a converging computation) iff DETP converges in $(A, \underline{a})$;
- if the relation P_A is a partial function, then $DETP_A = P_A$.

Sketch of the proof

The algorithm of DETP simulates successively all possible finite initial fragments of computations of P. Counters are used to generate the successive instructions. □

Theorem 2

If K is an apl such that $K \equiv KC$ then $L(K) \equiv L(NK)$.

Sketch of the proof

Use Theorem 1 to translate formulas of the form Halt(P), for $P \in NK$. □

The next theorem gives a motivation of the notion "semi-universal". It suffices to add counters to a semi-universal apl to obtain a language of universal power.

Theorem 3

Let K be a semi-universal apl. Then (N)KC $\equiv$ (N)FDSC, where (N)FDSC denotes the class of all (non)deterministic flow-diagrams with one stack and counters.

Sketch of the proof

The key point is that a program scheme with counters is able to compute value of a term of a given code. The program $SUCSEARCH_k$ (see Lemmas 1 and 2) provides a sequence of values of all terms, but not every term must be actually evaluated by $SUCSEARCH_k$. To overcome the difficulty, one uses a method introduced by Friedman ([F],Thm.1.6.) and Kfoury ([K1] ,Thm.II). The method is inductive. To compute the value of a term of a given code, the procedure finds values of all terms of smaller codes and remembers their positions in the order of $SUCSEARCH_k$. $\square$

Remark: The converse of Theorem 2 is not true in general. If σ consists of two unary function symbols, then FDC $\equiv$ FDSC, but FD is not semi-universal over σ (see [BHT], [ST], [U]).

Corollary 1

If K is a semi-universal apl over σ then (N)K is equivalent to (N)FDSC over all accessibly infinite interpretations, since counters may be simulated using Lemma 2. $\square$

In general, it is not known whether semi-universality suffices to eliminate nondeterminism. This may be a hard question, as shown by the following example: for σ consisting only one unary function symbol and no other symbols (except =), flow-diagrams form a semi-universal apl. For this signature, L(FD) $\equiv$ L(NFD) is equivalent to the well-known problem of nondeterminism in context-sensitive languages (see [TU]). However, for all semi-universal apl´s, if spec(NK) = spec(K) then L(K) $\equiv$ L(NK), which follows from the following:

Theorem 4

Let K_2 be a semi-universal apl. Then, for every (n)apl K_1, if spec(K_1) $\subseteq$ spec(K_2) then L(K_1) $\leq$ L(K_2).

Sketch of the proof

For $P \in K_1$, with help of $SUCSEARCH_k$, one constructs $Q \in K_2$, such that Halt(P) $\longleftrightarrow$ Halt(Q) holds in every interpretation. $\square$

An apl K is divergence closed iff, for every $P \in K$, there is $Q \in K$ such that Halt(P) $\longleftrightarrow$ Halt(Q) holds in all accessibly finite interpretations.

Theorem 5

Let K_1, K_2 be (n)apl´s. Assume that K_2 is semi-universal and divergence closed. Then the following conditions are equivalent:

i) $L(K_1) \leqslant L(K_2)$;

ii) spec$(K_1) \subseteq$ spec(K_2);

iii) for every $P \in K_1$, there is $Q \in K_2$ such that $\mathrm{Halt}(P) \leftrightarrow \mathrm{Halt}(Q)$ holds in every interpretation.

If both K_1 and K_2 are nondeterministic then the conditions (i-iii) are also equivalent to

iv) $K_1 \leqslant K_2$.

Remark: If both K_1 and K_2 are deterministic, then the equivalence (i)$\longleftrightarrow$(iv) holds under a stronger condition, namely:

- for every $P \in K_2$ there is $Q \in K_2$ such that, for every accessibly finite interpretation $(A,\underline{a})$, with at least two elements in $A(\underline{a})$,

$$A,\underline{a} \models \mathrm{Halt}(P) \quad \text{iff} \quad A,\underline{a} \models \langle Q \rangle\, x=y \;;$$
$$A,\underline{a} \models \neg\mathrm{Halt}(P) \quad \text{iff} \quad A,\underline{a} \models \langle Q \rangle\, x \neq y \;,$$

for some x,y not in Var(P).

Sketch of the proof

This is left to the reader. For (i)$\longrightarrow$(ii), observe that spec(K_2) = spec$L(K_2)$.

□

REFERENCES

[BHT] Berman,P.,Halpern,J.Y.,Tiuryn,J.,On the power of nondeterminism in Dynamic Logic, in: Proc of 9th ICALP (M.Nielsen and E.M.Schmidt,Eds.), Lecture Notes in Comp.Sci.,vol.140,Springer-Verlag,Berlin,1982,pp.48-60.

[CG] Constable,R.L.,Gries,D., On classes of program schemata, SIAM J.Comput. vol.1, no.1, 1972,66-118.

[CGH] Clarke,E.M.,German,S.M.,Halpern,J.Y., Effective axiomatizations of Hoare logics, Research Report, Harvard Univ.,1982.

[F] Friedman,H., Algorithmic procedures,generalized Turing algorithms, and elementary recursion theory, in: Logic Colloquium ´69 (R.O.Gandy and C.M.E.Yates, Eds.), North Holland, Amsterdam,1971,pp.361-389.

[H] Harel,D.,First-Order Dynamic Logic, Lecture Notes in Comp.Sci. vol.68,Springer-Verlag, Berlin, 1979.

[K1] Kfoury,A.J., Translatability of schemes over restricted interpretations, J.Comput.Sys.Sci., 8,1974,387-408.

[K2] Kfoury,A.J., Definability by programs in first-order structures, Theoret.Comput.Sci., 25,1983,1-66.

[L] Lipton,R.J., A necessary and sufficient condition for the existence of Hoare logics, in: Proc. 18th IEEE Symp.on FoCS, 1977.

[MT] Meyer,A.R.,Tiuryn,J., A note on equivalences among logics of programs, to appear in J.Comput.Sys.Sci.

[ST] Stolboushkin,A.P.,Taitslin,M.A., DDL is strictly weaker than DL, to appear in Inform.Control.

[T] Tiuryn,J., An introduction to first-order programming logics, a course given at Inter-University Centre for Postgraduate studies,Dubrovnik,1983, to appear.

[TU] Tiuryn,J.,Urzyczyn,P., Some relationships between logics of programs and complexity theory, in: Proc. 24th IEEE Symp. on FoCS, 1983.

[U] Urzyczyn,P., Nontrivial definability by flow-chart programs, to appear in Inform.Control.

THE COMPLEXITY OF PROBLEMS CONCERNING GRAPHS WITH REGULARITIES
(EXTENTED ABSTRACT)

K. Wagner
Sektion Mathematik
Friedrich-Schiller-Universität
DDR 6900 Jena, Universitätshochhaus

INTRODUCTION

In most cases finite graphs are described by a report of all vertices and edges. Especially in practically important cases (for example in VLSI design) graphs are of interest which have a lot of regularities. For such graphs it can be desirable and advantageous to use more compact descriptions such that the repeated description of similarly structured subgraphs can be avoided. Such compact descriptions have been investigated in [SM 73] for finite sets of natural numbers and in [BO 81] for finite sets of rectangles in the plane. The languages used there as well as a common generalization have been adapted in [Wa 84a] and [Wa 84c] to describe finite subsets of $\mathbb{N}^n$. In all these papers it has been studied how the complexity of problems increases when their instances are described in such a compact way. In the present paper we study this phenomenon for graph-theoretical problems where the languages defined in [Wa 84a] and [Wa 84c] are used to describe finite graphs. As an example, the graph accessibility problem being $\mathcal{NL}$-complete for usual descriptions becomes PSPACE-complete for compact descriptions.

COMPACT DESCRIPTIONS OF GRAPHS

A finite subset M of $\mathbb{N}^{2n}$ can be considered as an finite graph $G_M = (V_M, E_M)$ where

$$V_M = \{(a_1,\dots,a_n) : \text{there exists } (b_1,\dots,b_n) \text{ such that } (a_1,\dots,a_n,b_1,\dots,b_n) \in M \text{ or } (b_1,\dots,b_n,a_1,\dots,a_n) \in M\}$$

$E_M = \{((a_1,\ldots,a_n),(b_1,\ldots,b_n)): (a_1,\ldots,a_n,b_1,\ldots,b_n) \in M\}$.

Therefore languages to describe finite subsets of $\mathbb{N}^{2n}$ can be considered as languages to describe finite graphs. In [Wa 84a] and [Wa 84c] three languages to describe finite subsets of $\mathbb{N}^n$ are dealt with. We report their definitions and some of their properties.

For $n \geq 1$, the set HE^n of hierarchical expressions is defined as the smallest set fulfilling

1. $m \in \mathbb{N} \Rightarrow \langle \text{bin } m\rangle \in HE^n$, [1)]
   $m_1,\ldots,m_n \in \mathbb{N} \Rightarrow (\text{bin } m_1,\ldots,\text{bin } m_n) \in HE^n$,
2. $H,H' \in HE^n \Rightarrow (H \cup H'),(H+H') \in HE^n$.

Now we define the set HD^n of hierarchical descriptions. Such a hierarchical description is a sequence $D = (H_0,H_1,\ldots,H_r)$ where $r \geq 0$ and $H_i \in HE^n$ such that: if $\langle \text{bin } m\rangle$ occurs in H_i then $m < i$. The idea is the following: instead of writing down the subexpression H_m of H_i we replace it by $\langle \text{bin } m\rangle$ and describe H_m separately. Thus the subexpression H_m can be used several times but it must be described only once. This makes the hierarchical descriptions so powerful.

A description $D = (H_0,H_1,\ldots,H_r) \in HD^n$ describes the finite set $L(D) \subseteq \mathbb{N}^n$ which is defined as $L(H_r)$ where

1. $L(\langle \text{bin } m\rangle) = L(H_m)$, for $m = 0,1,\ldots,r-1$,
   $L((\text{bin } m_1,\ldots,\text{bin } m_n)) = \{(m_1,\ldots,m_n)\}$,
2. $L(H \cup H') = L(H) \cup L(H')$,
   $L(H+H') = \{(k_1+l_1,\ldots,k_n+l_n): (k_1,\ldots,k_n) \in L(H) \wedge (l_1,\ldots,l_n) \in L(H')\}$.

Example 1. For $r \geq 0$ the hierarchical description

$D_r \equiv ((0,0) \cup (0,1), \langle 0\rangle+\langle 0\rangle, \langle 1\rangle+\langle 1\rangle, \ldots, \langle r-1\rangle+\langle r-1\rangle)$ [2)]

describes $L(D_r) = \{(0,i): i = 0,1,\ldots,2^r\}$. Note that D_r is of lenth $\sim r \cdot \log r$.

Now we consider two restrictions of the class of hierarchical descriptions. First we do not allow the addition of two arbitrary expressions but only the addition of an arbitrary expression with an expression of form $(\text{bin } m_1,\ldots,\text{bin } m_n)$. By this restriction the class RHE^n of restricted hierarchical expressions and the class RHD^n of restricted hierarchical descriptions are defined. These descriptions can be considered as an analogue to the "general hierarchic input language" defined in [BO 81] to describe finite sets of rectangles in the plane.

Example 2. For $r \geq 1$, the restricted hierarchical description

$$E_r \equiv ((0,0) \cup (0,1), \langle 0\rangle \cup (\langle 0\rangle+(0,2)), \langle 1\rangle \cup (\langle 1\rangle+(0,4)), \ldots, \langle r-2\rangle \cup (\langle r-2\rangle+(0,2^{r-1})))$$

---

1) By bin m we denote the binary presentation of the natural number m.

2) All natural numbers are thought to be written in binary notation.

describes $L(E_r) = \{(0,i): i = 0,1,\ldots,2^r-1\}$. Note that E_r is of lengt $\sim r^2$.

Next we allow full addition but we do not allow the use of hierar chical means. Thus we obtain the class

$$IE^n = \{H: H \in HD^n \text{ and } \langle\ldots\rangle \text{ does not occur in } H\}$$

of integer-expressions defined in [SM 73] for $n = 1$.

Example 3. For $r \geq 1$, the integer-expression

$$H_r \equiv ((0,0)\vee(0,1))+((0,0)\vee(0,2))+((0,0)\vee(0,4))+\ldots+((0,0)\vee(0,2^{r-1}))$$

describes $L(H_r) = \{(0,i): i = 0,1,\ldots,2^r-1\}$. Note that H_r is of length $\sim r^2$.

Finally put $HD = \bigcup_{n\geq 1} HD^n$, $RHD = \bigcup_{n\geq 1} RHD^n$ and $IE = \bigcup_{n\geq 1} IE^n$.

Our examples have shown that there can be an exponential gap between the length of a description (of any type mentioned above) and the number of elements described by it. This gap cannot be enlarged.

Proposition 1. For every $H \in HD^n \cup IE^n$,

1. $\text{card } L(H) \leq 2^{|H|}$,
2. $x \in L(H)$ implies $|x| \leq |H|$, for all $x \in \mathbb{N}^n$. [1)]

Next we state that for a finite graph described by an $H \in X$ ($X \in \{HD, RHD, IE\}$) we can efficiently find an isomorphic graph described by an $H' \in X^2$ with essentially the same degree of compactness.

Proposition 2. Let $X \in \{HD, RHD, IE\}$. There exist logarithmic-space computable functins $g: \bigcup_{n\geq 1} X^{2n} \mapsto X^2$ and $h: \bigcup_{n\geq 1} X^{2n} \times \mathbb{N}^n \mapsto \mathbb{N}$ such that

1. $|g(H)| \leq c\cdot|H|^2$ for suitable $c > 0$,
2. $(a_1,\ldots,a_n) \in V_{L(H)} \Leftrightarrow h(H,a_1,\ldots,a_n) \in V_{L(g(H))}$ where for fixed H the function h is one-one,
3. $((a_1,\ldots,a_n),(b_1,\ldots,b_n)) \in E_{L(H)} \Leftrightarrow (h(H,a_1,\ldots,a_n),h(H,b_1,\ldots,b_n)) \in E_{L(g(H))}$.

Remark 1. Since all properties of graphs studied below do not depend on the naming of the vertices, Proposition 2 implies that it does not matter whether we consider graphs described by an H X^{2n} for arbitrary n or only graphs described by an $H \in X^2$, where $X \in \{HD, RHD, IE\}$. In other words: if X^n-PROB is a graph-theoretical problem based on descriptions from X^{2n} then $X^1\text{-PROB} \equiv_m^{\log} X^n\text{-PROB} \equiv_m^{\log} X\text{-PROB} =_{df} \bigcup_{i\geq 1} X^i\text{-PROB}$ ($\equiv_m^{\log}$ denotes the logarithmic-space m-reducibility).

1) Here and in what follows w denotes the length of the word w.

Now we compare the descriptional power of the integer-expressions on the one side and the restricted hierarchical descriptions on the other side. Though the reasons for which these descriptional languages are powerful seem to be different it can be shown that integer-expressions can be transformed efficiently into equivalent restricted hierarchical descriptions with essentially the same degree of compactness.

Theorem 1. There exist a logarithmic-space computable function f: IE $\longmapsto$ RHD such that for every H $\in$ IE,

1. L(f(H)) = L(H),
2. $|f(H)| \leq c \cdot |H| \cdot \log|H|$, for suitable $c > 0$.

Corollary 1. Let PROB be any problem concerning graphs described by the report of all vertices and edges. Let further X-PROB be the same problem but based on descriptions of X $\in$ {HD,RHD,IE}. Then we have

$$\text{PROB} \leq_m^{\log} \text{IE-PROB} \leq_m^{\log} \text{RHD-PROB} \leq_m^{\log} \text{HD-PROB}.$$

The question of whether restricted hierarchical descriptions can be transformed efficiently into equivalent integer- expressions and whether hierarchical descriptions can be transformed efficiently into restricted hierachical descriptions are still unresolved. We conjecture that both questions must be answered in the negative.

Finally we deal with the question about the maximum increase of the complexity of graph-theoretical problems when the graphs are described by our compact descriptions instead of the usual report of all vertices and edges. For $t: \mathbb{N} \longmapsto \mathbb{N}$ let DSPACE(t) (NSPACE(t)) be the class of all languages accepted by (non)deterministic multitape Turing machines within space t and let DTIME(t) (NTIME(t)) be the class of all languages accepted by (non)deterministic Turing machines within time t . Let Pol be the set of all polynomials. Furthermore, for a problem PROB involving natural numbers (as usual in binary notation) let PROB_{un} be the same problem but the natural numbers being represented in unary notation.

Theorem 2. Let X $\in$ {HD,RHD,IE} and Y $\in$ {D,N}.

1. For nondecreasing $s \geq \log$,
   $\text{PROB}_{un} \in \text{YSPACE}(s)$ implies $\text{X-PROB} \in \text{YSPACE}(s(2^n)+\text{Pol})$.
2. For nondecreasing $t \geq id$,
   $\text{PROB}_{un} \in \text{YTIME}(t)$ implies $\text{X-PROB} \in \text{YTIME}(t(2^{2n}) \cdot 2^{\text{Pol}})$.

Defining $\mathcal{L}$ = DSPACE(log), $\mathcal{NL}$ = NSPACE(log), $\mathcal{P}$ = DTIME(Pol), $\mathcal{NP}$ = NTIME(Pol) and PSPACE = DSPACE(Pol) = NSPACE(Pol) we obtain

Corollary 2. Let X $\in$ {HD,RHD,IE}.

1. $PROB_{un} \in \mathcal{NL}$ implies X-PROB $\in$ PSPACE.
2. PROB $\in \mathcal{NL}$ implies X-PROB $\in$ PSPACE.
3. $PROB_{un} \in \mathcal{P}$ implies X-PROB $\in$ DTIME(2^{Pol}).
4. PROB $\in \mathcal{P}$ implies X-PROB $\in$ DTIME(2^{Pol}).
5. PROB $\in \mathcal{NP}$ implies X-PROB $\in$ NTIME(2^{Pol}).

Theorem 2 and Corollary 2 show that using our compact descriptions the complexity of the problems can rise at most exponentially (in the argument of the bounding function). In the next section we give examples which really have this exponential gap (for instance, GAP and COLOUR). The behaviour of these problems can be described as follows: their complexity do not decrease when considering only graphs with extreme regularities. On the other hand, there are problems not having this exponential gap. As an example we have WGAP whose behaviour can be explained by the fact that natural numbers are involved in it and that its complexity decreases if only instances with small numbers are considered.

THE RESULTS

First we deal with some graph-theoretical problems which are in $\mathcal{L}$ when the graphs are described by the usual report of all vertices and edges. Let $X \in \{HD, RHD, IE\}$ and $n, k \geq 1$.

X^n-VERTEX = $\{(H,a): H \in X^{2n} \wedge a \in \mathbb{N}^n \wedge a \in V_{L(H)}\}$,

X^n-EDGE = $\{(H,a,b): H \in X^{2n} \wedge a,b \in \mathbb{N}^n \wedge (a,b) \in E_{L(H)}\}$,

X^n-k-INDEGREE = $\{H: H \in X^{2n} \wedge \bigwedge_a (a \in V_{L(H)} \rightarrow \text{indegree}(a) \leq k)\}$,

X^n-INDEGREE = $\{(H,k): H \in X^{2n} \wedge k \in \mathbb{N} \wedge \bigwedge_a (a \in V_{L(H)} \rightarrow \text{indegree}(a) \leq k)\}$,

X^n-TREE = $\{H: H \in X^{2n} \wedge G_{L(H)}$ is a tree$\}$,

X^n-TREE-CARD = $\{(H,k): H \in X^{2n} \wedge k \in \mathbb{N} \wedge G_{L(H)}$ is a forest having at most k trees$\}$,

X^n-k-CLIQUE = $\{H: H \in X^{2n} \wedge G_{L(H)}$ has a clique of size k$\}$,

X^n-k-INDEPENDENT-SET = $\{H: H \in X^{2n} \wedge G_{L(H)}$ has an independent set of size k$\}$.

The following theorem concerns only the languages RHD and IE. For HD we only know that all this problems are in PSPACE.
Let Σ_k^p and Π_k^p ($k \geq 0$) be the classes of the polynomial-time hierarchy (cf. [MS 72], [SM 73], [St 77]).

Theorem 3. Let $X \in \{RHD, IE\}$.

1. X-VERTEX, X-EDGE and X-k-CLIQUE are $\leq_m^{log}$-complete in $\mathcal{NP}$, for $k \geq 2$.
2. X-k-INDEGREE is $\leq_m^{log}$-complete in co$\mathcal{NP}$, for $k \geq 1$.
3. X-k-INDEPENDENT-SET is $\leq_m^{log}$-complete in Σ_2^p, for $k \geq 2$.
4. X-TREE is $\leq_m^{log}$-complete in Π_2^p.

Remark 2. The problems X-TREE-CARD and X-INDEGREE are $\leq_m^{log}$-complete in certain classes of the "counting polynomial-time hierarchy", an extension of the polynomial-time hierarchy, which has been defined in [Wa 84a] and [Wa 84b].

Remark 3. Though the problems CLIQUE and INDEPENDENT-SET seem to be essentially the same problems (cf. [GJ 79]) they are probably (i.e. if $\mathcal{NP} \neq \Sigma_2^p$) of different complexity when using our compact descriptions.

Remark 4. In [Wa 84a] and [Wa 84c] 20 further results of the same spirit can be found which concern properties of finite stes of natural numbers.

Next we deal with problems which are $\leq_m^{log}$-complete in PSPACE when their instances are described compactly. For $X \in \{HD, RHD, IE\}$ and $n \geq 1$ we define the graph accessibility problem for undirected graphs:

$$X^n\text{-UGAP} = \{(H,a,b): H \in X^{2n} \wedge a,b \in \mathbb{N}^n \wedge \text{there exist } r \in \mathbb{N},\ a_0,\dots,a_r \in \mathbb{N}^n \text{ such that } a_0 = a,\ a_r = b \text{ and } \{(a_i,a_{i+1}),(a_{i+1},a_i)\} \cap E_{L(H)} \neq 0 \text{ for } i = 0,\dots,r-1\},$$

for directed graphs:

$$X^n\text{-GAP} = \{(H,a,b): H \in X^{2n} \wedge a,b \in \mathbb{N}^n \wedge \text{there exist } r \in \mathbb{N},\ a_0,\dots,a_r \in \mathbb{N}^n \text{ such that } a_0 = a,\ a_r = b \text{ and } (a_i,a_{i+1}) \in E_{L(H)} \text{ for } i = 0,\dots,r-1\},$$

and for directed graphs with weighted edges:

$$X^n\text{-WGAP} = \{(H,a,b,c): H \in X^{2n+1} \wedge a,b \in \mathbb{N}^n \wedge c \in \mathbb{N} \wedge \text{there exist } r \in \mathbb{N},\ a_0,\dots,a_r \in \mathbb{N}^n,\ c_0,\dots,c_{r-1} \in \mathbb{N} \text{ such that } a_0 = a,\ a_r = b,\ \sum_{i=0}^{r-1} c_i = c \text{ and } (a_i,a_{i+1},c_i) \in E_{L(H)} \text{ for } i = 0,\dots,r-1\}$$

It is well known that (for graphs given by report of all vertices and edges) UGAP $\in \mathcal{NL}$ and GAP is $\leq_m^{log}$-complete in $\mathcal{NL}$, it is not known whether UGAP $\in \mathcal{L}$ or UGAP is $\leq_m^{log}$-complete in $\mathcal{NL}$, and it is easy to see that WGAP is $\leq_m^{log}$-complete in $\mathcal{NP}$ (by log-space reduction of the "sum of subst" problem to WGAP).

Theorem 4. For $X \in \{HD, RHD, IE\}$, X-UGAP, X-GAP and X-WGAP are $\leq_m^{log}$-complete in PSPACE.

The proof is made by reduction from B_ω (which has been shown to be $\leq_m^{log}$-complete in [SM 73], [St 77]) to B_{IE}^ω (which is an analogue to B_ω based on integer-expressions rather than on propositional formulas) and from B_{IE} toIE-UGAP. Since the graphs constructed in this reduction are acyclic and planar it is not hard to prove the following corollary. For $X \in \{HD, RHD, IE\}$ define

X-CYCLE = $\{H: H \in X \wedge G_{L(H)}$ is a directed graph having a cycle$\}$,

X-ACGAP = X-GAP $\cap$ $\overline{\text{X-CYCLE}}$ and

X-PLANAR = $\{H: H \in X \wedge G_{L(H)}$ is a planar graph$\}$

That CYCLE and ACGAP = GAP $\cap$ $\overline{\text{CYCLE}}$ are $\leq_m^{log}$-complete in $\mathcal{NL}$ has already been shown in [Su 75] and [Jo 75]. It is obvious that PLANAR $\in \mathcal{P}$.

Corollary 3. Let $X \in \{HD, RHD, IE\}$.

1. X-ACGAP and X-CYCLE are $\leq_m^{log}$-complete in PSPACE.
2. X-PLANAR is $\leq_m^{log}$-hard in PSPACE and X-PLANAR $\in$ DTIME(2^{Pol}).

Because of the close relationship between the graph accessibility problem for undirected graphs and the problem of whether a given graph is bipartite we can prove the following corollary. For $X \in \{HD, RHD, IE\}$ define

X-BIPARTITE = $\{H: H \in X \wedge G_{L(H)}$ is a bipartite graph$\}$.

Note that BIPARTITE $\leq_m^{log}$ UGAP (see [JLL 76]).

Corollary 4. For $X \in \{HD, RHD, IE\}$, BIPARTITE is $\leq_m^{log}$-complete in PSPACE.

Finally we deal with some problems which are $\leq_m^{log}$-complete in $\mathcal{NP}$ when the graphs are given by the report of all vertices and edges, and which are $\leq_m^{log}$-complete in NTIME(2^{Pol}) when using our compact descriptions. For $X \in \{HD, RHD, IE\}$ and $k \geq 1$ define

X-k-COLOUR = $\{H: H \in X \wedge$ the graph $G_{L(H)}$ can be coloured with k colours$\}$,

X-COLOUR = $\{(H,k): H \in X \wedge k \in \mathbb{N} \wedge$ the graph $G_{L(H)}$ can be coloured with k colours$\}$,

X-CLIQUE = $\{(H,k): H \in X \wedge k \in \mathbb{N} \wedge$ the graph $G_{L(H)}$ has a clique of size k$\}$,

X-INDEPENDENT-SET = $\{(H,k): H \in X \wedge k \in \mathbb{N} \wedge$ the graph $G_{L(H)}$ has an independent set of size k$\}$,

X-VERTEX-COVER = $\{(H,k): H \in X \wedge k \in \mathbb{N} \wedge \bigvee_{V \subseteq V_{L(H)}} (\text{card } V \leq k \wedge \bigwedge_{(a,b) \in E_{L(H)}} \{a,b\} \cap V \neq \emptyset)\}$.

Theorem 5. For $X \in \{HD, RHD, IE\}$ the following problems are $\leq_m^{log}$-complete in NTIME(2^{Pol}): X-k-COLOUR for $k \geq 3$, X-COLOUR, X-CLIQUE, X-INDEPENDENT-SET, X-VERTEX-COVER.

Remark 5. Corollary 4 shows that X-2-COLOUR = X-BIPARTITE is $\leq_m^{log}$-complete in PSPACE.

Remark 6. The usual reductions between $\mathcal{NP}$-complete problems do not work when the instances are given by our compact descriptions. This is caused (among other things) by the fact that simple set-theoretical operations like complementation and intersection can probably not be carried out efficiently on the base of our compact descriptions. Thus we have been forced to reduce arbitrary problems of NTIME(2^{Pol}) to each of the problems mentioned in Theorem 5. It is an interesting observation that the same reductions can be used to reduce arbitrary problems of $\mathcal{NP}$ to the problems k-COLOUR, COLOUR, CLIQUE, INDEPENDENT-SET and VERTEX-COVER. Thus we get new proofs for the $\mathcal{NP}$-completeness of these problems which are at least in the case of k-COLOUR and COLOUR considerably shorter than the usual proofs by reduction via the satifyability problem.

Remark 7. Contrary to the problems of Theorem 5, WGAP is an $\mathcal{NP}$-complete problem which does not become $\leq_m^{log}$-complete in NTIME(2^{Pol}) when the graphs are given by our compact descriptions (unless PSPACE = NTIME(2^{Pol})). This was caused by the fact that WGAP restricted to small weights is in $\mathcal{NL}$. Up till now we were unable to find a problem which remains $\mathcal{NP}$-complete for small numbers (i.e. which is strong $\mathcal{NP}$-complete, see [GJ 79]) but which also does not have the exponential complexity gap.

ACKNOWLEDGEMENTS

I am grateful to Burkard Monien, Gerd Wechsung and Jörg Vogel for some interesting discussions.

REFERENCES

[BO 81] Bentley, J.L., Ottmann, T., The complexity of manipulating hierarchically defined sets of rectangles, MFCS 1981, in Lecture Notes in Computer Science 118, 1-15.

[GJ 79] Garey, M.R., Johnson, D.S., Computers and Intractability, A Guide to the Theory of NP-Completeness, W.H. Freeman and Co., San Francisco 1979.

[JLL 76] Jones, N.D., Lien, Y.E., Laaser, W.T., New problems complete for nondeterministic log space, Mathematical Systems Theory 10(1976), 1–17.

[Jo 75] Jones, N.D., Space-bounded reducibility among combinatorial problems, Journal of Computer and System Sciences 11(1975), 68–85.

[MS 72] Meyer, A.R., Stockmeyer, L.J., The equivalence problem for regular expressions with squaring requires exponential space, Proc. of the 13th SWAT(1972), 125–129.

[SM 73] Stockmeyer, L.J., Meyer, A.R., Word problems requiring exponential time, Proc. of the 5th STOC(1973), 1–9.

[St 77] Stockmeyer, L.J., The polynomial-time hierarchy, Theoretical Computer Science 3(1977), 1–22.

[Su 75] Sudborough, I.H., On tape-bounded complexity classes and multihead finite automata, Journal of Computer and System Sciences 10(1975), 62–76.

[Wa 84a] Wagner, K., Compact descriptions and the counting polynomial-time hierarchy (extended abstract), to appear in the Proc. of the 2nd Frege Conf., Schwerin 1984.

[Wa 84b] Wagner, K., The counting polynomial-time hierarchy, preprint.

[Wa 84c] Wagner, K., The complexity of combinatorial problems with compactly described instances, preprint.

ON THE COMPLEXITY OF SLICE FUNCTIONS

I. Wegener
FB 20-Informatik, Johann Wolfgang
Goethe-Universität, 6000 Frankfurt a.M.,
Fed. Rep. of Germany

ABSTRACT

By a result of Berkowitz the monotone circuit complexity of slice functions cannot be much larger than the circuit (combinational) complexity of these functions for arbitrary complete bases. This result strengthens the importance of the theory of monotone circuits. We show in this paper that monotone circuits for slice functions can be understood as special circuits called set circuits. Here disjunction and conjunction are replaced by set union and set intersection. All known methods for proving lower bounds on the monotone complexity of Boolean functions do not work in their present form for slice functions. Furthermore we show that the canonical slice functions of the Boolean convolution, the Nechiporuk Boolean sums and the clique function can be computed with linear many gates.

1. INTRODUCTION

We investigate the complexity of slice functions. A function $f:\{0,1\}^n \to \{0,1\}^m$ is called a k-slice iff f(x) equals the 0-vector if x has less than k ones and f(x) equals the 1-vector if x has more than k ones. That means the interesting part of f happens on the k-slice of $\{0,1\}^n$.

For the computation of Boolean functions we consider Boolean circuits (for the definition and elementary properties see Savage [4]) and the circuit complexity (combinational complexity) either over the complete basis of all binary Boolean functions or over the monotone basis consisting of the binary conjunction and disjunction. These complexity measures are denoted by C and C_m. One knows that one may prove NP≠P by proving a non polynomial lower bound on the circuit complexity of a function in NP. In general we know only little about the relation between C and C_m. Berkowitz [2] (as cited in Valiant [5]) was able to show that these two complexity measures are closely connected for slice functions. In Chapter 2 we present these results in more detail.

In Chapter 3 we investigate the structure of monotone circuits for slice functions. If one is not interested in an additional additive $O(n \log n)$ term for the complexity one may replace conjunctions by set intersections and disjunctions by set unions. Set circuits are the heart of circuits and monotone circuits for slice functions.

Many important functions like the Boolean convolution, Boolean sums or the clique functions have the property that all prime implicants have the same length. If this length is k the proper k-slice may be called the canonical slice function. It has among some others the same prime implicants as the given function, while all other slices have only other prime implicants. We show in Chapter 4 that the canonical slice of the Boolean convolution has linear complexity and in Chapter 5 that the canonical slice of each clique function can be computed with a linear number of gates.

2. THE RESULTS OF BERKOWITZ

Since the results of Berkowitz [2] will be included in his Ph.D. thesis they are until now not published. Therefore we present here among perhaps some more results that result (Theorem 1) of Berkowitz which is most important for our purposes.

Definition 1: A Boolean function $f:\{0,1\}^n \to \{0,1\}^m$ is called a k-slice iff f(x) equals the 0-vector if x contains less than k ones and equals the 1-vector if x contains more than k ones.

Definition 2: Let T^n_m denote the threshold-m-function on n variables, computing 1 iff the input has at least m ones. Let E^n_m denote the exactly-m-function, computing 1 iff the input has exactly m ones.

Definition 3: For an arbitrary Boolean function f its k-slice $f_{s\ell-k}$ is defined by

$$f_{s\ell-k} := (f \wedge E^n_k) \vee T^n_{k+1} = (f \wedge T^n_k) \vee T^n_{k+1}.$$

Definition 4: If all prime implicants of a monotone function have length k we call $f_{s\ell-k}$ the canonical slice of f.

We notice that in this situation all prime implicants of f are prime implicants of its canonical slice and no prime implicant of f is a prime implicant of any other slice with the only exception of the (k-1)-slice which equals T^n_k. Since for monotone f with prime implicants of length k only $f \le T^n_k$ we can conclude that $f_{cs\ell} = f \vee T^n_{k+1}$ for the canonical slice $f_{cs\ell}$.

It is well known that the set of all threshold functions $T^n = (T^n_1, \dots, T^n_n)$ has linear complexity over a complete basis (Savage [4]) and has complexity O(n log n) over the monotone basis (Ajtai/Komlós/Szemerédi [1]). Furthermore $C_m(T^n_k)$ is O(n) if k is fixed. Thus the slices cannot be much harder than the given function f. But we can prove that not all slices of a given hard function can be easy.

<u>Proposition 1:</u> (i) $C(f) \leq \sum_{1 \leq k \leq n} C(f_{s\ell-k}) + O(n)$

(ii) $C(f_{s\ell-k}) \leq C(f) + O(n)$

(iii) $C_m(f_{s\ell-k}) \leq C_m(f) + O(n \log n)$

(iv) $C_m(f_{s\ell-k}) \leq C_m(f) + O(n)$ if k is fixed.

<u>Proof:</u> Obviously the first assertion follows from the facts that

$f = \bigvee_{1 \leq k \leq n} (f_{s\ell-k} \wedge E_k^n)$, $E_k^n = T_k^n \wedge \overline{T_{k+1}^n}$ and $C(T) = O(n)$. The second, third and fourth assertion follow from the definition of

$f_{s\ell-k}$, $C(T) = O(n)$, $C_m(T) = O(n \log n)$ and $C_m(T_k^n) = O(n)$ for fixed k. Q.E.D.

Proposition 1 states that in order to prove large lower bounds on the (monotone) circuit complexity of f it is sufficient to consider all slices of f. For a complete basis we even know, that some slice of a hard function has to be hard. This cannot be proved for the monotone basis directly. Since for k<k' obviously

$f_{s\ell-k'} \leq f_{s\ell-k}$ also $f_{s\ell-k'} \wedge f_{s\ell-k} = f_{s\ell-k'}$ and

$f_{s\ell-k'} \vee f_{s\ell-k} = f_{s\ell-k}$.

Thus we cannot compute in monotone circuits f from its slices.

Until now we can prove nonlinear lower bounds only for the monotone complexity of n-output functions (Wegener [6], Weiß [8]) and not for complete bases. It seems to be much easier to prove lower bounds for the monotone basis than for complete bases. The following result of Berkowitz shows that for slice functions the monotone complexity and the circuit complexity are closely related. Thus a hard Boolean function must have a slice whose <u>monotone</u> complexity is large. This proves the importance of the theory of monotone circuits.

<u>Theorem 1:</u> (Berkowitz [2]) For a slice function f it holds that $C_m(f) \leq O(C(f) + n^2 \log n)$. If f is a k-slice for fixed k the additive term may be reduced to $O(n^2)$.

<u>Proof:</u> Let f be a k-slice. An optimal circuit for f may be replaced by a circuit of conjunctions, disjunctions and negations only. The complexity increases only by a constant factor. By the rules of de Morgan we get a circuit where only the variables are negated.

The problem is how to compute $\overline{x_i}$ over the monotone basis. This is in general impossible. But we may use the following trick. Let $X:=\{x_1,\dots,x_n\}$ and $X_i:=X-\{x_i\}$. We compute $T_k^n(X)$, $T_{k+1}^n(X)$, $T_k^{n-1}(X_1),\dots,T_k^{n-1}(X_n)$ which can be done by $O(n^2\log n)$ monotone gates. Let us now consider only inputs with exactly k ones. For such an input $T_k^{n-1}(X_i)=0$ if $x_i=1$ and $T_k^{n-1}(X_i)=1$ if $x_i=0$, that means $T_k^{n-1}(X_i)=\overline{x_i}$.

By replacing in our circuit $\overline{x_i}$ by $T_k^{n-1}(X_i)$ we compute f correctly on all inputs with exactly k ones. Let f* be the function computed by our circuit. Since we may easily see by considering the cases where x has less than, exactly and more than k ones that $f=(f^* \wedge T_k^n) \vee T_{k+1}^n$ we are done. Q.E.D.

Altogether we have shown that a function f with a slice whose monotone complexity is large has a hard slice and is therefore hard itself.

3. MONOTONE CIRCUITS FOR SLICE FUNCTIONS VS. SET CIRCUITS

We have shown that it is important to investigate the monotone complexity of slice functions. Here we will show that the main structure of a monotone circuit for a slice function is given by a set circuit which we define later.

Let us consider a k-slice f. We are interested in monotone circuits for functions f' which compute 0 if the input has less than k ones, which equal f for k ones and which are arbitrary for inputs with more than k ones. Since $f=f' \vee T_{k+1}^n$ we get again a monotone circuit for f whose cost is at most by an additive term of $O(n \log n)$ larger than the cost of the circuit for f'. For constant k this additive term can be reduced to $O(n)$. At first we manipulate the inputs.

<u>Proposition 2:</u> If we replace in a monotone circuit for a k-slice f each variable input x_i by $x_i \wedge T_k^n$ the new circuit again computes f. The complexity of the new circuit is only by an additive term of $n+C_m(T_k^n)$ larger than the complexity of the given one.

<u>Proof:</u> The second assertion is obvious. Since $x_i \wedge T_k^n \le x_i$ and because of the monotonicity of the circuit the function f* computed by the new circuit has the property $f^* \le f$. Let us assume that for some input a $f^*(a)=0$ and $f(a)=1$. Then again by monotonicity $a_i=1$ and $a_i \wedge T_k^n(a)=0$ for some i. Thus $T_k^n(a)=0$ which implies by the definition of a k-slice $f(a)=0$, a contradiction. Q.E.D.

Investigating the main structure of monotone circuits for slice functions we may assume w.l.o.g. that we have changed the circuit in the way described in Proposition 2. The effect of this transformation is that afterwards all functions computed in the circuit have prime implicants of length at least k only. Let $f' := \bigvee_{t \in PI_k(f)} t$ where $PI_k(f)$ is the set of all prime implicants of f of length k. Then f' is one of the functions described at the beginning of this chapter where

$f = f' \vee T^n_{k+1}$. In our monotone circuit for f we use now supergates instead of monotone gates. A supergate (super $\wedge$-gate or super $\vee$-gate) works at first like a normal $\wedge$ or $\vee$ - gate and afterwards it destroys all prime implicants with more than k variables. By the first replacement rule of Mehlhorn/Galil [3] for monotone circuits the new circuit computes f' instead of f. This can be shown easily by induction on the topological order of the gates. We compute everywhere instead of some function g now $g' := \bigvee_{t \in PI_k(g)} t$.

Altogether we have now a monotone (super-) circuit for f' where all prime implicants of all computed functions have length k. Let us consider the effect of supergates. Let g_1 and g_2 be two functions of the described class. Let $g' := \text{super-}\vee(g_1, g_2)$ and $g'' := \text{super-}\wedge(g_1, g_2)$. We can conclude that $PI(g') = PI(g_1) \cup PI(g_2)$. The property "$\subseteq$" always holds. Here "$\supseteq$" holds too. The absorption rule cannot be applied since all prime implicants have the same length. Also

$PI(g'') = PI(g_1) \cap PI(g_2)$. For $g := g_1 \wedge g_2$ we have $g = (\bigvee_{t \in PI(g_1)} t) \wedge (\bigvee_{t' \in PI(g_2)} t')$. All t and t' have length k.

If $tt' \in PI(g_1) \cap PI(g_2)$ it is always a prime implicant of g too. Since it has length k it is a prime implicant even of g". All products tt' where $t \neq t'$ have length larger than k and become destroyed.

These observations motivate the following definition.

Definition 5: Let g be a monotone function whose prime implicants all have length k. A set circuit for g has inputs $x_i \wedge T^n_k$ for $1 \leq i \leq n$ and uses $\cap$ - and $\cup$ - gates. For two functions g_1 and g_2 whose prime implicants all have length k $g' := g_1 \cup g_2$ and $g'' := g_1 \cap g_2$ are defined by

$PI(g') := PI(g_1) \cup PI(g_2)$ and $PI(g'') := PI(g_1) \cap PI(g_2)$. The set complexity of g denoted by $SC_m(g)$ is the minimal number of gates in a set circuit computing g.

Theorem 2: Let f be a k-slice and $g := \bigvee_{t \in PI_k(f)} t$.

i) $C_m(f) \leq SC_m(g) + O(n \log n)$.

ii) If k is fixed even $C_m(f) \leq SC_m(g) + O(n)$.

iii) $SC_m(g) \leq C_m(f)$.

Proof: i)/ii) In order to compute f by a monotone circuit we compute

all $x_i \wedge T_k^n$. We use afterwards an optimal set circuit for g, where the inputs are already computed, and replace all $\cap$ by $\wedge$ and all $\cup$ by $\vee$ (cost $SC_m(g)$). We compute a function f' where all prime implicants have length at least k and where $PI_k(f')=PI_k(g)=PI_k(f)$. Thus $f=f' \vee T_{k+1}^n$.
iii) In order to compute g by a set circuit we use an optimal monotone circuit for f. We replace the inputs x_i by the inputs $x_i \wedge T_k^n$ which are given for free in set circuits. Furthermore we replace all $\vee$ by $\cup$ and al $\wedge$ by $\cap$. By our previous observations we obtain a set circuit for g. Q.E.

Theorem 2 shows that the heart of a monotone circuit for a slice function is a set circuit. The set circuit contains exactly all necessary information and it contains no unnecessary monoms. Many lower bounds on the monotone complexity of non slice functions use the fact that one has to destroy unnecessary monoms if one has computed them. For slice functions this work is easy by a disjunction with T_{k+1}^n. Thus the set circuit concentrates the view on the important part of the computation. Investigating the monotone complexity of slice functions one should therefore work in the model of set circuits.

To gain even more structure we make the following observations. Let us at first assume that we may partition the set of variables to k subsets and that each prime implicant of g contains exactly one variable of each subset. This property is fulfilled for the Boolean convolution or the Boolean matrix product for k=2 and for the generalized Boolean matrix product (Wegener [6]) for arbitrary k. Let x_j^i denote a variable of the i-th subclass. We can compute with O(n) gates all h_i, the disjunction of all variables in the i-th subclass. Let $g_i := \bigwedge_{j \neq i} h_j$ for $1 \le i \le k$. All g_i can be computed in the following way.
At first we compute all $h_1 \wedge h_2, h_3 \wedge h_4, h_5 \wedge h_6, \ldots$, afterwards all $h_1 \wedge \ldots \wedge h_4, h_5 \wedge \ldots \wedge h_8, \ldots$, and so on. All this can be done by O(k) gates. Each g_i can be expressed as the conjunction of $\lceil \log k \rceil$ of the computed terms. Altogether all g_i can be computed with O(n+k log k) gates. Proposition 2 and the appropriate version of Theorem 2 remain correct if we replace each x_j^i by $x_j^i \wedge g_i$. The advantage of this procedure is the following. The set of all possible prime implicants is the set of all combinations of one variable of each class, that means it is a k-dimensional discrete block or even a k-dimensional discrete cube if all classes have the same size. Each input is a (k-1)-dimensional subblock or subcube. The prime implicants form an arbitrary subset (pattern) of the block or cube.

The problem of constructing optimal set circuits for g (or optimal monotone circuits for slice functions f) is therefore equivalent to the geometric problem of constructing the subset of points of a k-dimensional block formed by the prime implicants of g by intersections and unions of its (k-1)-dimensional subblocks.

Arbitrary functions with prime implicants of length k only can be changed in the following way in order to apply the geometric approach. The set of variables $\{x_1,\ldots,x_n\}$ is replaced by the kn variables x_j^i $(1\le i\le k,\ 1\le j\le n)$. The prime implicants $x_{i_1}\ldots x_{i_k}$ where $i_1<\ldots<i_k$ are replaced by $x_{i_1}^1\ldots x_{i_k}^k$.

We combine our results. We are interested in the circuit complexity of f. Instead of that we may investigate the circuit complexity of the slices of f. For slice functions it is nearly equivalent to consider monotone circuits. At last we have shown that this problem can be replaced by the investigation of the set complexity of $g:=\bigvee_{t\in PI_k(f_{s\ell-k})} t$. On one hand this last problem turned out to be the key problem but on the other hand we will show the existence of set circuits for some slices whose efficiency is remarkable.

4. THE BOOLEAN CONVOLUTION

The canonical slice is that slice which at first sight is most similar to the given function. Here and in the following chapter we show that the canonical slice may be much easier than the given function.

Definition 6: The Boolean convolution $f:\{0,1\}^{2n}\to\{0,1\}^{2n-1}$ on the set of variables $\{x_1,\ldots,x_n,\ y_1,\ldots,y_n\}$ is given by $f_k:=\bigvee_{i+j=k} x_iy_j$ for $2\le k\le 2n$.

Weiß [8] has shown that $C_m(f)=\Omega(n^{3/2})$ and one conjectures that $C_m(f)=\Theta(n^2)$.

Theorem 3: The monotone complexity of the canonical slice of the Boolean convolution is linear.

Proof: By Theorem 2 it is sufficient to prove $SC_m(f)=O(n)$. We use the geometric approach and consider the square $\{1,\ldots,n\}^2$ where the input $A_i=x_i\wedge(y_1\vee\ldots\vee y_n)$ of the set circuit corresponds to the i-th row of the square and $B_j=y_j\wedge(x_1\vee\ldots\vee x_n)$ corresponds to the j-th column. The output f_k corresponds to the k-diagonal of all (i,j) where i+j=k. We assume that $n=m^2$ for some natural number m. Otherwise we could add some variables which we fix afterwards to 0. We use the following algorithm.

$D_\ell:=\bigcup_{1\le i\le m} A_{(\ell-1)m+i}$ and $E_\ell:=\bigcup_{1\le i\le m} B_{(\ell-1)m+i}$ $(1\le\ell\le m)$,

$F_{ij}:=D_i \cap E_j$ $(1\le i,j\le m)$, $G_\ell := \bigcup_{i+j=\ell} F_{ij}$ $(2\le \ell \le 2m)$,
$H_\ell := \bigcup_{1\le i\le m} A_{(i-1)m+\ell}$ and $I_\ell := \bigcup_{1\le i\le m} B_{(i-1)m+\ell}$ $(1\le \ell \le m)$,

$J_{ij}:=H_i \cap I_j$ $(1\le i,j\le m)$, $K_\ell := \bigcup_{i+j=\ell} J_{ij}$ $(2\le \ell \le 2m)$.

G_ℓ is the ℓ-diagonal of subsquares and K_ℓ is the union of all ℓ-diagonals of all subsquares. The cost up to now is $8m^2-8m+2$. T_k, the set corrponding to f_k is the k-diagonal of the whole square. T_k touches one or two diagonals of subsquares, say $G_{h(k)}$ and $G_{h(k)+1}$. The intersection of T_k and $G_{h(k)}$ (resp. $G_{h(k)+1}$) is some diagonal of the corresponding subsquares, say the diagonal $d_1(k)$ of the subsquares of $G_{h(k)}$ and the diagonal $d_2(k)$ of the subsquares of $G_{h(k)+1}$. Thus

$T_k := (G_{h(k)} \cap K_{d_1(k)}) \cup (G_{h(k)+1} \cap K_{d_2(k)})$ or $T_k := G_{h(k)} \cap K_{d(k)}$ $(2\le k\le 2n)$.

The cost altogether is less than $14n-8m-1$. Q.E.D.

5. CLIQUE FUNCTIONS

Definition 7: The k-clique function $f_k:\{0,1\}^N \to \{0,1\}$, $N=\binom{n}{2}$, has variables x_{ij} $(1\le i<j\le n)$ corresponding to the possible edges of an n-vertex graph G. f_k computes 1 iff the graph specified by the variables contains a clique of size k.

The clique problem is NP- complete. Therefore one may ask which slices are hard. We prove that the canonical slice f*, which is the $K=\binom{k}{2}$-slice, is easy to compute.

Theorem 4: The circuit complexity of the canonical slice of any clique-function is $O(N)$ and its monotone complexity is $O(N \log N)$. For fixed k even the monotone complexity is $O(N)$.

Proof: $f^*=(f \wedge E_K^N) \vee T_{K+1}^N = (f \wedge T_K^N) \vee T_{K+1}^N$. T_K^N and T_{K+1}^N can be computed with $O(N)$ gates. In the formula for f* above we may replace f by any function g which coincides with f on the set of inputs where $E_K^N(x)=1$. Let $X_i:=\{x_{1,i},\dots,x_{i-1,i}, x_{i,i+1},\dots,x_{i,n}\}$. We replace f by $g:=T_k^n(T_{k-1}^{n-1}(X_1),\dots,T_{k-1}^{n-1}(X_n))$. Each $T_{k-1}^{n-1}(X_i)$ and afterwards the function T_k^n may be computed with $O(n)$ gates, therefore we need only $O(n^2)=O(N)$ gates for g. The following graph theoretical claim proves the correctness of our replacement. Graphs with exactly K edges contain a k-clique if and only if at least k vertices have outdegree at least k-1. Q.E.D.

CONCLUSION

We have seen that monotone circuits and in particular set circuits are important tools for the proof of lower bounds on the circuit complexity of hard Boolean functions. The investigation of monotone circuits is sometimes easier than the investigation of circuits over a complete basis. But we have indicated that for all models it seems to be difficult to prove lower bounds for slice functions. Nevertheless the concept of considering the slice functions of a given function is an important new concept. The more detailed paper [7] contains also some additional results. The asymptotic complexity of the set of slice functions is investigated. Similar results to Chapter 4 are proved for a one-output function based on Boolean sums. Finally it is shown that one cannot apply directly the known methods for proving lower bounds on the monotone complexity to slice functions.

ACKNOWLEDGEMENTS

I like to thank Leslie G.Valiant who presented Theorem 1 of Steve Berkowitz at the Oberwolfach conference on complexity theory. Also thanks to Rüdiger Reischuk for suggesting the algorithm of Theorem 3 and to Friedhelm Meyer auf der Heide for discussing the ideas of this paper.

References:

[1] Ajtai,M./Komlós ,J./Szemerédi,E.: An O(n log n) sorting network, Proc. 15th STOC, 1-9, 1983

[2] Berkowitz,S.: Personal communication as cited in [5], 1982

[3] Mehlhorn,K./Galil,Z.: Monotone switching circuits and Boolean matrix product, Computing 16, 99-111, 1976

[4] Savage,J.E.: The complexity of computing, John Wiley, 1976

[5] Valiant,L.G.: Exponential lower bounds for restricted monotone circuits, Proc. 15th STOC, 110-117, 1983

[6] Wegener,I.: Boolean functions whose monotone complexity is of size $n^2/\log n$, Theoretical Computer Science 21, 213-224, 1982

[7] Wegener,I.: On the complexity of slice functions, Techn.Rep., Univ. Frankfurt, 1983 (submitted to Theoretical Computer Science)

[8] Weiß,J.: An $\Omega(n^{3/2})$ lower bound on the monotone complexity of Boolean convolution, to appear: Information and Control

AN EXPONENTIAL LOWER BOUND FOR ONE-TIME-ONLY BRANCHING PROGRAMS

Stanislav Žák
Institute for Computation Techniques
Technical University
Horská 3
128 00 Praha 2
Czechoslovakia

The bound is $2^{\sqrt{2n}/3 - \log(\sqrt{2n}/3)}$.

INTRODUCTION

We define branching programs following Borodin et al. [2] . Branching programs are acyclic labelled graphs with the following properties:

(i) There is exactly one source.

(ii) Every node has outdegree at most 2.

(iii) For every node v with outdegree 2, one of the edges leaving v is labelled by a Boolean variable x_i and the other by its complement $\bar{x}_i$.

(iv) Every sink is labelled by 0 or 1 .

Let P be a branching program with edges labelled by the Boolean variables, $x_1, \dots x_n$ and their complements. Given an input $a = (a_1, \dots ,a_n) \in \{0,1\}^n$, program P computes a function value $f_P(a)$ in the following way. The computation starts at the source. If the computation has reached a node v and if only one edge leaves v, then the computation proceeds via that edge. If 2 edges, with labels x_i and $\bar{x}_i$, leave v , then the computation proceeds via the edge labelled x_i if $a_i = 1$, and via the edge labelled $\bar{x}_i$ otherwise. Once the computation reaches a sink, the computation ends and $f_P(a)$ is defined to be the label of that sink. We call sinks accepting if they are labelled 1 and rejecting otherwise. By the complexity of a branching program we mean the number of its nodes.

Branching programs are a generalization on one hand of decision trees (Masek [4]), and on the other hand of Turing machines ([7], [8]). /Informally: the configurations of a Turing machines on words of a fixed length are , in fact, the nodes of a branching program; hence, the logarithm of a lower bound on branching programs is a lower bound on space complexity of TM's./

Aleliunas et al. [1] have proved that the reachability problem for undirected graphs is of polynomial complexity on branching programs. To prove a nonpolynomial lower bound in general case seems to be very difficult; there are results for some restricted cases. Borodin et al. [2] have proved an exponential lower bound for monotone width-two branching programs and $\Omega(n^2/\log n)$ for width-two branching programs. Chandra et al. [3] have proved nonlinear lower bound for constant-width branching programs. Pudlák [9] has proved a nonlinear lower bound for the general case without the assumption "constant-width". Another restriction is to consider only one-time-only branching programs. In this case, each computation looks for any variable at most once. Quadratic lower bounds were proved by Masek [4] and Wegener [5] . The present author proved an exponential lower bound [7] , [8]. Recently, Wegener independently proved the bound $\Omega(2^{\sqrt{2n}/4})$ [6] .

ONE-TIME-ONLY BRANCHING PROGRAMS

We say that a computation asks for a variable x_i iff it goes through such a node with out-degree 2 that its out-edges are labelled by $x_i, \bar{x}_i$. A branching program is called "one-time-only" iff each computation asks for each variable at most one time. It is a simple fact that a branching program is "one-time-only" iff each path from the source to a sink is a computation.

Now, let us present some technical definitions and lemmas. A computation on an input a will be considered as a sequence of nodes and will be denoted comp(a) . For a node $K \in$ comp(a) we write

$a_K = \{ i \mid$ before having reached K, comp(a) asks the variable $x_i \}$,

$a^K = \{ i \mid$ after having reached K, comp(a) asks the variable $x_i \}$.

The complement of $X \subseteq \{1, \dots ,n\}$ will be denoted $\bar{X}$. Clearly, $\overline{a_K} \supseteq a^K$ for any one-time-only branching program. Let $a = (a_1, \dots a_n)$, $b = (b_1, \dots ,b_n)$ and $S \subseteq \{1, \dots ,n\}$. We write $a =_S b$ iff $(\forall i \in S)(a_i = b_i)$.

Lemma 1. Let a one-time-only branching program be given. Let a,b be inputs and K be a node common to comp(a) and comp(b).

(a) If $a_K = b_K$, $c =_{a_K} a$ and $c =_{\overline{a_K}} b$ then c is accepted iff b is accepted.

(b) $b_K \subseteq \overline{a^K}$ /i.e. after K, b_K is inaccessible not only for comp(b) but also for comp(a) /.

(c) If $c =_{\overline{b_K - a_K}} a$ then comp(c) = comp(a).

Proof. (a) comp(c) follows comp(a) until K is reached, then it follows comp(b).

(b) Let p be the path which follows comp(b) to K, and which follows comp(a) after K. p must be a computation since we have a one-time-only branching program, /in fact, p = comp(c) where $c =_{b_K} b$, $c =_{\overline{b_K}} a$ /. Now suppose $b_K \cap a^K \neq \emptyset$. Then p asks twice an x_i^K, $i \in b_K$. A contradiction.

(c) comp(a) and comp(c) can branch only in $b_K - a_K$; but by (b) this is inaccessible for comp(a).

A finite graph $(\{v_i\}_{i=1}^m, E)$ may be given by a 0-1 matrix $(a_{ij})_{i,j=1}^m$ where $a_{ij} = 1$ iff $(v_i, v_j) \in E$. Assuming irreflexivity and symmetry of E, such a graph can be codded by the binary string $a_{12}a_{13}a_{14} \ldots a_{1,m}a_{23}a_{24} \ldots a_{2,m} \ldots a_{m-1,m}$.

By a half-clique we mean the code of any finite graph $G = (V_1 \cup V_2, E_1)$ where card V_1 = card V_2 and $E_1 = V_1 x V_1 - \{(v,v) \mid v \in V_1\}$. For simplicity, we shall use the same letter to denote both the code of G and V_1.

We say that a branching program accepts a set $S \subseteq \{0,1\}^n$ iff $(\forall a \in \{0,1\}^n)(f_P(a) = 1 \longleftrightarrow a \in S)$.

Lemma 2. Let a one-time-only branching program accepting the set of half-cliques be given. Let a be a half-clique, b be any input and K be a node common to comp(a) and to comp(b). Then $b_K - a_K = \emptyset$. /If b is also a half-clique then $b_K = a_K$./

Proof. Suppose $b_K - a_K \neq \emptyset$. By Lemma 1(c) there is an input c such that comp(c) = comp(a) and c is not a half-clique. A contradiction.

__Theorem 3.__ Each one-time-only branching program which accepts the set of codes of half-cliques of length n is of complexity at least $2^{\sqrt{2n}/3 - \log(\sqrt{2n}/3)}$.

The proof is based on the intuitive idea that, treating the last vertices of a half-clique , the computation "has to realize" what are the other vertices of this half-clique. Generally this is possible in two ways: either to remember all treated vertices, or to ask once again. For one-time-only branching programs the second possibility is forbidden. Now, in order to remember many half-cliques it is necessary to have many nodes.

Proof. Each $a \in \{0,1\}^n$ can be considered as a code of a graph with m vertices where m is maximal such that $(m^2 - m)/2 \leq n$, and therefore $m = \left[(1 + \sqrt{1+8n})/2\right]$. For a half-clique a, let F(a) be the first node in comp(a) such that the edges which are asked in nodes preceeding F(a) and in F(a) cover at least m/2 - 2 vertices of the clique of a. We choose a maximal set S of half-cliques such that each two half-cliques in S differ on at least 6 vertices. The cardinality of S is at least $\binom{m/3}{m/6} \geq \frac{2^{m/3}}{m/3 + 1} \geq 2^{\sqrt{2n}/3 - \log(\sqrt{2n}/3)}$. To see this, group the vertices into triads. Now it suffices to prove that our $F \restriction S$ is an injective mapping. Suppose $a,b \in S$, $a \neq b$ and $F(a) = F(b) = K$. By Lemma 2 , $a_K = b_K$ and we can choose an input c such that $c =_{a_K} a$, $c =_{\overline{a_K}} b$. By Lemma 1(a) , c is a half-clique too.
Let $v \in a - b$ be a vertex with at least one edge in a_K , hence $v \in c$. Let $u \in b$ be a vertex with no edge in b_K , let $w \in b - a$, $w \neq u$. Cleraly $w \in c$. But there is no edge between v and w in c, since in c there are only edges which are in a or in b. Therefore c is not a half--clique. A contradiction.

It is not difficult to construct a two-time-only branching program which computes half-cliques within a polynomial bound. We see that the number of asking is important since a small change of it dramatically reduces the complexity bound. There are many questions about hierarchies with respect to asking and the number of nodes.

References.

[1] R. Aleliunas, R.M. Karp, R.J. Lipton, L. Lovász, C. Rackoff Random Walks, Universal Traversal Sequences, and the Complexity of Maze Problems , Proc. 20-th IEEE Symp. on Foundations of Computer Science, 1979, 218-223.

[2] A. Borodin, D. Dolev, F.E. Fich, W.Paul Bounds for Width Two Branching Programs , 15th Annual ACM Symposium on Theory of Computing, 1983, 87-93.

[3] A.K. Chandra, M.L. Furst, R.J. Lipton , Multi-party Protocols , ibidem, 94-99.

[4] W. Masek , A Fast Algorithm for the String Editing Problem and Decision Graph Complexity , M.Sc. Thesis, M.I.T., May 1976 .

[5] I. Wegener , Optimal Decision Trees and One-time-only Branching Programs for Symmetric Boolean Functions , Interner Bericht 3/83 , Fachbereich Informatik, Universität Frankfurt .

[6] I. Wegener , private communication .

[7] P. Pudlák, S. Žák , Space complexity of computations , manuscript, 1982.

[8] S. Žák , Information in computation structures /preliminary version/ , to appear in Acta Polytechnica .

[9] P. Pudlák , A lower bound on complexity of branching programs , these proceedings.

and the analytic sets [K, M]. Several observations support this connection. NP sets are exactly those which are accepted by polynomial-size, nondeterministic circuits (ignoring uniformity issues). A Nondeterministic circuit is one with inputs that are nondeterministically set as well as ordinary inputs. By the addition of additional nondeterministic inputs these circuits may be converted to equivalent polynomial-size, depth-2, nondeterministic circuits. The infinitary analog to these, the countable, depth-2, nondeterministic circuits accept exactly the analytic sets.

This analogy suggests that the NP = co-NP question may be illuminated by the theorem stating that the class of analytic sets is not closed under complementation. The classical proof by diagonalization of this theorem does not seem to have a corresponding finitary argument. We give here a new purely combinatorial proof of this theorem.

PRELIMINARIES.

Let $\Sigma = \{0,1\}$ and Σ^{ω} be the set of infinite 0,1 sequences or <u>reals.</u> An <u>interval</u> is the set of reals extending a finite sequence. An <u>open set</u> is a union of intervals. Closing the open sets under countable union and intersection gives the <u>Borel sets.</u> An <u>analytic set</u> is a projection of a Borel set, (i.e., A is analytic if $A = \{\alpha\colon \langle\alpha,\beta\rangle \in B$ for some $\beta\}$ where B is Borel and $\langle\alpha,\beta\rangle$ is any pairing function).

Definition. A <u>literal</u> is a member of $\{x_1,\bar{x}_1,x_2,\bar{x}_2,\ldots\}$. A <u>$\vee_1$-circuit</u> is a collection of literals and an <u>$\wedge_2$-circuit</u> is a

A TOPOLOGICAL VIEW OF SOME PROBLEMS IN COMPLEXITY THEORY

Michael Sipser

Mathematics Department
Massachusetts Institute of Technology
Cambridge, Massachusetts 02139

ABSTRACT.

We present a new, combinatorial proof of the classical theorem that the analytic sets are not closed under complement. Possible connections with questions in complexity theory are discussed.

INTRODUCTION.

A number of recent results in circuit complexity theory have been stimulated by a new understanding of certain old theorems in descriptive set theory. The hierarchy theorem for polynomial-size, constant depth circuits is the finite counterpart to the Borel rank hierarchy theorem [S2]. The lower bound for circuits computing the parity function [FSS, A] in part stemmed from a result showing that infinite parity functions are not Borel definable [S1]. In both cases, the classical proofs do not exhibit enough combinatorial structure to yield insight into the finitary questions and new proofs were required.

The link between circuits and Borel sets stems from an analogy between polynomial growth and countability [S1]. In this paper, we propose a further link suggested by this analogy, one between NP

countable collection of V_1-circuits. These naturally represent functions from Σ^ω to Σ. A nondeterministic circuit has additional nondeterministic inputs represented by literals drawn from $\{y_i,\overline{y_i}\}$. It accepts a given real if there is some setting of the nondeterministic inputs which causes evaluation to 1. If a nondeterministic Λ_2-circuits accepts a real ρ, then a setting π of the x and y inputs causing evaluation to 1 is called a proof. If C is a member V_1-circuit then π satisfies C at j if the jth literal of C is 1 in π.

The nondeterministic circuits accept exactly the class of analytic sets.

Let N = {1,2,...} and N* be the set of finite sequences over N. A tree is a subset of N* closed under prefix. Let T be the set of all trees. We fix any enumeration of N* and obtain a natural correspondence between trees and reals. Hence we may speak of, say, an analytic set of trees. A tree is well-founded if it has no infinite branch, (i.e., tree τ is well-founded every $\alpha \in N^\omega$ contains a prefix $b \notin \tau$). Let W be the set of all well-founded trees. It is easy to verify that $\overline{W}$, the complement of W, is analytic. (nondeterministically guess the branch). We show that W itself is not.

We introduce some additional notation. If s,t are sequences in N* then st is the concatenation of s and t. If A is a set of sequences then sA = {st: t ∈ A}.

The Proof.

Theorem. There is an analytic set $\overline{W}$ whose complement is not analytic.

Proof. Let W be the set of all well-founded trees. We first establish the following Ramsey-like property of collections of trees.

Definition. For any tree τ, collection of trees A, and $s \in N^*$ we say the detail of τ at s, $\tau^s = \{t: st \in \tau\}$. The detail of A at s, $A^s = \{\tau^s: \tau \in A\}$. Say A is large at s if $W \subseteq A^s$ or simply large if it is large for some s. For example, W is large at e, the sequence of length o.

Claim. If A is large at s and is divided into a countable union of sets, $A = B_1 \cup B_2 \cup \ldots$ then for some i and j, B_i is large at sj.

Proof. Assume to the contrary that for each i,j B_i is not large at sj. So each detail of B_i at any sj lacks a tree $\tau_{i,j}$ in W. By pasting these together, one obtains the well-founded tree $\sigma = 1\tau_{1,1} \cup 2\tau_{2,2} \cup \cdots$ not in B_i^s for any i and therefore not in A^s. But $\sigma \in W$ contradicting the largeness of A at s.

□

To show that W is not analytic, we construct a sequence of large sets $W \supseteq A_1 \supseteq A_2 \supseteq \cdots$ containing trees which "converge" to one not in W.

Assume to the contrary that W is analytic, accepted by a nondeterministic Λ_2-circuited N containing V_1-circuits $C_1, C_2, \ldots$. Let $N^* = \{t_1, t_2, \ldots\}$. We perform a construction in stages. The goal of stage i is to construct $A_i \subseteq W$, $s_i \in N^*$, and $p_i \in N^*$ such that A_i is large at s_i, all $\sigma \in A_i$ agree on $t_1, \ldots, t_i$, and each $\sigma \in A_i$ has a proof which satisfies C_j at $p_i(j)$

(the jth position of p_i) for $j \le i$. Let $A_0 = W$, $s_i = e$, and $p_i = e$. Go to Stage 1.

Stage i. Let $B_m = \{\alpha \in A_{i-1}$: α has a proof which satisfies C_i at $m\}$. By the lemma, for some m and n, B_m is large at $s_{i-1}n$. Fix m and n. Let $D = \{\alpha \in B_m$: α contains $t_i\}$ and $E = B_m - D$. By the lemma, either D or E is large at a sequence $s_{i-1}nk$. Let A_i be this large set. Let $s_i = s_{i-1}nk$ and $p_i = p_{i-1}m$. Go to stage i+1.

It is straightforward to verify that upon completion of all stages there is exactly one tree α in every A_i. Furthermore α is not in W since it contains an infinite branch $S_1 \cup S_2 \cup \cdots$ and there is a proof $\pi = p_1 \cup p_2 \cup \cdots$ which satisfies every C_i. Therefore α is accepted by N, a contradiction.

CONCLUSION.

The links between topological notions such as open set, Borel set, and analytic set and their companions in circuit complexity bear further investigation. Ajtai's theorem [A] that every polynomial-size, depth-k definable set is well approximable by a union of cylinders is analogous to the theorem that all Borel sets are measurable, i.e., well approximable by open sets. There seems to be a parallel between Baire category theorem type constructions and constructions involving probabilistic methods. It is interesting to view these observations in the context of defining a notion of finite topological space.

ACKNOWLEDGEMENT.

I am deeply grateful to John Addison for having introduced me to descriptive set theory.

REFERENCES

[A] M. Ajtai, "Σ_1^1-formulae on finite structures", Annals of Pure and Applied Logic 24, 1983, 1-48.

[FSS] M. Furst, J.B. Saxe, M. Sipser, "Parity, circuits, and the polynomial time hierarchy", Proceedings of the 22nd Annual Symposium on Foundations of Computer Science, 1981, 260-270.

[K] K. Kuratowski, Topology, Academic Press, 1966.

[M] Y. Moschavakis, Descriptive Set Theory, North-Holland, 1980.

[S1] M. Sipser, "On polynomial versus exponential growth", unpublished report, 1981.

[S2] M. Sipser, "Borel sets and circuit complexity", Proceedings of the 15th Annual Symposium on Theory of Computing, 1983, 61-69.

PROPOSITIONAL DYNAMIC LOGIC WITH STRONG LOOP PREDICATE

Ryszard Danecki
Institute of Mathematics, Polish Acad. of Sci.
Mielżyńskiego 27/29, 61-725 Poznań, Poland

INTRODUCTION

By the loop-PDL we mean the Propositional Dynamic Logic of [FL] extended of a new formula-forming functor "loop" of one program variable. A formula loop(a) is satisfied at some state x if there exists an execution of the program a that starts and ends at x. The paper shows that the satisfiability problem for loop-PDL formulae is decidable in time $\exp(n^c)$, where n is the length of the formula and c is a constant, if nondeterministic models are allowed, and becomes undecidable (even Σ^1_1-hard) if restricted to models in which all atomic programs are deterministic.

We use the phrase "strong loop predicate" to underline the distinction from the delta functor $\Delta(a)$ of [S] expressing the existence of an infinite sequence of consecutive executions of the program a starting with the state at which $\Delta(a)$ is evaluated. It is lear that loop(a) implies $\Delta(a)$, but not vice versa. The delta-PDL of [S] is decidable in both nondeterministic and deterministic cases, and has the finite model property. The last fact is not true for loop-PDL. However, the decidability proof for loop-PDL is very similar to that of [S] and consists in a reduction to the emtiness problem for some restricted version of tree automata [R] . This involves a series of lemmas presented in three sections reflecting the main steps of the proof. Our undecidability result is obtained by a simple application of the recurring domino problem of [H] .

Acknowledgement: The loop-PDL was first presented to me by R. Knast who also claimed its undecidability in deterministic case. My thanks are also to Z. Habasiński for stimulating discussions.

DEFINITIONS AND RESULTS

Programs and formulae of the loop PDL are defined inductively as follows. A, B, C, ... are atomic programs, P, Q, R,... are atomic formulae. If a, b are programs and p, q formulae, then a;b, $a \cup b$, a^*, p? are programs and $\neg p$, $\langle a\rangle p$, loop(a) are formulae. By definition, $p\&q \equiv \langle p?\rangle q$, $[a]p \equiv \neg\langle a\rangle \neg p$, <u>true</u> $\equiv p \vee \neg p$. Formulae are interpreted in structures of the form $\mathcal{M}=(X, \models, \langle\ \rangle)$, where X is a nonempty set of nodes (often called states), $\models$ is a satisfiability relation for atomic formulae, i. e. $\models\ \subset X\times\{P, Q, R, \ldots\}$, $\langle\ \rangle \subset X\times\{A, B, \ldots\}\times X$ assigns a binary relation $\langle A\rangle$ to every atomic program A. $\mathcal{M}$ is said to be deterministic if for every atomic program A, $\langle A\rangle$ is a function, i. e. $x\langle A\rangle y$ and $x\langle A\rangle z$ imply $y=z$. Relations $\models$ and $\langle\ \rangle$ are extended to arbitrary formulae and programs as follows: $x\models\neg p$ iff not $x\models p$, $x\models\langle a\rangle p$ iff $\exists y\in X$: $x\langle a\rangle y$ and $y\models p$, $x\models$ loop(a) iff $x\langle a\rangle x$, $\langle a;b\rangle=\langle a\rangle\langle b\rangle$ (superposition of relations), $\langle a\cup b\rangle = \langle a\rangle\cup\langle b\rangle$, $\langle a^*\rangle=\langle a\rangle^*$ (transitive, reflexive closure). We say that $\mathcal{M}$ is a model for the formula p, $\mathcal{M}\models p$, if $x\models p$ for some node x of $\mathcal{M}$. A formula is satisfiable if it has a model.

The loop-PDL has no finite model property. For example, the formula $[A^*](\langle A\rangle$<u>true</u> $\&\neg$loop$(A;A^*))$ has a model, but not a finite one.

<u>Theorem 1.</u> The satisfiability problem for loop-PDL formulae is decidable in time $\exp(n^c)$, where n is the length of the formula and c some constant.

<u>Theorem 2.</u> The problem of whether or not a loop-PDL formula has a deterministic model is undecidable (Σ_1^1-hard).

The rest of the paper contains proofs of the above theorems. The Theorem 1 is proved in three major steps formulated by Lemmas 1, 2, and 4. First, the satisfiability of a formula is expressed in terms of the existence of a labelled graph in which paths and loops are, or are not, accepted by some finite automata called path acceptors. Then we show that the existence of such a graph is equivalent to the existence of a cactus with similar properties. Cactuses are not trees, but are very tree-like and can be easily represented by trees. In consequence, the satisfiability problem is reduced to the emptiness problem of some very restricted version of tree automata. The latter problem can be solved in time polynomial of the number of states. The proof of the

Theorem 2 is a short, direct reduction of a recurring domino problem of [H] to the existence of a deterministic model for loop-PDL formula.

PATH/LOOP CONDITIONS FOR LABELLED GRAPHS

By a Σ,Δ-graph we mean a triple $G=(X, E, v)$, where Σ,Δ are finite sets of labels, X is a set of nodes, $E \subset X \times \Delta \times X$ is a set of edges labelled with elements of Δ and $v: X \longrightarrow \Sigma$ is a node labelling function. By a path acceptor we mean a 5-tuple $\mathfrak{A}=(S, M, N, S_0, F)$, where S is a finite set of states, S_0, $F \subset S$ are the sets of initial and final states, respectively, $M \subset S \times \Delta \times S$ is a transition relation, and $N \subset S \times \Sigma$ is a testing relation. A finite path π in G defines a word $\sigma_0 \delta_1 \sigma_1 \ldots \delta_k \sigma_k$ belonging to $\Sigma(\Delta\Sigma)^*$. Here σ_0 is the label of the first node of the path, δ_1 is the label of the first edge, σ_1 is the label of the second node, and so on. We say that $\mathfrak{A}$ accepts π if there is a sequence of states $s_0 s_1 \ldots s_k$ such that $s_0 \in S_0$, $s_k \in F$, $(s_{i-1}, \delta_i, s_i) \in M$ for every $0 < i \leqslant k$, and $(s_i, \sigma_i) \in N$ for all $0 \leqslant i \leqslant k$. A loop is a path returning to its beginning. The notion " $\mathfrak{A}$ accepts a loop" is always relativized to some starting position, usually understood from the context.

Any pair $(\Gamma, \mathfrak{A})$, where $\Gamma \subset \Sigma$ and $\mathfrak{A}$ is a Σ,Δ-path acceptor, defines four conditions for graphs: positive path and loop conditions, and negative path and loop conditions. We say that a Σ,Δ-graph $G= =(X, E, v)$ satisfies the condition:

PATH$(\Gamma, \mathfrak{A})$ if every node x with v(x) in Γ is a beginning of some path accepted by $\mathfrak{A}$,

LOOP$(\Gamma, \mathfrak{A})$ if for every node x with v(x) in Γ there exists a loop from x to x accepted by $\mathfrak{A}$,

NO-PATH$(\Gamma, \mathfrak{A})$ if there is no path in G accepted by $\mathfrak{A}$ and beginning with a node x with v(x) in Γ,

NO-LOOP$(\Gamma, \mathfrak{A})$ if for any x with v(x) in Γ there is no loop from x to x accepted by $\mathfrak{A}$.

If $v(x) \in \Gamma$, then we say that x falls into the range of the condition $\Psi=\Psi(\Gamma, \mathfrak{A})$, and a path starting with x and accepted by $\mathfrak{A}$ is referred to as a path satisfying Ψ. We write $\Psi(\mathfrak{A})$ if $\Gamma=\Sigma$. Every negative condition can be put in such a form by redefining the testing relation N of $\mathfrak{A}$. Thus, any negative condition actually means: in the whole graph there is no path (loop) accepted by some $\mathfrak{A}$.

Lemma 1. For any loop-PDL formula p one can effectively construct finite sets Σ, Δ, $\Sigma' \subset \Sigma$ and a finite set of path/loop conditions $\Psi_1, \ldots, \Psi_k$ such that the formula p has a model iff there exists a Σ,Δ-graph $G=(X, E, v)$ satisfying all $\Psi_1, \ldots, \Psi_k$ and $v(x) \in \Sigma'$ for some $x \in X$.

Proof. Let p be a formula and Δ the set of atomic programs it contains. A structure $\mathcal{M}=(X, \models, \langle\rangle)$ can be considered together with a graph $G=(X, E, v)$ such that for every $x, y \in X$, $A \in \Delta$, $x\langle A\rangle y$ iff $(x, A, y) \in E$, and v(x) is a set of subformulae of p. $\mathcal{M} \models p$ iff there exists v such that for every subformula q of p, (a1): $x \models q$ iff $q \in v(x)$ for every $x \in X$, and (a2): $p \in v(x)$ for some $x \in X$. Now, the crucial observation is the following. A program b with tests $q_1?, \ldots, q_n?$ can be treated as a regular expression over the alphabet Δ and "$q_1?$", ... "$q_n?$" interpreted as additional symbols. It is not hard to construct a path acceptor $\mathfrak{A}(b)$ such that if $q_1, \ldots, q_n$ satisfy (a1), then some path from x to y in G is accepted by $\mathfrak{A}(b)$ iff $x\langle b\rangle y$ in $\mathcal{M}$. In this construction a letter from Δ means a transition through an edge with the same label, and a letter "$q_i?$" means checking if $q_i \in v(x)$.

Let Σ be a family of all sets U of subformulae of p, such that for every subformula q of p, $\neg q \in U$ iff $q \notin U$. Define $\Sigma(q)=\{U \in \Sigma: q \in U\}$, $\Sigma' = \Sigma(p)$. By structural induction (a1) and (a2) can be restated as follows. $\mathcal{M} \models p$ iff there exists v such that,

(b1): (a1) holds for every atomic subformula of p,

(b2): $v(x) \in \Sigma$ for every $x \in X$,

(b3): for every subformula $q=\langle a\rangle r$ of p, the graph G satisfies the conditions PATH($\Gamma, \mathfrak{A}$) and NO-PATH($\Sigma \setminus \Gamma, \mathfrak{A}$), where $\Gamma = \Sigma(q)$, $\mathfrak{A}=\mathfrak{A}(a;r?)$,

(b4): for every subformula $q=\text{loop}(a)$, G satisfies the conditions LOOP($\Gamma, \mathfrak{A}$) and NO-LOOP($\Sigma \setminus \Gamma, \mathfrak{A}$) with $\Gamma=\Sigma(q)$, $\mathfrak{A}=\mathfrak{A}(a)$,

(b5): $v(x) \in \Sigma'$ for some $x \in X$.

From this equivalence the Lemma 1 follows immediately. □

CACTUSES

Let G and G' be Σ,Δ-graphs with distinguished nodes x and x' referred to as the roots of G and G', respectively. The operation of grafting G' onto G in the node y of G can be done if y and x' have the same label. Its result is defined as the union of G and such a copy of G' which has exactly one node in common with G, namely the node y. The root of the new graph is x. If $x \neq y$, then we say that G is the ground of y, and G′ is a graft in y. If x=y, then both G and G′ are grafts in x. A path is simple if it contains no loop. A loop is simple if it has no proper subloop. Any finite simple path with the root at its beginning, or any simple loop with a distinguished node, is an elementary cactus. A Σ,Δ-cactus is a Σ,Δ-graph resulting from elementary cactuses by some, maybe infinite, number of grafting operations. The maximal number of elementary paths or loops grafted in one node is called the degree of a cactus. For any two different nodes x, y in a cactus, if there exists a path from x to y, then there exists exactly one simple path from x to y, denoted by $\overrightarrow{x,y}$.

Let Σ,Δ, $\Sigma' \subset \Sigma$ be finite sets and $\Psi_1, \ldots, \Psi_k$ path/loop conditions. We say that a Σ,Δ-cactus C is perfect for these conditions if (1): the label of the root of C belongs to Σ', (2): C satisfies all negative conditions among $\Psi_1, \ldots, \Psi_k$, (3): C satisfies all positive conditions by separate grafts. The last requirement means that for every node x of C the number of elementary grafts in it is equal to the number of positive conditions among Ψ_i, $i=1,\ldots, k$, whose ranges contain the label of x. Moreover, the conditions and grafts in x can be enumerated in such a way, that the i-th graft is a path or loop satisfying the i-th condition. Thus, the degree of C does not exceed k.

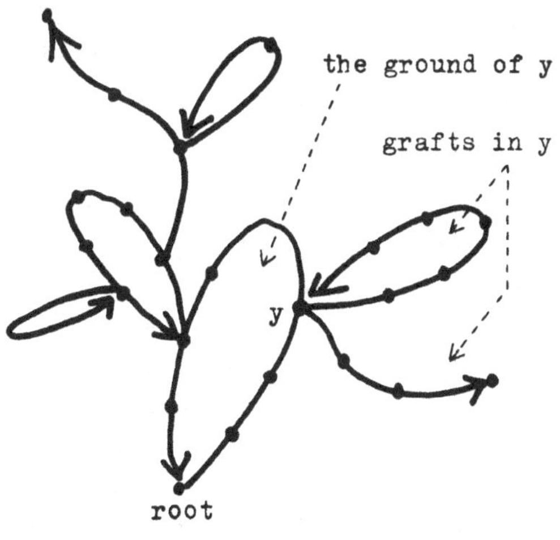

Fig. 1. A cactus

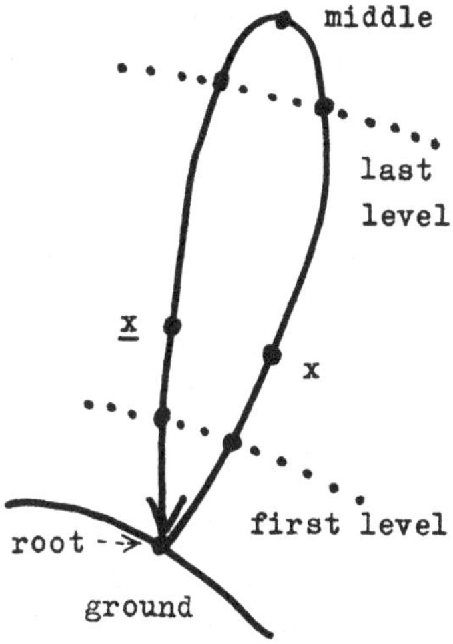

Fig. 2. Parts of a simple loop.

Lemma 2. If there exists a graph satisfying conditions from the second part of the Lemma 1, then there exists a cactus perfect for these conditions.

Hint to the proof. In a graph G satisfying the conditions we choose a node z with v(z) in Σ', and for every node x and every positive Ψ_i with v(x) in its range we choose a path (or loop) $\pi(xi)$ satisfying Ψ_i. Then, the perfect cactus C is constructed from the root z and copies of $\pi(xi)$'s as elementary grafts. □

TREES

By a n-ary Ω-tree we mean a function $t: T \longrightarrow \Omega$, where T is a full n-ary infinite tree, usually defined as $T=\{0, 1, \ldots, n-1\}^*$, and Ω is a finite set. The empty word λ is the root of T, and x0, x1, ..., x(n-1) are the sons of $x \in T$. A restricted tree automaton (RTA) over n-ary Ω-trees is a system $\mathcal{A} =(S, M, S_0, F)$, where S is a finite set of states, S_0, $F \subset S$ are the sets of initial and final states, respectively, and $M \subset S \times \Omega \times S^n$ is a tree transition relation. A tree $t: T \longrightarrow \Omega$ is accepted by $\mathcal{A}$ if there exists a mapping $r: T \longrightarrow S$, such that (1): $r(\lambda) \in S_0$, (2): $(r(x), t(x), r(x0), \ldots$ $\ldots, r(x(n-1))) \in M$ for every $x \in T$, and (3): for every $xi \in T$ with $i \neq 0$, $i \in \{1, \ldots, n-1\}$ there exists k such that $r(xi0^k) \in F$. RTAs are very particular tree automata [R] . The emtiness problem, i. e. the question of whether or not a given RTA accepts some tree, is decidable in time polynomial of $|S| \cdot |\Omega|$. The family of sets of trees recognizable by RTAs is closed under intersection and projection.

Consider a simple loop in some Σ,Δ-cactus C. The loop has a root, i. e. the node by which it was grafted, some number of levels, and a middle node (edge) if its length is even (odd), cf. Fig. 2. The notation is choosen in such a way that x and $\underline{x}$ always belong to one level of some simple loop, and that $\overrightarrow{x, \underline{x}}$ contains the middle, and $\overrightarrow{\underline{x}, x}$ the root of the loop. We will refer to $\overrightarrow{\underline{x}, x}$ (resp. $\overrightarrow{x, \underline{x}}$) as the lower (resp. upper) segment cut by the level of x. By convention, the levels of a loop do not contain the root and the middle node, thus the expression "the level of x" is unambiguous. Now, suppose that C is of degree k. We can transform C into a tree by folding loops, that is by considering any level x, $\underline{x}$ as a node with a double label and any pair of edges between neighbouring levels as a single edge with a double label. Such a tree can be further represented as a (2k+1)-ary tree

$t: T \longrightarrow \Omega$, where $T=I^*$, $I=\{0, -1, 1, \ldots, -k, k\}$, $|\Omega|=O(|\Delta|^2 \cdot |\Sigma|^2)$. The general idea is that if a level x, $\underline{x}$ of some loop π corresponds to $z \in T$, then the next level of π corresponds to z0, the first level of the i-th graft in x corresponds to zi, and the first level of the i-th graft in $\underline{x}$ corresponds to z(-i). Thus, any elementary graft in C corresponds to a sequence of the form zi, zi0, ..., $zi0^k$, where $i \neq 0$, and the root of the graft corresponds to z. The root of the whole C corresponds to λ. We omit tedious details of the definition of Ω. Generally, elements of Ω provide information about labels of nodes and edges of C in such a way, that any t defines no more than one cactus C(t), there exists a RTA which accepts t iff C(t) is defined, and a RTA working on t can easily perform transitions of a path acceptor working on C(t).

<u>Lemma 3.</u> For any negative path or loop condition Ψ one can effectively construct a restricted tree automaton $\mathfrak{B}$ such that, $\mathfrak{B}$ accepts a tree t iff the cactus corresponding to t satisfies Ψ.

<u>Proof.</u> Let $\mathfrak{A}=(S, M, N, S_0, F)$ be a Σ,Δ-path acceptor. In the case Ψ=NO-PATH($\mathfrak{A}$) a RTA $\mathfrak{B}$ puts at every node of C(t) all initial states of $\mathfrak{A}$ and computes all reachable states. A tree t is accepted if this can be done without reaching any final state of $\mathfrak{A}$. $\mathfrak{B}$ needs O(exp S) states.

The case Ψ =NO-LOOP($\mathfrak{A}$) is more interesting. Let C(t)=(X, E, v). For every mapping R: X $\longrightarrow \underline{P}(S\times S)$, /$\underline{P}$ means powerset/, and every path $\pi=x_1\delta_2x_2\ldots\delta_kx_k$ define a relation MNR(π) $\subset S\times S$ as follows: (s, s′) is in MNR(π) iff there exists a sequence of states from S, $s_1s_1's_2s_2'\ldots s_ks_k'$ such that (a): $s=s_1$, $s'=s_k$, (b): $(s_i', \delta_{i+1}, s_{i+1}) \in M$, for all $1\leqslant i<k$, (c): $(s_i, v(x_i))$ and $(s_i', v(x_i))$ belong to N for all $1\leqslant i\leqslant k$, (d): $s_i=s_i'$ or $(s_i, s_i') \in R(x_i)$ for all $1\leqslant i\leqslant k$.

We say that R contains MNR transitions through simple loops in C(t), if for every $x\in X$ and every simple path π from x to x, MNR(π)$\subset$R(x). Let X″ be the set of all levels x, $\underline{x}$ of C(t). For mappings L, U from X″ into $\underline{P}(S\times S)$ we say that L (resp. U) contains MNR transitions through lower (resp. upper) segments of C(t) if for every x, $\underline{x}$ from X″ , $MNR(\overrightarrow{\underline{x}, x}) \subset L(x, \underline{x})$, (resp. $MNR(\overrightarrow{x, \underline{x}}) \subset U(x, \underline{x})$).

<u>Fact A:</u> C(t) satisfies the negative loop condition Ψ iff there exists a mapping R: X $\longrightarrow \underline{P}(S\times S)$ such that, (A1): R contains MNR transitions through simple loops, (A2): R(x) is a transitive relation for every $x\in X$, and (A3): $(S_0\times F)\cap R(x)=\emptyset$, for every $x\in X$.

<u>Fact B:</u> A mapping R contains MNR transitions through simple loops iff there exist mappings L, U: $X'' \longrightarrow \underline{P}(S\times S)$ such that, (B1): L and U contain MNR transitions through lower and upper segments, respectively, (B2): $U(x, \underline{x})L(x, \underline{x}) \subset R(x)$ and $L(x, \underline{x})U(x, \underline{x}) \subset R(\underline{x})$ for every x, $\underline{x}$ (superposition of relations), (B3): if $x \in X$ is a root of some simple loop, y, $\underline{y}$ is its first level, and e, e′ are edges from x to y and from $\underline{y}$ to x, respectively, then $MNR(e)U(y, \underline{y})MNR(e') \subset R(x)$, (B4): if $x \in X$ is the middle node of some even simple loop, y, $\underline{y}$ is its last level, e is the edge from x to $\underline{y}$, and e′ is the edge from y to x, then $MNR(e)L(y, \underline{y})MNR(e') \subset R(x)$.

<u>Fact C:</u> Mappings L and U contain MNR transitions through lower and upper segments, respectively, iff (C1): for every first level x, $\underline{x}$ and every last level y, $\underline{y}$, $MNR(\overrightarrow{\underline{x}, x}) \subset L(x, \underline{x})$ and $MNR(\overrightarrow{y, \underline{y}}) \subset U(y, \underline{y})$, (C2): if y, $\underline{y}$ is the next level of x, $\underline{x}$ in some loop, and e is the edge from x to y, and e′ the edge from $\underline{y}$ to $\underline{x}$, then $MNR(e')L(x, \underline{x})MNR(e) \subset L(y, \underline{y})$, $MNR(e)U(y, \underline{y})MNR(e') \subset U(x, \underline{x})$.

By Facts A-C, C(t) satisfies Ψ iff there exist mappings R, L, U satisfying (A2-3), (B2-4), and (C1-2). All these conditions postulate the existence of some labelling on the input tree t, that satisfies some neighbourhood constraints. This can be easily checked by a RTA with $O(\exp|S|^c)$ states, for some constant c.□

For positive conditions $\Psi_1, \ldots, \Psi_n$ it is not hard to construct a RTA $\mathfrak{X}$ which accepts a tree t iff C(t) satisfies all of them by separate grafts. This, combined with the Lemma 3 and a RTA that checks if C(t) is defined, yields the following

<u>Lemma 4.</u> For finite sets Σ, Δ, $\Sigma' \subset \Sigma$ and path/loop conditions $\Psi_1, \ldots, \Psi_k$ one can effectively construct a restricted tree automaton $\mathfrak{X}$ which accepts a tree t iff the cactus C(t) is perfect for these conditions.□

Lemmas 1, 2 and 4 show that for every loop-PDL formula p one can effectively construct a RTA $\mathfrak{X}$ such that, p is satisfiable iff $\mathfrak{X}$ accepts some tree. Thus the problem is decidable. Now, few words about the size of $\mathfrak{X}$. If p is of the length n, then in Lemma 1: $|\Sigma| = \exp n$, $|\Delta| < n$, $k < 2n$, and any path acceptor involved has O(n) states. The RTA $\mathfrak{X}$ constructed in Lemma 4 works over k-ary Ω-trees with $|\Omega| = O(\exp cn)$. $\mathfrak{X}$ is a product of k+1 automata every of which in the worst case has $O(\exp n^c)$ states. Thus, $\mathfrak{X}$ has $O(\exp n^d)$ states, d = const., and its emptiness problem can be solved in time polynomial of the number of states. This ends the proof of Theorem 1.

PROOF OF THEOREM 2.

The following recurring domino problem is known to be undecidable (Σ_1^1-complete), [H]. Given: A finite set of domino types $D=\{0, 1, \dots, n\}$ and a relation $R \subset D\times D\times D$ of tiling rules. Question: Does there exist a function which to every pair (i, j) of natural numbers assigns some element f(i, j) from D in such a way, that (1): $(f(i, j), f(i+1, j), f(i, j+1)) \in R$ for every $i, j \in \mathcal{N}$, and (2): $f(0, j)=0$ for infinitely many $j \in \mathcal{N}$. For given $D=\{0, 1, \dots, n\}$ and R we construct a loop-PDL formula p such that, p has a deterministic model iff there exists f satisfying (1) and (2). Atomic programs of p are A, A′, B, B′, and atomic formulae are $P_0, P_1, \dots, P_n$. The formula p is the conjunction of the following five formulae. (The prefix $[(A\cup B)^*]\dots$ is written as ALL... .) p_1=ALL(⟨A⟩true & ⟨B⟩true), p_2=ALL($\&_{a\in I}$ loop(a)), where $I=\{AA', A'A, BB', B'B, ABA'B'\}$, p_3=ALL($\bigvee_{i\in D} P_i$)&ALL($\&_{i\neq j,\ i,j\in D}$ $\neg(P_i\&P_j)$), p_4=ALL($\bigvee_{(i, j, k)\in R} P_i\&\langle A\rangle P_j\&\langle B\rangle P_k$) , $p_5=[B^*]\langle B^*\rangle P_0$. The general idea is that in a structure nodes x, y, z with $x\langle A\rangle y$, $x\langle B\rangle z$, correspond to some (i, j), (i+1, j), (i, j+1), respectively, and $x\models P_k$ means $f(i, j)=k$. In a deterministic structure $p_1\&p_2$ defines some grid, p_3 defines f, and $p_4\&p_5$ saya that f satisfies (1) and (2).

R e f e r e n c e s:

[FL] M. J. Fischer, R. E. Ladner, Propositional dynamic logic of regular programs, JCSS 18:2 (1979) 194-211.

[H] D. Harel, Recurring dominoes: Making the highly undecidable highly understandable, Proc. FCT'83, LNCS 158, pp 177-194, (1983) Springer-Verlag.

[R] M. O. Rabin, Decidability of second-order theories and automata on infinite trees, Trans. AMS 141 (1969), 1-35.

[S] R. S. Streett, Propositional dynamic logic of looping and converse, Proc. 13th Ann. ACM Symp. on the Theory of Comput., (Milwaukee) 1981, 375-383.

Vol. 142: Problems and Methodologies in Mathematical Software Production. Proceedings, 1980. Edited by P.C. Messina and A. Murli. VII, 271 pages. 1982.

Vol. 143: Operating Systems Engineering. Proceedings, 1980. Edited by M. Maekawa and L.A. Belady. VII, 465 pages. 1982.

Vol. 144: Computer Algebra. Proceedings, 1982. Edited by J. Calmet. XIV, 301 pages. 1982.

Vol. 145: Theoretical Computer Science. Proceedings, 1983. Edited by A.B. Cremers and H.P. Kriegel. X, 367 pages. 1982.

Vol. 146: Research and Development in Information Retrieval. Proceedings, 1982. Edited by G. Salton and H.-J. Schneider. IX, 311 pages. 1983.

Vol. 147: RIMS Symposia on Software Science and Engineering. Proceedings, 1982. Edited by E. Goto, I. Nakata, K. Furukawa, R. Nakajima, and A. Yonezawa. V. 232 pages. 1983.

Vol. 148: Logics of Programs and Their Applications. Proceedings, 1980. Edited by A. Salwicki. VI, 324 pages. 1983.

Vol. 149: Cryptography. Proceedings, 1982. Edited by T. Beth. VIII, 402 pages. 1983.

Vol. 150: Enduser Systems and Their Human Factors. Proceedings, 1983. Edited by A. Blaser and M. Zoeppritz. III, 138 pages. 1983.

Vol. 151: R. Piloty, M. Barbacci, D. Borrione, D. Dietmeyer, F. Hill, and P. Skelly, CONLAN Report. XII, 174 pages. 1983.

Vol. 152: Specification and Design of Software Systems. Proceedings, 1982. Edited by E. Knuth and E. J. Neuhold. V, 152 pages. 1983.

Vol. 153: Graph-Grammars and Their Application to Computer Science. Proceedings, 1982. Edited by H. Ehrig, M. Nagl, and G. Rozenberg. VII, 452 pages. 1983.

Vol. 154: Automata, Languages and Programming. Proceedings, 1983. Edited by J. Díaz. VIII, 734 pages. 1983.

Vol. 155: The Programming Language Ada. Reference Manual. Approved 17 February 1983. American National Standards Institute, Inc. ANSI/MIL-STD-1815A-1983. IX, 331 pages. 1983.

Vol. 156: M.H. Overmars, The Design of Dynamic Data Structures. VII, 181 pages. 1983.

Vol. 157: O. Østerby, Z. Zlatev, Direct Methods for Sparse Matrices. VIII, 127 pages. 1983.

Vol. 158: Foundations of Computation Theory. Proceedings, 1983. Edited by M. Karpinski, XI, 517 pages. 1983.

Vol. 159: CAAP'83. Proceedings, 1983. Edited by G. Ausiello and M. Protasi. VI, 416 pages. 1983.

Vol. 160: The IOTA Programming System. Edited by R. Nakajima and T. Yuasa. VII, 217 pages. 1983.

Vol. 161: DIANA, An Intermediate Language for Ada. Edited by G. Goos, W.A. Wulf, A. Evans, Jr. and K.J. Butler. VII, 201 pages. 1983.

Vol. 162: Computer Algebra. Proceedings, 1983. Edited by J.A. van Hulzen. XIII, 305 pages. 1983.

Vol. 163: VLSI Engineering. Proceedings. Edited by T.L. Kunii. VIII, 308 pages. 1984.

Vol. 164: Logics of Programs. Proceedings, 1983. Edited by E. Clarke and D. Kozen. VI, 528 pages. 1984.

Vol. 165: T.F. Coleman, Large Sparse Numerical Optimization. V, 105 pages. 1984.

Vol. 166: STACS 84. Symposium of Theoretical Aspects of Computer Science. Proceedings, 1984. Edited by M. Fontet and K. Mehlhorn. VI, 338 pages. 1984.

Vol. 167: International Symposium on Programming. Proceedings, 1984. Edited by C. Girault and M. Paul. VI, 262 pages. 1984.

Vol. 168: Methods and Tools for Computer Integrated Manufacturing. Edited by R. Dillmann and U. Rembold. XVI, 528 pages. 1984.

Vol. 169: Ch. Ronse, Feedback Shift Registers. II, 1–2, 145 pages. 1984.

Vol. 171: Logic and Machines: Decision Problems and Complexity. Proceedings, 1983. Edited by E. Börger, G. Hasenjaeger and D. Rödding. VI, 456 pages. 1984.

Vol. 172: Automata, Languages and Programming. Proceedings, 1984. Edited by J. Paredaens. VIII, 527 pages. 1984.

Vol. 173: Semantics of Data Types. Proceedings, 1984. Edited by G. Kahn, D.B. MacQueen and G. Plotkin. VI, 391 pages. 1984.

Vol. 174: EUROSAM 84. Proceedings, 1984. Edited by J. Fitch. XI, 396 pages. 1984.

Vol. 175: A. Thayse, P-Functions and Boolean Matrix Factorization, VII, 248 pages. 1984.

Vol. 176: Mathematical Foundations of Computer Science 1984. Proceedings, 1984. Edited by M.P. Chytil and V. Koubek. XI, 581 pages. 1984.